AF361782

Novel Strategies in the Development of New Therapies, Drug Substances and Drug Carriers 2.0

Novel Strategies in the Development of New Therapies, Drug Substances and Drug Carriers 2.0

Editors

Andrzej Kutner
Geoffrey Brown
Enikö Kallay

Basel • Beijing • Wuhan • Barcelona • Belgrade • Novi Sad • Cluj • Manchester

Editors

Andrzej Kutner
Medical University of
Warsaw
Warsaw, Poland

Geoffrey Brown
University of Birmingham
Edgbaston, UK

Enikö Kallay
Medical University of Vienna
Vienna, Austria

Editorial Office
MDPI
St. Alban-Anlage 66
4052 Basel, Switzerland

This is a reprint of articles from the Special Issue published online in the open access journal *International Journal of Molecular Sciences* (ISSN 1422-0067) (available at: https://www.mdpi.com/journal/ijms/special_issues/ACCORD2022).

For citation purposes, cite each article independently as indicated on the article page online and as indicated below:

Lastname, A.A.; Lastname, B.B. Article Title. *Journal Name* **Year**, *Volume Number*, Page Range.

ISBN 978-3-0365-8656-4 (Hbk)
ISBN 978-3-0365-8657-1 (PDF)
doi.org/10.3390/books978-3-0365-8657-1

Cover image courtesy of Teresa Zolek

Contents

About the Editors

Andrzej Kutner

Andrzej Kutner received his PhD in organic chemistry from the Chemistry Department of the University of Warsaw, Poland, his DSc (habilitation) in pharmaceutical sciences from the Faculty of Pharmacy of the Medical University of Warsaw, Poland, and has been awarded the title of Professor of Pharmaceutical Sciences. He is a trainee in US academia (University of Wisconsin—Madison, New York University, University of Minnesota, Duluth, University of California, Riverside). He works at the Department of Drug Chemistry, Faculty of Pharmacy of the Medical University of Warsaw. He also lectures on instrumental analysis in drug development at the Faculty of Pharmacy of the Medical University of Warsaw and on strategies for pharmaceutical syntheses at the Chemistry Department of the University of Warsaw. His main area of interest is medicinal chemistry, the synthesis of active substances, and the analysis of the correlation between biological activity and molecular structure.

Geoffrey Brown

Geoffrey Brown received his BSc from Queen Elizabeth College, London, and his PhD from University College (with Prof Sir Mel Greaves), London. His postdoctoral research took place at the MRC Immunochemistry Unit (with Alan Williams and Prof. Rodney Porter) and the Nuffield Department of Clinical Medicine (in Prof. Sir David Weatherall's Department), Oxford, where he was also an IBM Fellow, University of Oxford, and a Research Lecturer, Christ Church College. He is now a Reader in Cellular Immunology (Honorary) at the College of Medical and Dental Sciences, University of Birmingham. His research concerns the development of blood cells and the nature of leukemia stem cells.

Enikö Kallay

Enikö Kallay received her PhD in biochemistry in 1999 from the University of Vienna and her habilitation in experimental pathophysiology in 2004 from the Medical University of Vienna. She has been a trainee at several US universities (Harvard Medical School, John Hopkins University, MD, Strang Cancer Prevention Center, Cornell University, NY) and was a Marie Curie European Research Fellow at the University of Oxford, UK. She currently works at the Dept. Pathophysiology and Allergy Research, Center of Pathophysiology, Infectiology, and Immunology Medical University of Vienna, and lectures in cancer biology at the Karl Landsteiner University of Health Sciences, Krems, Austria. Her main area of interest is the unveiling of the biochemical and molecular mechanisms of the anti-tumorigenic and anti-inflammatory effects of vitamin D and dietary calcium in inflammatory bowel diseases and in cancers of the colon and ovaries.

Preface

The highly successful Volume 1.0 of the Special Issue "Novel Strategies in the Development of New Therapies, Drug Substances, and Drug Carriers", in the *International Journal of Molecular Sciences* (IJMS), edited also as an MDPI reprint, comprised 21 papers. We, therefore, reopened this topic as Volume 2.0 of the Special Issue in *IJMS*. The interest remained high, leading to the publication of 24 original research articles and reviews. As for Volume 1.0, Special Issue Volume 2.0 was also conceived to address a wide range of the processes involved in drug development at the molecular level, complying with the scope of the journal.

Despite several decades of intensive research and the development of several generations of increasingly selective and effective drugs, cancer is still the leading cause of death. For this reason, anticancer therapy, specifically against leukemia and lymphoma, and solid tumors, including prostate, melanoma, ovarian, lung, urinary bladder, and breast cancers, remains a consistently leading area of interest for researchers worldwide. The interest was shared also by the participants of the Interdisciplinary Conference on Drug Sciences, ACCORD 2022, held at the Medical University of Warsaw, Poland. Most of the papers published in this Special Issue were presented, at least in part, at this conference, as invited lectures, scientific communications, and posters. The main subjects of the communications were manifold: markers in neurodegenerative disorders, mediators of inflammation, non-coding micro-RNAs, enzymes regulating cell division, measures to mitigate the post-operative side effects of cancer treatment, genetic disorders such as cystic fibrosis, skin disorders, multifunctional biomaterials, plant-derived topical preparations, crystal forms of active pharmaceutical ingredients, innovative drug product formulations, the quality control of drug substances and drug products, including the chemical stability of drug substances and impurity profile, drug membrane permeability and the identification of drug candidates via the virtual screening of a compound library.

The papers within the Special Issue Volume 2.0, edited also as this reprint, address a wide range of topics regarding the development of new drug substances, new drug carriers, and drug products. As conceived, the topics in this reprint address the identification of the molecular target of new drugs, docking supported by molecular dynamics simulation, quantum chemical simulations, studies of protein–drug interactions, the modeling and optimization of the structure of small molecules with drug-like activity, the determination of the functional profile, pre-formulation studies, new pharmaceutical formulations, preclinical development to the design, and prediction of the efficacy of agents in clinical trials. Elucidation of the molecular mechanism of action is now an integral prerequisite for transforming an active substance into a drug candidate.

Andrzej Kutner, Geoffrey Brown, and Enikö Kallay

Editors

International Journal of
Molecular Sciences

Editorial

Novel Strategies in the Development of New Therapies, Drug Substances, and Drug Carriers Volume II

Andrzej Kutner [1,*], Geoffrey Brown [2] and Enikö Kallay [3]

[1] Department of Drug Chemistry, Faculty of Pharmacy, Medical University of Warsaw, 1 Banacha, 02-097 Warsaw, Poland
[2] School of Biomedical Sciences, Institute of Clinical Sciences, College of Medical and Dental Sciences, University of Birmingham, Birmingham B15 2TT, UK
[3] Department of Pathophysiology and Allergy Research, Center of Pathophysiology, Infectiology & Immunology, Medical University of Vienna, Währinger Gürtel 18-20, A-1090 Vienna, Austria
* Correspondence: andrzej.kutner@wum.edu.pl

Citation: Kutner, A.; Brown, G.; Kallay, E. Novel Strategies in the Development of New Therapies, Drug Substances, and Drug Carriers Volume II. *Int. J. Mol. Sci.* **2023**, 24, 5621. https://doi.org/10.3390/ijms24065621

Received: 1 February 2023
Revised: 24 February 2023
Accepted: 24 February 2023
Published: 15 March 2023

The highly successful previous Volume 1.0 of the Special Issue "Novel Strategies in the Development of New Therapies, Drug Substances, and Drug Carriers" [1], in the *International Journal of Molecular Sciences* (IJMS), edited also as a book [2], comprised 21 papers. We, therefore, reopened this topic as Volume 2.0 of the Special Issue in *IJMS*. The interest remained high, leading to the publication of 24 original research articles and reviews. As for Volume I, this Special Issue was also conceived to cover a wide range of processes of drug development at the molecular level, complying with the scope of the journal.

Despite several decades of intensive research and the development of several generations of increasingly selective and effective drugs, cancer is still the leading cause of death. For this reason, anticancer therapy, specifically against leukemia and lymphoma, and solid tumors, including prostate, melanoma, ovarian, lung, urinary bladder, and breast cancers, remains a consistently leading area of interest for researchers worldwide. The interest was shared also by the participants of the Interdisciplinary Conference on Drug Sciences, ACCORD 2022, held at the Medical University of Warsaw. Most of the papers published in this Issue were presented, at least in part, at this Conference, as invited lectures, scientific communications, and posters. The main subjects of the communications were manifold: markers in neurodegenerative disorders, mediators of inflammation, non-coding micro-RNAs, enzymes regulating cell division, measures to mitigate post-operative side effects of cancer treatment, genetic disorders such as cystic fibrosis, skin disorders, multifunctional biomaterials, plant-derived topical preparations, crystal forms of active pharmaceutical ingredients, innovative drug product formulations, quality control of drug substances and drug products, including the chemical stability of drug substances and impurity profile, drug membrane permeability and identification of drug candidates by virtual screening a compound library.

Most chemotherapeutics are effective against the bulk of cancer cells but usually leave cancer stem cells intact. Therapeutics that target cancer stem cells may also provide a cure for cancer. Brown [3] reviewed the two rationales for targeting the retinoic acid receptor γ (RARγ), which is expressed selectively within primitive cells. RARγ is a putative oncogene for several cancers, and prostate cancer cells depend on RARγ for their survival. Antagonizing all RAR-caused necroptosis of prostate cancer stem cell-like cells, and antagonizing RARγ was sufficient to drive necroptosis. The normal prostate epithelium was less sensitive to the RARγ antagonist and pan-RAR antagonist than prostate cancer cells, while fibroblasts and blood mononuclear cells were insensitive. The author concluded that the RARγ-selective antagonist and pan-RAR antagonist are potential therapeutics in cancer treatment.

Bexarotene, a drug that is active against cutaneous T-cell lymphoma, interferes with retinoid X-receptor (RXR)-dependent pathways and might cause serious side effects such as hypothyroidism. Therefore, Jurutka et al. [4] synthesized analogs of this drug, aiming to retain its key function and avoid harmful side effects. The bexarotene alicyclic ring, aliphatic linker, and benzoic acid moiety were substituted with an isochroman ring and nitrogen heterocycles, respectively. The ability of the new analogs to agonize RXR compared to bexarotene was evaluated. Analogs were modeled for RXR binding affinity, and EC_{50} and IC_{50} values were determined for a leukemia cell line. The analogs were tested for their ability to activate liver-X receptor (LXR) and increase the level of triglycerides, as well as LXRE-mediated transcription of brain ApoE expression as a marker of neurodegenerative disorders. While some RXR agonists cross-signaled the retinoic acid receptor (RAR), interestingly, many of the new analogs presented in this paper showed reduced activity against RAR. The authors demonstrated that selective modifications of rexinoids, such as bexarotene, can lead to analogs with increased RXR selectivity, decreased cross-signaling, and improved anti-proliferative characteristics against a leukemia cell line.

To evaluate the correlation between the synergy of an MDM2 inhibitor (siremadlin) combined with a MEK inhibitor (trametinib) in vitro and in vivo, Witkowski et al. [5] evaluated the interaction between these anticancer agents at the level of pharmacokinetics (PK) and pharmacodynamics (PD). The authors examined the cytotoxicity of a siremadlin and trametinib combination against melanoma A375 cells. Calculated drug interaction showed high synergy between siremadlin and trametinib and an increase in the potency of the drug combination. The authors recommended coupling physiologically based PK/PD (PBPK/PD) modeling, with further studies using cancer xenograft models to select the optimal PD interaction parameter to translate in vitro synergy into an animal model.

To identify highly effective therapeutic strategies, Witkowski et al. [6] developed in vitro/in vivo translational methods for synergistic drug combinations. The authors modeled PBPK/PD for siremadlin, trametinib, and a combination of these agents at various dose levels and dosing schedules for an A375 melanoma cell xenograft mouse model. The modeling was based on in vitro absorption, distribution, metabolism, and excretion, and in vivo, PK/PD reported data or estimated by the Simcyp Animal simulator (V21). The developed PBPK/PD models revealed the interactions between siremadlin and trametinib at the PK and PD levels. The interaction at the PK level is presented by an interplay between absorption and tumor disposition levels, whereas the PD interaction is based entirely on the vitro data. From these studies, and based on in vitro and in vivo extrapolation, the synergistic and most efficacious dose levels and dosing schedules for combinations of siremadlin and trametinib in mice were effectively developed.

To extend their previous studies that aimed to provide an optimal therapeutic scheme, Witkowski et al. [7] elaborated on PBPK/PD modeling and clinical trial simulation for siremadlin and trametinib combination for use in melanoma patients. Clinical information was obtained from reported data or the Simcyp simulator. The PBPK/PD models accounted for the interactions between siremadlin and trametinib at the PK and PD levels. PK characteristics were predicted based on animal studies, while PD was based on in vitro cytotoxicity. These data, combined with virtual clinical trials, allowed for the estimation of PK/PD profiles and identified melanoma patients for whom this therapy might be non-inferior to the standard dabrafenib and trametinib combination. The authors demonstrated that PBPK/PD modeling, combined with clinical trial simulation, allows for the design of clinical trials and for predicting the clinical effectiveness of anticancer drug combinations.

Ibuprofen, an aryl propionic acid, is used to treat rheumatoid arthritis and is an over-the-counter drug for use to treat minor pains. Chemically, it is a racemate, and only (S)-ibuprofen shows therapeutic function. The chiral metabolic inversion may cause the accumulation of one of the enantiomers, leading to toxicity. To overcome the side effects of dextrorotary ibuprofen (DXI), Thiruchenthooran et al. [8] designed nanostructured lipid carriers (NLCs) of DXI to be used in anticancer therapy. The formulation was optimized by a two-level factorial design. The spherical shape of the NLCs prolonged DXI release. Moreover, DXI-NLCs were more stable than the parent drug. DXI-NLCs showed in vitro cytotoxicity and anticancer potential against breast cancer cells. Therefore, the authors suggested that DXI-NLCs might become a promising antiproliferative therapy that is especially effective against breast cancer.

Platinum (II) complexes are still used to treat almost half of all cancer patients, and they exhibit anticancer activity by interacting with DNA and inducing programmed cell death. They undergo several bio-transformations and can form reactive transient species, which can complex macromolecules such as DNA. Szefler et al. [9] investigated the interactions of oxaliplatin with vitamin B, as compared to native purines. The authors carried out quantum-chemical simulations, with the set representing atomic orbitals of the platinum atom, using the Polarizable Continuum Model in a water environment. Additionally, time-dependent density functional theory (TD-DFT) was employed to study molecular properties in the electronically excited state. Interactions of vitamin B and oxaliplatin were investigated using UV-VIS spectroscopy. The free-energy values indicated spontaneous reactions with mono-aqua and the preferred di-aqua derivatives of oxaliplatin. The free energy values obtained for vitamin B indicated a lower affinity of oxaliplatin compared with the respective values obtained for guanine, adenine, and cytosine. The exception is the mono-aqua form of vitamin B1 (thiamine) at the MN15/def2-TZVP levels of calculations. An application of atoms in molecules (AIM) theory revealed non-covalent interactions that were present in the studied complexes. The data computed by the authors showed a good agreement with the experimental spectroscopic properties of the complexes.

By interacting with nucleobases of DNA, platinum complexes form mono- and di-aqua products. These products are further complexed with guanine or adenine, leading to the death of cancer cells. The structures of vitamin B resemble the structures of nucleobases. Therefore, Szefler and Czeleń [10] reviewed theoretical and experimental studies of the interactions of vitamins B with Pt (II) complexes, as compared to guanine. Two levels of simulations were implemented at the theoretical level as well as the polarizable continuum model in an aqueous environment. Free-energy values showed spontaneous reactions with mono- and di-aqua derivatives of cisplatin and oxaliplatin. However, interactions with di-aqua derivatives are preferred. The strength of these interactions was also compared. Carboplatin products had the weakest interaction with the studied structures. The authors demonstrated the presence of non-covalent interactions in the complexes. Contrary to expectations, vitamin B formed weaker complexes with the products of hydrolysis of chemotherapeutics compared to nucleobases.

Overexpression of the enzymes that regulate cell division, such as cyclin-dependent kinases (CDK), is a key factor that contributes to carcinogenesis. Czeleń et al. [11] selected an indole derivative, isatin, as a freely available starting point for developing inhibitors for CDKs. Based on computational chemistry, docking, and molecular dynamics, the authors designed a series of potential inhibitors of the CDK2 enzyme, using isatin and benzoyl hydrazine as structure templates to synthesize newly selected analogs. The physicochemical properties of the analogs were well characterized. The authors successfully evaluated some new analogs as potential inhibitors of the CDK2 enzyme.

Metabolic resistance determines the lifetime and, therefore, the applicability of a compound as a drug substance candidate. For this reason, Żołek et al. [12] performed rigid docking supported by molecular dynamics simulation using the known structure of the native CYP3A4 enzyme to predict the metabolic resistance of analogs of 1,25-dihydroxyvitamin D_2 (1,25D2) in vivo. The microsomal cytochrome P450 3A4 (CYP3A4) and mitochondrial cytochrome P450 24A1 (CYP24A1) hydroxylating enzymes both metabolize vitamin D and its analogs. The three-dimensional (3D) structure of the full-length native human CYP3A4 was solved, but the respective structure of the main vitamin D hydroxylating CYP24A1 enzyme remains unknown. Modeling using 3D data from human CYP3A4 provided a rationale for the substantial differences in the metabolic conversion of the side-chain geometric analogs of 1,25D2. The calculated free enthalpy of the binding of an analog of 1,25D2 to CYP3A4 agreed with the experimentally observed conversion of the analog by CYP24A1. This way, the authors proved that the metabolic resistance of an analog of 1,25D2 against the main vitamin D hydroxylating enzyme CYP24A1 can be predicted and explained by the binding strength of the analog to the known 3D structure of the CYP3A4 enzyme.

Invasive breast cancer therapy requires the development and selection of appropriate measures to mitigate post-operative side effects. Pawlicka et al. [13] evaluated the safety and potency of genistein (GE) using breast cancer MCF-7 cells and BJ skin fibroblasts. In a concentration-dependent manner, genistein affected both healthy dermal BJ fibroblasts and MCF-7 cells. Genistein at lower concentrations of 10 and 20 µM increased the abundance of dermal fibroblasts. However, genistein at a higher concentration, above 50 µM, was detrimental to fibroblasts at longer exposure times. Genistein shows high potential regarding the treatment of skin injuries, wounds, and surgical scars in women during and after breast cancer treatment. Even so, the authors recommended the need to be very cautious in selecting a concentration of isoflavonoids for use in treatment.

Given that isoflavonoids, such as genistein (GE), are effective antioxidants with antitumor activity, Stolarczyk et al. [14] characterized oxidation products from structurally related thiogenistein (TGE), both in solution and on the 2D surface of the Au electrode as a self-assembling TGE monolayer, using electrospray ionization mass spectrometry and Fourier-transform infrared spectroscopy. Density functional theory was used to support the experimental results for the estimation of the antioxidant activity of TGE, as well as for the molecular modeling of oxidation products. TGE showed high cytotoxic activity against human breast cancer cells and neutralized the production of lipopolysaccharide (LPS)-induced reactive oxygen species (ROS) more efficiently than GE. TGE also exhibited (2,2′-azino-bis-3-(ethylbenzothiazoline-6-sulphonic acid (ABTS) radical scavenging ability. TGE redox properties were related to its pharmacological activities. Most importantly, the authors demonstrated that the cytotoxic activity of TGE against human breast cancer cells was almost twice as high as that of GE.

Human neutrophil elastase (HNE), a serine protease of the chymotrypsin family, can hydrolyze extracellular matrix proteins. Therefore, HNE is a major mediator of inflammation and has become a therapeutic target for small-molecule inhibitors. Donarska et al. [15] designed and synthesized a series of thiazoles, based on 3,3-diethylazetidine-2,4-dione, as new HNE inhibitors in the nanomolar range. The molecular docking study revealed a good correlation between the binding energies, indicating that the inhibition was largely dependent on the ligand alignment in the binding cavity. Most of the active compounds were stable and had antiproliferative activity against human leukemia, lung carcinoma, breast adenocarcinoma, and urinary bladder carcinoma cells, with IC_{50} values in the micromolar range. Additionally, some of the compounds induced growth arrest at the G2/M phase of the cell cycle and apoptosis via caspase-3 activation. The authors suggested that these new compounds might be effective against cancer and other diseases in which immunoreactive HNE is released.

Non-coding micro-RNAs (miRNAs) play an important role in the response of cancer cells to drug treatment by regulating the protein expression responsible for cell growth and proliferation. Gleba et al. [16] investigated the correlation between selected miRNAs and the proteins that they regulate in response to the active endogenous forms of vitamin D_3, calcitriol, and tacalcitol by using several leukemia and lymphoma cell lines. Five miRNAs were selected as well as the respective regulated proteins. The authors concluded that the level of selected miRNAs correlated well with the levels of the proteins. They also identified some miRNAs that were most likely responsible for the anticancer activity of the most active forms of vitamin D_3 in human leukemias and lymphomas.

4,6,4′-Trimethylangelicin (TMA) is a promising agent for the treatment of cystic fibrosis, a genetic disorder that is caused by mutations in the gene for transmembrane conductance regulator (CFTR) protein. However, TMA shows serious disadvantages, such as poor solubility, phototoxicity, and mutagenicity. To overcome these side effects, Vaccarin et al. [17] designed and synthesized a library of TMA analogs. The authors showed that bulky aromatic substituents at C-4 of TMA impaired DNA intercalation and prevented the photoreaction of the lactone or furan double bond of the furocoumarin scaffold with the nucleic acid. Based on this finding, the authors obtained 4-phenyl- and 6-phenyl- analogs showing an F508del CFTR rescue ability. Interdisciplinary studies confirmed that the analogs lacked side effects. Pharmacokinetic studies revealed a favorable profile, especially after the incorporation of the analogs into lipid formulations. The authors concluded that the analogs are good candidates for novel CFTR correctors based on the angelicin scaffold.

The increasing number of orthopaedical surgeries and the need to replace bone tissue have stimulated interest in developing multifunctional biomaterials to be used in bone diseases. Hence, Pajor et al. [18] searched for synthetic materials as carriers for delivering drugs, such as antibiotics, to combat surgical site infections. The authors studied the physicochemical properties and biological activity of different types of porous granules containing silver or gallium ions. Hydroxyapatite powders were doped with Ga^{3+} or Ag^+. Then, the powders were used to fabricate ceramic micro granules and alginate/hydroxyapatite composite granules (AgT and GaT). The AgT and GaT granules showed high porosity and a large specific surface, whereas the micro granules contained very fine and numerous micropores. As expected, the granules released the incorporated ions slowly. All the granules except AgT were found to be non-cytotoxic. The authors subjected the granules to various antibacterial tests against key bacterial strains, which revealed that the new material had high antibacterial potency.

Skin disorders of different etiology, such as dermatitis, atopic dermatitis, eczema, psoriasis, wounds, burns, and others, are widespread in the population. For this reason, Melnyk et al. [19] first reviewed data on the plant-derived topical preparations used in Poland and Ukraine and then indicated future studies based on a recent understanding of the etiology of skin diseases. In severe cases, skin diseases require the topical application of antibiotics, steroids, and calcineurin inhibitors. Milder symptoms are treated with other medications, dietary supplements, and cosmetic products of plant origin. These skin care products were applied in various pharmaceutical formulations, including raw infusions, tinctures, creams, and ointments. The mechanisms of the beneficial effects of these treatments are often unclear. Recent developments in the role of the skin microbiota in maintaining skin homeostasis have helped researchers to understand the function of topically applied products of plant origin. As the authors concluded, knowledge of the medications' interaction with the skin microbial ecosystem network remains to be upgraded.

The existence of multiple crystal forms of an active pharmaceutical ingredient (API) is a serious challenge to the selection of the most suitable solid for drug development. It is also important for intellectual property protection. Braga et al. [20] reviewed the scientific, commercial, and ethical importance of investigating the crystalline forms (polymorphs, hydrates, and co-crystals) of active pharmaceutical ingredients. The physicochemical properties, such as solubility, dissolution rate, thermal stability, processability, therapeutic efficacy, etc., of the solid active pharmaceutical ingredient, usually depend on the crystal form and influence the efficacy of the drug product. The authors suggest that in addition to improving the understanding and control of the crystal form of the API, the study of hydrates and co-crystals may afford the alternative, often improved, innovative properties of the drug. The author provided examples of handling multiple crystal forms and showed how to exploit the intentional preparation of new crystalline forms as innovation.

The quality control of drug products and the appropriate physicochemical properties of drug substances, including chemical stability and the right impurity profile, were of interest to several contributors. Since the worldwide crisis of contamination of medicinal products with nitrosamine, effective trace-level analysis of these contaminants is of continuing importance to the quality control of medicines. Witkowska et al. [21] developed a novel GC-MS method with electron ionization and microextraction for the simultaneous determination of main carcinogenic nitrosamines in a range of representative active pharmaceutical ingredients. The method demonstrated good linearity and wide limits of detection. The trace-level GC-MS method was specific, accurate, and precise. Nitrosamines were not detected in most of the APIs screened. However, in some of them, one nitrosamine was detected at the limit level of quantitation. Therefore, the authors stated that the novel method can be effectively applied for tracing nitrosamines in a range of APIs and can be used for the routine quality control of APIs to ensure the safety and high quality of medicinal products.

The chemical stability of drug substances is one of the major challenges in drug development. In this respect, Żyżyńska-Granica et al. [22] evaluated the stability of two biologically active peptide drug candidates that combine both opioid and neurotensin pharmacophores in one hybrid compound. The hybrids are structurally similar and differ only in an amino acid at position 9 of the peptide chain. Isoleucine in the parent compound was replaced with its isomer, *tert*-leucine. The study aimed to assess and compare the chemical stability of the hybrids. Contrary to the change in biological activity, substituting tert-leucine with isoleucine did not affect chemical stability. Neither alkaline nor acidic hydrolysis and oxidative degradation resulted in stability differences between the hybrids. However, the number of degradation products under acidic conditions increased. In summary, the authors showed that the introduced modification reduced the compound's half-life.

Gurba-Bryśkiewicz et al. [23] carried out a complete Analytical Quality by Design (AQbD) approach, including screening, optimization, and validation, for the development of a new method for the quantitative determination of the complete impurity profile of an innovative pharmaceutical substance with structure-based pre-development. The authors demonstrated a novel approach to the development of an ultra HPLC method using the AQbD with the Design of an Experiment (DOE). The method was applied for the quantitative determination of the impurity profile of a JAK/ROCK inhibitor (CPL409116) during preclinical drug discovery. The critical method parameters (CMPs), including the stationary phase of the columns, pH of the aqueous mobile phase, and composition of the organic mobile phase, were tested extensively. The authors selected the resolution between the peaks and peak symmetry of analytes as the critical method attributes (CMAs). The influence of various levels of CMPs on the CMAs was evaluated based on a full fractional design. The robustness tests were established from the knowledge space of the screening and completed by fractional factorial design. Method-operable design region (MODR) was created. Monte-Carlo simulations provided the probability of complying with the specifications. The authors documented that the optimized method is specific, linear, precise, and robust.

In the field of innovative agents used for biomedical applications, glutaraldehyde is one of the most popular synthetic cross-linkers for drug product formulations. However, the unreacted cross-linker can be released from the drug product and cause unexpected side effects. To overcome this difficulty, Wegrzynowska-Drzymalska et al. [24] obtained dialdehyde starch nanocrystals (NDASs) as an alternative to the commonly used cross-linking agents. Then, NDASs were used for chemical cross-linking of native chitosan (CS), gelatin (Gel), and a mixture of these two biopolymers (CS-Gel). The thin films of the materials were characterized by Attenuated Total Reflectance FTIR, Scanning Electron Microscopy (SEM), and X-ray diffraction (XRD), and toxicity was tested by the Microtox test. Thermal and mechanical properties were determined by thermal gravimetric analysis (TGA) and tensile testing. Moreover, all cross-linked biopolymers were characterized by hydrophilic character, swelling ability, and protein absorption. The authors revealed that dialdehyde starch nanocrystals form beneficial plant-derived cross-linking agents, leading to cross-linked biopolymers that are advantageous to biomedical applications.

Membrane permeability is important in the development of therapeutic peptides of large molecular size. Therefore, Mazzanti et al. [25] computed the permeability coefficient of a cyclic decapeptide by comparing two physical models. The authors applied the inhomogeneous solubility–diffusion model (ISDM) requiring umbrella–sampling simulations. A chemical kinetics model was also used including multiple unconstrained simulations (MSM). The authors compared the accuracy of the approaches and concluded that the MSM method requires much more computer resources than the ISDM method while giving comparable predictions of the membrane permeability of peptides.

Papain-like protease (PL^{pro}) is a significant target for anti-COVID-19 drugs. Therefore, Wu et al. [26] virtually screened a large library of compounds against this protein and identified several drug candidates. The binding energy of the candidates was estimated as higher than proposed in previous studies. By analyzing the docking results, the authors demonstrated that the critical interactions between the candidates and PL^{pro} revealed by the computations were consistent with the biological experiments. The predicted binding energies among selected candidates and absorption, distribution, metabolism, and excretion (ADME) characteristics were consistent with their IC$_{50}$ values.

The papers within this Special Issue cover a wide range of topics regarding the development of new drug substances, new drug carriers, and drug products. As conceived, the topics in the Issue cover the identification of the molecular target of new drugs, docking supported by molecular dynamics simulation, quantum-chemical simulations, studies of protein–drug interactions, modeling and optimization of the structure of small molecules with drug-like activity, determination of the functional profile, pre-formulation studies, new pharmaceutical formulations, preclinical development to the design, and prediction of the efficacy of agents in clinical trials. Elucidation of the molecular mechanism of action is now an integral prerequisite for transforming an active substance into a drug candidate.

Author Contributions: A.K. wrote the original draft that was completed and revised by G.B. and E.K. All authors have read and agreed to the published version of the manuscript.

Conflicts of Interest: The authors declare no conflict of interest.

References

1. Kutner, A.; Brown, G.; Kallay, E. (Eds.) International Journal of Molecular Sciences, Special Issue "Novel Strategies in the Development of New Therapies, Drug Substances, and Drug Carriers". Available online: https://www.mdpi.com/journal/ijms/special_issues/ACCORD2021 (accessed on 3 March 2023).
2. Kutner, A.; Brown, G.; Kallay, E. (Eds.) *Novel Strategies in the Development of New Therapies, Drug Substances, and Drug Carriers*; MDPI: Basel, Switzerland, 2022; Available online: https://www.mdpi.com/books/book/5818-novel-strategies-in-the-development-of-new-therapies-drug-substances-and-drug-carriers (accessed on 3 March 2023).
3. Brown, G. Antagonizing RARγ Drives Necroptosis of Cancer Stem Cells. *Int. J. Mol. Sci.* **2022**, *23*, 4814. [CrossRef] [PubMed]
4. Jurutka, P.W.; di Martino, O.; Reshi, S.; Mallick, S.; Sausedo, M.A.; Moen, G.A.; Lee, I.J.; Ivan, D.J.; Krall, T.D.; Peoples, S.J.; et al. An Isochroman Analog of CD3254 and Allyl-, Isochroman-Analogs of NEt-TMN Prove to Be More Potent Retinoid-X-Receptor (RXR) Selective Agonists Than Bexarotene. *Int. J. Mol. Sci.* **2022**, *23*, 16213. [CrossRef] [PubMed]
5. Witkowski, J.; Polak, S.; Rogulski, Z.; Pawelec, D. In Vitro/In Vivo Translation of Synergistic Combination of MDM2 and MEK Inhibitors in Melanoma Using PBPK/PD Modelling: Part I. *Int. J. Mol. Sci.* **2022**, *23*, 12984. [CrossRef] [PubMed]
6. Witkowski, J.; Polak, S.; Rogulski, Z.; Pawelec, D. In Vitro/In Vivo Translation of Synergistic Combination of MDM2 and MEK Inhibitors in Melanoma Using PBPK/PD Modelling: Part II. *Int. J. Mol. Sci.* **2022**, *23*, 11939. [CrossRef] [PubMed]
7. Witkowski, J.; Polak, S.; Pawelec, D.; Rogulski, Z. In Vitro/In Vivo Translation of Synergistic Combination of MDM2 and MEK Inhibitors in Melanoma Using PBPK/PD Modelling: Part III. *Int. J. Mol. Sci.* **2023**, *24*, 2239. [CrossRef]
8. Thiruchenthooran, V.; Świtalska, M.; Bonilla, L.; Espina, M.; García, M.L.; Wietrzyk, J.; Sánchez-López, E.; Gliszczyńska, A. Novel Strategies against Cancer: Dexibuprofen-Loaded Nanostructured Lipid Carriers. *Int. J. Mol. Sci.* **2022**, *23*, 11310. [CrossRef]
9. Szefler, B.; Czeleń, P.; Wojtkowiak, K.; Jezierska, A. Affinities to Oxaliplatin: Vitamins from B Group vs. Nucleobases. *Int. J. Mol. Sci.* **2022**, *23*, 10567. [CrossRef]
10. Szefler, B.; Czeleń, P. Will the Interactions of Some Platinum (II)-Based Drugs with B-Vitamins Reduce Their Therapeutic Effect on Cancer Patients? Comparison of Chemotherapeutic Agents such as Cisplatin, Carboplatin, and Oxaliplatin—A Review. *Int. J. Mol. Sci.* **2023**, *24*, 1548. [CrossRef]
11. Czeleń, P.; Skotnicka, A.; Szefler, B. Designing and Synthesis of New Isatin Derivatives as Potential CDK2 Inhibitors. *Int. J. Mol. Sci.* **2022**, *23*, 046. [CrossRef]
12. Żołek, T.; Yasuda, K.; Brown, G.; Sakaki, T.; Kutner, A. In Silico Prediction of the Metabolic Resistance of Vitamin D Analogs against CYP3A4 Metabolizing Enzyme. *Int. J. Mol. Sci.* **2022**, *23*, 7845. [CrossRef]
13. Pawlicka, M.A.; Zmorzyński, S.; Popek-Marciniec, S.; Filip, A.A. The Effects of Genistein at Different Concentrations on *MCF-7* Breast Cancer Cells and *BJ* Dermal Fibroblasts. *Int. J. Mol. Sci.* **2022**, *23*, 12360. [CrossRef] [PubMed]
14. Stolarczyk, E.U.; Strzempek, W.; Łaszcz, M.; Leś, A.; Menaszek, E.; Stolarczyk, K. Thiogenistein—Antioxidant Chemistry, Antitumor Activity, and Structure Elucidation of New Oxidation Products. *Int. J. Mol. Sci.* **2022**, *23*, 7816. [CrossRef] [PubMed]
15. Donarska, B.; Świtalska, M.; Wietrzyk, J.; Płaziński, W.; Mizerska-Kowalska, M.; Zdzisińska, B.; Łączkowski, K.Z. Discovery of New 3,3-Diethylazetidine-2,4-dione Based Thiazoles as Nanomolar Human Neutrophil Elastase Inhibitors with Broad-Spectrum Antiproliferative Activity. *Int. J. Mol. Sci.* **2022**, *23*, 7566. [CrossRef]

16. Gleba, J.J.; Kłopotowska, D.; Banach, J.; Mielko, K.A.; Turlej, E.; Maciejewska, M.; Kutner, A.; Wietrzyk, J. Micro-RNAs in Response to Active Forms of Vitamin D_3 in Human Leukemia and Lymphoma Cells. *Int. J. Mol. Sci.* **2022**, *23*, 5019. [CrossRef]
17. Vaccarin, C.; Gabbia, D.; Franceschinis, E.; De Martin, S.; Roverso, M.; Bogialli, S.; Sacchetti, G.; Tupini, C.; Lampronti, I.; Gambari, R.; et al. Improved Trimethylangelicin Analogs for Cystic Fibrosis: Design, Synthesis and Preliminary Screening. *Int. J. Mol. Sci.* **2022**, *23*, 11528. [CrossRef]
18. Pajor, K.; Michalicha, A.; Belcarz, A.; Pajchel, L.; Zgadzaj, A.; Wojas, F.; Kolmas, J. Antibacterial and Cytotoxicity Evaluation of New Hydroxyapatite-Based Granules Containing Silver or Gallium Ions with Potential Use as Bone Substitutes. *Int. J. Mol. Sci.* **2022**, *23*, 7102. [CrossRef]
19. Melnyk, N.; Vlasova, I.; Skowrońska, W.; Bazylko, A.; Piwowarski, J.P.; Granica, S. Current Knowledge on Interactions of Plant Materials Traditionally Used in Skin Diseases in Poland and Ukraine with Human Skin Microbiota. *Int. J. Mol. Sci.* **2022**, *23*, 9644. [CrossRef] [PubMed]
20. Braga, D.; Casali, L.; Grepioni, F. The Relevance of Crystal Forms in the Pharmaceutical Field: Sword of Damocles or Innovation Tools? *Int. J. Mol. Sci.* **2022**, *23*, 9013. [CrossRef]
21. Witkowska, A.B.; Giebułtowicz, J.; Dąbrowska, M.; Stolarczyk, E.U. Development of a Sensitive Screening Method for Simultaneous Determination of Nine Genotoxic Nitrosamines in Active Pharmaceutical Ingredients by GC-MS. *Int. J. Mol. Sci.* **2022**, *23*, 12125. [CrossRef]
22. Żyżyńska-Granica, B.; Mollica, A.; Stefanucci, A.; Granica, S.; Kleczkowska, P. Comparative Study of Chemical Stability of a PK20 Opioid–Neurotensin Hybrid Peptide and Its Analogue [Ile^9]PK20—The Effect of Isomerism of a Single Amino Acid. *Int. J. Mol. Sci.* **2022**, *23*, 10839. [CrossRef]
23. Gurba-Bryśkiewicz, L.; Dawid, U.; Smuga, D.A.; Maruszak, W.; Delis, M.; Szymczak, K.; Stypik, B.; Moroz, A.; Błocka, A.; Mroczkiewicz, M.; et al. Implementation of QbD Approach to the Development of Chromatographic Methods for the Determination of Complete Impurity Profile of Substance on the Preclinical and Clinical Step of Drug Discovery Studies. *Int. J. Mol. Sci.* **2022**, *23*, 10720. [CrossRef]
24. Wegrzynowska-Drzymalska, K.; Mylkie, K.; Nowak, P.; Mlynarczyk, D.T.; Chelminiak-Dudkiewicz, D.; Kaczmarek, H.; Goslinski, T.; Ziegler-Borowska, M. Dialdehyde Starch Nanocrystals as a Novel Cross-Linker for Biomaterials Able to Interact with Human Serum Proteins. *Int. J. Mol. Sci.* **2022**, *23*, 7652. [CrossRef] [PubMed]
25. Mazzanti, L.; Ha-Duong, T. Understanding Passive Membrane Permeation of Peptides: Physical Models and Sampling Methods Compared. *Int. J. Mol. Sci.* **2023**, *24*, 5021. [CrossRef] [PubMed]
26. Wu, Y.; Pegan, S.D.; Crich, D.; Lou, L.; Nicole Mullininx, L.; Starling, E.B.; Booth, C.; Chishom, A.E.; Chang, K.Y.; Xie, Z.-R. Identifying Drug Candidates for COVID-19 with Large-Scale Drug Screening. *Int. J. Mol. Sci.* **2023**, *24*, 4397. [CrossRef] [PubMed]

International Journal of
Molecular Sciences

MDPI

Review

Antagonizing RARγ Drives Necroptosis of Cancer Stem Cells

Geoffrey Brown

School of Biomedical Sciences, Institute of Clinical Sciences, College of Medical and Dental Sciences, University of Birmingham, Edgbaston, Birmingham B15 2TT, UK; g.brown@bham.ac.uk; Tel.: +44-(0)121-414-4082

Abstract: There is a need for agents that eliminate cancer stem cells, which sustain cancer and are also largely responsible for disease relapse and metastasis. Conventional chemotherapeutics and radiotherapy are often highly effective against the bulk of cancer cells, which are proliferating, but spare cancer stem cells. Therapeutics that target cancer stem cells may also provide a *bona fide* cure for cancer. There are two rationales for targeting the retinoic acid receptor (RAR)γ. First, RARγ is expressed selectively within primitive cells. Second, RARγ is a putative oncogene for a number of human cancers, including cases of acute myeloid leukemia, cholangiocarcinoma, and colorectal, renal and hepatocellular carcinomas. Prostate cancer cells depend on active RARγ for their survival. Antagonizing all RARs caused necroptosis of prostate and breast cancer stem cell-like cells, and the cancer stem cells that gave rise to neurospheres from pediatric patients' primitive neuroectodermal tumors and an astrocytoma. As tested for prostate cancer, antagonizing RARγ was sufficient to drive necroptosis. Achieving cancer-selectively is a longstanding paradigm for developing new treatments. The normal prostate epithelium was less sensitive to the RARγ antagonist and pan-RAR antagonist than prostate cancer cells, and fibroblasts and blood mononuclear cells were insensitive. The RARγ antagonist and pan-RAR antagonist are promising new cancer therapeutics.

Keywords: cancer stem cells; oncogenes; retinoic acid receptors; prostate cancer; necroptosis

Citation: Brown, G. Antagonizing RARγ Drives Necroptosis of Cancer Stem Cells. *Int. J. Mol. Sci.* **2022**, *23*, 4814. https://doi.org/10.3390/ijms23094814

Academic Editor: Gianpaolo Papaccio

Received: 15 April 2022
Accepted: 25 April 2022
Published: 27 April 2022

Publisher's Note: MDPI stays neutral with regard to jurisdictional claims in published maps and institutional affiliations.

1. Introduction

An often-asked question is will there ever be a cure for cancer? Cancer is a group of diseases and moreover, and for every cancer, variant clones evolve in a Darwinian manner as the disease progresses, including clones that are resistant to chemotherapeutics [1]. Potentially, chemotherapy provides a selective pressure that leads to the expansion of drug-resistant variants [2]. These matters lie at the heart of the failure to cure many cancers. The challenge is to either sustain control of disease for the rest of a patient's life and/or to increase the chances of achieving a cure at disease presentation and the onset of treatment. Regarding both of these options, the cells that sustain a cancer are cancer stem cells (CSCs) which produce the hierarchy of cells for cancer. Their frequency within cancer varies from exceedingly rare, as for human acute myeloblastic leukemia (AML) [3], to up to 25%, as for human melanoma [4]. For many solid cancers, the nature and frequency of CSCs are still uncertain. CSCs are also largely responsible for disease relapse and metastasis, and the treatment of metastasized cancers has not advanced significantly and often they are beyond successful treatment. To increase the chances of providing a *bona fide* cure for cancer there is the need to develop a strategy to control and/or eliminate CSCs.

The failure of conventional treatments to eliminate CSCs is well illustrated by studies of chronic myeloid leukemia (CML). This leukemia arises from the transformation of a hematopoietic stem cell (HSC) [5], and it has been known since 1999 that CML leukemia stem cells (LSCs) are insensitive to high doses of chemotherapeutic agents that target the cell cycle [6]. A subpopulation of primitive CML leukemia cells that were highly quiescent LSCs was isolated from patients with chronic-phase CML and were insensitive to high doses of chemotherapeutic agents that are efficacious against the dividing leukemia cells. Additionally, tyrosine kinase inhibitors, such as imatinib, are used to treat CML and many different types of cancer. In vitro, CML LSCs were observed to be insensitive to imatinib, and, are presumed to be insensitive in vivo [7,8]. The challenge to curing CML is one of eliminating the LSCs that are spared by current treatments. CML LSCs can also cause disease relapse even after allogeneic transplantation [9].

There are substantial efforts to develop the means to eradicate CSCs [10,11]. The approaches include chimeric antigen receptor T cells (CAR T cells), antibodies, and small molecules that target CSCs. For example, the premise in CAR T cell targeting is that the molecules recognized are over-expressed on cancer cells and at a low level on normal cells. As yet, successes from the use of CAR T cells are rare, because surface antigen expression by cancer cells is highly heterogeneous and the need to identify a proper target for each patient, including even for patients with the same type of cancer [12]. Heterogeneity of antigen expression by cancer cells and whether the antigen is expressed by normal cells are also concerns regarding antibody targeting of CSCs. A general approach to developing anticancer drugs is to interfere with the intracellular events that regulate the survival of cancer cells. Moreover, it is important to eliminate CSCs without too much damage to normal stem cells. This review focuses on targeting RARγ to eliminate CSCs by the use of an antagonist to switch-off RARγ.

2. Why Target RARγ to Eliminate CSCs?

The three main isotypes of RAR are RARα, RARβ, and RARγ. They form dimers with members of the retinoid X receptor subfamily which bind to response elements to act as ligand-regulated transcription factors. The ligand for RARs is all-*trans* retinoic acid (ATRA), which is the major bioactive metabolite of retinol or vitamin A. Disruption of ATRA signaling is thought to play a role in the etiology of many cancers. The list includes leukemias, breast cancer, glioblastoma, head and neck cancer, liver cancer, lung cancer, ovarian cancer, neuroblastoma, pancreatic cancer, prostate cancer (PCa), renal cell cancer, and skin cancer [13]. ATRA is a potent pro-differentiation, anti-proliferation, pro-apoptosis agent and its therapeutic use has provided a cure for acute promyelocytic leukemia (APL) [14]. APL accounts for 5–15% of cases of AML and is classified as AML M3 under the French-American-British (FAB) system. The hallmark signature of APL is the t(15;17)(q24;q21) translocation that fuses the *PML* and *RARA* genes leading to the expression of the oncogenic PML-RARα protein. There is evidence to support the view that APL arises in an HSC from the presence of the PML-RARα protein in patients' LSCs [15]. ATRA targeting of both PML-RARα and wild type RARα results in the dissociation of transcriptional corepressors, proteolytic degradation of PML-RARα and wild type RARα, and differentiation and apoptosis of APL cells. The therapeutic use of a combination of ATRA and arsenic trioxide has led to long-lasting disease remission, and in this regard, greater demethylation of genes may be important [16]. Unfortunately, the success of ATRA in providing differentiation therapy for APL has not translated to other cancers.

2.1. The Role of RARγ within Stem Cells

There are three rationales for targeting RARγ to eliminate CSCs. RARγ is selectively expressed within stem cells and their immediate offspring, it plays a crucial role in the survival of these cells, and is an oncogene for a number of cancers. The selective expression of RARγ and its role are well described for HSC development and RARγ and RARα have discrete physiological roles. Active RARγ promotes HSC survival and self-renewal, whereas active RARα promotes differentiation, and this balance is critical to the proper conduct of hematopoiesis. RARγ expression is restricted to primitive hematopoietic cells and ATRA-activated RARγ supports HSC self-renewal as knockout mice had a reduced number of HSCs [17]. It is well established that ATRA-activated RARα drives the terminal maturation of committed granulocyte/monocyte progenitors [18].

Studies of the early development of zebra fish embryos have provided further support to the fact that RARγ plays a crucial role in stem cells. Zebra fish embryos were treated in vitro with a RARγ-selective agonist. RARγ activation blocked stem cell development, preventing fin, bone and neural ganglia development. Stem cell numbers were unaffected because wash-out or the use of a RARγ antagonist to reverse the action of the RARγ agonist restored fin formation. In this case and in the absence of ATRA, RARγ functions to maintain stem cells [19]. Studies of RARγ knockout embryonic stem (ES) cells have also revealed the importance of RARγ to stem cells. For ES cells, ATRA-regulated transcripts are dependent on a functional RARγ and, therefore, RARγ is essential for transcriptional activation in ES cells. The studies also revealed that RARγ is essential for chromatin remodeling and DNA epigenetic marks [20]. It is also important to bear in mind that RAR/retinoid X receptor dimers bind to genomic regions that are characterized by the binding of pluripotency-associated factors [21].

2.2. RARγ Is an Oncogene for a Number of Cancers

The need is to spare normal stem cells whilst eliminating CSCs. In this regard, RARγ is an oncogene for a number of cancers. As mentioned above, APL is characterized by *RARA*-associated gene rearrangements. There is dysregulation of the RARG gene in APL-like leukemia. Fusions have been identified between the *RARG* gene and the genes for PML, CPSF6 (a subunit of the RNA binding protein cleavage factor 1), NPM1 (nucleophosmin), and NuP98 (nucleoporin) [22–24]. These patients did not respond to ATRA treatment. For an APL-like patient that lacked a *RARA* rearrangement, a reciprocal fusion involving *RARG* and *HNRPC3* (heterogeneous nuclear ribonucleoprotein C) has been reported. The patient was treated with ATRA and arsenious acid, arsenious acid was withdrawn because a *RARA* rearrangement was lacking, and there was no response to ATRA [25]. The potential impact of RARγ expression on disease is illustrated by a patient with relapsed AML who died from rapid disease progression after ATRA treatment. An increase in the level of nuclear RARγ was observed when primary cells from this patient were treated with ATRA in vitro, which may explain the rapid disease progression in response to ATRA therapy [26].

From quantitative PCR and Western blotting studies, RARγ mRNA and protein are frequently overexpressed in human colorectal cancer (CRC) tissue versus the surrounding non-tumorous colorectal tissue. Similarly, expression of RARγ is increased in the CRC cell lines HT29, HCT116, RKO and SW480 as compared with the HCoEpiC normal colonic epithelial cells. For the HT29, HCT116, and RKO cell lines, knockdown of RARγ enhanced their sensitivity to 5-fluorouracil, oxaliplatin, and vincristine. This was found to be related to a decreased expression of the multi-drug resistance 1 protein. RARγ is, therefore, a potential therapeutic target for chemotherapeutic resistance CRC [27]. Similarly, overexpression of RARγ in the bile duct carcinoma cholangiocarcinoma (CCA) is associated with a poor prognosis and resistance to 5-fluorouracil. Knockdown of RARγ expression in the three human CCA cell lines QBC939, SK-ChA-1, and MZ-ChA-1, by siRARγ, resulted in the suppression of cell proliferation. Colony formation and xenograft tumor growth in nude mice were reduced in the case of QBC939 cells that were stably transfected [28]. RARγ appears, therefore, to be important for CCA tumorigenesis. The majority of primary tissue samples from patients with hepatocellular carcinoma (HCC) overexpress RARγ, as do HCC cell lines. For the HCC cell line HepG2, colony formation and xenograft engraftment were promoted by overexpression of RARγ [29]. From the use of qPCR and bioinformatics analyses, around 50% of tissues from patients with clear cell renal cell carcinoma were observed to overexpress RARγ [30].

2.3. PCa Cells Are Dependent on Active RARγ for Their Survival

Patients' PCa cells and normal prostate epithelium express RARα and RARγ. The key finding that supports the view that RARγ is an oncogene in PCa is that PCa cells depend on activated RARγ for their survival as follows. Patients' PCa cells survive and grow in an abnormally low level of ATRA (Figure 1) because the level in patients' tissue was observed to be very close to the limit of detection, at around 1 ng/gram tissue. The level in the surrounding normal tissue and benign prostate hyperplasia is up to 8 times higher [31]. It has been reported that RARγ has a constitutive closed helix 12 conformation that interferes with corepressor recruitment and that there is a level of target gene activation in the absence of ATRA [32]. Even so, transactivation studies have shown that 0.24 nM ATRA (equivalent to 1 ng/gram tissue) transactivates RARγ, whereas RARα transactivation requires a much higher concentration of 19.3 nM ATRA [33]. The low level of ATRA in patients' PCa tissue is important because it is likely that just RARγ is transactivated in PCa cells which are, therefore, reliant on active RARγ for survival and proliferation.

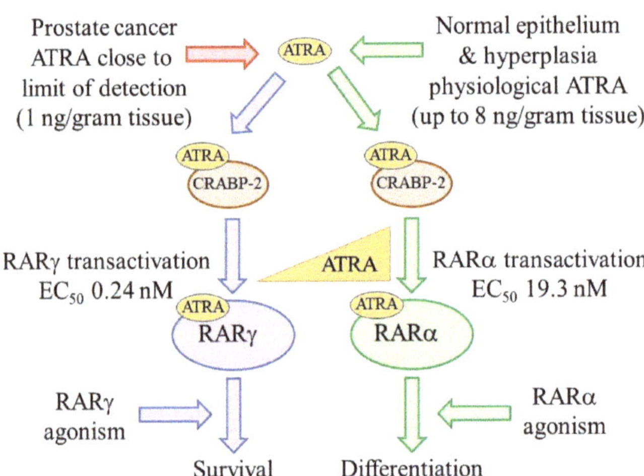

Figure 1. Patients PCa cells survive and grow in an abnormally low level of ATRA. The low level of ATRA within PCa tissue is important because 0.24 nM ATRA transactivates RARγ, whereas RARα transactivation requires a much higher concentration of 19.3 nM ATRA. Patients' PCa cells are, therefore, reliant on RARγ transactivation for survival and proliferation. From various studies, active RARγ is pro-survival whereas active RARα is pro-differentiation.

There are a number of possibilities regarding the abnormally low level of ATRA within PCa tissue. There is decreased expression of the aldehyde dehydrogenase isoforms that convert retinol to ATRA within malignant prostate cancer tissue [34] and the PCa cell line LNCaP [35]. Dietary vitamin A is transported in the bloodstream as hydrophobic retinol and solely by the retinol-binding protein (RBP). The integral membrane receptor STR6 recognizes RBP-retinol and mediates cellular retinol uptake by triggering the release and internalization of retinol [36]. Induction of the expression of STR6 by retinol and ATRA has been shown to be defective with PC3 cells, compared to normal prostate epithelium. This might reduce the efficiency by which PCa cells can sequester retinol from the environment. Retinol is also metabolized by PCa cells to retinyl esters for storage by lecithin:retinol acyltransferase. The activity of this enzyme is required for the uptake of an appropriate amount of retinol by cells and the level of expression of mRNA was found to be reduced in PC3 cells [37].

3. Agonists and Antagonists of RARs

Antagonists of RARs were developed in the late 1990s [38], and the synthetic retinoids developed included antagonists, and agonists, that are highly selective for RAR subtypes. A pharmacological level of ATRA (10^{-6} M) is often used to reveal activity against carcinoma cell lines. Hence, there was interest in whether RAR antagonists are more effective than ATRA against carcinoma cells. As determined by the [^{3}H]-ATRA displacement method [39], the equilibrium binding affinities for the new retinoids for their receptor are in the nM range [40], Table 1. They were used at down to nM levels to test their biological activities. Carcinoma cell lines had been adapted to grow long-term in a serum-free medium to avoid any positive or negative effects of the ATRA that is present in serum.

Table 1. Binding affinities (ED$_{50}$ in nM) of selected retinoids against RARs. Nuclear extracts were prepared from baculovirus infected Sf21 insect cells engineered to express either human RARα, -β or -γ. The equilibrium binding affinities of each retinoid analog were estimated by the [^{3}H]-ATRA displacement method. ND, not conducted.

Retinoids	RARα	RARβ	RARγ	Classification
RAR Agonists—Equilibrium Binding Affinities in nM				
ATRA	ND	ND	ND	RARαβγ
AGN195183	20.1	>5000	>5000	RARα
AGN190168	>1000	14.2	135	RARβγ
AGN205327	3700	734	32	RARγ
RAR Antagonists—Equilibrium Binding Affinities in nM				
AGN194310	4.3	5	2	RARαβγ
AGN196996	3.9	4036	>10,000	RARα
AGN194431	300	6	20	RARβγ
AGN205728	2400	4248	3	RARγ

The pan-RAR and RARγ antagonists are of particular interest regarding the killing of CSCs and their structures, alongside that of a RARα antagonist which does not kill CSCs, are shown in Figure 2. Agonists are shown for comparison.

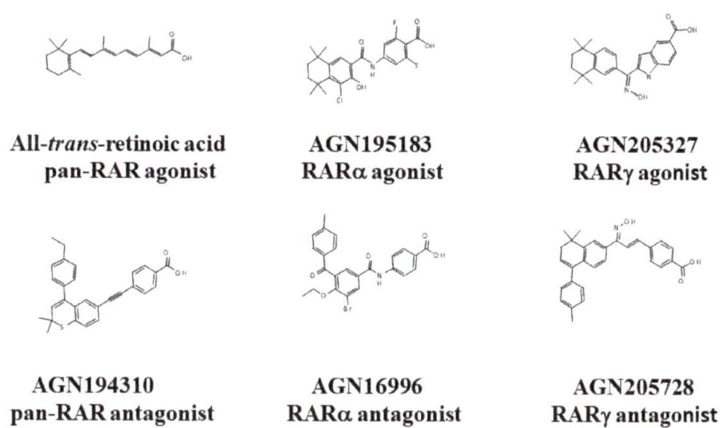

Figure 2. Structures of the RAR agonists and antagonists.

4. Antagonizing RARγ Kills CSCs

PCa was a focus of attention in testing the potential therapeutic use of RAR antagonists. It is a complex disease whereby in the first instance there is aberrant differentiation and hyperproliferation of the prostate epithelium. This is followed by malignant transformation, an alteration of the cellular mechanisms to favor an increased survival of malignant cells, and the eventual appearance of genetically diverse malignant clones which are metastatic [41]. At an early stage of the disease, the majority of patients have excessive androgen production which is also often associated with aberrant androgen receptor (AR) signaling [42]. These events are the major drivers of disease progression. Patients with localized disease, a low/intermediate risk of recurrence, and who are diagnosed and treated at an early stage have a 99% overall survival for 10 years [43]. A major pharmacological treatment for prostate cancer is androgen ablation, using androgen synthesis inhibitors and/or AR antagonists [44]. However, and in the majority of cases, an aggressive disease develops that does not respond to androgen ablation, termed castration refractory PCa. These patients often respond poorly to chemotherapy and other pharmacological interventions [45]. There is a need to develop new treatments for PCa.

The human DU-145, LNCaP, and PC-3 cell lines are widely used in PCa research and were derived from metastatic disease [46]. AR variants have been identified in the human PCa cell line LNCaP [47] and the human PCa cell lines DU-145 and PC-3 have an apparently normal AR gene [48,49]. These three cell lines express AR protein, the levels are lower in DU-145 and PC-3 than in LNCaP, and increases in the levels of AR protein were observed in response to dihydrotestosterone [50].

Table 2 shows the potency of pan-RAR and RARγ antagonists against the PCa cell lines and patients' cells when grown as flask cultures [51–53]. The pan-RAR antagonists AGN194310, AGN19309, and AGN193776 were highly effective in inhibiting the growth of the three PCa cell lines (IC_{50} values from 3.5 to 6.8×10^{-7} M), AGN194310 was equally effective against patients' PCa cells (IC_{50} value 4.7×10^{-7} M), and normal prostate epithelium was less sensitive to the AGN194310 antagonist than the PCa lines (IC_{50} value 1.0×10^{-6} M) [51–53]. The pan-RAR antagonist LG100815 is not structurally related to the AGN antagonists. It was a little less effective than the AGN antagonists against the PCa lines, and normal prostate epithelium was less sensitive than the PCa lines. Antagonism of RARγ, by AGN205728, was sufficient to growth arrest the PCa cell lines (IC_{50} values from 3.0 to 6.0×10^{-7} M) and primary cells from a PCa patient (IC_{50} value 3.0×10^{-7} M) [53]. Normal prostate epithelial cells and the non-malignant RWPE-1 cells were less sensitive to the RARγ antagonist (IC_{50} values of 7.2×10^{-7} M and 2.3×10^{-6} M, respectively). The RARα antagonist AGN196996 and RARβ antagonist LE135 [54] did not affect the growth of the PCa lines. ATRA and the non-hydrolysable pan-RAR agonist TTNPB [55] inhibited the growth of the three PCa lines at concentrations $>10^{-8}$ M. The proliferation of LNCaP cells was markedly increased by exposure to 10^{-10} to 10^{-8} M ATRA, TTNPB, and the RARγ agonist AGN205327. The RARγ agonist inhibited adipogenic differentiation of PCa cells [53]. From all of the above, RARγ agonism stimulates growth and inhibits the differentiation of PCa cells, and RARγ antagonism drives growth arrest.

Table 2. Pan-RAR and RARγ antagonists are potent inhibitors of the growth of flask cultures of PCa cells * mean of the IC_{50} values obtained for the AGN194310 pan-RAR antagonist when tested against primary cells from 14 patients.

Cells	AGN194310 pan-RAR Antagonist IC_{50} Values	AGN193109 pan-RAR Antagonist IC_{50} Values	AGN193776 pan-RAR Antagonist IC_{50} Values	LG100815 pan-RAR Antagonist IC_{50} Values	AGN205728 RARγ Antagonist IC_{50} Values
PCa cells					
DU-145	5.0×10^{-7} M			1.8×10^{-6} M	6.0×10^{-7} M
LNCaP	4.0×10^{-7} M	4.2×10^{-7} M	3.9×10^{-7} M	5.2×10^{-7} M	4.5×10^{-7} M
PC-3	3.5×10^{-7} M	6.8×10^{-7} M	5.7×10^{-7} M	1.0×10^{-6} M	4.7×10^{-7} M
Patients' cells	$4.7 \pm 2.1 \times 10^{-7}$ M *				3.0×10^{-7} M
Non-malignant prostate cells					
Prostate epithelial	1.0×10^{-6} M	1.4×10^{-6} M	1.1×10^{-6} M	$>1 \times 10^{-5}$ M	7.2×10^{-7} M
RWPE-1					2.3×10^{-6} M

The cells that give rise to large colonies when PCa cell line cells are dispersed in a petri dish are CSC-like cells. Table 3 shows the potency of pan-RAR and RARγ antagonists against these cells. The pan-RAR antagonist AGN194310 and the RARγ antagonist AGN205728 were potent inhibitors of colony formation by the DU-145, LNCaP, and PC-3 cell lines [51–53]. The IC_{50} values obtained for the pan-RAR antagonist AGN194310 and the RARγ antagonist AGN205728 were between 16 to 34×10^{-9} M and 50 to 60×10^{-9} M, respectively. The RARα antagonist AGN196996 did not affect colony formation, and the IC_{50} values obtained for ATRA were between 3.2 to 4.2×10^{-7} M. The pan-RAR antagonist AGN194310 was equally effective in preventing colony formation by the breast cancer cell lines MCF7 and MDA-MB-231 which were derived from metastatic disease. Other workers have also shown that pharmacological or genetic blockage of RARγ signaling drives growth arrest, differentiation, and cell death of breast cancer cells [56,57].

Table 3. Pan-RAR and RARγ antagonists are potent inhibitors of colony formation by PCa cell lines.

PCa Lines	AGN194310 pan-RAR Antagonist IC_{50} Values	AGN205728 RARγ Antagonist IC_{50} Values	AGN196996 RARα Antagonist IC_{50} Values	ATRA pan-RAR Agonist IC_{50} Values
DU-145	34×10^{-9} M	60×10^{-9} M	$>1 \times 10^{-5}$ M	4.0×10^{-7} M
LNCaP	16×10^{-9} M	55×10^{-9} M	$>1 \times 10^{-5}$ M	3.2×10^{-7} M
PC-3	18×10^{-9} M	50×10^{-9} M	$>1 \times 10^{-5}$ M	4.2×10^{-7} M

For LNCaP cells plated in the absence of ATRA, around 70% of the colonies had a large and holoclone/merclone morphology, and these colonies contain stem cells [58]. The remaining colonies had a small and differentiated paraclone morphology. Treatment of LNCaP cells with 10^{-10} M ATRA, to activate RARγ, increased the proportion of stem cell-like colonies to 85%, whereas 10^{-6} M ATRA increased the proportion of differentiation-committed and paraclone colonies to 60% [53]. The low level of ATRA seen within PCa tissue via transactivating RARγ would, therefore, increase the population of clonogenic and stem cell-like cells or block differentiation of these cells.

Treatment of flask-grown cultures of the PCa cell lines with the pan-RAR antagonist AGN194310 and the RARγ antagonist AGN205728 led to growth arrest in G_1 of the cell cycle followed by cell death. The cell death was mitochondria depolarization-dependent, and involved cellular DNA fragmentation, but was caspase-independent [52,53]. This form of cell death is termed necroptosis and was seen for Jurkat T leukemia cells that were deprived of retinoids. This led to the activation of the poly(ADP-ribose) polymerase PARP-1 which ribosylates a wide variety of proteins, including those involved in transcription and cell cycle, to change their function [59]. Caspase-independent cell death has also been seen for ischemia-reperfusion injury, diabetes, inflammatory-mediated injury, and neurotoxicity [60]. Cell stress activation of PARP-1 has been associated with mitochondrial dysfunction with the release of ATP, NAD^+ and the caspase-independent nucleases AIF and endonuclease G. They fragment DNA. Inhibition of PARP-1, by 1,5-dihydroisoquinoline (at 1×10^{-4} M), blocked the actions of RAR antagonists on the PCa cell lines.

5. Antagonizing All RARs Is Effective against Pediatric Brain Tumors

The most common cause of cancer mortality in children are tumors of the peripheral and central nervous system. Post-treatment, a significant proportion of patients have life-long and induced neurological, cognitive and endocrine disturbances [61]. Several of the peripheral and central nervous system tumors share an embryological origin in the neuroectoderm and have been grouped as primitive neuroectodermal tumors (PNETs). They include neuroblastoma, Ewing's sarcoma, retinoblastoma, medulloblastoma, and supratentorial primitive neuroectodermal tumors (stPNETs) [62]. ATRA has been implicated in the development of the central nervous system [63]. Oral 13-*cis* retinoic acid is a key component of the therapy for neuroblastoma as a consolidation treatment [64], and ATRA has been shown to inhibit the proliferation of human PNET cells [65]. Exploration of the effectiveness of the pan-RAR antagonist AGN194310 was driven by the use of 13-*cis* retinoic acid to treat neuroblastoma requires a high serum level and is severely limited by toxicity.

When cells from PNET patient biopsies were cultured in a serum-free neural stem cell medium (Neurocult) supplemented with 20 ng/mL epidermal growth factor they generated neurospheres which produced differentiated cells that migrated (Figure 3). The cells that give rise to neurospheres are CSCs. The activity of the pan-RAR antagonist AGN194310 was examined against two pediatric PNETs and a pediatric astrocytoma. These cells were plated into wells, treated with AGN194310, and cellular ATP levels were measured on day 5. AGN194310 was highly effective against the two pediatric PNETS and the pediatric astrocytoma. Neurospheres and their progeny were completely ablated by 10^{-6} M AGN194310. ATRA and the RARα antagonist AGN195183 were somewhat ineffective [53].

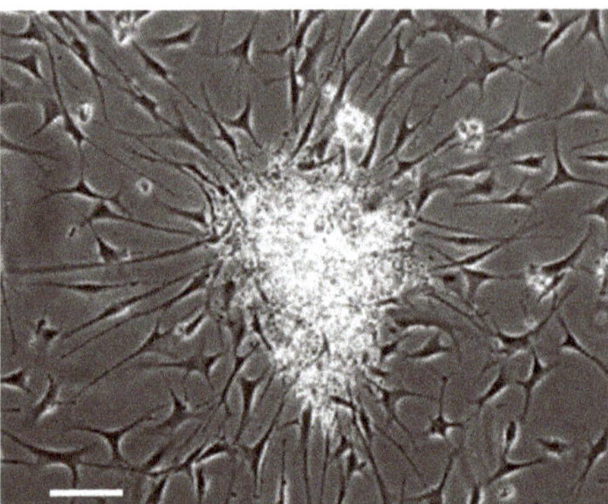

Figure 3. Primary culture of cells from PNET biopsies generates neurospheres with differentiating cells. The cells that give rise to neurospheres are CSCs. Scale bar = 100 μm.

6. The Effect of Antagonizing RARs on Normal Cells

Selectivity is all important to killing CSCs and sparing normal tissue cells as much as is possible. As mentioned above, normal prostate epithelium cells were less sensitive to the pan-RAR antagonist AGN194310 and RARγ antagonist AGN205728 than PCa cell lines and patients' cells. The antagonists did not have an effect on human peripheral blood lymphocytes and primary cultures of human fibroblasts.

An important matter is that we expected the pan-RAR to influence the differentiation of HSCs and/or hematopoietic progenitor cells (HPCs) because active RARα is required for the differentiation of neutrophil/monocyte progenitors [18]. We investigated this by using flask cultures of purified human bone marrow HSCs/HPCs (CD34$^+$). Indeed, treatment of these cells with the pan-RAR antagonist AGN194310 and the RARα antagonist AGN196996 prolonged the lifespan of cultures, up to 55 days, and there was a substantial increase in the production of neutrophils and monocytes. This was not related to cell differentiation slowing down, and instead there was an expansion of the number of HSCs and HPCs [33]. From these findings, the use of the pan-RAR antagonist AGN194310 to drive an increase in neutrophil production may be of benefit to patients with neutropenia, including those with chemotherapy-provoked neutropenia.

7. Concluding Remarks

PCa cell line CSC-like cells and primary cultures of cells from PCa patient biopsies were ablated by treatment with the RARγ antagonist AGN205728 or the pan-RAR antagonist AGN194310. Normal prostate epithelium cells were less sensitive to the actions of the two antagonists, and human peripheral blood lymphocytes and primary human fibroblast were unaffected. The two compounds were also equally active against breast cancer cell line CSC-like cells and the pan-RAR antagonist AGN194310 was effective against PNET CSCs. The use of a metastatic model of epidermal growth factor receptor-mutant lung cancer has shown that a pan-RAR antagonist dramatically reduces lung cancer metastasis to the brain [66]. Cancer patients receiving intensive chemotherapy often develop neutropenia. The pan-RAR antagonist AGN194310 as used as an adjunct to chemotherapy may provide additional therapeutic benefit because it increases the production of neutrophils by HSCs/HPCs.

Treatment of the PCa cell lines with the RAR antagonists led to necroptosis of CSC-like cells. Necroptosis is mediated by active PARP-1, and the PARP-1 inhibitor 1,5-dihydroisoquinoline blocked the actions of RAR antagonists. A significant aspect of the oncogenic action of active RARγ is that it blocks necroptosis. Necroptosis is viewed as a fail-safe cell death pathway for apoptosis-resistant cells [67] and one that defends against cancer [68]. The ability of the antagonists to drive necroptosis emphasizes an important avenue to treating cancer.

8. Perspectives

The findings from in vitro studies of PCa and breast cancer cells and in vivo studies of lung cancer support the further development of the RARγ antagonist AGN205728 and/or the pan-RAR antagonist AGN194310 for use to treat these cancers. The activity of AGN205728 and AGN194310 may extend to other solid tumor CSCs because RARγ is an oncogene for CRC, CCA, and HCC. However, there is need to test whether the antagonists are active against these cancer cells. To move the antagonists forward as drug candidates, there is the need to undertake preclinical studies using the 3-dimensional models that are available for PCa, human organoids, and xenograft models of, for example, breast and CRC cancer. These preclinical studies will reveal whether a projected therapeutic dose (from in vitro studies) eliminates CSCs, and is safe to use. Safety tests, including Ames mutagenicity, Chinese hamster ovary chromosomal aberration, and mouse micronucleus, are important to showing that there is no genotoxicity.

The antagonists are highly selective for RARs. However, we do not as yet know the precise mode of action of RAR© in CSCs, nor how RARγ antagonism triggers necroptosis. There are two possible ways. Analysis of F9 embryonal stem cells, by integrative genomics, has revealed that RARγ regulates a large network of genes within stem cells [69]. RARγ is also essential for ATRA induced chromatin remodeling and the activation of transcription in embryonic stem cells [20]. Disruption to these processes may be the cause of RARγ antagonist driven necroptosis. RARγ antagonism may influence the cellular location of RARγ and this is important because cytosolic RARγ plays a role in controlling Riptosome (RIPK1/RIPK2)-mediated DNA damage-induced necroptosis when the cellular inhibitor of apoptosis is blocked [70,71]. Therefore, a more complete understanding of the precise mode of action of RAR© within CSCs and how RAR antagonism triggers necroptosis is needed.

Funding: G.B. received funding from the European Union's Seventh Framework Programme for research, technological development and demonstration under grant agreement no 315902. G.B. was the coordinator of the Marie Curie Initial Training Network DECIDE.

Institutional Review Board Statement: Not applicable.

Informed Consent Statement: Not applicable.

Data Availability Statement: Not applicable.

Conflicts of Interest: The author declares no conflict of interest.

References

1. Greaves, M. Evolutionary determinants of cancer. *Cancer Discov.* **2015**, *5*, 806–820. [CrossRef]
2. Greaves, M.; Maley, C.C. Clonal evolution in cancer. *Nature* **2012**, *481*, 306–313. [CrossRef] [PubMed]
3. Dick, J.E. Stem cell concepts renew cancer research. *Blood* **2008**, *112*, 4793–4807. [CrossRef] [PubMed]
4. Quintana, E.; Shackleton, M.; Sabel, M.S.; Fullen, D.R.; Johnson, T.M.; Morrison, S.J. Efficient tumour formation by single human melanoma cells. *Nature* **2008**, *456*, 593–598. [CrossRef] [PubMed]
5. Fialkow, P.J.; Denman, A.M.; Jacobson, G.J.; Lowenthal, M.N. Chronic myelocytic leukaemia: Origin of some lymphocytes from leukaemic stem cell. *J. Clin. Investig.* **1978**, *62*, 815–823. [CrossRef]
6. Holyoake, T.; Jiang, X.; Eaves, C.; Eaves, A. Isolation of a highly quiescent subpopulation of primitive leukemic cells in chronic myeloid leukemia. *Blood* **1999**, *94*, 2056–2064. [CrossRef]
7. Graham, S.M.; Jorgensen, H.G.; Allan, E.; Pearson, C.; Alcorn, M.J.; Richmond, L.; Holyoake, T.L. Primitive, quiescent, Philadelphia-positive stem cells from patients with chronic myeloid leukemia are insensitive to STI571 in vitro. *Blood* **2002**, *99*, 319–325. [CrossRef]
8. Corbin, A.S.; Agarwal, A.; Loriaux, M.; Cortes, J.; Deininger, M.W.; Druker, B.J. Human chronic myeloid leukemia stem cells are insensitive to imatinib despite inhibition of BCR-ABL activity. *J. Clin. Investig.* **2011**, *121*, 396–409. [CrossRef]

9. Lim, Z.; Brand, R.; Martino, R.; van Biezen, A.; Finke, J.; Bacigalupo, A.; Beelen, D.; Devergie, A.; Alessandrino, E.; Willemze, R.; et al. Allogeneic hematopoietic stem-cell transplantation for patients 50 years or older with myelodysplastic syndromes or secondary acute myeloid leukemia. *J. Clin. Oncol.* **2010**, *28*, 405–411. [CrossRef]

10. Yang, L.; Shi, P.; Zhao, G.; Xu, J.; Peng, W.; Zhang, J.; Zhang, G.; Wang, X.; Dong, Z.; Chen, F.; et al. Targeting cancer stem cell pathways for cancer therapy. *Signal Transduct. Target. Ther.* **2020**, *5*, 8. [CrossRef]

11. Chen, K.; Huang, Y.-H.; Chen, J.-L. Understanding and targeting cancer stem cells: Therapeutic implications. *Acta Pharmacol. Sin* **2013**, *34*, 732–740. [CrossRef] [PubMed]

12. Liu, B.; Yan, L.; Zhou, M. Targeted selection of CAR T cell therapy in accordance with the TME for solid cancers. *Am. J. Cancer Res.* **2019**, *9*, 228–241. [PubMed]

13. di Masi, A.; Leboffe, L.; De Marinis, E.; Pagano, F.; Cicconi, L.; Rochette-Egly, C.; Le-Coco, F.; Ascenzi, P.; Nervi, C. Retinoic acid receptors: From molecular mechanisms to cancer therapy. *Mol. Asp. Med.* **2015**, *41*, 1–115. [CrossRef] [PubMed]

14. Sanz, M.A.; Fenaux, P.; Tallman, M.S.; Estey, E.H.; Lowenberg, B.; Naoe, T.; Lengfelder, E.; Dohner, H.; Burnet, A.K.; Chen, S.J.; et al. Management of acute promyelocytic leukema: Updated recommendations from an expert panel of the European Leukemia. *Net. Blood* **2019**, *133*, 1630–1643. [CrossRef]

15. Edwards, R.H.; Wasik, M.A.; Finan, J.; Rodriquez, R.; Moore, J.; Kamoun, M.; Rennert, H.; Bird, J.; Novell, P.C.; Salhany, K.E. Evidence for early progenitor cell involvement in promyelocytic leukemia. *Am. J. Clin. Pathol.* **1999**, *112*, 819–827. [CrossRef]

16. Huynh, T.T.; Sultan, M.; Vidovic, D.; Dean, C.A.; Cruichshank, B.M.; Lee, K.; Loung, C.Y.; Holloway, R.W.; Hoskin, D.W.; Waisman, D.M.; et al. Retinoic acid and arsenic trioxide induce lasting differentiation and demethylation of target genes in APL cells. *Sci. Rep.* **2019**, *9*, 9414. [CrossRef]

17. Purton, L.E.; Dworkin, S.; Olsen, G.M.; Walkley, C.; Fabb, S.A.; Collins, S.J.; Chambon, P. RAR© is critical for maintaining a balance between hematopoietic stem cell self-renewal and differentiation. *J. Exp. Med.* **2006**, *203*, 1283–1293. [CrossRef]

18. Purton, L.E.; Bernstein, I.D.; Collins, S.J. All-trans retinoic acid delays the differentiation of primitive hematopoietic precursors (lin−c-kit+Sca-1(+)) while enhancing the terminal maturation of committed granulocyte/monocyte progenitors. *Blood* **1999**, *94*, 483–495. [CrossRef]

19. Wai, H.A.; Kawakami, K.; Wada, H.; Muller, F.; Vernalis, A.B.; Brown, G.; Johnson, W.E.B. The development and growth of tissues derived from cranial neural crest and primitive mesoderm is dependent on the ligation status of retinoic acid receptor©: Evidence that retinoic acid receptor γ functions to maintain stem/progenitor cells in the absence of retinoic acid. *Stem. Cells Dev.* **2015**, *24*, 507–519.

20. Kashyap, V.; Laursen, K.B.; Brenet, F.; Viale, A.J.; Scandura, J.M.; Gudas, L.J. RAR© is essential for retinoic acid induced chromatin remodelling and transcriptional activation in embryonic stem cells. *J. Cell Sci.* **2012**, *126*, 999–1008.

21. Chatagon, A.; Veber, P.; Morin, V.; Bedo, J.; Triqueneaux, G.; Semon, M.; Laudet, V.; d'Alche-Buc, F.; Benoit, G. RAR/RXR binding dynamics distinguish pluripotency from differentiation associated cis-regulatory elements. *Nucl. Acids Res.* **2015**, *43*, 4833–4854. [CrossRef] [PubMed]

22. Such, E.; Cervera, J.; Valencia, A. A novel NUP98/RARG gene fusion in acute myeloid leukemia resembling acute promyelocytic leukemia. *Blood* **2011**, *117*, 242–245. [CrossRef] [PubMed]

23. Qin, Y.Z.; Huang, X.J.; Zhu, H.H. Identification of a novel CPSF6-RARG fusion transcript in acute myeloid leukemia resembling acute promyelocytic leukemia. *Leukemia* **2018**, *32*, 2285–2287. [CrossRef] [PubMed]

24. Conserva, M.R.; Redavid, I.; Anelli, L.; Zagaria, A.; Specchia, G.; Albano, F. RARG gene dysregulation in acute myeloid leukemia. *Front. Mol. Biosci.* **2019**, *6*, 114. [CrossRef] [PubMed]

25. Su, Z.; Liu, X.; Xu, Y.; Zhao, C.; Zhao, H.; Feng, X.; Zhang, S.; Yang, S.; Yang, J.; Shi, X.; et al. Novel reciprocal fusion genes involving HNRPC1 and RARG in acute promyelocytic leukemia lacking RARA rearrangement. *Haematologica* **2020**, *105*, e376. [CrossRef] [PubMed]

26. Watts, J.M.; Perez, A.; Pereira, L.; Fan, Y.S.; Brown, G.; Vega, F.; Petrie, K.; Swords, R.T.; Zelent, A.A. A case of AML characterised by a novel t(4;15)(q31;q22) translocation that confers a growth-stimulatory response to retinoid-based therapy. *Int. J. Mol. Sci.* **2017**, *18*, 1492. [CrossRef] [PubMed]

27. Huang, G.L.; Song, W.; Zhou, P.; Fu, Q.R.; Lin, C.L.; Chen, Q.X.; Shen, D.Y. Oncogenic retinoic acid receptor gamma knockdown reverses multi-drug resistance of human colorectal cancer via Wnt/beta-catenin pathway. *Cell Cycle* **2017**, *16*, 685–692. [CrossRef]

28. Huang, G.L.; Luo, Q.; Rui, G.; Zhang, W.; Zhang, Q.Y.; Chen, Q.X.; Shen, D.Y. Oncogenic activity of retinoic acid receptor gamma is exhibited through activation of the Akt/NF-kappaB and Wnt/beta-catenin pathways in cholangiocarcinoma. *Mol. Cell. Biol.* **2013**, *33*, 3416–3425. [CrossRef]

29. Yan, T.D.; Wu, H.; Zhang, H.P.; Lu, N.; Ye, P.; Yu, F.H.; Zhou, H.; Li, W.G.; Cao, X.; Lin, Y.Y.; et al. Oncogenic potential of retinoic acid receptor-gamma in hepatocellular carcinoma. *Cancer Res.* **2010**, *70*, 2285–2295. [CrossRef]

30. Kudryavtseva, A.V.; Nyushko, K.M.; Zaretsky, A.R.; Shagin, D.A.; Kaprin, A.D.; Alekseev, B.Y.; Snezhkina, A.V. Upregulation of Rarb, Rarg, and Rorc Genes in Clear Cell Renal Cell Carcinoma. *Biomed. Pharmacol. J.* **2016**, *9*, 967–975. [CrossRef]

31. Pasquali, D.; Thaller, C.; Eichele, G. Abnormal level of retinoic acid in prostate cancer tissues. *J. Clin. Endocrinol. Metab.* **1996**, *81*, 2186–2191. [PubMed]

32. Farboud, B.; Hauksdotter, H.; Wu, Y.; Privalsky, M.L. Isotype-restricted corepressor recruitment: A constitutively closed helix 12 conformation in retinoic acid receptors beta and gamma interferes with corepressor recruitment and prevents transcriptional repression. *Mol. Cell. Biol.* **2003**, *23*, 2844–2858. [CrossRef] [PubMed]

33. Brown, G.; Marchwicka, A.; Cunningham, A.; Toellner, K.-M.; Marcinkowska, E. Antagonizing retinoic acid receptors increases myeloid cell production by cultured human hematopoietic stem cells. *Arch. Immunol. Ther. Exp.* **2017**, *65*, 69–81. [CrossRef] [PubMed]
34. Ryzlak, M.T.; Ambroziak, W.; Schaffer, C.P. Humam prostatic aldehyde dehydrogenase of healthy controls and diseased prostates. *Biochim. Biophys. Acta* **1992**, *1139*, 287–294. [CrossRef]
35. Trasino, S.E.; Harrison, E.H.; Wang, T.T. Androgen regulation of aldehyde dehydrogenase 1A3 (ALDH1A3) in the androgen-responsive human prostate cancer cell line LNCaP. *Exp. Biol. Med.* **2007**, *232*, 762–771.
36. Chen, Y.; Clarke, O.B.; Kim, J.; Stowe, S.; Kim, Y.-K.; Assur, Z.; Cavalier, M.; Goday-Ruiz, R.; von Alpen, D.C.; Manzini, C.; et al. Structure of the STRA6 receptor for retinol uptake. *Science* **2016**, *353*, aad8266. [CrossRef]
37. Cai, K.; Gudas, L.J. Retinoic acid receptors and GATA transcription factors activate the transcription of the human lecithin: Retinol acyltransferase gene. *Int. J. Biochem. Cell Biol.* **2009**, *41*, 546–553. [CrossRef]
38. Johnson, A.T.; Wang, L.; Standeven, A.M.; Escobar, M.; Chandraratna, R.A. Synthesis and biological activity of high-affinity retinoic acid receptor antagonists. *Bioorg. Med. Chem.* **1999**, *7*, 1321–1338. [CrossRef]
39. Heyman, R.; Mangelsdorf, D.; Dyck, J.; Stein, R.; Eichele, G.; Evans, R.; Thaller, C. 9-cis-retinoic acid is a high affinity ligand for the retinoid-X-receptor. *Cell* **1992**, *68*, 392–406. [CrossRef]
40. Hughes, P.J.; Zhao, Y.; Chandraratna, R.A.; Brown, G. Retinoid-mediated stimulation of steroid sulfatase activity in myeloid leukemic cells requires RARα and RXR and involves the phosphoinositide 3-kinase and ERK-MAP pathways. *J. Cell. Biochem.* **2006**, *97*, 327–350. [CrossRef]
41. Berges, R.R.; Vukanovic, J.; Epstein, J.L.; Carmichel, M.; Cisek, L.; Johnson, D.E.; Veltri, R.W.; Walsh, P.C.; Isaaccs, J.T. Implication of cell kinetic changes during the progression of human prostate cancer. *Clin. Cancer Res.* **1995**, *1*, 473–480.
42. Knudsen, K.E.; Penning, T.M. Partners in crime: Deregulation of AR activity and androgen synthesis in prostate cancer. *Trends Endocrinol. Metab.* **2010**, *21*, 315–324. [CrossRef] [PubMed]
43. Rebello, R.J.; Ding, C.; Knudsen, K.; Loeb, S.; Johnson, D.C.; Reiter, R.E.; Gillessen, S.; Van der Kwast, T.; Bristow, R.G. Prostate cancer. *Nat. Rev. Dis. Prim.* **2021**, *7*, 9. [CrossRef]
44. Quan, H.; Loblaw, D.A. Androgen deprivation for prostate cancer–review of indications in 2010. *Curr. Oncol.* **2010**, *17*, S38–S44. [CrossRef] [PubMed]
45. Attard, G.; Reid, A.H.; Olmos, D.; de Bono, J.S. Antitumor activity with CYP17 blockade indicates that castration-resistant prostate cancer frequently remains hormone driven. *Cancer Res.* **2009**, *69*, 4937–4940. [CrossRef]
46. Namekawa, T.; Ikeda, K.; Harie-Inoue, K.; Inoue, S. Application of prostate cancer models for preclinical studies: Advantages and limitations of cell lines, patient derived xenografts, and three-dimensional culture of patient-derived cells. *Cells* **2019**, *8*, 74. [CrossRef]
47. Linja, M.J.; Vinsakorpi, T. Alternations of androgen receptor in prostate cancer. *J. Steroid Biochem. Mol. Biol.* **2004**, *92*, 255–264. [CrossRef]
48. Culig, Z.; Klocker, K.; Eberle, J.; Kaspar, F.; Habish, A.; Cronauer, M.V.; Bartsch, G. DNA sequence of the androgen receptor in prostate tumor cell lines and tissue specimens assessed by means of the polymerase chain reaction. *Prostate* **1993**, *22*, 11–22. [CrossRef]
49. Tilley, W.D.; Bentel, J.M.; Aspinall, J.O.; Hall, R.E.; Horsfall, D.J. Evidence for a novel mechanism of androgen resistance in the human prostate cancer cell line PC-3. *Steroids* **1995**, *60*, 180–186. [CrossRef]
50. Alimireh, F.; Chen, J.; Basrawala, Z.; Xin, H.; Choubey, D. DU-145 and PC-3 human prostate cancer cell lines express androgen receptor: Implications for the androgen receptor functions and regulation. *FEBS Lett.* **2006**, *580*, 2294–2300. [CrossRef] [PubMed]
51. Hammond, L.A.; Krinks, C.H.V.; Durham, J.; Tomkins, S.E.; Burnett, R.D.; Jones, E.L.; Chandraratna, R.A.S.; Brown, G. Antagonists of retinoic acid receptors (RARs) are potent growth inhibitors of prostate carcinoma cells. *Br. J. Cancer* **2001**, *85*, 453–462. [CrossRef] [PubMed]
52. Keedwell, R.G.; Zhao, Y.; Hammond, L.A.; Wen, K.; Qin, S.; Atangan, L.I.; Shurland, D.-L.; Wallace, D.M.A.; Bird, R.; Reitmair, A.; et al. An antagonist of retinoic acid receptors more effectively inhibits growth of human prostate cancer cells than normal prostate epithelium. *Br. J. Cancer* **2004**, *91*, 580–588. [CrossRef] [PubMed]
53. Petrie, K.; Urban, Z.; Sbirkov, Y.; Graham, A.; Hamann, A.; Brown, G. Retinoic acid receptor-© is a therapeutically targetable driver of growth and survival in prostate cancer. *Cancer Rep.* **2020**, *3*, e1284. [CrossRef] [PubMed]
54. Li, Y.; Hashomoto, Y.; Agadir, A.; Kagechika, H.; Zhang, X.K. Identification of a novel class of retinoic acid receptor beta-selective retinol antagonists and their inhibitory effects on AP-1 activity and retinoic acid-induced apoptosis in huam breast cancer cells. *J. Biol. Chem.* **1999**, *274*, 15360–15366. [CrossRef] [PubMed]
55. Lemaire, G.; Balaguer, P.; Michel, S.; Rahmani, R. Activation of retinoic acid receptor-dependent transcription by organochlorine pesticides. *Toxicol. Appl. Pharmacol.* **2005**, *207*, 38–49. [CrossRef]
56. Lu, Y.; Bertan, S.; Samuels, T.A.; Mira-y-Lopez, R.; Farias, E.F. Mechanism of inhibition of MMTV-neu and MMTV-wnt induced mammary oncogenesis by RARα agonist Am580. *Oncogene* **2010**, *29*, 3554–3576. [CrossRef]
57. Bosch, A.; Bertran, S.P.; Lu, Y.; Garcia, A.; Jones, A.M.; Dawson, M.I.; Farias, E.F. Reversal by RARα agonist Am580 of c-Myc-induced imbalance in RARα/RAR© expression during MMTV-Myc tumorigenesis. *Breast Cancer Res.* **2012**, *14*, R121. [CrossRef]

58. Beaver, M.; Ahmed, A.; Masters, J.R. Clonogenic holoclones and merclones contain stem cells. *PLoS ONE* **2014**, *9*, e89834. [CrossRef]
59. Chiu, H.J.; Fishman, D.A.; Hammerling, U. Vitamin A depletion causes oxidative stress, mitochondrial dysfunction, and PARP-1-dependent energy deprevation. *FASEB J.* **2008**, *22*, 3878–3887. [CrossRef]
60. Luo, X.; Kraus, W.L. On PAR with PARP: Cellular stress signaling through poly(ADP-ribose) and PARP-1. *Genes Dev.* **2012**, *26*, 417–432. [CrossRef]
61. Pomeroy, S.L.; Ullrich, N.L. Pediatric brain tumors. *Neurol. Clin.* **2003**, *4*, 897–913.
62. Kleihues, P.; Loius, D.; Scheithauer, B.W.; Rourke, L.R.; Reifenberger, G.; Burges, P.C.; Cavanee, W.K. The WHO classification of tumors of the nervous stsyem. *Neurpathol. Exp. Neurol.* **2002**, *61*, 215–235. [CrossRef] [PubMed]
63. Maden, M. Role and distribution of retonic acid during CNS development. *Int. Rev. Cytol.* **2001**, *209*, 1–77. [PubMed]
64. Mattay, K.K.; Maris, J.M.; Schleirmacher, G.; Nagagawara, A.; Mackell, C.L.; Diller, L.; Weiss, W.A. Neuroblastoma. *Nat. Rev. Des. Prim.* **2016**, *2*, 16078. [CrossRef]
65. Rosolen, A.; Favaretto, G.; Masarotto, G.; Cavazzana, A.; Zanesco, L.; Franscella, E. Effect of all-trans retinoic acid and interferon α in peripheral neuroectodermal tumour cell cultures and xenografts. *Int. J. Oncol.* **1998**, *13*, 943–949.
66. Biswas, A.K.; Han, S.; Tai, Y.; Ma, W.; Coker, C.; Quinn, S.A.; Shakri, A.R.; Zhong, T.J.; Scholze, H.; Lagos, G.G.; et al. Targeting S100A9-ALDH1AI-retinoic acid signaling to suppress brain relapse in EGFR-mutant lung cancer. *Cancer Discov.* **2022**, *12*, 1002–1021. [CrossRef]
67. Gong, Y.; Fan, Z.; Luo, G.; Yang, C.; Huang, Q.; Fan, K.; Cheng, H.; Jin, K.; Ni, Q.; Yu, X.; et al. The role of necroptosis in cancer biology and therapy. *Mol. Cancer* **2019**, *18*, 100. [CrossRef]
68. Chen, D.; Yu, J.; Zhang, L. Necroptosis: An alternative cell death program defending against cancer. *Biochim. Biophys. Acta* **2016**, *1865*, 228–236. [CrossRef]
69. Mendoza-Parra, M.A.; Walia, M.; Sankar, M.; Gronemeyer, H. Dissecting the retinoid-induced differentiation of F9 embryonal stem cells by integrative genomics. *Mol. Syst. Biol.* **2011**, *7*, 538. [CrossRef]
70. Xu, Q.; Jitkaew, S.; Choksi, S.; Kadigamuwa, C.; Choe, M.; Jang, J.; Liu, C.; Liu, Z.-G. The cytoplasmic nuclear RAR© controls RIPI initiated cell death when cIAP activity is inhibited. *Nat. Commun.* **2017**, *8*, 425. [CrossRef]
71. Kadigamuwa, C.; Choksi, S.; Xu, Q.; Cataisson, C.; Greenbaum, S.S.; Yuspa, S.H.; Liu, Z.-G. Role of retinoic acid receptor-© in DNA dmaage-induced necroptosis. *IScience* **2019**, *17*, 74–86. [CrossRef] [PubMed]

International Journal of
Molecular Sciences

Article

An Isochroman Analog of CD3254 and Allyl-, Isochroman-Analogs of NEt-TMN Prove to Be More Potent Retinoid-X-Receptor (RXR) Selective Agonists Than Bexarotene

Peter W. Jurutka [1,2], Orsola di Martino [3], Sabeeha Reshi [1], Sanchita Mallick [1], Michael A. Sausedo [1], Grant A. Moen [1], Isaac J. Lee [1], Dominic J. Ivan [1], Tyler D. Krall [1], Samuel J. Peoples [1,4], Anthony Perez [1], Lucas Tromba [1], Anh Le [1], Iraj Khadka [1], Ryan Petros [1], Brianna M. Savage [1], Eleine Salama [1], Jakline Salama [1], Joseph W. Ziller [5], Youngbin Noh [4], Ming-Yue Lee [6], Wei Liu [7], John S. Welch [8], Pamela A. Marshall [1] and Carl E. Wagner [1,*]

[1]   School of Mathematical and Natural Sciences, Arizona State University, Glendale, AZ 85306, USA
[2]   Department of Basic Medical Sciences, University of Arizona College of Medicine, Phoenix, AZ 85004, USA
[3]   Department of Internal Medicine, Washington University, St. Louis, MO 63110, USA
[4]   School of Molecular Sciences, Arizona State University, Tempe, AZ 85281, USA
[5]   Department of Chemistry, University of California, Irvine, CA 92697, USA
[6]   Poseida Therapeutics, Inc. 9390 Towne Centre Drive, Suite 200, San Diego, CA 92121, USA
[7]   Cancer Center and Department of Pharmacology and Toxicology, Medical College of Wisconsin, Milwaukee, WI 53226, USA
[8]   A2 Biotherapeutics, 30301 Agoura Rd, Agoura Hills, CA 91301, USA
*   Correspondence: carl.wagner@asu.edu; Tel.: +1-(602)-543-6937

Citation: Jurutka, P.W.; di Martino, O.; Reshi, S.; Mallick, S.; Sausedo, M.A.; Moen, G.A.; Lee, I.J.; Ivan, D.J.; Krall, T.D.; Peoples, S.J.; et al. An Isochroman Analog of CD3254 and Allyl-, Isochroman-Analogs of NEt-TMN Prove to Be More Potent Retinoid-X-Receptor (RXR) Selective Agonists Than Bexarotene. *Int. J. Mol. Sci.* **2022**, *23*, 16213. https://doi.org/10.3390/ijms232416213

Academic Editors: Andrzej Kutner, Geoffrey Brown and Enikö Kallay

Received: 12 November 2022
Accepted: 14 December 2022
Published: 19 December 2022

Publisher's Note: MDPI stays neutral with regard to jurisdictional claims in published maps and institutional affiliations.

Abstract: Bexarotene is an FDA-approved drug for the treatment of cutaneous T-cell lymphoma (CTCL); however, its use provokes or disrupts other retinoid-X-receptor (RXR)-dependent nuclear receptor pathways and thereby incites side effects including hypothyroidism and raised triglycerides. Two novel bexarotene analogs, as well as three unique CD3254 analogs and thirteen novel NEt-TMN analogs, were synthesized and characterized for their ability to induce RXR agonism in comparison to bexarotene (**1**). Several analogs in all three groups possessed an isochroman ring substitution for the bexarotene aliphatic group. Analogs were modeled for RXR binding affinity, and EC_{50} as well as IC_{50} values were established for all analogs in a KMT2A-MLLT3 leukemia cell line. All analogs were assessed for liver-X-receptor (LXR) activity in an LXRE system to gauge the potential for the compounds to provoke raised triglycerides by increasing LXR activity, as well as to drive LXRE-mediated transcription of brain ApoE expression as a marker for potential therapeutic use in neurodegenerative disorders. Preliminary results suggest these compounds display a broad spectrum of off-target activities. However, many of the novel compounds were observed to be more potent than **1**. While some RXR agonists cross-signal the retinoic acid receptor (RAR), many of the rexinoids in this work displayed reduced RAR activity. The isochroman group did not appear to substantially reduce RXR activity on its own. The results of this study reveal that modifying potent, selective rexinoids like bexarotene, CD3254, and NEt-TMN can provide rexinoids with increased RXR selectivity, decreased potential for cross-signaling, and improved anti-proliferative characteristics in leukemia models compared to **1**.

Keywords: retinoid-x-receptor; retinoid; rexinoid; leukemia; small molecule therapeutic; structure-activity-relationship

1. Introduction

Nuclear receptors (NRs) are ligand-dependent transcription factors that bind DNA sequence-specific motifs in enhancers and promoters to transactivate their target genes [1]. The retinoid X receptors (RXRs) are ligand-activated NRs that have pleiotropic effects including the control of hematopoietic stem cell self-renewal and differentiation. There are three different isoforms of each receptor (α, β, and γ) that are differently expressed in mouse and human tissues [2,3]. The RXRs, often working in concert with other NRs regulate gene transcription through receptor-specific molecular signals. The RXRs are remarkably versatile compared to other NRs, since they partner with many of the NRs to form heterodimers that modulate cell differentiation, migration, proliferation, and metabolic pathways. Several critical NR pathways that are RXR-dependent include those regulated by the retinoic acid receptor (RAR), the vitamin D receptor (VDR), the peroxisome proliferator-activated receptor (PPAR), the thyroid hormone receptor (TR), the farnesoid X receptor (FXR), and the liver-X-receptor (LXR), to cite just a few. All NRs function as transcriptional modulators, most often promoting transcription as a result of the presence of corresponding receptor ligand in addition to any obligate receptor partner. The receptor ligands are often endogenous molecules that bind to a ligand-binding domain (LBD) in the receptor. This subsequently forces the receptor into a new conformation more conducive towards dimerizing with another receptor, recruiting associated co-factors, and finally binding to a high affinity hormone responsive element (HRE) specific to the genes the receptor controls in the DNA. While several HREs have been located proximal to or inside the promoter region of the regulated genes, HREs are increasingly being observed a considerable distance down- or upstream from the regulated genes. The HREs display a shared sequence specificity that includes two repeat hexads enclosing a specific quantity of spacers that separate those inverted, everted, or direct repeats [4]. RAR, TR, and VDR's HREs comprise half-sites enclosing, respectively, five, four, and three nucleotide spacers [5,6].

Before RXR was well known, TR, RAR, and VDR were believed to assemble into homodimers [7] in order to bind their respective HREs, though heterodimerizing with RXR was later discovered to be the prerequisite for them to bind and activate their HREs [8]. The naturally occurring 9-cis-retinoic acid (9-cis-RA)—a geometric isomer of the all-trans-retinoic acid (ATRA)—was identified as an RXR-specific agonist (a rexinoid) by Zhang and coworkers, who documented that its binding to RXR's LBD triggers RXR homodimerization and the subsequent association of the homodimer to the RXR responsive elements (RXREs) [9]. In other NR heterodimers where RXR is involved, the LBD of RXR may not necessarily need to possess a rexinoid. For instance, the VDR-RXR heterodimer functions absent a ligand or rexinoid binding to RXR [10]. Conversely, there are examples of RXR heterodimers where a rexinoid bound to RXR enhances that heterodimer's activity, such as for the LXR-RXR heterodimer [11]. This remarkable versatility—where RXR can partner with many other NRs with and without rexinoids—has led to RXR's classification as the indispensable, master receptor [12].

A great number of studies reported in the literature concerning RXR partnering with other NRs and comprising many rexinoids have been distilled to yield two primary classifications for RXR heterodimers—permissive and nonpermissive. For purely nonpermissive heterodimers of RXR, only the other NR's agonist can activate the heterodimer, whereas permissive RXR heterodimers can be activated by either a rexinoid or the partnering NR's agonists [13]. The RAR-RXR, TR-RXR, and VDR-RXR heterodimers are generally nonpermissive. In the majority, but not every instance, the partnering RXR receptor is "silent" in the TR and VDR heterodimers. The RAR-RXR heterodimer, meanwhile, displays increased activity in the presence of certain rexinoids as well as agonists specific for RAR. The presence of certain rexinoids have been shown to activate RAR-RXR despite the absence of agonists specific for RAR [14]. Thus, the classical idea of purely nonpermissive RXR heterodimers has evolved towards a spectrum of conditions for permissibility, such that some RXR heterodimers, such as RAR-RXR, could be more accurately described as conditionally nonpermissive. The LXR-RXRs, FXR-RXRs, and the PPAR-RXRs, however, are fully permissive.

Potent rexinoids can disrupt the proper functioning of both types of RXR heterodimers, giving rise to pleiotropy by stimulating activity in the permissive RXR heterodimers or by removing RXR from participation in the nonpermissive heterodimers. The tendency to exert pleotropic effects has blocked clinical development of many rexinoids for various therapeutic applications. Rexinoids like 9-cis-RA, for example, arrest activity for VDR-RXR [15–17] and TR-RXR [18]. Similarly, the 1,25-dihydroxyvitamin D_3 (1,25D) and T3 promote formation of VDR-RXR and TR-RXR, respectively, and thus deplete RXR availability for other RXR-dependent pathways. This effect has been termed cross-receptor squelching and is manifested in the loss of TR function via (1,25D)-VDR-RXR-modulated inhibition [18,19], or similarly, in the loss of VDR activity paralleling T_3-TR-RXR-activation [20,21]. However, more than just RXR depletion may account for this crosstalk inhibition. Accordingly, the potency and selectivity are the primary characteristics that must be considered in the development of novel rexinoids to minimize side effects and maximize therapeutic potential. Hence, the approach of slightly modifying the structural features of a parent rexinoid's structure could impact both characteristics and generate rexinoids with less severe pleiotropy and greater specificity, resulting in specific NR modulators (SNuRMs) [22].

A number of rexinoid SNuRMs can be found in advanced stages of pre-clinical or clinical investigation as therapeutics, especially as preventative or treatment regimens for various cancers where selective RXR versus RAR activation exerts therapeutic effects and avoids RAR toxicities [23] in treating many human cancers. After multiple studies [24,25] that used 9-cis-RA as a starting point to model and design new compounds for RXR-selectivity, 4-[1-(3,5,5,8,8,-pentamethyltetralin-2-yl)ethynyl]benzoic acid (**1**) [26] was identified as a lead compound in terms of its stability and RXR-selectivity, though there were many additional candidates with equally promising profiles. After Ligand Pharmaceuticals Inc. earned FDA approval for **1** as a treatment for cutaneous T-cell lymphoma, the tradename of "bexarotene"—the more common name by which **1** is known—was assigned. Many studies making very minor modifications of the structure of bexarotene have identified structurally similar analogs of **1** that display similar profiles of activity in RXR-binding and activity—disilabexarotene (**2**) [27] is a representative example.

Even though bexarotene (**1**) is an approved treatment for CTCL, there are an increasing number of studies in other human cancers and cancer models such as lung [28], breast [29], and colon cancer [30]. In fact, bexarotene is prescribed in certain cases of non-small cell lung cancer off-label, since a proof-of-concept (POC) clinical trial showed benefits of **1** as a treatment [31,32]. Increasingly, studies are linking suppression of cell-proliferation and apoptosis synergy for combination-chemotherapeutic approaches that target RXR-controlled pathways. There are numerous studies in the literature for bexarotene (**1**) and analogous synthetic rexinoids exhibiting therapeutic effects for non-insulin-dependent diabetes mellitus murine models likely tied to metabolic impacts from RXR:PPAR activity [33]. While the selective activation of RXR by bexarotene avoids toxicities associated with RAR activity, humans dosed with **1** often suffer hyperlipidemia and hypothyroidism [34], and sometimes cutaneous toxicity, as the most significant side effects. These side effects arise because, much like 9-cis-RA, bexarotene (**1**) disrupts nonpermissive heterodimers—such as TR-RXR to incite hypothyroidism [35]—and concurrently stimulates permissive heterodimers—such as LXR-RXR to provoke hyperlipidemia [36,37] or cutaneous toxicity [38] via activating RAR at raised dose concentrations. It is difficult to dissociate the potential of a potent synthetic rexinoid to provoke increased activity in RXR's permissive heterodimers from its selectivity for RXR alone, though many research groups are exploring this area. Developing potent rexinoids that mitigate permissive RXR heterodimer activity is a timely objective since **1** has exhibited potential to stem neurodegenerative progression in models of Alzheimer's disease [39] (AD) as well as Parkinson's disease (PD) [40]. A limited POC clinical trial for moderate AD patients treated with **1** or placebo revealed that non-apoE4 genotypes treated with **1** showed a reduction of soluble amyloid beta from cerebrospinal fluid that was statistically significant [41].

Many groups in this field have successfully used modeling and the structures of potent rexinoids in the literature as the basis to develop new rexinoids that display unique profiles. One well-known rexinoid that is in phase II clinical trials for prostate cancer [42] and pre-clinical trials for PD, AD, and multiple sclerosis is IRX-4204 (**3**) [43], a chiral rexinoid shown to be more selective and potent for RXR than its enantiomer. The 9cUAB30 (**4**) [44] is another rexinoid in clinical trials for breast cancer [45–47], and studies of its methylated analogs [48,49] have helped elucidate the reasons that **4** does not provoke hyperlipidemia via LXR-RXR agonism. Boehm's group reported unbranched trienes terminating in carboxylic acids [50], along with analogs incorporating one [51] or fused aryl ring-systems [52]—in the case of the latter, compound **5** [52] is an example. When our group first entered the field, we reported a fluorinated bexarotene analog (**6**) [53] followed by other halogenated, and even a difluorinated bexarotene analog (**7**) [54]. The pyridine bexarotene analog (**8**) [55] and the pyrimidine bexarotene analog (**9**) [56], as well as LGD100268 (**10**) [55] and the LGD100268 pyrimidine analog (**11**) [56] all showed enhanced RXR activity compared to **1** and superior therapeutic effects in mouse models of cancer [28,57]. Installing an unsaturation in bexarotene's aliphatic ring results in compound **12** [58,59], and the unsaturated-fluorinated bexarotene (**13**) [56] also activates LXR [60]. CD3254 (**14**) [61] and CD2915 (**15**) [62] are two synthetic rexinoids with activity comparable to **1**. Our group utilized **14** and **15** to design analogous rexinoids **16–19** [56]. Kakuta and colleagues reported compound **20** (NEt-TMN) [63–66] as well as its analogous compounds **21** [67–69] and **22** [67,68]—all of which showed high potency and selectivity for RXR alongside many other NEt-TMN derivatives that our group reported [70]. Even replacing the ethyl group with a methyl group on the linking nitrogen atom of NEt-TMN leads to potent rexinoids such as **23** [71], **24** [71], and **25** [71] (Figure 1).

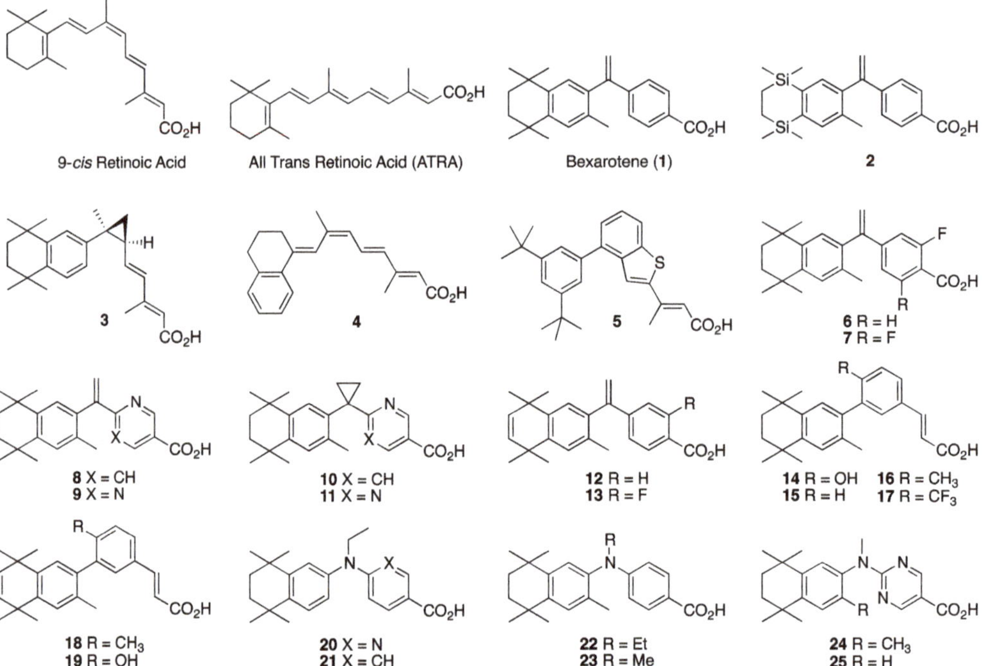

Figure 1. Structures of 9-*cis* Retinoic Acid, ATRA, Bexarotene (**1**), and Rexinoids **2–25**.

The current study uses many of the above compounds cited above as starting points to investigate how introducing new modifications result in changes to the compounds' activities. For example, we were interested in substituting an isochroman group for the aliphatic ring system in bexarotene and some of the CD3254 analogs that have been reported, as well as a multiple fused aryl-ring system, and hence we targeted the synthesis of rexinoids **26–30**. We were also interested in exploring a pyridine aromatic ring substitution from reported analogs **23–25**, so we targeted the synthesis of **31**. Due to the potency of

NEt-TMN (**20**) and its analogs, we were curious about the impacts that substituting an allyl group, varying aromatic rings, and adding a methyl group would have on RXR activity for the new rexinoids, so we targeted the synthesis of **32–35**. Finally, we were interested in substituting the isochroman group for the aliphatic ring system of NEt-TMN and then varying the *N*-alkyl chains—including methy, ethyl and allyl—along with different aromatic acid ring systems, and so we targeted **36–44** for synthesis (Figure 2). Interestingly, compound **34** [67] was previously made and disclosed by Kagechika and co-workers, so we were eager to synthesize several possible analogs of it (**32, 33, 35, 40, 41**, and **44**) and compare their activities.

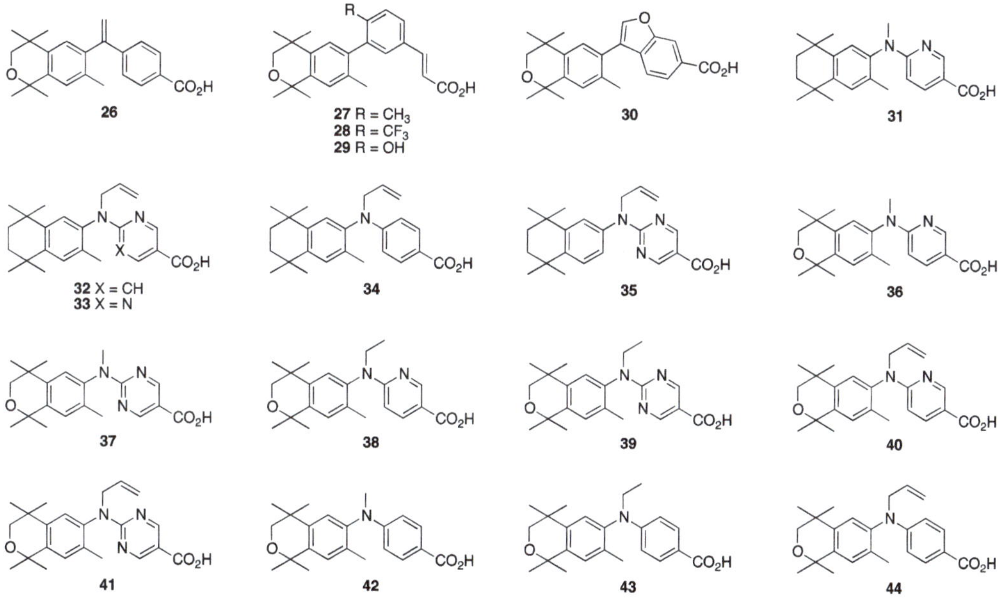

Figure 2. Structures of Reported Rexinoids **3–27**.

While the isochroman group places a fairly polar, hydrogen-bonding oxygen atom in the nonpolar aliphatic ring system for the known parent rexinoids, we hypothesized that it would not disrupt the largely non-polar binding interaction in the RXR LBD. Further, we hypothesized that the isochroman group would make the rexinoids possessing it more resistant to metabolic oxidation. If the isochroman rexinoids were as potent and selective as their non-isochroman counterparts, we also hypothesized that they would possess similar in vitro activities as their parent compounds. Importantly, we wanted to assess these specific hypotheses about the incorporation of an isochroman group across multiple parent compound structures that include bexarotene (**1**), CD3254 (**14**), and various NEt-TMN (**20**) analogs coupled with other potential modifications to assess similarities or differences in the activities of the resulting analogs and their predecessors. Hence, we proceeded with the synthesis and testing of these compounds.

2. Results: Molecular Modeling

The binding affinity, predicted using AutoDock Vina [72], of human-RXR for each ligand is output as an energy unit (in kcal/mol, Table 1). These predicted ligand-bound RXR complexes were then visually inspected using PyMol (version 2.3, Shrodinger, LLC) (Figure 3). To further analyze and illustrate the interactions between RXR protein residue sidechains with the ligands, PoseView (BioSolvIT [73,74]) was used to generate the detailed two-dimensional depictions (Figure 3B,C). In these two-dimensional depictions, hydrogen bonds are presented as dashed lines between interaction partners, and hydrophobic interactions are depicted as smooth contour lines.

Table 1. Tabe of Auto-Dock Vina Binding Energies, EC_{50} (nM) values, 96h IC_{50} (nM) + 100 nM ATRA values, LXRE Activities (% of Bex), LHS Score (vs. Bex), and RARE Activity (%ATRA at 10 nM).

Compound	Auto-Dock Vina Scores (kcal/mol)	EC_{50} (nM) +/− (SD)	96 h IC_{50} (nM) + 100 nM ATRA +/− SD	LXRE Activity (% of Bex)	LHS Score (vs. Bex)	RARE Activity (%ATRA at 10 nM)
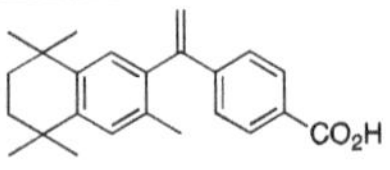 1	−12.7	17.8 (1.0)	7.8 (1.1)	100	1	37.16
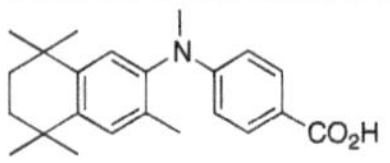 23	−11.5	8.6 (0.2)	3.7 (1.1)	99.5	0.59	24.33
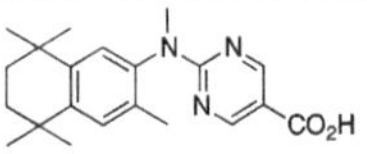 24	−10.9	6.2 (0.1)	3.9 (1.1)	105.98	0.95	32.04
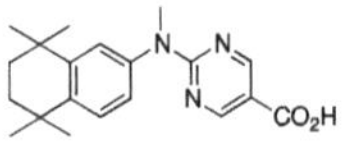 25	−10.6	17.8 (0.2)	12.9 (1.2)	80.98	0.7	14.32
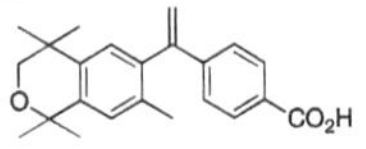 26	−12.4	51.0 (0.1)	60.0 (1.2)	213.84	2.81	10.1
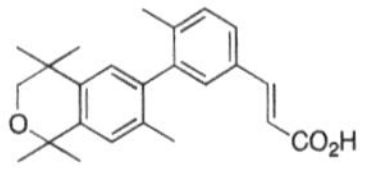 27	−9.3	65.3 (0.1)	27.6 (1.2)	253.64	1.76	19.45
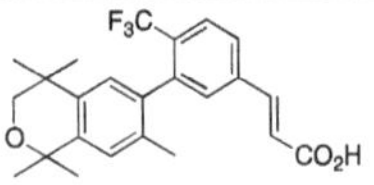 28	−11.7	59.2 (0.1)	38.4 (1.3)	208.05	1.06	15.85
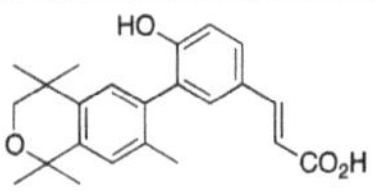 29	−11.5	3.9 (0.5)	3.7 (1.2)	221.39	1.29	54.64
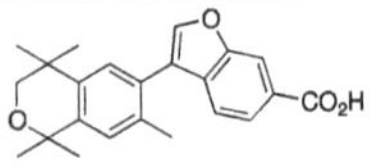 30	−11.9	>1000	>1000	161.38	4.03	7.05
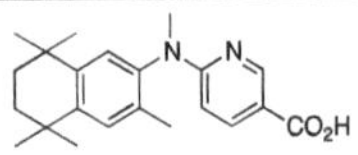 31	−11.5	5.4 (0.1)	2.1 (1.1)	114.37	1.04	48.61

Table 1. *Cont.*

Compound	Auto-Dock Vina Scores (kcal/mol)	EC$_{50}$ (nM) +/− (SD)	96 h IC$_{50}$ (nM) + 100 nM ATRA +/− SD	LXRE Activity (% of Bex)	LHS Score (vs. Bex)	RARE Activity (%ATRA at 10 nM)
32	−11.4	1.8 (1.1)	2.0 (1.6)	92.2	0.61	61.79
33	−11.0	5.3 (0.2)	1.3 (1.1)	103.99	0.72	39.94
34	−11.6	11.5 (0.2)	4.1 (1.2)	77.44	0.53	18.74
35	−10.6	0.54 (0.11)	4.68 (1.91)	94.07	0.79	28.76
36	−11.0	75.0 (0.1)	42.3 (1.2)	51.57	0.69	6.25
37	−10.8	404.9 (0.1)	>1000	62.72	1.36	4.02
38	−11.0	23.7 (0.2)	11.6 (1.1)	79.69	0.69	12.59
39	−10.7	151.6 (0.1)	>1000	61.52	0.92	4.96
40	−11.1	10.7 (0.2)	16.6 (1.2)	73.76	0.63	11.13
41	−10.7	143.4 (0.3)	105 (1.6)	75.66	0.71	5.16
42	−11.4	87.7 (0.1)	136.7 (1.6)	61.86	0.56	7.99

Table 1. *Cont.*

Compound	Auto-Dock Vina Scores (kcal/mol)	EC$_{50}$ (nM) +/− (SD)	96 h IC$_{50}$ (nM) + 100 nM ATRA +/− SD	LXRE Activity (% of Bex)	LHS Score (vs. Bex)	RARE Activity (%ATRA at 10 nM)
43	−11.5	23.1 (0.1)	14.8 (1.2)	71.09	0.55	9.6
44	−10.7	116.5 (0.1)	~1000	68.04	0.68	11.13

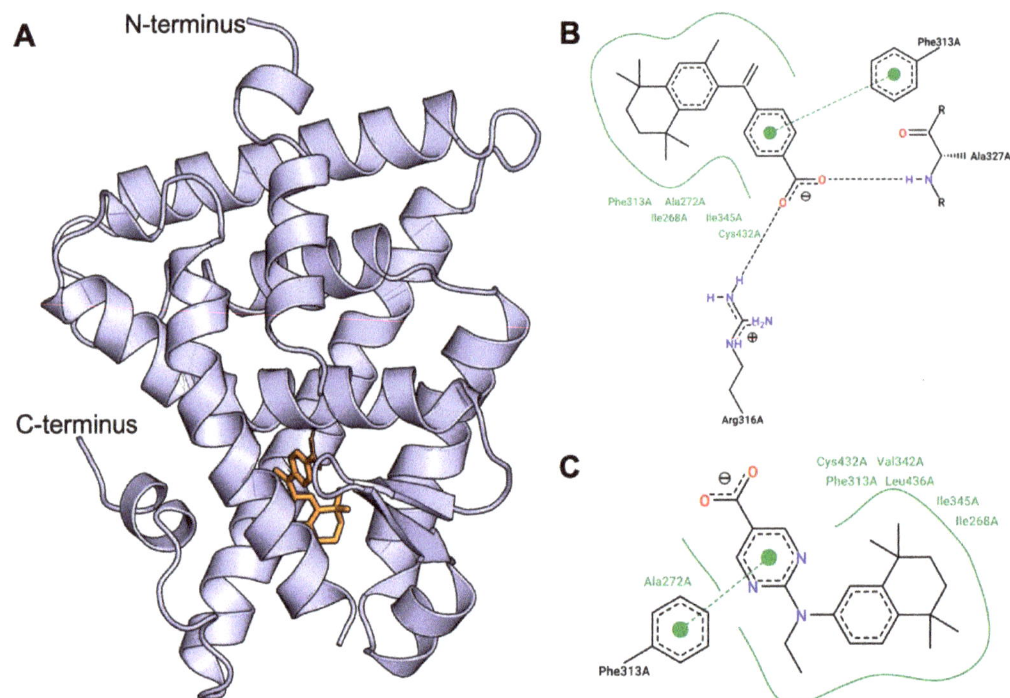

Figure 3. AutoDock Vina simulation of bexarotene bound to human RXR protein. (**A**) Cartoon representation of the human RXR alpha ligand binding domain (PDB:1FBY, blue) and the docked compound bexarotene (orange). N- and C-termini are labeled. (**B,C**) Two-dimensional depiction of the interactions between protein residue sidechains with bexarotene (**B**) and compound **26** (**C**) using PoseView (BioSolvIT [73,74]). In both (**B**) and (**C**), hydrogen bonds are presented as dashed lines between interaction partners, and hydrophobic interactions are depicted as smooth contour lines.

The AutoDock Vina docking results showed that the standard compound bexarotene (**1**), with a score of −12.7 kcal/mol, was the most potent among all compounds. Compound **26** has a comparable score of −12.4 kcal/mol. Additionally, similar interactions with RXR residue sidechains were observed between **1** and **26** such as direct interactions between Phe313 and the aromatic cores of either ligand, as well as hydrophobic residues Ile268, Ile345, and Ala272 bearing common interactions (Figure 3B,C). Several other compounds have similar but slightly lower scores of −11.9 kcal/mol (**30**), −11.7 kcal/mol (**28**),

−11.6 kcal/mol (**34**), and −11.5 kcal/mol (**23, 29, 31, 32, 42, 43,** and **44**) (Table 1). Therefore, we can conclude that RXR compounds with a docking score within 10% that of bexarotene could potentially possess comparable, or better, EC_{50} and IC_{50} profiles for further study.

3. Results: Chemistry

The synthesis of compound **26** begins with the bromination of commercially available 1,1,4,4,7-pentamethylisochroman (**45**) with bromine in dichloromethane and aluminum chloride catalyst to give 6-bromo-1,1,4,4,7-pentamethylisochroman (**46**) in 96% yield (Scheme 1) [75].

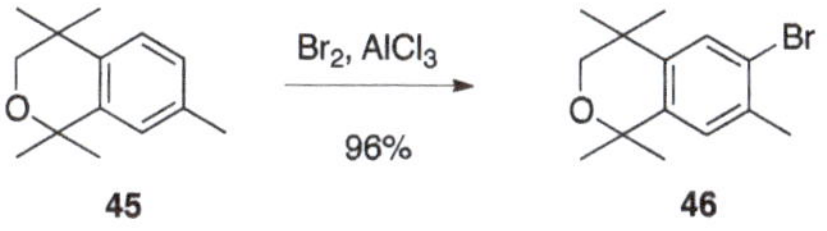

Scheme 1. Synthesis of bromo-isochroman **46**.

Compound **46** was lithiated by treatment with n-butyllithium, and the resulting aryl lithium reagent (**47**) was transferred to a solution of the reported amide **48** [59] in THF to provide ketone (**49**) in 56% purified yield whose nitrile group is then hydrolyzed to a carboxylic acid (**50**) in quantitative yield. Compound **50**, when treated with more than two equivalents of methylmagnesium chloride is converted to alcohol **51** that is then dehydrated to compound **26** in 58.3% over two steps (Scheme 2).

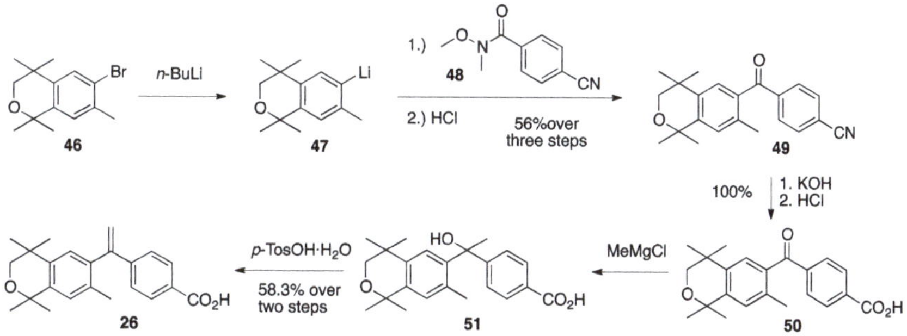

Scheme 2. Conversion of **46** to lithium-isochroman **47**, its reaction with amide-nitrile **48** to give ketone-nitrile **49** that was subsequently hydrolyzed to ketone-acid **50** which was converted to alcohol-acid **51** that was finally dehydrated to isochroman acid **26**.

Acid **26** formed single, transparent crystals, so an X-ray diffraction study was conducted to confirm the structure of **26** (Figure 4).

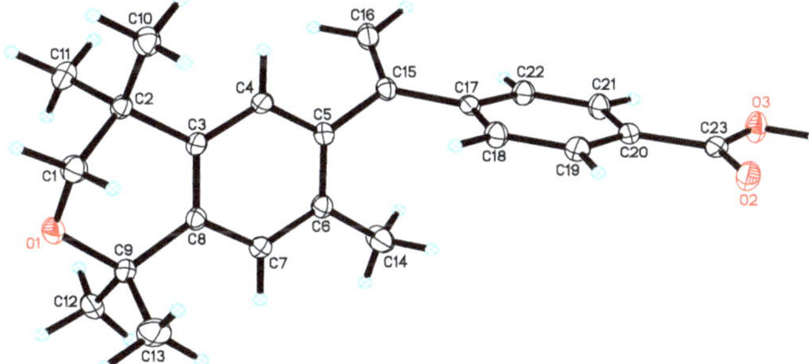

Figure 4. X-ray crystal structure of a 0.114 × 0.141 × 0.380 mm^3 crystal of **26**. Thermal ellipsoids are shown at 50% probability level.

To synthesize compound **27**, compound **46** was treated with n-butyllithium and then triisopropylborate to give **52** in 73% yield (Scheme 3).

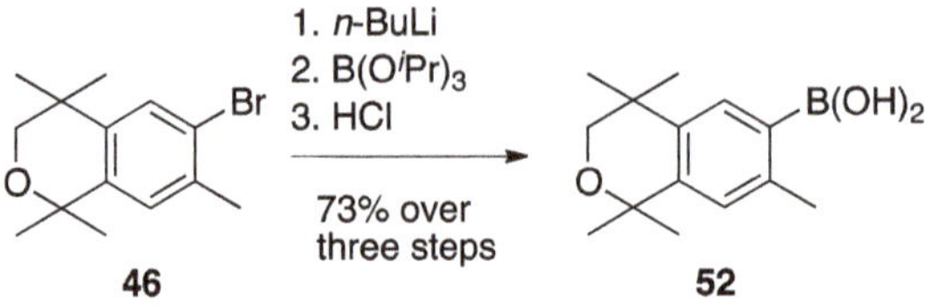

Scheme 3. Synthesis of boronic acid **52**.

The known ethyl ester **53** [56] was combined with boronic acid **53** along with Pd(II) diacetate, TBAB, and sodium carbonate in water and heated to boiling for 5 min to give **54** in 64% yield, and ethyl ester **54** was saponified with KOH in methanol to give **27** in 81.8% yield after acidification with 1N HCl, filtration, and purification by column chromatography (Scheme 4).

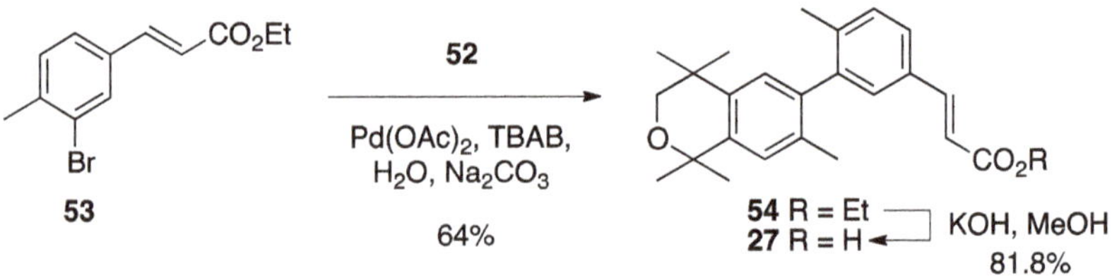

Scheme 4. Synthesis of biphenyl **54** and its saponification to give acid **27**.

A similar procedure was used to make compounds **28**, **29**, and **30**. For compound **28**, the known ethyl ester **55** [56] was combined with boronic acid **52** along with Pd(II) diacetate, TBAB, and sodium carbonate in water and heated to boiling for 5 min to give **56** in 67% yield, and ethyl ester **56** was saponified with KOH in methanol to give **28** in 71.9% yield after acidification with 1N HCl, filtration, and purification by column chromatography (Scheme 5).

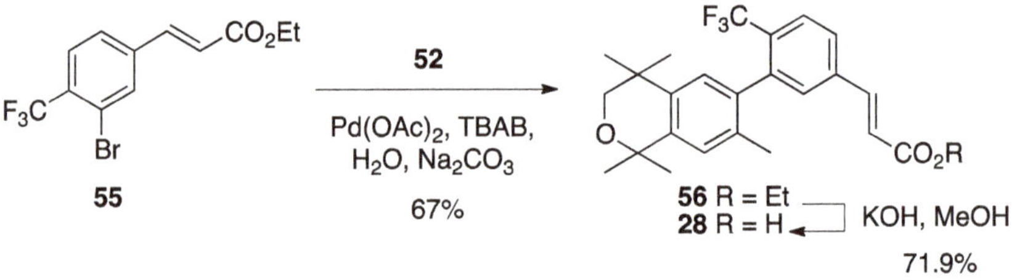

Scheme 5. Synthesis of biphenyl **56** and its saponification to give acid **28**.

For compound **5**, the known ethyl ester **57** [56] was combined with boronic acid **52** along with Pd(II) diacetate, TBAB, and sodium carbonate in water and heated to boiling for 5 min to give **58** in 41.7% yield, and ethyl ester **58** was saponified with KOH in methanol to give **29** in 75% yield after acidification with 1N HCl, filtration, and purification by column chromatography (Scheme 6).

For compound **30**, the commercially available methyl ester **59** was combined with boronic acid **52** along with Pd(II) diacetate, TBAB, and sodium carbonate in water and heated to boiling for 5 min to give **60** in 48% yield, and methyl ester **60** was saponified with KOH in methanol to give **30** in 83% yield after acidification with 1N HCl, filtration, and purification by column chromatography (Scheme 7).

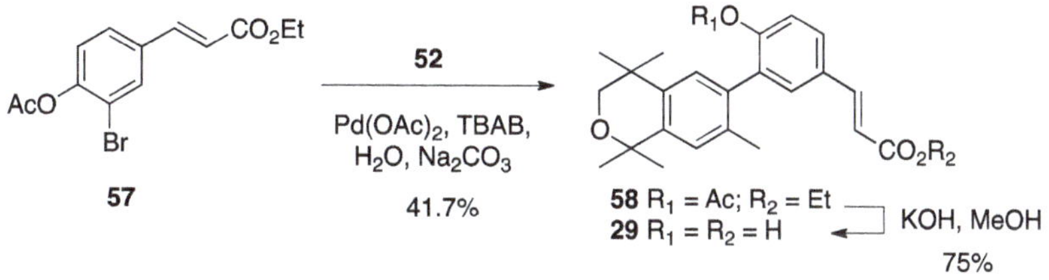

Scheme 6. Synthesis of biphenyl **58** and its saponification to give acid **29**.

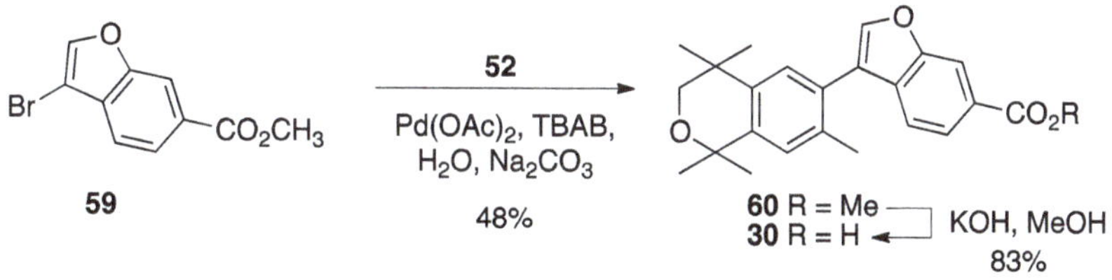

Scheme 7. Synthesis of methyl ester **60** and its saponification to give acid **30**.

The synthesis of known compounds **23**, **24**, **25**, and novel compound **31** begins with the conversion of known compounds **61–64** [70] to compounds **65–68** by treatment with sodium hydride followed by methyl iodide in DMF with stirring at room temperature in good yields. The methyl esters **65–68** were saponified by treatment with KOH in methanol followed by acidification and purification to give compounds **23–25** and **31**, respectively, in moderate to good yields (Scheme 8).

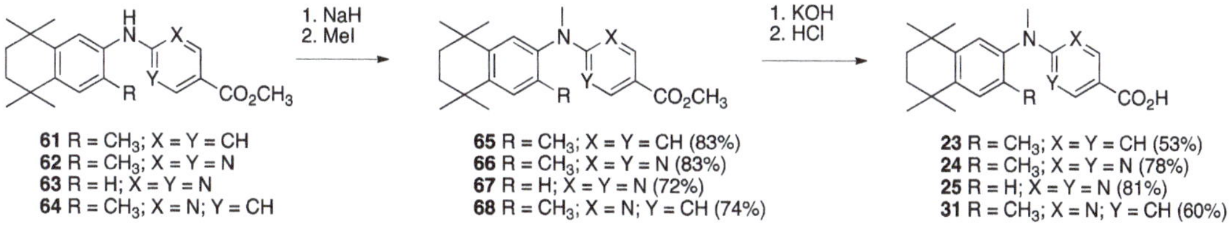

Scheme 8. Synthesis of methyl esters **65–68** and their saponification to give acids **23–25** and **31**.

The synthesis of compounds **32–35** begins with the conversion of known compounds [70] **64**, **62**, **61**, and **63** to compounds **69–72** by treatment with sodium hydride followed by allyl bromide in DMF with stirring at room temperature in good yields. The methyl esters **69–72** were saponified by treatment with KOH in methanol followed by acidification and purification to give compounds **32–35** in moderate to good yields (Scheme 9).

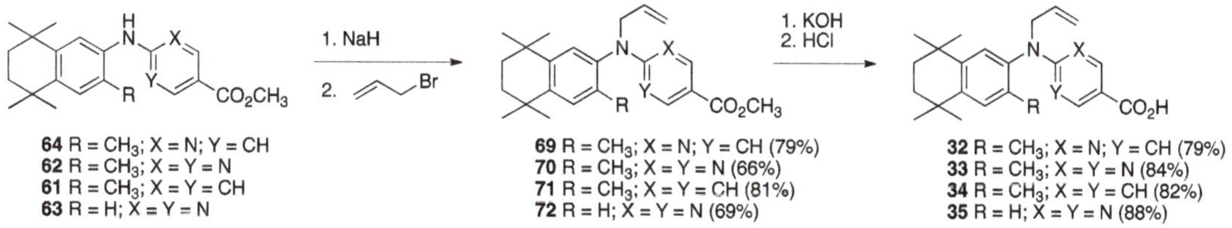

Scheme 9. Synthesis of methyl esters **69–72** and their saponification to give acids **32–35**.

The synthesis of compounds **36–41** begins with the palladium-catalyzed reaction of the known aryl bromide **46** with methyl 6-aminonicotinate (**73**) or methyl 2-aminopyrimidine-5-carboxylate (**74**) to give diaryl amines **75** or **76** in moderate yields of 55% and 40%, respectively. The diaryl amines **75** and **76** were treated with sodium hydride followed by methyl iodide, ethyl iodide, or allyl bromide to give methyl esters **77–82** in moderate to good yields which were then saponified by treatment with KOH in methanol followed by acidification and purification to give compounds **36–41** in moderate to good yields (Scheme 10).

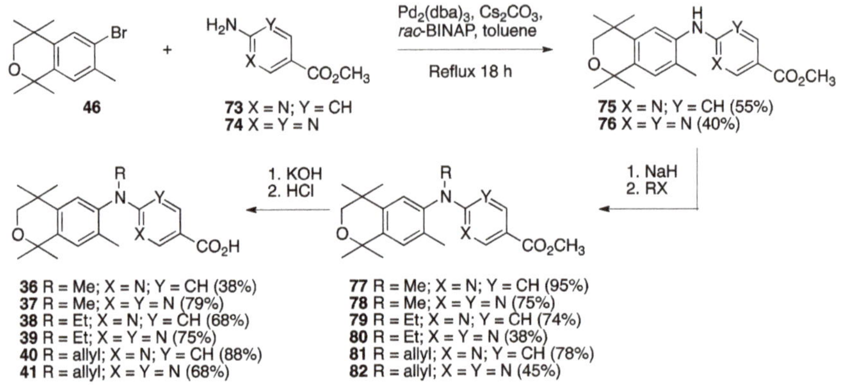

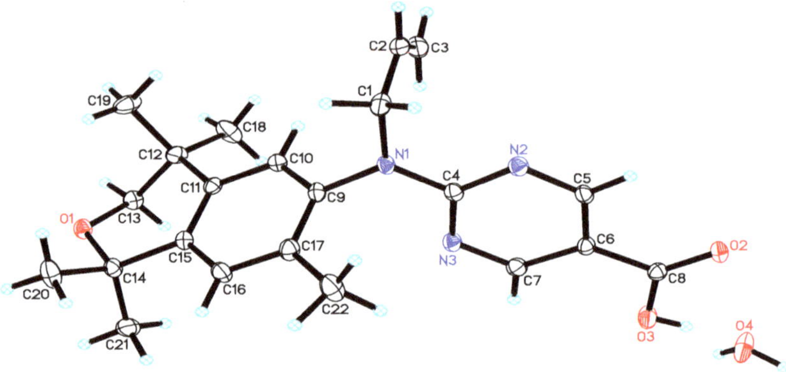

Scheme 10. Synthesis of diaryl amines **75–76**, their alkylation to give methyl esters **77–82** and their saponification to give acids **36–41**.

Acid **41** formed single, transparent crystals from d6-DMSO, so an X-ray diffraction study was undertaken to confirm the structure for **41** (Figure 5).

Figure 5. X-ray crystal structure of a 0.184 × 0.217 × 0.371 mm^3 crystal of compound **41** as a monohydrate. Thermal ellipsoids are shown at 50% probability level.

Finally, the synthesis of novel rexinoids **42–44** begins with the nitration of isochroman (**45**) to give **83** in 98% yield, followed by the hydrogenation of **83** to give aryl amine **84** in quantitative yield. Aryl amine **84** is then combined with 4-iodomethylbenzoate (**85**) in a palladium-catalyzed reaction to give **86** in 24% yield. Diaryl amine **86** is then treated with sodium hydride followed by methyl iodide, ethyl iodide, or allyl bromide to give methyl esters **87–89** in moderate to good yields and the methyl esters are saponified with potassium hydroxide in methanol followed by acid work-up to give rexinoids **42–44** in good yields after purification (Scheme 11).

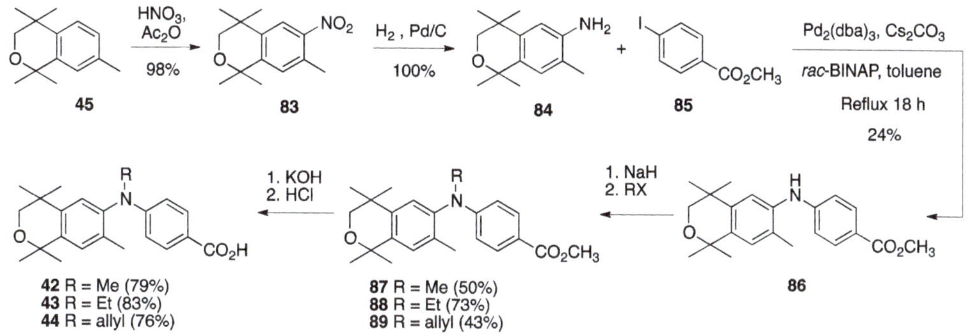

Scheme 11. Synthesis of nitro-isochroman **83**, its reduction to aniline **84** and subsequent coupling with **85** to provide **86** and its subsequent alkylation to give methyl esters **87–89** and their saponification to yield acids **42–44**.

4. Results and Discussion: Biological Assays

Bexarotene (**1**) and the isochroman and other analogs of bexarotene, CD3254, and NEt-TMN **26–44** were tested in a KMT2A-MLLT3 cell line to determine RXRα activation EC50 values by both a Luc– and GFP-assay, and then IC50 values for a 96 h cell viability assay were determined both in the presence and absence of 100 nM ATRA and the resulting data is tabulated in Table 1. Compounds **1** and **26–44** were examined for cytotoxicity and mutagenicity in Saccharomyces cerevisiae [76]. None of the compounds were mutagenic or cytotoxic.

We also evaluated the analogs for their capacity to bind and activate the liver-X-receptor (LXR) using a liver-X-receptor responsive element (LXRE)-based assay, and assessed the effects in the presence vs. absence of an authentic LXR compound (TO901317), as well as with bexarotene alone. LXR is known to regulate inflammatory responses and lipid metabolism in multiple tissues, including the central nervous system, and there is sufficient evidence that strong cholesterol and lipid metabolism in the brain, along with enhanced ApoE expression, is paramount to reducing the risk of human dementias. The "activity assessment" of our novel RXR agonists for their ability to enhance gene expression using a DNA responsive element sequence (LXRE) that is found to occur in the natural human promoter of LXR-RXR controlled genes (such as ApoE) was carried out in U87 glial cells with bexarotene (**1**) as a comparative control. Using the above assay system, the activation from this natural LXRE was probed with either 100 nM RXR analogs (or bexarotene) alone or in combination with 100 nM of both the RXR agonist and an LXR ligand T0901317 (TO). The employment of both LXR and RXR agonists collectively was anticipated to demonstrate a more vigorous response in LXRE transactivation due to cumulative and/or synergistic effects of dual ligand activation of the RXR-LXR heterodimer. The results demonstrated that in comparison to the parent bexarotene (**1**) compound alone, separate dosing of the cells with analogs **26**, **27**, **28**, **29**, and **30** displayed more LXR/LXRE activity (Figure 6A, $p < 0.05$).

Specifically, the analogs displayed activities ranging from 52% to 254% of the bexarotene control (set to 100%; Table 1). Furthermore, when the LXR synthetic ligand (TO) was used in combination with bexarotene (**1**) or analogs, similar results were observed (Figure 6A, Figure 7A, and Figure 8A, stippled bars). Although some of the analogs exhibit lower LXR activation when compared to bexarotene (**1**), it is imperative to evaluate this activity in the background of the RXR-RXR homodimer activity of each analog, and to thus "normalize" the LXR/LXRE heterodimer activation results. The LXRE Heterodimer Specificity (LHS) score (Figure 6B, Figure 7B, and Figure 8B) is thus calculated based on both the RXR homodimer activity, as well as the LXR heterodimer activation. The results of this LHS analysis (Table 1) reveal that many of our novel compounds (e.g., **26**, **27**, **29**, **30**, **31**, and **37**, $p < 0.05$) display enhanced LXR/LXRE activity via increased heterodimer specificity compared to the parent bexarotene (**1**).

Finally, even though compound **1** is selective for binding to RXR, it does include some "residual" RAR activity. We assessed the potential of our analogs to induce transcription via the retinoic acid response element (RARE) and retinoic acid receptor. For these studies, we employed human embryonic cells (HEK293) that were transfected with human RARα and dosed with 10 nM of either all-trans retinoic acid (ATRA), the endogenous ligand for RARα, compound **1**, or our panel of analogs. The results of the assay revealed that compound **1** displayed an average of 37% of the activity of the ATRA control (set to 100%, Table 1). Analog **32** exhibited the greatest RARE activation at 62% of ATRA, while analog **37** showed the lowest RARE activity at 4%, which is indistinguishable from the ethanol control (Table 1). Importantly, most of our novel analogs (14 out of 18) possess attenuated "cross-over" onto RAR-RARE signaling compared to bexarotene (**1**).

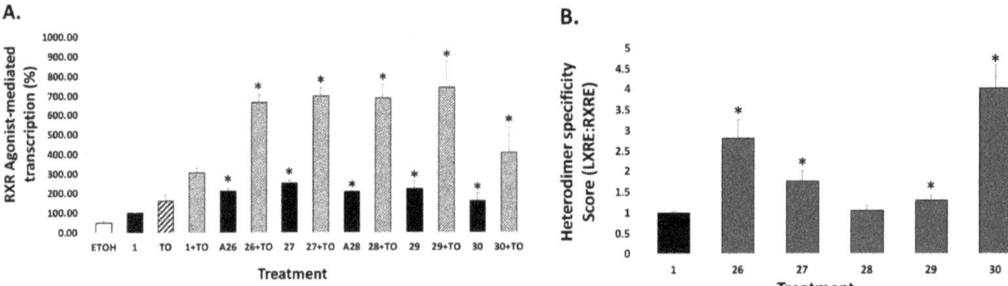

Figure 6. RXR agonist potentiation of LXRE-regulated transactivation with or without the T0901317 LXR agonist. (**A**) U87 glial cells were transfected with an expression vector for human LXRα, an LXRE-luciferase reporter gene with three tandem copies of the LXRE from the human ApoE gene, and a Renilla control plasmid. Cells were transfected for 24 h utilizing a liposome-mediated transfection protocol and then treated with ethanol vehicle, or 100 nM of the indicated compound alone or in combination with 100 nM TO901317 (TO). LXRE-directed activity was compared to compound 1 (Bexarotene), set to 100%. (**B**) The "Heterodimer Specificity Score" was determined by the LXRE:RXRE ratio with compound 1 set to 1.0. All error bars represent standard deviations; the data are representative of at least three independent experiments with six replicates in each treatment group. * $p < 0.05$ versus control compound **1**.

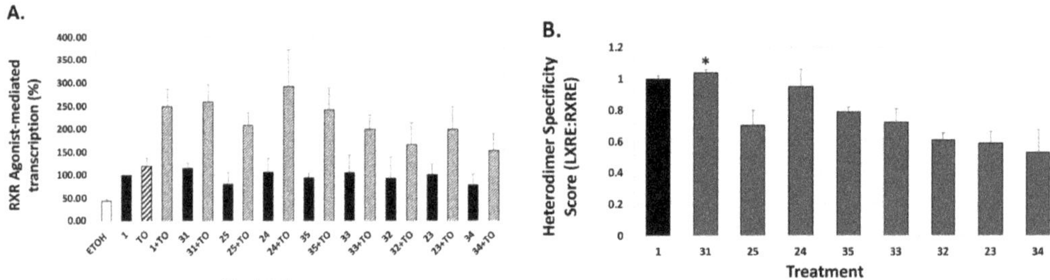

Figure 7. RXR agonist potentiation of LXRE-regulated transactivation with or without the T0901317 LXR agonist. (**A**) U87 glial cells were transfected with an expression vector for human LXRα, an LXRE-luciferase reporter gene with three tandem copies of the LXRE from the human ApoE gene, and a Renilla control plasmid. Cells were transfected for 24 h utilizing a liposome-mediated transfection protocol and then treated with ethanol vehicle, or 100 nM of the indicated compound alone or in combination with 100 nM TO901317 (TO). LXRE-directed activity was compared to compound 1 (Bexarotene), set to 100%. (**B**) The "Heterodimer Specificity Score" was determined by the LXRE:RXRE ratio with compound 1 set to 1.0. All error bars represent standard deviations; the data are representative of at least three independent experiments with six replicates in each treatment group. * $p < 0.05$ versus control compound **1**.

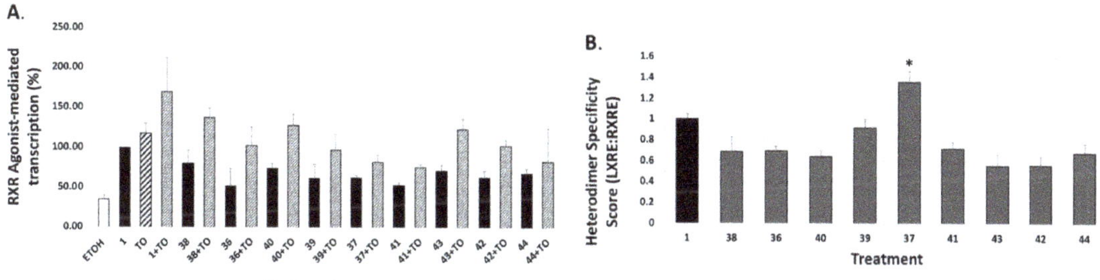

Figure 8. **Evaluation of RXR agonists to potentiate LXRE-mediated transactivation in the absence and presence of LXR ligand T0901317.** (**A**) U87 glial cells were transfected with an expression vector for human LXRα, an LXRE-luciferase reporter gene with three tandem copies of the LXRE from the human ApoE gene, and a Renilla control plasmid. Cells were transfected for 24 h utilizing a liposome-mediated transfection protocol and then treated with ethanol vehicle, or 100 nM of the indicated compound alone or in combination with 100 nM TO901317 (TO). LXRE-directed activity was compared to compound 1 (Bexarotene), set to 100%. (**B**) The "Heterodimer Specificity Score" was determined by the LXRE:RXRE ratio with compound 1 set to 1.0. All error bars represent standard deviations; the data are representative of at least three independent experiments with six replicates in each treatment group. * $p < 0.05$ versus control compound 1.

5. Conclusions

While the substitution of the saturated, pentamethyl-tetrahydronaphthyl ring system of bexarotene, CD3254, or Net-TMN-analog parent compound frameworks with a pentamethyl-isochroman ring system tended to decrease both modeled binding scores as well as RXR EC50 values, there were nevertheless some examples of very potent compounds possessing the pentamethyl-isochroman ring system. Several reported and novel compounds displayed lower or equivalent EC50 values than bexarotene (e.g., reported compounds **23–25** and novel compounds **29, 31–33, 35,** and **40**), and several reported and novel compounds exhibited lower IC50 values along with 100 nM ATRA (e.g., reported compounds **23** and **24**, and novel compounds **29** and **31–35**) than bexarotene. It should be noted that compound **29** is an isochroman analog of CD3254 and compound **40** is an isochroman analog of NEt-TMN, so while the isochroman substituted analog tends to reduce potency compared to the parent compounds, rexinoids possessing the isochroman group can still be synthesized that exhibit greater potency than bexarotene. Notably, some of the compounds possessing the greatest RARE activity at 10 nM vs. ATRA also exhibited lower IC50 values along with 100 nM ATRA (e.g., **29** and **31–33**). Additionally, the range of LXRE activity vs. bexarotene demonstrated that several compounds exhibit increased LXR activity and specificity vs. bexarotene. In particular, compounds **29** and **31–35** appear to be good candidates for further investigation, and this study suggests that modification of known, potent rexinoids results in novel compounds with unique and promising therapeutic potential meriting further examination. We are actively pursuing additional studies towards these ends.

6. Materials and Methods

Molecular Modeling. The three-dimensional structures of the compounds reported herein were generated using ChemDraw 3D (PerkinElmer Informatics, Waltham, MA, USA), energy minimized, and exported in the Protein Data Bank (PDB) format. The human RXR alpha ligand binding domain structure model was obtained from the PDB (PDB code: 1FBY [77]). The crystallized ligand, 9-cis retinoic acid, was removed from the protein model prior to docking simulations. Further, 9-cis retinoic acid was also used as a positive control in the docking studies presented here. Both the protein and ligand models were prepared using MGLTools (version 1.5.7) [78] and screened virtually using

AutoDock Vina [72]. The search space volume (4032 Å3) was determined using MGLTools (center_x = 12.848, center_y = 29.174, center_z = 50.269, size_x = 16, size_y = 14, size_z = 18). The exhaustiveness was set to 8.

Hematopoietic cell culture. Murine bone marrow Kit$^+$ cells were isolated using an Automacs Pro (Miltenyi Biotec, San Diego, CA, USA) according to the manufacturer's protocol. Kit$^+$ cells were plated in progenitor expansion medium (RPMI 1640 medium, 15% fetal bovine serum, stem factor [50 ng/mL], interleukin 3 [10 ng/mL], Flt3L [25 ng/mL], thrombopoietin [10 ng/mL], L-glutamine [2 mM], sodium pyruvate [1 mM], HEPES buffer [10 mM], penicillin/streptomycin [100 units/mL], and b-mercaptoethanol [50 mM]) overnight and transduced with MSCV-KMT2A-MLLT3 retrovirus by spinfection with 10 µg/mL polybrene and 10 mM HEPES at 2400 rpm, 30 °C for 90 min in an Eppendorf 5810R centrifuge. Cells were transplanted into sublethally irradiated mice and subsequent leukemia harvested 4 to 6 months later. KMT2A-MLLT3 leukemia cells were cultured in vitro using similar media, but without Flt3L or thrombopoietin.

UAS/Gal4 assay. Bone marrow cells from UAS-GFP mice [79] were transduced with retroviruses MSCV-Gal4 (DNA binding domain, DBD)—RXRα (ligand binding domain, LBD)—IRES—mCherry. Gal4 is a yeast transcription factor and the UAS sequence is not recognized by mammalian transcription factors. Cells were treated, and after 48 h, GFP was measured by flow cytometry.

Luciferase detection. HEK293T cells were transfected using Lipofectamine 2000 (Invitrogen) [80]. Six hours after transfection, the cells were collected and plated into a 48-well plate in 1% BSA media in biological triplicates and treated with compounds. After 40 h incubation, the cells were harvested and assayed for luciferase (Luc Assay System with Reporter Lysis Buffer, Promega) in a Beckman Coulter LD400 plate reader.

LXRE assay. Human glioblastoma cells (U87) were purchased from American Type Culture Collection (ATCC, Manassas, VA, USA) and used to perform the LXRE-mediated assays. Authentication and validation of these master stocks included both testing for mycoplasma contamination (via Universal Mycoplasma Detection Kit; ATCC, Cat. #30-1012K), as well as short tandem repeat (STR) analysis to confirm cell line identity. For LXRE experiments, cells were seeded at a density of 80,000 cells/well in a 24-well plate and maintained in DMEM (Hyclone) supplemented with 10% fetal bovine serum, 100 µg/mL streptomycin, 100 U/mL penicillin (Invitrogen, Carlsbad, CA, USA) at 37 degrees Celsius, 5% CO2 for 24 h. The cells were transiently transfected in individual wells using Polyethylenimine (PEI) (Polysciences, Inc., Warrington, PA, USA) according to the manufacturer's protocol. The cells in each well received 250 ng of an LXRE-luciferase reporter gene, 50 ng of CMX-h-LXRα (an expression vector for human LXRα), 50 ng of pSG5-human RXRα (an expression vector for human RXRα), and 20 ng of Renilla control plasmid was used along with 1.25 µL of PEI reagent. After 22–24 h of transfection, the cells were treated with either vehicle control ethanol, reference compound bexarotene (100 nM) alone, or in combination with 100 nM T0901317 (an LXR ligand), or 100 nM of the indicated bexarotene analog either alone or in combination with T0901317, as indicated. All compounds were solubilized in ethanol. After 24 h of treatment, the cells were lysed in 1X passive lysis buffer (Promega, Madison, WI USA) and the amount of reporter gene product (luciferase) was quantified using the Dual-Luciferase Reporter Assay System based on the manufacturer's protocol (Promega) in a Sirius FB12 luminometer (Berthold Detection Systems, Pforzheim, Germany). Luminescence resulting from the inducible firefly luciferase was divided by luminescence from the constitutively expressed Renilla luciferase in order to normalize for transfection efficacy, cell death, and cellular toxicity from ligand exposure. The data are a compilation of between six to eight independent assays with each treatment group dosed in triplicate for each independent assay. The LXRE-directed transcriptional activation of the reporter gene was measured in comparison to the reference compound bexarotene (**1**) set to 100%. Error bars on all graphs indicate the standard deviation of the replicate experiments.

RARE assay. Human embryonic kidney cells (HEK293) were plated at 60,000 cells per well in a 24-well plate and maintained as described above. After 22–24 h, the cells were transiently transfected with 250 ng pTK-DR5(X2)-Luc, 25 ng pSG5-human RXRα, and 20 ng of Renilla control plasmid using 1.25 μL polyethylenimine (PEI) per well for 24 h. The sequence of the double DR5 RARE is: 5'-AA<u>AGGTCA</u>CCGAA<u>AGGTCA</u>CCATCCCGGG <u>AGGTCA</u>CCGAA<u>AGGTCA</u>CC-3' (DR5 responsive elements underlined). After 22–24 h of transfection, the cells were treated with ethanol vehicle (0.1%), all-trans-retinoic acid (ATRA, the endogenous ligand for RAR), or the indicated rexinoid analog at a final concentration of 10 nM. After 24 h of treatment, the cells were lysed and the retinoid activity was measured as described above (dual luciferase assay). The activity of compound **1** (or analog) divided by the activity of ATRA (expressed as a percentage) represents the RARE activity. The data (Table 1) are a compilation of between three to four independent assays with each treatment group dosed in triplicate for each independent experiment. The value for the positive control ATRA was set to 100%.

Data analysis. Statistical analysis was performed using Microsoft Excel (Microsoft, Redmond, WA, USA) software. *t*-tests were performed, as appropriate. All error bars represent the standard deviation. For Figures 6–8, data are expressed as means ± SD. Statistical differences between two groups (generally the bexarotene (**1**) control group versus bexarotene analog group) were determined by a two-sided Student's *t*-test. A *p*-value of less than or equal to 0.05 was considered significant.

Mutagenicity and Toxicity Assay. Mutagenicity and toxicity were assessed in a eukaryotic Saccharomyces cerevisiae model as described previously [76]. Each compound was solubilized in DMSO at increasing concentrations. D7 *S. cerevisiae* indicator cells were incubated with the compounds for 3 h before plating on selective media or YPD. Growth on plates was used to determine mutagenicity and cytotoxicity. Growth of colonies on the full nutrient YPD plate for each treatment was compared to the DMSO only control. Eleven micrograms per microliter was the highest concentration tested.

X-ray Data Collection, Structure Solution and Refinement. A colorless crystal of indicated dimensions was mounted on a glass fiber and transferred to a Bruker SMART APEX II diffractometer system. The APEX2 (APEX2 Version 2014.11-0, Bruker AXS, Inc.; Madison, WI 2014 USA) program package was used to determine the unit-cell parameters and for data collection (60 sec/frame scan time). The raw frame data were processed using SAINT (SAINT Version 8.34a, Bruker AXS, Inc.; Madison, WI 2013) and SADABS (Sheldrick, G.M. SADABS, Version 2014/5, Bruker AXS, Inc.; Madison, WI 2014 USA) to yield the reflection data file. Subsequent calculations were carried out using the SHELXTL (Sheldrick, G.M. SHELXTL, Version 2014/7, Bruker AXS, Inc.; Madison, WI 2014 USA) program package. For compound **26**, the diffraction symmetry was 2/m and the systematic absences were consistent with the monoclinic space group $P2_1/n$ that was later determined to be correct. The structure for **26** was solved by direct methods and refined on F^2 by full-matrix least-squares techniques, and the analytical scattering factors [81] for neutral atoms were used throughout the analysis. Hydrogen atoms in **26**, except those associated with C(14), were located from a difference-Fourier map and refined (x,y,z and U_{iso}), and H(14A), H(14B), and H(14C) were included using a riding model. Least-squares analysis for **26** yielded wR2 = 0.1064 and Goof = 1.025 for 328 variables refined against 3880 data (0.80 Å), R1 = 0.0426 for those 3293 data with I > 2.0σ(I). For compound **41**, the diffraction symmetry was 2/m and the systematic absences were consistent with the monoclinic space groups Cc and C2/c, and it was later determined that space group Cc was correct. The structure for **41** was solved by direct methods and refined on F^2 by full-matrix least-squares techniques, and again, the analytical scattering factors for neutral atoms were used throughout the analysis. Hydrogen atoms in **41** were located from a difference-Fourier map and refined (x,y,z and U_{iso}), and there was one water solvent molecule present. Least-squares analysis for **41** yielded wR2 = 0.0792 and Goof = 1.022 for 378 variables refined against 5116 data (0.73 Å), R1 = 0.0322 for those 4861 data with I > 2.0σ(I), and the absolute structure could not be assigned by refinement of the Flack parameter [82].

The wR2 = $[\Sigma[w(F_o{}^2 - F_c{}^2)^2]/\Sigma[w(F_o{}^2)^2]\,]^{1/2}$ and R1 = $\Sigma||F_o| - |F_c||/\Sigma|F_o|$. Goof = S = $[\Sigma[w(F_o{}^2 - F_c{}^2)^2]/(n - p)]^{1/2}$ where n is the number of reflections and *p* is the total number of parameters defined.

Supplementary Materials: The following are available online at https://www.mdpi.com/article/10.3390/ijms232416213/s1.

Author Contributions: Conceptualization, C.E.W., P.W.J., P.A.M., O.d.M. and J.S.W.; methodology, C.E.W., P.W.J., P.A.M., O.d.M., S.R., S.M., W.L., M.-Y.L., Y.N. and J.S.W.; X-ray Diffraction Analysis, J.W.Z.; formal analysis, O.d.M., S.R., M.A.S., M.A.S., G.A.M., I.J.L., D.J.I., T.D.K., S.J.P., A.P., L.T., A.L., I.K., R.P., B.M.S., E.S. and J.S.; writing—original draft preparation, C.E.W., P.W.J., P.A.M., O.d.M., M.-Y.L., Y.N. and J.S.W.; writing—review and editing, C.E.W., P.W.J., P.A.M., O.d.M., M.-Y.L., W.L. and J.S.W.; funding acquisition, C.E.W., P.W.J., P.A.M. and J.S.W. All authors have read and agreed to the published version of the manuscript.

Funding: This research was funded by US National Institutes of Health, grant number 1 R15 CA139364-01A2 and 1R15CA249617-01 to C.E.W., P.W.J. and P.A.M., by R01 HL128447 to J.S.W., and by the Siteman Investment Program to J.S.W.

Institutional Review Board Statement: Not applicable.

Informed Consent Statement: Not applicable.

Data Availability Statement: CCDC numbers 2191083 and 2191084 contain the supplementary crystallographic data for this paper. These data can be obtained free of charge from The Cambridge Crystallographic Data Centre via www.ccdc.cam.ac.uk/data_request/cif, accessed on 15 December 2022.

Acknowledgments: This work was also supported by The Rotary Coins for Alzheimer's Research Trust Fund. We also thank Arizona State University New College Undergraduate Inquiry and Research Experiences (NCUIRE) Program for supporting undergraduate student co-authors. Thanks are given to Felix Grun of the High-Resolution Mass Spectrometry Laboratory at University of California, Irvine (UCI). We thank Gayla Hadwiger and Anh Vu for technical support. We thank the Alvin J. Siteman Cancer Center at Washington University School of Medicine for the use of the Flow Cytometry Core. The Siteman Cancer Center is supported in part by an NCI Cancer Center Support Grant P30 CA91842.

Conflicts of Interest: The authors declare no conflict of interest. Patent applications covering the technologies described in this work have been applied for on behalf of the Arizona Board of Regents.

Abbreviations

ATRA	all-*trans*-retinoic acid
AD	Alzheimer's disease
9-*cis*-RA	9-*cis*-retinoic acid
CTCL	cutaneous T-cell lymphoma
DMF	dimethylformamide
DMSO	dimethylsulfoxide
DNA	deoxyribonucleic acid
FXR	farnesoid-X-receptor
HCl	hydrochloric acid
HRE	hormone responsive element
KOH	potassium hydroxide
LBD	ligand binding domain
LHS	LXR Heterodimer Specificity
LR	lipid risk assessment index
LXR	liver-X-receptor
LXRE	liver-X-receptor element
NaBu	sodium butyrate
NR	nuclear receptor
POC	proof of concept

PPAR	peroxisome proliferator activating receptor
RAR	retinoic-acid-receptor
RARE	retinoic acid receptor element
RXR	retinoid-X-receptor
RXRE	retinoid-X-receptor element
SNuRMs	specific nuclear receptor modulators
SREBP	sterol regulatory element binding protein
TR	thyroid hormone receptor
VDR	vitamin D receptor

References

1. Evans, R.M.; Mangelsdorf, D.J. Nuclear Receptors, RXR, and the Big Bang. *Cell* **2014**, *157*, 255–266. [CrossRef] [PubMed]
2. Mangelsdorf, D.J.; Umesono, K.; Evans, R.M. *The Retinoids*; Academic Press: Orlando, FL, USA, 1994; pp. 319–349.
3. Leid, M.; Kastner, P.; Chambon, P. Multiplicity generates diversity in the retinoic acid signalling pathways. *Trends Biochem. Sci.* **1992**, *17*, 427–433. [CrossRef] [PubMed]
4. Olefsky, J.M. Nuclear Receptor Minireview Series. *J. Biol. Chem.* **2001**, *276*, 36863–36864. [CrossRef] [PubMed]
5. Perlmann, T.; Rangarajan, P.N.; Umesono, K.; Evans, R.M. Determinants for selective RAR and TR recognition of direct repeat HREs. *Genes Dev.* **1993**, *7*, 1411–1422. [CrossRef]
6. Phan, T.Q.; Jow, M.M.; Privalsky, M.L. DNA recognition by thyroid hormone and retinoic acid receptors: 3,4,5 rule modified. *Mol. Cell. Endocrinol.* **2010**, *319*, 88–98. [CrossRef]
7. Forman, B.M.; Yang, C.-R.; Au, M.; Casanova, J.; Ghysdael, J.; Samuels, H.H. A Domain Containing Leucine-Zipper-Like Motifs Mediate Novel in Vivo Interactions between the Thyroid Hormone and Retinoic Acid Receptors. *Mol. Endocrinol.* **1989**, *3*, 1610–1626. [CrossRef]
8. Mangelsdorf, D.J.; Evans, R.M. The RXR heterodimers and orphan receptors. *Cell* **1995**, *83*, 841–850. [CrossRef]
9. Zhang, X.-K.; Lehmann, J.; Hoffmann, B.; Dawson, M.I.; Cameron, J.; Graupner, G.; Hermann, T.; Tran, P.; Pfahl, M. Homodimer formation of retinoid X receptor induced by 9-cis retinoic acid. *Nature* **1992**, *358*, 587–591. [CrossRef]
10. Thompson, P.D.; Remus, L.S.; Hsieh, J.C.; Jurutka, P.W.; Whitfield, G.K.; Galligan, M.A.; Dominguez, C.E.; Haussler, C.A.; Haussler, M.R. Distinct retinoid X receptor activation function-2 residues mediate transactivation in homodimeric and vitamin D receptor heterodimeric contexts. *J. Mol. Endocrinol.* **2001**, *27*, 211–227. [CrossRef]
11. Svensson, S.; Östberg, T.; Jacobsson, M.; Norström, C.; Stefansson, K.; Hallén, D.; Johansson, I.C.; Zachrisson, K.; Ogg, D.; Jendeberg, L. Crystal structure of the heterodimeric complex of LXRα and RXRβ ligand-binding domains in a fully agonistic conformation. *EMBO J.* **2003**, *22*, 4625–4633. [CrossRef]
12. Nahoum, V.; Pérez, E.; Germain, P.; Rodríguez-Barrios, F.; Manzo, F.; Kammerer, S.; Lemaire, G.; Hirsch, O.; Royer, C.A.; Gronemeyer, H.; et al. Modulators of the structural dynamics of the retinoid X receptor to reveal receptor function. *Proc. Natl. Acad. Sci. USA* **2007**, *104*, 17323–17328. [CrossRef] [PubMed]
13. Forman, B.M.; Umesono, K.; Chen, J.; Evans, R.M. Unique response pathways are established by allosteric interactions among nuclear hormone receptors. *Cell* **1995**, *81*, 541–550. [CrossRef]
14. Lala, D.S.; Mukherjee, R.; Schulman, I.G.; Koch, S.S.C.; Dardashti, L.J.; Nadzan, A.M.; Croston, G.E.; Evans, R.M.; Heyman, R.A. Activation of specific RXR heterodimers by an antagonist of RXR homodimers. *Nature* **1996**, *383*, 450–453. [CrossRef] [PubMed]
15. Lemon, B.D.; Freedman, L.P. Selective effects of ligands on vitamin D3 receptor- and retinoid X receptor-mediated gene activation in vivo. *Mol. Cell. Biol.* **1996**, *16*, 1006–1016. [CrossRef]
16. MacDonald, P.N.; Dowd, D.R.; Nakajima, S.; Galligan, M.A.; Reeder, M.C.; Haussler, C.A.; Ozato, K.; Haussler, M.R. Retinoid X receptors stimulate and 9-cis retinoic acid inhibits 1,25-dihydroxyvitamin D3-activated expression of the rat osteocalcin gene. *Mol. Cell. Biol.* **1993**, *13*, 5907–5917. [CrossRef] [PubMed]
17. Thompson, P.D.; Jurutka, P.W.; Haussler, C.A.; Whitfield, G.K.; Haussler, M.R. Heterodimeric DNA Binding by the Vitamin D Receptor and Retinoid X Receptors Is Enhanced by 1,25-Dihydroxyvitamin D3 and Inhibited by 9-cis-Retinoic Acid: Evidence for Allosteric Receptor Interactions. *J. Biol. Chem.* **1998**, *273*, 8483–8491. [CrossRef] [PubMed]
18. Lehmann, J.M.; Zhang, X.K.; Graupner, G.; Lee, M.O.; Hermann, T.; Hoffmann, B.; Pfahl, M. Formation of retinoid X receptor homodimers leads to repression of T3 response: Hormonal cross talk by ligand-induced squelching. *Mol. Cell. Biol.* **1993**, *13*, 7698–7707. [CrossRef] [PubMed]
19. Yen, P.M.; Liu, Y.; Sugawara, A.; Chin, W.W. Vitamin D receptors repress basal transcription and exert dominant negative activity on triiodothyronine-mediated transcriptional activity. *J. Biol. Chem.* **1996**, *271*, 10910–10916. [CrossRef]
20. Raval-Pandya, M.; Freedman, L.P.; Li, H.; Christakos, S. Thyroid hormone receptor does not heterodimerize with the vitamin D receptor but represses vitamin D receptor-mediated transactivation. *Mol. Endocrinol.* **1998**, *12*, 1367–1379. [CrossRef]
21. Thompson, P.D.; Hsieh, J.C.; Whitfield, G.K.; Haussler, C.A.; Jurutka, P.W.; Galligan, M.A.; Tillman, J.B.; Spindler, S.R.; Haussler, M.R. Vitamin D receptor displays DNA binding and transactivation as a heterodimer with the retinoid X receptor, but not with the thyroid hormone receptor. *J. Cell. Biochem.* **1999**, *75*, 462–480. [CrossRef]
22. Altucci, L.; Leibowitz, M.D.; Ogilvie, K.M.; de Lera, A.R.; Gronemeyer, H. RAR and RXR modulation in cancer and metabolic disease. *Nat. Rev. Drug Discov.* **2007**, *6*, 793–810. [CrossRef]

23. Lehmann, J.M.; Jong, L.; Fanjul, A.; Cameron, J.F.; Lu, X.P.; Haefner, P.; Dawson, M.I.; Pfahl, M. Retinoids selective for retinoid X receptor response pathways. *Science* **1992**, *258*, 1944–1946. [CrossRef] [PubMed]

24. Jong, L.; Lehmann, J.M.; Hobbs, P.D.; Harlev, E.; Huffman, J.C.; Pfahl, M.; Dawson, M.I. Conformational effects on retinoid receptor selectivity. 1. Effect of 9-double bond geometry on retinoid X receptor activity. *J. Med. Chem.* **1993**, *36*, 2605–2613. [CrossRef] [PubMed]

25. Dawson, M.I.; Jong, L.; Hobbs, P.D.; Cameron, J.F.; Chao, W.-R.; Pfahl, M.; Lee, M.-O.; Shroot, B.; Pfahl, M. Conformational Effects on Retinoid Receptor Selectivity. 2. Effects of Retinoid Bridging Group on Retinoid X Receptor Activity and Selectivity. *J. Med. Chem.* **1995**, *38*, 3368–3383. [CrossRef]

26. Boehm, M.F.; Zhang, L.; Badea, B.A.; White, S.K.; Mais, D.E.; Berger, E.; Suto, C.M.; Goldman, M.E.; Heyman, R.A. Synthesis and Structure-Activity Relationships of Novel Retinoid X Receptor-Selective Retinoids. *J. Med. Chem.* **1994**, *37*, 2930–2941. [CrossRef] [PubMed]

27. Daiss, J.O.; Burschka, C.; Mills, J.S.; Montana, J.G.; Showell, G.A.; Fleming, I.; Gaudon, C.; Ivanova, D.; Gronemeyer, H.; Tacke, R. Synthesis, Crystal Structure Analysis, and Pharmacological Characterization of Disila-bexarotene, a Disila-Analogue of the RXR-Selective Retinoid Agonist Bexarotene. *Organometallics* **2005**, *24*, 3192–3199. [CrossRef]

28. Zhang, D.; Leal, A.S.; Carapellucci, S.; Shahani, P.H.; Bhogal, J.S.; Ibrahim, S.; Raban, S.; Jurutka, P.W.; Marshall, P.A.; Sporn, M.B.; et al. Testing Novel Pyrimidinyl Rexinoids: A New Paradigm for Evaluating Rexinoids for Cancer Prevention. *Cancer Prev. Res.* **2019**, *12*, 211–224. [CrossRef] [PubMed]

29. Yen, W.-C.; Prudente, R.Y.; Lamph, W.W. Synergistic effect of a retinoid X receptor-selective ligand bexarotene (LGD1069, Targretin) and paclitaxel (Taxol) in mammary carcinoma. *Breast Cancer Res. Treat.* **2004**, *88*, 141–148. [CrossRef]

30. Cesario, R.M.; Stone, J.; Yen, W.-C.; Bissonnette, R.P.; Lamph, W.W. Differentiation and growth inhibition mediated via the RXR:PPARγ heterodimer in colon cancer. *Cancer Lett.* **2006**, *240*, 225–233. [CrossRef]

31. Yen, W.-C.; Corpuz, M.R.; Prudente, R.Y.; Cooke, T.A.; Bissonnette, R.P.; Negro-Vilar, A.; Lamph, W.W. A Selective Retinoid X Receptor Agonist Bexarotene (Targretin) Prevents and Overcomes Acquired Paclitaxel (Taxol) Resistance in Human Non–Small Cell Lung Cancer. *Clin. Cancer Res.* **2004**, *10*, 8656–8664. [CrossRef]

32. Dragnev, K.H.; Petty, W.J.; Shah, S.J.; Lewis, L.D.; Black, C.C.; Memoli, V.; Nugent, W.C.; Hermann, T.; Negro-Vilar, A.; Rigas, J.R.; et al. A Proof-of-Principle Clinical Trial of Bexarotene in Patients with Non–Small Cell Lung Cancer. *Clin. Cancer Res.* **2007**, *13*, 1794–1800. [CrossRef] [PubMed]

33. Mukherjee, R.; Davies, P.J.A.; Crombie, D.L.; Bischoff, E.D.; Cesario, R.M.; Jow, L.; Hamann, L.G.; Boehm, M.F.; Mondon, C.E.; Nadzan, A.M.; et al. Sensitization of diabetic and obese mice to insulin by retinoid X receptor agonists. *Nature* **1997**, *386*, 407–410. [CrossRef] [PubMed]

34. Sherman, S.I.; Gopal, J.; Haugen, B.R.; Chiu, A.C.; Whaley, K.; Nowlakha, P.; Duvic, M. Central Hypothyroidism Associated with Retinoid X Receptor–Selective Ligands. *N. Engl. J. Med.* **1999**, *340*, 1075–1079. [CrossRef] [PubMed]

35. Li, D.; Li, T.; Wang, F.; Tian, H.; Samuels, H.H. Functional Evidence for Retinoid X Receptor (RXR) as a Nonsilent Partner in the Thyroid Hormone Receptor/RXR Heterodimer. *Mol. Cell. Biol.* **2002**, *22*, 5782–5792. [CrossRef] [PubMed]

36. Field, F.J.; Born, E.; Mathur, S.N. LXR/RXR ligand activation enhances basolateral efflux of beta-sitosterol in CaCo-2 cells. *J. Lipid Res.* **2004**, *45*, 905–913. [CrossRef]

37. Murthy, S.; Born, E.; Mathur, S.N.; Field, F.J. LXR/RXR activation enhances basolateral efflux of cholesterol in CaCo-2 cells. *J. Lipid Res.* **2002**, *43*, 1054–1064. [CrossRef]

38. Thacher, S.M.; Standeven, A.M.; Athanikar, J.; Kopper, S.; Castilleja, O.; Escobar, M.; Beard, R.L.; Chandraratna, R.A.S. Receptor Specificity of Retinoid-Induced Epidermal Hyperplasia: Effect of RXR-Selective Agonists and Correlation with Topical Irritation. *J. Pharmacol. Exp. Ther.* **1997**, *282*, 528–534.

39. Cramer, P.E.; Cirrito, J.R.; Wesson, D.W.; Lee, C.Y.D.; Karlo, J.C.; Zinn, A.E.; Casali, B.T.; Restivo, J.L.; Goebel, W.D.; James, M.J.; et al. ApoE-Directed Therapeutics Rapidly Clear β-Amyloid and Reverse Deficits in AD Mouse Models. *Science* **2012**, *335*, 1503. [CrossRef]

40. McFarland, K.; Spalding, T.A.; Hubbard, D.; Ma, J.-N.; Olsson, R.; Burstein, E.S. Low Dose Bexarotene Treatment Rescues Dopamine Neurons and Restores Behavioral Function in Models of Parkinson's Disease. *ACS Chem. Neurosci.* **2013**, *4*, 1430–1438. [CrossRef]

41. Cummings, J.L.; Zhong, K.; Kinney, J.W.; Heaney, C.; Moll-Tudla, J.; Joshi, A.; Pontecorvo, M.; Devous, M.; Tang, A.; Bena, J. Double-blind, placebo-controlled, proof-of-concept trial of bexarotene Xin moderate Alzheimer's disease. *Alzheimers Res. Ther.* **2016**, *8*, 4. [CrossRef]

42. Kabbinavar, F.F.; Zomorodian, N.; Rettig, M.; Khan, F.; Greenwald, D.R.; Davidson, S.J.; DiCarlo, B.A.; Patel, R.; Pandit, L.; Chandraratna, R.; et al. An open-label phase II clinical trial of the RXR agonist IRX4204 in taxane-resistant, castration-resistant metastatic prostate cancer (CRPC). *J. Clin. Oncol.* **2014**, *32*, 5073. [CrossRef]

43. Vuligonda, V.; Thacher, S.M.; Chandraratna, R.A. Enantioselective syntheses of potent retinoid X receptor ligands: Differential biological activities of individual antipodes. *J. Med. Chem.* **2001**, *44*, 2298–2303. [CrossRef] [PubMed]

44. Muccio, D.D.; Brouillette, W.J.; Breitman, T.R.; Taimi, M.; Emanuel, P.D.; Zhang, X.-k.; Chen, G.-q.; Sani, B.P.; Venepally, P.; Reddy, L.; et al. Conformationally Defined Retinoic Acid Analogues. 4. Potential New Agents for Acute Promyelocytic and Juvenile Myelomonocytic Leukemias. *J. Med. Chem.* **1998**, *41*, 1679–1687. [CrossRef] [PubMed]

45. Atigadda, V.R.; Vines, K.K.; Grubbs, C.J.; Hill, D.L.; Beenken, S.L.; Bland, K.I.; Brouillette, W.J.; Muccio, D.D. Conformationally Defined Retinoic Acid Analogues. 5. Large-Scale Synthesis and Mammary Cancer Chemopreventive Activity for (2E,4E,6Z,8E)-8-(3′,4′-Dihydro-1′(2′H)-naphthalen-1′-ylidene)-3,7-dimethyl-2,4,6-octatrienoic Acid (9cUAB30). *J. Med. Chem.* **2003**, *46*, 3766–3769. [CrossRef]

46. Kolesar, J.M.; Hoel, R.; Pomplun, M.; Havighurst, T.; Stublaski, J.; Wollmer, B.; Krontiras, H.; Brouillette, W.; Muccio, D.; Kim, K.; et al. A Pilot, First-in-Human, Pharmacokinetic Study of 9cUAB30 in Healthy Volunteers. *Cancer Prev. Res.* **2010**, *3*, 1565–1570. [CrossRef]

47. Hansen, N.J.; Wylie, R.C.; Phipps, S.M.; Love, W.K.; Andrews, L.G.; Tollefsbol, T.O. The low-toxicity 9-cis UAB30 novel retinoid down-regulates the DNA methyltransferases and has anti-telomerase activity in human breast cancer cells. *Int. J. Oncol.* **2007**, *30*, 641–650. [CrossRef]

48. Atigadda, V.R.; Xia, G.; Desphande, A.; Boerma, L.J.; Lobo-Ruppert, S.; Grubbs, C.J.; Smith, C.D.; Brouillette, W.J.; Muccio, D.D. Methyl substitution of a rexinoid agonist improves potency and reveals site of lipid toxicity. *J. Med. Chem.* **2014**, *57*, 5370–5380. [CrossRef] [PubMed]

49. Desphande, A.; Xia, G.; Boerma, L.J.; Vines, K.K.; Atigadda, V.R.; Lobo-Ruppert, S.; Grubbs, C.J.; Moeinpour, F.L.; Smith, C.D.; Christov, K.; et al. Methyl-substituted conformationally constrained rexinoid agonists for the retinoid X receptors demonstrate improved efficacy for cancer therapy and prevention. *Bioorganic Med. Chem.* **2014**, *22*, 178–185. [CrossRef]

50. Michellys, P.Y.; Ardecky, R.J.; Chen, J.H.; Crombie, D.L.; Etgen, G.J.; Faul, M.M.; Faulkner, A.L.; Grese, T.A.; Heyman, R.A.; Karanewsky, D.S.; et al. Novel (2E,4E,6Z)-7-(2-alkoxy-3,5-dialkylbenzene)-3-methylocta-2,4,6-trienoic acid retinoid X receptor modulators are active in models of type 2 diabetes. *J. Med. Chem.* **2003**, *46*, 2683–2696. [CrossRef]

51. Michellys, P.Y.; Ardecky, R.J.; Chen, J.H.; D'Arrigo, J.; Grese, T.A.; Karanewsky, D.S.; Leibowitz, M.D.; Liu, S.; Mais, D.A.; Mapes, C.M.; et al. Design, synthesis, and structure-activity relationship studies of novel 6,7-locked-[7-(2-alkoxy-3,5-dialkylbenzene)-3-methylocta]-2,4,6-trienoic acids. *J. Med. Chem.* **2003**, *46*, 4087–4103. [CrossRef]

52. Michellys, P.Y.; D'Arrigo, J.; Grese, T.A.; Karanewsky, D.S.; Leibowitz, M.D.; Mais, D.A.; Mapes, C.M.; Reifel-Miller, A.; Rungta, D.; Boehm, M.F. Design, synthesis and structure-activity relationship of novel RXR-selective modulators. *Bioorganic Med. Chem. Lett.* **2004**, *14*, 1593–1598. [CrossRef] [PubMed]

53. Wagner, C.E.; Jurutka, P.W.; Marshall, P.A.; Groy, T.L.; van der Vaart, A.; Ziller, J.W.; Furmick, J.K.; Graeber, M.E.; Matro, E.; Miguel, B.V.; et al. Modeling, Synthesis and Biological Evaluation of Potential Retinoid X Receptor (RXR) Selective Agonists: Novel Analogues of 4-[1-(3,5,5,8,8-Pentamethyl-5,6,7,8-tetrahydro-2-naphthyl)ethynyl]benzoic Acid (Bexarotene). *J. Med. Chem.* **2009**, *52*, 5950–5966. [CrossRef] [PubMed]

54. Furmick, J.K.; Kaneko, I.; Walsh, A.N.; Yang, J.; Bhogal, J.S.; Gray, G.M.; Baso, J.C.; Browder, D.O.; Prentice, J.L.S.; Montano, L.A.; et al. Modeling, Synthesis and Biological Evaluation of Potential Retinoid X Receptor-Selective Agonists: Novel Halogenated Analogues of 4-[1-(3,5,5,8,8-Pentamethyl-5,6,7,8-tetrahydro-2-naphthyl)ethynyl]benzoic Acid (Bexarotene). *ChemMedChem* **2012**, *7*, 1551–1566. [CrossRef] [PubMed]

55. Boehm, M.F.; Zhang, L.; Zhi, L.; McClurg, M.R.; Berger, E.; Wagoner, M.; Mais, D.E.; Suto, C.M.; Davies, P.J.A.; Heyman, R.A.; et al. Design and Synthesis of Potent Retinoid X Receptor Selective Ligands That Induce Apoptosis in Leukemia Cells. *J. Med. Chem.* **1995**, *38*, 3146–3155. [CrossRef] [PubMed]

56. Jurutka, P.W.; Kaneko, I.; Yang, J.; Bhogal, J.S.; Swierski, J.C.; Tabacaru, C.R.; Montano, L.A.; Huynh, C.C.; Jama, R.A.; Mahelona, R.D.; et al. Modeling, Synthesis, and Biological Evaluation of Potential Retinoid X Receptor (RXR) Selective Agonists: Novel Analogues of 4-[1-(3,5,5,8,8-Pentamethyl-5,6,7,8-tetrahydro-2-naphthyl)ethynyl]benzoic Acid (Bexarotene) and (E)-3-(3-(1,2,3,4-tetrahydro-1,1,4,4,6-pentamethylnaphthalen-7-yl)-4-hydroxyphenyl)acrylic Acid (CD3254). *J. Med. Chem.* **2013**, *56*, 8432–8454. [CrossRef]

57. Liby, K.; Rendi, M.; Suh, N.; Royce, D.B.; Risingsong, R.; Williams, C.R.; Lamph, W.; Labrie, F.; Krajewski, S.; Xu, X.; et al. The Combination of the Rexinoid, LG100268, and a Selective Estrogen Receptor Modulator, Either Arzoxifene or Acolbifene, Synergizes in the Prevention and Treatment of Mammary Tumors in an Estrogen Receptor–Negative Model of Breast Cancer. *Clin. Cancer Res.* **2006**, *12*, 5902–5909. [CrossRef]

58. Zhang, L.; Badea, B.A.; Enyeart, D.; Berger, E.M.; Mais, D.E.; Boehm, M.F. Syntheses of isotopically labeled 4-[1-(3,5,5,8,8-pentamethyl-5,6,7,8-tetrahydro-2-naphthyl)ethenyl]benzoic acid (LGD1069), a potent retinoid x receptor-selective ligand. *J. Label. Comp. Radiopharm.* **1995**, *36*, 701–712. [CrossRef]

59. Faul, M.M.; Ratz, A.M.; Sullivan, K.A.; Trankle, W.G.; Winneroski, L.L. Synthesis of Novel Retinoid X Receptor-Selective Retinoids. *J. Org. Chem.* **2001**, *66*, 5772–5782. [CrossRef]

60. Marshall, P.A.; Jurutka, P.W.; Wagner, C.E.; van der Vaart, A.; Kaneko, I.; Chavez, P.I.; Ma, N.; Bhogal, J.S.; Shahani, P.; Swierski, J.C.; et al. Analysis of differential secondary effects of novel rexinoids: Select rexinoid X receptor ligands demonstrate differentiated side effect profiles. *Pharm. Res. Perspect.* **2015**, *3*, e00122. [CrossRef]

61. Santín, E.P.; Germain, P.; Quillard, F.; Khanwalkar, H.; Rodríguez-Barrios, F.; Gronemeyer, H.; de Lera, Á.R.; Bourguet, W. Modulating Retinoid X Receptor with a Series of (E)-3-[4-Hydroxy-3-(3-alkoxy-5,5,8,8-tetramethyl-5,6,7,8-tetrahydronaphthalen-2-yl)phenyl]acrylic Acids and Their 4-Alkoxy Isomers. *J. Med. Chem.* **2009**, *52*, 3150–3158. [CrossRef]

62. Gianni, M.; Ponzanelli, I.; Mologni, L.; Reichert, U.; Rambaldi, A.; Terao, M.; Garattini, E. Retinoid-dependent growth inhibition, differentiation and apoptosis in acute promyelocytic leukemia cells. Expression and activation of caspases. *Cell Death Differ.* **2000**, *7*, 447–460. [CrossRef] [PubMed]

63. Fujii, S.; Ohsawa, F.; Yamada, S.; Shinozaki, R.; Fukai, R.; Makishima, M.; Enomoto, S.; Tai, A.; Kakuta, H. Modification at the acidic domain of RXR agonists has little effect on permissive RXR-heterodimer activation. *Bioorg. Med. Chem. Lett.* **2010**, *20*, 5139–5142. [CrossRef] [PubMed]

64. Ohsawa, F.; Morishita, K.-I.; Yamada, S.; Makishima, M.; Kakuta, H. Modification at the Lipophilic Domain of RXR Agonists Differentially Influences Activation of RXR Heterodimers. *ACS Med. Chem. Lett.* **2010**, *1*, 521–525. [CrossRef]

65. Kakuta, H.; Yakushiji, N.; Shinozaki, R.; Ohsawa, F.; Yamada, S.; Ohta, Y.; Kawata, K.; Nakayama, M.; Hagaya, M.; Fujiwara, C.; et al. RXR Partial Agonist CBt-PMN Exerts Therapeutic Effects on Type 2 Diabetes without the Side Effects of RXR Full Agonists. *ACS Med. Chem. Lett.* **2012**, *3*, 427–432. [CrossRef]

66. Ohsawa, F.; Yamada, S.; Yakushiji, N.; Shinozaki, R.; Nakayama, M.; Kawata, K.; Hagaya, M.; Kobayashi, T.; Kohara, K.; Furusawa, Y.; et al. Mechanism of Retinoid X Receptor Partial Agonistic Action of 1-(3,5,5,8,8-Pentamethyl-5,6,7,8-tetrahydro-2-naphthyl)-1H-benzotriazole-5-carboxylic Acid and Structural Development To Increase Potency. *J. Med. Chem.* **2013**, *56*, 1865–1877. [CrossRef] [PubMed]

67. Kagechika, H.; Koichi, S.; Sugioka, T.; Sotome, T.; Nakayama, Y.; Doi, K. Retinoid Activity Regulators. European Patent WO9845242A1, 15 October 1998.

68. Ohta, K.; Tsuji, M.; Kawachi, E.; Fukasawa, H.; Hashimoto, Y.; Shudo, K.; Kagechika, H. Potent Retinoid Synergists with a Diphenylamine Skeleton. *Biol. Pharm. Bull.* **1998**, *21*, 544–546. [CrossRef]

69. Ohta, K.; Kawachi, E.; Fukasawa, H.; Shudo, K.; Kagechika, H. Diphenylamine-based retinoid antagonists: Regulation of RAR and RXR function depending on the N-substituent. *Bioorg. Med. Chem.* **2011**, *19*, 2501–2507. [CrossRef]

70. Heck, M.C.; Wagner, C.E.; Shahani, P.H.; MacNeill, M.; Grozic, A.; Darwaiz, T.; Shimabuku, M.; Deans, D.G.; Robinson, N.M.; Salama, S.H.; et al. Modeling, Synthesis, and Biological Evaluation of Potential Retinoid X Receptor (RXR)-Selective Agonists: Analogues of 4-[1-(3,5,5,8,8-Pentamethyl-5,6,7,8-tetrahydro-2-naphthyl)ethynyl]benzoic Acid (Bexarotene) and 6-(Ethyl(5,5,8,8-tetrahydronaphthalen-2-yl)amino)nicotinic Acid (NEt-TMN). *J. Med. Chem.* **2016**, *59*, 8924–8940. [CrossRef]

71. Ohta, K.; Kawachi, E.; Inoue, N.; Fukasawa, H.; Hashimoto, Y.; Itai, A.; Kagechika, H. Retinoidal pyrimidinecarboxylic acids. Unexpected diaza-substituent effects in retinobenzoic acids. *Chem. Pharm. Bull.* **2000**, *48*, 1504–1513. [CrossRef]

72. Trott, O.; Olson, A.J. AutoDock Vina: Improving the speed and accuracy of docking with a new scoring function, efficient optimization, and multithreading. *J. Comput. Chem.* **2010**, *31*, 455–461. [CrossRef]

73. Stierand, K.; Rarey, M. Drawing the PDB: Protein−Ligand Complexes in Two Dimensions. *ACS Med. Chem. Lett.* **2010**, *1*, 540–545. [CrossRef] [PubMed]

74. Stierand, K.; Rarey, M. From Modeling to Medicinal Chemistry: Automatic Generation of Two-Dimensional Complex Diagrams. *ChemMedChem* **2007**, *2*, 853–860. [CrossRef] [PubMed]

75. Tachdjian, C.; Guo, J.; Boudjelal, M.; Al-Shamma, H.A.; Giachino, A.F.; Jakubowicz-Jaillardon, K.; Chen, Q.; Zapf, J.W.; Pfahl, M. Preparation of Substituted Isochroman Compounds for the Treatment of Metabolic Disorders, Cancer and Other Diseases. Patent CN1774246A, 17 May 2004.

76. Jurutka, P.W.; di Martino, O.; Reshi, S.; Mallick, S.; Sabir, Z.L.; Staniszewski, L.J.P.; Warda, A.; Maiorella, E.L.; Minasian, A.; Davidson, J.; et al. Modeling, Synthesis, and Biological Evaluation of Potential Retinoid-X-Receptor (RXR) Selective Agonists: Analogs of 4-[1-(3,5,5,8,8-Pentamethyl-5,6,7,8-tetrahyro-2-naphthyl)ethynyl]benzoic Acid (Bexarotene) and 6-(Ethyl(4-isobutoxy-3-isopropylphenyl)amino)nicotinic Acid (NEt-4IB). *Int. J. Mol. Sci.* **2021**, *22*, 1237. [CrossRef]

77. Egea, P.F.; Mitschler, A.; Rochel, N.; Ruff, M.; Chambon, P.; Moras, D. Crystal structure of the human RXRalpha ligand-binding domain bound to its natural ligand: 9-cis retinoic acid. *EMBO J.* **2000**, *19*, 2592–2601. [CrossRef] [PubMed]

78. Harris, R.; Olson, A.J.; Goodsell, D.S. Automated prediction of ligand-binding sites in proteins. *Proteins Struct. Funct. Bioinf.* **2008**, *70*, 1506–1517. [CrossRef] [PubMed]

79. Di Martino, O.; Niu, H.; Hadwiger, G.; Kuusanmaki, H.; Ferris, M.A.; Vu, A.; Beales, J.; Wagner, C.; Menéndez-Gutiérrez, M.P.; Ricote, M.; et al. Endogenous and combination retinoids are active in myelomonocytic leukemias. *Haematologica* **2021**, *106*, 1008–1021. [CrossRef]

80. Di Martino, O.; Ferris, M.A.; Hadwiger, G.; Sarkar, S.; Vu, A.; Menéndez-Gutiérrez, M.P.; Ricote, M.; Welch, J.S. RXRA DT448/9PP generates a dominant active variant capable of inducing maturation in acute myeloid leukemia cells. *Haematologica* **2021**, *107*, 417–426. [CrossRef]

81. Wilson, A.J.C.; Hahn, T.; Wilson, A.J.C.; Shmueli, U.; Crystallography, I.U.O. *International Tables for Crystallography, Volume C: Mathematical, Physical and Chemical Tables*; Springer: Dordrecht, The Netherland, 1992.

82. Parsons, S.; Flack, H.D.; Wagner, T. Use of intensity quotients and differences in absolute structure refinement. *Acta Crystallogr. Sect. B Struct. Sci. Cryst. Eng. Mater.* **2013**, *69*, 249–259. [CrossRef]

International Journal of
Molecular Sciences

Article

In Vitro/In Vivo Translation of Synergistic Combination of MDM2 and MEK Inhibitors in Melanoma Using PBPK/PD Modelling: Part I

Jakub Witkowski [1,2,*], Sebastian Polak [3,4], Zbigniew Rogulski [1] and Dariusz Pawelec [2]

[1]   Faculty of Chemistry, University of Warsaw, Pasteura 1, 02-093 Warsaw, Poland
[2]   Adamed Pharma S.A., Adamkiewicza 6a, 05-152 Czosnów, Poland
[3]   Faculty of Pharmacy, Jagiellonian University, Medyczna 9, 30-688 Kraków, Poland
[4]   Simcyp Division, Certara UK Limited, Level 2-Acero, 1 Concourse Way, Sheffield S1 2BJ, UK
*   Correspondence: j.witkowski6@uw.edu.pl

Abstract: Translation of the synergy between the Siremadlin (MDM2 inhibitor) and Trametinib (MEK inhibitor) combination observed in vitro into in vivo synergistic efficacy in melanoma requires estimation of the interaction between these molecules at the pharmacokinetic (PK) and pharmacodynamic (PD) levels. The cytotoxicity of the Siremadlin and Trametinib combination was evaluated in vitro in melanoma A375 cells with MTS and RealTime-Glo assays. Analysis of the drug combination matrix was performed using Synergy and Synergyfinder packages. Calculated drug interaction metrics showed high synergy between Siremadlin and Trametinib: 23.12%, or a 7.48% increase of combined drug efficacy (concentration-independent parameter β from Synergy package analysis and concentration-dependent δ parameter from Synergyfinder analysis, respectively). In order to select the optimal PD interaction parameter which may translate observed in vitro synergy metrics into the in vivo setting, further PK/PD studies on cancer xenograft animal models coupled with PBPK/PD modelling are needed.

Keywords: anticancer drugs; preclinical study; pharmacokinetics; pharmacodynamics; drug combination; PBPK/PD modelling; MDM2 inhibitor; MEK inhibitor

Citation: Witkowski, J.; Polak, S.; Rogulski, Z.; Pawelec, D. In Vitro/In Vivo Translation of Synergistic Combination of MDM2 and MEK Inhibitors in Melanoma Using PBPK/PD Modelling: Part I. *Int. J. Mol. Sci.* **2022**, *23*, 12984. https://doi.org/10.3390/ijms232112984

Academic Editor: Takeo Nakanishi

Received: 10 October 2022
Accepted: 24 October 2022
Published: 26 October 2022

Publisher's Note: MDPI stays neutral with regard to jurisdictional claims in published maps and institutional affiliations.

1. Introduction

One of the first targeted therapies approved was a drug combination targeting the MAPK (mitogen-activated protein kinase family) signaling pathway of Dabrafenib (BRAF inhibitor) and Trametinib (MEK inhibitor) [1]. The Dabrafenib and Trametinib combination showed synergistic efficacy and significantly increased the overall survival of melanoma patients. Preclinical evidence suggests that the drug combination with MEK and MDM2 (mouse double minute 2) inhibitors may also act synergistically in the treatment of melanoma [2]. This pharmacodynamic interaction has been characterized at a molecular level, and may be explained by the DUSP6 mechanism (DUSP6 suppression followed by increased p53 phosphorylation) in BRAFV600E and p53WT melanoma cells, which leads to synergistic induction of the expression of genes encoding PUMA and BIM that increase apoptosis ratio and growth inhibition of melanoma cells [3]. There is also in vivo evidence in animals with melanoma tumour xenografts that this synergistic efficacy may be efficient in the treatment of skin cancer [4]. However, long-term administration of MDM2 and MEK inhibitors can lead to acquired resistance caused by the mechanism of spontaneous p53 and MAP2K1 (MEK1) mutations or expression of BRAF-V600E splice variants [5–7]. Thus, combining these two classes of drugs can bring benefit to the patients by restoring anticancer activity (salvage therapy if one acquires resistance against either drug [8]) or delaying development of resistance to the treatment. It should be noted that combining the drugs may increase the adverse effects observed in patients. Therefore, the selection of drugs with a different profile of adverse effects for drug combination is very important for the sake of patient safety. Examination of this particular drug combination on healthy cells was not in the scope of this publication, but previously published data for MDM2 and MEK inhibitors indicate very low toxicity in healthy cells [9,10]. Moreover, since such a drug combination is currently the subject of many studies in clinical trials (ClinicalTrials.gov identifiers: NCT02110355, NCT03714958, NCT02016729, NCT01985191, NCT03566485) it is assumed that this combination is generally safe and not excessively toxic to healthy cells. This drug combination utility was confirmed in the clinical setting with moderately active MDM2 inhibitor AMG232 [11], but it is believed that the next generation of more potent MDM2 inhibitors, such as Siremadlin (HDM201), may further enhance this synergistic drug interaction.

In order to assess how the addition of Siremadlin to Trametinib could improve anti-cancer response, performance of preclinical translational studies and the development of in vitro/in vivo translational methods are truly essential. A bench-to-bedside approach for drug combination may be possible only when PK/PD data for both drugs are available because it must account for the interaction between two (or even more) drugs at two different levels: pharmacokinetic (PK) and pharmacodynamic (PD).

Analyses of the interactions between two or more drugs at the PD level are impeded by a lack of consensus on which method/theoretical model should be used to describe drug interaction. The quantification of the interaction between drugs is based on the comparison of the observed combination response to the expected effect predicted by a reference model under the assumption of non-interaction of those drugs. Depending on whether the combination response is greater or less than what is expected, the drug combinations can be classified as synergistic or antagonistic, respectively. In a case where the drug combination response equals the expected effect, it can be classified as additive (for some authors, such lack of drug interaction is also referred to as independence or noninteraction [12]). Historically, there were also other types of drug interactions, such as inertism or coalism, described by Greco et al., and Roell et al. [12,13]. Over the past years, many drug interaction frameworks have been developed, including the most recognized models, the Loewe additivity [14], Bliss independence [15], or highest single agent (HSA) [16] models, and the most recent but less recognized models that overcome many limitations of already existing models, such as zero interaction potency (ZIP) [17] or multi-dimensional synergy of combinations (MuSyC) [18].

Choosing the model to evaluate combination data became very problematic in the light of scientific discussion over the past years. There has often been a dilemma when a drug combination is classified as synergistic according to one model but antagonistic by the other [16,19,20]. According to Tang et al. [19], this phenomenon can be explained by consistency between models, which can be indicative of the degree of drug interaction; e.g., if both the Bliss model and the Loewe model classify a drug combination as synergistic, then it may be described as a strong synergy. On the other hand, if the drug combination interaction is classified as synergistic according to one model only, then it may be described as a weak synergy or additivity. Synergy and efficacy concepts are highly related, but it is very important to not treat them as the same. Synergy is a type of drug interaction and a measure of its degree, while efficacy is the magnitude of the phenotypic response of a drug combination. It can be observed that a combination of drugs can be highly synergistic, whereas its response may not be sufficient to achieve therapeutic efficacy. It is also possible that a drug combination can show a strong response which is not related to a synergistic interaction (e.g., only one drug is responsible for the observed response) [18,19,21].

Unlike the models mentioned above, the MuSyC model addresses those two concepts of drug interaction by describing their interaction with three different metrics that decouple synergy or antagonism based on increased/decreased efficacy (parameter β), potency (parameter α), and cooperativity (parameter γ), which may be advantageous due to disease type. Moreover, those drug interaction parameters are dose-independent and, thus, observed synergistic interaction may not lead to ambiguous results (for a given combination drug interaction may be synergistic or not, as both drugs can synergize at some concentrations, and antagonize at others) [18].

Due to the nature of the drug combination studies which are typically conducted in the high throughput screening (HTS) format, the most practical solution would be to use software capable of calculating multiple drug interaction metrics for large datasets. Unfortunately, there are only a few software packages for which such features are available: Synergyfinder [22], Synergy [23], and Combenefit software [24]. Because the Combenefit software has not been developed since 2016 and provides analysis with the same models as Synergyfinder (Loewe, Bliss and HSA), analysis involving Combenefit was not considered in this work.

The Synergyfinder package allows for calculating delta score (δ) value (which corresponds to the percentage of drug combination response beyond expectation) for Loewe, Bliss, HSA and ZIP models. For example, a delta of 10 would indicate that the drug combination will produce, on average, 10% more response compared to the expected effect predicted by given theoretical drug interaction model (Loewe, Bliss, HSA and ZIP models), which we would refer here as synergistic drugs interaction, while a delta of -10 would indicate an antagonistic drug interaction with the same level of magnitude in this case. Applying a threshold of 5% response (delta score $\delta = 5$), which is the typical noise level in large-scale drug combination experiments, minimizes the rate of false-positive results [17,25]. Therefore, in this context, and according to Tang's group's experience, classification of drug combinations was formulated based on delta score value, as shown in Table 1 [26].

Table 1. Drug interaction classification based on delta score (Synergyfinder package).

Description	δ Score Value
Antagonism	≤ -5
Additivity	$(-5; 5)$
Synergism	≥ 5

Despite the fact that the Synergy package permits use of several drug interaction models, such as concentration (dose)-independent parametric models (MuSyC, Zimmer, and BRAID) and concentration (dose)-dependent nonparametric models (Loewe, Bliss, HSA, CombinationIndex, Schindler, and ZIP), its main focus was on the MuSyC model to determine which drug interaction metric among the efficacy (parameter β), potency (parameter α), and cooperativity (parameter γ) parameters would be the most significant in the context of in vitro/in vivo translation of the drug combination interaction at the pharmacodynamics level. Parameter β may be interpreted as the percent increase in maximal efficacy of the combination over the most efficacious single agent. Parameter α quantifies, in fold, how the effectiveness of one drug is altered by the presence of the other (fold change in the potency of combined drugs). Gamma parameter provides information about the change of a drug's Hill slope (cooperativity) due to the other drug. There are two values for α and γ because each drug can independently modulate the potency and cooperativity of the other. Classification of drug combination interaction in the MuSyC model based on α, β, and γ score values is shown in Table 2 [27,28].

Table 2. Drug interaction classification based on alpha, beta, and gamma scores (Synergy package).

Description	α_{12}/α_{21} Score Value	β Score Value	γ_{12}/γ_{21} Score Value
Antagonism	<1	<0	<1
Additivity	1	0	1
Synergism	>1	>0	>1

Translation of the in vitro into in vivo synergistic efficacy must cover interactions at the PK and PD levels. One of the approaches allowing for the incorporation of such interactions is physiologically based pharmacokinetic/pharmacodynamic (PBPK/PD) modelling.

The main goal of this work is focused on estimation of the PD interaction parameter, which may serve for translatability of in vitro drug combination results into in vivo settings. Such an approach, involving results from in vitro cytotoxicity studies coupled with PBPK/PD modelling, may facilitate the determination of the most synergistic and efficacious schedule and dose levels for Siremadlin and Trametinib in mice, in vivo, and also may be the basis for better estimation of drug combination efficacy in melanoma patients.

2. Results

2.1. In Vitro Cytotoxicity

Siremadlin and Trametinib efficacy in monotherapy and in combination was studied in in vitro A375 human melanoma cells with the use of an MTS assay. One of the limitations of the preclinical drug combinations is the reproducibility of the measured drug interaction metrics [29–31]; thus, additional in vitro efficacy study was performed with the use of RealTime-Glo assay to confirm the efficacy and observed drug interaction metrics. Results indicate that both drugs are very active against A375 melanoma cells (see Table 3).

Table 3. Comparison of observed Siremadlin and Trametinib IC50 with data from the literature in A375 cells (72 h incubation). For MTS assay, $n = 4$, and for RealTime-Glo assay, $n = 3$.

Compound	MTS IC50 ± SD (nM)	RealTime-Glo IC50 ± SD (nM)	Literature IC50 (nM)
Siremadlin (HDM201)	65.7 ± 4.7	260.1 ± 170.5	764.1 [32] [1]
Trametinib	0.58 ± 0.03	0.8 ± 0.4	1.0 [33] [2]

[1] Cell count was measured using an amount of ATP (CellTiter-Glo assay). [2] Cell count was measured using 4′,6-diamidino-2-phenylindole (DAPI) nuclei staining.

Despite the high efficacy of tested compounds, some populations of cells remained resistant even at high concentrations (see Table S1). RealTime-Glo assay revealed that the killing effect and its intensity for both drugs was concentration- and time-dependent (see Figures 1 and 2).

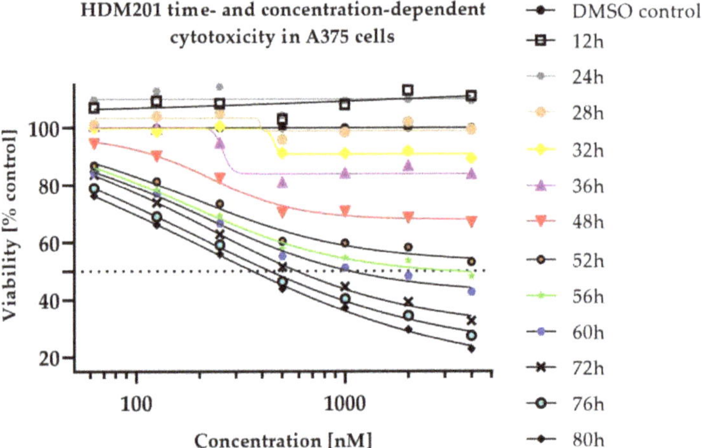

Figure 1. Time- and dose-dependent cytotoxicity of HDM201 in A375 cells. Mean from $n = 3$ (RealTime-Glo assay).

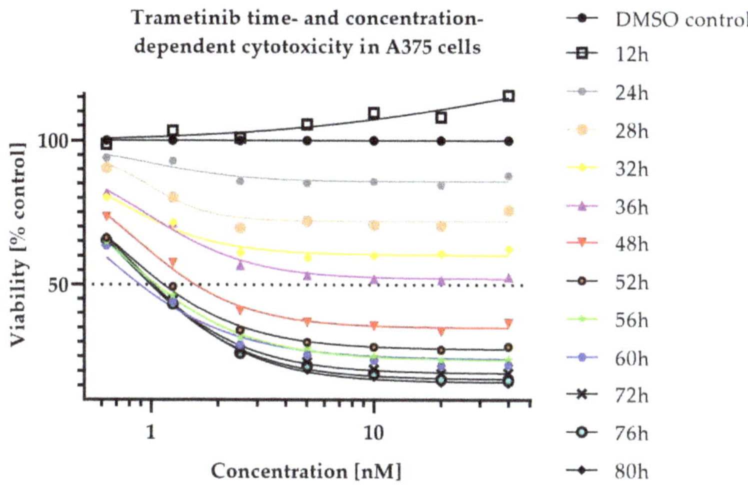

Figure 2. Time and dose-dependent cytotoxicity of Trametinib in A375 cells. Mean from $n = 3$ (RealTime-Glo assay).

The combination of those two compounds caused a profound increase in cytotoxicity, demonstrating an increase in potency and efficacy against A375 cells. For survival curve-shift, see Figures 3 and 4.

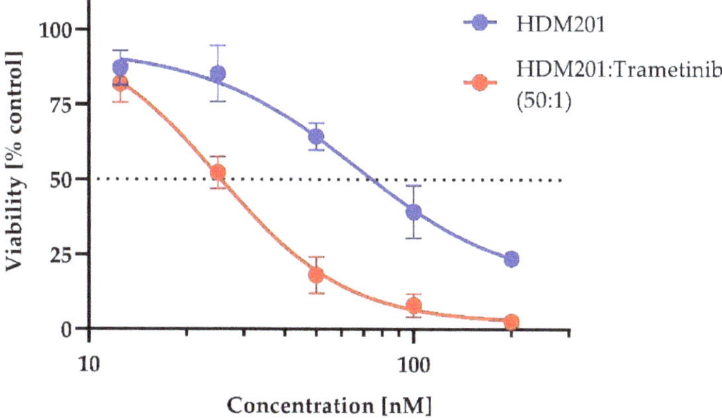

Figure 3. Curve shift for HDM201 combination with Trametinib. MTS assay, mean from $n = 4$.

2.2. In Vitro Drug Combination Analysis

Results from both assay methods (MTS and RealTime-Glo) with analysis involving use of Synergyfinder package (δ score from ZIP, Loewe, has, and Bliss models) were generally consistent, comparable, and indicated a synergistic interaction between studied drugs. Two exceptions were the ZIP model (RealTime-Glo assay) and Bliss model (MTS assay); however, calculated δ scores were very close to the synergistic threshold. Synergistic delta score is also reflected in calculated mean across the methods and models (see Table 4 and Table S2).

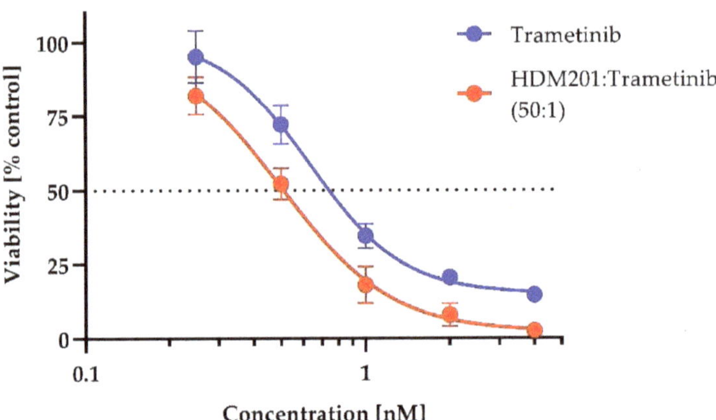

Figure 4. Curve shift for Trametinib combination with HDM201. MTS assay, mean from $n = 4$.

Table 4. Drug interaction between Siremadlin and Trametinib in A375 cells. For MTS assay, $n = 4$, and for RealTime-Glo assay, $n = 3$. Synergyfinder package analysis.

Assay	Timepoints (h)	ZIP $\delta \pm$ SD	Loewe $\delta \pm$ SD	HSA $\delta \pm$ SD	Bliss $\delta \pm$ SD	Mean across Models $\delta \pm$ SD
MTS	72	5.353 ± 2.613	5.111 ± 1.926	12.394 ± 2.085	4.881 ± 3.117	6.935 ± 2.420
RealTime-Glo	28–80	4.858 ± 1.346	7.113 ± 4.355	13.513 ± 3.111	5.540 ± 1.957	7.756 ± 1.614
Mean	-	5.023 ± 1.768	6.446 ± 3.546	13.140 ± 2.769	5.321 ± 2.344	7.482 ± 1.883

Results from both assay methods with analysis using the Synergy package (MuSyC model) were comparable for α_{21}, β, and γ_{21} parameters (especially in the 48–80 h interval for α_{21} and γ_{21} parameters, as shown in Table S3), indicating synergistic interaction in terms of increased potency (α parameter) and efficacy (β parameter) between Siremadlin and Trametinib, but not in terms of increased cooperativity (γ parameter); however, it seems that synergistic cooperativity is more important in neurological disorders than in treating cancer; thus, lack of synergy in this metric is clinically not relevant [22] (see Tabels 5 and S3).

Table 5. Drug interaction between Siremadlin and Trametinib in A375 cells. For MTS assay, $n = 4$, and for RealTime-Glo assay, $n = 3$. Synergy package analysis.

Assay	Timepoints (h)	$\alpha_{12}/\alpha_{21} \pm$ SD	$\beta \pm$ SD	$\gamma_{12}/\gamma_{21} \pm$ SD
MTS	72	$2.229 \pm 1.065/$ 1.498 ± 0.351	0.217 ± 0.045	$0.402 \pm 0.102/$ 0.710 ± 0.286
RealTime-Glo	28–80	$2.095 \pm 0.780/$ $12{,}507 \pm 26{,}999$	0.244 ± 0.050	$0.901 \pm 0.136/$ $6878 \pm 21{,}748$
Mean		$2.162 \pm 0.923/$ $6254 \pm 13{,}500$	0.231 ± 0.048	$0.652 \pm 0.119/$ $3440 \pm 10{,}874$

2.3. Siremadlin and Trametinib Pharmacokinetics (PK)

Pharmacokinetic profiles were determined after single oral administration of HDM201 (100 mg/kg) and Trametinib (1 mg/kg) in vehicle formulation in CD-1 nude mice. Initial analysis of compound concentrations included plasma and A375 tumour tissue homogenates. Additionally, analysis included HDM201 and Trametinib administered in combination (100 + 1 mg/kg, respectively) to determine whether there were interactions at the PK level, as shown in Figures 5–8.

The values of calculated PK parameters suggest that HDM201 and Trametinib are absorbed relatively quickly (typical Tmax are in the 1.5–4 h range) and maintain high exposure in plasma within 24 h. Data from A375 tumour tissue indicate that both compounds are well distributed in the tumour achieving higher maximal concentrations and exposure (Cmax and AUC, respectively) as shown in Table 6. PK analysis revealed that plasma and tumour Cmax and AUC for HDM201 have higher values when co-administered with Trametinib, while for Trametinib those parameters are higher only in A375 tumour. Interestingly, in plasma, Cmax and AUC for Trametinib are significantly lower when co-administered with HDM201, as depicted in Figure 6.

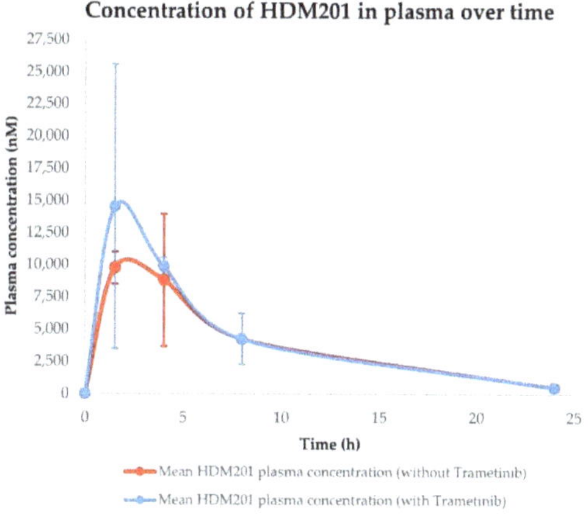

Figure 5. Pharmacokinetic profiles of HDM201 (Siremadlin) in plasma after administration with and without Trametinib. Observed data are means ± standard deviation (SD) from $n = 3$.

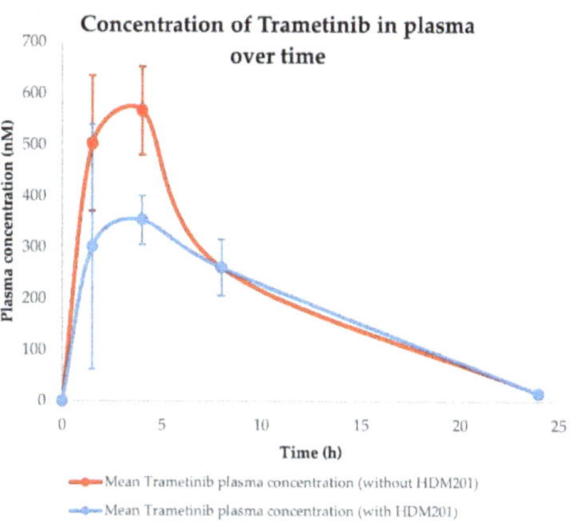

Figure 6. Pharmacokinetic profiles of Trametinib in plasma after administration with and without HDM201. Observed data are means ± standard deviation (SD) from $n = 3$.

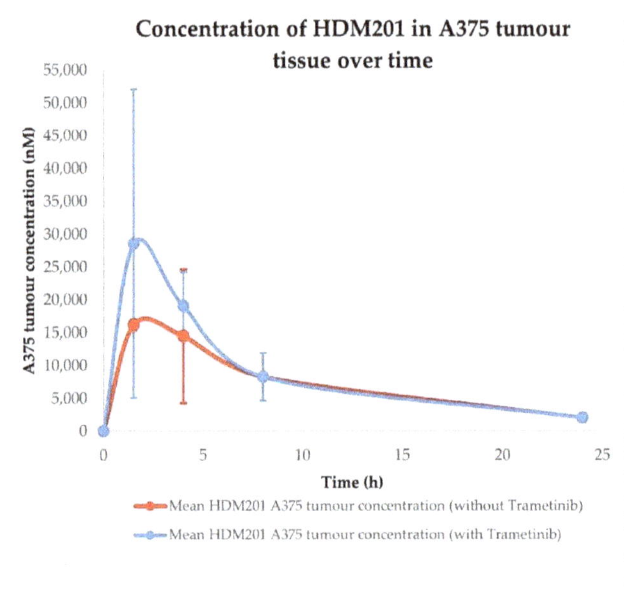

Figure 7. Pharmacokinetic profiles of HDM201 (Siremadlin) in A375 tumour tissue after administration with and without Trametinib. Observed data are means ± standard deviation (SD) from $n = 3$.

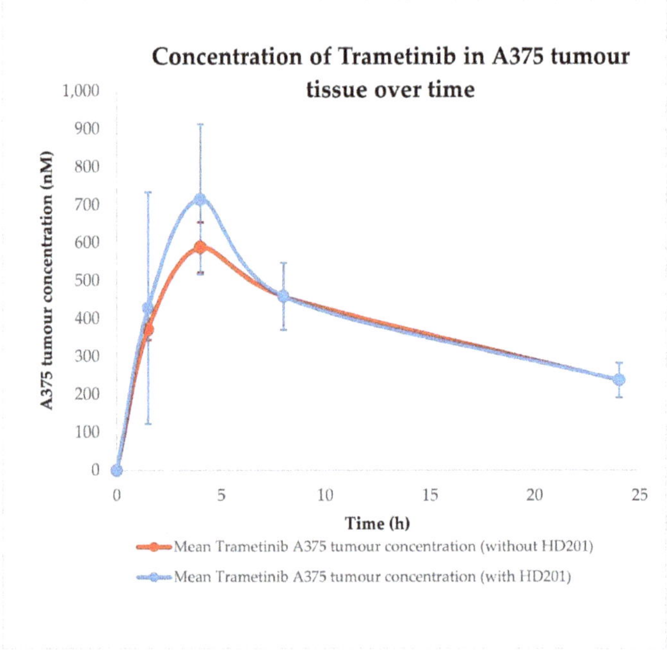

Figure 8. Pharmacokinetic profiles of Trametinib in A375 tumour tissue after administration with and without HDM201. Observed data are means ± standard deviation (SD) from $n = 3$.

Table 6. Calculated PK parameters for HDM201 and Trametinib in plasma and A375 tumour tissue.

Conditions	Tissue	AUC$_{0-24h}$ ± SD (nM × h)	Cmax ± SD (nM)	Tmax ± SD (h)
HDM201 without Trametinib	Plasma	95,092.97 ± 34,215.83	9777.67 ± 2976.84	1.50 ± 1.44
HDM201 with Trametinib	Plasma	107,993.98 ± 26,303.00	14,559.95 ± 7433.26	1.50 ± 1.44
Trametinib without HDM201	Plasma	5580.83 ± 566.66	567.02 ± 49.38	4.00 ± 1.44
Trametinib with HDM201	Plasma	4484.99 ± 1171.06	353.65 ± 105.55	4.00 ± 1.44
HDM201 without Trametinib	A375 tumour	179,026.48 ± 65,901.61	16,214.30 ± 5459.78	1.50 ± 1.44
HDM201 with Trametinib	A375 tumour	218,677.07 ± 91,168.31	28,613.74 ± 16,751.20	1.50 ± 1.44
Trametinib without HDM201	A375 tumour	9131.17 ± 1296.84	587.25 ± 66.35	4.00 ± 0.00
Trametinib with HDM201	A375 tumour	9656.67 ± 1393.80	714.53 ± 197.48	4.00 ± 0.00

2.4. Siremadlin and Trametinib Pharmacodynamics (PD)

The studied compounds, namely HDM201 and Trametinib, were tested separately and in combination as a therapy against an A375 melanoma tumour model. Tested compounds were administrated in three and six doses per schedule. All tested compounds decreased tumour volume in comparison to the group treated by formulation. Data from this efficacy study indicate that HDM201 and Trametinib are more efficacious when they are used in combination, compared to their efficacy when administered separately (please refer to maximal tumour growth inhibition (TGI) percentage, presented in Table 7, and compounds efficacy in single and combined administrations, shown in Figure 9).

Table 7. Mean tumour growth inhibition (TGI) values with standard error of mean (SEM) as metrics of in vivo efficacy of HDM201, Trametinib, and their combination in A375-inoculated CD-1 nude mice $n = 6$.

Group	Max TGI (%) ± SEM
HDM201 40 mg/kg qdx3	33.39 ± 13.90
HDM201 100 mg/kg qdx3	76.94 ± 5.38
Trametinib 0.3 mg/kg qdx6	65.47 ± 21.29
Trametinib 1 mg/kg qdx6	90.05 ± 1.13
HDM201 + Trametinib 40 + 0.3 mg/kg qdx3/qdx6	91.83 ± 1.37
HDM201 + Trametinib 40 + 1 mg/kg qdx3/qdx6	93.68 ± 1.63
HDM201 + Trametinib 100 + 0.3 mg/kg qdx3/qdx6	94.56 ± 1.77
HDM201 + Trametinib 100 + 1 mg/kg qdx3/qdx6	95.99 ± 0.84

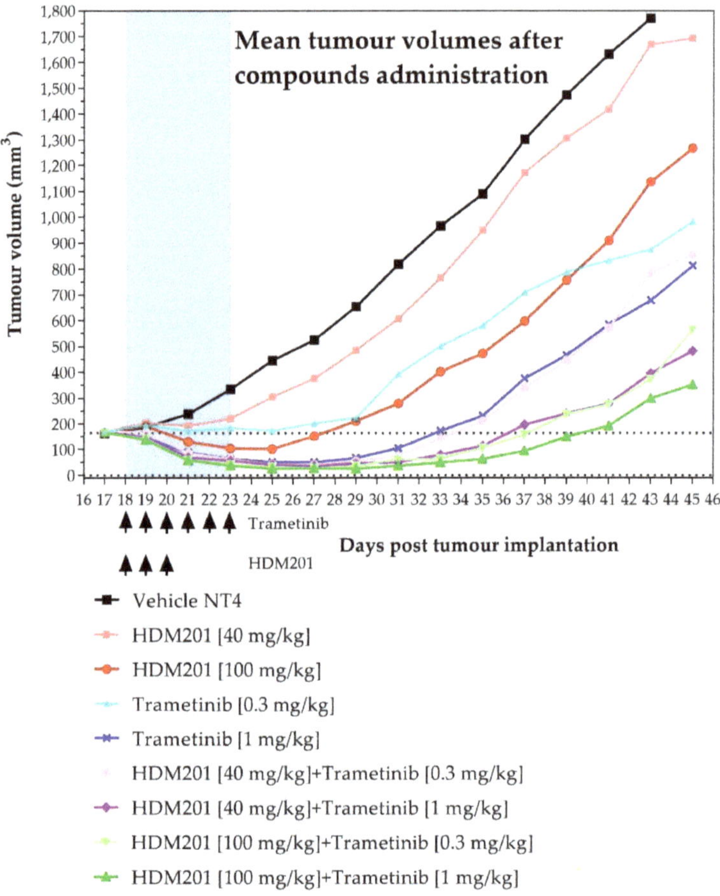

Figure 9. In vivo efficacy of HDM201 and Trametinib administered separately and in combination against a mouse model of human melanoma tumour (A375). HDM201 was dosed in a qdx3 schedule and Trametinib in a qdx6 schedule. Observed data are means from *n* = 6.

3. Discussion

In vitro cytotoxicity data demonstrated high efficacy of Siremadlin and Trametinib against A375 cells, which justifies further studies in in vivo models. Calculated Trametinib IC50 values were very similar to the values reported previously in the literature [32,33]; however, greater difference was observed for Siremadlin IC50. Differences in IC50 values may be explained by dissimilarities in methods used for counting living cells (different assays) or by the influence of cell-seeding density on cytotoxic sensitivity [34]. Performed experiments revealed that cytotoxicity of the studied compounds is concentration- and time-dependent with an initial delay of response. The delay in response to these drugs is most likely related to the duration of signal transduction cascade associated with the activation of the p53-MDM2 and MAPK pathways, resulting in cell death.

Resistance is an inherent part of anticancer treatment; therefore, the population of resistant cells was assessed for both drugs in the performed study, the description of which may play a critical role in predicting and optimizing treatment response and may improve therapy scheduling [35,36]. However, further in vitro studies on drug-resistant A375 melanoma sublines with the Siremadlin and Trametinib combination would be needed for in-depth analysis of resistance mechanisms, and to test if such combination treatment would be suitable for prolonged treatment, which is often characterized by increased resistance [7,37–42]. Results from drug interaction analysis indicated synergistic interaction between the drugs studied in an A375 melanoma model. The use of two different cytotoxicity assays (MTS and RealTime-Glo) indicated consistency in the drug interaction metrics obtained. Regarding analysis at particular timepoints, it can be noted that in analysis with use of the MuSyC model, the values of the α_{21} and γ_{21} parameters up to 36 h were very high and irregular. This may be caused by different initial responses of cells, since therapy started when cells were at different stages of the cell cycle. It can be hypothesized that those parameters began to stabilize when the majority of those cells reached the stasis/apoptosis stage.

Synergistic interaction was identified by using a number of the most commonly used theoretical drug interaction models with the use of Synergyfinder and Synergy software. Calculated synergy metrics were different in terms of their magnitude and foundations: a 7.48% increase in drug combination efficacy in the case of Synergyfinder (for the mean concentration-dependent δ parameter, see Table 4) and a 23.12% of increase in efficacy in the case of the Synergy software (for the mean concentration-independent β parameter, see Table 5). Those calculated synergy metrics will serve as a PD interaction parameter for further translational PBPK/PD modelling. Additional information obtained from this in vitro study about the delay of the response and the percentage of resistant cells, in terms of the total cell population, is very useful and will be also incorporated into PD model development.

Initial pharmacokinetic analysis performed on mice plasma indicated that HDM201 and Trametinib undergo fast absorption, as the Tmax values were in the 1.5–4 h range; however, due to a limited quantity of available timepoints (sparse sampling), these values may be not accurately determined. HDM201 and Trametinib in plasma and A375 tumours are characterized by high exposure within 24 h.

The pharmacokinetic profile of HDM201 co-administered with Trametinib in plasma is characterized by higher maximal concentration and exposure (Cmax and AUC, respectively) than after single drug administration. Interestingly, in the case of Trametinib co-administered with HDM201, the situation is the opposite. Observed Cmax and AUC are lower than after Trametinib administration alone. This may be related to the occurrence of PK interaction; however, further PK studies combined with PBPK modelling are required to prove the existence of this interaction and the mechanism of its formation. Higher Cmax and AUC in A375 tumour tissue than in plasma were observed for both drugs. Such observation is very favourably in context of potential combination therapy using HDM201 and Trametinib. Nevertheless, further PK studies combined with PBPK modelling are required to explain the mechanisms of pharmacokinetic interactions in the tumour compartment. More detailed pharmacokinetic analysis on heart, liver, spleen, muscle, brain, kidney, lung, gut, and skin tissue homogenates, combined with PBPK modelling and simulation, is the main topic of the second part (part II) of this publication cycle [43].

Efficacy data showed that all tested compounds decreased tumour volume in comparison to the group treated by vehicle formulation. Results from this study indicated that HDM201 and Trametinib were much more efficient when they are used in combination, compared to their efficacy when the those two compounds are administered separately. This observation should be supported by the results of subsequent in vivo studies using a higher number of animals per group. Moreover, it remains unclear how the PK interaction combined with PD interaction influences the observed anticancer efficacy for this drug combination. Therefore, further PK/PD studies on mice, coupled with PBPK/PD modelling, are needed in order to determine the mechanism of the PK interaction formation and to select the optimal PD interaction parameter which will translate observed in vitro synergy metrics into in vivo settings. Such an approach may facilitate the determination of the most synergistic and efficacious schedule and dose levels for Siremadlin and Trametinib in in vivo models, and may provide a basis for better estimation of drug combination efficacy in melanoma patients.

4. Materials and Methods

4.1. Materials

HDM201 (catalog number HY-18658) and Trametinib (catalog number HY-10999) used in this study were obtained from MedChemExpress. RealTime-Glo™ MT Cell Viability Assay kit and CellTiter 96® AQueous Non-Radioactive Cell Proliferation Assay (MTS) were provided by Promega. The A375 cell line used in the in vitro studies was obtained from American Type Culture Collection (CRL-1619). PEG 400 (catalog number 81172) and Cremophor RH40 (catalog number 07076) were provided by Merck (formerly Sigma-Aldrich), EtOH (catalog number 1016/12/19) was provided by POCH, and Labrafil M1944CS (catalog number 178290) was provided by Gattefosse. For drug combination in vivo studies, A375 cell line was provided by European Collection of Authenticated Cell Cultures (88113005).

4.2. Software

For in vitro drug combination studies, raw data processing was performed in Microsoft Excel 2016 with the use of Excel Visual Basic for Applications (VBA) macros. PK parameters and TGI values were estimated with Microsoft Excel (Excel version 2016, Microsoft Corporation, Redmond, WA, USA, 2016, https://www.office.com). For in vitro drug combination analysis, Synergyfinder (version 2.5.1 compiled on 01.12.2020 from Github resource https://github.com/shuyuzheng/synergyfinder) and Synergy (version 0.4.5 compiled on 13.02.2021 from Github resource https://github.com/djwooten/synergy) packages were used. Visualization of the in vivo efficacy and calculation of the IC_{50} values by curve fitting were performed using GraphPad Prism version 9.3.1 for Windows, GraphPad Software, San Diego, CA, USA, 2022, www.graphpad.com.

4.3. In Vitro Drug Combination Studies

Inhibition of tumour cell viability after single and combination drug treatment was measured with the use of MTS assay (CellTiter 96® AQueous Non-Radioactive Cell Proliferation Assay) and RealTime-Glo assay (RealTime-Glo™ MT Cell Viability Assay) in A375 cell line using standard manufacturer protocols. Briefly, cells were plated at optimized seeding density (0.5×10^3 cells/well) in a 96-well culture plate in an appropriate cell culture medium (DMEM 4.5 g/L glucose supplemented with 10% v/v FBS), cell cultures were stimulated with compounds 24 h after cell seeding with MDM2 and MEK inhibitors. In the MTS assay, compounds were added at 5 concentrations, ranging from 12.5 nmol/L to 200 nmol/L and 0.25 nmol/L to 4 nmol/L (for HDM201 and Trametinib, respectively) along with a dimethyl sulfoxide (DMSO) control. In the RealTime-Glo assay, compounds were added at 7 concentrations, ranging from 62.5 nmol/L to 4000 nmol/L for HDM201 and 0.625 nmol/L to 40 nmol/L for Trametinib, with a dimethyl sulfoxide (DMSO) control. Drug combinations were tested in the matrix layout with the increasing concentrations of both drugs (please see Figures S1 and S2). Tumour cells' viability was measured after 72 h of cell incubation in the presence of tested compounds in the MTS assay and at 0,12, 24, 28, 32, 36, 48, 52, 56, 60, 72, 76, 80 h timepoints in the RealTime-Glo assay (with exception in first experiment, performed with RealTime-Glo assay, on which 28 and 32 h timepoints were not measured). Several independent assay repetitions were performed for MTS ($n = 4$) and RealTime-Glo assays ($n = 3$).

4.4. Drug Combination Interaction Analysis

Drug combination interaction analysis was performed on in vitro cytotoxicity data from MTS and RealTime-Glo assays. For RealTime-Glo data, all further calculations of drug interaction parameters were performed in the 28–80 h time range, due to the lack of significant efficacy of single Siremadlin and Trametinib at 12 and 24 h timepoints and the impossibility of finding proper drug interaction model fit ($R^2 < 0.8$). Analysis with the use of the Synergyfinder package (version 2.5.1 compiled on 01.12.2020 from Github resource https://github.com/shuyuzheng/synergyfinder) was performed with a script written in the R language in RStudio Version 1.2.5001 Build 1468 (see Code S1 in Supplementary Materials). For Synergyfinder analysis parameters, see Table S4. For a more detailed description of all Synergyfinder functions, please see its documentation and user instructions [44].

Analysis with the use of the Synergy package (version 0.4.5 compiled on 13.02.2021 from Github resource https://github.com/djwooten/synergy was performed with a script written in Python language in Python 3.9.12 (see Code S2 in Supplementary Materials). For Synergy analysis parameters, see Table S5.

4.5. Studies Involving Animals

Crl:CD-1-*Foxn1^{nu}* female mice, 4–5 weeks old, from Charles River Germany, inoculated subcutaneously with A375 cells, were used for in vivo studies. Determination of compound concentrations in plasma and A375 tumour tissue homogenates were performed with the use of quantitative LC-MS/MS system. Plasma and tissue samples were resected at the following timepoints: 1.5, 4, 8, 24 h ($n = 3$ per timepoint) after oral administration. Pharmacokinetic parameters AUC, C_{max}, and T_{max} were calculated using MS Excel 2016. Area under the concentration versus time curve was calculated using the linear trapezoidal rule [45].

Determination of tumour growth was performed after oral gavage of Vehicle (60% PEG 400 (v/v), 10% Cremophor RH40 (v/v), 10% EtOH (v/v) and 20% Labrafil M1944CS (v/v)), Siremadlin, Trametinib or their combination in Vehicle. The volume of the administered formulation (10 mL/kg) of the compounds was always adjusted to the mice body weight. Initial tumour volumes, doses, dose schedules, and number of animals in particular in vivo studies are summarized in Table 8.

Table 8. Summary of performed in vivo studies on CD-1 nude mice xenografted with A375 tumour.

Compound	Initial Tumour Volume (mm^3)	Doses (mg/kg)	Dose Schedule	N	Comments
Vehicle	~162	-	qdx6	11	Efficacy
Siremadlin	~163–172	40/100	qdx3	6	Efficacy
Trametinib	~167–180	0.3/1	qdx6	6	Efficacy
Siremadlin+ Trametinib	~165–169	40 + 0.3/40 + 1/ 100 + 0.3/100 + 1	qdx3/qdx6	6	Efficacy
Siremadlin	~300	100	qdx1	12	PK
Trametinib	~300	1	qdx1	12	PK
Siremadlin+ Trametinib	~300	100 + 1	qdx1	12	PK

Tumour volume (V) was recorded with an electronic calliper 2–3 times a week and was calculated based on its length and width, using the prolate ellipsoid equation [46] (Equation (1)):

$$V \ (\text{mm}^3) = d^2 \times D/2 \tag{1}$$

where d is the tumour width (mm) and D is the tumour length (mm).

Tumour growth inhibition (TGI) value was calculated using Equation (2) [47]:

$$\text{TGI} \ (\%) = (100 - (T/C \times 100)) \tag{2}$$

where C is the mean tumour size in control group (mm^3) and T is the mean tumour size in treated group (mm^3).

Supplementary Materials: The following supporting information can be downloaded at: https://www.mdpi.com/article/10.3390/ijms232112984/s1.

Author Contributions: Conceptualization, J.W.; methodology, J.W.; software, J.W.; validation, J.W., S.P. and D.P.; formal analysis, J.W.; investigation, J.W.; resources, J.W. and Z.R.; data curation, J.W.; writing—original draft preparation, J.W.; writing—review and editing, S.P., Z.R. and D.P.; visualization, J.W.; supervision, S.P., Z.R. and D.P.; project administration, J.W.; funding acquisition, Z.R. and D.P. All authors have read and agreed to the published version of the manuscript.

Funding: This work has been performed under the Ministry of Science and Higher Education's program "Applied Ph.D." at the Faculty of Chemistry of the University of Warsaw and at Adamed Pharma S.A., based on contract no. 0058/DW/2018/01/1. Data obtained from Adamed Pharma S.A. were performed under the Project "Preclinical development of an innovative anti-cancer drug using the mechanism of reactivation of p53 protein" (POIR/01.02.00-14-31/15), co-financed by the European Union funds under Measure 1.2 Sectoral R & D programs of the Operational Programme Smart Growth 2014–2020 (INNOMED—Program of scientific research and development works for economic sector in the field of innovative medicine).

Institutional Review Board Statement: The animal study protocols were approved by the Ethics Committee of II Local Ethics Committee for Animal Experiments in Warsaw (permission for the experimental use of animals WAW2_6/2016 approved 16.03.2016 and WAW2_19/2016 approved 18.05.2016) for studies involving animals.

Informed Consent Statement: Not applicable.

Data Availability Statement: The data presented in this study are available in the article or Supplementary Materials. Raw data from MTS and RealTime-Glo assays and PK and PD studies are available on request from the corresponding author.

Acknowledgments: The authors would like to thank: Maria Mazur and Katarzyna Jastrzębska-Mazur for their helpful advice and reviewing this manuscript. Shuyu Zheng, Christian Meyer, and David Wooten for very valuable discussions about drug combinations and advice regarding interpretation of the obtained results. Małgorzata Drabczyk-Pluta, Sabina, Jeleń, Grażyna Peszyńska-Sularz, Anna Żyłko, Agnieszka Jakubiak, and Grzegorz Stasiłojć for support in performing PK/PD studies. Special thanks to Michal Łęcicki for his help with R and Python script development used for in vitro drug combination analysis.

Conflicts of Interest: Jakub Witkowski is Adamed Pharma employee and co-inventor of the patent WO2019141549 related to MDM2 inhibitor developed by Adamed Pharma. Dariusz Pawelec is a Adamed Pharma employee. Sebastian Polak is a Certara UK employee. No other conflicts of interest, financial or otherwise, were disclosed by the authors. The funders had no role in the design of the study; in the collection, analyses, or interpretation of data; in the writing of the manuscript, or in the decision to publish the results.

References

1. Eroglu, Z.; Ribas, A. Combination Therapy with BRAF and MEK Inhibitors for Melanoma: Latest Evidence and Place in Therapy. *Adv. Med. Oncol.* **2016**, *8*, 48–56. [CrossRef] [PubMed]
2. Saiki, A.Y.; Caenepeel, S.; Yu, D.; Lofgren, J.A.; Osgood, T.; Robertson, R.; Canon, J.; Su, C.; Jones, A.; Zhao, X.; et al. MDM2 Antagonists Synergize Broadly and Robustly with Compounds Targeting Fundamental Oncogenic Signaling Pathways. *Oncotarget* **2014**, *5*, 2030–2043. [CrossRef] [PubMed]
3. Wu, C.-E.; Koay, T.S.; Esfandiari, A.; Ho, Y.-H.; Lovat, P.; Lunec, J. ATM Dependent DUSP6 Modulation of P53 Involved in Synergistic Targeting of MAPK and P53 Pathways with Trametinib and MDM2 Inhibitors in Cutaneous Melanoma. *Cancers* **2018**, *11*, 3. [CrossRef] [PubMed]
4. Caenepeel, S.; CANON, J.; Hughes, P.; Oliner, J.D.; Rickles, R.J.; Saiki, A.Y. Combination Therapy Including an MDM2 Inhibitor and One or More Additional Pharmaceutically Active Agents for the Treatment of Cancers. U.S. Patent 10,881,648, 5 January 2021.
5. Cinatl, J.; Speidel, D.; Hardcastle, I.; Michaelis, M. Resistance Acquisition to MDM2 Inhibitors. *Biochem. Soc. Trans.* **2014**, *42*, 752–757. [CrossRef] [PubMed]
6. Hoffman-Luca, C.G.; Yang, C.-Y.; Lu, J.; Ziazadeh, D.; McEachern, D.; Debussche, L.; Wang, S. Significant Differences in the Development of Acquired Resistance to the MDM2 Inhibitor SAR405838 between In Vitro and In Vivo Drug Treatment. *PLoS ONE* **2015**, *10*, e0128807. [CrossRef]
7. Tripathi, R.; Liu, Z.; Jain, A.; Lyon, A.; Meeks, C.; Richards, D.; Liu, J.; He, D.; Wang, C.; Nespi, M.; et al. Combating Acquired Resistance to MAPK Inhibitors in Melanoma by Targeting Abl1/2-Mediated Reactivation of MEK/ERK/MYC Signaling. *Nat. Commun.* **2020**, *11*, 5463. [CrossRef]

8. Berberich, A.; Kessler, T.; Thomé, C.M.; Pusch, S.; Hielscher, T.; Sahm, F.; Oezen, I.; Schmitt, L.-M.; Ciprut, S.; Hucke, N.; et al. Targeting Resistance against the MDM2 Inhibitor RG7388 in Glioblastoma Cells by the MEK Inhibitor Trametinib. *Clin. Cancer Res.* **2019**, *25*, 253–265. [CrossRef]

9. Shangary, S.; Qin, D.; McEachern, D.; Liu, M.; Miller, R.S.; Qiu, S.; Nikolovska-Coleska, Z.; Ding, K.; Wang, G.; Chen, J.; et al. Temporal Activation of P53 by a Specific MDM2 Inhibitor Is Selectively Toxic to Tumours and Leads to Complete Tumour Growth Inhibition. *Proc. Natl. Acad. Sci. USA* **2008**, *105*, 3933–3938. [CrossRef]

10. De la Puente, P.; Muz, B.; Jin, A.; Azab, F.; Luderer, M.; Salama, N.N.; Azab, A.K. MEK Inhibitor, TAK-733 Reduces Proliferation, Affects Cell Cycle and Apoptosis, and Synergizes with Other Targeted Therapies in Multiple Myeloma. *Blood Cancer J.* **2016**, *6*, e399. [CrossRef]

11. Moschos, S.J.; Sandhu, S.K.; Lewis, K.D.; Sullivan, R.J.; Johnson, D.B.; Zhang, Y.; Rasmussen, E.; Henary, H.A.; Long, G.V. Phase 1 Study of the P53-MDM2 Inhibitor AMG 232 Combined with Trametinib plus Dabrafenib or Trametinib in Patients (Pts) with TP53 Wild Type (TP53WT) Metastatic Cutaneous Melanoma (MCM). *J. Clin. Oncol.* **2017**, *35*, 2575. [CrossRef]

12. Greco, W.; Unkelbach, H.-D.; Pöch, G.; Sühnel, J.; Kundi, M.; Bödeker, W. Consensus on Concepts and Terminology for Combined-Action Assessment: The Saariselkä Agreement. *Arch. Complex Environ. Stud.* **1992**, *4*, 65–69.

13. Roell, K.R.; Reif, D.M.; Motsinger-Reif, A.A. An Introduction to Terminology and Methodology of Chemical Synergy—Perspectives from Across Disciplines. *Front. Pharm.* **2017**, *8*, 158. [CrossRef]

14. Loewe, S. The Problem of Synergism and Antagonism of Combined Drugs. *Arzneimittelforschung* **1953**, *3*, 285–290.

15. Bliss, C.I. The Toxicity of Poisons Applied Jointly1. *Ann. Appl. Biol.* **1939**, *26*, 585–615. [CrossRef]

16. Berenbaum, M.C. What Is Synergy? *Pharm. Rev.* **1989**, *41*, 93–141.

17. Yadav, B.; Wennerberg, K.; Aittokallio, T.; Tang, J. Searching for Drug Synergy in Complex Dose–Response Landscapes Using an Interaction Potency Model. *Comput. Struct. Biotechnol. J.* **2015**, *13*, 504–513. [CrossRef]

18. Meyer, C.T.; Wooten, D.J.; Paudel, B.B.; Bauer, J.; Hardeman, K.N.; Westover, D.; Lovly, C.M.; Harris, L.A.; Tyson, D.R.; Quaranta, V. Quantifying Drug Combination Synergy along Potency and Efficacy Axes. *Cell Syst.* **2019**, *8*, 97–108.e16. [CrossRef]

19. Tang, J.; Wennerberg, K.; Aittokallio, T. What Is Synergy? The Saariselkä Agreement Revisited. *Front. Pharm.* **2015**, *6*, 181. [CrossRef]

20. Cokol, M.; Chua, H.N.; Tasan, M.; Mutlu, B.; Weinstein, Z.B.; Suzuki, Y.; Nergiz, M.E.; Costanzo, M.; Baryshnikova, A.; Giaever, G.; et al. Systematic Exploration of Synergistic Drug Pairs. *Mol. Syst. Biol.* **2011**, *7*, 544. [CrossRef]

21. Sen, P.; Saha, A.; Dixit, N.M. You Cannot Have Your Synergy and Efficacy Too. *Trends Pharmacol. Sci.* **2019**, *40*, 811–817. [CrossRef]

22. Ianevski, A.; He, L.; Aittokallio, T.; Tang, J. SynergyFinder: A Web Application for Analyzing Drug Combination Dose–Response Matrix Data. *Bioinformatics* **2017**, *33*, 2413–2415. [CrossRef] [PubMed]

23. Wooten, D.J.; Albert, R. Synergy: A Python Library for Calculating, Analyzing and Visualizing Drug Combination Synergy. *Bioinformatics* **2021**, *37*, 1473–1474. [CrossRef] [PubMed]

24. Di Veroli, G.Y.; Fornari, C.; Wang, D.; Mollard, S.; Bramhall, J.L.; Richards, F.M.; Jodrell, D.I. Combenefit: An Interactive Platform for the Analysis and Visualization of Drug Combinations. *Bioinformatics* **2016**, *32*, 2866–2868. [CrossRef] [PubMed]

25. Lehàr, J.; Krueger, A.S.; Avery, W.; Heilbut, A.M.; Johansen, L.M.; Price, E.R.; Rickles, R.J.; Short, G.F.; Staunton, J.E.; Jin, X.; et al. Synergistic Drug Combinations Improve Therapeutic Selectivity. *Nat. Biotechnol.* **2009**, *27*, 659–666. [CrossRef] [PubMed]

26. Malyutina, A.; Majumder, M.M.; Wang, W.; Pessia, A.; Heckman, C.A.; Tang, J. Drug Combination Sensitivity Scoring Facilitates the Discovery of Synergistic and Efficacious Drug Combinations in Cancer. *PLoS Comput. Biol.* **2019**, *15*, e1006752. [CrossRef] [PubMed]

27. Wooten, D.J.; Meyer, C.T.; Lubbock, A.L.R.; Quaranta, V.; Lopez, C.F. MuSyC Is a Consensus Framework That Unifies Multi-Drug Synergy Metrics for Combinatorial Drug Discovery. *Nat. Commun.* **2021**, *12*, 4607. [CrossRef]

28. Wooten, D. MuSyC Model Description (Github). Available online: https://github.com/djwooten/synergy/blob/b13e2ac961f8738d58c01f79e5b93ec2dd78f555/src/synergy/combination/musyc.py (accessed on 27 May 2022).

29. Meyer, C.T.; Wooten, D.J.; Lopez, C.F.; Quaranta, V. Charting the Fragmented Landscape of Drug Synergy. *Trends Pharmacol. Sci.* **2020**, *41*, 266–280. [CrossRef]

30. Day, D.; Siu, L.L. Approaches to Modernize the Combination Drug Development Paradigm. *Genome Med.* **2016**, *8*, 115. [CrossRef]

31. Zagidullin, B.; Aldahdooh, J.; Zheng, S.; Wang, W.; Wang, Y.; Saad, J.; Malyutina, A.; Jafari, M.; Tanoli, Z.; Pessia, A.; et al. DrugComb: An Integrative Cancer Drug Combination Data Portal. *Nucleic Acids Res.* **2019**, *47*, W43–W51. [CrossRef]

32. Jeay, S.; Ferretti, S.; Holzer, P.; Fuchs, J.; Chapeau, E.A.; Wartmann, M.; Sterker, D.; Romanet, V.; Murakami, M.; Kerr, G.; et al. Dose and Schedule Determine Distinct Molecular Mechanisms Underlying the Efficacy of the P53–MDM2 Inhibitor HDM201. *Cancer Res.* **2018**, *78*, 6257–6267. [CrossRef]

33. Jing, J.; Greshock, J.; Holbrook, J.D.; Gilmartin, A.; Zhang, X.; McNeil, E.; Conway, T.; Moy, C.; Laquerre, S.; Bachman, K.; et al. Comprehensive Predictive Biomarker Analysis for MEK Inhibitor GSK1120212. *Mol. Cancer Ther.* **2012**, *11*, 720–729. [CrossRef]

34. He, Y.; Zhu, Q.; Chen, M.; Huang, Q.; Wang, W.; Li, Q.; Huang, Y.; Di, W. The Changing 50% Inhibitory Concentration (IC 50) of Cisplatin: A Pilot Study on the Artifacts of the MTT Assay and the Precise Measurement of Density-Dependent Chemoresistance in Ovarian Cancer. *Oncotarget* **2016**, *7*, 70803–70821. [CrossRef]

35. Howard, G.R.; Johnson, K.E.; Rodriguez Ayala, A.; Yankeelov, T.E.; Brock, A. A Multi-State Model of Chemoresistance to Characterize Phenotypic Dynamics in Breast Cancer. *Sci. Rep.* **2018**, *8*, 12058. [CrossRef]

36. Yoon, N.; Vander Velde, R.; Marusyk, A.; Scott, J.G. Optimal Therapy Scheduling Based on a Pair of Collaterally Sensitive Drugs. *Bull. Math. Biol.* **2018**, *80*, 1776–1809. [CrossRef]

37. Drummond, C.J.; Esfandiari, A.; Liu, J.; Lu, X.; Hutton, C.; Jackson, J.; Bennaceur, K.; Xu, Q.; Makimanejavali, A.R.; Bello, F.D.; et al. TP53 Mutant MDM2-Amplified Cell Lines Selected for Resistance to MDM2-P53 Binding Antagonists Retain Sensitivity to Ionizing Radiation. *Oncotarget* **2016**, *7*, 46203–46218. [CrossRef]

38. Wu, C.-E.; Koay, T.S.; Ho, Y.-H.; Lovat, P.; Lunec, J. Abstract 3034: TP53 Mutant Cell Lines Selected for Resistance to MDM2 Inhibitors Retain Growth Inhibition by MAPK Pathway Inhibitors but a Reduced Apoptotic Response. *Cancer Res.* **2019**, *79*, 3034. [CrossRef]

39. Deben, C.; Boullosa, L.F.; Domen, A.; Wouters, A.; Cuypers, B.; Laukens, K.; Lardon, F.; Pauwels, P. Characterization of Acquired Nutlin-3 Resistant Non-Small Cell Lung Cancer Cells. *Cancer Drug Resist.* **2021**, *4*, 233–243. [CrossRef]

40. Michaelis, M.; Rothweiler, F.; Schneider, C.; Rothenburger, T.; Mernberger, M.; Nist, A.; Stiewe, T.; Cinatl, J.; Nevels, M. Long-Term Cultivation Using Ineffective MDM2 Inhibitor Concentrations Alters the Drug Sensitivity Profiles of PL21 Leukaemia Cells. *Exp. Results* **2020**, *1*, e5. [CrossRef]

41. Poulikakos, P.I.; Persaud, Y.; Janakiraman, M.; Kong, X.; Ng, C.; Moriceau, G.; Shi, H.; Atefi, M.; Titz, B.; Gabay, M.T.; et al. RAF Inhibitor Resistance Is Mediated by Dimerization of Aberrantly Spliced BRAF(V600E). *Nature* **2011**, *480*, 387–390. [CrossRef]

42. Basile, K.J.; Abel, E.V.; Dadpey, N.; Hartsough, E.J.; Fortina, P.; Aplin, A.E. In Vivo MAPK Reporting Reveals the Heterogeneity in Tumoural Selection of Resistance to RAF Inhibitors. *Cancer Res.* **2013**, *73*, 7101–7110. [CrossRef]

43. Witkowski, J.; Polak, S.; Rogulski, Z.; Pawelec, D. In Vitro/In Vivo Translation of Synergistic Combination of MDM2 and MEK Inhibitors in Melanoma Using PBPK/PD Modelling: Part II. *Int. J. Mol. Sci.* **2022**, *23*, 11939. [CrossRef] [PubMed]

44. Zheng, S.; Tang, J. Synergyfinder Package Documentation. Available online: https://bioconductor.org/packages/release/bioc/manuals/synergyfinder/man/synergyfinder.pdf (accessed on 13 June 2022).

45. Curry, S.H.; Whelpton, R. Appendix 1: Mathematical Concepts and the Trapezoidal Method. In *Introduction to Drug Disposition and Pharmacokinetics*; John Wiley & Sons: Hoboken, NJ, USA, 2016; pp. 293–299. ISBN 978-1-119-26108-7.

46. Jensen, M.M.; Jørgensen, J.T.; Binderup, T.; Kjær, A. Tumour Volume in Subcutaneous Mouse Xenografts Measured by MicroCT Is More Accurate and Reproducible than Determined by 18F-FDG-MicroPET or External Caliper. *BMC Med. Imaging* **2008**, *8*, 16. [CrossRef] [PubMed]

47. Hather, G.; Liu, R.; Bandi, S.; Mettetal, J.; Manfredi, M.; Shyu, W.-C.; Donelan, J.; Chakravarty, A. Growth Rate Analysis and Efficient Experimental Design for Tumour Xenograft Studies. *Cancer Inf.* **2014**, *13*, 65–72. [CrossRef]

International Journal of
Molecular Sciences

Article

In Vitro/In Vivo Translation of Synergistic Combination of MDM2 and MEK Inhibitors in Melanoma Using PBPK/PD Modelling: Part II

Jakub Witkowski [1,2,*], Sebastian Polak [3,4], Zbigniew Rogulski [1] and Dariusz Pawelec [2]

1   Faculty of Chemistry, University of Warsaw, Pasteura 1, 02-093 Warsaw, Poland
2   Adamed Pharma S.A., Adamkiewicza 6a, 05-152 Czosnów, Poland
3   Faculty of Pharmacy, Jagiellonian University, Medyczna 9, 30-688 Krakow, Poland
4   Simcyp Division, Certara UK Limited, Level 2-Acero, 1 Concourse Way, Sheffield S1 2BJ, UK
*   Correspondence: j.witkowski6@uw.edu.pl

Abstract: The development of in vitro/in vivo translational methods for synergistically acting drug combinations is needed to identify the most effective therapeutic strategies. We performed PBPK/PD modelling for siremadlin, trametinib, and their combination at various dose levels and dosing schedules in an A375 xenografted mouse model (melanoma cells). In this study, we built models based on in vitro ADME and in vivo PK/PD data determined from the literature or estimated by the Simcyp Animal simulator (V21). The developed PBPK/PD models allowed us to account for the interactions between siremadlin and trametinib at PK and PD levels. The interaction at the PK level was described by an interplay between absorption and tumour disposition levels, whereas the PD interaction was based on the in vitro results. This approach allowed us to reasonably estimate the most synergistic and efficacious dosing schedules and dose levels for combinations of siremadlin and trametinib in mice. PBPK/PD modelling is a powerful tool that allows researchers to properly estimate the in vivo efficacy of the anticancer drug combination based on the results of in vitro studies. Such an approach based on in vitro and in vivo extrapolation may help researchers determine the most efficacious dosing strategies and will allow for the extrapolation of animal PBPK/PD models into clinical settings.

Keywords: anticancer drugs; preclinical study; pharmacokinetics; pharmacodynamics; drug combination; PBPK/PD modelling; MDM2 inhibitor; MEK inhibitor

Citation: Witkowski, J.; Polak, S.; Rogulski, Z.; Pawelec, D. In Vitro/In Vivo Translation of Synergistic Combination of MDM2 and MEK Inhibitors in Melanoma Using PBPK/PD Modelling: Part II. *Int. J. Mol. Sci.* **2022**, *23*, 11939. https://doi.org/10.3390/ijms231911939

Academic Editors: Andrzej Kutner, Geoffrey Brown and Enikö Kallay

Received: 24 August 2022
Accepted: 30 September 2022
Published: 8 October 2022

Publisher's Note: MDPI stays neutral with regard to jurisdictional claims in published maps and institutional affiliations.

1. Introduction

Metastatic melanoma is a cancer condition that is dangerous and difficult to treat due to its ability to spread early and aggressively. Before the development of new therapeutic strategies, the median survival of patients with metastatic melanoma was only 6–9 months [1] and the 10-year survival rate was less than 10% [2]. Although recent therapeutic advances for metastatic melanoma have considerably increased the overall survival of patients with melanoma, a subset of patients do not respond to immunotherapy or targeted therapies [3,4]. Such limited responses may be explained by arising resistance. Drug combinations targeting multiple signalling pathways in cancer cells may provide a remedy for emerging resistance development [5], and this is why new anticancer drug combinations and therapies are so important and urgently required. One of the novel therapeutic options is the drug combination of siremadlin (MDM2 inhibitor) and trametinib (MEK inhibitor). Preclinical evidence suggests that siremadlin (previously known as HDM201) and trametinib synergistically act in melanoma treatment [6,7]. To assess how polytherapy may improve the anticancer response, the performance of preclinical translational studies and the development of in vitro/in vivo translational methods are highly needed. A bench-to-bedside approach is much more challenging for drug combinations than for a single drug because it must combine assumptions regarding the interaction between two (or more) drugs at both the pharmacokinetic (PK) and pharmacodynamic (PD) levels [8,9]. One of the solutions allowing for such a prediction is physiologically based pharmacokinetic–pharmacodynamic (PBPK/PD) modelling. The PBPK/PD modelling approach allows researchers to combine information on the drug characteristics with their knowledge of physiology and biology at the organ and whole-organism levels. Such an approach allows researchers to achieve a representation of the drug in a biological system and the simulation of drug concentration–time profiles (pharmacokinetic profiles) and to link it to the drug's efficacy (its pharmacodynamic effect). This modelling approach offers an advantage over the traditional PK/PD modelling approach because it potentially allows for extrapolation into conditions for which pharmacokinetic studies have not been conducted. PBPK models consider different organs and tissues (whole-body PBPK model) that are the most relevant to the absorption, distribution, metabolism, and excretion (ADME) of the drug. Thus, the drug concentration–time (pharmacokinetic) profile can be accurately simulated in particular organs and tissues. Such a prediction is of high pharmacological relevance because it enables the estimation of drug exposure at the site of its action (for example, in a tumour), which may be difficult or impossible to experimentally measure in animals or humans.

Analysis of the interactions between two or more drugs at the PD level is difficult because of a lack of consensus on which theoretical model should be used to describe the drug interaction type. As previously discussed, this issue is usually related to a dilemma when a drug combination is classified as synergistic according to one model but antagonistic in the other [6,10–12]. In this study, we chose the previously proposed synergy metrics δ score (from the Synergyfinder package analysis) and β parameter (from the Synergy package analysis) to be tested as translational in vitro/in vivo PD interaction parameters [6].

Our main goal in this study was to develop and optimise a PBPK/PD model which could allow the translation of the in vitro drug combination results to an in vivo situation. An approach that involves using the results from in vitro studies coupled with PBPK/PD modelling may accurately describe the observed tumour growth inhibition (TGI) data and may suggest the most synergistic and efficacious schedules and dose levels for siremadlin and trametinib in mice in vivo. This modus operandi may lead to more accurate estimations of drug combination efficacy in virtual clinical trials (VCTs), which might be performed on a virtual representation of cancer patients and ultimately provide the rationale for using this drug combination in clinical trials on real patients with melanoma cancer.

2. Results

2.1. PBPK Models (with and without PK Interaction)

The developed PBPK models properly described the observed concentration–time data for siremadlin, trametinib, and their combination in plasma, A375 tumour, and other tissues (the muscle, spleen, brain, heart, kidneys, skin, lungs, gut, and liver), as shown in Figures 1–7. These results are in line with those of a numerical analysis of fold errors (the predicted and observed values) for the most important pharmacokinetic parameters $AUC_{0–24h}$, C_{max}, and T_{max}, which, in most cases, were within the 2-fold error range 0.5–2.0 (Tables S3–S6).

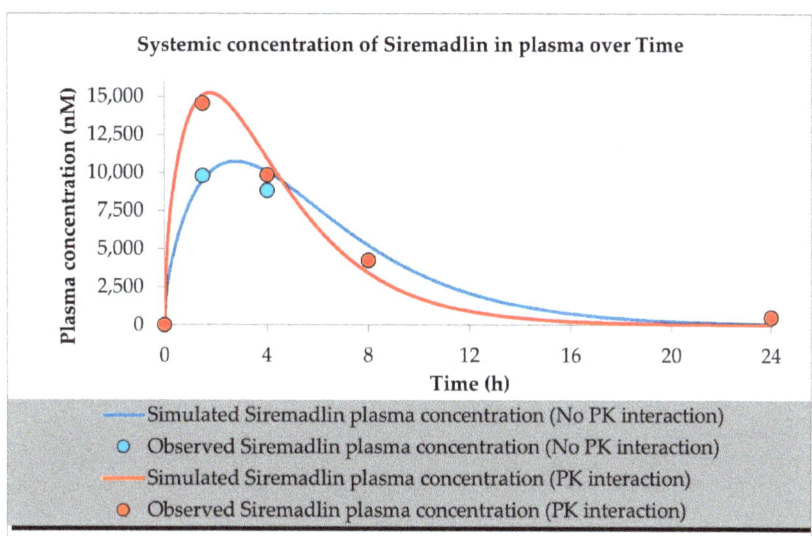

Figure 1. PBPK model of siremadlin with and without PK interaction. Observed data are means from *n* = 3.

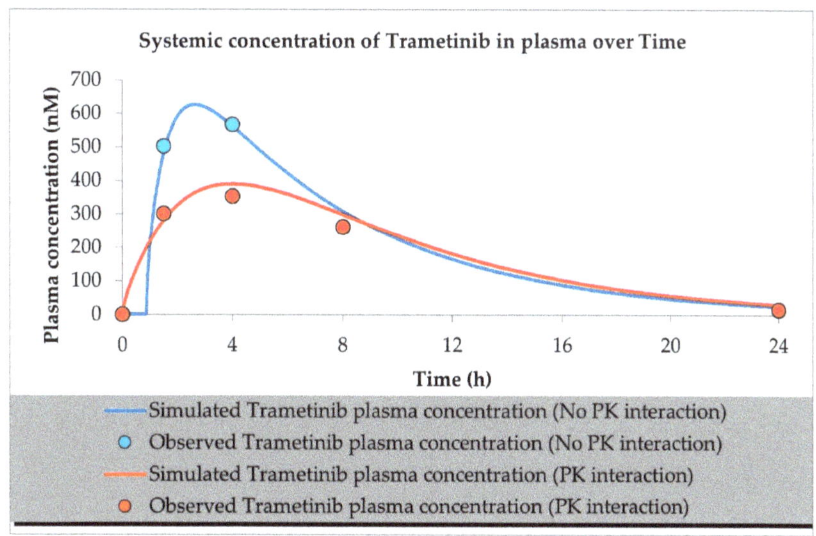

Figure 2. PBPK model of trametinib with and without PK interaction. Observed data are means from *n* = 3.

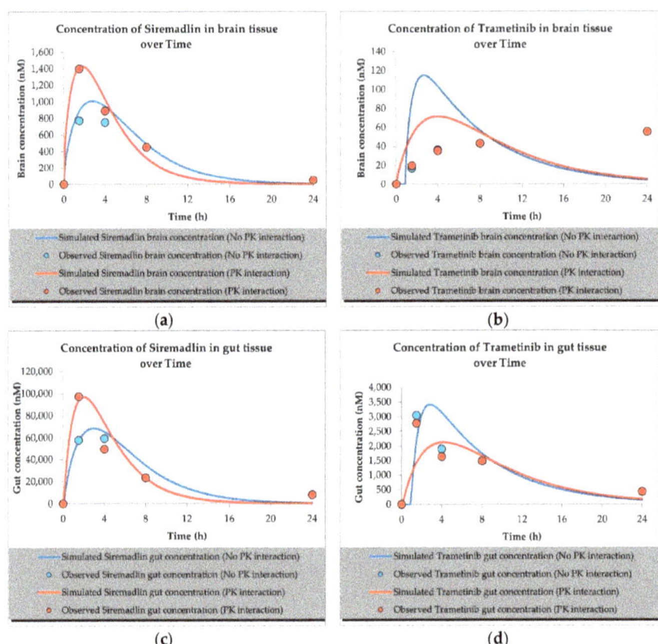

Figure 3. Values of predicted concentrations with and without PK interaction for (**a**) siremadlin and (**b**) trametinib in brain tissue; (**c**) siremadlin and (**d**) trametinib in gut tissue; observed data are means from $n = 3$.

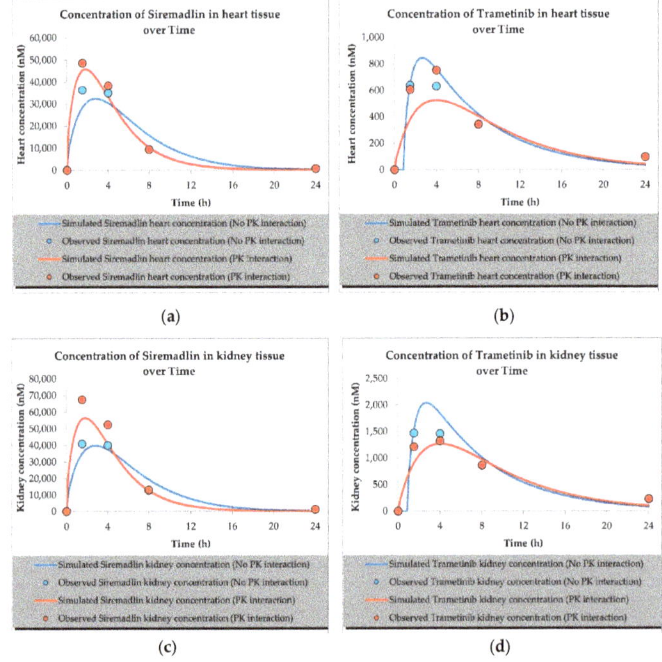

Figure 4. Values of predicted concentrations with and without PK interaction for (**a**) siremadlin and (**b**) trametinib in heart tissue; (**c**) siremadlin and (**d**) trametinib in kidney tissue; observed data are means from $n = 3$.

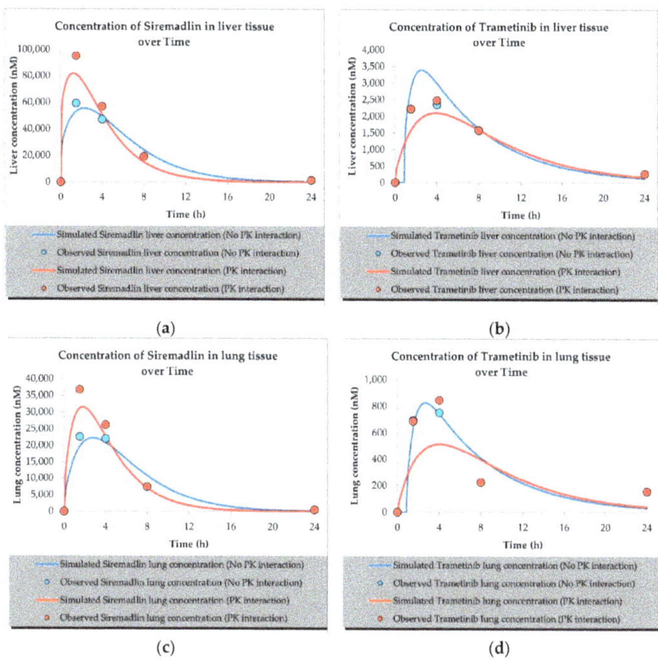

Figure 5. Values of predicted concentrations with and without PK interaction for (**a**) siremadlin and (**b**) trametinib in liver tissue; (**c**) siremadlin and (**d**) trametinib in lung tissue; observed data are means from $n = 3$.

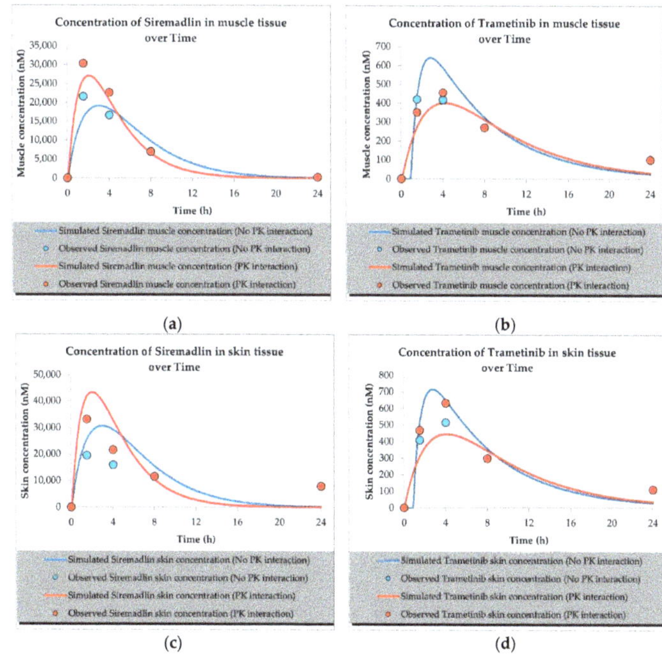

Figure 6. Values of predicted concentrations with and without PK interaction for (**a**) siremadlin and (**b**) trametinib in muscle tissue; (**c**) siremadlin and (**d**) trametinib in skin tissue; observed data are means from $n = 3$.

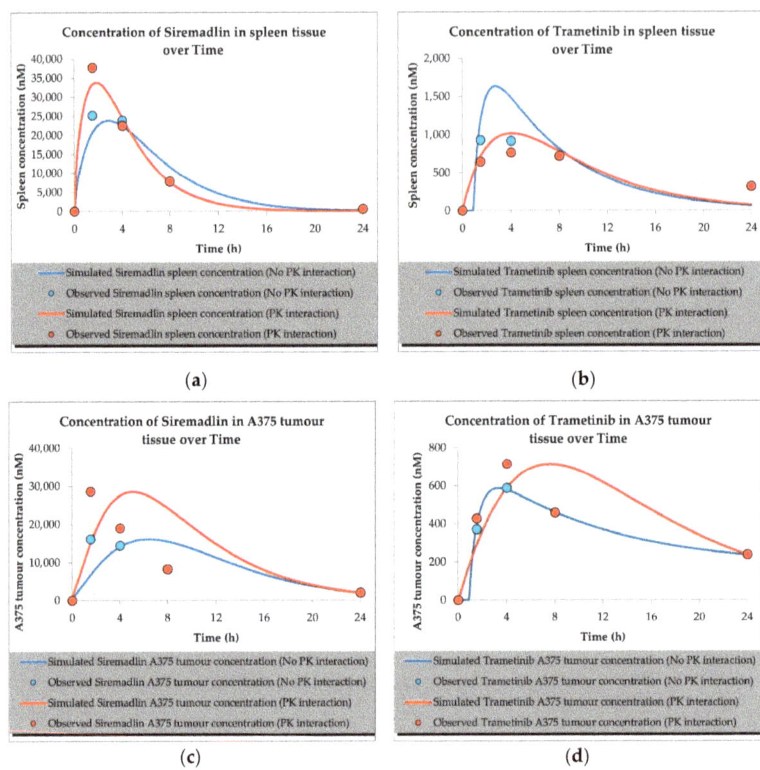

Figure 7. Values of predicted concentrations with and without PK interaction for (**a**) siremadlin and (**b**) trametinib in spleen tissue; (**c**) siremadlin and (**d**) trametinib in A375 tumour tissue; observed data are means from $n = 3$.

Regarding the estimated AUC_{0-24h}, the PBPK simulation for siremadlin indicated that this parameter was accurately predicted for all tissues within a fold error of 0.93, except for the A375 tumour tissue, which was slightly overestimated (1.22), as shown in Table S3. We obtained a similar observation for the estimated trametinib AUC_{0-24h}. The estimated fold errors were also close to unity (0.93) in all tissues except for the A375 tumour, for which we noted a 0.96-fold error (Table S4).

Concerning the predicted C_{max} for siremadlin, this parameter was overpredicted in the following tissues: the plasma (1.10), A375 tumour (1.00), brain (1.31), skin (1.57), and gut (1.15); please refer to Figures 1, 3a,c, 6c and 7c and Table S3. Compared with the tissues that were well supplied with blood, the C_{max} was somewhat underpredicted in the muscles (0.89), spleen (0.95), heart (0.89), kidney (0.97), lungs (0.99), and liver (0.94), as shown in Figures 4a,c, 5a,c, 6a and 7a and Table S3. Regarding trametinib, the estimated fold errors for this PK parameter were somewhat overpredicted in all the tissues (range: 1.00–2.08), as shown in Figures 2, 3b,d, 4b,d, 5b,d, 6b,d and 7b,d and Table S4. We noted a significant fold error (2.08) in the brain tissue, which could have been caused by the suboptimal blood perfusion for this tissue in the constructed PBPK model (Figure 3b).

Regarding the T_{max} parameter for siremadlin, the estimations for this parameter were generally overpredicted for all the tissues (1 < fold error < 2). The exceptions were the gut and A375 tumour tissue, for which this parameter was slightly underpredicted (0.74) and remarkably overpredicted (4.26), respectively. This may have been caused by unoptimised tumour blood perfusion in the mouse model; alternatively, the observed T_{max} values might not have been not properly determined due to sparse sampling, as depicted in Figures 1, 3a,c, 4a,c, 5a,c, 6a,c and 7a,c and Table S3. Concerning the predicted

T_{max} for trametinib, the calculated fold errors for most of the tissues were within a 2-fold error range. The only exception was for the brain tissue (fold error 0.11), which may have been caused by suboptimal blood perfusion for this tissue in the constructed PBPK model (Figures 2, 3b,d, 4b,d, 5b,d, 6b,d and 7b,d and Table S4).

Basic PBPK models were further improved by introducing PK interactions as a result of the coadministration of the two studied drugs. The PK interactions could be explained by the altered absorption process and distribution in the tumour compartment as well (Table S2). The modification of those parameters allowed us to accurately fit the models to the data with PK interaction in the plasma and other organs (Figures 1–7).

In the developed models assuming PK interactions, we estimated the most important pharmacokinetic parameters (AUC_{0-24h}, C_{max}, and T_{max}). In most cases, those parameters were within the 2-fold error range (0.5–2.0). Regarding the estimated AUC_{0-24h} for the siremadlin model assuming PK interaction, this parameter was slightly underestimated for all the tissues (within a fold error range of 0.76–0.86), except for the A375 tumour tissue, for which this parameter was slightly overestimated (1.54), as shown in Table S5. Our observation was similar for the estimated trametinib model assuming PK interaction. The estimated fold errors for the AUC_{0-24h} were also slightly underestimated (fold error range: 0.76–0.88) in all the tissues except for the A375 tumour and plasma, for which we noted 1.22- and 1.02-fold errors, respectively (Table S6).

Concerning the predicted C_{max} for the siremadlin model assuming PK interaction, this parameter was overpredicted in the following tissues: the plasma (1.15), A375 tumour (1.00), brain (1.02), skin (1.31), and gut (1.00); please refer to Figures 1, 3a,c, Figures 6c and 7c and Table S5. Compared with the tissues that were well supplied with blood, the C_{max} was some-what underpredicted in the studied tissues, including the muscle (0.89), spleen (0.90), heart (0.94), kidney (0.84), lungs (0.86), and liver (0.86), as shown in Figures 4a,c, 5a,c, 6a and 7a and Table S5. Regarding the trametinib model considering PK interaction, the estimated fold errors for this parameter were slightly overpredicted in the following tissues: the plasma (1.10), A375 tumour (1.00), spleen (1.33), and brain (1.29). In the well-perfused tissues, the C_{max} was slightly underestimated in the muscle (0.88), heart (0.70), kidney (0.96), lungs (0.61), gut (0.77), and liver (0.85). However, in this model, we also noted that the C_{max} value of the skin was underestimated (Figures 2, 3b,d, 4b,d, 5b,d, 6b,d and 7b,d and Table S6).

Regarding the T_{max} parameter for the siremadlin model assuming PK interaction, estima-tions of this parameter was generally overpredicted for all the tissues (1 < fold error < 2), except for the liver and A375 tumour tissues, for which this parameter was slightly underpredicted (0.81) and considerably overpredicted (3.38), respectively. This may have been caused by unoptimised tumour blood perfusion in the mouse model, or the observed T_{max} values might not have been not properly determined due to sparse sampling, as depicted in Figures 1, 3a,c, 4a,c, 5a,c, 6a,c, 7a,c and Table S5. Concerning the predicted T_{max} for the trametinib model accounting for PK interaction, the calculated fold errors in most of the tissues were within a 2-fold error range. The only exceptions were for the brain tissue (fold error 0.17), which may have been caused by suboptimal blood perfusion for this tissue, and the gut (fold error 2.79), which may have been due to one or more of the following reasons: the combined effect of the high modification of the ka parameter, sparse sampling of the gut tissue homogenate, or homogenisation of the initial fragment of the intestine where the trametinib concentrations more quickly appeared (Figures 2, 3b,d, 4b,d, 5b,d, 6b,d and 7b,d and Table S6).

Generally, T_{max} was the PK parameter most often mispredicted. Such a discrepancy between the predicted and observed values might have been related to the fact that we report the observed T_{max} values on the highest observed C_{max}, which might not have been properly determined due to the sparse sampling of the observed data.

In the final stage of the siremadlin and trametinib PBPK model development, we compared the models' predictions with external PK data digitised from the literature [13,14]. We assumed that differences in the used formulations between our study and the already-published data would only be associated with the absorption process (Table S2). As shown in Figures S1 and S2, the developed PBPK models were able to effectively capture the plasma concentration–time data observed in the external studies, which therefore validated those models. The final PBPK models' parameters with and without PK interactions are summarised in Tables S2, S7 and S8.

2.2. PD (TGI) Models

To determine the characteristics of unperturbed tumour growth, we plotted the log tumour volume versus time for the mean tumour volume in vehicle-treated animals, as shown in Figure S3. As a result, the growth curve in the vehicle control group was initially characterised by a fast-growing exponential phase that ultimately approached a plateau once a certain tumour volume was reached, indicating the saturation of the tumour growth. According to our current knowledge, this may have been caused by rapid tumour growth, which leads to limited oxygen and nutrient supply [15]. The selected logistic growth model best described the unperturbed tumour growth in terms of the model score.

The final perturbed TGI models for siremadlin and trametinib assumed logistic tumour growth, the Skipper–Schabel–Wilcox (log-kill) tumour-cell-killing hypothesis, drug effects described by the exponential drug-killing model, acquired resistance to the therapy, and treatment effect delay described by the signal distribution model with four transit compartments. For drugs administered in monotherapy, those models were able to accurately capture the changes in tumour volume in time, with a mean relative error (RE) < 20%, as shown in Figure 8 and Table S9.

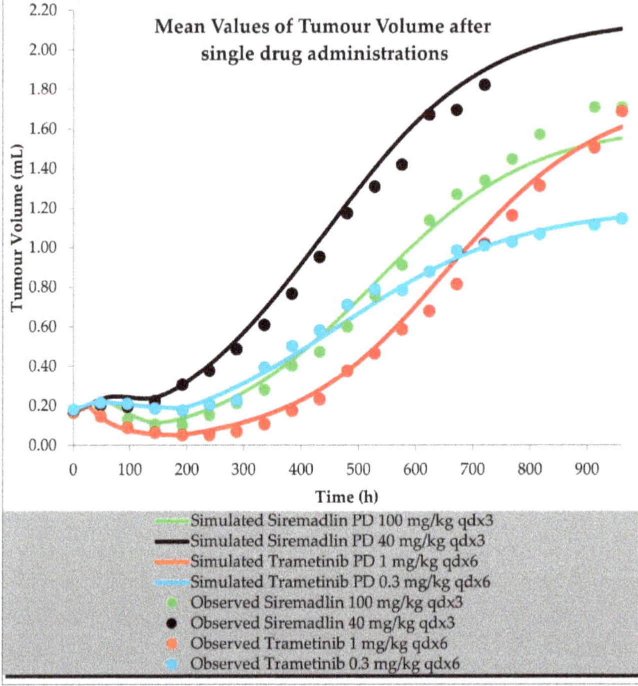

Figure 8. Tumour volume simulations after single drug administration of siremadlin and trametinib. Observed data are means from *n* = 6.

To properly fit the tumour volume data, TGI modelling for drug combinations required additional parameters for the drug interactions at the PK and PD levels. We tested two different parameters that were determined to be drug interaction parameters at the PD level during the in vitro data analysis (Table S1). We selected the β parameter from the

MuSyC drug interaction model as the translational in vitro/in vivo PD parameter. We predicted that the interactions at the PK level (the AUC ratio parameters for siremadlin and trametinib) would be dose-dependent in all the tested drug combination arms. The predicted tumour volumes for the drug combination were within a mean RE of <20%, as shown in Figure 9 and Table S9.

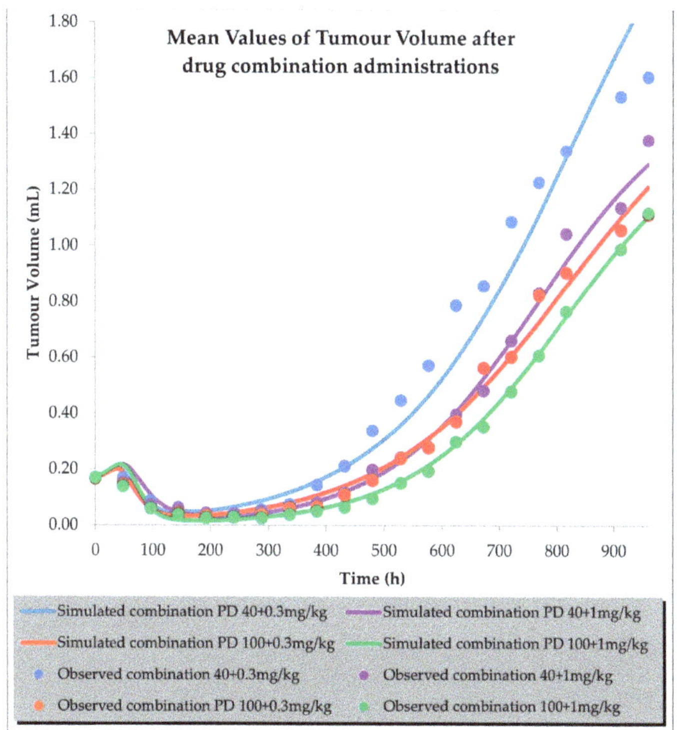

Figure 9. Tumour volume simulations after combined drug administration of siremadlin and trametinib. Observed data are means from *n* = 6.

We successfully verified the TGI models with external efficacy data, as described in Section 4.5.2 and as shown in Figures S5 and S6 and Tables S10 and S11.

The key input parameters for the final TGI models for the single drug and drug combination are shown in Tables S9–S11.

2.3. PBPK/PD Estimation with Universal Model for Drug Combination at Human Equivalent Doses (HEDs)

We created a universal model based on a visual inspection of the data from the current study, validation studies, and score analysis. Determining the relationships between the TGI model parameters allowed us to carefully extrapolate the values of the model parameters for different doses and dosing regimens (Tables S12, S15–S17 and Figures S7–S14). As shown in Figure 10, the results of the tumour volume simulations after two cycles of the combined therapy at HEDs revealed that at least 40 days (960 h) of continuous dosing of trametinib with siremadlin (regardless of the siremadlin dosing regimen) is needed for complete tumour regression (tumour volume $\leq$ 32 mm^3). Additionally, the results from simulations for two therapy cycles suggested that the synergistic efficacy of the siremadlin and trametinib drug combination reduces the number of trametinib doses needed to achieve tumour stasis (tumour volume $\leq$ 170 mm^3) to only 21 doses in a treatment cycle. In general simulations of combinations accounting for continuous siremadlin dosing, the results indicated that qdx7 or qdx14 had a higher efficacy than intermittent dosing schedules qwx2 or qdx1, as shown in Figures S15, S16 and S19. Simulations of siremadlin, trametinib, and their combination efficacy after one and two cycles of therapy at HEDs are depicted in Figures S15–S19.

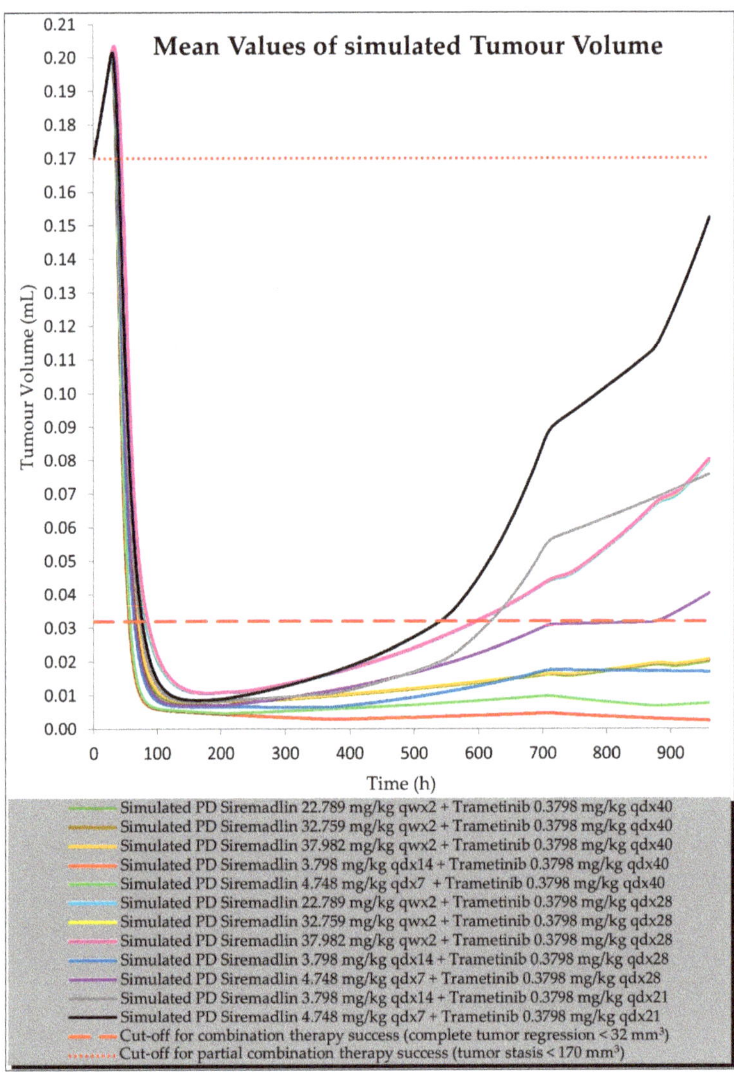

Figure 10. Simulation of siremadlin and trametinib combination efficacy at HEDs (2 therapy cycles) using universal TGI model.

3. Discussion

The developed mouse PBPK/PD models for the MDM2 inhibitor siremadlin, the MEK inhibitor trametinib, and their therapeutic combination were able to describe both the pharmacokinetic and pharmacodynamic profiles of those drugs. The models consider the oral (P.O.) administration of siremadlin and trametinib, full-body distribution model, hepatic metabolism for siremadlin, and intravenous clearance for trametinib. Additionally, the models implement a permeability-limited tumour distribution model and drug interactions at absorption and tumour distribution levels, which allowed us to capture changes in siremadlin and trametinib concentrations in plasma and other tissues (e.g., the heart, liver, spleen, muscle, brain, kidney, A375 tumour, lung, gut, and skin) when we separately or simultaneously administered both compounds. The interactions at the absorption level include changes in the absorption rate constant (ka) for siremadlin and alterations in the absorption rate constant (ka), fraction absorbed (fa), and lag time (tlag) in the case of trametinib. We hypothesised that those two drugs are competing for intestinal transporters related to absorption. Whereas siremadlin might be preferentially transported, a saturation of absorption and transport mechanisms may decrease trametinib absorption. In turn, in tumour interaction depending on the shift in passive permeability and efflux P-gp transporter clearances, an increased disposition of both compounds may occur, which may additionally abolish tumour resistance, which is often related to increased efflux transporter abundance and activity [16–21]. However, further PK studies are needed to confirm such observations in plasma and tumour compartments. The application of PBPK models for siremadlin and trametinib allowed us to calculate key the pharmacokinetic parameters AUC_{0-24h}, C_{max}, and T_{max}, which were mainly within the 2-fold error range for all tissues except the A375 tumour and brain tissues. This may have been caused by suboptimal blood perfusion in those tissues. In the case of the A375 tumour, a slower distribution and, consequently, a higher predicted T_{max} than that observed was probably related to insufficient blood perfusion. Applying higher blood perfusion would be possible, but it might be out of the range of measured perfusion in human melanoma xenografts [22–24]. In models assuming coadministration and associated PK interaction, we also characterised the AUC_{0-24h}, C_{max}, and T_{max} mainly within the 2-fold range for all the tissues except for the A375 tumour, brain, and gut. Regarding the A375 tumour and brain tissues, the high T_{max} fold error might be related to unoptimised blood perfusion in this model. The higher calculated trametinib T_{max} for the gut may have been related to the combined effect of the considerable modification of the ka parameter with the sparse sampling of the gut tissue homogenate, or to the homogenisation of the initial fragment of the intestine (where the concentrations more quickly appeared), which could have impacted the predicted T_{max} for this tissue. The T_{max} was the PK parameter most often mispredicted. Such a discrepancy between the predicted and observed values for this parameter might have been related to the fact that we used the observed T_{max} values on the highest observed C_{max}, which might not have been accurately determined because of the sparse sampling of the observed data in this study. We successfully verified the obtained PBPK models with external PD data extracted from the literature [13,14].

The developed TGI models describe the time course of siremadlin and trametinib efficacy when administered separately and together to mice xenografted with A375 melanoma cells. The logistic growth model best described the unperturbed tumour growth using data from the current study and validation sets.

Based on previously performed in vitro studies, it is known that compounds' killing effect is concentration- and time-dependent with an initial delay in the response, and resistance that might arise. These initial assumptions led us to develop final perturbed TGI models for siremadlin and trametinib. The models assume logistic tumour growth, the Skipper–Schabel–Wilcox (log-kill) tumour-cell-killing hypothesis, drug effect described by the exponential drug killing model, acquired resistance to the therapy, and treatment effect delay described by the signal distribution model with four transit compartments. The delay in the effect of these drugs is most likely related to the duration of the signal transduction associated with the activation of the p53–MDM2 and MAPK pathways, resulting in cell death. Resistance is an inherent part of anticancer treatment; therefore, a population of resistant cells was assessed for both drugs separately and when combined in the performed study, and its description may play a critical role in predicting and optimising the treatment response and may improve therapy scheduling [25,26]. Modelling the drug combination approach required the selection of a PD interaction parameter that could be translational in in vitro/in vivo extrapolation. The selected β parameter from the MuSyC drug interaction model was the most optimal solution for the modelling data obtained from the present study; however, further PK/PD studies on mice coupled with PBPK/PD modelling are needed to validate the optimal PD interaction parameter choice. We successfully verified the created PBPK/PD models with different dosing and scheduling regimens from external PD data. The final TGI models for the unperturbed and perturbed groups properly predicted tumour volume within 20% of the mean relative error acceptance criteria for the data from the present study and external PD data.

Among the most relevant limitations of the present work is that the current models are restricted only to the experimental data available from a single mouse study with a limited number of animals ($n = 6$); additionally, no data from other mouse melanoma xenografts were available. Furthermore, differences in the exposure ratios (AUC ratio) in the models using external PD data may have been caused by the different formulations used in those studies, but they could also be explained by the variability in the resistance parameter lambda. However, because the used cell lines were authenticated, the possibility that the genomic drift impacted the arising resistance is unlikely. The selected TGI model for the MDM2 inhibitor is similar to already reported models for RG7388 (idasanutlin) [27] and HDM201 (siremadlin) [28] molecules from the same class. Even though those models were built on data from osteosarcoma SJSA-1 derived mouse xenografts, common mechanisms such as delayed drug effects and arising resistance can be concluded for this class of small-molecule inhibitors.

The results of an in-depth analysis of the TGI models' parameter dependencies allowed us to extrapolate PD model predictions for different doses and dosing frequencies of the studied drugs after one or two cycles of the therapy, which allowed us to select the most optimal therapeutic strategies that ensure a high efficacy. The modelling and simulation approach suggested several effective dosing schedules; however, the most important are those assuming tumour stasis and complete tumour regression. The simulations of the tumour volume after two cycles of the combined therapy at HEDs revealed that at least 40 days of continuous dosing of trametinib combined with siremadlin (regardless of the siremadlin dosing regimen) is needed for the tumour to completely regress (tumour volume ≤ 32 mm^3). Moreover, these simulations suggested that the synergistic efficacy of the siremadlin and trametinib drug combination reduces the number of successive trametinib doses needed to achieve tumour stasis (tumour volume ≤ 170 mm^3) to only 21 doses in the treatment cycle. This might be especially useful for patients who may develop hypersensitivity, a serious skin rash, or other adverse events grade 2 or higher after using trametinib [29]. Such suggestions may play a role in the development of a potential clinical trial protocol to study melanoma-bearing patients that will be treated with a siremadlin and trametinib combination.

In the simulations of the combination, trametinib seemed to be less effective than in monotherapy, which might be explained by the assumed higher resistance to the therapy when both drugs are simultaneously administered (but smaller parts of the cell population will be resistant). Additionally, by comparing the results from the efficacy simulations after 40 doses of trametinib (Figure S17) with data from the administration of 36 doses (Figure S6, external PD data from the literature), the simulation results suggested a higher efficiency, which might be related to the variability in the lambda parameter or exposure ratio. The presented PBPK/PD translational approach also has other associated limitations, such as not considering the influence of MDM2 inhibition in stromal or immune microenvironments [30–32]. Nonetheless, despite these many limitations, the developed PBPK/PD models reasonably accurately described the PK and time course of the tumour growth across all doses and dosing schedules.

Further analyses are encouraged to externally validate the developed PBPK/PD models for siremadlin, trametinib, and their combination toward predicting tumour volume after human equivalent dose administrations.

The in vitro/in vivo translational approach presented in this study facilitated the determination of the most synergistic and efficacious schedules and dose levels for the siremadlin and trametinib combination in mice and may provide a rationale for planned translational modelling between mice and melanoma-bearing patients.

Recently published clinical data on siremadlin [28,33] and trametinib [29,34] combined with the findings of this study may support the extrapolation of animal PBPK/PD data into a clinical situation. Nonetheless, due to the limited amount of in vivo drug combination data available from this study, such extrapolation may only predict the initial efficacy in patients. The performance of additional in vivo efficacy studies with a larger number of animals and different melanoma xenografts is warranted to improve simulation predictions, although the performance of virtual clinical trials (VCTs) may also facilitate simulation prediction improvement regarding the melanoma patient subpopulation. Further development of the clinical PBPK/PD models for siremadlin and trametinib is needed to construct a proper drug combination model in clinical settings.

PBPK/PD modelling is a powerful tool that allows researchers to properly estimate the in vivo efficacy of anticancer drug combinations based on the results of in vitro studies. Such an approach may indicate the most efficacious dosing strategies. This method allows for better planning of the clinical trials and estimation of drug combination efficacy in such trials on the virtual representation of cancer patients.

4. Materials and Methods

4.1. Materials

Siremadlin (catalogue number HY-18658) and trametinib (catalogue number HY-10999) used in this study were obtained from MedChemExpress. PEG 400 (catalogue number 81172) and Cremophor RH40 (catalogue number 07076) were provided by Merck (formerly Sigma-Aldrich), EtOH (catalogue number 1016/12/19) was provided by POCH, and Labrafil M1944CS (catalogue number 178290) was provided by Gattefosse. The A375 cell line used in the single drug administration in vivo studies was obtained from American Type Culture Collection (CRL-1619). For drug combination in vivo studies, the A375 cell line was provided by European Collection of Authenticated Cell Cultures (88113005).

4.2. Software

PK parameters were estimated with Microsoft Excel (Excel version 2016, Microsoft Corporation, Redmond, WA, USA, 2016, https://www.office.com). Digitalisation of the literature-derived data was performed with the use of WebPlotDigitizer software (version 4.4, Ankit Rohatgi, Pacifica, CA, USA, 2021, https://automeris.io/WebPlotDigitizer). PD modelling was performed with Monolix software (Monolix version 2021R1, Lixoft SAS, Antony, France, 2022, http://lixoft.com/products/monolix/). Monolix custom PD model in Mlxtran can be found in Code S1. PBPK/PD modelling was performed in Simcyp simulator software (Simcyp Animal V21, Certara UK Limited, Sheffield, UK, 2022, https://www.certara.com/software/simcyp-pbpk). Custom PK interaction and PD models in Lua can be found in Codes S2 and S3. The relationship between PD parameters in the universal PD model was determined by 2D curve fitting in Microsoft Excel and 3D curve fitting in Python ZunZun3 tool (ZunZunSite3, James R. Phillips, Birmingham, AL, USA, 2016, http://findcurves.com).

4.3. Studies Involving Animals

Crl:CD-1-*Foxn1*[nu] female 4–5-week-old mice from Charles River Germany inoculated subcutaneously with A375 cells were used for in vivo studies. Determination of compound concentrations in plasma, heart, liver, spleen, muscle, brain, kidney, A375 tumour, lung, gut, and skin tissue homogenates were performed with the use of a quantitative LC-MS/MS system. Tissues were resected at the following timepoints: 1.5, 4, 8, 24 h ($n = 3$ per timepoint). Pharmacokinetic parameters (AUC, C_{max}, relative AUC ratio, and T_{max}) and tissue:plasma partition coefficients (Kp) were calculated using MS Excel 2016. Area under the concentration versus time curve was calculated using the linear trapezoidal rule. Tissue:plasma partition coefficients (Kp) were calculated as proposed by Rodgers and Rowland [35] with an assumption of linear pharmacokinetics [36] (Equation (1)).

$$Kp\ (K_{tissue:plasma}) = C_{tissue,ss}/C_{plasma,ss} = AUC_{tissue}/tau/AUC_{plasma}/tau = AUC_{tissue}/AUC_{plasma}, \tag{1}$$

where C is concentration in particular tissue or plasma at steady state (ss), AUC is observed area under the curve for particular tissue/plasma, and tau is the dosing interval.

Determination of tumour growth was performed after oral gavage of vehicle (60% PEG 400 (v/v), 10% Cremophor RH40 (v/v), 10% EtOH (v/v), and 20% Labrafil M1944CS (v/v)), siremadlin, trametinib, or their combination in vehicle. The volume of the administration (10 mL/kg) of the compounds was always adjusted to the mouse body weight. Initial tumour volumes, doses, dose schedules, and numbers of animals in particular in vivo studies are summarised in Table 1.

Table 1. Summary of performed in vivo studies on CD-1 nude mice xenografted with A375 tumour.

Compound	Initial Tumour Volume (mm^3)	Doses (mg/kg)	Dose Schedule	N	Comments
Vehicle (Adamed)	~135	-	q1dx5/q7dx2	10	Adamed reference
Siremadlin	~137	25/50	q1dx5	10	Adamed reference
Siremadlin	~137	50/100	q7dx2	10	Adamed reference
Vehicle (current study)	~162	-	qdx6	11	Efficacy in current study
Siremadlin	~163–172	40/100	qdx3	6	Efficacy in current study
Trametinib	~167–180	0.3/1	qdx6	6	Efficacy in current study
Siremadlin + Trametinib	~165–169	40 + 0.3/40 + 1/100 + 0.3/100 + 1	qdx3/qdx6	6	Efficacy in current study
Siremadlin	~300	100	qdx1	12	PK in current study
Trametinib	~300	1	qdx1	12	PK in current study
Siremadlin + Trametinib	~300	100 + 1	qdx1	12	PK in current study

Tumour volume (V) was recorded with an electronic calliper twice or thrice a week and was calculated based on its length and width using the prolate ellipsoid equation (Equation (2)).

$$V = d^2 \times D/2, \tag{2}$$

where d is tumour width (mm) and D is tumour length (mm). For tumour volume simulation, it was assumed that 1 cm^3 = 1 mL.

4.4. Physiologically Based Pharmacokinetic Models

4.4.1. General PBPK Modelling Strategy

The modelling strategy was based on "middle-out" approach combining advantages of "bottom-up" and "top-down" approaches, whereby some parameters were fixed (such as in vitro determined or literature-derived data for siremadlin and trametinib [37–40]) and others were estimated. Parameter estimation (PE) was performed using the PE Module of the Simcyp Animal V21 using the Nelder–Mead method, weighted least squares by the reciprocal of the square of the maximum observed value as the objective function, and the termination criterion defined as the improvement of less than 1% of the objective function value. Optimisation was performed manually to fit the observed data. Due to the limitations of the mouse model in the current V21 version of Simcyp Animal software, estimation of inter-individual variability was not possible; therefore, models were fitted to the averaged values of PK and PD data at particular timepoints. PBPK model performance was evaluated based on the "2-fold" criterion for maximum concentration (C$_{max}$) and area under the concentration vs. time curve (AUC) and T$_{max}$ [41,42]. A graphical representation of PBPK model development is presented in Figures S20–S22.

4.4.2. Mouse Population

The physiological parameters of the Simcyp mouse population were modified to reproduce the CD-1 nude mouse population used in the experimental procedure. With regard to this, body weight, cardiac output [43], tissue volumes, blood flows, and tumour properties (including tumour tissue volume, blood flow, composition, and pH) were adapted to the studied population. Since no raw data for blood flow in the A375 xenograft were available [44], blood flow was set as in human melanoma xenografts [22–24].

4.4.3. PBPK Model Verification

The final siremadlin and trametinib PBPK models were compared with external PK data [13,14] through both visual check and numerical analysis. Experimental and predicted longitudinal plasma concentration (Cp) and tissue concentration (Ct) profiles were generated, including the mean predicted concentrations. Local sensitivity analysis (parameter scanning) was performed to evaluate the relative impact of fa, HepCL, and fu_inc in the plasma PK parameters (AUC_{0-24h} and Cmax) for siremadlin and fa and CL_{iv} for trametinib in the ranges presented in Tables S7 and S8. The performance of the siremadlin and trametinib PBPK models was assessed by the fold error for each tissue, which referred to the ratio of the predicted AUC_{0-24h}, C_{max}, or T_{max} to the observed AUC_{0-24h}, C_{max}, or T_{max} respectively (Equation (3)). AUC_{0-24h} was calculated by the linear trapezoidal rule. Both visual checks and numerical analyses were performed in Microsoft Excel 2016.

$$\text{Fold Error PK parameter} = \text{Predicted PK parameter}/\text{Observed PK parameter}, \quad (3)$$

4.5. Pharmacodynamic Modelling

4.5.1. General PD Modelling Strategy

Optimal PD models used further in PBPK/PD modelling were established in Monolix software. Due to the comprehensive library of Monolix PD (TGI) models [45], the first step in proper PD selection was based on the screening of various PD models with the use of an automatic model initialisation (auto-init) function. Auto-init allows finding the good initial values of parameters before starting the population modelling approach. This custom optimisation method is performed on the pooled data, without inter-individual variability, and as result finds only a local minimum of the fitted PD model (the global minimum has yet to be set). Model selections were based on visual inspection of individual observed vs. predicted data and comparisons of resulting values of model score (Equation (4)).

$$\text{Model score} = -2 \times \text{log-likelihood} \ (-2LL, \text{called also objective function value—OFV}) + \text{corrected Bayesian Information Criteria (BICc)}, \quad (4)$$

−2LL and BICc were estimated by linearisation method to accelerate calculations. For TGI models further developed in Simcyp Animal, the goodness of TGI model fit was evaluated based on the mean relative error (RE) value (Equation (5)) being < 20% as proposed in [46]:

$$\text{RE (\%)} = 100 \times (\text{Predicted Tumour Volume} - \text{Observed Tumour Volume})/\text{Observed Tumour Volume}, \quad (5)$$

4.5.2. PD (TGI) Model Development and Verification

The first stage of TGI model development was focused on the selection of a mathematical model describing properly unperturbed tumour growth of A375 xenografts. Different tumour growth models assuming growth saturation such as logistic, generalised logistic, hybrid Simeoni–logistic, Gompertz, exponential Gompertz, and von Bertalanffy were evaluated and fitted to the data from the vehicle group in the Monolix software. The selected unperturbed logistic growth model assumes an exponential growth rate (kge) which decelerates linearly with respect to the initial tumour size (TS0) and is described by the following Equation (6):

$$dTS/dt = kge \times TS0 \times (1 - (TotalTS/TSmax)), \tag{6}$$

where kge is tumour growth (1/day), TS0 is initial tumour size (mL), TotalTS is total tumour size (mL), and TSmax is maximal tumour size (mL).

After the characterisation of tumour growth, the tumour growth inhibition models (perturbed models) were developed for siremadlin, trametinib, and their combination separately based on observed tumour volume data. Several structural TGIs were compared in terms of model score in order to select the best fit to the averaged tumour volume data.

A screen of multiple models incorporating introduction of the type of killing hypothesis (Norton–Simon or Skipper–Schabel–Wilcox log-kill), dynamics of treatment effect (linear, Emax, Emax–Hill, exponential kill), type of treatment effect delay (cell distribution model [47] or signal distribution model with 3 or 4 transit compartments [48]), type of resistance arising (claret exponential [49] or 2-population model [48,50–52]), and value of gamma (PD interaction parameter) (Synergyfinder-derived δ score or β parameter calculated with synergy package [6]) was performed in Monolix software.

The most important features of the selected perturbed TGI model characterising single siremadlin-, single trametinib-, and combination-treated groups are written in the following equations and initial conditions:

$$TotalTS(t) = TS(t) + TSr(t), \tag{7}$$

$$TS\_0 = TS0, \tag{8}$$

$$TSr\_0 = TSr0, \tag{9}$$

$$TSmax = crck \times (TS0 + TSr0), \tag{10}$$

$$kkill\_Siremadlin = a \times Siremadlin\ dose \times number\ of\ doses, \tag{11}$$

$$kkill\_Trametinib = b \times Trametinib\ dose \times number\ of\ doses, \tag{12}$$

$$kkill\_combination = (kkill\_Siremadlin \times AUC\_ratio\_Siremadlin + kkill\_Trametinib \times AUC\_ratio\_Trametinib) \times gamma, \tag{13}$$

$$kkill = kkill\_Siremadlin/kkill\_Trametinib/kkill\_combination, \tag{14}$$

$$C(t) = C\_Siremadlin/C\_Trametinib/(C\_Siremadlin + C\_Trametinib), \tag{15}$$

$$K(t) = kkill \times (1 - e\hat{\ }(-s \times C(t))), \tag{16}$$

$$K1\_0 = 0, \tag{17}$$

$$K2\_0 = 0, \tag{18}$$

$$K3\_0 = 0, \tag{19}$$

$$K4\_0 = 0, \tag{20}$$

$$dK1/dt = (dK - K1)/tau, \tag{21}$$

$$dK2/dt = (K1 - K2)/tau, \tag{22}$$

$$dK3/dt = (K2 - K3)/tau, \tag{23}$$

$$dK4/dt = (K3 - K4)/tau, \tag{24}$$

$$dTS/dt = 0, \tag{25}$$

$$dTSr/dt = 0, \tag{26}$$

$$dTS/dt = (kge*TS*(1 - (TotalTS/TSmax))) - (K4 + ksr*K4)*TS, \tag{27}$$

$$dTSr/dt = (kge*TSr*(1 - (TotalTS/TSmax))) + (ksr*K4*TS) - (K4/lambda*TSr), \quad (28)$$

where resistance is defined by the introduction of sensitive (TS—tumour size) and resistant population (TSr—resistant tumour size) of cancer cells. It was assumed that at time 0, the total tumour size (TotalTS) is represented by the sum of TS and TSr cells (Equation (7)). TotalTS is assumed to be ~100% comprising TS cells because TSr0 = ~0 at t = 0. Initial volumes of TS and TSr at time = 0 are denoted as TS0 and TSr0 (Equations (8) and (9)). Correlation between initial tumour size (for TS0 and TSr0, respectively) and maximal tumour size (TSmax) was parametrised as the crck parameter as proposed in [28] (Equation (10)). The tumour-cell-killing constant of siremadlin or trametinib was determined to be dependent on the dose and number of doses, after introducing compound-specific killing constants—a and b (Equations (11) and (12)). The killing constant for combination (kkill_combination) was characterised as the sum of the siremadlin and trametinib killing constants adjusted with PK interaction parameters (AUC ratio parameter which was calculated for siremadlin + trametinib 100 + 1 mg/kg dose and estimated for the other doses) as well as gamma (PD interaction parameter)—β parameter determined from analysis of in vitro data (Equation (13)). Depending on the treated group, the killing constant could be assigned to the killing constants of siremadlin, trametinib, or their combination (Equation (14)). Total plasma concentration of siremadlin, trametinib, or their combination was used as the input for the drug effect (Equation (15)). The tumour growth inhibition model uses a log-kill killing hypothesis with the treatment dynamics following exponential kill kinetics: kkill is the killing constant, and s is the killing constant coefficient (Equations (16), (27), and (28)). A delay of killing effect (K) has been implemented by the introduction of 4 signal transit compartments (K1, K2, K3, K4), as suggested by [48]. The duration of this delay is determined by the parameter tau (Equations (21–24)). It was assumed that transit compartments equal 0 in time = 0 (Equations (17–20)). It was also assumed that initial change of tumour volume for sensitive and resistant cells populations equal 0 (Equations (25) and (26)). The tumour logistic growth model and growth rate (kge) were assumed to be the same for the sensitive and treatment-resistant cell populations. Acquired resistance to the therapy also assumes that part of sensitive cells will convert into resistant ones. The conversion rate from sensitive to resistant population is regulated by a rate constant denoted as ksr as previously proposed [28,53] (Equations (27) and (28)). It was assumed that studied drugs are also inducing a killing effect on the resistant cell population but with a reduced potency (Equation (28)). The parameter lambda denotes the fold-change loss in drug potency on resistant cells relative to sensitive cells. Units for particular parameters are summarised in Table S9.

In the next stage of TGI model development, models for single drug administration were compared with external efficacy data: siremadlin efficacy from a previously performed study on mice xenografted with A375 cells (unpublished data, courtesy of Adamed Pharma) and trametinib efficacy data digitised from the literature [54,55] (studies carried out on mice xenografted with A375 cells). The applied TGI models allow the tumour volume data in external efficacy studies to be fitted properly, therefore validating those models (Figures S5 and S6 and Tables S10 and S11). Due to differences in exposure (AUC_{0-24h}) in external data, the value of the killing-effect-related parameter (kkill) was adjusted with the AUC ratio parameter which was calculated for siremadlin based on previously performed PK studies (unpublished data, courtesy of Adamed Pharma as shown in Table S13) and was estimated in particular studies for trametinib.

In the last step, unperturbed and perturbed tumour growth inhibition models for siremadlin, trametinib, and their combination previously developed in Monolix were translated into Lua programming language and applied within the Simcyp Animal V21 for further development to achieve a mean relative error (RE) value of < 20% (Equation (5)). A graphical summary of TGI model development and verification is presented in Figures S20–S22.

4.5.3. TGI Model Parameter Dependence Estimations (Universal Model Development)

After the construction of the final PD models, relationships between input parameters were established in order to construct universal TGI models which allowed the proper prediction of tumour volume even after using different doses and dosing frequencies of the studied drugs.

Relationships between input parameters were developed for siremadlin, trametinib, and combination TGI models separately and only for selected input parameters (crck, kge, exposure ratio, kkill, and tau), while the other ones determined experimentally or estimated through TGI model development (TS0, TSr0, s, lambda, ksr, and gamma) were fixed at specific values (Table S12).

Development of the dependency of the parameters was based on the initial assumption that values of input parameters scaled with the dose and were summed up in the combination TGI model. Various mathematical equations were screened in the ZunZun3 standard 3D equations library (including over 300 equations from bioscience, enzyme kinetics, exponential, logarithmic, polynomial, power, rational, sigmoidal, trigonometric, and many more miscellaneous equations) in order to find the best fit describing dependencies between input parameter data. Then, the best models (in terms of the lowest sum of squared absolute error) were verified on the siremadlin and trametinib verification datasets.

4.5.4. Tumour Volume Simulation for Drug Combination at Human Equivalent Doses

Tumour volume for studied drugs and their combination was estimated using universal PBPK/PD model within a 0–960 h simulation timeframe at the human equivalent doses and clinically examined dosing regimens for each drug (Table S14). For simulation purposes, the initial tumour size (TS0) was assumed to be 170 mm^3 (0.17 mL); therefore, the cut-off for tumour stasis was set to be $\leq$ 170 mm^3 at the end of the simulation (960 h). Following complete eradication of the tumour, scarring often occurs at the site of tumour implantation, leaving behind connective tissue that may be mistaken for a small tumour; thus, tumour width or length below the limit of detection (4 mm) resulting in tumour volume of 32 mm^3 was selected as the cut-off for complete tumour regression at the end of the simulation (960 h). The chosen cut-off for complete response (tumour volume $\leq$ 32 mm^3) is in line with reported values [56–58]. TGI model parameters for simulations of tumour volume at HEDs for studied compounds after 1 or 2 cycles of therapy are summarised in Tables S15–S17.

Animal doses equivalent to human doses were calculated according to Equation (29) [59].

$$\text{Animal equivalent dose (AED) [mg/kg]} = \text{Human equivalent dose (HED) [mg/kg]}/(\text{Weight}_{\text{animal}} \text{ [kg]}/ \text{Weight}_{\text{human}} \text{ [kg]})^{(1-0.67)}, \tag{29}$$

where the mean weight of mice was 0.02755 kg (mean mouse weight in current study) and human weight selected as typical patient weight was 70 kg.

Supplementary Materials: The following supporting information can be downloaded at: https://www.mdpi.com/article/10.3390/ijms231911939/s1.

Author Contributions: Conceptualization, J.W.; methodology, J.W.; software, J.W. and S.P.; validation, J.W., S.P. and D.P.; formal analysis, J.W.; investigation, J.W.; resources, J.W. and Z.R.; data curation, J.W.; writing—original draft preparation, J.W.; writing—review and editing, S.P., Z.R. and D.P.; visualization, J.W.; supervision, S.P., Z.R. and D.P.; project administration, J.W.; funding acquisition, Z.R. and D.P. All authors have read and agreed to the published version of the manuscript.

Funding: This work has been performed under the Ministry of Science and Higher Education's programme "Applied PhD", at the Faculty of Chemistry of the University of Warsaw and the Adamed Pharma S.A., based on contract No. 0058/DW/2018/01/1. Data obtained from Adamed Pharma S.A. were collected under the project "Preclinical development of an innovative anti-cancer drug using the mechanism of reactivation of p53 protein" (POIR/01.02.00-14-31/15) co-financed by the European Union funds under Measure 1.2 Sectoral R&D Programmes of the Operational Programme

Smart Growth 2014–2020 (INNOMED—programme of scientific research and development works for the economic sector in the field of innovative medicine).

Institutional Review Board Statement: The animal study protocols were approved by the Ethics Committee of II Local Ethics Committee for Animal Experiments in Warsaw (permission for the experimental use of animals WAW2_6/2016 approved 16 March 2016 and WAW2_19/2016 approved 18 May 2016) for studies involving animals.

Informed Consent Statement: Not applicable.

Data Availability Statement: The data presented in this study are available in the article or supplementary material. Raw data from PK and PD studies are available on request from the corresponding author.

Acknowledgments: The authors would like to thank Maria Mazur and Katarzyna Jastrzębska-Mazur for helpful advice and reviewing this manuscript and Małgorzata Drabczyk-Pluta, Damian Matyśniak, Magdalena Oślislok, Michał Krystosik, Sabina, Jeleń, Małgorzata Choroś, Emilia Mróz, Katarzyna Woś-Latosi, Urszula Połaska, Joanna Jaszczewska, Piotr Syga, Daniel Nosal, Marlena Gałązka, Katarzyna Bukato, Piotr Rózga, Grażyna Peszyńska-Sularz, Anna Żyłko, Agnieszka Jakubiak, and Grzegorz Stasiłojć for support in performing PK/PD studies. The authors give special thanks to Tariq Abdulla for his valuable advice and help with Lua script development used for custom PK interaction and PD models used in Simcyp simulations and Géraldine Cellière for help with the development of the Mlxtran PD model. The authors would like to also thank Certara UK (Simcyp Division) and Lixoft SAS for granting free access to the Simcyp Simulators and Monolix Suite software through an academic license.

Conflicts of Interest: Jakub Witkowski is an Adamed Pharma employee and co-inventor of the patent WO2019141549 related to the MDM2 inhibitor developed by Adamed Pharma. Dariusz Pawelec is an Adamed Pharma employee. Sebastian Polak is a Certara UK employee. No other conflicts of interest, financial or otherwise, were disclosed by the authors. The funders had no role in the design of the study; in the collection, analyses, or interpretation of data; in the writing of the manuscript; or in the decision to publish the results.

References

1. Steininger, J.; Gellrich, F.F.; Schulz, A.; Westphal, D.; Beissert, S.; Meier, F. Systemic Therapy of Metastatic Melanoma: On the Road to Cure. *Cancers* **2021**, *13*, 1430. [CrossRef] [PubMed]
2. Bhatia, S.; Tykodi, S.S.; Thompson, J.A. Treatment of Metastatic Melanoma: An Overview. *Oncology* **2009**, *23*, 488–496. [PubMed]
3. Morrison, C.; Pabla, S.; Conroy, J.M.; Nesline, M.K.; Glenn, S.T.; Dressman, D.; Papanicolau-Sengos, A.; Burgher, B.; Andreas, J.; Giamo, V.; et al. Predicting Response to Checkpoint Inhibitors in Melanoma beyond PD-L1 and Mutational Burden. *J. Immunother. Cancer* **2018**, *6*, 32. [CrossRef] [PubMed]
4. Atkinson, V.; Batty, K.; Long, G.V.; Carlino, M.S.; Peters, G.D.; Bhave, P.; Moore, M.A.; Xu, W.; Brown, L.J.; Arneil, M.; et al. Activity and Safety of Third-Line BRAF-Targeted Therapy (TT) Following First-Line TT and Second-Line Immunotherapy (IT) in Advanced Melanoma. *J Clin Oncol.* **2020**, *38*, 10049. [CrossRef]
5. Chatterjee, N.; Bivona, T.G. Polytherapy and Targeted Cancer Drug Resistance. *Trends Cancer* **2019**, *5*, 170–182. [CrossRef] [PubMed]
6. Witkowski, J.; Polak, S.; Rogulski, Z.; Pawelec, D. In Vitro/In Vivo Translation of Synergistic Combination of MDM2 and MEK Inhibitors in Melanoma Using PBPK/PD Modelling: Part I. *Int. J. Mol. Sci.* **2022**, *23*, 11939. [CrossRef]
7. Wu, C.-E.; Koay, T.S.; Esfandiari, A.; Ho, Y.-H.; Lovat, P.; Lunec, J. ATM Dependent DUSP6 Modulation of P53 Involved in Synergistic Targeting of MAPK and P53 Pathways with Trametinib and MDM2 Inhibitors in Cutaneous Melanoma. *Cancers* **2018**, *11*, 3. [CrossRef]
8. Goldblatt, E.M.; Lee, W.-H. From Bench to Bedside: The Growing Use of Translational Research in Cancer Medicine. *Am. J. Transl Res.* **2010**, *2*, 1–18.
9. Niu, J.; Straubinger, R.M.; Mager, D.E. Pharmacodynamic Drug–Drug Interactions. *Clin. Pharmacol. Ther.* **2019**, *105*, 1395–1406. [CrossRef]
10. Berenbaum, M.C. What Is Synergy? *Pharmacol. Rev.* **1989**, *41*, 93–141.
11. Tang, J.; Wennerberg, K.; Aittokallio, T. What Is Synergy? The Saariselkä Agreement Revisited. *Front. Pharmacol.* **2015**, *6*, 181. [CrossRef] [PubMed]
12. Cokol, M.; Chua, H.N.; Tasan, M.; Mutlu, B.; Weinstein, Z.B.; Suzuki, Y.; Nergiz, M.E.; Costanzo, M.; Baryshnikova, A.; Giaever, G.; et al. Systematic Exploration of Synergistic Drug Pairs. *Mol. Syst. Biol.* **2011**, *7*, 544. [CrossRef]
13. Jeay, S.; Ferretti, S.; Holzer, P.; Fuchs, J.; Chapeau, E.A.; Wartmann, M.; Sterker, D.; Romanet, V.; Murakami, M.; Kerr, G.; et al. Dose and Schedule Determine Distinct Molecular Mechanisms Underlying the Efficacy of the P53–MDM2 Inhibitor HDM201. *Cancer Res.* **2018**, *78*, 6257–6267. [CrossRef] [PubMed]

14. Yoo, J.Y.; Swanner, J.; Otani, Y.; Nair, M.; Park, F.; Banasavadi-Siddegowda, Y.; Liu, J.; Jaime-Ramirez, A.C.; Hong, B.; Geng, F.; et al. Oncolytic HSV Therapy Increases Trametinib Access to Brain Tumors and Sensitizes Them in Vivo. *Neuro Oncol.* **2019**, *21*, 1131–1140. [CrossRef] [PubMed]

15. Wek, R.C.; Staschke, K.A. How Do Tumours Adapt to Nutrient Stress? *EMBO J.* **2010**, *29*, 1946–1947. [CrossRef] [PubMed]

16. Theile, D.; Wizgall, P. Acquired ABC-Transporter Overexpression in Cancer Cells: Transcriptional Induction or Darwinian Selection? *Naunyn-Schmiedeberg's Arch. Pharmacol.* **2021**, *394*, 1621–1632. [CrossRef]

17. Huang, Q.; Cai, T.; Bai, L.; Huang, Y.; Li, Q.; Wang, Q.; Chiba, P.; Cai, Y. State of the Art of Overcoming Efflux Transporter Mediated Multidrug Resistance of Breast Cancer. *Transl. Cancer Res.* **2019**, *8*, 319–329. [CrossRef]

18. Chen, J.; Wang, Z.; Gao, S.; Wu, K.; Bai, F.; Zhang, Q.; Wang, H.; Ye, Q.; Xu, F.; Sun, H.; et al. Human Drug Efflux Transporter ABCC5 Confers Acquired Resistance to Pemetrexed in Breast Cancer. *Cancer Cell Int.* **2021**, *21*, 136. [CrossRef]

19. Samimi, G.; Safaei, R.; Katano, K.; Holzer, A.K.; Rochdi, M.; Tomioka, M.; Goodman, M.; Howell, S.B. Increased Expression of the Copper Efflux Transporter ATP7A Mediates Resistance to Cisplatin, Carboplatin, and Oxaliplatin in Ovarian Cancer Cells. *Clin. Cancer Res.* **2004**, *10*, 4661–4669. [CrossRef]

20. Lotz, C.; Kelleher, D.K.; Gassner, B.; Gekle, M.; Vaupel, P.; Thews, O. Role of the Tumor Microenvironment in the Activity and Expression of the P-Glycoprotein in Human Colon Carcinoma Cells. *Oncol. Rep.* **2007**, *17*, 239–244. [CrossRef]

21. Legrand, O.; Simonin, G.; Beauchamp-Nicoud, A.; Zittoun, R.; Marie, J.P. Simultaneous Activity of MRP1 and Pgp Is Correlated with in Vitro Resistance to Daunorubicin and with in Vivo Resistance in Adult Acute Myeloid Leukemia. *Blood* **1999**, *94*, 1046–1056. [CrossRef] [PubMed]

22. Kallinowski, F.; Schlenger, K.H.; Runkel, S.; Kloes, M.; Stohrer, M.; Okunieff, P.; Vaupel, P. Blood Flow, Metabolism, Cellular Microenvironment, and Growth Rate of Human Tumor Xenografts. *Cancer Res.* **1989**, *49*, 3759–3764. [PubMed]

23. Benjaminsen, I.C.; Graff, B.A.; Brurberg, K.G.; Rofstad, E.K. Assessment of Tumor Blood Perfusion by High-Resolution Dynamic Contrast-Enhanced MRI: A Preclinical Study of Human Melanoma Xenografts. *Magn. Reason. Med.* **2004**, *52*, 269–276. [CrossRef] [PubMed]

24. Graff, B.A.; Benjaminsen, I.C.; Melås, E.A.; Brurberg, K.G.; Rofstad, E.K. Changes in Intratumor Heterogeneity in Blood Perfusion in Intradermal Human Melanoma Xenografts during Tumor Growth Assessed by DCE-MRI. *Magn. Reason. Imaging* **2005**, *23*, 961–966. [CrossRef]

25. Howard, G.R.; Johnson, K.E.; Rodriguez Ayala, A.; Yankeelov, T.E.; Brock, A. A Multi-State Model of Chemoresistance to Characterize Phenotypic Dynamics in Breast Cancer. *Sci. Rep.* **2018**, *8*, 12058. [CrossRef]

26. Yoon, N.; Vander Velde, R.; Marusyk, A.; Scott, J.G. Optimal Therapy Scheduling Based on a Pair of Collaterally Sensitive Drugs. *Bull. Math. Biol.* **2018**, *80*, 1776–1809. [CrossRef]

27. Higgins, B.; Glenn, K.; Walz, A.; Tovar, C.; Filipovic, Z.; Hussain, S.; Lee, E.; Kolinsky, K.; Tannu, S.; Adames, V.; et al. Preclinical Optimization of MDM2 Antagonist Scheduling for Cancer Treatment by Using a Model-Based Approach. *Clin. Cancer Res.* **2014**, *20*, 3742–3752. [CrossRef]

28. Guerreiro, N.; Jullion, A.; Ferretti, S.; Fabre, C.; Meille, C. Translational Modeling of Anticancer Efficacy to Predict Clinical Outcomes in a First-in-Human Phase 1 Study of MDM2 Inhibitor HDM201. *AAPS J.* **2021**, *23*, 28. [CrossRef]

29. Flaherty, K.T.; Robert, C.; Hersey, P.; Nathan, P.; Garbe, C.; Milhem, M.; Demidov, L.V.; Hassel, J.C.; Rutkowski, P.; Mohr, P.; et al. Improved Survival with MEK Inhibition in BRAF-Mutated Melanoma. *N. Engl. J. Med.* **2012**, *367*, 107–114. [CrossRef]

30. Wang, H.Q.; Mulford, I.J.; Sharp, F.; Liang, J.; Kurtulus, S.; Trabucco, G.; Quinn, D.S.; Longmire, T.A.; Patel, N.; Patil, R.; et al. Inhibition of MDM2 Promotes Antitumor Responses in P53 Wild-Type Cancer Cells through Their Interaction with the Immune and Stromal Microenvironment. *Cancer Res.* **2021**, *81*, 3079–3091. [CrossRef]

31. Fang, D.D.; Tang, Q.; Kong, Y.; Wang, Q.; Gu, J.; Fang, X.; Zou, P.; Rong, T.; Wang, J.; Yang, D.; et al. MDM2 Inhibitor APG-115 Synergizes with PD-1 Blockade through Enhancing Antitumor Immunity in the Tumor Microenvironment. *J. Immuno Ther. Cancer* **2019**, *7*, 327. [CrossRef] [PubMed]

32. Zhou, J.; Kryczek, I.; Li, S.; Li, X.; Aguilar, A.; Wei, S.; Grove, S.; Vatan, L.; Yu, J.; Yan, Y.; et al. The Ubiquitin Ligase MDM2 Sustains STAT5 Stability to Control T Cell-Mediated Antitumor Immunity. *Nat. Immunol.* **2021**, *22*, 460–470. [CrossRef] [PubMed]

33. Stein, E.M.; DeAngelo, D.J.; Chromik, J.; Chatterjee, M.; Bauer, S.; Lin, C.-C.; Suarez, C.; de Vos, F.; Steeghs, N.; Cassier, P.A.; et al. Results from a First-in-Human Phase I Study of Siremadlin (HDM201) in Patients with Advanced Wild-Type TP53 Solid Tumors and Acute Leukemia. *Clin. Cancer Res.* **2022**, *28*, 870–881. [CrossRef] [PubMed]

34. Mistry, H.B.; Orrell, D.; Eftimie, R. Model Based Analysis of the Heterogeneity in the Tumour Size Dynamics Differentiates Vemurafenib, Dabrafenib and Trametinib in Metastatic Melanoma. *Cancer Chemother. Pharm.* **2018**, *81*, 325–332. [CrossRef] [PubMed]

35. Rodgers, T.; Leahy, D.; Rowland, M. Physiologically Based Pharmacokinetic Modeling 1: Predicting the Tissue Distribution of Moderate-to-Strong Bases. *J. Pharm. Sci.* **2005**, *94*, 1259–1276. [CrossRef]

36. Jeong, Y.-S.; Jusko, W.J. Consideration of Fractional Distribution Parameter Fd in the Chen and Gross Method for Tissue-to-Plasma Partition Coefficients: Comparison of Several Methods. *Pharm Res.* **2022**, *39*, 463–479. [CrossRef]

37. Lei, Y.; Zhenglin, Y.; Heng, L. MDM2 Inhibitors. U.S. Patent 11,339,171, 24 May 2022.

38. Hofmann, F. Small Molecule HDM201 Inhibitor HDM201. In Proceedings of the AACR Annual Meeting 2016, New Orleans, LA, USA, 16–20 April 2016.

39. Pharmaceuticals and Medical Devices Agency (PMDA), Mekinist Initial Approval. Available online: https://www.pmda.go.jp/files/000233741.pdf (accessed on 19 August 2022).

40. Food and Drug Administration (FDA), Trametinib Pharmacology Review. Available online: https://www.accessdata.fda.gov/drugsatfda_docs/nda/2013/204114Orig1s000PharmR.pdf (accessed on 19 August 2022).

41. Tsakalozou, E.; Alam, K.; Babiskin, A.; Zhao, L. Physiologically-Based Pharmacokinetic Modeling to Support Determination of Bioequivalence for Dermatological Drug Products: Scientific and Regulatory Considerations. *Clin. Pharmacol. Ther.* **2022**, *111*, 1036–1049. [CrossRef]

42. Shebley, M.; Sandhu, P.; Emami Riedmaier, A.; Jamei, M.; Narayanan, R.; Patel, A.; Peters, S.A.; Reddy, V.P.; Zheng, M.; de Zwart, L.; et al. Physiologically Based Pharmacokinetic Model Qualification and Reporting Procedures for Regulatory Submissions: A Consortium Perspective. *Clin. Pharmacol. Ther.* **2018**, *104*, 88–110. [CrossRef]

43. Stypmann, J.; Engelen, M.A.; Epping, C.; van Rijen, H.V.M.; Milberg, P.; Bruch, C.; Breithardt, G.; Tiemann, K.; Eckardt, L. Age and Gender Related Reference Values for Transthoracic Doppler-Echocardiography in the Anesthetized CD1 Mouse. *Int. J. Cardiovasc. Imaging* **2006**, *22*, 353–362. [CrossRef]

44. Jeanne, A.; Boulagnon-Rombi, C.; Devy, J.; Théret, L.; Fichel, C.; Bouland, N.; Diebold, M.-D.; Martiny, L.; Schneider, C.; Dedieu, S. Matricellular TSP-1 as a Target of Interest for Impeding Melanoma Spreading: Towards a Therapeutic Use for TAX2 Peptide. *Clin. Exp. Metastasis* **2016**, *33*, 637–649. [CrossRef]

45. Traynard, P. A Library of Tumor Growth and Tumor Growth Inhibition Models for the MonolixSuite. In Proceedings of the Tumor Growth Inhibition Modeling with Monolixsuite, ACoP 11 Virtual Conference, 13 November 2020.

46. Reig-López, J.; Maldonado, M. del M.; Merino-Sanjuan, M.; Cruz-Collazo, A.M.; Ruiz-Calderón, J.F.; Mangas-Sanjuán, V.; Dharmawardhane, S.; Duconge, J. Physiologically-Based Pharmacokinetic/Pharmacodynamic Model of MBQ-167 to Predict Tumor Growth Inhibition in Mice. *Pharmaceutics* **2020**, *12*, 975. [CrossRef] [PubMed]

47. Simeoni, M.; Magni, P.; Cammia, C.; De Nicolao, G.; Croci, V.; Pesenti, E.; Germani, M.; Poggesi, I.; Rocchetti, M. Predictive Pharmacokinetic-Pharmacodynamic Modeling of Tumor Growth Kinetics in Xenograft Models after Administration of Anticancer Agents. *Cancer Res.* **2004**, *64*, 1094–1101. [CrossRef] [PubMed]

48. Lobo, E.D.; Balthasar, J.P. Pharmacodynamic Modeling of Chemotherapeutic Effects: Application of a Transit Compartment Model to Characterize Methotrexate Effects in Vitro. *AAPS J.* **2002**, *4*, 212–222. [CrossRef] [PubMed]

49. Claret, L.; Girard, P.; Hoff, P.M.; Van Cutsem, E.; Zuideveld, K.P.; Jorga, K.; Fagerberg, J.; Bruno, R. Model-Based Prediction of Phase III Overall Survival in Colorectal Cancer on the Basis of Phase II Tumor Dynamics. *J Clin Oncol.* **2009**, *27*, 4103–4108. [CrossRef] [PubMed]

50. Hahnfeldt, P.; Folkman, J.; Hlatky, L. Minimizing Long-Term Tumor Burden: The Logic for Metronomic Chemotherapeutic Dosing and Its Antiangiogenic Basis. *J. Theor. Biol.* **2003**, *220*, 545–554. [CrossRef]

51. Desmée, S.; Mentré, F.; Veyrat-Follet, C.; Guedj, J. Nonlinear Mixed-Effect Models for Prostate-Specific Antigen Kinetics and Link with Survival in the Context of Metastatic Prostate Cancer: A Comparison by Simulation of Two-Stage and Joint Approaches. *AAPS J.* **2015**, *17*, 691–699. [CrossRef] [PubMed]

52. Jusko, W.J. A Pharmacodynamic Model for Cell-Cycle-Specific Chemotherapeutic Agents. *J. Pharmacokinet. Biopharm.* **1973**, *1*, 175–200. [CrossRef]

53. Mould, D.; Walz, A.-C.; Lave, T.; Gibbs, J.; Frame, B. Developing Exposure/Response Models for Anticancer Drug Treatment: Special Considerations. *CPT Pharmacomet. Syst. Pharmacol.* **2015**, *4*, e00016. [CrossRef] [PubMed]

54. King, A.J.; Arnone, M.R.; Bleam, M.R.; Moss, K.G.; Yang, J.; Fedorowicz, K.E.; Smitheman, K.N.; Erhardt, J.A.; Hughes-Earle, A.; Kane-Carson, L.S.; et al. Dabrafenib; Preclinical Characterization, Increased Efficacy When Combined with Trametinib, While BRAF/MEK Tool Combination Reduced Skin Lesions. *PLoS ONE* **2013**, *8*, e67583. [CrossRef]

55. Gilmartin, A.G.; Bleam, M.R.; Groy, A.; Moss, K.G.; Minthorn, E.A.; Kulkarni, S.G.; Rominger, C.M.; Erskine, S.; Fisher, K.E.; Yang, J.; et al. GSK1120212 (JTP-74057) Is an Inhibitor of MEK Activity and Activation with Favorable Pharmacokinetic Properties for Sustained in Vivo Pathway Inhibition. *Clin. Cancer Res.* **2011**, *17*, 989–1000. [CrossRef]

56. Cubas, R.; Moskalenko, M.; Cheung, J.; Yang, M.; McNamara, E.; Xiong, H.; Hoves, S.; Ries, C.H.; Kim, J.; Gould, S. Chemotherapy Combines Effectively with Anti–PD-L1 Treatment and Can Augment Antitumor Responses. *J. Immunol.* **2018**, *201*, 2273–2286. [CrossRef]

57. Jahchan, N.; Ramoth, H.; Juric, V.; Mayes, E.; Mankikar, S.; Mehta, R.; Binnewies, M.; Dash, S.; Palmer, R.; Pollack, J.; et al. 859 Tuning the Tumor Microenvironment by Reprogramming TREM1+ Myeloid Cells to Unleash Anti-Tumor Immunity in Solid Tumors. *J. Immunother. Cancer* **2021**, *9*, A900. [CrossRef]

58. Houghton, P.J.; Morton, C.L.; Tucker, C.; Payne, D.; Favours, E.; Cole, C.; Gorlick, R.; Kolb, E.A.; Zhang, W.; Lock, R.; et al. The Pediatric Preclinical Testing Program: Description of Models and Early Testing Results. *Pediatric Blood Cancer* **2007**, *49*, 928–940. [CrossRef]

59. Nair, A.B.; Jacob, S. A Simple Practice Guide for Dose Conversion between Animals and Human. *J. Basic Clin. Pharm* **2016**, *7*, 27–31. [CrossRef]

International Journal of
Molecular Sciences

Article

In Vitro/In Vivo Translation of Synergistic Combination of MDM2 and MEK Inhibitors in Melanoma Using PBPK/PD Modelling: Part III

Jakub Witkowski [1,2,*], Sebastian Polak [3,4], Dariusz Pawelec [2] and Zbigniew Rogulski [1]

[1] Faculty of Chemistry, University of Warsaw, Pasteura 1, 02-093 Warsaw, Poland
[2] Adamed Pharma S.A., Adamkiewicza 6a, 05-152 Czosnów, Poland
[3] Faculty of Pharmacy, Jagiellonian University, Medyczna 9, 30-688 Krakow, Poland
[4] Simcyp Division, Certara UK Limited, Level 2-Acero, 1 Concourse Way, Sheffield S1 2BJ, UK
*   Correspondence: j.witkowski6@uw.edu.pl

Abstract: The development of in vitro/in vivo translational methods and a clinical trial framework for synergistically acting drug combinations are needed to identify optimal therapeutic conditions with the most effective therapeutic strategies. We performed physiologically based pharmacokinetic–pharmacodynamic (PBPK/PD) modelling and virtual clinical trial simulations for siremadlin, trametinib, and their combination in a virtual representation of melanoma patients. In this study, we built PBPK/PD models based on data from in vitro absorption, distribution, metabolism, and excretion (ADME), and in vivo animals' pharmacokinetic–pharmacodynamic (PK/PD) and clinical data determined from the literature or estimated by the Simcyp simulator (version V21). The developed PBPK/PD models account for interactions between siremadlin and trametinib at the PK and PD levels. Interaction at the PK level was predicted at the absorption level based on findings from animal studies, whereas PD interaction was based on the in vitro cytotoxicity results. This approach, combined with virtual clinical trials, allowed for the estimation of PK/PD profiles, as well as melanoma patient characteristics in which this therapy may be noninferior to the dabrafenib and trametinib drug combination. PBPK/PD modelling, combined with virtual clinical trial simulation, can be a powerful tool that allows for proper estimation of the clinical effect of the above-mentioned anticancer drug combination based on the results of in vitro studies. This approach based on in vitro/in vivo extrapolation may help in the design of potential clinical trials using siremadlin and trametinib and provide a rationale for their use in patients with melanoma.

Keywords: anticancer drugs; virtual clinical trials; pharmacokinetics; pharmacodynamics; drug combination; PBPK/PD modelling; MDM2 inhibitor; MEK inhibitor

Citation: Witkowski, J.; Polak, S.; Pawelec, D.; Rogulski, Z. In Vitro/In Vivo Translation of Synergistic Combination of MDM2 and MEK Inhibitors in Melanoma Using PBPK/PD Modelling: Part III. *Int. J. Mol. Sci.* **2023**, *24*, 2239. https://doi.org/10.3390/ijms24032239

Academic Editors: Andrzej Kutner, Geoffrey Brown and Enikö Kallay

Received: 15 December 2022
Revised: 17 January 2023
Accepted: 19 January 2023
Published: 23 January 2023

1. Introduction

Conducting well-designed studies to establish the pharmacokinetic–pharmacodynamic (PK/PD) relationships in animal models in a way that allows us to scale the results to humans is a crucial element of preclinical drug development. Determining and understanding the relationships (or lack of them) between PK and PD significantly improves the interpretation of drug-related data and facilitates successful translation to human conditions [1].

The translational sciences aim to transfer results from basic science to the animal or patient level (bench to bedside). Currently, in vitro/in vivo extrapolation (IVIVE) methods combined with physiologically based pharmacokinetic (PBPK) modelling are often an inherent element of pharmacokinetic (PK) and pharmacodynamic (PD) properties estimation in the drug discovery and development process [2–4]. Such an approach can potentially help implement the 3R principles (replacement, reduction, and refinement) for the ethical use of animals by replacing living animals and reduce the number required to obtain meaningful PK/PD data [5].

Despite substantial progress in the development of in silico methods, the challenge of in vitro model verification and in vitro result extrapolation to animals or humans in vivo remains [6]. That is why studies involving animals are still irreplaceable at the preclinical drug development level, especially in the field of oncology [7]. This situation is also caused by the fact that some preclinical results may not directly translate to the clinic because some in vitro systems do not fully mimic the in vivo environment. This is why further development of in vitro/in vivo translational methods is crucial to better characterise a clinically observed drug's efficacy and safety and to increase the ethical usage of animals and decrease the number of animals needed in preclinical testing programs. One alternative way to avoid the excessive use of animals is the concept of bidirectional translational research. Clinical and in vivo animal studies inform basic science and in vitro research, and vice versa [8]. The translational concept of the bidirectional "learn, confirm and refine" paradigm is adopted in the modelling and simulation (M&S) approach [9]. One of the M&S applications that bridges the gap between in vitro and in vivo research is PBPK/PD modelling [10,11]. The selection of sensitive cancer types usually requires preclinical in vitro and in vivo efficacy data using indication-related cell lines or patient-derived xenografts (PDX). Such a process demands translational studies and data modelling to estimate effective clinical doses and dosing schedules. This is especially important for difficult-to-treat cancer indications requiring drug combinations. The modelling of such data is additionally complicated by the fact that there are a lack of IVIVE solutions for drug combinations. Admittedly, there have been some attempts [12,13]; however, the proposed clinical translational solutions were based only on the in vitro data and neglected the animal in vivo verification context. It is also worth mentioning the concepts of tumour static concentration and tumour static exposure developed by Cardilin et al. [14,15]. This methodology may also aid in translational efforts for drug combinations utilising the Chou–Talalay combination index (CI) theorem [16] and in vivo-derived data. The same group also recently published an interesting scaling technique to predict the clinical efficacy of drug combinations based on their efficacy in mouse xenografts [17].

Metastatic melanoma is a cancer condition that is life-threatening and difficult to treat due to its ability to spread early and aggressively. The treatment of aggressive and fast-spreading cancer usually requires a combination of various therapeutic options to stop the cancer from developing further. One of the newly proposed therapeutic options is the drug combination of siremadlin MDM2 (mouse double minute 2) inhibitor and trametinib MEK (mitogen-activated protein kinase kinase) inhibitor. Drug combinations utilising this class of inhibitor are currently the subject of many studies in clinical trials (clinicaltrials.gov (accessed on 14 December 2022) identifiers: NCT02110355, NCT03714958, NCT02016729, NCT01985191 and NCT03566485). Preclinical evidence suggests that siremadlin (previously known as HDM201) and trametinib act synergistically in melanoma [18–22]. The mechanisms of action of both drugs (including the mechanisms influencing melanin production [23,24], which may prevent melanoma metastasis [25,26]) are depicted in Figure S1.

Moreover, previous in vitro and in vivo data suggest that these two compounds might synergistically interact with each other at the PK and PD levels; thus, co-administration may result in synergistic drug–drug interaction (DDI) at the PK and PD levels, which provides the basics for further consideration as a therapeutic option.

Recent studies showed that a tumour size change of at least 10% is a prognostic factor and is highly correlated with the improvement of overall survival in melanoma [27,28]. Thus, it is believed that the increased antitumour activity of the siremadlin and trametinib drug combination may contribute to increased overall survival of melanoma patients.

The main aim of this study was to establish an accurate simulation methodology for drug combination using available in vitro/in vivo absorption, distribution, metabolism, and excretion (ADME), and PK and PD data, to develop and optimise a PBPK/PD model. Such a model could potentially allow for the translation of previously presented in vitro/in vivo findings into an in vivo human situation. This model was used to estimate the clinical effect (tumour size reduction) of an MDM2 and MEK inhibitor combination.

The performance of virtual clinical trials (VCTs), which might be carried out on a virtual representation of cancer patients, may lead to accurate estimations of the drug combination's efficacy. This can ultimately provide the rationale for using siremadlin and trametinib in combination in clinical trials with melanoma cancer patients.

2. Results

2.1. PBPK Models (with and without PK Interaction)

The two PBPK models were developed for siremadlin. The first model with a mixed zero- and first-order absorption mechanism (hereafter referred to as the MO model) was originally proposed in the literature [29], and the second one was simplified to a first-order absorption mechanism only (referred to as the FO model). Both models reasonably well described the plasma concentration–time data for siremadlin administered daily and in the intermittent regimens in cancer patients' representatives, as shown in the example in Figures 1 and S2–S4. PBPK simulations for siremadlin for all the study participants are depicted in Figures S5–S19.

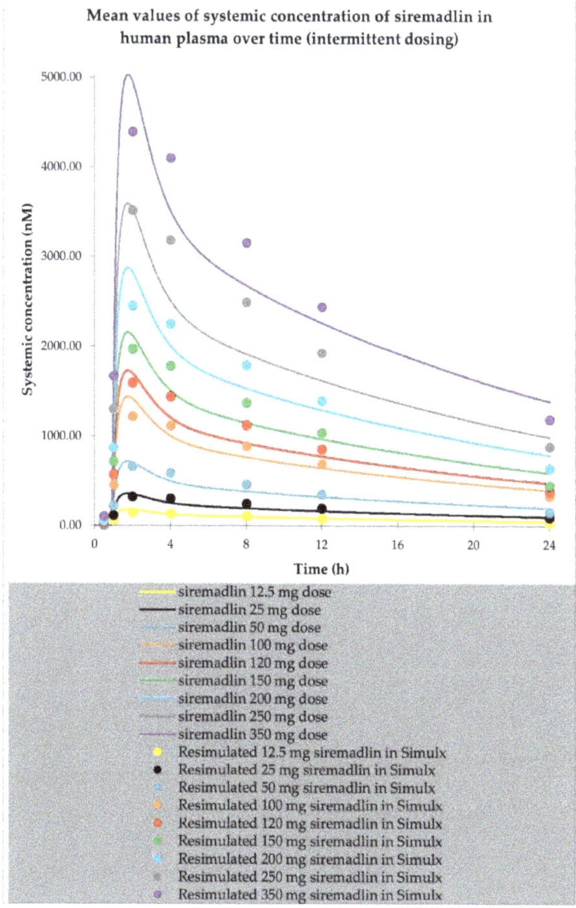

Figure 1. PBPK model of siremadlin administered in an intermittent regimen in cancer patient representatives using first-order absorption mechanism. Resimulated data are presented as a geometric mean from number of study participants × 10 (see Table 7 in Section 4).

The PBPK models developed for siremadlin indicated that the estimated area under the curve ($AUC_{0\text{-}inf}$, hereafter also referred to as exposure) and maximal concentration (C_{max}) values were accurately predicted only with the FO model. The numerical analysis of errors (fold difference between the predicted and clinically observed values) for the most important pharmacokinetic parameters, $AUC_{0\text{-}inf}$ and C_{max}, showed that values predicted by the FO model were within the 2-fold error range (0.5–2.0) of the observed values, except the exposure predicted for a 4 mg dose (Tables 1 and 2). The exception for a 4 mg dose was most probably caused by the fact that the observed exposure for a 4 mg dose was not dose-proportional. The observed exposure ratio at 4 mg and 2 mg doses (AUC 4 mg dose/AUC 2 mg dose) was 1.27, whereas the predicted exposure ratio at those doses was dose-proportional (the AUC ratio equalled 1.96). Further analyses also showed that the fold errors of $AUC_{0\text{-}inf}$ and C_{max} predicted by the FO model were closer to unity compared to those predicted by the MO model (Tables S1 and S2). Moreover, the introduction of the FO model decreased over a $50\times$ computation time and allowed for model simplification with increased precision. This is why the first-order model was selected as the absorption mechanism in the final PBPK model and was further used in the estimation of PK interaction and in drug combination.

Table 1. Comparison of predicted vs. observed $AUC_{0\text{-}inf}$ for siremadlin. Population representative and total population-derived $AUC_{0\text{-}inf}$ parameters were generated via first-order absorption mechanism and are presented as geometric means (%CV).

Dose (mg)	Representative $AUC_{0\text{-}inf}$ Predicted (nM × h)	Population $AUC_{0\text{-}inf}$ Predicted (nM × h)	$AUC_{0\text{-}inf}$ Observed (nM × h)	Representative AUC Predicted/ Observed	Population AUC Predicted/ Observed
1	257.90	286.40 (44.8%)	241.80 (-%)	1.07	1.18
2	515.80	537.83 (42.5%)	304.46 (31.5%)	1.69	1.77
4	1031.60	1073.86 (36.5%)	387.64 (22.0%)	2.66	2.77
7.5	1934.26	2013.49 (36.5%)	1076.86 (49.8%)	1.80	1.87
12.5	3223.76	3580.00 (44.8%)	2670.28 (-%)	1.21	1.34
15	3868.52	3816.39 (40.0%)	2343.49 (71.0%)	1.65	1.63
20	5158.02	5119.24 (36.9%)	4122.18 (29.6%)	1.25	1.24
25	6447.53	6399.05 (36.9%)	4803.12 (28.3%)	1.34	1.33
50	12,895.05	13,423.28 (36.5%)	14,455.63 (25.6%)	0.89	0.93
100	25,790.10	26,846.56 (36.5%)	25,723.34 (58.5%)	1.00	1.04
120	30,948.13	29,889.60 (38.8%)	33,275.78 (62.7%)	0.93	0.90
150	38,685.16	37,968.46 (41.8%)	42,719.97 (43.2%)	0.91	0.89
200	51,580.21	51,192.34 (36.9%)	47,271.75 (56.2%)	1.09	1.08
250	64,475.26	63,698.83 (43.2%)	74,579.68 (71.2%)	0.86	0.85
350	90,265.34	89,586.57 (36.9%)	99,211.21 (34.4%)	0.91	0.90

Table 2. Comparison of predicted vs. observed C_{max} for siremadlin. Population representative and total population- derived C_{max} parameters were generated via first-order absorption mechanism and are presented as geometric means (%CV).

Dose (mg)	Representative C_{max} Predicted (nM)	Population C_{max} Predicted (nM)	C_{max} Observed (nM)	Representative C_{max} Predicted/ Observed	Population C_{max} Predicted/ Observed
1	14.37	14.05 (33.6%)	14.22 (-%)	1.01	0.99
2	28.75	28.77 (27.5%)	21.61 (23.7%)	1.33	1.33
4	57.49	58.14 (27.0%)	31.69 (22.8%)	1.81	1.83
7.5	107.79	109.01 (27.0%)	70.22 (43.9%)	1.54	1.55

Table 2. *Cont.*

Dose (mg)	Representative C$_{max}$ Predicted (nM)	Population C$_{max}$ Predicted (nM)	C$_{max}$ Observed (nM)	Representative C$_{max}$ Predicted/ Observed	Population C$_{max}$ Predicted/ Observed
12.5	179.64	175.62 (33.6%)	212.46 (-%)	0.85	0.83
15	215.60	209.79 (26.2%)	164.74 (56.9%)	1.31	1.27
20	266.36	278.51 (28.6%)	269.17 (20.3%)	0.99	1.03
25	359.32	348.15 (28.6%)	422.57 (27.5%)	0.85	0.82
50	718.63	726.77 (27.0%)	840.82 (13.0%)	0.85	0.86
100	1437.19	1453.48 (27.0%)	1194.25 (30.1%)	1.20	1.22
120	1724.76	1596.25 (26.2)	1871.59 (51.5%)	0.92	0.85
150	2156.00	2093.20 (25.4%)	2600.42 (27.8%)	0.83	0.80
200	2874.55	2785.12 (28.6%)	2104.39 (43.7%)	1.37	1.32
250	3593.23	3482.48 (27.4%)	3629.21 (69.5%)	0.99	0.96
350	5030.18	4873.72 (28.6%)	4066.91 (56.9%)	1.24	1.20

The PBPK model developed for trametinib properly described the plasma concentration–time profiles in cancer patients after a single 2 mg dose and multiple (15) doses, as shown in Figures 2 and 3. The PBPK simulations for trametinib for all study participants are depicted in Figures S20 and S21.

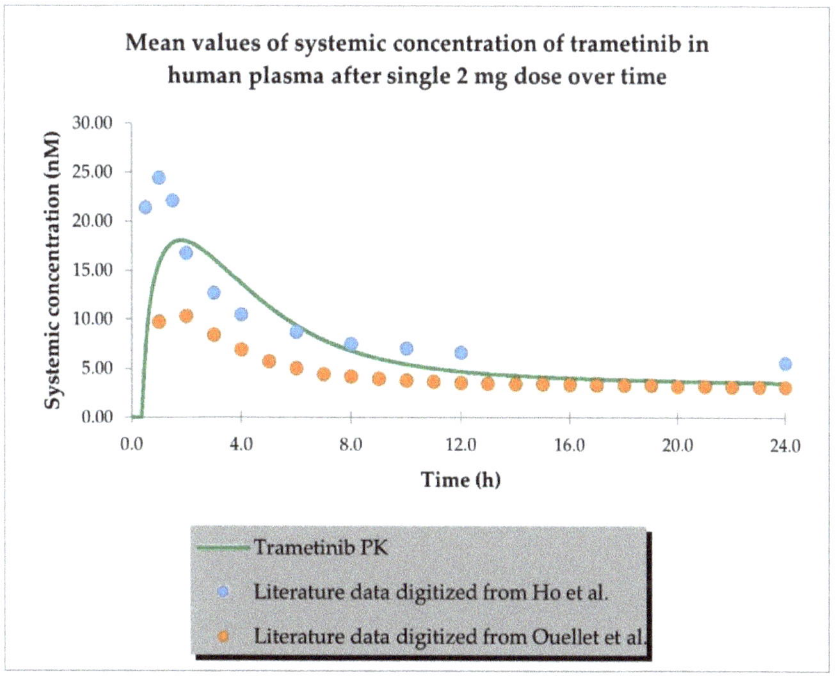

Figure 2. PBPK model of trametinib after a single dose in cancer patient representative. Observed data are presented as means from literature data (digitised from Ho et al. [30] and Ouellet et al. [31] as indicated in Table 3).

The numerical analysis of errors (fold difference between the predicted and observed values) for the most important pharmacokinetic parameters, AUC$_{0-24\,h}$ and C$_{max}$, showed that the predicted data met the 2-fold error (range: 0.5–2.0) acceptance criteria (Table 3).

The last step of PBPK model development for siremadlin and trametinib was the estimation of possible PK interaction at the absorption level using the AUC ratio relationships established previously in animals [19]. Plasma concentration–time profiles for siremadlin and trametinib with and without estimated PK interaction are shown in Figures 4 and 5.

Table 3. Comparison of predicted vs. observed key PK parameters ($AUC_{0-24\,h}$ and C_{max}) for trametinib. Population representative and population cohort-derived $AUC_{0-24\,h}$ and C_{max} parameters were generated via first-order absorption mechanism and are presented as geometric means (%CV).

Drug	Trametinib	Trametinib	Trametinib	Trametinib
Representative $AUC_{0-24\,h}$ (day 1)	165.39	-	-	-
Population $AUC_{0-24\,h}$ (day 1)	170.70 (29%)	-	200.76 (20%) *	109.60 (2%) **
Representative $AUC_{0-24\,h}$ (day 15)	570.91	-	-	-
Population $AUC_{0-24\,h}$ (day 15)	656.45 (50%)	601.24 (22%)	-	586.28 (16%) **
Population AUC ratio (day 1)	-	-	0.85	1.56
Population AUC ratio (day 15)	-	1.09	-	1.12
Representative C_{max} (day 1)	18.08	-	-	-
Population C_{max} (day 1)	18.50 (19%)	-	24.44 (3%) *	10.45 (0.3%) **
Representative C_{max} (day 15)	36.44	-	-	-
Population C_{max} (day 15)	40.93 (36%)	36.07 (28%)	-	31.67 (12%) **
Population C_{max} ratio (day 1)	-	-	0.76	1.77
Population C_{max} ratio (day 15)	-	1.13	-	1.29
Source	Current study	Digitised from Infante et al. [32]	Digitised from Ho et al. [30]	Digitised from Ouellet et al. [31]

* Geometric mean from digitised data. ** Geometric mean from digitised data from typical male and female with a median body weight of 79 kg.

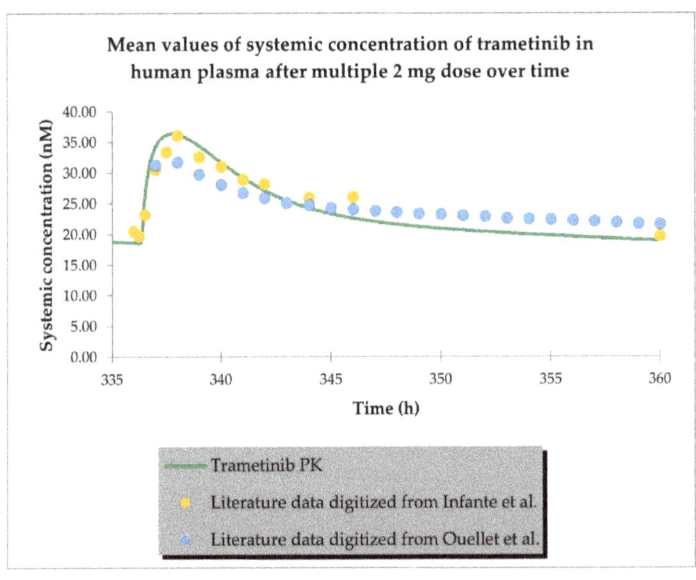

Figure 3. PBPK model of trametinib after a repeated dose in cancer patient representative. Observed data are presented as means from literature data (digitised from Infante et al. [32] and Ouellet et al. [31] as indicated in Table 3).

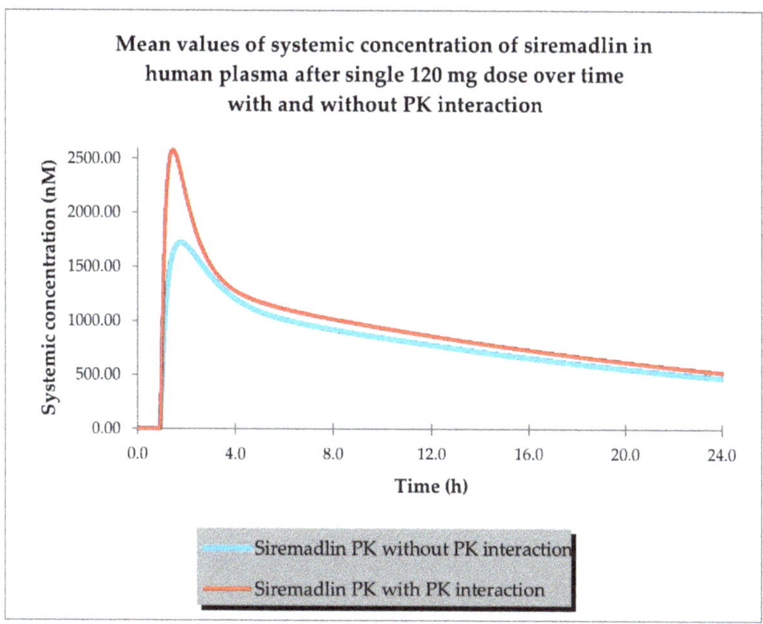

Figure 4. PBPK model of siremadlin with and without estimated PK interaction in cancer patient representative (*n* = 1).

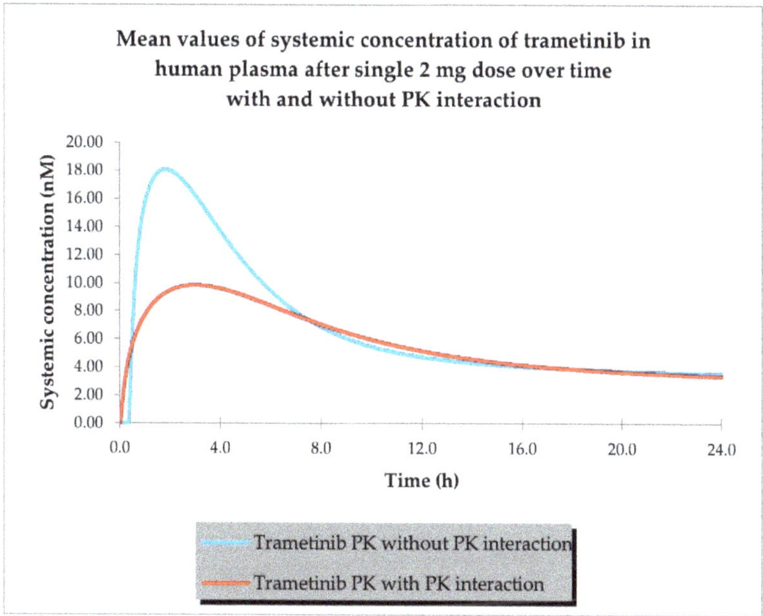

Figure 5. PBPK model of trametinib with and without estimated PK interaction in cancer patient representative (*n* = 1).

The predicted changes in $AUC_{0-24\,h}$ and C_{max} for siremadlin and trametinib with and without PK interaction for population representative are shown in Table 4. The final siremadlin and trametinib PBPK models' input parameters with and without PK interactions are summarised in Tables S3–S5.

Table 4. Comparison of predicted key PK parameters ($AUC_{0-24\,h}$ and C_{max}) with and without PK interaction for population representative for siremadlin and trametinib at the 120 and 2 mg doses (with an initial tumour size of 6.4 cm).

Drug	Siremadlin	Trametinib
Dose (mg)	120	2
AUC no PK DDI	19,240.73	174.01
AUC PK DDI	21,858.34	133.15
AUC ratio (PK DDI)	1.1360	0.7652
Cmax no PK DDI	1724.96	19.02
Cmax PK DDI	2580.39	9.61
Cmax ratio (PK DDI)	1.4959	0.5051

2.2. PD (TGI) Models

The final perturbed tumour growth inhibition (TGI) model for siremadlin and trametinib assumed exponential tumour growth, the Skipper–Schabel–Wilcox (log-kill) tumour-cell-killing hypothesis, the drug effect described by the linear drug-killing model, primary resistance to the therapy, and treatment effect delay described by the signal distribution model with four transit compartments. Changes in the sum of the longest diameters (SLD, hereafter also referred to as tumour size) over time in cancer population representatives after treatment with siremadlin (various doses and dosing regimens) and trametinib (2 mg dose administered daily) are presented as examples in Figures 6, 7, S22 and S23. TGI models for siremadlin and trametinib administered in monotherapy in the cancer patient population are depicted in Figures S24–S44. Since there was no available information about tumour growth in the trametinib study, two TGI model variants were developed. The first one had high tumour growth of kgh = 0.00028 1/h (Figure 8) and low tumour growth of kgh = 0.0000261 1/h (Figure 9). For drugs administered in monotherapy, this model was able to accurately capture the changes in tumour size over time, with a mean relative error (RE) < 20% (except for siremadlin administered in regimen 1A with a 25 mg dose, which was slightly over the threshold—20.06%), as shown in Table S6. The key input parameters (with corresponding %CV calculated based on equations from [33]) for the final TGI models for siremadlin and trametinib monotherapy are shown in Table S6.

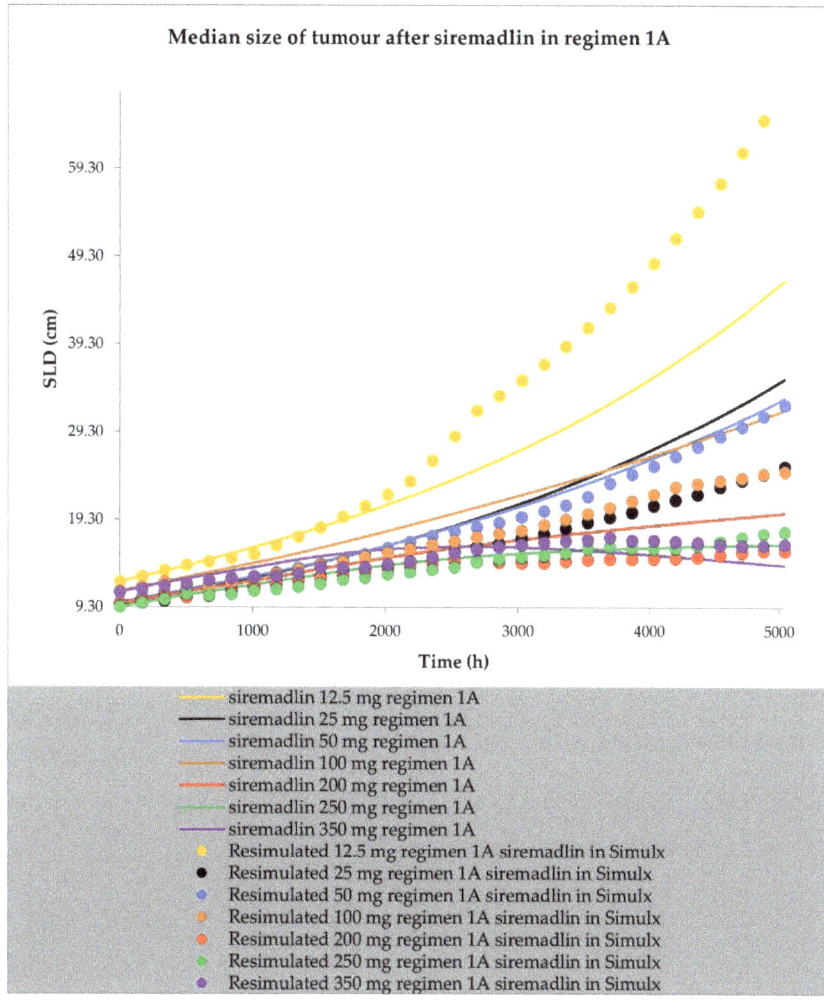

Figure 6. TGI model of siremadlin administered in regimen 1A in cancer patient representatives ($n = 1$ per treatment arm). Resimulated data are presented as medians from number of study participants $\times$ 10 (see Table 7 in Section 4).

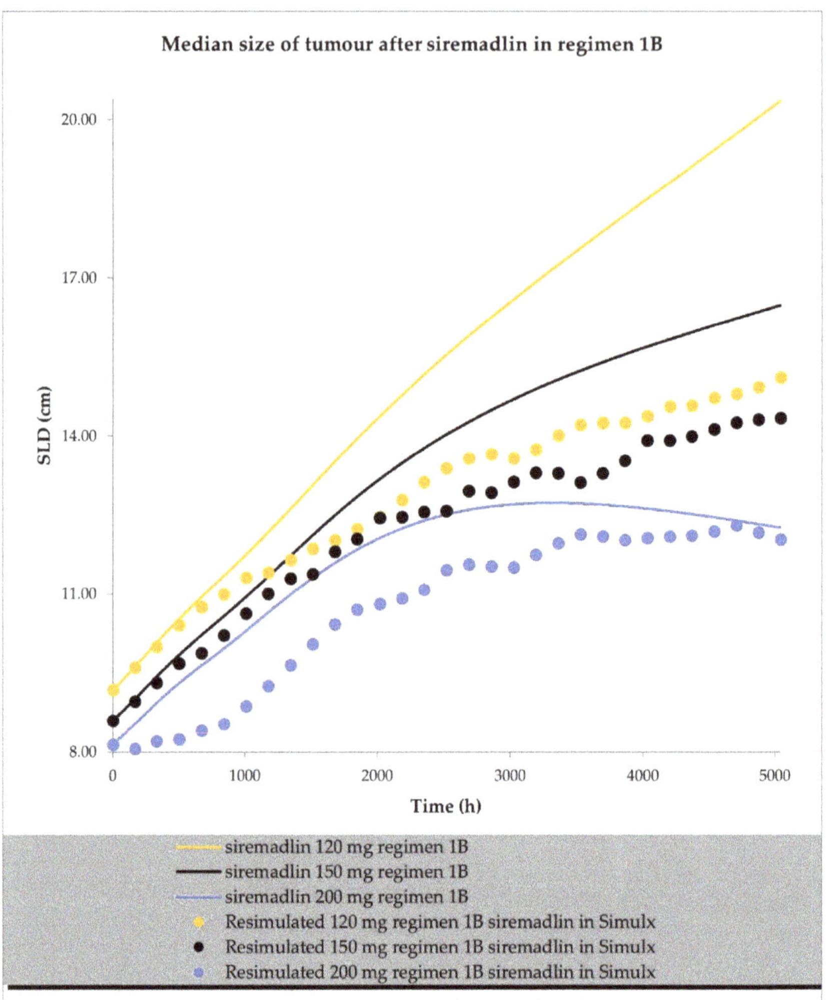

Figure 7. TGI model of siremadlin administered in regimen 1B in cancer patient representatives (n = 1 per treatment arm). Resimulated data are presented as medians from number of study participants $\times$ 10 (see Table 7 in Section 4).

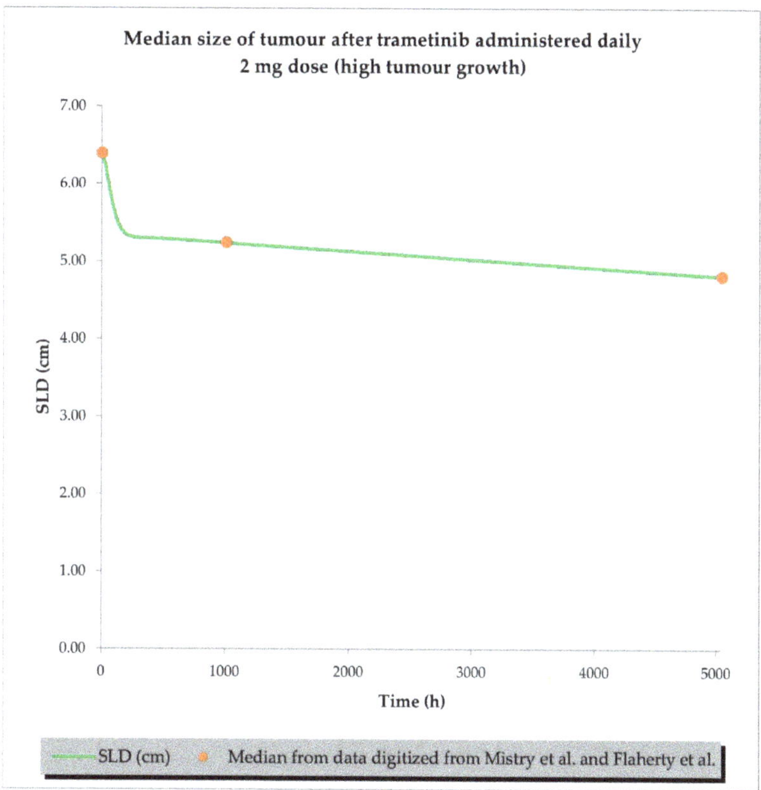

Figure 8. TGI model of trametinib administered daily in cancer patient representatives (*n* = 1) with assumption of high tumour growth (kgh = 0.00028 1/h). Observed data are presented as medians from literature data (data digitised from Mistry et al. [34] and Flaherty et al. [35]).

2.3. PBPK/PD Simulations for Drug Combination

The drug combination model was built based on universal TGI model principles previously established in animals [19] and a developed TGI model for siremadlin and trametinib. The relationships between TGI parameters for the drug combination are shown in Tables S5 and S6. For the drug combination simulations, siremadlin's recommended dose for expansion of 120 mg (in regimen 1B) [36] and an approved dose of 2 mg (daily dosing) [37] of trametinib were selected. Several scenarios were simulated for the combination of siremadlin and trametinib, including the assumption that the drugs will interact with each other at the PK or PD level. Additionally, several parameters were tested to examine how they influence the combination treatment outcome:

- Case A assumed high tumour growth (kgh), high initial tumour size (SLD0), and a high fraction of sensitive cells.
- Case B assumed a similar approach to that described above, but assumed low tumour growth (kgh).
- Case C assumed high tumour growth (kgh), high initial tumour size (SLD0), and a low fraction of sensitive cells.
- Case D assumed high tumour growth (kgh), low initial tumour size (SLD0), and a high fraction of sensitive cells.

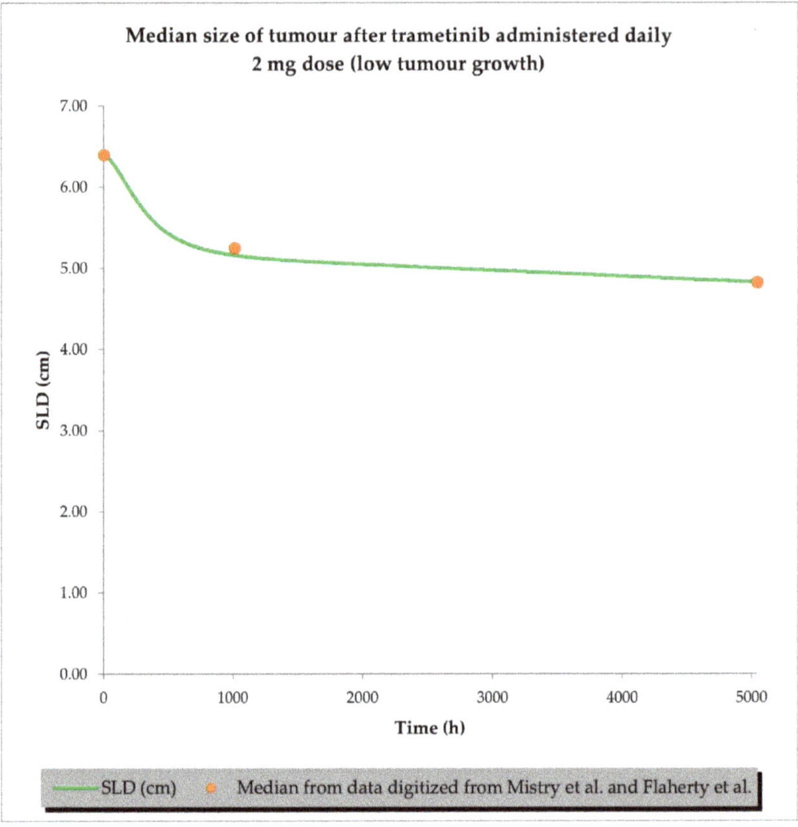

Figure 9. TGI model of trametinib administered daily in cancer patient representatives (*n* = 1) with assumption of low tumour growth (kgh = 0.0000261 1/h). Observed data are presented as medians from literature data (data digitised from Mistry et al. [34] and Flaherty et al. [35]).

Several sub-scenarios were also included in which the potential influence of interactions at the PK and PD levels between siremadlin and trametinib were studied:

- Scenario 1—without PK and PD drug interactions
- Scenario 2—without PK but with PD drug interactions
- Scenario 3—with PK and without PD drug interactions
- Scenario 4—with PK and PD drug interactions

The previously proposed universal TGI model for the siremadlin and trametinib drug combination allowed us to extrapolate the values of the clinical model parameters in a melanoma cancer population. The key input parameters, relationships between input parameters, and simulation results for these scenarios are shown in Figures 10–13 and Tables S7–S12.

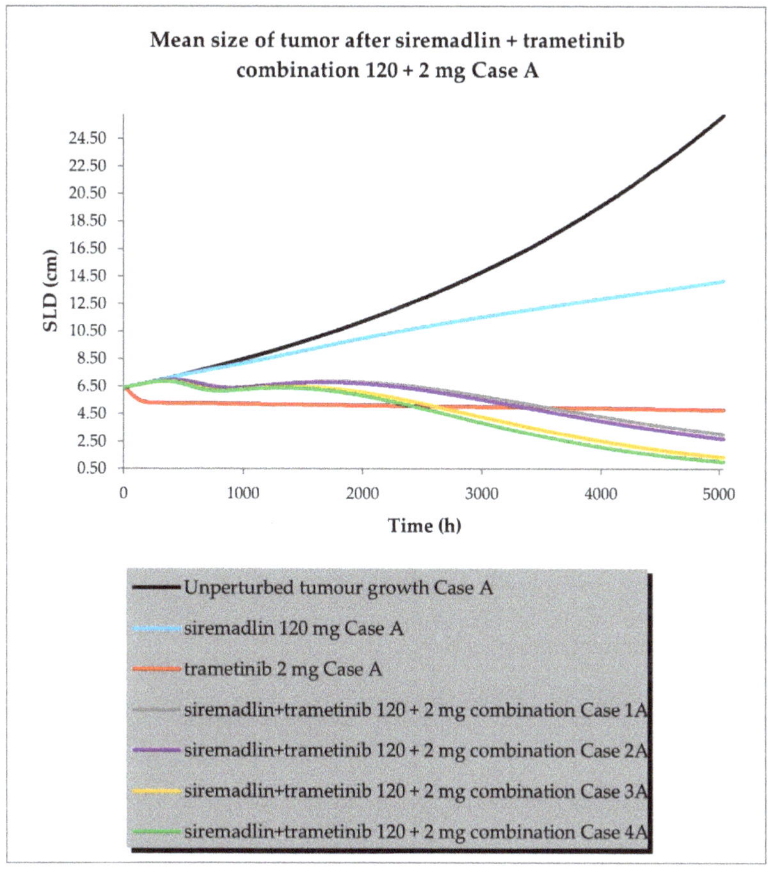

Figure 10. Siremadlin and trametinib combination's 120 + 2 mg efficacy in population representative of melanoma cancer (Case A—high tumour growth, high initial tumour size, and high fraction of sensitive cells).

The simulations of tumour growth patterns after administration of the siremadlin and trametinib combination indicated that there was a substantial tumour size reduction, regardless of the occurrence of PK or PD interaction. The simulations of tumour growth pointed out a biphasic response to the drug combination with initial and subsequent delayed tumour size reduction (cases A and D), while a monophasic response was observed with low tumour growth and a low fraction of sensitive cells (cases B and C). The simultaneous introduction of PK and PD interactions had the greatest impact on the tumour growth reduction (scenario 4) while a lack of such interactions resulted in only a limited increase in tumour growth reduction over trametinib monotherapy (scenario 1). The introduction of interaction only at the PD level had a lower effect than the introduction of interaction at the PK level in terms of tumour size reduction (scenarios 2 and 3, respectively). The simulated tumour growth dynamics affected the final tumour size (size at the end of the 30-week simulation: 5040 h) for the most efficacious scenario in which PK and PD interactions were employed. For highly proliferating tumours, the simulated neoplasm size was reduced to around 1 cm, while for low tumour growth, it was approximately four-times higher (4 cm).

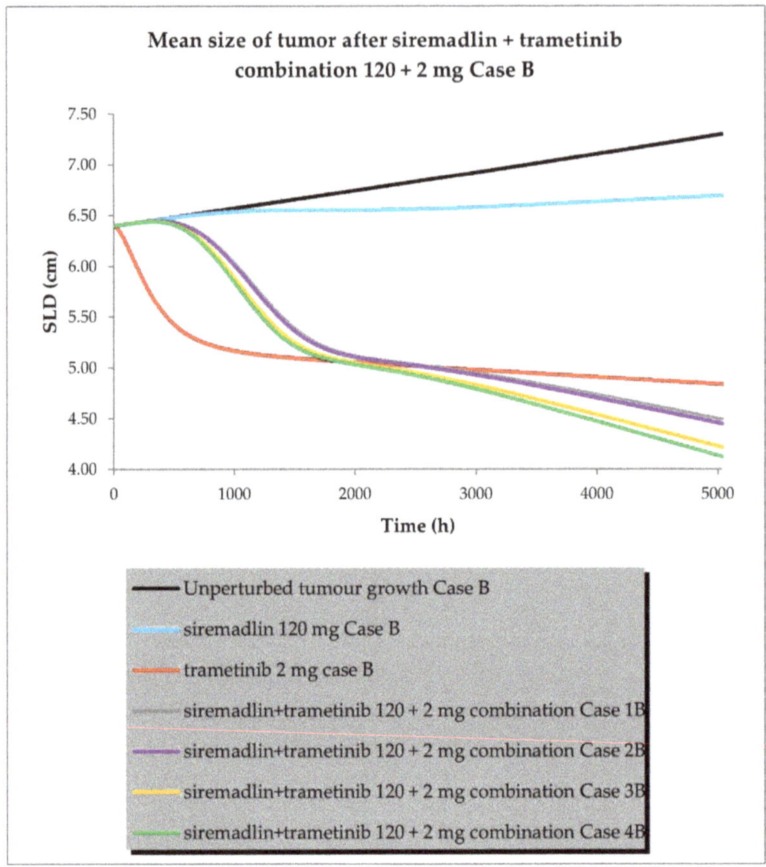

Figure 11. Siremadlin and trametinib combination's 120 + 2 mg efficacy in population representative of melanoma cancer (Case B—low tumour growth, high initial tumour size, and high fraction of sensitive cells).

The increased combination efficacy as a mean objective response rate (ORR—percentage of patients with a complete or partial response at any time during the study) was statistically significant in each drug combination scenario compared to the most efficacious drug—trametinib (Table 5).

Table 5. Comparison of estimated efficacy of siremadlin and trametinib combination at the 120 and 2 mg doses assuming various case scenarios.

Trial	n	No. of Trials	Difference (Drug Combination vs. Trametinib)	%ORR (Mean from 10 Trials)	Statistical Significance (p Value)
Siremadlin Case A	29	10	51.64%	0.00%	<0.0001
Trametinib Case A	214	10	0.00%	51.64%	<0.0001
Case 1a	243	10	26.51%	78.15%	<0.0001
Case 2a	243	10	31.16%	82.80%	<0.0001
Case 3a	243	10	45.81%	97.45%	<0.0001
Case 4a	243	10	47.29%	98.93%	<0.0001
Siremadlin Case B	29	10	25.84%	0.00%	<0.0001
Trametinib Case B	214	10	0.00%	25.84%	<0.0001

Table 5. *Cont.*

Trial	n	No. of Trials	Difference (Drug Combination vs. Trametinib)	%ORR (Mean from 10 Trials)	Statistical Significance (*p* Value)
Case 1b	243	10	26.67%	52.51%	<0.0001
Case 2b	243	10	30.70%	56.54%	<0.0001
Case 3b	243	10	52.22%	78.07%	<0.0001
Case 4b	243	10	59.14%	84.98%	<0.0001
Siremadlin Case C	29	10	51.64%	0.00%	<0.0001
Trametinib Case C	214	10	0.00%	51.64%	<0.0001
Case 1c	243	10	14.41%	66.05%	<0.0001
Case 2c	243	10	20.83%	72.47%	<0.0001
Case 3c	243	10	42.52%	94.16%	<0.0001
Case 4c	243	10	45.81%	97.45%	<0.0001
Siremadlin Case D	29	10	51.64%	0.00%	<0.0001
Trametinib Case D	214	10	0.00%	51.64%	<0.0001
Case 1d	243	10	26.51%	78.15%	<0.0001
Case 2d	243	10	31.16%	82.80%	<0.0001
Case 3d	243	10	45.81%	97.45%	<0.0001
Case 4d	243	10	47.29%	98.93%	<0.0001

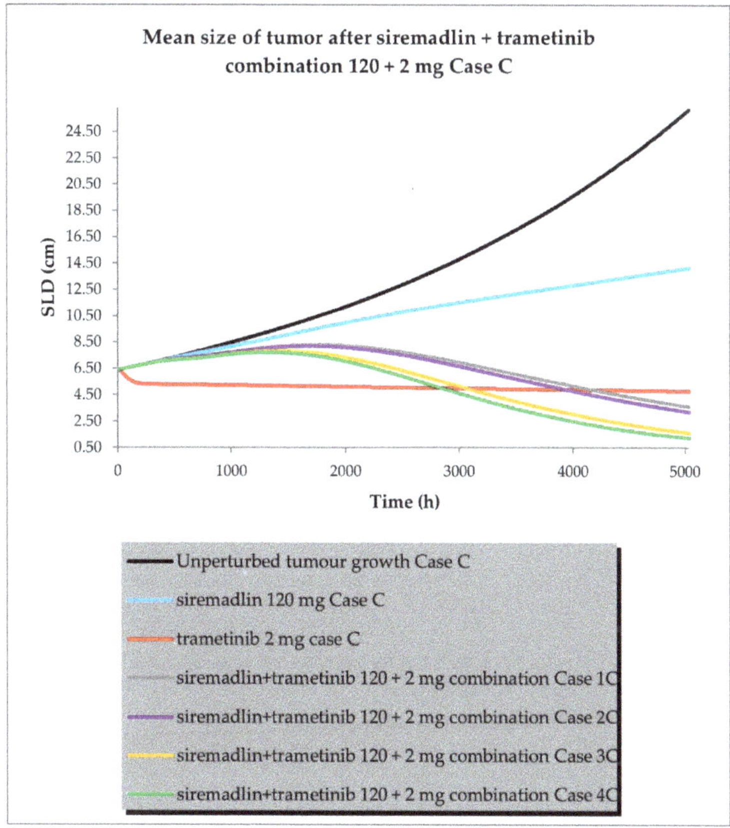

Figure 12. Siremadlin and trametinib combination's 120 + 2 mg efficacy in population representative if melanoma cancer (Case C—high tumour growth, high initial tumour size, and low fraction of sensitive cells).

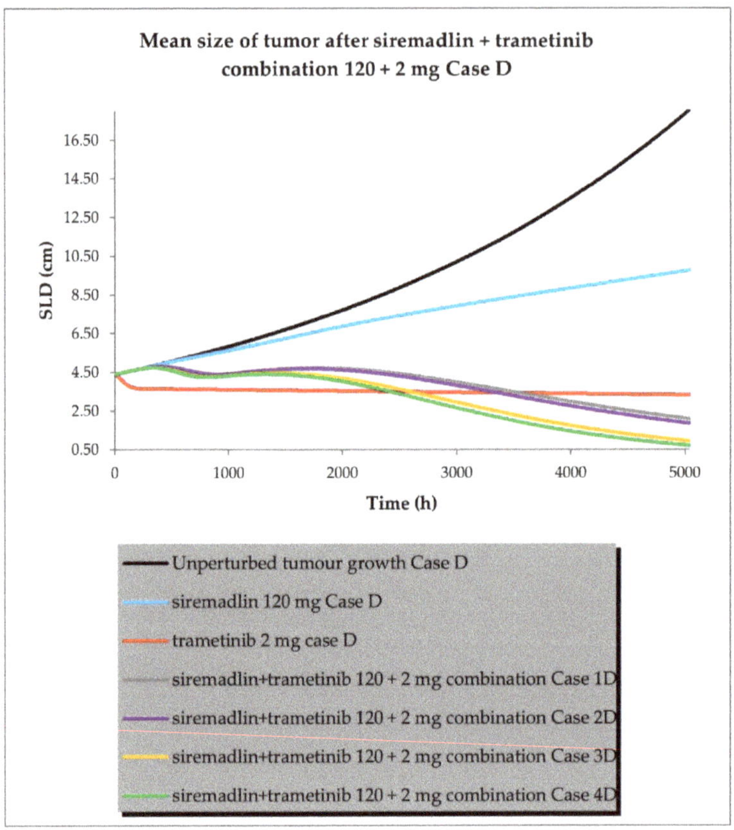

Figure 13. Siremadlin and trametinib combination's 120 + 2 mg efficacy in population representative of melanoma cancer (Case D—high tumour growth, low initial tumour size, and high fraction of sensitive cells).

3. Discussion

The developed PBPK/PD models for the MDM2 inhibitor siremadlin and the MEK inhibitor trametinib were able to describe both the pharmacokinetic and pharmacodynamic profiles of these drugs. The models consider the oral (p.o.) administration of siremadlin and trametinib, the first-order absorption mechanism, a full-body distribution model, hepatic metabolism and renal excretion for siremadlin (renal excretion calculated using the FCIM method [38]), intravenous clearance for trametinib, and a permeability-limited tumour distribution model.

The first developed PBPK model for siremadlin assumes a mixed zero- and first-order absorption mechanism, as originally suggested in the literature [29], and the second model includes a first-order absorption mechanism only. The analysis of the fold difference between the predicted and clinically observed values for key PK parameters ($AUC_{0\text{-inf}}$ and C_{max}) indicated that those values were accurately predicted only with the model assuming a first-order absorption mechanism. In our opinion, the model with mixed zero- and first-order absorption is more sensitive to the initial tumour size (SLD0) than the model with first-order absorption. Therefore, the predicted values may be more likely to deviate from those observed in the clinical trial. The exposure and maximal concentration values predicted by the model assuming first-order absorption were within the 2-fold error range (0.5–2.0) of the observed values, except the $AUC_{0\text{-inf}}$ predicted for a 4 mg dose. The exception for the 4 mg dose was most probably caused by the fact that the observed exposure for a 4 mg dose was not dose-proportional compared, for example, with exposure for a 2 mg dose (AUC 4 mg dose/AUC 2 mg dose: 1.27), whereas the predicted exposure was nearly dose-proportional (AUC 4 mg dose/AUC 2 mg dose: 1.96). The model tended to slightly overpredict exposure in the 2–25 mg dose range, while for higher doses, the fold error was close to unity. This may have been caused by several factors: different sampling of data in the simulation and in the clinical protocol, or an unexplained variability of nearly 17%, which was assumed during the generation of resimulated PK data [29], while in the PBPK model, no unexplained PK variability was assumed. Since the PBPK models consider various compartments, rapid changes in tumour size may affect the volume of distribution, which could also explain the variability in the estimated PK parameters. The final PBPK model was successfully verified with resimulated data from the PK model published by Guerreiro et al. [29] and external PK data digitised from the literature [39].

The PBPK model developed for trametinib properly described the plasma concentration–time profiles in cancer patients after a 2 mg dose administered once and after reaching a steady state (after approximately 15 doses [40]). The predicted data met the 2-fold error range (0.5–2.0) acceptance criteria for $AUC_{0-24\,h}$ and C_{max}, as well. The developed PBPK model was successfully verified with external PK data extracted from the literature [30–32].

The above-mentioned models were further extended with the introduction of PK interaction at the absorption level. Estimations of the exposure changes for both drugs were made based on relationships between the AUC ratios and drug doses previously established in animals [19]. This interaction included changes in the absorption rate constant (ka) and fraction absorbed (fa) for siremadlin and alterations in the absorption rate constant (ka), fraction absorbed (fa), and lag time (tlag) in the case of trametinib.

The translation of PK drug interaction from animals to patients is very challenging. This is due to differences in digestive tract anatomy and physiology, transporter abundance, intestinal pH, internal organ blood flow, metabolising enzymes, and potential drug absorption mechanisms [41,42]. It appears that ex vivo or cadaver studies in an Ussing chamber might be one of the methods that allows for precise estimation of the fraction absorbed for single drugs and drug combinations. However, further studies are needed to prove its utility [43–45]. Therefore, the occurrence of PK interaction between siremadlin and trametinib estimated in this work should be taken with caution. Only results from the currently ongoing siremadlin and trametinib clinical trial (NCT03714958) or following trials may confirm or exclude the occurrence of such PK interaction.

Previously performed in vitro and in vivo studies have shown that the killing effect of siremadlin and trametinib compounds is concentration- and time-dependent, and an initial delay in the response and resistance to the treatment might arise [18,19]. These findings allowed us to develop a perturbed TGI model for siremadlin and trametinib. The developed model assumed exponential tumour growth, the Skipper–Schabel–Wilcox (log-kill) tumour-cell-killing hypothesis, the drug effect described by the linear drug-killing model, and acquired resistance to the therapy and treatment effect delay described by the signal distribution model with four transit compartments. The delay in the effect of these drugs is most likely related to the duration of the signal transduction associated with the activation of the p53–MDM2 and MAPK pathways, resulting in cell death. Resistance is an inherent part of anticancer treatment; therefore, its description may play a critical role in predicting and optimising the treatment response and may improve therapy scheduling [46,47]. Clinically predicted resistance in siremadlin and trametinib models was much higher (13–26× higher) than estimated in animal models, which further warrants the need for drug combination usage to overcome it. The applied model assumed that two distinct tumour growth patterns might occur as originally proposed for solid tumours [29], the first with high tumour growth 0.00028 cm/h and the second with low tumour growth 0.0000261 cm/h. These assumptions were in line with the reported tumour growth range observed in melanoma cancer (0.00030 cm/h and 0.000015 cm/h for high and low melanoma growth, respectively [48]) and a previously performed study with melanoma-derived A375 xenografts in mice (0.00024–0.00031 1/h range [19]). Therefore,

0.00028 cm/h and 0.0000261 cm/h were used as estimates for high and low tumour growth in the melanoma population, as well.

The final TGI models for the perturbed groups properly predicted the tumour size within 20% of the mean relative error (%RE) between the predicted and resimulated data in the case of siremadlin (except for a 25 mg dose of siremadlin administered in regimen 1A), and the predicted versus the literature data for trametinib. Most of the observed differences were related to the TGI model developed for siremadlin. This may have been due to the fact that when generating the resimulated PD data, nearly 4% of unexplained variability [29] was assumed in contrast to the PBPK/PD model, where no unexplained PD variability was assumed. Moreover, differences in the predicted PK between the PK model in Simulx and the PBPK model developed in Simcyp may additionally deepen these differences. We believe that further modelling work is warranted to optimise this TGI model to ensure that all doses and dosing regimens meet the 20% acceptance criteria.

No clinical data from the siremadlin and trametinib combination are available yet. Simulation of this combination required us to test virtual "what if" scenarios related to the influence of various model parameters and the introduction of potential interactions at the PK and PD levels. The results of an in-depth analysis of the TGI model parameter dependencies and experiences from mouse species allowed us to extrapolate PD model predictions for this combination into a clinical context. The developed PK interaction model for siremadlin and trametinib was used in simulations, assuming PK interaction during co-administration and that the previously selected β parameter from the MuSyC drug interaction model was used as the PD interaction parameter.

The simulations covered the influence of the tumour growth pattern (kgh parameter), fraction of sensitive cells (fs parameter) and initial tumour size (SLD0 parameter) on the siremadlin recommended dose for expansion of 120 mg dose (in regimen 1B) [36] and the approved 2 mg dose (daily dosing) [37] of trametinib. As mentioned earlier, two distinct tumour growth estimates (kgh high and kgh low), which correspond to the values observed in human melanoma [48], were tested in simulations of efficacy in patients with melanoma. The fraction of sensitive cells for the drug combination was unknown; thus, two assumptions were made: first, that the estimated sensitive fractions will be summed up for drug combination (fs high), and second, that there will only be some parts of the cells that respond (fs low). The assumed low fraction of sensitive cells was equal to the inter-individual variability of the fs parameter for siremadlin ($0.0321 \times 1.93 = 0.0620$), which was similar to the reported values for pembrolizumab in melanoma (0.0628) [49]. Since no clinical data are available, these assumptions should also be taken with caution. Moreover, this model did not assume conversion from sensitive to resistant cells, which might occur in the clinical trials [50–53]. However, this issue is out of the scope of this work because access to raw clinical data would be needed to verify such a modelling hypothesis. According to the pooled analysis from four randomised clinical trials [54], the baseline SLD value was a prognostic factor of the overall survival of patients treated with MEK and BRAF inhibitors if the initial SLD (SLD0) was lower than or equal to 4.4 cm; therefore, the influence of this parameter on the combination's efficacy was also tested (Case D).

The simulations for the drug combination showed that high tumour growth (khg high) results in a biphasic response to the drug combination, with an initial and subsequent delayed tumour size reduction, regardless of the initial tumour size (SLD0), as shown in cases A and D. In turn, a monophasic response was observed in simulations with low tumour growth (kgh low) and a low fraction of sensitive cells (fs low), as may be observed in cases B and C. These simulation results suggest that the tumour characteristics (such as tumour growth or sensitivity to combined drugs) may have a considerable impact on the onset of the observable therapeutic effect. The simultaneous introduction of PK and PD interactions had the greatest impact on tumour growth reduction (scenario 4), while a lack of such interactions resulted in only a limited increase in tumour growth reduction over trametinib monotherapy (scenario 1). This may be simply explained by the fact that all predicted PK and PD interactions were acting synergistically, amplifying the predicted response. The introduction of the interaction only at the PD level had a lower effect than the introduction of the interaction at the PK level in terms of tumour size reduction (scenarios 2 and 3, respectively). The simulated tumour growth dynamics affected the final tumour size (size at the end of the 30-week (5040 h) simulation) for the most efficacious scenario in which PK and PD interactions were employed. For highly proliferating tumours (kgh high), the simulated neoplasm size was reduced to ~1 cm, while for low tumour growth (kgh low), it was four-times higher (~4 cm). Interestingly, alteration of the initial tumour size (Case D) did not increase the calculated ORR; however, further modelling with different doses and dosing regimens is needed to confirm such an observation.

The simulation results indicated that this drug combination significantly increases the probability of achieving at least a partial response in melanoma patients regardless of the occurrence of PK and PD interaction. For Cases A, C and D, the calculated ORR was in the 61–98% range, which seems to be comparable to the reported ORR for the already approved dabrafenib (BRAF inhibitor) and trametinib combination (ORR range: 50–76% [55–59]). Nevertheless, the estimation of the superiority or noninferiority of the siremadlin and trametinib combination over the already approved dabrafenib and trametinib combination was not possible with such limited availability of data. Hence, without access to raw PK/PD clinical data coupled with the development of population PBPK models for those molecules, such conclusions cannot be made.

The calculated ORR for trametinib monotherapy in Case B was twice as low as in other scenarios, and was the most comparable with the observed ORR in the clinic (26 vs. 22%) [35]. Likewise, in Case B for drug combination scenarios, a lower ORR was calculated (ORR range: 48–81%), which also seems to be comparable with the data reported for the approved dabrafenib and trametinib combination. This finding supports the hypothesis that all patients suffering from metastatic melanoma, regardless of baseline tumour size, tumour growth dynamics, and the fraction of sensitive cells, may potentially benefit from this drug combination therapy. It should also be noted that other clinical factors can affect the clinical response assessment, such as shrinkage in nontarget tumours such as pathologic lymph nodes or the appearance of malignant lesions indicating cancer progression. Due to the fact that the developed models accounted only for the tumour size of the target lesion without a distinction between lymph nodes or tumour metastases, such issues described in the Response Evaluation Criteria in Solid Tumors (RECIST) guidelines [60] cannot be fully accounted for in such simulations. Therefore, the interpretation of results and drawing of direct comparisons with the actual clinical response should always be approached with caution.

The developed PBPK/PD models also have other associated limitations which should be discussed. Synergistic pharmacodynamic interaction in the current TGI model was based on results from in vitro studies that were further validated on mouse xenografts from the same A375 melanoma cells. One of the limitations of such an approach is that cell-line-derived xenograft models cannot fully simulate the microenvironment of tumours in humans, such as the vascular, lymphatic, and immune environments; moreover, such models often lose genetic heterogeneity compared to primary tumours [61]. In recent years, the patient-derived xenograft model (PDX model) has emerged as a promising tool that provides translational value with better mimicking of the tumour microenvironment; however, one of the disadvantages of this model is the loss of the human tumour stroma, which is entirely replaced by the murine stroma [62,63]. There is evidence that both the tumour stroma and the corresponding microenvironment may affect the drug response [64–66]. This may be important based on the reports that MDM2 inhibition may affect stromal or immune microenvironments, which was not considered in this work [67–69]. The results from ongoing clinical trials (NCT03611868, NCT03787602) may shed light on the utility of MDM2 inhibitors in skin cancer.

Another troubling problem regarding the developed PBPK/PD model is the repetitiveness of the pharmacodynamic interaction parameter (β) in the heterogeneous melanoma patient population. Additional in vitro studies, followed by in vivo verification with melanoma-derived xenografts or patient-derived xenografts (PDX), are needed to validate the magnitude of the estimated pharmacodynamic synergy between siremadlin and trametinib and to improve the translational value of the present study.

The current TGI model did not account for the cell transition from a sensitive to a resistant state (defined in the literature as the ksr parameter [29,51]), which might be useful for the estimation of long-term therapies such as cancer treatment. Further model development with raw clinical study data would be needed to assess whether the introduction of such a feature would improve the fit to clinically observed data for siremadlin and trametinib.

Due to the limited availability of response data for trametinib in the literature, the developed model may be not optimal and does not cover delayed resistance, resulting in tumour recurrence in patients with partials and complete responses. Thereby, the developed TGI model for trametinib predicts a higher percentage of responders than clinically observed (especially in the scenario in which high tumour growth is assumed) [35,70,71].

Nonetheless, despite these many limitations, the developed PBPK/PD models reasonably accurately described the PK and time course of tumour growth across all doses and dosing schedules. Further analyses and modelling work with different doses and dose regimens are encouraged to externally validate the developed PBPK/PD models for siremadlin, trametinib, and their combination toward predicting tumour size reduction. Although modelling work with different doses and dosing schedules is still ongoing, the results from the in vitro/in vivo translational approach presented in this study are promising. This study shows that the currently examined 120 mg dose of siremadlin in regimen 1B with daily dosed trametinib may act with sufficient synergy to elicit a significant clinical response.

Further performance of virtual clinical trial simulations with different doses and dosing regimens of the siremadlin and trametinib combination are ongoing to help select the most synergistic, efficacious, and safe dose levels and dosing regimens for melanoma-bearing patients. Further development of the presented clinical PBPK/PD models for siremadlin and trametinib is needed to develop a proper drug combination model in clinical settings and aid in the design of potential clinical trials using siremadlin and trametinib in patients with melanoma.

4. Materials and Methods

4.1. Clinical Studies Used

Data modelling usually requires two types of dataset: a training dataset, used in the development of the model, and a verification dataset, used for independent verification of the developed model. When early clinical data are available (e.g., from phase I clinical studies) a PK model is developed using only data from such studies and validated using data from later phases of clinical studies (phases 2 and 3). For siremadlin, resimulated data from Guerreiro et al. [29] were used as a training dataset and data from Stein et al. [29] and Jeay et al. [39] as verification datasets.

The details of data from clinical studies used for PBPK and PD (TGI) model development and verification are summarised in Table 6.

4.2. Software

The PK parameters were estimated using Microsoft Excel (Excel version 2016, Microsoft Corporation, Redmond, WA, USA, 2016, https://www.office.com (accessed on 14 December 2022)). Digitisation of the literature-derived data was performed with the use of WebPlotDigitizer software (version 4.4, Ankit Rohatgi, Pacifica, CA, USA, 2021, https://automeris.io/WebPlotDigitizer (accessed on 14 December 2022)). TGI modelling for trametinib was performed using Monolix software (Monolix version 2021R1, Lixoft SAS, a Simulations Plus company, Antony, France, 2022, http://lixoft.com/products/monolix/ (accessed on 14 December 2022)). The simulation of clinical PK/PD data for siremadlin was performed using Simulx software (Simulx version 2021R1, Lixoft SAS, a Simulations Plus company, Antony, France, 2022, https://lixoft.com/products/simulx/ (accessed on 14 December 2022)).

Table 6. Details of the clinical studies used for the development and verification of the siremadlin and trametinib PBPK model.

Drug	Siremadlin	Trametinib	Trametinib	Trametinib	Trametinib
NCT number	NCT02143635	NCT00687622	NCT01387204	NCT01245062	NCT00687622/ NCT01037127/ NCT01245062
Phase	1/2	1/2	1	3	1/2/3
Doses (mg)	1–350	0.125–10	2	2	0.125–10/2/2
Administration	Oral	Oral	Oral	Oral	Oral/oral/oral
n	115	206	2	214	206/97/214
Women (%)	44	46	0	44	46/30/44
Age (Years)	18–80	19–92	54–66	23–85	19–92/23–79/ 23–85
Dataset purpose	PK/PD training/ verification	PK training	PK training	PD training/ verification	PK verification
Reference	Guerreiro et al. [29] Stein et al. [36] Jeay et al. [39]	Infante et al. [32]	Ho et al. [30]	Flaherty et al. [35] Mistry et al. [34]	Ouellet et al. [31]

The Mlxtran model used for PK/PD resimulation in Simulx can be found in Code S1. PBPK/PD modelling was performed using Simcyp simulator software (Simcyp simulator V21, Certara UK Limited, Sheffield, UK, 2022, https://www.certara.com/software/simcyp-pbpk (accessed on 14 December 2022)). The custom mixed zero- and first-order absorption PK for siremadlin, PK drug interaction for trametinib, and PD (TGI) models in Lua can be found in Codes S2–S4. Fisher's test was performed using GraphPad Prism version 9.4.1 for Windows, GraphPad Software, San Diego, CA, USA, 2022, www.graphpad.com (accessed on 14 December 2022).

4.3. Statistical Methods

According to RECIST v1.1 metric [60], patients whose calculated maximal % reduction in change from baseline tumour size was ≥30% at any time during study were scored as responders (PR—partial response and CR—complete response), and the others as non-responders (SD—stable disease and PD—progressive disease). The percentage of responders represents the overall response rate (ORR). The ORRs were tabulated based on the number and percentage of subjects attaining an overall best response of complete or partial response in the melanoma patient population. Fisher's exact test was used to evaluate the ORR statistically between trametinib and the estimated trametinib plus siremadlin drug combination.

From the available sample size—n = 457: 214 patients from the trametinib study and 243 patients (214 patients from the trametinib study plus 29 patients from the siremadlin study) from the trametinib/siremadlin drug combination study—we calculated the precision of the effect (%ORR) estimate between trametinib and trametinib plus siremadlin to be 12.7% at a statistical power corresponding to 80% using a two-sided test with a significance level of α = 0.05. The precision of the effect was calculated using a sample size calculator [72]. Microsoft Excel was used to calculate the %ORR and GraphPad Prism was used to perform Fisher's test.

4.4. Resimulation of Clinical PK and PD Data for Siremadlin

In the literature, PK profiles for siremadlin were available only for 46 individuals [39] instead of the whole cohort involved in clinical trial (n = 115). Additionally for already published PD data [29], the assignment of particular dosing levels and dosing schedules was not possible. For these reasons, patients' siremadlin PK and PD profiles were resimulated in Simulx using the PK/PD model from [29].

The PK and PD model input parameters used for resimulation in Simulx are shown in Table S13. Descriptions of the dose levels, dosing schedules, and the number of patients in each resimulated group are shown in Table 7.

Table 7. Descriptions of resimulated groups treated with siremadlin.

Dose (mg)	Regimen	Dosing Schedule	No. of Patients *	No. of Trials	Notes
1	2A	qdx14 in 28-day cycle	1	10	
2	2A	qdx14 in 28-day cycle	2	10	
4	2A	qdx14 in 28-day cycle	4	10	
7.5	2A	qdx14 in 28-day cycle	4	10	
15	2A	qdx14 in 28-day cycle	4	10	
20	2A	qdx14 in 28-day cycle	5	10	
15	2C	qdx7 in 28-day cycle	8	10	
20	2C	qdx7 in 28-day cycle	6	10	
25	2C	qdx7 in 28-day cycle	5	10	
12.5	1A	qdx1 in 21-day cycle	1	10	
25	1A	qdx1 in 21-day cycle	1	10	
50	1A	qdx1 in 21-day cycle	4	10	
100	1A	qdx1 in 21-day cycle	4	10	
200	1A	qdx1 in 21-day cycle	5	10	
250	1A	qdx1 in 21-day cycle	9	10	Including patients from eltrombopag group ($n = 3$)
350	1A	qdx1 in 21-day cycle	5	10	
120	1B	qwx2 (day 1/8) in 28-day cycle	29	10	
150	1B	qwx2 (day 1/8) in 28-day cycle	15	10	Including patients from eltrombopag group ($n = 7$)
200	1B	qwx2 (day 1/8) in 28-day cycle	3	10	

* Numbers of patients were obtained from siremadlin clinical trial protocol [73].

4.5. Physiologically Based Pharmacokinetic Models

4.5.1. General PBPK Modelling Strategy

The modelling strategy was based on the "middle-out" approach combining the advantages of the "bottom-up" and "top-down" approaches [74]. In our case, some parameters were fixed (such as in vitro-determined or literature-derived data for siremadlin and trametinib [18,19,29,37,39,40,75–80]) and others were estimated. Parameter estimation was performed using the PE Module of the Simcyp Simulator V21 using the Nelder–Mead method, the weighted least squares by the reciprocal of the square of the maximum observed value as the objective function, and a termination criterion defined as improvement of less than 1% of the objective function value. Optimisation was performed manually to fit the observed or resimulated data in the case of siremadlin. PBPK model performance was evaluated based on the "2-fold" criterion for maximum concentration (C_{max}) and area under the concentration–time curve (AUC) [81,82].

4.5.2. Virtual Population Characteristics (System Data)/Patient Population

As both studied drugs are intended for use in patients with cancer, the standard cancer population (Sim-Cancer) which is included within the Simcyp simulator was used. This software allows us to perform simulations on a typical cancer population patient (population representative) or on the whole cancer population. This special population has many adjustments (age distribution, height–age–weight relationships, prediction of glomerular filtration rate, and changes in plasma protein concentrations, e.g., alpha-1-acid glycoprotein, human serum albumin, and serum creatinine), which are made to better account for the specific changes that are expected to be found in the physiological parameters of such a population [83,84]. Some of the physiological parameters of the Simcyp cancer population were modified to specifically mimic the melanoma cancer population. Concerning this, tumour properties including tumour tissue size, blood flow (blood flow was set as in human melanoma xenografts [75,79,80]), composition (skin-derived), and pH were adapted to the studied population.

4.5.3. PBPK Model Verification

The final siremadlin and trametinib PBPK models were compared with external observed or resimulated PK data [30–32,36,39] through both visual checks and numerical analysis. Predicted longitudinal plasma concentration profiles were generated, including the geometric mean predicted concentrations. Local sensitivity analysis (parameter scanning) was performed to evaluate the relative impacts of fa, ka, tlag, Hep intrinsic CL, fu_inc, and additional renal CL on the plasma PK parameters ($AUC_{0-24\,h}/AUC_{0\text{-inf}}$ and C_{max}) for siremadlin, and fa, ka, tlag, and CL_{iv} for trametinib. The performance of the siremadlin and trametinib PBPK models was assessed using the fold error in plasma, which referred to the ratio of the predicted $AUC_{0-24\,h}/AUC_{0\text{-inf}}$ and C_{max} to the observed values (Equation (1)). $AUC_{0-24\,h}/AUC_{0\text{-inf}}$ was calculated using the linear trapezoidal rule. Both visual checks and numerical analyses were performed in Microsoft Excel 2016. Changes in PBPK parameters related to PK DDI at the absorption level are shown in Table S3.

$$\text{Fold Error PK parameter} = \text{Predicted PK parameter}/\text{Observed PK parameter}, \tag{1}$$

4.6. Pharmacodynamic Modelling

4.6.1. General PD (TGI) Modelling Strategy

The initial PD model used further in PBPK/PD modelling for siremadlin and trametinib was established using Monolix software. Model selection was based on visual inspection of the individual observed vs. predicted data and a comparison of the resulting values of the model score (Equation (2)).

$$\begin{aligned}\text{Model score} = -2 \times \text{log-likelihood } (-2LL, \text{called also objective function value—OFV})\\ + \text{ corrected Bayesian Information Criteria (BICc)},\end{aligned} \tag{2}$$

−2LL and BICc were estimated using the linearisation method to accelerate the calculations. For the TGI model further developed in the Simcyp simulator, the goodness of TGI model fit was evaluated based on the mean relative error (RE) value (Equation (3)) being < 20%, as proposed in [85]:

$$\text{RE (\%)} = 100 \times (\text{Predicted Tumour Volume} - \text{Observed Tumour Volume})/\text{Observed Tumour Volume}, \tag{3}$$

4.6.2. PD (TGI) Model Development and Verification

The TGI model for siremadlin and trametinib was developed based on a model previously proposed by Guerreiro et al. [29]. In this model, unperturbed tumour growth is assumed to be exponential, as proposed for melanoma [86] (Equation (4)):

$$dTS/dt = kgh \times SLD0, \tag{4}$$

where kgh is tumour growth (1/h) and SLD0 is the initial sum of the longest diameters (cm).

After the characterisation of tumour growth, the tumour growth inhibition models (perturbed models) were developed for siremadlin, trametinib, and their combination based on the observed tumour size for trametinib and the resimulated tumour size data for siremadlin.

The final selected perturbed TGI model characterising single siremadlin-, single trametinib-, and combination-treated groups is written in the following equations and under the initial conditions:

$$\text{TotalSLD}(t) = Ts(t) + Tsr(t), \tag{5}$$

$$Ts\_0 = SLD0 \times fs, \tag{6}$$

$$Tsr\_0 = SLD0 \times (1 - fs), \tag{7}$$

$$TSCs\_combination = (TSCs\_siremadlin/AUC\_ratio\_siremadlin—TSCs\_trametinib/ AUC\_ratio\_trametinib)/gamma, \tag{8}$$

$$TSCs = TSCs\_siremadlin \mid TSCs\_trametinib \mid TSCs\_combination, \tag{9}$$

$$C(t) = C\_siremadlin \mid C\_trametinib \mid (C\_siremadlin + C\_trametinib), \tag{10}$$

$$TK(t) = kgh/TSCs \times C(t), \tag{11}$$

$$K1\_0 = 0, \tag{12}$$

$$K2\_0 = 0, \tag{13}$$

$$K3\_0 = 0, \tag{14}$$

$$K4\_0 = 0, \tag{15}$$

$$dK1/dt = (dTK - K1)/tau, \tag{16}$$

$$dK2/dt = (K1 - K2)/tau, \tag{17}$$

$$dK3/dt = (K2 - K3)/tau, \tag{18}$$

$$dK4/dt = (K3 - K4)/tau, \tag{19}$$

$$dTs/dt = 0, \tag{20}$$

$$dTsr/dt = 0, \tag{21}$$

$$dTs/dt = kgh \times Ts - K4 \times Ts, \tag{22}$$

$$dTsr/dt = kgh \times Tsr - K4/(1 + lambda) \times Tsr, \tag{23}$$

where pre-existing resistance is defined by the introduction of the sensitive (Ts—tumour size) and resistant population (Tsr—resistant tumour size) of cancer cells. It was assumed that at time 0, the total sum of the longest diameters of the tumour (TotalSLD) is represented by the sum of Ts and Tsr (Equation (5)). The initial Ts and Tsr size at time = 0 are calculated with the fraction of sensitive cells (fs) and the initial sum of the longest diameters (SLD0), as in Equations (6) and (7). The tumour static concentration for the sensitive fraction (TSCs) for the combination (TSCs_combination) was characterised as the subtraction of the siremadlin and trametinib TSCs adjusted with the PK interaction parameters (the AUC ratio parameters which were calculated for siremadlin and trametinib based on findings from the preclinical model (see Table S8)) as well as gamma (the PD interaction parameter) and β parameter determined from the analysis of in vitro data [18] (Equation (8)). Depending on the treated group, the TSCs constant could be assigned to the TSCs constants of siremadlin, trametinib, or their combination (Equation (9)). The total plasma concentration of siremadlin, trametinib, or their combination was used as the input for the drug effect (Equation (10)). The tumour-killing effect (TK) assumes linear dynamics of the treatment effect, which is defined by tumour growth (kgh), tumour static concentration for sensitive cells (TSCs), and total plasma concentration (C) (Equation (11)). A delay of the killing effect (TK) was implemented through the introduction of 4 signal-transit compartments (K1, K2, K3, and K4), as suggested by [87]. The duration of this delay is determined by the parameter tau (Equations (16)–(19)). It was assumed that transit compartments equalled 0 at time = 0 (Equations (12)–(15)). It was also assumed that the initial change in tumour size for sensitive and resistant cell populations was equal to 0 (Equations (20) and (21)). The tumour growth rate (kgh) was assumed to be the same for the sensitive and treatment-resistant cell populations. The Skipper–Schabel–Wilcox log-kill hypothesis was applied to describe tumour size changes over time for sensitive and resistant cell populations (Equations (22) and (23)). Additionally, it was assumed that the studied drugs also induce a killing effect on the resistant cell population but with reduced potency (Equation (23)). The parameter lambda denotes the fold-change loss in drug potency in resistant cells relative to sensitive cells. The units for particular parameters are summarised in Table S6.

In the last step, unperturbed and perturbed tumour growth inhibition models for siremadlin, trametinib, and their combination, previously developed in Monolix, were translated into Lua programming language and applied within the Simcyp simulator V21 for further development to achieve a mean relative error (RE) value of <20% (Equation (3)).

4.6.3. Tumour Size Simulation for the Drug Combination

Tumour size for the studied drugs and their combination was estimated using the PBPK/PD model within a 0–5040 h (210 days) simulation timeframe at clinically relevant doses (120 mg for siremadlin and 2 mg for trametinib) and dosing regimens for each drug (qwx2 for siremadlin and qd for trametinib). The trial design assumed a cancer population in a 20–80-year age range, 50% of patients of either sex, and tumour blood flow typical for melanoma. For simulation purposes, the influence of the initial (baseline) sum of the longest diameters (SLD0) was tested. Each simulated patient response was classified according to RECIST criteria [60]; therefore, all patients whose tumour size decreased by at least 30% were classified as responders. The overall response rate (defined as the percentage of subjects who achieved CR or PR at any time during the study) was calculated for each simulated scenario: drugs administered in monotherapy and drug combinations.

Additionally, according to the RECIST guidelines, the minimum measurable tumour size is 10 mm; therefore, if a patient's tumour size was lower than 1 cm during the simulation timeframe (5040 h), that patient was also counted as a responding patient.

Supplementary Materials: The following supporting information can be downloaded at: https://www.mdpi.com/article/10.3390/ijms24032239/s1.

Author Contributions: Conceptualisation, J.W.; methodology, J.W.; software, J.W. and S.P.; validation, J.W., S.P. and D.P.; formal analysis, J.W.; investigation, J.W.; resources, J.W. and Z.R.; data curation, J.W.; writing—original draft preparation, J.W.; writing—review and editing, S.P., Z.R. and D.P.; visualisation, J.W.; supervision, S.P., Z.R. and D.P.; project administration, J.W.; funding acquisition, Z.R. and D.P. All authors have read and agreed to the published version of the manuscript.

Funding: This work was performed under the Ministry of Science and Higher Education's programme "Applied PhD", at the Faculty of Chemistry of the University of Warsaw and Adamed Pharma S.A., based on contract No. 0058/DW/2018/01/1. Data obtained from Adamed Pharma S.A. were collected under the project "Preclinical development of an innovative anti-cancer drug using the mechanism of reactivation of p53 protein" (POIR/01.02.00-14-31/15), co-financed by European Union funds under Measure 1.2: Sectoral R&D Programmes of the Operational Programme Smart Growth 2014–2020 (INNOMED—programme of scientific research and development works for the economic sector in the field of innovative medicine).

Institutional Review Board Statement: Not applicable.

Informed Consent Statement: Not applicable.

Data Availability Statement: The data presented in this study are available in the article or supplementary material. Raw data from the PK and PD simulations are available on request from the corresponding author.

Acknowledgments: The authors would like to give special thanks to: Tariq Abdulla for his valuable advice and help with development of the Lua script used for the custom PK interaction and PD models in the Simcyp simulations; Géraldine Cellière for her help with the development of the Mlxtran PD model. The authors would like to also thank Certara UK (Simcyp division) and Lixoft SAS for granting us free access to the Simcyp simulators and Monolix suite software through an academic license.

Conflicts of Interest: Jakub Witkowski is an Adamed Pharma employee and co-inventor of the patent WO2019141549 related to the MDM2 inhibitor developed by Adamed Pharma. Dariusz Pawelec is an Adamed Pharma employee. Sebastian Polak is a Certara UK employee. No other conflicts of interest, financial or otherwise, are disclosed by the authors. The funders had no role in the design of the study; in the collection, analyses, or interpretation of the data; in the writing of the manuscript; or in the decision to publish the results.

References

1. Yadav, J.; El Hassani, M.; Sodhi, J.; Lauschke, V.M.; Hartman, J.H.; Russell, L.E. Recent Developments in in Vitro and in Vivo Models for Improved Translation of Preclinical Pharmacokinetics and Pharmacodynamics Data. *Drug Metab. Rev.* **2021**, *53*, 207–233. [CrossRef] [PubMed]
2. Jamei, M. Recent Advances in Development and Application of Physiologically-Based Pharmacokinetic (PBPK) Models: A Transition from Academic Curiosity to Regulatory Acceptance. *Curr. Pharmacol. Rep.* **2016**, *2*, 161–169. [CrossRef] [PubMed]
3. Zhuang, X.; Lu, C. PBPK Modeling and Simulation in Drug Research and Development. *Acta Pharm. Sin. B* **2016**, *6*, 430–440. [CrossRef]
4. Jones, H.M.; Dickins, M.; Youdim, K.; Gosset, J.R.; Attkins, N.J.; Hay, T.L.; Gurrell, I.K.; Logan, Y.R.; Bungay, P.J.; Jones, B.C.; et al. Application of PBPK Modelling in Drug Discovery and Development at Pfizer. *Xenobiotica* **2012**, *42*, 94–106. [CrossRef] [PubMed]
5. Yuan, Y.; He, Q.; Zhang, S.; Li, M.; Tang, Z.; Zhu, X.; Jiao, Z.; Cai, W.; Xiang, X. Application of Physiologically Based Pharmacokinetic Modeling in Preclinical Studies: A Feasible Strategy to Practice the Principles of 3Rs. *Front. Pharmacol.* **2022**, *13*, 895556. [CrossRef] [PubMed]
6. Mattes, W.B. In Vitro to In Vivo Translation. *Curr. Opin. Toxicol.* **2020**, *23–24*, 114–118. [CrossRef]
7. Ireson, C.R.; Alavijeh, M.S.; Palmer, A.M.; Fowler, E.R.; Jones, H.J. The Role of Mouse Tumour Models in the Discovery and Development of Anticancer Drugs. *Br. J. Cancer* **2019**, *121*, 101–108. [CrossRef] [PubMed]
8. Hiemstra, P.S.; Sterk, P.J. Translation of In Vitro Findings to Patients with Asthma: A Timely and Compelling Challenge. *Eur. Respir. J.* **2019**, *54*, 1901759. [CrossRef]
9. Jones, H.; Rowland-Yeo, K. Basic Concepts in Physiologically Based Pharmacokinetic Modeling in Drug Discovery and Development. *CPT Pharmacomet. Syst. Pharmacol.* **2013**, *2*, e63. [CrossRef]
10. Zhou, Z.; Zhu, J.; Jiang, M.; Sang, L.; Hao, K.; He, H. The Combination of Cell Cultured Technology and in Silico Model to Inform the Drug Development. *Pharmaceutics* **2021**, *13*, 704. [CrossRef]
11. Imaoka, T.; Huang, W.; Shum, S.; Hailey, D.W.; Chang, S.-Y.; Chapron, A.; Yeung, C.K.; Himmelfarb, J.; Isoherranen, N.; Kelly, E.J. Bridging the Gap between in Silico and in Vivo by Modeling Opioid Disposition in a Kidney Proximal Tubule Microphysiological System. *Sci. Rep.* **2021**, *11*, 21356. [CrossRef] [PubMed]
12. Susanto, B.O.; Wicha, S.G.; Hu, Y.; Coates, A.R.M.; Simonsson, U.S.H. Translational Model-Informed Approach for Selection of Tuberculosis Drug Combination Regimens in Early Clinical Development. *Clin. Pharmacol. Ther.* **2020**, *108*, 274–286. [CrossRef] [PubMed]
13. Fleisher, B.; Lezeau, J.; Werkman, C.; Jacobs, B.; Ait-Oudhia, S. In Vitro to Clinical Translation of Combinatorial Effects of Doxorubicin and Abemaciclib in Rb-Positive Triple Negative Breast Cancer: A Systems-Based Pharmacokinetic/Pharmacodynamic Modeling Approach. *Breast Cancer (Dove Med Press).* **2021**, *13*, 87–105. [CrossRef] [PubMed]
14. Cardilin, T.; Almquist, J.; Jirstrand, M.; Sostelly, A.; Amendt, C.; El Bawab, S.; Gabrielsson, J. Tumor Static Concentration Curves in Combination Therapy. *AAPS J.* **2017**, *19*, 456–467. [CrossRef]
15. Cardilin, T.; Almquist, J.; Jirstrand, M.; Gabrielsson, J. Evaluation and Translation of Combination Therapies in Oncology—A Quantitative Approach. *Eur. J. Pharmacol.* **2018**, *834*, 327–336. [CrossRef] [PubMed]
16. Chou, T.-C. Drug Combination Studies and Their Synergy Quantification Using the Chou-Talalay Method. *Cancer Res.* **2010**, *70*, 440–446. [CrossRef]
17. Baaz, M.; Cardilin, T.; Lignet, F.; Jirstrand, M. Optimized Scaling of Translational Factors in Oncology: From Xenografts to RECIST. *Cancer Chemother. Pharmacol.* **2022**, *90*, 239–250. [CrossRef]
18. Witkowski, J.; Polak, S.; Rogulski, Z.; Pawelec, D. In Vitro/In Vivo Translation of Synergistic Combination of MDM2 and MEK Inhibitors in Melanoma Using PBPK/PD Modelling: Part I. *Int. J. Mol. Sci.* **2022**, *23*, 12984. [CrossRef]
19. Witkowski, J.; Polak, S.; Rogulski, Z.; Pawelec, D. In Vitro/In Vivo Translation of Synergistic Combination of MDM2 and MEK Inhibitors in Melanoma Using PBPK/PD Modelling: Part II. *Int. J. Mol. Sci.* **2022**, *23*, 11939. [CrossRef]
20. Wu, C.-E.; Koay, T.S.; Esfandiari, A.; Ho, Y.-H.; Lovat, P.; Lunec, J. ATM Dependent DUSP6 Modulation of P53 Involved in Synergistic Targeting of MAPK and P53 Pathways with Trametinib and MDM2 Inhibitors in Cutaneous Melanoma. *Cancers* **2018**, *11*, 3. [CrossRef]
21. Caenepeel, S.; Canon, J.; Hughes, P.; Oliner, J.D.; Rickles, R.J.; Saiki, A.Y. Combination Therapy Including an MDM2 Inhibitor and One or More Additional Pharmaceutically Active Agents for the Treatment of Cancers. U.S. Patent 10,881,648, 5 January 2021.
22. Decaudin, D.; Frisch Dit Leitz, E.; Nemati, F.; Tarin, M.; Naguez, A.; Zerara, M.; Marande, B.; Vivet-Noguer, R.; Halilovic, E.; Fabre, C.; et al. Preclinical Evaluation of Drug Combinations Identifies Co-Inhibition of Bcl-2/XL/W and MDM2 as a Potential Therapy in Uveal Melanoma. *Eur. J. Cancer* **2020**, *126*, 93–103. [CrossRef] [PubMed]

23. Koo, H.-M.; VanBrocklin, M.; McWilliams, M.J.; Leppla, S.H.; Duesbery, N.S.; Woude, G.F.V. Apoptosis and Melanogenesis in Human Melanoma Cells Induced by Anthrax Lethal Factor Inactivation of Mitogen-Activated Protein Kinase Kinase. *Proc. Natl. Acad. Sci. USA* **2002**, *99*, 3052–3057. [CrossRef] [PubMed]

24. Alesiani, D.; Cicconi, R.; Mattei, M.; Bei, R.; Canini, A. Inhibition of Mek 1/2 Kinase Activity and Stimulation of Melanogenesis by 5,7-Dimethoxycoumarin Treatment of Melanoma Cells. *Int. J. Oncol.* **2009**, *34*, 1727–1735. [CrossRef]

25. Saud, A.; Sagineedu, S.R.; Ng, H.-S.; Stanslas, J.; Lim, J.C.W. Melanoma Metastasis: What Role Does Melanin Play? (Review). *Oncol. Rep.* **2022**, *48*, 217. [CrossRef]

26. Sarna, M.; Krzykawska-Serda, M.; Jakubowska, M.; Zadlo, A.; Urbanska, K. Melanin Presence Inhibits Melanoma Cell Spread in Mice in a Unique Mechanical Fashion. *Sci. Rep.* **2019**, *9*, 9280. [CrossRef]

27. Almansour, H.; Afat, S.; Serna-Higuita, L.M.; Amaral, T.; Schraag, A.; Peisen, F.; Brendlin, A.; Seith, F.; Klumpp, B.; Eigentler, T.K.; et al. Early Tumor Size Reduction of at Least 10% at the First Follow-Up Computed Tomography Can Predict Survival in the Setting of Advanced Melanoma and Immunotherapy. *Acad. Radiol.* **2022**, *29*, 514–522. [CrossRef]

28. Jain, R.K.; Lee, J.J.; Ng, C.; Hong, D.; Gong, J.; Naing, A.; Wheler, J.; Kurzrock, R. Change in Tumor Size by RECIST Correlates Linearly with Overall Survival in Phase I Oncology Studies. *J. Clin. Oncol.* **2012**, *30*, 2684–2690. [CrossRef] [PubMed]

29. Guerreiro, N.; Jullion, A.; Ferretti, S.; Fabre, C.; Meille, C. Translational Modeling of Anticancer Efficacy to Predict Clinical Outcomes in a First-in-Human Phase 1 Study of MDM2 Inhibitor HDM201. *AAPS J.* **2021**, *23*, 28. [CrossRef] [PubMed]

30. Ho, M.Y.K.; Morris, M.J.; Pirhalla, J.L.; Bauman, J.W.; Pendry, C.B.; Orford, K.W.; Morrison, R.A.; Cox, D.S. Trametinib, a First-in-Class Oral MEK Inhibitor Mass Balance Study with Limited Enrollment of Two Male Subjects with Advanced Cancers. *Xenobiotica* **2014**, *44*, 352–368. [CrossRef] [PubMed]

31. Ouellet, D.; Kassir, N.; Chiu, J.; Mouksassi, M.-S.; Leonowens, C.; Cox, D.; DeMarini, D.J.; Gardner, O.; Crist, W.; Patel, K. Population Pharmacokinetics and Exposure-Response of Trametinib, a MEK Inhibitor, in Patients with BRAF V600 Mutation-Positive Melanoma. *Cancer Chemother. Pharmacol.* **2016**, *77*, 807–817. [CrossRef]

32. Infante, J.R.; Fecher, L.A.; Falchook, G.S.; Nallapareddy, S.; Gordon, M.S.; Becerra, C.; DeMarini, D.J.; Cox, D.S.; Xu, Y.; Morris, S.R.; et al. Safety, Pharmacokinetic, Pharmacodynamic, and Efficacy Data for the Oral MEK Inhibitor Trametinib: A Phase 1 Dose-Escalation Trial. *Lancet Oncol.* **2012**, *13*, 773–781. [CrossRef] [PubMed]

33. Elassaiss-Schaap, J.; Heisterkamp, S. Variability as Constant Coefficient of Variation: Can We Right Two Decades in Error? Available online: https://www.page-meeting.org/pdf_assets/4964-Elassaiss-Schaap%20-%20Equations%20variability%20reporting%20PK-PD%20-%20Final.pdf (accessed on 15 December 2022).

34. Mistry, H.B.; Orrell, D.; Eftimie, R. Model Based Analysis of the Heterogeneity in the Tumour Size Dynamics Differentiates Vemurafenib, Dabrafenib and Trametinib in Metastatic Melanoma. *Cancer Chemother. Pharmacol.* **2018**, *81*, 325–332. [CrossRef] [PubMed]

35. Flaherty, K.T.; Robert, C.; Hersey, P.; Nathan, P.; Garbe, C.; Milhem, M.; Demidov, L.V.; Hassel, J.C.; Rutkowski, P.; Mohr, P.; et al. Improved Survival with MEK Inhibition in BRAF-Mutated Melanoma. *N. Engl. J. Med.* **2012**, *367*, 107–114. [CrossRef] [PubMed]

36. Stein, E.M.; DeAngelo, D.J.; Chromik, J.; Chatterjee, M.; Bauer, S.; Lin, C.-C.; Suarez, C.; de Vos, F.; Steeghs, N.; Cassier, P.A.; et al. Results from a First-in-Human Phase I Study of Siremadlin (HDM201) in Patients with Advanced Wild-Type TP53 Solid Tumors and Acute Leukemia. *Clin. Cancer Res.* **2022**, *28*, 870–881. [CrossRef]

37. Pharmaceuticals and Medical Devices Agency (PMDA). Mekinist Initial Approval. Available online: https://www.pmda.go.jp/files/000233741.pdf (accessed on 15 December 2022).

38. Paine, S.W.; Ménochet, K.; Denton, R.; McGinnity, D.F.; Riley, R.J. Prediction of Human Renal Clearance from Preclinical Species for a Diverse Set of Drugs That Exhibit Both Active Secretion and Net Reabsorption. *Drug Metab. Dispos.* **2011**, *39*, 1008–1013. [CrossRef] [PubMed]

39. Jeay, S.; Ferretti, S.; Holzer, P.; Fuchs, J.; Chapeau, E.A.; Wartmann, M.; Sterker, D.; Romanet, V.; Murakami, M.; Kerr, G.; et al. Dose and Schedule Determine Distinct Molecular Mechanisms Underlying the Efficacy of the P53-MDM2 Inhibitor HDM201. *Cancer Res.* **2018**, *78*, 6257–6267. [CrossRef] [PubMed]

40. Leonowens, C.; Pendry, C.; Bauman, J.; Young, G.C.; Ho, M.; Henriquez, F.; Fang, L.; Morrison, R.A.; Orford, K.; Ouellet, D. Concomitant Oral and Intravenous Pharmacokinetics of Trametinib, a MEK Inhibitor, in Subjects with Solid Tumours. *Br. J. Clin. Pharmacol.* **2014**, *78*, 524–532. [CrossRef]

41. Tang, C.; Prueksaritanont, T. Use of In Vivo Animal Models to Assess Pharmacokinetic Drug-Drug Interactions. *Pharm. Res.* **2010**, *27*, 1772–1787. [CrossRef]

42. Jaiswal, S.; Sharma, A.; Shukla, M.; Vaghasiya, K.; Rangaraj, N.; Lal, J. Novel Pre-Clinical Methodologies for Pharmacokinetic Drug–Drug Interaction Studies: Spotlight on "Humanized" Animal Models. *Drug Metab. Rev.* **2014**, *46*, 475–493. [CrossRef]

43. Sjöberg, Å.; Lutz, M.; Tannergren, C.; Wingolf, C.; Borde, A.; Ungell, A.-L. Comprehensive Study on Regional Human Intestinal Permeability and Prediction of Fraction Absorbed of Drugs Using the Ussing Chamber Technique. *Eur. J. Pharm. Sci.* **2013**, *48*, 166–180. [CrossRef]

44. Li, M.; de Graaf, I.A.M.; Groothuis, G.M.M. Precision-Cut Intestinal Slices: Alternative Model for Drug Transport, Metabolism, and Toxicology Research. *Expert Opin. Drug Metab. Toxicol.* **2016**, *12*, 175–190. [CrossRef] [PubMed]

45. Hashimoto, S.; Honda, K.; Fujita, K.; Miyachi, Y.; Isoda, K.; Misaka, K.; Suga, Y.; Kato, S.; Tsuchiya, H.; Kato, Y.; et al. Effect of Coadministration of Rifampicin on the Pharmacokinetics of Linezolid: Clinical and Animal Studies. *J. Pharm. Health Care Sci.* **2018**, *4*, 27. [CrossRef] [PubMed]

46. Howard, G.R.; Johnson, K.E.; Rodriguez Ayala, A.; Yankeelov, T.E.; Brock, A. A Multi-State Model of Chemoresistance to Characterize Phenotypic Dynamics in Breast Cancer. *Sci. Rep.* **2018**, *8*, 12058. [CrossRef] [PubMed]

47. Yoon, N.; Vander Velde, R.; Marusyk, A.; Scott, J.G. Optimal Therapy Scheduling Based on a Pair of Collaterally Sensitive Drugs. *Bull. Math. Biol.* **2018**, *80*, 1776–1809. [CrossRef]

48. Liu, W.; Dowling, J.P.; Murray, W.K.; McArthur, G.A.; Thompson, J.F.; Wolfe, R.; Kelly, J.W. Rate of Growth in Melanomas: Characteristics and Associations of Rapidly Growing Melanomas. *Arch. Dermatol.* **2006**, *142*, 1551–1558. [CrossRef]

49. Chatterjee, M.; Elassaiss-Schaap, J.; Lindauer, A.; Turner, D.; Sostelly, A.; Freshwater, T.; Mayawala, K.; Ahamadi, M.; Stone, J.; de Greef, R.; et al. Population Pharmacokinetic/Pharmacodynamic Modeling of Tumor Size Dynamics in Pembrolizumab-Treated Advanced Melanoma. *CPT Pharmacomet. Syst. Pharmacol.* **2017**, *6*, 29–39. [CrossRef]

50. Settleman, J.; Neto, J.M.F.; Bernards, R. Thinking Differently about Cancer Treatment Regimens. *Cancer Discov.* **2021**, *11*, 1016–1023. [CrossRef]

51. Mould, D.; Walz, A.-C.; Lave, T.; Gibbs, J.; Frame, B. Developing Exposure/Response Models for Anticancer Drug Treatment: Special Considerations. *CPT Pharmacomet. Syst. Pharmacol.* **2015**, *4*, e00016. [CrossRef]

52. Wang, Q.; Shen, X.; Chen, G.; Du, J. Drug Resistance in Colorectal Cancer: From Mechanism to Clinic. *Cancers* **2022**, *14*, 2928. [CrossRef]

53. Wang, X.; Zhang, H.; Chen, X. Drug Resistance and Combating Drug Resistance in Cancer. *Cancer Drug Resist.* **2019**, *2*, 141–160. [CrossRef]

54. Hauschild, A.; Larkin, J.; Ribas, A.; Dréno, B.; Flaherty, K.T.; Ascierto, P.A.; Lewis, K.D.; McKenna, E.; Zhu, Q.; Mun, Y.; et al. Modeled Prognostic Subgroups for Survival and Treatment Outcomes in BRAF V600–Mutated Metastatic Melanoma. *JAMA Oncol.* **2018**, *4*, 1382–1388. [CrossRef]

55. Saiag, P.; Robert, C.; Grob, J.-J.; Mortier, L.; Dereure, O.; Lebbe, C.; Mansard, S.; Grange, F.; Neidhardt, E.-M.; Lesimple, T.; et al. Efficacy, Safety and Factors Associated with Disease Progression in Patients with Unresectable (Stage III) or Distant Metastatic (Stage IV) BRAF V600-Mutant Melanoma: An Open Label, Non-Randomized, Phase IIIb Study of Trametinib in Combination with Dabrafenib. *Eur. J. Cancer* **2021**, *154*, 57–65. [CrossRef]

56. Clinical Trials Results (Study 113220/NCT01072175): Novartis Clinical Trial Results Database. Available online: https://www.novctrd.com/ctrdweb/trialresult/trialresults/pdf?trialResultId=17378 (accessed on 15 December 2022).

57. Clinical Trials Results (Study 115306/NCT01584648) Novartis Clinical Trial Results Database. Available online: https://www.novctrd.com/ctrdweb/trialresult/trialresults/pdf?trialResultId=17627 (accessed on 15 December 2022).

58. Clinical Trials Results (Study BRF117277/NCT02039947) Novartis Clinical Trial Results Database. Available online: https://www.novctrd.com/ctrdweb/trialresult/trialresults/pdf?trialResultId=17326 (accessed on 15 December 2022).

59. Long, G.V.; Eroglu, Z.; Infante, J.; Patel, S.; Daud, A.; Johnson, D.B.; Gonzalez, R.; Kefford, R.; Hamid, O.; Schuchter, L.; et al. Long-Term Outcomes in Patients With BRAF V600-Mutant Metastatic Melanoma Who Received Dabrafenib Combined With Trametinib. *J. Clin. Oncol.* **2018**, *36*, 667–673. [CrossRef]

60. Eisenhauer, E.A.; Therasse, P.; Bogaerts, J.; Schwartz, L.H.; Sargent, D.; Ford, R.; Dancey, J.; Arbuck, S.; Gwyther, S.; Mooney, M.; et al. New Response Evaluation Criteria in Solid Tumours: Revised RECIST Guideline (Version 1.1). *Eur. J. Cancer* **2009**, *45*, 228–247. [CrossRef]

61. Pan, B.; Wei, X.; Xu, X. Patient-Derived Xenograft Models in Hepatopancreatobiliary Cancer. *Cancer Cell Int.* **2022**, *22*, 41. [CrossRef] [PubMed]

62. Blomme, A.; Van Simaeys, G.; Doumont, G.; Costanza, B.; Bellier, J.; Otaka, Y.; Sherer, F.; Lovinfosse, P.; Boutry, S.; Palacios, A.P.; et al. Murine Stroma Adopts a Human-like Metabolic Phenotype in the PDX Model of Colorectal Cancer and Liver Metastases. *Oncogene* **2018**, *37*, 1237–1250. [CrossRef]

63. Hidalgo, M.; Amant, F.; Biankin, A.V.; Budinská, E.; Byrne, A.T.; Caldas, C.; Clarke, R.B.; de Jong, S.; Jonkers, J.; Mælandsmo, G.M.; et al. Patient Derived Xenograft Models: An Emerging Platform for Translational Cancer Research. *Cancer Discov.* **2014**, *4*, 998–1013. [CrossRef]

64. Hirata, E.; Sahai, E. Tumor Microenvironment and Differential Responses to Therapy. *Cold Spring Harb. Perspect. Med.* **2017**, *7*, a026781. [CrossRef]

65. Ni, Y.; Zhou, X.; Yang, J.; Shi, H.; Li, H.; Zhao, X.; Ma, X. The Role of Tumor-Stroma Interactions in Drug Resistance Within Tumor Microenvironment. *Front. Cell Dev. Biol.* **2021**, *9*, 637675. [CrossRef]

66. McMillin, D.W.; Negri, J.M.; Mitsiades, C.S. The Role of Tumour–Stromal Interactions in Modifying Drug Response: Challenges and Opportunities. *Nat. Rev. Drug Discov.* **2013**, *12*, 217–228. [CrossRef]

67. Wang, H.Q.; Mulford, I.J.; Sharp, F.; Liang, J.; Kurtulus, S.; Trabucco, G.; Quinn, D.S.; Longmire, T.A.; Patel, N.; Patil, R.; et al. Inhibition of MDM2 Promotes Antitumor Responses in P53 Wild-Type Cancer Cells through Their Interaction with the Immune and Stromal Microenvironment. *Cancer Res.* **2021**, *81*, 3079–3091. [CrossRef]

68. Fang, D.D.; Tang, Q.; Kong, Y.; Wang, Q.; Gu, J.; Fang, X.; Zou, P.; Rong, T.; Wang, J.; Yang, D.; et al. MDM2 Inhibitor APG-115 Synergizes with PD-1 Blockade through Enhancing Antitumor Immunity in the Tumor Microenvironment. *J. Immunother. Cancer* **2019**, *7*, 327. [CrossRef]

69. Zhou, J.; Kryczek, I.; Li, S.; Li, X.; Aguilar, A.; Wei, S.; Grove, S.; Vatan, L.; Yu, J.; Yan, Y.; et al. The Ubiquitin Ligase MDM2 Sustains STAT5 Stability to Control T Cell-Mediated Antitumor Immunity. *Nat. Immunol.* **2021**, *22*, 460–470. [CrossRef]

70. Clinical Trials Results (Study 114267/NCT01245062) GlaxoSmithKline Clinical Trial Results Database. Available online: https://www.gsk-studyregister.com/en/trial-details/?id=114267 (accessed on 15 December 2022).

71. Clinical Trials Results (Study CTMT212AUS55/NCT05611229) Novartis Clinical Trial Results Database. Available online: https://www.novctrd.com/ctrdweb/trialresult/trialresults/pdf?trialResultId=17991 (accessed on 15 December 2022).

72. Ristl, R. Sample Size Calculator. Available online: https://homepage.univie.ac.at/robin.ristl/samplesize.php?test=fishertest (accessed on 24 November 2022).

73. Clinical Trial Results (Study CHDM201X2101/NCT02143635) Novartis Clinical Trial Results Database. Available online: https://www.novctrd.com/ctrdweb/trialresult/trialresults/pdf?trialResultId=17828 (accessed on 15 December 2022).

74. Tylutki, Z.; Polak, S.; Wiśniowska, B. Top-down, Bottom-up and Middle-out Strategies for Drug Cardiac Safety Assessment via Modeling and Simulations. *Curr. Pharmacol. Rep.* **2016**, *2*, 171–177. [CrossRef]

75. Kallinowski, F.; Schlenger, K.H.; Runkel, S.; Kloes, M.; Stohrer, M.; Okunieff, P.; Vaupel, P. Blood Flow, Metabolism, Cellular Microenvironment, and Growth Rate of Human Tumor Xenografts. *Cancer Res.* **1989**, *49*, 3759–3764. [PubMed]

76. Hofmann, F. Small Molecule HDM201 Inhibitor HDM201. Presented at the AACR Annual Meeting 2016, New Orleans, LA, USA, 16–20 April 2016.

77. Lei, Y.; Zhenglin, Y.; Heng, L. MDM2 Inhibitors. U.S. Patent 11,339,171, 24 May 2022.

78. Food and Drug Administration (FDA). Trametinib Pharmacology Review. Available online: https://www.accessdata.fda.gov/drugsatfda_docs/nda/2013/204114Orig1s000PharmR.pdf (accessed on 15 December 2022).

79. Benjaminsen, I.C.; Graff, B.A.; Brurberg, K.G.; Rofstad, E.K. Assessment of Tumor Blood Perfusion by High-Resolution Dynamic Contrast-Enhanced MRI: A Preclinical Study of Human Melanoma Xenografts. *Magn. Reson. Med.* **2004**, *52*, 269–276. [CrossRef]

80. Graff, B.A.; Benjaminsen, I.C.; Melås, E.A.; Brurberg, K.G.; Rofstad, E.K. Changes in Intratumor Heterogeneity in Blood Perfusion in Intradermal Human Melanoma Xenografts during Tumor Growth Assessed by DCE-MRI. *Magn. Reson. Imaging* **2005**, *23*, 961–966. [CrossRef]

81. Shebley, M.; Sandhu, P.; Emami Riedmaier, A.; Jamei, M.; Narayanan, R.; Patel, A.; Peters, S.A.; Reddy, V.P.; Zheng, M.; de Zwart, L.; et al. Physiologically Based Pharmacokinetic Model Qualification and Reporting Procedures for Regulatory Submissions: A Consortium Perspective. *Clin. Pharmacol. Ther.* **2018**, *104*, 88–110. [CrossRef]

82. Tsakalozou, E.; Alam, K.; Babiskin, A.; Zhao, L. Physiologically-Based Pharmacokinetic Modeling to Support Determination of Bioequivalence for Dermatological Drug Products: Scientific and Regulatory Considerations. *Clin. Pharmacol. Ther.* **2022**, *111*, 1036–1049. [CrossRef]

83. Khoshaein, N.; Ezuruike, U.; Hatley, O.; Gill, K.; Gardner, I. Performance Verification and Application of a Cancer Population for Use in Physiologically Based Pharmacokinetic Modelling. Available online: https://www.certara.com/app/uploads/2019/09/Khoshaein_2019_JRC_cancer.pdf (accessed on 2 December 2022).

84. Hatley, O. Predicting Drug Exposure in Cancer Patients Using a PBPK Oncology Population. Available online: https://www.certara.com/blog/predicting-drug-exposure-in-cancer-patients-using-a-pbpk-oncology-population/ (accessed on 2 December 2022).

85. Reig-López, J.; Maldonado, M.D.M.; Merino-Sanjuan, M.; Cruz-Collazo, A.M.; Ruiz-Calderón, J.F.; Mangas-Sanjuán, V.; Dharmawardhane, S.; Duconge, J. Physiologically-Based Pharmacokinetic/Pharmacodynamic Model of MBQ-167 to Predict Tumor Growth Inhibition in Mice. *Pharmaceutics* **2020**, *12*, 975. [CrossRef]

86. Tejera-Vaquerizo, A.; Cañueto, J.; Toll, A.; Santos-Juanes, J.; Jaka, A.; Ferrandiz, C.; Sanmartín, O.; Ribero, S.; Moreno-Ramírez, D.; Almazán, F.; et al. Estimated Effect of COVID-19 Lockdown on Skin Tumor Size and Survival: An Exponential Growth Model. *Actas Dermo-Sifiliográficas (Engl. Ed.)* **2020**, *111*, 629–638. [CrossRef]

87. Lobo, E.D.; Balthasar, J.P. Pharmacodynamic Modeling of Chemotherapeutic Effects: Application of a Transit Compartment Model to Characterize Methotrexate Effects in Vitro. *AAPS J.* **2002**, *4*, 212–222. [CrossRef] [PubMed]

International Journal of
Molecular Sciences

MDPI

Article

Novel Strategies against Cancer: Dexibuprofen-Loaded Nanostructured Lipid Carriers

Vaikunthavasan Thiruchenthooran [1], Marta Świtalska [2], Lorena Bonilla [3,4], Marta Espina [3,4], Maria Luisa García [3,4], Joanna Wietrzyk [2], Elena Sánchez-López [3,4,5,*] and Anna Gliszczyńska [1,*]

[1] Department of Food Chemistry and Biocatalysis, Wrocław University of Environmental and Life Sciences, Norwida 25, 50-375 Wrocław, Poland
[2] Department of Experimental Onclogy, Ludwik Hirszfeld Institute of Immunology and Experimental Therapy, Polish Academy of Sciences, Weigla 12, 53-114 Wrocław, Poland
[3] Department of Pharmacy, Pharmaceutical Technology and Physical Chemistry, University of Barcelona, 08028 Barcelona, Spain
[4] Institute of Nanoscience and Nanotechnology (IN2UB), University of Barcelona, 08028 Barcelona, Spain
[5] Unit of Synthesis and Biomedical Applications of Peptides, IQAC-CSIC, 08034 Barcelona, Spain
* Correspondence: esanchezlopez@ub.edu (E.S.-L.); anna.gliszczynska@upwr.edu.pl (A.G.)

Abstract: The aim of this work was to design innovative nanostructured lipid carriers (NLCs) for the delivery of dexibuprofen (DXI) as an antiproliferative therapy against tumoral processes, and overcome its side effects. DXI-NLC samples were prepared with beeswax, Miglyol 812 and Tween 80 using high-pressure homogenization. A two-level factorial design 2^4 was applied to optimize the formulation, and physicochemical properties such as particle size, zeta potential, polydispersity index and entrapment efficiency were measured. Optimized parameters of DXI-NLCs exhibited a mean particle size of 152.3 nm, a polydispersity index below 0.2, and high DXI entrapment efficiency (higher than 99%). Moreover, DXI-NLCs provided a prolonged drug release, slower than the free DXI. DXI-NLCs were stable for 2 months and their morphology revealed that they possess a spherical shape. In vitro cytotoxicity and anticancer potential studies were performed towards prostate (PC-3) and breast (MDA-MB-468) cancer cell lines. The highest activity of DXI-NLCs was observed towards breast cancer cells, which were effectively inhibited at 3.4 μM. Therefore, DXI-NLCs constitute a promising antiproliferative therapy that has proven to be especially effective against breast cancer.

Keywords: dexibuprofen; anticancer activity; nanostructured lipid carriers (NLCs); factorial design; particle size; zeta potential; cytotoxicity; drug delivery

Citation: Thiruchenthooran, V.; Świtalska, M.; Bonilla, L.; Espina, M.; García, M.L.; Wietrzyk, J.; Sánchez-López, E.; Gliszczyńska, A. Novel Strategies against Cancer: Dexibuprofen-Loaded Nanostructured Lipid Carriers. *Int. J. Mol. Sci.* **2022**, *23*, 11310. https://doi.org/10.3390/ ijms231911310

Academic Editor: Antonella Piozzi

Received: 30 August 2022
Accepted: 21 September 2022
Published: 25 September 2022

Publisher's Note: MDPI stays neutral with regard to jurisdictional claims in published maps and institutional affiliations.

1. Introduction

Cancerous diseases are the second leading cause of death worldwide, representing a serious health concern. According to the World Health Organization (WHO) report, by 2040 the number of cancer patients will increase by 47% compared to 2020, and will reach more than 28 million cases and 16 million deaths [1]. Hence, many research centers are looking for effective compounds in the fight against cancerous diseases. In 2020 and 2021, the Food and Drug Administration (FDA) approved 53 and 50 new drugs based on 40 and 36 new chemical entities (NCEs), respectively [2,3]. In 2020, 18 new anticancer drugs were released on the market, whereas in 2021 only 15 were found [4].

Developing and bringing a new drug to the market constitutes a tedious and expensive procedure that takes, on average, twelve to sixteen years and requires USD 2.558 billion [5]. The pharmacologist and Nobel laureate James Black used to say that "the most fruitful basis for the discovery of new drug is to start with an old drug". Applying his principle, it is possible to bypass almost 40% of the overall cost of bringing a drug to market [6]. This is because drugs approved by regulatory agencies with known pharmacokinetics and safety profiles can be rapidly evaluated in phase II clinical trials where their novel application is confirmed.

Turning to drugs that have already been recognized as safe by the FDA is being increasingly recognized as an attractive strategy to obtain effective treatments. Consequently, many efforts have been made for repurposing drugs that already exist by finding new targets, delivery methods and formulations.

In the area of tumoral processes, great hopes are placed on NSAIDs as effective antiproliferative agents, as these have exhibited the ability to reduce the incidence and mortality of many types of cancer in numerous epidemiological and animal studies [7]. It is suggested that anti-inflammatory drugs prevent malignant cell formation and tumor progression. They may also promote the antitumoral effects of chemotherapy and radiotherapy [8]. The first mention of the use of this group of drugs as anti-tumoral compounds appeared during the 1980s [9]. Harris and co-workers reported that their regular use for 5 to 9 years caused a 21% reduction in the incidence of breast cancer [10]. However, despite promising in vitro results and clinical trials assessing their anticancer activity, the practical application of these drugs is still impossible because of individual adverse effects, such as increased gastrointestinal ulceration arising after their long-term administration [11]. Therefore, in recent decades, many attempts have been made to find the effective solution for their delivery to overcome these side effects.

One of the NSAID drugs with proved anticancer potential is ibuprofen (IBU), the mechanism of action of which has been extensively studied. Greenspan et al. suggested that IBU can effectively target both the Wnt/β-catenin and NF-κB pathways, indicating a novel mechanism of chemopreventive efficacy of NSAIDs [12]. Another research group published that this drug regulates expression and function in epithelial cells [13].

Dexibuprofen (DXI) is the dextrorotatory isomer of ibuprofen (IBU), being almost 160 times more active as an anti-inflammatory molecule and less toxic than its enantiomer *R* [14]. DXI is better tolerated in comparison with other NSAIDs, but still exhibits some side effects characteristic of this group of drugs. Therefore, encapsulation and methods of fabrication of the lipid nanostructures of IBU have been studied [15–17].

The aim of this work was to develop a controlled-release DXI delivery system based on lipid nanocarriers. Among them, nanostructured lipid carriers (NLCs) were selected due to their excellent pharmaceutical and drug delivery properties, in addition to their biocompatibility and biodegradability [18]. NLCs are able to encapsulate lipophilic drugs and are formed by mixing solid and liquid lipids with aqueous surfactant dispersion [19]. NLCs provide some advantages over traditional solid lipid nanoparticles (SLNs) and other colloidal lipid carriers. On one hand, both NLCs and SLNs are known to provide a prolonged drug release due to the presence of a solid lipid matrix [20]. However, SLNs possess a restricted stability due to the fact that they tend to expel the drug from the lipid matrix. Thus, NLCs overcome SLN problems due to the incorporation of a liquid lipid that offers higher drug loading capacity, as well as increased stability. Using NLCs, it would be possible to obtain a formulation providing prolonged-release DXI, and to produce the formulation at an industrial scale [21]. Therefore, during this work, the preparation, physicochemical characterization and therapeutic efficacy of DXI-loaded NLCs will be assessed.

2. Results and Discussion

2.1. Preformulation Studies

The selection of solid lipid used in the preparation of NLCs was made based on the literature data indicating the antibacterial properties of beeswax against Gram-positive bacteria and its anti-inflammatory activity, which could contribute to enhancing the antiproliferative properties of DXI [22,23].

Knowledge about interactions between the solid lipid and the drug is crucial in the process of designing nanocarriers for further drug entrapment [24]. First, we examined the interactions between beeswax and DXI by preparing two different mixtures of beeswax:DXI in ratios 70:20 and 70:5 (%, *w/w*). Samples of DXI and beeswax, as well as their mixtures, were heated at 90 °C, cooled down to room temperature, and finally visualized under a light microscope. Under the microscope, a noncomplete mixture of DXI with solid lipid at a ratio of 70:20 was observed. The observable dark opaque areas denote the area of recrystallized DXI. In contrast, the 70:5 ratio had a reduced exposure of dark opaque DXI recrystallisation under light microscopy. Based on this observation, the ratio 70:5 of beeswax: DXI was selected and examined by differential scanning calorimetry (DSC). DSC constitutes an effective method to investigate the melting and crystallization behavior of lipid molecules and drugs. Thermograms were obtained by heating the samples (beeswax, DXI and beeswax:DXI) from 25 °C to 100 °C (Figure 1B). The data obtained indicate that the melting point of DXI was 53.6 °C, the melting point of beeswax was 65.94 °C, and the melting point of their mixture was 62.4 °C.

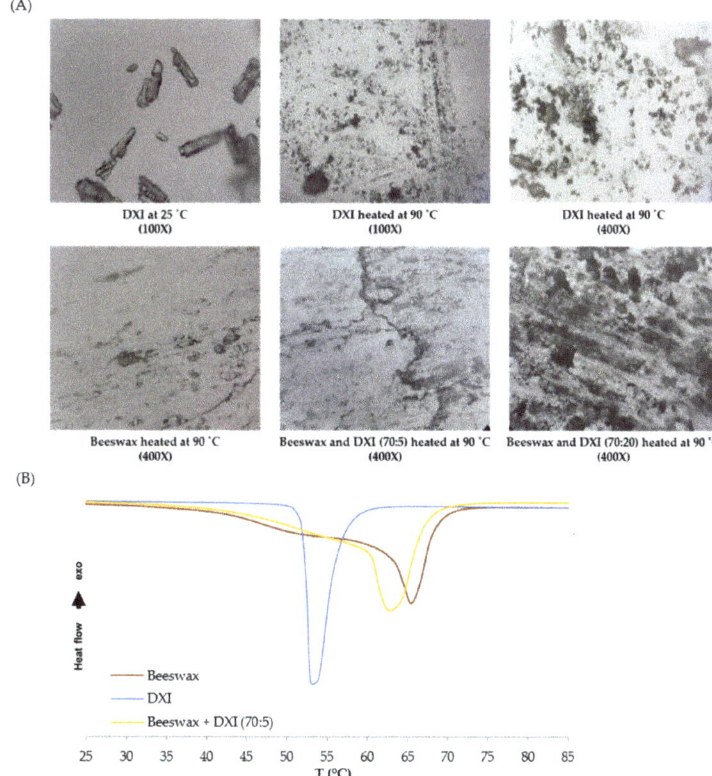

Figure 1. Optical microscopy images (**A**) and differential scanning calorimetry (**B**) of physical mixture of beeswax and DXI.

This increase in the melting point of the mixture against the DXI melting point, broadening the endothermic curves, corresponds to the solubilization of the lipid compound with the drug. Using DSC studies, we confirmed microscopic results in order to ensure that DXI could be completely dissolved into the solid lipid [25]. Furthermore, we also observed a slight reduction in the heating enthalpy of beeswax (ΔH 134.14 J/g) during its mixture with DXI (ΔH 119.93 J/g). Applying heterogenic solid lipids with a higher melting point has the capacity to disorder the recrystallization of drugs, and during this study it is evidenced that beeswax prevents DXI from undergoing recrystallization, and could help improve the entrapment of DXI in the beeswax matrix during DXI-NLC formulation [24].

Afterwards, liquid lipids were selected by investigating castor oil and Miglyol 812. Castor oil is a natural oil with high viscosity. Its main component is a ricinoleic acid: glyceride [26]. Miglyol 812 is a mixture of triglycerides of saturated and unbranched C8 to C12 fatty acids, and is characterized by high chemical stability and low viscosity [27]. Both products are water-insoluble and non-ionic lipids with different HLB values (14 and 15.36, respectively). Miglyol 812 has been described to possess antiproliferative activity and it has been widely used as liquid lipid in NLCs with a broad range of antitumoral applications [28,29]. Castor oil is considered an important plant product in herbal medicine. It is known to penetrate well through human skin and eradicate tumors near the breast skin surface [30].

Tween 80, an oleate ester of sorbitol, was used as a surfactant. Tween is a water-soluble and non-ionic synthetic product with an HLB of 15.

With the selected excipients, preformulations A and B were performed using the concentrations given in Table 1 and the high-pressure homogenizer (HPH). The HPH, in which tension is generated by high pressure, is the main method to produce lipid nanoparticles due to its feasibility for scale-up, short production time, uniform particle size reduction, and the avoidance of organic solvents [31]. Despite this, the number of cycles and the pressure applied should be optimized, since these are the critical parameters in HPH.

Table 1. Concentrations corresponding to the preformulation studies of NLCs encapsulating DXI.

Compounds	Concentrations of Selected Excipients	
	Preformulation A	Preformulation B
DXI (mg/mL)	1	1
Beeswax (%)	4	4
Castor oil (%)	2	0
Miglyol 812 (%)	0	2
Tween 80 (%)	3	3

Physicochemical parameters of both formulations were evaluated. Based on the results summarized in Table 2, Miglyol 812 was selected for further studies instead of castor oil because it showed better properties in terms of polydispersity index (PDI, obtaining values below 0.2) and mean particle size (Z_{ave} around 100 nm), while in terms of entrapment efficiency (EE) and zeta potential (ZP), the results were similar in both cases, obtaining values higher than 95% and lower than −14 mV.

Table 2. Physicochemical characterization of the synthesis of nonoptimized DXI-NLCs. (A) NLCs with beeswax and castor oil; (B) NLCs with beeswax and Miglyol 812.

Preformulation	Z_{ave} (nm)	PDI	ZP (mV)	EE (%)
A	129.8 ± 1.3	0.232 ± 0.025	−18.9 ± 0.5	96.94
	123.1 ± 0.6	0.223 ± 0.021	−17.7 ± 0.3	96.89
B	103.8 ± 1.1	0.171 ± 0.019	−15.9 ± 0.2	97.02
	92.35 ± 2.32	0.188 ± 0.012	−14.8 ± 0.3	95.69

Results are presented as mean ± SD (n = 3); Z_{ave}: mean average size; PDI: polydispersity index; ZP: zeta potential; EE: encapsulation efficiency.

2.2. Optimization of Dexibuprofen–Nanostructured Lipid Carriers

In order to obtain the final desired product with optimal physiochemical properties and maximal DXI encapsulation, the design of experiment (DoE) approach was used. The effects of the formulation variables (independent variables) on the response parameters (dependent variables) were evaluated. According to the matrix generated,

26 experiments (16 factorial points, 8 axial points, and 2 replicated center points) were performed (Table 3) [32].

Table 3. Values of the experimental factors according to the matrix designed by 2^4 + star central composite rotatable factorial design parameters and measured responses.

	DXI		Beeswax		Miglyol 812		Tween 80		Parameters			
	Coded Level	mg/mL	Coded Level	%	Coded Level	%	Coded Level	%	Z_{ave} (nm)	PDI	ZP (mV)	EE (%)
Factorial Points												
F1	1	1.25	1	5	1	2.5	1	4	89.1 ± 1.09	0.17 ± 0.01	−17.6 ± 2.91	98.96
F2	−1	0.75	1	5	1	2.5	1	4	103.9 ± 1.96	0.18 ± 0.03	−16.5 ± 0.61	97.72
F3	1	1.25	−1	3	1	2.5	1	4	72.52 ± 0.84	0.18 ± 0.01	−14.9 ± 2.01	98.87
F4	−1	0.75	−1	3	1	2.5	1	4	77.43 ± 1.46	0.2 ± 0.01	−15.6 ± 1.57	96.01
F5	1	1.25	1	5	−1	1.5	1	4	95.51 ± 0.06	0.19 ± 0.01	−16.7 ± 1.65	97.14
F6	−1	0.75	1	5	−1	1.5	1	4	98.25 ± 1.93	0.19 ± 0.05	−18.7 ± 0.82	96.52
F7	1	1.25	−1	3	−1	1.5	1	4	71.73 ± 0.59	0.13 ± 0.01	−13.1 ± 2.46	97.91
F8	−1	0.75	−1	3	−1	1.5	1	4	70.25 ± 0.72	0.22 ± 0.01	−16.7 ± 0.7	95.7
F9	−1	0.75	1	5	1	2.5	−1	2	154.3 ± 2.44	0.15 ± 0.04	−19.5 ± 0.51	98.96
F10	−1	0.75	−1	3	1	2.5	−1	2	126.9 ± 1.3	0.17 ± 0.01	−20 ± 2.01	99.11
F11	−1	0.75	1	5	−1	1.5	−1	2	145.8 ± 0.58	0.12 ± 0.03	−20.6 ± 0.4	99.48
F12	−1	0.75	−1	3	−1	1.5	−1	2	112.6 ± 1.23	0.18 ± 0.01	−17.8 ± 1.47	96.28
F13	1	1.25	1	5	1	2.5	−1	2	160.4 ± 1.43	0.12 ± 0.02	−21.8 ± 2.66	96.09
F14	1	1.25	−1	3	1	2.5	−1	2	129.7 ± 2.35	0.15 ± 0.01	−18.1 ± 0.54	98.12
F15	1	1.25	1	5	−1	1.5	−1	2	146.9 ± 0.98	0.09 ± 0.04	−20.6 ± 0.76	99.88
F16	1	1.25	−1	3	−1	1.5	−1	2	112.7 ± 0.78	0.17 ± 0.03	−18.1 ± 0.7	97.46
Axial Points												
F17	−2	0.5	0	4	0	2	0	3	108.70 ± 1.82	0.17 ± 0.01	−21.7 ± 1.48	99.29
F18	2	1.5	0	4	0	2	0	3	106.7 ± 0.89	0.17 ± 0.02	−15.9 ± 0.53	97.87
F19	0	1	0	4	0	2	2	5	78.94 ± 1.56	0.18 ± 0.02	−14.1 ± 0.51	96.3
F20	0	1	0	4	2	3	0	3	116.7 ± 1.01	0.18 ± 0.01	−17.7 ± 0.38	98.52
F21	0	1	0	4	−2	1	0	3	100 ± 3.07	0.2 ± 0.03	−19.4 ± 0.06	97.65
F22	0	1	2	6	0	2	0	3	112.1 ± 2.12	0.17 ± 0.02	−20.3 ± 0.38	98.89
F23	0	1	−2	2	0	2	0	3	71.86 ± 0.41	0.16 ± 0.08	−16.8 ± 0.22	98.37
F24	0	1	0	4	0	2	−2	1	182.8 ± 0.27	0.16 ± 0.02	−20.8 ± 0.27	98.54
Central Points												
F25	0	1	0	4	0	2	0	3	110.4 ± 2.03	0.16 ± 0.03	−15.6 ± 0.31	97.2
F26	0	1	0	4	0	2	0	3	91.75 ± 1.77	0.18 ± 0.01	−17 ± 0.72	98.3

Results are presented as mean ± SD (n = 3); F1-F26: formulations of DXI-NLCs; Z_{ave}: mean average size; PDI: polydispersity index; ZP: zeta potential; EE: encapsulation efficiency.

The results obtained were analyzed and, as can be observed in Figure 2, on the Pareto's chart of Z_{ave}, it is clearly visible that the statistically significant variables corresponded to Tween 80®, beeswax and Miglyol (Figure 2).

Regarding the mean particle size, site-specific NLCs aiming to deliver chemotherapeutic agents should have a diameter range of 50–300 nm for increased cellular uptake [33]. In our studies, we observed that the average size of the DXI-NLC strongly depended on the concentrations of Tween 80® and beeswax, and was reduced with the increase in the surfactant (Figure 2A,F), whereas the opposite tendency was reported for beeswax. At concentrations of Tween 80® lower than 2%, the average size of DXI-NLC decreases. On the other hand, it was observed that the concentrations of DXI do not influence this parameter in a significant manner (Figure 2A,E).

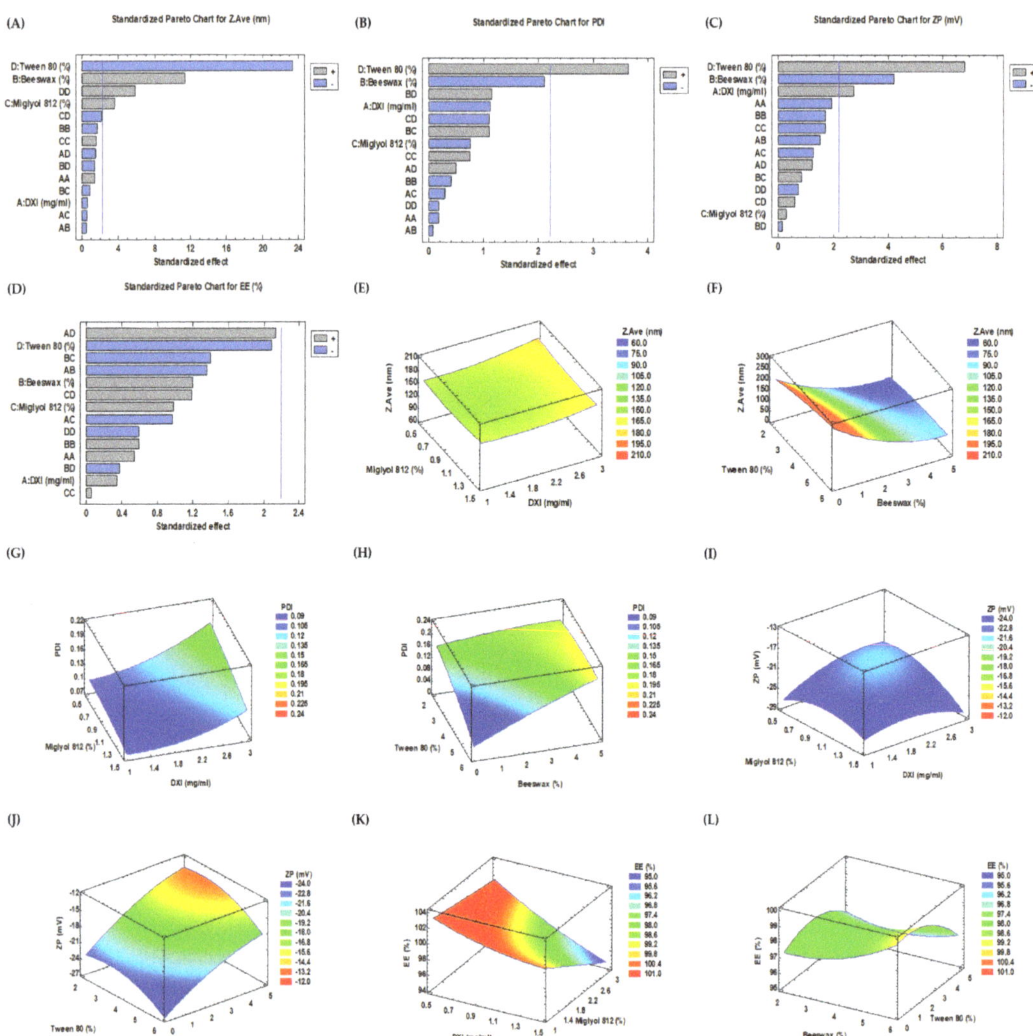

Figure 2. (**A–D**) Pareto chart shows positive and negative influence of dexibuprofen (DXI), Miglyol 812, Tween 80 and beeswax on the particle size (Z_{ave}), polydispersity index (PDI), zeta potential (ZP) and encapsulation efficiency (EE). (**E–L**) Surface response plots denoting changes in Z_{ave}, PDI, ZP and EE during different concentrations of DXI, Miglyol 812, Tween 80 and beeswax. (**A**) Pareto chart for Z_{Ave}; (**B**) Pareto chart for PDI; (**C**) Pareto chart for ZP; (**D**) Pareto chart for EE. (**E,F**) Surface response plot for Z_{Ave}; (**G,H**) surface response plot for PDI; (**I,J**) surface response plot for ZP; (**K,L**) surface response plot for EE. In the Pareto chart, a bar crossing the blue line is considered a significant effect.

The PDI has an Important effect on the physical stability and uniformity of NLC. The values of this parameter should be as low as possible to ensure the long-term stability of the formulation. PDI values in the range between 0.1 and 0.25 show a narrow size distribution, while PDI values greater than 0.5 indicate a very broad distribution [34]. In the conditions studied, the concentrations of Tween 80[®] and beeswax have a strong impact on the PDI (Figure 2B,H). Since the homogeneity of the formulation is a key issue, low PI is sought. Therefore, according to our data, the concentration of surfactant should be high while the solid lipid should be maintained as low as possible. In the case of DXI, a nonsignificant impact of its concentration was observed.

All obtained formulations had a negative surface charge ranging from −13.1 to −21.8 mV (Table 3), predicting a suitable short-term stability regardless of DXI concentration, and thus suggesting that DXI did not significantly alter the ZP of the formulation. This is in accordance with previous studies which point out that when beeswax is used the formulations possess anionic character [35]. These negatively charged NLCs have higher half-life during circulation and are less likely show toxicity because of their relatively low interaction with the mononuclear phagocytic system [36]. In addition, in the blood stream, the majority of plasma proteins are negatively charged, and cationic NLCs would bind to these proteins at a higher rate [37,38]. Taking all this into account and aiming for suitable short-term stability, highly negative ZP values should be obtained. Therefore, high beeswax amounts would be suitable since values around −20 mV can be obtained (Figure 2C,J).

The concentrations of DXI, beeswax and surfactants had no significant impact on the encapsulation efficiency (EE) of DXI (Figure 2D). As it can be seen in Table 1, a large amount of DXI (EE > 99%) was incorporated in some DXI-NLC formulations, suggesting its preferential partition into the lipid matrix of the NLCs. From the surface response chart (Figure 2L), we can find that at a lower concentration of Tween 80[®] and a higher concentration of beeswax, the EE parameter is higher (Figure 2L).

The DoE approach allowed us to develop an optimized DXI-NLC. Based on the analyzed data, 10 mL samples were prepared using 600 mg of beeswax, 200 mg of Miglyol, 200 mg of Tween 80[®], and 15 mg of DXI for their subsequent characterization (Table 4). Samples were prepared in triplicate to verify their reproducibility and stored under the same conditions. As shown in Table 4, predicted and experimental values were highly similar, except for the EE, which was higher than expected.

Table 4. Predicted and expected physiochemical parameters and EE of DXI-NLC.

Parameters	Predicted Values	Experimental Values
Z_{ave} (nm)	~150	152.3 ± 1.6
PDI	<0.18	0.149 ± 0.03
ZP (mV)	−20	−19.8 ± 0.764
EE (%)	>95	99.17

Experimental values are presented as mean ± SD (n = 3); Z-Ave: mean size; PDI: polydispersity index; ZP: zeta potential; EE (%): encapsulation efficacy.

2.3. DXI-NLC Characterization Studies

In order to study the thermal profile of DXI-NLC, DSC of dried DXI-NLC, as well as the individual compounds, was carried out. Figure 3 shows that DXI exhibits a melting point of 53.6 °C, while the melting point of the empty NLC is 63.03 °C. The DXI-NLC melting point is similar to the empty NLC (63.06 °C). These data confirm that DXI is in a disordered crystalline state without exhibiting its melting peak in DXI-NLC [39].

An X-ray diffractogram (XRD) was performed to characterize the DXI-NLC crystalline in its amorphous state (Figure 4). The XRD profiles of DXI present sharp peaks indicating crystalline structure, while the empty NLC shows two high diffraction peaks (2θ: 21.4 and 23.8) corresponding to orthorhombic subcell diffraction of waxes [40]. Interestingly, in DXI-NLC, the intensity of these two peaks slightly dropped.

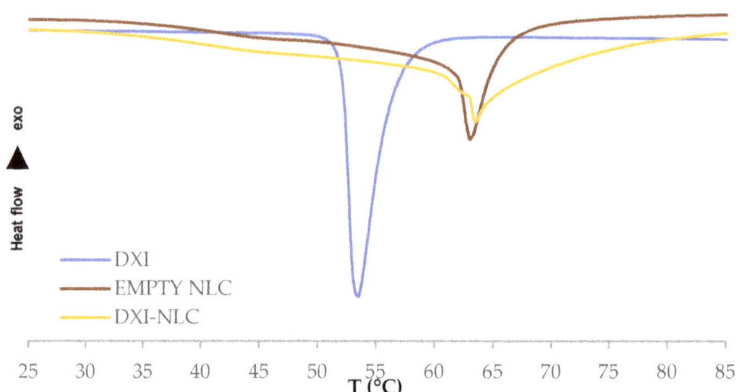

Figure 3. Differential scanning calorimetry (DSC) of DXI-NLC, DXI and empty NLC.

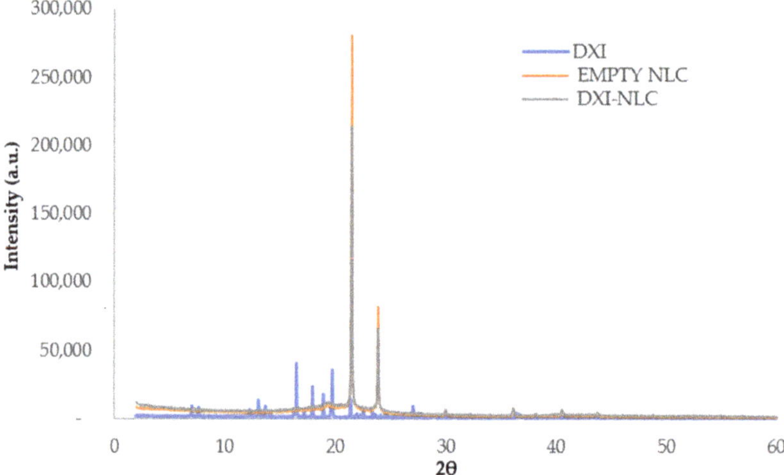

Figure 4. X-ray diffractogram pattern of DXI-NLC, DXI and empty NLC.

Beeswax has a heterogeneous chemical composition that makes it less crystalline, and the reduction in sharpness of these peaks in DXI-NLC shows the changes in the packing β-modification of beeswax [40]. The lack of peaks, corresponding to DXI not being visible in the drug-loaded particles, indicates that DXI is present in the dissolved state (molecular dispersion). Moreover, no peaks of DXI can be observed in DXI-NLC, thus indicating that DXI is encapsulated in the NLC matrix.

Fourier-transform infrared spectroscopy (FTIR) is a useful tool to confirm the analysis of bonds between molecules. Thus, FTIR was performed to identify the interaction between DXI and the lipid matrix. In Figure 5, it can be observed that there are no new covalent bonds formed between DXI and NLC. The main peak characteristic for DXI from the C=O bond-stretching vibration is presented at 1701 cm^{-1}. Other small peaks from C-C bonds (1508 cm^{-1}, 1466 cm^{-1} and 1417 cm^{-1}), C-O stretching (1282 cm^{-1}) and O-H (778 cm^{-1}) were also observed [41].

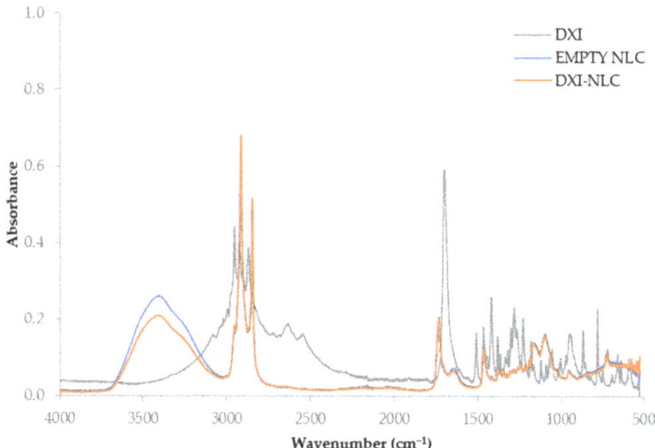

Figure 5. Fourier-transform infrared spectroscopy (FTIR) of DXI-NLC, DXI and empty NLC.

In the cases of the empty NLC and DXI-NLC, similar vibrations attributed to bonds appeared. There were several vibrations observed at the beeswax fingerprint region (719–1098 cm^{-1}). The bond vibrations corresponded to the CH$_2$ rocking (719 cm^{-1}), C-H bending (1098 cm^{-1}), and C=O stretching (1737 cm^{-1}) of hydrocarbons, esters and free fatty acids in beeswax [42]. The major vibrations between 2848 and 2915 cm^{-1} were attributed to the symmetric and asymmetric CH$_2$ stretching vibrations of hydrocarbon compounds present in beeswax, respectively.

The notable change during DXI loading is shown in the reduction in the water molecular O-H bond-stretching vibration at 3411 cm^{-1}. This is an indication of the dispersant being present in the interfacial or pocketed regions of beeswax, and being incompletely removed during DXI loading because of its hydrophobicity. There were no reportable additional bond vibrations observed between the empty NLC and DXI-NLC.

The morphology of DXI-NLC was observed using transmission electron microscopy (TEM). The images presented in Figure 6 confirm that there was no aggregation between the nanoparticles. DXI-NLCs were spherical and uniform in shape, with smooth surfaces. This is relevant because the shape of the colloidal dispersion system is known to affect the delivery and response in medicinal applications, and spherically shaped formulations are known to have a better immune response [33]. Moreover, for the effective delivery of active molecules in the body, the size of nanoparticles is also largely responsible. NLC populations with Z$_{ave}$ above 200 nm are known to compromise the human complement system, whereas Z$_{ave}$ values below 10 nm are known to be excreted by the human kidney [33]. Therefore, the Z$_{ave}$ measured using DLS was similar to that obtained with TEM, thus corroborating the results obtained where NLC measured less than 200 nm.

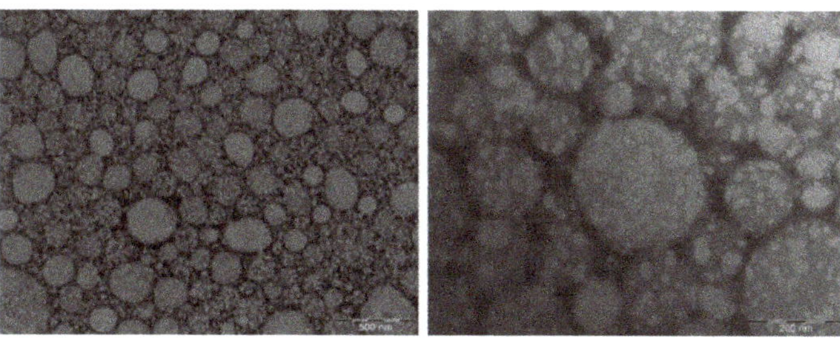

Figure 6. Transmission electron microscopic images of DXI-loaded nanostructured lipid carriers (DXI-NLC) with 500 nm and 200 nm scale.

2.4. In Vitro Drug Release

The release profiles of free DXI and DXI encapsulated in NLCs are shown in Figure 7. The release medium had been prepared at pH7 to contain sodium citrate and urea at concentrations of 1.74 M and 0.18 M, respectively, to ensure sink conditions throughout the experiment. During this study, the faster release kinetics of free DXI in comparison to DXI-NLC can be observed. After 10 h, the drug was almost fully released, whereas after this time only 73% of the initial amount of drug was released from the NLCs. Moreover, significant differences were obtained by comparing the release data one hour after release ($p < 0.05$). Before one hour, no significant differences between free DXI and DXI NLC were found. This could be due to the fact that, during the first hour, the DXI that was located on the NLCs' surface was released in a similar manner to free DXI due to a burst effect [43,44]. After the first hour, the DXI that was encapsulated in the inner core started being released at a slower rate than the free DXI, causing significant differences. These two phases confirm that DXI-NLC possesses a kinetic profile characteristic of prolonged drug-release formulations [45].

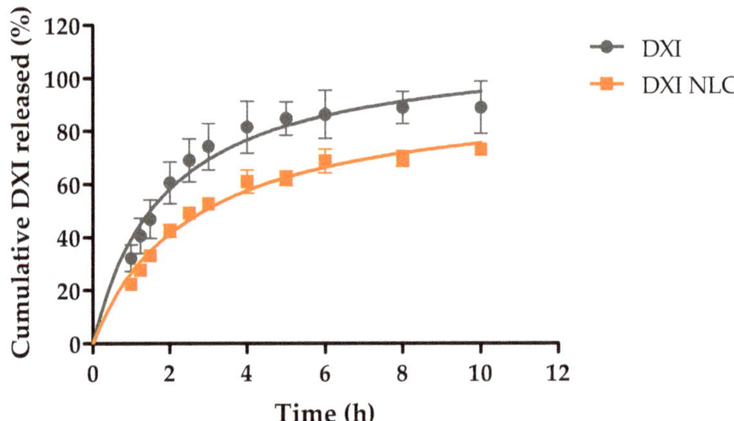

Figure 7. Released profile of free DXI and DXI-loaded nanostructured lipid carriers (DXI-NLC) after 10 h of incubation, adjusting the data to a hyperbola equation.

Drug release is affected by many other factors, such as morphometric and morphological paraments, the type of surfactant, and preparation methodology [46]. In the literature, it is reported that a faster release occurs due to drug diffusion, and could not be caused by matrix degradation, as is the case in DXI-NLC [47]. However, this burst release is one of the limitations among SLN and NLC, as they tend to show an initial fast release [48]. To further understand how DXI is released from formulations, the results were fitted to the most common biopharmaceutical equations and the best fit corresponded to the hyperbola equation [49]. Comparatively, a higher equilibrium dissociation constant (K_d), as is the case of DXI-NLC, reports a slower drug release from the NLC solid lipid matrix, thus confirming the slower DXI release from the particle matrix (Table 5) [39].

Table 5. Results of hyperbola equation for cumulative DXI release vs. time.

Parameters	DXI	DXI-NLC
$B_{max} \pm$ SD (%)	118.30 ± 5.75	91.09 ± 3.14
$K_d \pm$ SD (h)	2.24 ± 0.28	2.90 ± 0.22
R^2 value	0.9264	0.9763

B_{max}: maximum binding capacity; K_d: equilibrium dissociation constant.

2.5. Short-Term Storage Stability

Although NLCs are known to be more stable than SLNs, still one of the major limitations of NLCs is their restricted storage stability [20]. In order to address this issue, the optimized DXI-NLCs were stored at three different temperatures (4, 25 and 37 °C) for up to 60 days, and the resulting modifications in the physiochemical parameters were measured (Table 6). In addition, NLCs are limited to lipophilic drugs such as DXI, and are less suitable for hydrophilic drugs [20].

Table 6. Results of stability studies of DXI-loaded nanostructured lipid carriers (NLCs).

Day	T (°C)	Z_{ave} (nm)	PDI	ZP (mV)
0	4	152.3 ± 1.6	0.149 ± 0.03	−19.8 ± 0.76
	25	152.3 ± 1.6	0.149 ± 0.03	−19.8 ± 0.76
	37	152.3 ± 1.6	0.149 ± 0.03	−19.8 ± 0.76
30	4	152.6 ± 2.1	0.145 ± 0.01	−19.2 ± 0.53
	25	150.7 ± 2.4	0.143 ± 0.02	−19.6 ± 0.4
	37	152.4 ± 2.9	0.143 ± 0.02	−18 ± 0.17
60	4	149.8 ± 0.8	0.158 ± 0.01	−22.5 ± 0.15 *
	25	150.9 ± 1.9	0.140 ± 0.04	−21.6 ± 0.17 *
	37	150.4 ± 1	0.158 ± 0.01	−17.4 ± 0.4 *

Results are presented as mean ± SD (n = 3); T (°C): temperature; Zave: mean size; PDI: polydispersity index; ZP: zeta potential; * significant difference.

Although the HPH technique provides a uniform rigid NLC population, during storage, Z_{ave}, PDI and ZP could be altered by multiple factors [43]. Analyzing the components of fabricated formulation, it can be pointed out that Tween 80® is known to provide suitable storage stability against other surfactants [50,51]. However, waxes and glycerides are sometimes avoided for NLC preparation due to low encapsulation and drug expulsion [40]. In our case, beeswax at 30 days releases a low percentage of DXI, and this is lower at 4 than at 25 °C (2.35% and 3.25%, respectively). From the obtained results, initial and 60-day storage showed nonsignificant physicochemical modifications also denoting that DXI-NLC could be better stored at 4 °C than 25 °C in the long term. As explained by Freitas and Müller, to retain size, formulations should be stored at a low temperature between 2 and 8 °C [52]. However, the 4 °C temperature should be studied for long-term periods to confirm these data. Furthermore, storage at 37 °C causes a decrease in ZP values, probably due to alterations in the formulation, as has been previously reported in other studies [53].

2.6. Effect of Gamma Radiation

Before assessing the antiproliferative activity of DXI-NLC, sterilization using gamma radiation was performed. Gamma radiation is performed at the recommended dose of 25KGy, which is known to kill microbial contaminants [54]. This sterilization method is preferable due to NLCs' heat-sensitive properties, and methods using high temperatures such as ethylene oxide or autoclaving have been reported to cause significant adverse impacts on the physicochemical properties of NLCs [50].

Data on the physicochemical parameters after gamma sterilization are shown in Figure 8. As can be observed, NLC Z_{ave} was not significantly affected by sterilization (Figure 8A). However, the ZP of both the empty NLC and DXI-NLC was affected significantly (n = 3, t statistics: 9.986, p value: 0.001), as was PDI in the case of DXI-NLC (n = 3, t statistics: 8.302, p value: 0.001) (Figure 8B,C). Despite this, ZP was still negative and PDI was below 0.2. DXI EE remained higher than 99%. Small changes in chemical structural parameters were found in previous NLC preparations, but did not show any impact on the medicinal properties. After sterilization, the empty NLC had a Z_{ave}, PDI and ZP of

155.3 ± 3.21 nm, 0.16 ± 0.01, and -12.3 ± 0.64 mV, respectively, whereas the DXI-NLC properties were 150.3 ± 2.98 nm, 0.13, and -22.1 ± 0.75 mV, respectively.

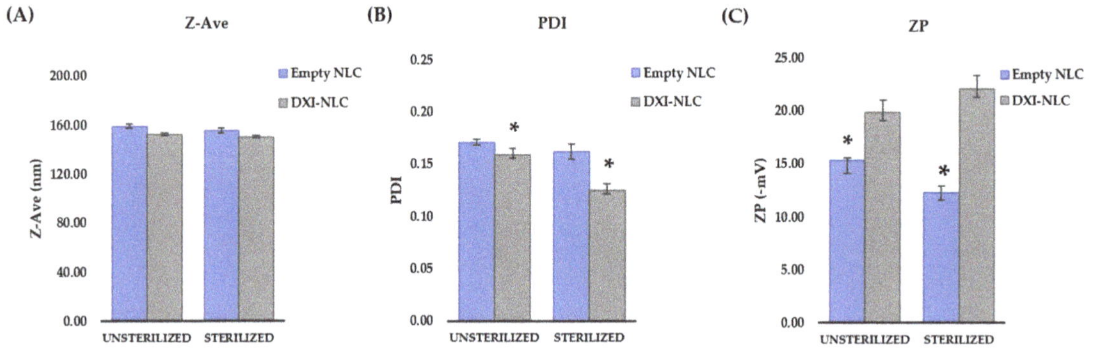

Figure 8. Effect of gamma sterilization on DXI-loaded nanostructured lipid carriers (NLC). (**A**): bar chart for sterilization effect on size (Z-Ave) of empty NLC and DXI-NLC; (**B**): bar chart for sterilization effect on polydispersity index (PDI) of empty NLC and DXI-NLC; (**C**): bar chart for sterilization effect on zeta potential (ZP) of empty NLC and DXI-NLC; * significant difference.

2.7. Cytotoxicity of Fabricated Formulation towards Selected Cancer Cell Lines

The antiproliferative activity of DXI and DXI-NLC towards prostate (PC-3) and breast (MDA-MB-468) cancer cell lines was evaluated after 24, 48 and 72 h. The results were obtained with an cellular viability assessment (SRB colorimetric assay). The data on in vitro anticancer activity are reported in Table 7, and are expressed as the IC_{50} concentration of the compound (in µM) that inhibits the proliferation of the cells by 50%, compared to the untreated control cells.

Table 7. IC_{50} values of DXI-NLC and empty NLC at several incubation times.

| Formulation | IC_{50} (µM) | | | | | |
| | PC-3 | | | MDA-MB-468 | | |
	24h	48h	72h	24h	48h	72h
Empty NLC	28.5 ± 4.3	14.9 ± 5.6	12.9 ± 1.3	42.5 ± 11.8	69.6 ± 27.5	11.6 ± 6.7
DXI-NLC	10.1 ± 3.1	72.9 ± 21.3	60.8 ± 10.3	3.4 ± 0.4	82.4 ± 65.2	66.1 ± 22.4

Results are presented as mean $\pm$ SD ($n = 3$); IC_{50}: 50% inhibitory concentration; PC-3: human prostate cancer cell line; MDA-MB-468: human breast cancer cell line 3.

After 24 h of incubation, the activity of DXI-NLC was almost three- and twelve-fold higher than the activity of the empty NLC (DXI was not assessed due to its low water solubility) toward PC-3 and MDA-MB-468, respectively. During this time, the activity of the encapsulated DXI was getting lower; as listed in Table 7, DXI-NLC exhibited much lower activity after 72 h than after 24 h. This is in correlation with the results presented on the release profile, since after 24 h it is predicted that around 90% of the encapsulated DXI will be released. This resulted in a higher antiproliferative activity of DXI-NLC after 24 h than after 48 or 72 h. Interestingly, empty NLCs also pose antitumoral properties by themselves, probably due to the beeswax effect. In this sense, several studies claim that honey-derived products have the ability to kill cancer cell lines [55,56]. Therefore, the combined properties of NLC and DXI have been demonstrated in vitro.

3. Materials and Methods

3.1. Chemicals and Reagents

S-(+)-ibuprofen (dexibuprofen/DXI) was purchased from Amadis Chemical® (Hangzhou, Zhejiang, China). Beeswax, Tween® 80 (polysorbate 80) and ortho-phosphoric acid (85%) were purchased from Sigma-Aldrich (Madrid, Spain). Miglyol 812 (caprylic/capric triglycerides) was purchased from Roig Farma SA (Terrassa, Spain). Sodium citrate tri-base and urea were purchased from Montplet & Esteban SA (Barcelona, Spain). Methanol (99.8%) was a product of Fisher Scientific (Geel, Belgium). All other reagents including ethanol (99.8%, PanReac AppliChem, Barcelona, Spain) were of analytical grade. Millipore® Milli® Q system (Darmstadt, Germany) purified water was used across all experiments and formulations.

3.2. Methods

3.2.1. Preparation of Nanostructured Lipid Carriers

DXI-NLC was produced using the high-shear hot homogenization technique [57]. In brief, the lipid phase was prepared by melting DXI, beeswax and Miglyol 812. The aqueous phase was prepared by dissolving Tween 80. After a 30 min water bath (~90 °C), both phases were mixed and the pre-emulsion was homogenized with an IKA® T10 basic ULTRA-TURRAX® (Staufen, Germany) at a speed of 8000 rpm for 30 s. Subsequently, the mixture was exposed to high-pressure homogenization (HPH) at a pressure of 800 mbar and 85 °C for 3 homogenization cycles using a Stansted pressure cell homogenizer FPG12800 (Stansted Fluid Power, Harlow, UK). The prepared DXI-NLC samples were allowed to settle down for 12 h at room temperature before further experiments. Empty NLCs were prepared using the same procedure without adding DXI.

3.2.2. Physicochemical Characterization of Nanostructured Lipid Carriers

Z_{ave} and PDI were analyzed using dynamic light scattering (DLS) [39]. NLCs were diluted in water (1:10) and Z_{ave} and PDI values were measured using Zetasizer Nano ZS (Malvern, UK). ZP values of DXI-NLC were measured with an electrophoretic light scattering (ELS) technique using Zetasizer Nano ZS (Malvern, UK) equipped with an electrophoresis laser Doppler [39]. NLCs were diluted in water (1:20) and loaded in a folded capillary cell (Malvern, UK) for ZP measurements. All measurements were carried out in triplicate at 25 °C and values are presented as mean ± standard deviation [44,45].

The EE of DXI in DXI-NLC was determined indirectly by measuring the free DXI in the aqueous phase (Equation (1)). Prepared NLCs were diluted in water (1:10) and filtered by centrifugation through Amicon® 0.5 ml Ultracel 100 K filters (Merck KGaA, Darmstadt, Germany) for 15 min at 14,000 rpm to separate the free DXI from the NLCs. Free DXI in the aqueous phase was quantified by reverse-phase high-performance liquid chromatography (RP-HPLC) [39]. In the HPLC system, samples were analyzed using a Waters™ 600 Controller (Waters, Milford, MA, USA), Waters™ 717 plus autosampler (Waters, Milford, MA, USA), and a Kromasil® Classic C18 (5 μm × 4.6 × 150 mm) column (Nouryon, Amsterdam, The Netherlands). The mobile phase constituted an aqueous phase containing 0.05% ortho-phosphoric acid in water and an organic phase containing methanol in a water: methanol ratio of 20:80, and at a 1 mL/min flow rate. A Waters™ 2996 Photodiode array detector (Waters, Milford, MA, USA) was used at 220 nm wavelength [58,59]. Data were processed using Empower® version 3 (Waters, Milford, MA, USA). The calibration curve was calculated using the DXI concentration range between 0.25 and 500 μg/mL against the area of the corresponding detected peaks.

$$EE\ (\%) = \frac{\text{total amount of DXI} - \text{free DXI}}{\text{total amount of DXI}} \times 100 \tag{1}$$

3.2.3. Optimization of Nanostructured Lipid Carrier Parameters

The optimization of the formulation was carried out by studying the influence of the independent variables (DXI, beeswax, Miglyol 812, polysorbate 80) on the dependent variables (Z_{ave}, PDI, ZP and EE) [32]. A two-level factorial design (2^4) was carried out using Statgraphics 18. The matrix of the design used and the levels of the independent parameters are given in Table 8.

Table 8. Chemical concentrations employed for optimization.

Compounds	Levels				
	−2	−1	0	1	2
DXI (mg/mL)	0.5	0.75	1	1.25	1.5
Beeswax (%)	2	3	4	5	6
Miglyol (%)	1	1.5	2	2.5	3
Tween 80 (%)	1	2	3	4	5

3.2.4. Interaction Studies and Nanosphere Characterization

Differential Scanning Calorimetry

Thermograms of DXI-NLC, DXI and empty NLCs were obtained using a DSC system (Mettler TA 4000 system, Mettler, Greifensee, Switzerland) equipped with a DSC-25 cell. Dried samples were transferred to perforated aluminum pans and weighed on Mettler M3 Microbalance (Mettler, Greifensee, Switzerland). A pan with indium (purity $\geq$99.95%, Fluka, Switzerland) was used for the calibration of the calorimetric system. An empty identical aluminum pan was used as a reference. Samples were melted under nitrogen flow at 10 °C/min to measure the melting transition point by heating the ramp from 25 to 100 °C [39].

Fourier-Transform Infrared Spectroscopy

The Fourier-transform infrared (FTIR) spectra of DXI-NLC, DXI, and the empty NLC were analyzed. The scan was performed between the range of 525 and 4000 cm^{-1} on a Thermo Scientific Nicolet iZ10 spectrometer equipped with an ATR diamond and a DTGS detector [39].

X-ray Diffraction

X-ray diffraction (XRD) was carried on DXI-NLC, DXI, and the empty NLC. XRD was performed to analyze the amphous or crystalline state of the formulations with and without DXI. Dried formulations and DXI were placed in between the polyester films and exposed to CuK″ radiation (45 kV, 40 mA, λ = 1.5418 Å). Measurements were carried at 2θ from 2° to 60° with a step size and time interval of 0.026° and 200 s, respectively [39].

Transmission Electron Microscopy

Morphological observations of DXI-NLC were carried out under transmission electron microscopy (TEM) on a Jeol 1010 microscope (Akishima, Japan). DXI-NLC were diluted (1:5) with water and placed on a surface of copper grids activated with UV light. Subsequently, negative staining was performed using 2% uranyl acetate [58].

3.2.5. In Vitro Drug Release

In vitro drug release was examined with the direct dialysis technique [60]. The release of DXI from DXI-NLC (n = 3) was estimated against the same concentration of DXI (n = 3) by placing the samples in a 500 µL dialysis cassette (10K MWCO, Slide-A-LyzerTM, Thermo Scientific, Rockford, IL, USA). The release medium was prepared with sodium citrate and urea to improve the hydrophilicity of DXI in a proportion of DXI:S sodium citrate: urea 1:3:7, as explained elsewhere [41]. The release medium was placed under magnetic stirring and incubated at 37 °C. At various time intervals, samples of 400 μL of release medium were collected and replaced by 400 μL of clean release medium. Collected samples were analyzed with RP-HPLC [39]. The amount of DXI released was plotted as cumulative drug release against time (h) in Graph Pad Prism version 5.0, and fitted to the most common biopharmaceutical equations (Higuchi, Korsmeyer–Peppas, hyperbola and first-order equation).

3.2.6. Short-Term Stability of Solid Lipid Nanoparticles

Short-term stability was assessed by preparing DXI-NLC samples and storing them at three different temperatures (4, 25 and 37 °C). The physicochemical parameters (Z_{ave}, PDI, ZP and EE) were analyzed monthly for 60 days.

3.2.7. Sterilization by Gamma Radiation

Gama radiation was carried out in order to sterilize the samples to develop further cellular assays. DXI-NLC samples were sterilized using a dose of 25 KGy of 60 Co as the gamma irradiation source (Aragogamma, Barcelona, Spain) [44]. Physicochemical parameters were measured before and after gamma radiation.

3.2.8. In Vitro Cytotoxicity Assay
Cell Lines

Human prostate cancer PC-3 cell line was obtained from the European Collection of Authenticated Cell Cultures (UK); human breast cancer MDA-MB-468 cell line cells were obtained from the Leibniz Institute DSMZ—German Collection of Microorganisms and Cell Cultures (Germany). All the cell lines were maintained at the Hirszfeld Institute of Immunology and Experimental Therapy, PAS, Wroclaw, Poland. PC-3 and MDA-MB-468 cells were cultured in RPMI 1640 +GlutaMAX medium (Gibco, UK) with 10% (PC-3) or 20% (MDA-MB-468) fetal bovine serum (FBS) (HyClone, Cytiva). All culture media were supplemented with 100 units/mL penicillin, (Polfa Tarchomin S.A., Poland) and 100 μg/mL streptomycin (Merck, Germany). All cell lines were grown at 37 °C with a 5% CO_2 humidified atmosphere.

Determination of Antiproliferative Activity

DXI-NLC and the empty NLCs were assayed between 0.8 and 100 μM. Formulations were diluted based on the total encapsulated DXI concentration (1.49 mg/ml) in DXI-NLC. Then, 24 h before adding NLCs, the cells were placed in 96-well plates (Sarstedt, Germany) at a density of 1×10^4 cells per well. The antiproliferative activity assay was performed after 24, 48 and 72 h of exposure to NLCs. The in vitro cytotoxic effect was examined using the SRB assay, as previously described [48]. The results were calculated as an IC_{50}: the concentration of NLCs which was cytotoxic for 50% of the cancer cells. IC_{50} values were calculated for each experiment separately using the Prolab-3 system based on Cheburator 0.4 software [49], and mean values $\pm$ SD are presented in Table 4. Each compound in each concentration was tested in triplicate in a single experiment which was repeated 3–5 times.

3.2.9. Statistics

Student's t-tests were carried for two-group comparisons to explain the significant differences between treatments using Statgraphics 18. A statistically significant difference was considered at $p \leq 0.05$.

4. Conclusions

In the present study, a novel drug delivery system based on nanostructured lipid carriers encapsulating DXI has been developed and optimized for the treatment of tumoral diseases. DXI-NLC were prepared using the high-shear homogenization procedure using HPH, which offers an easy industrial scale-up process. Moreover, DXI-NLC was confirmed to be stable for 2 months at 4 and 25 °C and to provide a prolonged drug release, slower than free DXI. Furthermore, DXI-NLC showed suitable physicochemical properties that allow it to act as an effective antitumoral agent, as demonstrated in breast and prostate cancer cell lines. Therefore, DXI-NLC constitutes a promising formulation to be used as antitumoral agent, and is especially effective against breast cancer cells.

Author Contributions: Conceptualization, E.S.-L. and A.G.; methodology, V.T., M.Ś., M.E., L.B. and E.S.-L.; investigation, V.T., L.B., M.E. and M.Ś.; validation, V.T., M.Ś., J.W., M.L.G., E.S.-L. and A.G.; formal analysis, A.G., E.S.-L., M.L.G. and J.W.; visualization, V.T., A.G., E.S.-L.; supervision, A.G. and E.S.-L.; writing—original draft, V.T.; writing—review and editing, A.G. and E.S.-L. All authors have read and agreed to the published version of the manuscript.

Funding: This work was supported by Wrocław University of Environmental and Life Sciences (Poland) under the Leading Research Groups support project from the subsidy increased for the period 2020–2025 in the amount of 2% of the subsidy referred to Art. 387 (3) of the Law of 20 July 2018 on Higher Education and Science, obtained in 2019. E.S.-L. want to acknowledge the support of the Grants for the Requalification of the Spanish University System. The authors M.E., M.L.G. and E.S.-L. want to acknowledge the Spanish Ministry of Science and Innovation (project reference PID2021-122187NB-C32).

Institutional Review Board Statement: Not applicable.

Informed Consent Statement: Not applicable.

Conflicts of Interest: The authors declare no conflict of interest.

References

1. Sung, H.; Ferlay, J.; Siegel, R.L.; Laversanne, M.; Soerjomataram, I.; Jemal, A.; Bray, F. Global Cancer Statistics 2020: GLOBOCAN Estimates of Incidence and Mortality Worldwide for 36 Cancers in 185 Countries. *CA. Cancer J. Clin.* **2021**, *71*, 209–249. [CrossRef]
2. FDA. *Novel Drug Approvals for 2021*; FDA: Silver Spring, MD, USA, 2021.
3. FDA. *Novel Drug Approvals for 2020*; FDA: Silver Spring, MD, USA, 2020.
4. Meegan, M.J.; O'Boyle, N.M. Special Issue "Anticancer Drugs 2021". *Pharmaceuticals* **2022**, *15*, 479. [CrossRef] [PubMed]
5. Brown, K. Repurposing Old Drugs for New Uses. *DePaul J. Art Technol. Intellect. Prop. Law* **2019**, *28*, 1.
6. DiMasi, J.A.; Hansen, R.W.; Grabowski, H.G. The price of innovation: New estimates of drug development costs. *J. Health Econ.* **2003**, *22*, 151–185. [CrossRef]
7. Wong, R.S.Y. Role of Nonsteroidal Anti-Inflammatory Drugs (NSAIDs) in Cancer Prevention and Cancer Promotion. *Adv. Pharmacol. Sci.* **2019**, *2019*, 3418975. [CrossRef] [PubMed]
8. Manrique-Moreno, M.; Heinbockel, L.; Suwalsky, M.; Garidel, P.; Brandenburg, K. Biophysical study of the non-steroidal anti-inflammatory drugs (NSAID) ibuprofen, naproxen and diclofenac with phosphatidylserine bilayer membranes. *Biochim. Biophys. Acta* **2016**, *1858*, 2123–2131. [CrossRef]
9. Pollard, M.; Luckert, P.H.; Schmidt, M.A. The suppressive effect of piroxicam on autochthonous intestinal tumors in the rat. *Cancer Lett.* **1983**, *21*, 57–61. [CrossRef]
10. Harris, R.E.; Chlebowski, R.T.; Jackson, R.D.; Frid, D.J.; Ascenseo, J.L.; Anderson, G.; Loar, A.; Rodabough, R.J.; White, E.; Mctiernan, A. Breast Cancer and Nonsteroidal Anti-Inflammatory Drugs: Prospective Results from the Women's Health Initiative 1. *CANCER Res.* **2003**, *63*, 6096–6101.
11. Makins, R.; Ballinger, A. Gastrointestinal side effects of drugs. *Expert Opin. Drug Saf.* **2003**, *2*, 421–429. [CrossRef]
12. Greenspan, E.J.; Madigan, J.P.; Boardman, L.A.; Rosenberg, D.W. Ibuprofen inhibits activation of nuclear β-catenin in human colon adenomas and induces the phosphorylation of GSK-3β. *Cancer Prev. Res.* **2011**, *4*, 161–171. [CrossRef]
13. Chai, A.C.; Robinson, A.L.; Chai, K.X.; Chen, L.M. Ibuprofen regulates the expression and function of membrane-associated serine proteases prostasin and matriptase. *BMC Cancer* **2015**, *15*, 1025. [CrossRef] [PubMed]
14. Gliszczyńska, A.; Sánchez-López, E. Dexibuprofen Therapeutic Advances: Prodrugs and Nanotechnological Formulations. *Pharmaceutics* **2021**, *13*, 414. [CrossRef] [PubMed]
15. Almeida, H.; Lobão, P.; Frigerio, C.; Fonseca, J.; Silva, R.; Sousa Lobo, J.M.; Amaral, M.H. Preparation, characterization and biocompatibility studies of thermoresponsive eyedrops based on the combination of nanostructured lipid carriers (NLC) and the polymer Pluronic F-127 for controlled delivery of ibuprofen. *Pharm. Dev. Technol.* **2017**, *22*, 336–349. [CrossRef] [PubMed]
16. Sütő, B.; Berkó, S.; Kozma, G.; Kukovecz, Á.; Budai-Szucs, M.; Erös, G.; Kemény, L.; Sztojkov-Ivanov, A.; Gáspár, R.; Csányi, E. Development of ibuprofen-loaded nanostructured lipid carrier-based gels: Characterization and investigation of in vitro and in vivo penetration through the skin. *Int. J. Nanomed.* **2016**, *11*, 1201–1212. [CrossRef]
17. Süto, B.; Weber, S.; Zimmer, A.; Farkas, G.; Kelemen, A.; Budai-Szucs, M.; Berkó, S.; Szabó-Révész, P.; Csányi, E. Optimization and design of an ibuprofen-loaded nanostructured lipid carrier with a 23 full factorial design. *Chem. Eng. Res. Des.* **2015**, *104*, 488–496. [CrossRef]

18. Sánchez-López, E.; Espina, M.; Doktorovova, S.; Souto, E.B.; García, M.L. Lipid nanoparticles (SLN, NLC): Overcoming the anatomical and physiological barriers of the eye—Part II—Ocular drug-loaded lipid nanoparticles. *Eur. J. Pharm. Biopharm.* **2017**, *110*, 58–69. [CrossRef]
19. Souto, E.B. *Lipid Nanocarriers in Cancer Diagnosis and Therapy*; Smithers Rapra: Shrewsbury, UK, 2011; ISBN 9781847354785.
20. Martins, S.; Sarmento, B.; Ferreira, D.C.; Souto, E.B. Lipid-based colloidal carriers for peptide and protein delivery—Liposomes versus lipid nanoparticles. *Int. J. Nanomed.* **2007**, *2*, 595.
21. Müller, R.H.; Mäder, K.; Gohla, S. Solid lipid nanoparticles (SLN) for controlled drug delivery—A review of the state of the art. *Eur. J. Pharm. Biopharm.* **2000**, *50*, 161–177. [CrossRef]
22. Fratini, F.; Cilia, G.; Turchi, B.; Felicioli, A. Beeswax: A minireview of its antimicrobial activity and its application in medicine. *Asian Pac. J. Trop. Med.* **2016**, *9*, 839–843. [CrossRef]
23. Ravelo, Y.; Molina, V.; Carbajal, D.; Fernández, L.; Fernández, J.C.; Arruzazabala, M.L.; Más, R. Evaluation of anti-inflammatory and antinociceptive effects of D-002 (beeswax alcohols). *J. Nat. Med.* **2011**, *65*, 330–335. [CrossRef]
24. Zur Mühlen, A.; Schwarz, C.; Mehnert, W. Solid lipid nanoparticles (SLN) for controlled drug delivery—Drug release and release mechanism. *Eur. J. Pharm. Biopharm.* **1998**, *45*, 149–155. [CrossRef]
25. Zur Mühlen, A.; Zur Mühlen, E.; Niehus, H.; Mehnert, W. Atomic force microscopy studies of Solid Lipid Nanoparticles. *Pharm. Res.* **1996**, *13*, 1411–1416. [CrossRef] [PubMed]
26. Gonzalez-Mira, E.; Egea, M.A.; Garcia, M.L.; Souto, E.B. Design and ocular tolerance of flurbiprofen loaded ultrasound-engineered NLC. *Colloids Surf. B Biointerfaces* **2010**, *81*, 412–421. [CrossRef] [PubMed]
27. Souto, E.B.; Doktorovova, S.; Gonzalez-Mira, E.; Egea, M.A.; Garcia, M.L. Feasibility of lipid nanoparticles for ocular delivery of anti-inflammatory drugs. *Curr. Eye Res.* **2010**, *35*, 537–552. [CrossRef] [PubMed]
28. Cao, C.; Wang, Q.; Liu, Y. Lung cancer combination therapy: Doxorubicin and β-elemene co-loaded, pH-sensitive nanostructured lipid carriers. *Drug Des. Dev. Ther.* **2019**, *13*, 1087–1098. [CrossRef]
29. Bondì, M.L.; Craparo, E.F.; Giammona, G.; Cervello, M.; Azzolina, A.; Diana, P.; Martorana, A.; Cirrincione, G. Nanostructured Lipid Carriers-Containing Anticancer Compounds: Preparation, Characterization, and Cytotoxicity Studies. *Drug. Deliv.* **2008**, *14*, 61–67. [CrossRef]
30. Levitsky, D.O.; Dembitsky, V.M. Anti-breast Cancer Agents Derived from Plants. *Nat. Products Bioprospect.* **2015**, *5*, 1–16. [CrossRef]
31. Amasya, G.; Aksu, B.; Badilli, U.; Onay-Besikci, A.; Tarimci, N. QbD guided early pharmaceutical development study: Production of lipid nanoparticles by high pressure homogenization for skin cancer treatment. *Int. J. Pharm.* **2019**, *563*, 110–121. [CrossRef]
32. Fangueiro, J.F.; Andreani, T.; Egea, M.A.; Garcia, M.L.; Souto, S.B.; Silva, A.M.; Souto, E.B. Design of cationic lipid nanoparticles for ocular delivery: Development, characterization and cytotoxicity. *Int. J. Pharm.* **2014**, *461*, 64–73. [CrossRef]
33. Mitchell, M.J.; Billingsley, M.M.; Haley, R.M.; Wechsler, M.E.; Peppas, N.A.; Langer, R. Engineering precision nanoparticles for drug delivery. *Nat. Rev. Drug Discov.* **2020**, *20*, 101–124. [CrossRef]
34. Subramaniam, B.; Siddik, Z.H.; Nagoor, N.H. Optimization of nanostructured lipid carriers: Understanding the types, designs, and parameters in the process of formulations. *J. Nanopart. Res.* **2020**, *22*, 141. [CrossRef]
35. Qushawy, M. Effect of the Surfactant and Liquid Lipid Type in the Physico-Chemical Characteristics of Beeswax-Based Nanostructured Lipid Carrier (NLC) of Metformin. *Pharm. Nanotechnol.* **2021**, *9*, 200–209. [CrossRef] [PubMed]
36. Blanco, E.; Shen, H.; Ferrari, M. Principles of nanoparticle design for overcoming biological barriers to drug delivery. *Nat. Biotechnol.* **2015**, *33*, 941–951. [CrossRef] [PubMed]
37. Li, X.; Jia, X.; Niu, H. Nanostructured lipid carriers co-delivering lapachone and doxorubicin for overcoming multidrug resistance in breast cancer therapy. *Int. J. Nanomed.* **2018**, *13*, 4107. [CrossRef]
38. Fleischer, C.C.; Kumar, U.; Payne, C.K. Cellular Binding of Anionic Nanoparticles is Inhibited by Serum Proteins Independent of Nanoparticle Composition. *Biomater. Sci.* **2013**, *1*, 975. [CrossRef]
39. Sánchez-López, E.; Egea, M.A.A.; Cano, A.; Espina, M.; Calpena, A.C.C.; Ettcheto, M.; Camins, A.; Souto, E.B.B.; Silva, A.M.M.; García, M.L.L. PEGylated PLGA nanospheres optimized by design of experiments for ocular administration of dexibuprofen—In vitro, ex vivo and in vivo characterization. *Colloids Surf. B Biointerfaces* **2016**, *145*, 241–250. [CrossRef]
40. Jenning, V.; Gohla, S. Comparison of wax and glyceride solid lipid nanoparticles (SLN®). *Int. J. Pharm.* **2000**, *196*, 219–222. [CrossRef]
41. El-Houssieny, B.M.; El-Dein, E.Z.; El-Messiry, H.M. Enhancement of solubility of dexibuprofen applying mixed hydrotropic solubilization technique. *Drug Discov. Ther.* **2014**, *8*, 178–184. [CrossRef]
42. Svečnjak, L.; Baranović, G.; Vinceković, M.; Prđun, S.; Bubalo, D.; Gajger, I.T. An approach for routine analytical detection of beeswax adulteration using FTIR-ATR spectroscopy. *J. Apic. Sci.* **2015**, *59*, 37–49. [CrossRef]
43. Jenning, V.; Thünemann, A.F.; Gohla, S.H. Characterisation of a novel solid lipid nanoparticle carrier system based on binary mixtures of liquid and solid lipids. *Int. J. Pharm.* **2000**, *199*, 167–177. [CrossRef]
44. López-Machado, A.; Díaz, N.; Cano, A.; Espina, M.; Badía, J.; Baldomà, L.; Calpena, A.C.; Biancardi, M.; Souto, E.B.; García, M.L.; et al. Development of topical eye-drops of lactoferrin-loaded biodegradable nanoparticles for the treatment of anterior segment inflammatory processes. *Int. J. Pharm.* **2021**, *609*, 121188. [CrossRef] [PubMed]

45. López-Machado, A.; Díaz-Garrido, N.; Cano, A.; Espina, M.; Badia, J.; Baldomà, L.; Calpena, A.C.; Souto, E.B.; García, M.L.; Sánchez-López, E. Development of Lactoferrin-Loaded Liposomes for the Management of Dry Eye Disease and Ocular Inflammation. *Pharmaceutics* **2021**, *13*, 1698. [CrossRef] [PubMed]

46. Hu, F.Q.; Jiang, S.P.; Du, Y.Z.; Yuan, H.; Ye, Y.Q.; Zeng, S. Preparation and characteristics of monostearin nanostructured lipid carriers. *Int. J. Pharm.* **2006**, *314*, 83–89. [CrossRef] [PubMed]

47. Son, G.H.; Lee, B.J.; Cho, C.W. Mechanisms of drug release from advanced drug formulations such as polymeric-based drug-delivery systems and lipid nanoparticles. *J. Pharm. Investig.* **2017**, *47*, 287–296. [CrossRef]

48. Ghasemiyeh, P.; Mohammadi-Samani, S. Solid lipid nanoparticles and nanostructured lipid carriers as novel drug delivery systems: Applications, advantages and disadvantages. *Res. Pharm. Sci.* **2018**, *13*, 288. [CrossRef] [PubMed]

49. Sánchez-López, E.; Ettcheto, M.; Egea, M.A.; Espina, M.; Cano, A.; Calpena, A.C.; Camins, A.; Carmona, N.; Silva, A.M.; Souto, E.B.; et al. Memantine loaded PLGA PEGylated nanoparticles for Alzheimer's disease: In vitro and in vivo characterization. *J. Nanobiotechnol.* **2018**, *16*, 32. [CrossRef] [PubMed]

50. Youshia, J.; Kamel, A.O.; El Shamy, A.; Mansour, S. Gamma sterilization and in vivo evaluation of cationic nanostructured lipid carriers as potential ocular delivery systems for antiglaucoma drugs. *Eur. J. Pharm. Sci.* **2021**, *163*, 105887. [CrossRef]

51. Cunha, S.; Costa, C.P.; Loureiro, J.A.; Alves, J.; Peixoto, A.F.; Forbes, B.; Lobo, J.M.S.; Silva, A.C. Double Optimization of Rivastigmine-Loaded Nanostructured Lipid Carriers (NLC) for Nose-to-Brain Delivery Using the Quality by Design (QbD) Approach: Formulation Variables and Instrumental Parameters. *Pharmaceutics* **2020**, *12*, 599. [CrossRef] [PubMed]

52. Freitas, C.; Müller, R.H. Effect of light and temperature on zeta potential and physical stability in solid lipid nanoparticle (SLN™) dispersions. *Int. J. Pharm.* **1998**, *168*, 221–229. [CrossRef]

53. Sánchez-López, E.; Egea, M.A.; Davis, B.M.; Guo, L.; Espina, M.; Silva, A.M.; Calpena, A.C.; Souto, E.M.B.; Ravindran, N.; Ettcheto, M.; et al. Memantine-Loaded PEGylated Biodegradable Nanoparticles for the Treatment of Glaucoma. *Small* **2018**, *14*, 1701808. [CrossRef]

54. Gainza, G.; Pastor, M.; Aguirre, J.J.; Villullas, S.; Pedraz, J.L.; Hernandez, R.M.; Igartua, M. A novel strategy for the treatment of chronic wounds based on the topical administration of rhEGF-loaded lipid nanoparticles: In vitro bioactivity and in vivo effectiveness in healing-impaired db/db mice. *J. Control. Release* **2014**, *185*, 51–61. [CrossRef] [PubMed]

55. Nainu, F.; Masyita, A.; Bahar, M.A.; Raihan, M.; Prova, S.R.; Mitra, S.; Emran, T.B.; Simal-Gandara, J. Pharmaceutical prospects of bee products: Special focus on anticancer, antibacterial, antiviral, and antiparasitic properties. *Antibiotics* **2021**, *10*, 822. [CrossRef] [PubMed]

56. Giampieri, F.; Quiles, J.L.; Orantes-Bermejo, F.J.; Gasparrini, M.; Forbes-Hernandez, T.Y.; Sánchez-González, C.; Llopis, J.; Rivas-García, L.; Afrin, S.; Varela-López, A.; et al. Are by-products from beeswax recycling process a new promising source of bioactive compounds with biomedical properties? *Food Chem. Toxicol.* **2018**, *112*, 126–133. [CrossRef]

57. Lason, E.; Sikora, E.; Miastkowska, M.; Escribano, E.; Garcia-Celma, M.J.; Solans, C.; Llinas, M.; Ogonowski, J. NLC as a potential carrier system for transdermal delivery of forskolin. *Acta Biochim. Pol.* **2018**, *65*, 437–442. [CrossRef] [PubMed]

58. Sánchez-López, E.; Ettcheto, M.; Egea, M.A.; Espina, M.; Calpena, A.C.; Folch, J.; Camins, A.; García, M.L. New Potential Strategies for Alzheimer's Disease Prevention: Pegylated Biodegradable Dexibuprofen Nanospheres Administration to APPswe/PS1dE9. *Nanomed. Nanotechnol. Biol. Med.* **2017**, *13*, 1171–1182. [CrossRef] [PubMed]

59. Sánchez-López, E.; Esteruelas, G.; Ortiz, A.; Espina, M.; Prat, J.; Muñoz, M.; Cano, A.; Calpena, A.C.; Ettcheto, M.; Camins, A.; et al. Dexibuprofen biodegradable nanoparticles: One step closer towards a better ocular interaction study. *Nanomaterials* **2020**, *10*, 720. [CrossRef]

60. Abla, M.J.; Banga, A.K. Formulation of tocopherol nanocarriers and in vitro delivery into human skin. *Int. J. Cosmet. Sci.* **2014**, *36*, 239–246. [CrossRef] [PubMed]

International Journal of
Molecular Sciences

MDPI

Article

Affinities to Oxaliplatin: Vitamins from B Group vs. Nucleobases

Beata Szefler [1,*], Przemysław Czeleń [1], Kamil Wojtkowiak [2] and Aneta Jezierska [2]

[1] Department of Physical Chemistry, Faculty of Pharmacy, Collegium Medicum, Nicolaus Copernicus University, Kurpińskiego 5, 85-096 Bydgoszcz, Poland
[2] Faculty of Chemistry, University of Wrocław, F. Joliot-Curie 14, 50-383 Wrocław, Poland
* Correspondence: beatas@cm.umk.pl

Abstract: Oxaliplatin, similar to Cisplatin, exhibits anticancer activity by interacting with DNA and inducing programmed cell death. It is biotransformed through a number of spontaneous and non-enzymatic processes. In this way, several transient reactive species are formed, including dichloro-, monochloro-, and diaqua-DACH platin, which can complex with DNA and other macromolecules. The molecular level suggests that such interactions can also take place with vitamins containing aromatic rings with lone pair orbitals. Theoretical and experimental studies were performed to investigate interactions of vitamins from the B group with Oxaliplatin, and the results were compared with values characterizing native purines. Quantum-chemical simulations were carried out at the B3LYP/6-31G(d,p) level, with the LANL2DZ basis set representing atomic orbitals of platinum atom, and at the MN15/def2-TZVP levels of theory with the use of Polarizable Continuum Model (IEF-PCM formulation) and water as a solvent. Additionally, time-dependent density functional theory (TD-DFT) was employed to study molecular properties in the electronic excited state. Interactions of vitamins and Oxaliplatin were investigated using UV-Vis spectroscopy. Values of the free energy (ΔG_r) indicate spontaneous reactions with monoaqua [PtH$_2$OClDACH]$^+$ and diaqua [Pt(H$_2$O)$_2$DACH]$^{2+}$ derivatives of Oxaliplatin. However, diaqua derivatives were found to be preferable. The free energy (ΔG_r) values obtained for vitamins from the B group indicate lower affinity of Oxaliplatin compared with values characterizing complexes formed by guanine, adenine, and cytosine. The exception is the monoaqua form of vitamin B1 (thiamine) at the MN15/def2-TZVP levels of calculations. An application of atoms in molecules (AIM) theory revealed non-covalent interactions present in the complexes studied. The comparison of computed and experimental spectroscopic properties showed a good agreement.

Keywords: Oxaliplatin; platinum-based drugs; colorectal cancer; gastric cancer; esophageal cancer; cancer treatment; vitamins from B group; UV-Vis; DFT; TD-DFT; PCM

Citation: Szefler, B.; Czeleń, P.; Wojtkowiak, K.; Jezierska, A. Affinities to Oxaliplatin: Vitamins from B Group vs. Nucleobases. *Int. J. Mol. Sci.* **2022**, *23*, 10567. https://doi.org/10.3390/ijms231810567

Academic Editors: Andrzej Kutner, Geoffrey Brown and Enikö Kallay

Received: 8 August 2022
Accepted: 8 September 2022
Published: 12 September 2022

Publisher's Note: MDPI stays neutral with regard to jurisdictional claims in published maps and institutional affiliations.

1. Introduction

Oxaliplatin (Figure 1) is a complex compound of platinum in the second oxidation state with 1,2-diaminocyclohexane and an oxalate group (DACH). It is a cytostatic and alkylating drug used in chemotherapy of malignant tumors, mainly in the treatment of colorectal cancer, in combination with fluoropyrimidines [1–13]. A randomized study suggests the equivalence of Cisplatin and Oxaliplatin in the treatment of gastric or esophageal cancer [14,15]. In the preclinical studies, Oxaliplatin has demonstrated efficacy against a broad spectrum of investigated tumors, including some Cisplatin- and Carboplatin-resistant cell lines [16–23].

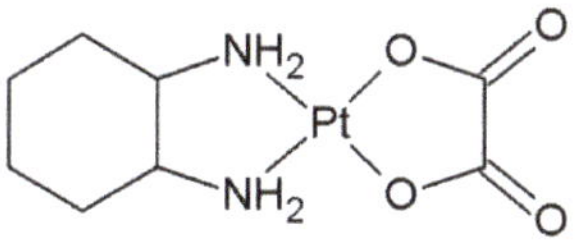

Figure 1. Structure of Oxaliplatin.

The main mechanism of action of the drug is based on the DNA damage (Figure 2), [24,25]. Similar to Cisplatin, Oxaliplatin binds to DNA, leading to GG intra-strand crosslinks [1]. Cancer cell apoptosis is caused by the DNA damage and blockage of its synthesis. After entering the cell, the drug approaches the cell nucleus, where it is directed to the sites rich in guanine, adenine and cytosine. Next, it interacts with the nitrogen atom (N_7) forming a monoadduct, and then a diadduct (Figures 2 and 3) [26]. This results in the formation of intra-chain and inter-chain cross-link bonds, and bonds between the DNA and the proteins (Figure 2). The DNA replication and transcription are stopped, followed by the cell apoptosis [27].

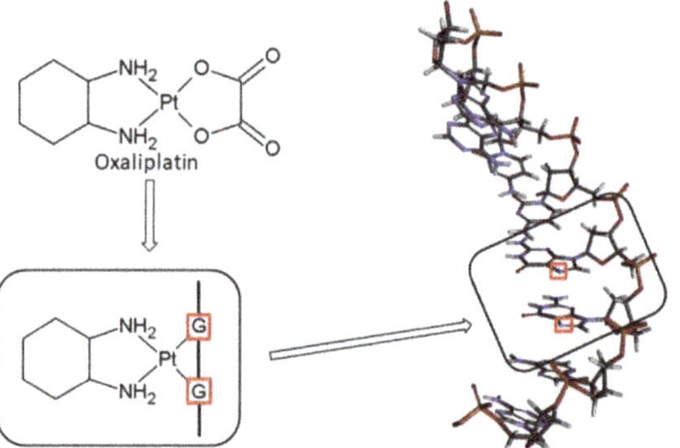

Figure 2. Pathway of Oxaliplatin and anticancer activity.

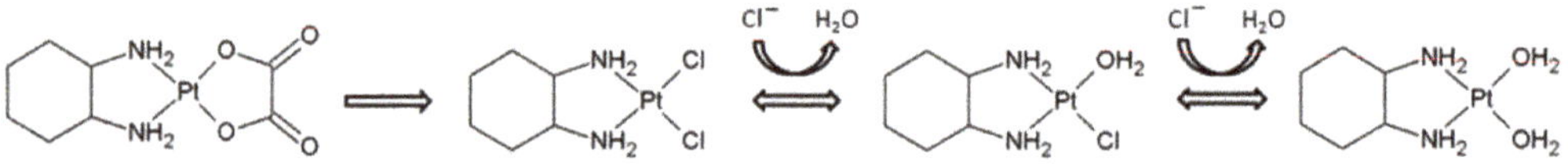

Figure 3. Biotransformation pathway of Oxaliplatin.

Oxaliplatin is biotransformed through a number of spontaneous and non-enzymatic processes [28–31]. Nucleophiles, such as Cl^-, HCO_3^-, and H_2O molecules, participate in these reactions (Figure 3) [30,32,33]. In this way, several transient reactive species are formed, including dichloro-, monochloro-, and diaqua-DACH platin, which can complex with DNA and other macromolecules [30,32,33].

Oxaliplatin belongs to the same group of Pt (II) derivatives used in cancer treatment, and it has a similar mechanism of action when compared to Cisplatin. For the activation of relatively inert platinum (II) complexes, the hydrolysis of Oxaliplatin to monoaqua form and diaqua complexes is necessary for the replacement of chloride ligands with water molecules (Figure 3) [34]. In this way, Oxaliplatin can interact with DNA and induce programmed cell death (Figure 2). Therefore, it can be assumed (analogously to Cisplatin) that the presence of other compounds possessing aromatic rings with electron lone pairs, such as N_7 or/and N_1 atoms, influences the therapeutic effect of the chemotherapeutic [24,25]. During chemotherapy, patients often consume juices rich in niacin (vitamin B3), pyridoxal phosphate (vitamin B6), riboflavin (vitamin B2), and thiamine (vitamin B1) [24,25]. The structures of selected vitamins B are presented in Figure 4. Therefore, on the basis of our previous findings, which clearly showed the possibility of forming complexes of B vitamins with nucleobases [24,25], it can be assumed that the presence of B vitamins could reduce the therapeutic effect of the Oxaliplatin [24].

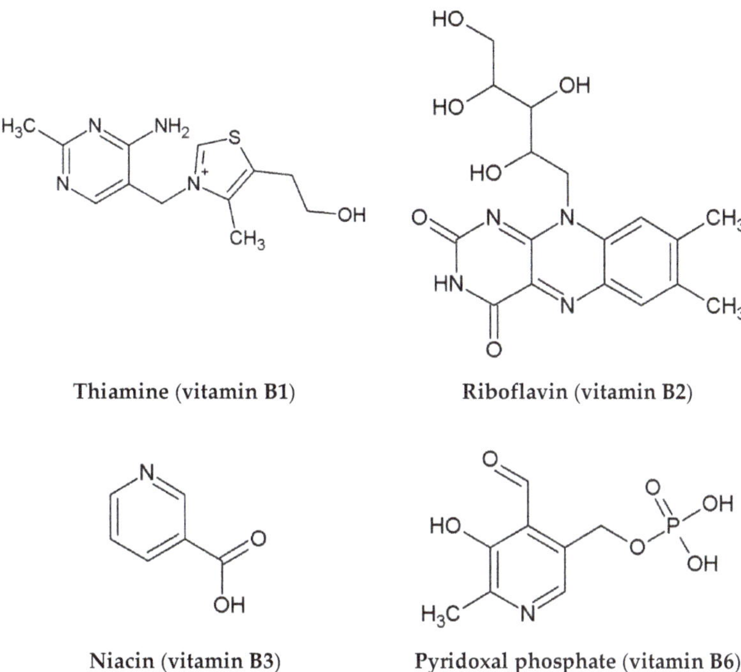

Thiamine (vitamin B1)   Riboflavin (vitamin B2)

Niacin (vitamin B3)   Pyridoxal phosphate (vitamin B6)

Figure 4. B vitamins chosen for the current study. The N_1 and N_7 atoms are indicated.

The theoretical and experimental studies of the interactions of the N_7 nitrogen atom of B vitamins (and, in addition, the N_1 nitrogen atom in the case of thiamine) with Oxaliplatin have been performed. The comparison of interactions of the studied drug with the nucleobases, such as adenine, guanine, and cytosine, which occur during chemotherapy, have been carried out. Finally, our results were compared with earlier obtained data for Cisplatin and Carboplatin [24,25].

2. Results and Discussion

2.1. Energetic and Electronic Structure Characterization in the Electronic Ground and Excited States (DFT and TD-DFT Study)

Monoaqua $[PtH_2OClDACH]^+$ and diaqua $[Pt(H_2O)_2DACH]^{2+}$ derivatives of Oxaliplatin, with vitamins from the B group such as thiamine (vitamin B1), riboflavin (vitamin B2), niacin (vitamin B3), and pyridoxal phosphate (vitamin B6) were studied (see Figures 4 and 5). All these vitamins are present in carrot and beet juices, which are often consumed by patients during chemotherapy.

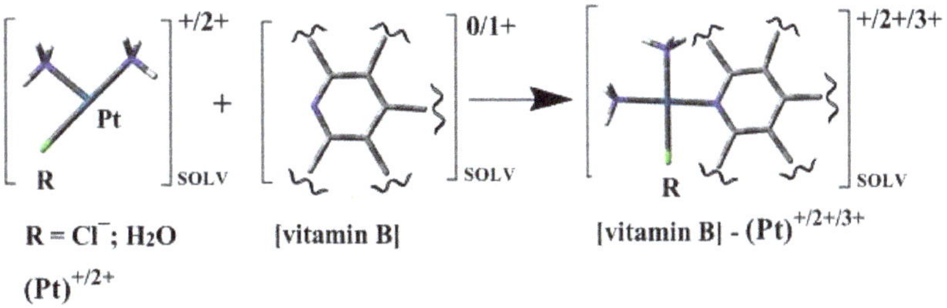

Figure 5. Scheme of complexes of B vitamins with Pt-mono and Pt-diaqua complexes.

Because vitamins from the B group have a purine and/or pyrimidine rings with lone pair orbitals analogous to N_7 in purine, they have a structure similar to native DNA purines and can freely react with Oxaliplatin. The vitamin B1 (thiamine), in contrast to other vitamins, has the second nitrogen atom in the position N_1, which could also act as a possible reaction center (Figure 4).

As is shown in Tables 1–4, the estimated values of Gibbs free energy of reaction ΔG_r are negative, not only for guanine, adenine, and cytosine, but also for all chloroaqua- and diaqua-platinumated complexes with vitamins studied from the B group. However, the affinities of all analyzed vitamins to platinum derivatives are smaller if compared to guanine (GUA), adenine (ADE), and cytosine (CYT). In Tables 1 and 2 the energetic characteristics of the complexes with vitamins from the B group and nucleobases with Pt-Chloroaqua are presented, while in Tables 3 and 4, those with diaqua-platinumated complexes are shown. The results were obtained based on the structures computed at the B3LYP/6-31G(d,p)/LANL2DZ level of theory (Tables 1–3). The estimated values of ΔG_r were further compared with the results obtained at the MN15/def2-TZVP level of theory (Tables 2 and 4). Furthermore, the obtained values of ΔG_r for Oxaliplatin were compared with the results of Cisplatin [24] and Carboplatin [25].

Table 1. The energetic characteristics of the Pt-Chloroaqua reactions with vitamin B and nucleobases obtained at the B3LYP/6-31G(d,p)/LANL2DZ level of theory. All energies are given in kcal/mol. Symbol Pt* stands for PtClDACH.

Number of Reaction	ΔG_r	Reaction
1	−13.87	$[Pt^*]^+ + [B1_{(N7)}]^+ \rightarrow [Pt^*\text{-}B1_{(N7)}]^{2+}$
2	−16.24	$[Pt^*]^+ + [B1_{(N1)}]^+ \rightarrow [Pt^*\text{-}B1_{(N1)}]^{2+}$
3	−1.46	$[Pt^*]^+ + B2 \rightarrow [Pt^*\text{-}B2]^+$
4	−22.32	$[Pt^*]^+ + B3 \rightarrow [Pt^*\text{-}B3]^+$
5	−17.06	$[Pt^*]^+ + B6 \rightarrow [Pt^*\text{-}B6]^+$
6	−33.46	$[Pt^*]^+ + GUA \rightarrow [Pt^*\text{-}GUA]^+$
7	−28.80	$[Pt^*]^+ + ADE \rightarrow [Pt^*\text{-}ADE]^+$
8	−26.89	$[Pt^*]^+ + CYT \rightarrow [Pt^*\text{-}CYT]^+$

Table 2. The energetic characteristics of the Pt-Chloroaqua reactions with vitamin B and nucleobases obtained at the MN15/def2-TZVP level of theory. All energies are given in kcal/mol. Symbol Pt* stands for PtClDACH.

Number of Reaction	ΔG_r	Reaction
1	−46.48	$[Pt^*]^+ + [B1_{(N7)}]^+ \rightarrow [Pt^*\text{-}B1_{(N7)}]^{2+}$
2	−48.91	$[Pt^*]^+ + [B1_{(N1)}]^+ \rightarrow [Pt^*\text{-}B1_{(N1)}]^{2+}$
3	−41.55	$[Pt^*]^+ + B2 \rightarrow [Pt^*\text{-}B2]^+$
4	−39.37	$[Pt^*]^+ + B3 \rightarrow [Pt^*\text{-}B3]^+$
5	−45.93	$[Pt^*]^+ + B6 \rightarrow [Pt^*\text{-}B6]^+$
6	−46.10	$[Pt^*]^+ + GUA \rightarrow [Pt^*\text{-}GUA]^+$
7	−42.59	$[Pt^*]^+ + ADE \rightarrow [Pt^*\text{-}ADE]^+$
8	−45.23	$[Pt^*]^+ + CYT \rightarrow [Pt^*\text{-}CYT]^+$

Oxaliplatin has the best affinity for vitamin B3 (niacin) with value of the Gibbs free energy equal −22.32 kcal/mol (Table 1). It means that niacin easily forms complexes, not only with cis-Pt-Chloroaqua, cis~$[Pt(NH_3)_2H_2O]^+$ [24], but also with $[PtH_2OClDACH]^+$. The second highest value of the ΔG_r equal −17.06 kcal/mol was obtained for vitamin B6 (pyridoxal phosphate). For vitamin B1 (thiamine), the ΔG_r value is equal −16.24 kcal/mol in the reaction with lone pair orbitals analogous to N_1 in purine (Table 1). In the case of reaction, where lone pair orbitals analogous to N_7 in purine of vitamin B1, the value of affinity is −13.87 kcal/mol (Table 1). The vitamin B2 (riboflavin) has the worst affinity, with slightly negative values of the Gibbs free energy of the reaction, approaching zero (−1.46 kcal/mol). In general, it can be concluded that the order of affinity of Oxaliplatin for

B vitamins is similar to Cisplatin [24], and quite different compared with Carboplatin [25] at the B3LYP/6-31G(d,p)/LANL2DZ level of theory.

Table 3. The energetic characteristics of the Pt-diaqua reactions with vitamin B and nucleobases obtained at the B3LYP/6-31G(d,p)/LANL2DZ level of theory. All energies are given in kcal/mol.

Number of Reaction	ΔG_r	Reaction
1	-25.10	$[Pt^{**}]^{2+} + [B1_{(N7)}]^+ \rightarrow [Pt^{**}\text{-}B1_{(N7)}]^{3+}$
2	-30.15	$[Pt^{**}]^{2+} + [B1_{(N1)}]^+ \rightarrow [Pt^{**}\text{-}B1_{(N1)}]^{3+}$
3	-13.64	$[Pt^{**}]^{2+} + B2 \rightarrow [Pt^{**}\text{-}B2]^{2+}$
4	-35.53	$[Pt^{**}]^{2+} + B3 \rightarrow [Pt^{**}\text{-}B3]^{2+}$
5	-30.60	$[Pt^{**}]^{2+} + B6 \rightarrow [Pt^{**}\text{-}B6]^{2+}$
6	-50.03	$[Pt^{**}]^{2+} + GUA \rightarrow [Pt^{**}\text{-}GUA]^{2+}$
7	-41.37	$[Pt^{**}]^{2+} + ADE \rightarrow [Pt^{**}\text{-}ADE]^{2+}$
8	-29.07	$[Pt^{**}]^{2+} + CYT \rightarrow [Pt^{**}\text{-}CYT]^{2+}$

where Pt^{**} symbol is PtH_2ODACH.

Table 4. The energetic characteristics of the Pt-diaqua reactions with vitamin B and nucleobases obtained at the MN15/def2-TZVP level of theory. All energies are given in kcal/mol.

Number of Reaction	ΔG_r	Reaction
1	-46.45	$[Pt^{**}]^{2+} + [B1_{(N7)}]^+ \rightarrow [Pt^{**}\text{-}B1_{(N7)}]^{3+}$
2	-50.58	$[Pt^{**}]^{2+} + [B1_{(N1)}]^+ \rightarrow [Pt^{**}\text{-}B1_{(N1)}]^{3+}$
3	-42.96	$[Pt^{**}]^{2+} + B2 \rightarrow [Pt^{**}\text{-}B2]^{2+}$
4	-40.61	$[Pt^{**}]^{2+} + B3 \rightarrow [Pt^{**}\text{-}B3]^{2+}$
5	-47.11	$[Pt^{**}]^{2+} + B6 \rightarrow [Pt^{**}\text{-}B6]^{2+}$
6	-51.01	$[Pt^{**}]^{2+} + GUA \rightarrow [Pt^{**}\text{-}GUA]^{2+}$
7	-43.67	$[Pt^{**}]^{2+} + ADE \rightarrow [Pt^{**}\text{-}ADE]^{2+}$
8	-47.89	$[Pt^{**}]^{2+} + CYT \rightarrow [Pt^{**}\text{-}CYT]^{2+}$

where Pt^{**} symbol is PtH_2ODACH.

The results obtained at the MN15/def2-TZVP (Table 2) show a completely different tendency when compared to the results derived from the lower level of theory. At this higher level of calculations, the best affinity to Oxaliplatin was showed by vitamin B1 (thiamine), in reaction with lone pair orbitals analogous to N_1 in purine, equal -48.91 kcal/mol. Compared to B3LYP/6-31G(d,p)/LANL2DZ level, the vitamin B2 (riboflavin) showed a much greater affinity for the drug, because the ΔG_r is equal -41.55 kcal/mol, at the MN15/def2-TZVP level of theory. At this higher level of calculations, the vitamin B3 (niacin) has the worst affinity, equal -39.37 kcal/mol.

The energetic characteristics of the Pt-diaqua reactions with derivatives of vitamin B show that the values of ΔG_r are more favorable when compared with Pt-Chloroaqua reactions (Tables 1 and 3), and are found to be in the range from -35.53 kcal/mol to -13.64 kcal/mol (Table 3) at the B3LYP/6-31G(d,p)/LANL2DZ level of theory, and in the range from -50.58 kcal/mol to -40.61 kcal/mol at the MN15/def2-TZVP level of theory.

The affinity of Pt-diaqua reactions with guanine, adenine, and cytosine is higher comparing with Pt-Chloroaqua (Table 3), as could be expected from the previous studies for Cisplatin [24]. As in the case of Pt-Chloroaqua reactions, the highest value of ΔG_r is observed for vitamin B3 (niacin), -35.53 kcal/mol. The Pt-diaqua also easily forms bonds with vitamin B6 (pyridoxal phosphate) and vitamin B1 (N_1, thiamine), which is analogous not only to Pt-Chloroaqua, but also to Cisplatin [25], with values of ΔG_r equal -30.60 kcal/mol and -30.15 kcal/mol, respectively (Table 3). The worst affinity is observed for vitamin B2 (riboflavin), -13.64 kcal/mol.

Again, it can be noticed that in the case of Pt-diaqua reactions, the computational results obtained at both levels of theory differ (B3LYP/6-31G(d,p)/LANL2DZ vs MN15/def2-TZVP); for details see Tables 3 and 4. It was found that, on the basis of the results obtained at the MN15/def2-TZVP level of theory, the best affinity to Oxaliplatin was shown by the vitamin B1 (thiamine) in reaction with lone pair orbitals analogous to N_1 in purine, and with value of Gibbs Free Energy equal -50.58 kcal/mol. The vitamin B2 (riboflavin) exhibits a much greater affinity for the drug than in the case of application of lower level of theory with value -42.96 kcal/mol (Table 3). The values of the Gibbs free energy of nucleobases are higher compared with the affinity values obtained at the B3LYP/6-31G(d,p)/LANL2DZ (Table 3). It can be concluded that the behavior of Oxaliplatin is similar to Cisplatin in relation to nucleobases (adenine, guanine, and cytosine) and all vitamins from the B group [24] at the B3LYP/6-31G(d, p)/LANL2DZ level of calculations. This level of theory was based on Baik's studies [35]. However, the Gibbs free energy values were obtained at two levels of theory and compared to each other. The higher level of theory, MN15/def2-TZVP [36,37], was chosen because it describes better possible non-covalent interactions present in the complexes studied. Unfortunately, the use of a higher level of calculations changes the order of the obtained values of the affinity of nucleobases and B vitamins to the tested drug, Oxaliplatin. However, regardless of the level of computation, the order of affinities of the studied vitamins to the drug for the Pt-Chloroaqua and Pt-diaqua derivatives are the same, at the B3LYP/6-31G(d, p)/LANL2DZ level of theory for vitamins B3→B6→B1(N_1)→B1(N_7)→B2 and for purines GUA→ADE→CYT, and at the MN15/def2-TZVP level of calculations for vitamins B1(N_1)→B1(N_7)→B6→B2→B3 and for purines GUA→CYT→ADE. Regardless of the level of calculations and the forms of the studied structures (chloro- and –diaqua), guanine has the highest affinity for Oxaliplatin. An exception is Pt-Chloroaqua derivatives with the vitamin B1, which showed a greater affinity compared to guanine at the MN15/def2-TZVP level of calculations.

The molar volume was computed for six exemplary complexes with nucleobases to estimate quantitatively changes associated with the replacement of a chlorine atom with a water molecule. The data is presented in Table 5. As is shown, the exchange of the species is associated with conformation changes. This further affects the molar volume of the complexes and their size. Such a study gave us a general overview of possible interactions with macromolecules (e.g., nucleic acid).

Table 5. Molar volume (cm^3/mol) of selected complexes computed at the MN15/def2-TZVP level of theory. Molecular structures of the complexes are shown in Figure 6.

Volume	PtClDACH-Ade (A1)	PtH2ODACH-Ade (A2)	PtClDACH-Cyt (B1)	PtH2ODACH-Cyt (B2)	PtClDACH-Gua (C1)	PtH2ODACH-Gua (C2)
	205.392	183.606	188.870	181.336	198.370	210.671

The results of the AIM analysis are presented in Figure 6 and Table 6. The goal was to investigate topological and electronic structure changes upon the exchange of a chlorine atom with a water molecule. As is shown in Figure 6, the BCPs have been found for covalent and non-covalent interactions. It is worth underlining that, depending on the complex, the network of chemical bonds or non-covalent interactions differs. Therefore, our attention was mostly focused on the Pt interactions with the closest chemical environment. Concerning the complex (A1), it is clear that Pt interacts with the chlorine atom and three nitrogen atoms (one from the adenine ring, two from the NH_2 groups). There is also a BCP between Pt and a hydrogen atom of the NH_2 group of adenine. In the case of the complex (A2), the network of interactions is similar. There are BCPs between Pt and nitrogen atoms, the oxygen atom from the water molecule, and the hydrogen atom. Interestingly, there is a BCP between nitrogen and hydrogen atoms from the NH_2 groups. For PtClDACH-Cyt complex (B1), the following BCPs were found: Pt interacting with a chlorine atom, nitrogen atoms, and a hydrogen atom from the NH_2 group. The exchange of a chlorine atom with a water molecule (complex (B2)) did not change the binding mode of the Pt atom. However, here, as well as in the case of other complexes, the Cl-Pt bond is characterized by significantly larger electronic density at the BCP than its H_2O-Pt counterpart. The Pt atom interacts with three nitrogen atoms, an oxygen atom from the water molecule, and with the hydrogen from NH_2 group. In the complex (C1), BCPs between Pt and chlorine atom and three nitrogen atoms have been observed. There is an intramolecular interaction between the oxygen atom from guanine and the hydrogen atom from the NH_2 group of Oxaliplatin moiety. In the last analyzed complex (C2), the interactions between Pt and three nitrogen atoms and an oxygen atom from the water molecule were found. There is a BCP between an oxygen atom from guanine and a hydrogen atom from the NH_2 group of Oxaliplatin moiety. The topological analysis revealed the network of interactions stabilizing the complexes. In addition, it could be useful in the design of new complexes where some of the intramolecular interactions are not preferable.

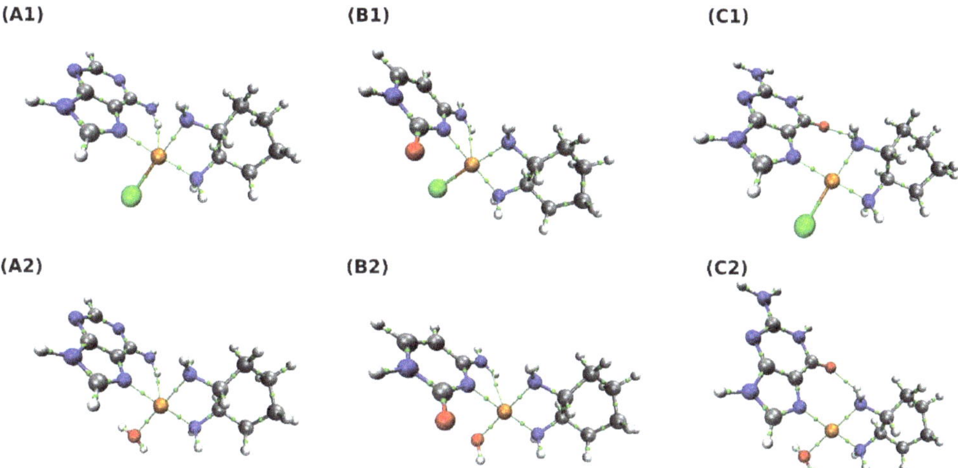

Figure 6. Atoms in molecules (AIM) molecular graphs of the studied complexes. The simulations were performed at the MN15/def2-TZVP level of theory. The BCPs of covalent and non-covalent interactions are presented and marked as small green spheres along bond paths. The designations in Figure are as follows: (**A1**) PtClDACH-Ade, (**A2**) PtH$_2$ODACH-Ade, (**B1**) PtClDACH-Cyt, (**B2**) PtH$_2$ODACH-Cyt, (**C1**) PtClDACH-Gua, (**C2**) PtH$_2$ODACH-Gua.

In Table 6, the values of electron density and its Laplacian at BCPs are presented. In addition, the potential electronic energy density was calculated to provide a general picture of the strength of the interactions. The data were collected for the BCPs found between Pt and fragments of nucleobases and the Oxaliplatin moiety. Some details of the intramolecular and intermolecular hydrogen bonds are reported. As it is shown, depending on the nucleobase and the presence of the chlorine atom or water molecule in the vicinity of the Pt atom of the Oxaliplatin moiety, the electron density and its Laplacian values were changing. This is also visible in the potential electronic energy density. The values of the electron density and its Laplacian at the BCPs for Pt-Cl and Pt-H$_2$O are similar. However, the potential energy density values differ (Table 6).

Time-dependent density functional theory (TD-DFT) was applied to investigate spectroscopic properties of sixteen selected complexes for which experimental UV-Vis measurements were done (see Section 2.2. below). Table S1 presents excitation energies for singlet-singlet transitions in eV as well as in nm. In addition, the oscillator strength is provided. It was found that the most intensive transitions are at 243.56 nm for PtClDACH-Ade, 254.22 nm for PtClDACH-Gua, 250.34 nm for PtClDACH-Cyt, 243.06 nm for PtH$_2$ODACH-Ade, 260.50 nm for PtH$_2$ODACH-Gua, 250.54 nm for PtH$_2$ODACH-Cyt, 252.64 nm for PtClDACH-vitB1(N$_1$), 255.04 nm for PtClDACH-vitB1(N$_7$), 248.69 nm for PtH$_2$ODACH-vitB1(N$_1$), 255.04 nm for PtH$_2$ODACH-vitB1(N$_7$), 443.05 nm for PtClDACH-vitB2, 447.71 nm for PtH$_2$ODACH-vitB2, 301.58 nm for PtClDACH-vitB3, 250.47 nm for PtH$_2$ODACH-vitB3, 258.68 nm for PtClDACH-vitB6, and 258.55 nm for PtH$_2$ODACH-vitB6. The comparison with experimental findings is provided in the end of Section 2.2.

Table 6. Bond Critical Points (BCPs) and potential energy density obtained at the MN15/def2-TZVP level of theory for six selected complexes. Electron density ρ_{BCP} is given in $e*a_0{}^{-3}$ atomic units, and its Laplacian $\nabla^2\rho_{BCP}$ in $e*a_0{}^{-5}$ units.

BCP	PtClDACH-Ade (A1)			BCP	PtH$_2$ODACH-Ade (A2)		
	ρ_{BCP}	$\nabla^2\rho_{BCP}$	V(r)		ρ_{BCP}	$\nabla^2\rho_{BCP}$	V(r)
H$_2$N-Pt	1.23×10^{-1}	4.16×10^{-1}	-1.94×10^{-1}	H$_2$O-Pt	9.63×10^{-2}	4.65×10^{-1}	-1.66×10^{-1}
N-Pt	1.27×10^{-1}	4.41×10^{-1}	-2.06×10^{-1}	H$_2$N-Pt	1.39×10^{-1}	3.89×10^{-1}	-2.14×10^{-1}
NH-Pt	1.39×10^{-2}	3.93×10^{-2}	-7.94×10^{-3}	N-Pt	1.29×10^{-1}	4.16×10^{-1}	-2.06×10^{-1}
H$_2$N-Pt	1.26×10^{-1}	4.15×10^{-1}	-2.00×10^{-1}	H$_2$N-Pt	1.26×10^{-1}	3.99×10^{-1}	-1.97×10^{-1}
Cl-Pt	1.04×10^{-1}	1.89×10^{-1}	-1.28×10^{-1}	NH-Pt	1.38×10^{-2}	4.00×10^{-2}	-8.12×10^{-3}

BCP	PtClDACH-Cyt (B1)			BCP	PtH$_2$ODACH-Cyt (B2)		
	ρ_{BCP}	$\nabla^2\rho_{BCP}$	V(r)		ρ_{BCP}	$\nabla^2\rho_{BCP}$	V(r)
Cl-Pt	1.02×10^{-1}	1.85×10^{-1}	-1.25×10^{-1}	H$_2$O-Pt	9.61×10^{-2}	4.59×10^{-1}	-1.65×10^{-1}
NH-Pt	1.88×10^{-2}	6.01×10^{-2}	-1.32×10^{-2}	H$_2$N-Pt	1.38×10^{-1}	3.95×10^{-1}	-2.14×10^{-1}
H$_2$N-Pt	1.23×10^{-1}	4.24×10^{-1}	-1.97×10^{-1}	N-Pt	1.24×10^{-1}	3.88×10^{-1}	-1.92×10^{-1}
N-Pt	1.21×10^{-1}	4.12×10^{-1}	-1.92×10^{-1}	H$_2$N-Pt	1.26×10^{-1}	4.03×10^{-1}	-1.98×10^{-1}
H$_2$N-Pt	1.25×10^{-1}	4.14×10^{-1}	-1.98×10^{-1}	NH-Pt	1.74×10^{-2}	5.83×10^{-2}	-1.24×10^{-2}

BCP	PtClDACH-Gua (C1)			BCP	PtH$_2$ODACH-Gua (C2)		
	ρ_{BCP}	$\nabla^2\rho_{BCP}$	V(r)		ρ_{BCP}	$\nabla^2\rho_{BCP}$	V(r)
Cl-Pt	1.02×10^{-1}	1.97×10^{-1}	-1.27×10^{-1}	H$_2$O-Pt	9.54×10^{-2}	4.69×10^{-1}	-1.65×10^{-1}
N-Pt	1.25×10^{-1}	4.53×10^{-1}	-2.06×10^{-1}	N-Pt	1.28×10^{-1}	4.31×10^{-1}	-2.07×10^{-1}
H$_2$N-Pt	1.25×10^{-1}	4.12×10^{-1}	-1.96×10^{-1}	H$_2$N-Pt	1.39×10^{-1}	3.77×10^{-1}	-2.12×10^{-1}
H$_2$N-Pt	1.26×10^{-1}	4.15×10^{-1}	-1.99×10^{-1}	H$_2$N-Pt	1.27×10^{-1}	4.04×10^{-1}	-1.99×10^{-1}
O...HN	1.84×10^{-2}	7.27×10^{-2}	-1.23×10^{-2}	O...HN	1.90×10^{-2}	7.36×10^{-2}	-1.27×10^{-2}

2.2. UV-Vis Experimental Study

Physico-chemical characterization of the interactions of vitamins from the B group and nucleobases with Oxaliplatin were experimentally measured by using the UV-Vis spectroscopic method, in the wavelength range from 190 nm to 500 nm (Figure 7). The UV-Vis measurement was carried out at specific time intervals of 1 h, 2 h, 3 h, 24 h, 48 h, 72 h, 96 h, and 168 h. The mixtures of nucleobases or vitamins and Oxaliplatin (synonyms: [SP-4-2-(1R-trans)]-(1,2-Cyclohexanediamine-N,N′)[ethanedioata(2-)-O,O]platinum; Oxaliplatinum; (SP-4-2)-[(1R,2R)-Cyclohexane-1,2-diamine-κN,κN′]-[ethanedioato(2-)-κO1,κO2]platinum) were incubated in buffer: 1 mmol L^{-1} phosphate buffer, 4 mmol L^{-1} sodiumchloride, pH 7.4. Absorption maxima were determined for all mixtures.

Some of the mixtures showed more than one absorption maximum (Figure 7). The results of the experimental studies confirmed the in silico findings. The mixtures with nucleobases (adenine, guanine, cytosine) showed a significant decrease in the absorbance maximum in relation to the baseline (0 h). In the case of complexation of Oxaliplatin with adenine, where maximum absorption was observed at λ = 260 nm, a significant decrease in baseline is observed after 72 h, approximately 3.5 times the initial concentration of adenine in the mixture, and over 14 times after 168 h. Slight decrease in the concentration of adenine in the mixture, one time, is observed after 1 h, 2 h, 3 h. After 48 h, the concentration of adenine in the mixture is reduced twofold from the initial concentration. The decrease in the absorbance of baseline is caused by the formation of adenine-Oxaliplatin complexes. Guanine has two absorbance maxima, at λ = 218 nm and λ = 270 nm. Only the latter one decreases after the complexation of guanine with Oxaliplatin (Figure 7). After the complexation of guanine with Oxaliplatin, a shift in the absorbance maximum of guanine towards shorter wavelengths, λ = 260 nm, was noticed. First, a significant drop in baseline was observed after 24 h. Even so, the greatest decrease in the absorbance maximum, of about 1.5 times, occurred after 72 h. Extending the incubation time of the mixture of guanine with Oxaliplatin did not lead to any further changes in the amount of the absorbance peak.

Cytosine showed two maxima of absorption, at $\lambda = 227$ nm and $\lambda = 266$ nm. Both of these significantly decreased after 168 h: by 2.7 times and 4.5 times, respectively. After 72 h, the complexation of cytosine–Oxaliplatin causes a decrease in the initial concentration of this nucleobase in the mixture by only onefold. It was observed that the time range 1–3 h is not enough for complexation in this case. The complexation of Oxaliplatin with B vitamins is much slower than that of nucleobases. Up to 3 h, the decrease in concentrations of vitamins in the mixture are insignificant. Meaningful drops in the concentration of vitamin B in the mixture, greater than onefold each time, are observed after 72 h and 168 h. The pyridoxal phosphate (vitamin B6) interacts best with the chemotherapeutic, among all vitamins studied. In this case, after 168 h a decrease in the concentration of about 1.5 times is observed. In the case of the vitamins B1 (thiamine) and B2 (riboflavin), after 72 h a decrease of greater than 1 time is observed in the baseline and, respectively, after a little more after 168 h. In the case of vitamins B3 (niacin) and B6 (pyridoxal phosphate), a significant decrease in the baseline is visible after 96 h, while after 168 h there is no further reduction in the concentration of the tested vitamins. The size of the reduction in the concentration of the tested vitamins depends little on the wavelength at which the maximum absorbance occurs.

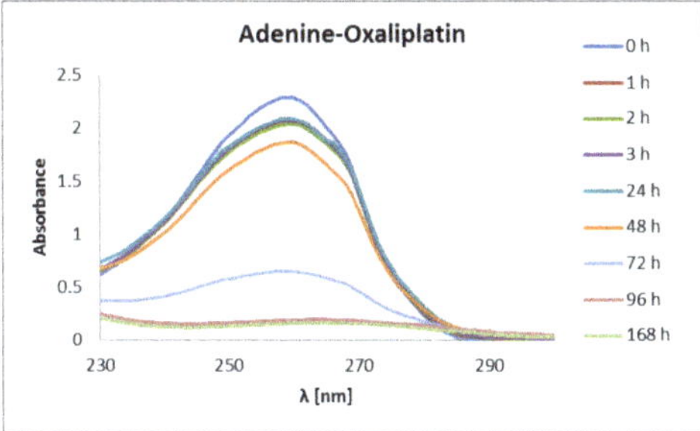

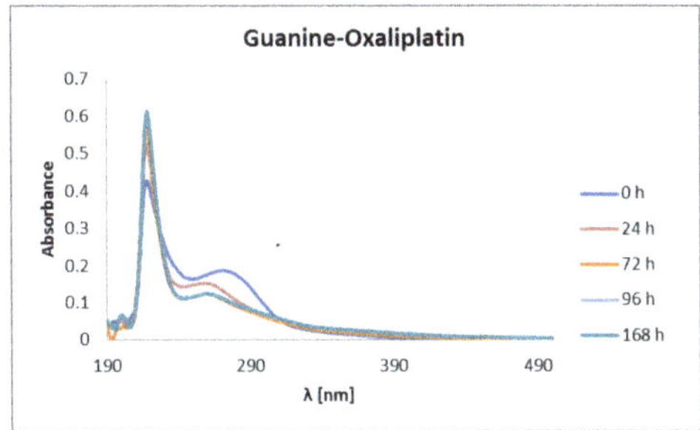

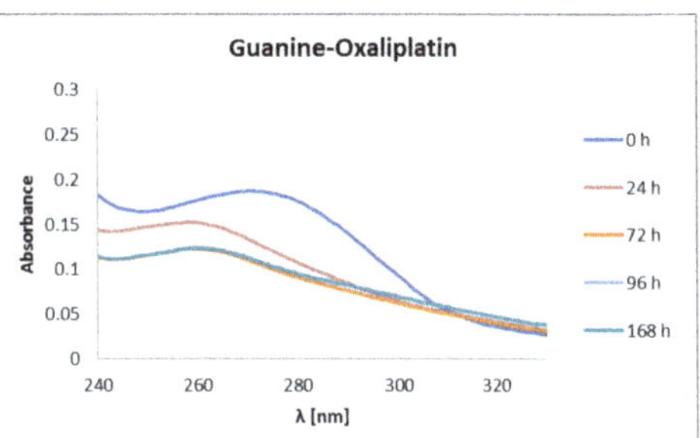

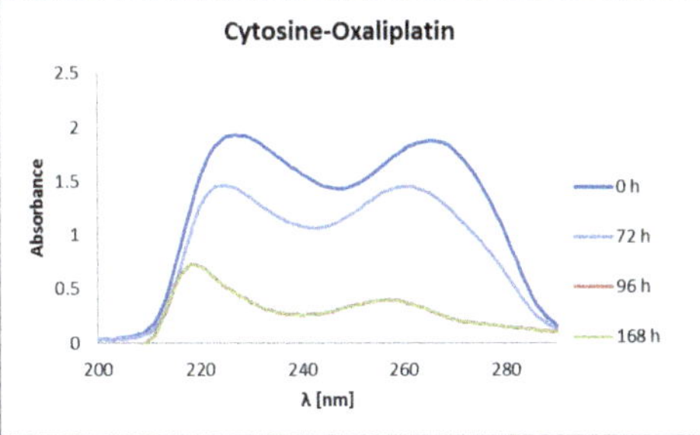

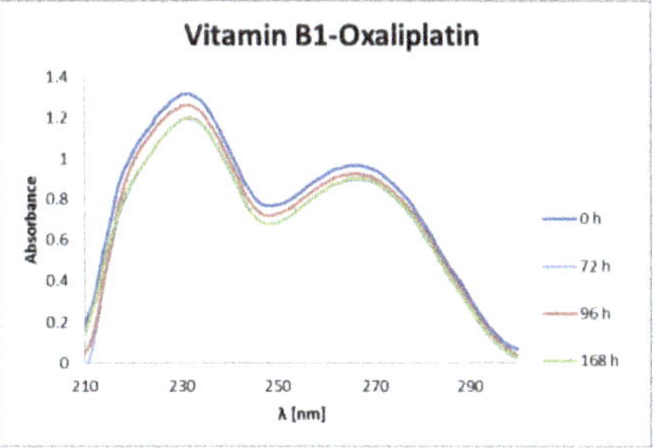

Figure 7. *Cont.*

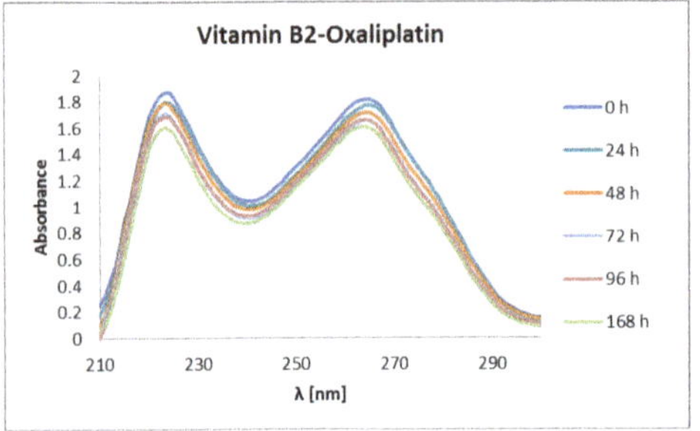
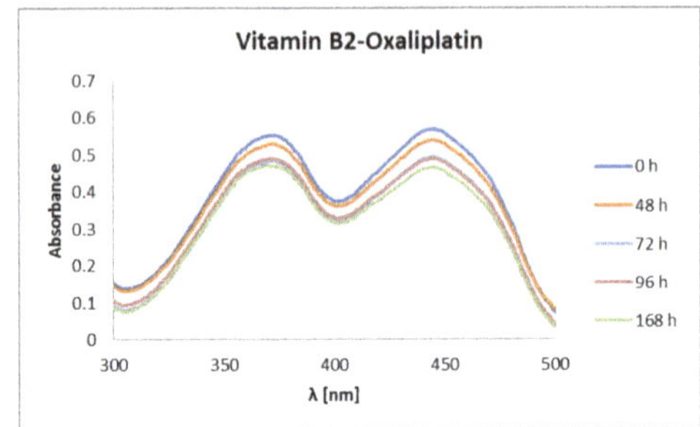
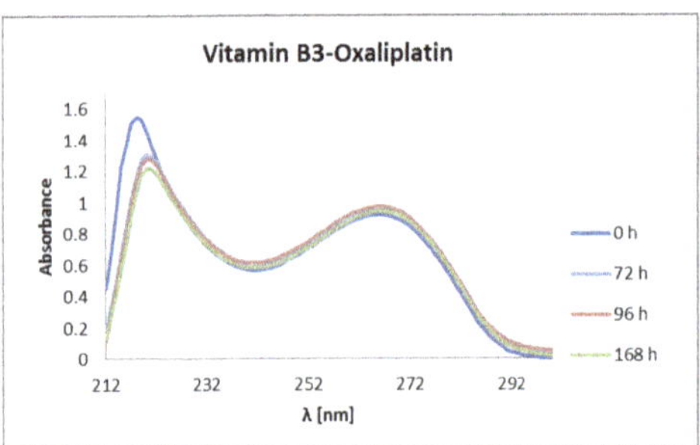
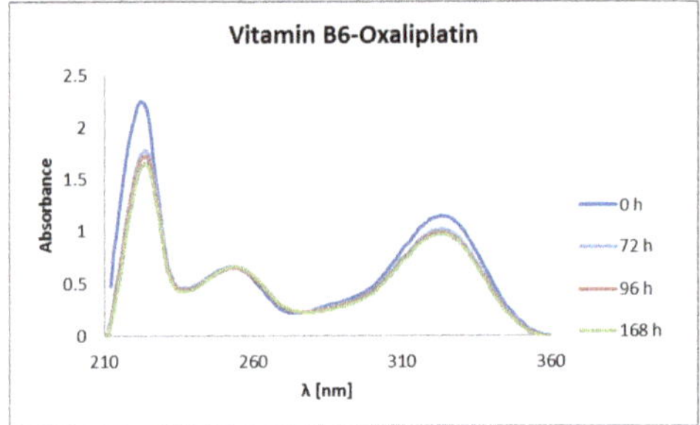
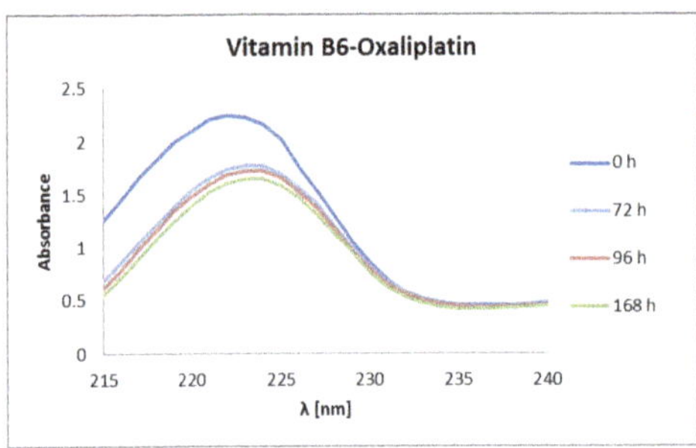
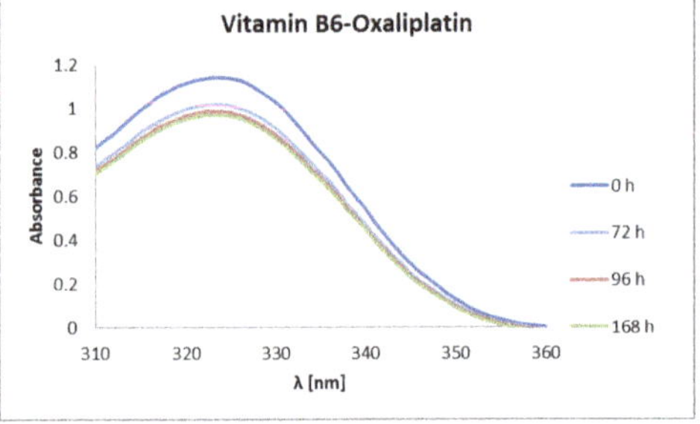

Figure 7. UV-Vis absorption spectra of vitamins from the B group and Oxaliplatin mixture in an incubation buffer (1 mmol L^{-1} phosphate buffer, 4 mmol L^{-1} sodium-chloride, pH 7.4). 0h shows initial concentrations of nucleobases or B vitamins before the Oxaliplatin is added to the mixture. The nucleobases and the B vitamins were incubated at 37 °C with Oxaliplatin in a ratio of 2:1.

Let us compare the theoretical and experimental excitation energies for the selected complexes. The theoretical findings concerning the most intensive singlet-singlet transitions for PtClDACH-Ade and PtH$_2$ODACH-Ade are 243.56 nm and 243.06 nm, respectively. The experimental data showed that the most intensive band occurred at 260 nm. Concerning the complex with guanine, the computed values are 254.22 nm for PtClDACH-Gua and 260.50 nm for PtH$_2$ODACH-Gua, while the experimental spectra showed two absorbance maxima, at 218 nm and 270 nm, respectively. The last nucleic base taken into consideration in the study was cytosine. For the complexes with a chlorine atom and water molecule, the most intensive transitions were noticed at 250.34 nm (for PtClDACH-Cyt) and 250.54 nm (for PtH$_2$ODACH-Cyt), whereas the experimental findings showed two absorption maxima, at 227 nm and 266 nm. The computed values for vitamin B1 are as follows: 252.64 nm for PtClDACH-vit. B1(N$_1$), 248.69 nm for PtH$_2$ODACH-vit. B1(N$_1$), 255.04 nm for PtClDACH-vit. B1(N$_7$), and 255.04 nm for PtH$_2$ODACH-vit. B1(N$_7$), while the experimental data showed the absorption maximum at 231 nm and 266 nm, respectively. For the complexes with vitamin B2, we obtained 443.05 nm for PtClDACH-vitB2 and

447.71 nm for PtH$_2$ODACH-vitB2, based on TD-DFT, which is in a good agreement with experimental data locating the maximum absorbance band at 445 nm (as well as at 221 nm, 266 nm, and 372 nm). Concerning vitamin B3, the computational results showed the most intensive singlet-singlet transitions at 301.58 nm for PtClDACH-vitB3 and 250.47 nm for PtH$_2$ODACH-vitB3. In the experimental spectrum, the absorption maximum was found at 218 nm and 266 nm. Finally, we have computed absorption spectra for vitamin B6. It was found that the absorbance maxima occurred at 258.68 nm for PtClDACH-vitB6 and at 258.55 nm for PtH$_2$ODACH-vitB6. The experimental spectrum showed the absorbance maximum at 222 nm, 252 nm, and 323 nm. The computed UV-Vis spectra are shown in Figure S1. The comparison showed a very good agreement between that computed at the MN15/def2-TZVP level of theory with application of the Polarizable Continuum Model (PCM) and water as a solvent with experimental spectra.

3. Materials and Methods

Since B vitamins are similar in structure to nucleobases, possessing aromatic rings with the nitrogen atom N$_7$/N$_1$, they could easily react with Oxaliplatin (as has been shown in earlier studies of Cisplatin) [24]. Such reaction could decrease the therapeutic effect of the anticancer drug. This is the main reason for the chemical affinity studies of Oxaliplatin to vitamins from the B group. Two possible reactions were investigated in which monoaqua or diaqua complexes play an electrophile role [38], namely [PtH$_2$OClDACH]$^+$ and [Pt(H$_2$O)$_2$DACH]$^{2+}$. In this respect, theoretical and experimental studies were performed.

3.1. Theoretical Study

Density functional theory (DFT) [39,40] was applied to develop theoretical models describing the geometric, energetic, and electronic structure parameters of the compounds studied. The models for quantum-chemical simulations were prepared manually using the Molden program [41]. The energy minimization was carried out at the B3LYP/6-31G(d, p) [42] and MN15/def2-TZVP [36,37] levels of theory. However, the atomic orbitals of the platinum atom were represented by the LANL2DZ basis set, which includes relativistic effective core potentials needed in the case of heavy atoms concerning the simulations with application of the B3LYP functional [43]. Such methodology was applied by Baik and co-workers [35]. The quantum-chemical computations were performed using the Gaussian 09 Rev. A.02 [44] and Gaussian 16 Rev. C.01 [45,46] suite of programs. The harmonic vibrational frequencies calculations were carried out to confirm that the obtained structures correspond with minima on the potential energy surface (PES) as well as to derive the zero-point energy (ZPE). The same level of theory was used in the case of vibrational entropy corrections at room temperature. A self-consistent reaction field (SCRF) approach was applied to reproduce the polar environment influence and to estimate the solvation free energies [47,48]. The polarizable continuum model (IEF-PCM formulation) was used to describe the aqueous solution with a dielectric constant of 78 for water [49]. By adding thermal corrections to the enthalpy and entropy terms, contributions of ZPE corrections and continuum solvation free energies, the chemical affinity was computed. Finally, for six selected complexes (denoted as: PtClDACH-Ade, PtH$_2$ODACH-Ade, PtClDACH-Cyt, PtH$_2$ODACH-Cyt, PtClDACH-Gua, PtH$_2$ODACH-Gua) molar volume was computed, and the electronic structure was analyzed on the basis of atoms in molecules (AIM) theory [50]. The electron density and its Laplacian were estimated at bond critical points (BCPs) and the potential electronic energy density (V$_r$) was computed. The AIM analysis was carried out with assistance of the Multiwfn 3.7 program [51]. Finally, time-dependent density functional theory (TD-DFT) [47] was applied for selected complexes investigated experimentally (for more details see the text below). We have applied the same computational setup as described above (MN15/def2-TZVP, PCM) concerning the DFT calculations. We have carried out calculations of the vertical excitation energies from the ground state (S$_0$) to the first five excited states (S$_1$, S$_2$, S$_3$, S$_4$, and S$_5$) for the optimized complexes. The TD-

DFT study was performed with the Gaussian 16 Rev. C.01 [45,46] suite of programs. The graphical presentation of the obtained results was prepared with the VMD 1.9.3 program [52].

3.2. Experimental Study (UV-Vis Measurements)

The spectrophotometrical measurements were performed on a Biosens UV-6000 spectrophotometer (Biosens, Warsaw, Poland). Oxaliplatin (synonims: [SP-4-2-(1R-trans)]-(1,2-Cyclohexanediamine-N,N′)[ethanedioata(2–)-O,O]platinum; Oxaliplatinum; (SP-4-2)-[(1R,2R)-Cyclohexane-1,2-diamine-κN,κN′]-[ethanedioato(2-)-κO1,κO2]platinum), and vitamins from B group were dissolved in the incubation buffer (1 mmol L^{-1} phosphate buffer, 4 mmol L^{-1} sodiumchloride, pH 7.4). During the measurements, of the nucleobases and the vitamins from B group were incubated at 37 °C with Oxaliplatin in a ratio of 2:1. The initial concentrations of study structures are 0.352 mmol L^{-1} for adenine, 0.614 mmol L^{-1} for cytosine, 0.503 mmol L^{-1} for guanine, 0.178 mmol L^{-1} for vitamin B1 (thiamine), 0.107 mmol L^{-1} for vitamin B2 (riboflavin), 0.712 mmol L^{-1} for vitamin B3 (niacin), and 0.365 mmol L^{-1} for vitamin B6 (pyridoxal phosphate). Physico-chemical characteristics of the interaction of nucleobases (adenine, guanine, cytosine) or vitamins from B group with Oxaliplatin was performed by using UV-Vis spectroscopic technique, in wavelength range from 190 nm to 500 nm. Depending on the type of nucleobases and vitamins, the tested compounds showed maximum absorptions at different wavelengths (Figures 7 and S2). The cytosine and vitamins from B group showed more than one absorption maximum (Figures 7 and S2). The samples of mixture of nucleobases or vitamins B and Oxaliplatin were taken after 0 h, 1 h, 2 h, 3 h, 24 h, 72 h, and 168 h incubation.

4. Conclusions

Theoretical studies based on DFT, TD-DFT, and experimental (UV-Vis) investigations were carried out on complexes formed with nucleobases/vitamins from the B group with Oxaliplatin. The affinity of the complexes was estimated to shed more light on the interactions and their consequences on cancer therapy. The order of affinity of Oxaliplatin for B vitamins for both monoaqua [PtH$_2$OClDACH]$^+$ and diaqua [Pt(H$_2$O)$_2$DACH]$^{2+}$ derivatives is niacin (vitamin B3), pyridoxal phosphate (vitamin B6), thiamine (vitamin B1), and riboflavin (vitamin B2), as obtained at the B3LYP/6-31G(d,p) level of theory. However, in the case of the MN15/def2-TZVP level of theory, the order of the affinity of Oxaliplatin for the vitamins studied was different: from thiamine (vitamin B1), to pyridoxal phosphate (vitamin B6), riboflavin (vitamin B2), and up to niacin (vitamin B3). Theoretical studies confirmed the clinical observations and suggest high therapeutic effectiveness of Oxaliplatin in cancer treatment. The AIM theory application revealed the existence of intramolecular non-covalent interactions, which could stabilize the conformation and further provided an impact on the interactions with macromolecules. Qualitative agreement between the computed and experimental absorption spectra was observed. The UV-Vis measurements confirmed the findings obtained based on theoretical approaches. A significant decrease of the concentrations of nucleobases in the mixtures were observed after the complexation with Oxaliplatin. It was noted that the vitamins from the B group form weaker and more time-consuming complexes with the chemotherapeutic studied. It can be stated that the B vitamins from the juice the patients drink are less likely to interfere with cancer treatment involving Oxaliplatin, based on the data presented in the study.

Supplementary Materials: The following supporting information can be downloaded at: https://www.mdpi.com/article/10.3390/ijms231810567/s1.

Author Contributions: Conceptualization, B.S.; methodology, B.S. and A.J.; software, B.S. and K.W.; validation, B.S. and A.J.; formal analysis, B.S. and A.J.; investigation, B.S. and K.W.; resources, B.S. and K.W.; data curation, B.S. and A.J.; writing—original draft preparation, B.S. and A.J.; writing—review and editing, B.S., A.J., P.C. and K.W.; visualization, B.S., K.W. and P.C.; supervision, B.S.; project administration, B.S.; funding acquisition, B.S. All authors have read and agreed to the published version of the manuscript.

Funding: Publication of this article (article processing charge) was financially supported by the Excellence Initiative Research University (IDUB) program for the Nicolaus Copernicus University (https://www.umk.pl/idub accessed on 8 August 2022).

Institutional Review Board Statement: Not applicable.

Informed Consent Statement: All authors have read and agreed to the published version of the manuscript.

Data Availability Statement: The data presented in the current study are available in the article and in the associated Supplementary Material.

Acknowledgments: The authors gratefully acknowledge the generous grants of CPU time, facilities, and support provided by the Wrocław Centre for Networking and Supercomputing (WCSS), the Poznań Supercomputing and Networking Center (PSNC), the Academic Computing Centre Cyfronet-Kraków (Prometheus supercomputer, part of the PL-Grid infrastructure), and the Centre of Informatics Tricity Academic Supercomputer and networK (CI TASK).

Conflicts of Interest: The authors declare no conflict of interest.

References

1. Seetharam, R.N. Oxaliplatin: Preclinical perspectives on the mechanisms of action, response and resistance. *Ecancermedicalscience* **2010**, *3*, 153. [CrossRef] [PubMed]
2. Martinez-Balibrea, E.; Martínez-Cardús, A.; Ginés, A.; Ruiz de Porras, V.; Moutinho, C.; Layos, L.; Manzano, J.L.; Bugés, C.; Bystrup, S.; Esteller, M.; et al. Tumor-Related Molecular Mechanisms of Oxaliplatin Resistance. *Mol. Cancer Ther.* **2015**, *14*, 1767–1776. [CrossRef] [PubMed]
3. Cvitkovic, E. Ongoing and unsaid on oxaliplatin: The hope. *Br. J. Cancer* **1998**, *77* (Suppl. S4), 8–11. [CrossRef] [PubMed]
4. Wang, D.; Lippard, S.J. Cellular processing of platinum anticancer drugs. *Nat. Rev. Drug Discov.* **2005**, *4*, 307–320. [CrossRef]
5. Di Francesco, A.M.; Ruggiero, A.; Riccardi, R. Cellular and molecular aspects of drugs of the future: Oxaliplatin. *Cell. Mol. Life Sci.* **2002**, *59*, 1914–1927. [CrossRef]
6. Rivory, L.P. Experimental and Clinical Pharmacology: New drugs for colorectal cancer-mechanisms of action. *Aust. Prescr.* **2002**, *25*, 108–110. [CrossRef]
7. Farrell, N.P. Preclinical perspectives on the use of platinum compounds in cancer chemotherapy. *Semin. Oncol.* **2004**, *31*, 1–9. [CrossRef]
8. Sharma, S.; Gong, P.; Temple, B.; Bhattacharyya, D.; Dokholyan, N.V.; Chaney, S.G. Molecular dynamic simulations of cisplatin- and oxaliplatin-d(GG) intrastand cross-links reveal differences in their conformational dynamics. *J. Mol. Biol.* **2007**, *373*, 1123–1140. [CrossRef]
9. Chaney, S.G.; Campbell, S.L.; Bassett, E.; Wu, Y. Recognition and processing of cisplatin- and oxaliplatin-DNA adducts. *Crit. Rev. Oncol. Hematol.* **2005**, *53*, 3–11. [CrossRef]
10. Raymond, E.; Chaney, S.G.; Taamma, A.; Cvitkovic, E. Oxaliplatin: A review of preclinical and clinical studies. *Ann. Oncol.* **1998**, *9*, 1053–1071. [CrossRef]
11. Rixe, O.; Ortuzar, W.; Alvarez, M.; Parker, R.; Reed, E.; Paull, K.; Fojo, T. Oxaliplatin, tetraplatin, cisplatin, and carboplatin: Spectrum of activity in drug-resistant cell lines and in the cell lines of the national cancer institute's anticancer drug screen panel. *Biochem. Pharmacol.* **1996**, *52*, 1855–1865. [CrossRef]
12. Martin, L.P.; Hamilton, T.C.; Schilder, R.J. Platinum resistance: The role of DNA repair pathways. *Clin. Cancer Res.* **2008**, *14*, 1291–1295. [CrossRef]
13. Meyerhardt, J.A.; Mayer, R.J. Systemic therapy for colorectal cancer. *N. Engl. J. Med.* **2005**, *352*, 476–487. [CrossRef]
14. Webster, R.G.; Brain, K.L.; Wilson, R.H.; Grem, J.L.; Vincent, A. Oxaliplatin induces hyperexcitability at motor and autonomic neuromuscular junctions through effects on voltage-gated sodium channels. *Br. J. Pharmacol.* **2005**, *7*, 1027–1039. [CrossRef]
15. Gebremedhn, E.G.; Shortland, P.J.; Mahns, D.A. The incidence of acute oxaliplatin-induced neuropathy and its impact on treatment in the first cycle: A systematic review. *BMC Cancer* **2018**, *1*, 410. [CrossRef]
16. Fukuda, M.; Ohe, Y.; Kanzawa, F.; Oka, M.; Hara, K.; Saijo, N. Evaluation of novel platinum complexes, inhibitors of topoisomerase I and II in non-small cell lung cancer (NSCLC) sublines resistant to cisplatin. *Anticancer Res.* **1995**, *2*, 393–398. [CrossRef]

17. Kidani, Y.; Inagaki, K.; Saito, R.; Tsukagoshi, S. Synthesis and anti tumor activities of platinum(II) complexes of 1,2 diaminocyclohexane isomers and their related derivatives. *Wadley Med. Bull.* **1976**, *7*, 197–209.
18. Göschl, S.; Schreiber-Brynzak, E.; Pichler, V.; Cseh, K.; Heffeter, P.; Jungwirth, U.; Jakupec, M.A.; Berger, W.; Keppler, B.K. Comparative studies of oxaliplatin-based platinum(iv) complexes in different in vitro and in vivo tumor models. *Metallomics* **2017**, *9*, 309–322. [CrossRef]
19. Riccardi, A.; Ferlini, C.; Meco, D.; Mastrangelo, R.; Scambia, G.; Riccardi, R. Antitumour activity of oxaliplatin in neuroblastoma cell lines. *Eur. J. Cancer* **1999**, *1*, 86–90. [CrossRef]
20. Dunn, T.A.; Schmoll, H.J.; Grünwald, V.; Bokemeyer, C.; Casper, J. Comparative cytotoxicity of oxaliplatin and cisplatin in non-seminomatous germ cell cancer cell lines. *Investig. New Drugs* **1997**, *15*, 109–114. [CrossRef]
21. Pendyala, L.; Creaven, P.J. In Vitro Cytotoxicity, Protein Binding, Red Blood Cell Partitioning, and Biotransformation of Oxaliplatin. *Cancer Res.* **1993**, *53*, 5970–5976.
22. Kraker, A.J.; Moore, C.W. Accumulation of cis-Diamminedichloroplatinum(II) and Platinum Analogues by Platinum-resistant Murine Leukemia Cells in Vitro. *Cancer Res.* **1988**, *48*, 9–13.
23. Monti, D.M.; Loreto, D.; Iacobucci, I.; Ferraro, G.; Pratesi, A.; D'Elia, L.; Monti, M.; Merlino, A. Protein-based delivery systems for anticancer metallodrugs: Structure and biological activity of the oxaliplatin/β-lactoglobulin adduct. *Pharmaceuticals* **2022**, *15*, 425. [CrossRef]
24. Szefler, B.; Czeleń, P.; Szczepanik, A.; Cysewski, P. Does the Affinity of Cisplatin to B-Vitamins Impair the Therapeutic Effect in the Case of Patients with Lung Cancer-consuming Carrot or Beet Juice? *Anticancer Agents Med. Chem.* **2019**, *19*, 1775–1783. [CrossRef] [PubMed]
25. Szefler, B.; Czeleń, P.; Krawczyk, P. The Affinity of Carboplatin to B-Vitamins and Nucleobases. *Int. J. Mol. Sci.* **2021**, *22*, 3634. [CrossRef] [PubMed]
26. Hodgkinson, E.; Neville-Webbe, H.L.; Coleman, R.E. Magnesium Depletion in Patients Receiving Cisplatin-based Chemotherapy. *Clin. Oncol.* **2006**, *18*, 710–718. [CrossRef] [PubMed]
27. Fischer, J.; Robin Ganellin, C. Analogue-based Drug Discovery. *Chem. Int.—Newsmag. IUPAC* **2006**, *32*, 12–15.
28. American Association for Cancer Research; Lockwood, G.F.; Greenslade, D.; Brienza, S.; Bayssas, M.; Gamelin, E. *Clinical Cancer Research: An Official Journal of the American Association for Cancer Research*; Association for Cancer Research: Philadelphia, PA, USA, 1995; Volume 6.
29. Puisset, F.; Schmitt, A.; Chatelut, E. Standardization of Chemotherapy and Individual Dosing of Platinum Compounds. *Anticancer Res.* **2014**, *34*, 465–470.
30. Kurter, H.; Yeşil, J.; Daskin, E.; Çalibaşi Koçal, G.; Ellidokuz, H.; Başbinar, Y. Drug Resistance Mechanisms on Colorectal Cancer. *J. Basic Clin. Health Sci.* **2021**, *5*, 88–93. [CrossRef]
31. Han, C.H.; Khwaounjoo, P.; Hill, A.G.; Miskelly, G.M.; McKeage, M.J. Predicting effects on oxaliplatin clearance: In vitro, kinetic and clinical studies of calcium- and magnesium-mediated oxaliplatin degradation. *Sci. Rep.* **2017**, *7*, 4073. [CrossRef]
32. Graham, M.A.; Lockwood, G.F.; Greenslade, D.; Brienza, S.; Bayssas, M.; Gamelin, E. Clinical Pharmacokinetics of Oxaliplatin: A Critical Review. *Clin. Cancer Res.* **2000**, *6*, 1205–1218.
33. Alcindor, T.; Beauger, N. Oxaliplatin: A review in the era of molecularly targeted therapy. *Curr. Oncol.* **2011**, *18*, 18–25. [CrossRef]
34. World Cancer Research Fund/American Institute for Cancer. Diet, nutrition, physical activity and breast cancer. *Contin. Updat. Proj. Expert Rep.* **2018**, 1–116.
35. Baik, M.-H.; Friesner, R.A.; Lippard, S.J. Theoretical Study of Cisplatin Binding to Purine Bases: Why Does Cisplatin Prefer Guanine over Adenine? *J. Am. Chem. Soc.* **2003**, *125*, 14082–14092. [CrossRef]
36. Yu, H.S.; He, X.; Li, S.L.; Truhlar, D.G. MN15: A Kohn–Sham global-hybrid exchange–correlation density functional with broad accuracy for multi-reference and single-reference systems and noncovalent interactions. *Chem. Sci.* **2016**, *7*, 5032–5051. [CrossRef]
37. Weigend, F.; Ahlrichs, R. Balanced basis sets of split valence, triple zeta valence and quadruple zeta valence quality for H to Rn: Design and assessment of accuracy. *Phys. Chem. Chem. Phys.* **2005**, *7*, 3297. [CrossRef]
38. Legendre, F.; Bas, V.; Kozelka, J.; Chottard, J.C. A complete kinetic study of GG versus AG platination suggests that the doubly aquated derivatives of cisplatin are the actual DNA binding species. *Chem.-A Eur. J.* **2000**, *6*, 2002–2010. [CrossRef]
39. Kohn, W.; Sham, L.J. Self-consistent equations including exchange and correlation effects. *Phys. Rev.* **1965**, *140*, A1133. [CrossRef]
40. Hohenberg, P.; Kohn, W. Inhomogeneous Electron Gas. *Phys. Rev.* **1964**, *136*, B864–B871. [CrossRef]
41. Schaftenaar, G.; Noordik, J.H. Molden: A pre- and post-processing program for molecular and electronic structures *. *J. Comput. Aided. Mol. Des.* **2000**, *14*, 123–134. [CrossRef]
42. Becke, A.D. Densityfunctional thermochemistry. III. The role of exact exchange Density-functional thermochemistry. III. The role of exact exchange. *J. Chem. Phys. Addit. Inf. J. Chem. Phys. J. Homepage* **1993**, *98*, 5848. [CrossRef]
43. Hay, P.J.; Wadt, W.R. Ab initio effective core potentials for molecular calculations. Potentials for the transition metal atoms Sc to Hg. *J. Chem. Phys.* **1985**, *82*, 270–283. [CrossRef]
44. Frisch, M.J.; Trucks, G.W.; Schlegel, H.B.; Scuseria, G.E.; Robb, M.A.; Cheeseman, J.R.; Scalmani, G.; Barone, V.; Mennucci, B.; Petersson, G.A.; et al. Gaussian 09 Citation | Gaussian.com. Available online: http://gaussian.com/g09citation/ (accessed on 7 December 2017).
45. Ortiz Quantum Chemistry Group: Collaboration with Gaussian Inc. Available online: https://www.auburn.edu/cosam/faculty/chemistry/ortiz/research/info_gaussian.html (accessed on 5 July 2022).

46. Gaussian 16 | Gaussian.com. Available online: https://gaussian.com/gaussian16/ (accessed on 5 July 2022).
47. Cramer, C.J.; Truhlar, D.G. Implicit Solvation Models: Equilibria, Structure, Spectra, and Dynamics. *Chem. Rev.* **1999**, *99*, 2161–2200. [CrossRef]
48. Marten, B.; Kim, K.; Cortis, C.; Friesner, R.A.; Murphy, R.B.; Ringnalda, M.N.; Sitkoff, D.; Honig, B. New Model for Calculation of Solvation Free Energies: Correction of Self-Consistent Reaction Field Continuum Dielectric Theory for Short-Range Hydrogen-Bonding Effects. *J. Phys. Chem.* **1996**, *100*, 11775–11788. [CrossRef]
49. Barone, V.; Cossi, M.; Tomasi, J. A new definition of cavities for the computation of solvation free energies by the polarizable continuum model. *J. Chem. Phys.* **1998**, *107*, 3210–3221. [CrossRef]
50. Bader, R.F.W. *Atoms in Molecules: A Quantum Theory*; Clarendon Press: Oxford, UK, 1990; ISBN 9780198558651.
51. Lu, T.; Chen, F. Multiwfn: A multifunctional wavefunction analyzer. *J. Comput. Chem.* **2012**, *33*, 580–592. [CrossRef]
52. Humphrey, W.; Dalke, A.; Schulten, K. VMD: Visual molecular dynamics. *J. Mol. Graph.* **1996**, *14*, 33–38. [CrossRef]

International Journal of
Molecular Sciences

Review

Will the Interactions of Some Platinum (II)-Based Drugs with B-Vitamins Reduce Their Therapeutic Effect in Cancer Patients? Comparison of Chemotherapeutic Agents such as Cisplatin, Carboplatin and Oxaliplatin—A Review

Beata Szefler * and Przemysław Czeleń

Department of Physical Chemistry, Faculty of Pharmacy, Collegium Medicum, Nicolaus Copernicus University, Kurpińskiego 5, 85-096 Bydgoszcz, Poland
* Correspondence: beatas@cm.umk.pl

Abstract: Pt (II) derivatives show anti-cancer activity by interacting with nucleobases of DNA, thus causing some spontaneous and non-spontaneous reactions. As a result, mono- and diaqua products are formed which further undergo complexation with guanine or adenine. Consequently, many processes are triggered, which lead to the death of the cancer cell. The theoretical and experimental studies confirm that such types of interactions can also occur with other chemical compounds. The vitamins from B group have a similar structure to the nucleobases of DNA and have aromatic rings with single-pair orbitals. Theoretical and experimental studies were performed to describe the interactions of B vitamins with Pt (II) derivatives such as cisplatin, oxaliplatin and carboplatin. The obtained results were compared with the values for guanine. Two levels of simulations were implemented at the theoretical level, namely, B3LYP/6-31G(d,p) with LANL2DZ bases set for platinum atoms and MN15/def2-TZVP. The polarizable continuum model (IEF–PCM preparation) and water as a solvent were used. UV-Vis spectroscopy was used to describe the drug–nucleobase and drug–B vitamin interactions. Values of the free energy (ΔG_r) show spontaneous reactions with mono- and diaqua derivatives of cisplatin and oxaliplatin; however, interactions with diaqua derivatives are more preferable. The strength of these interactions was also compared. Carboplatin products have the weakest interaction with the studied structures. The presence of non-covalent interactions was demonstrated in the tested complexes. A good agreement between theory and experiment was also demonstrated.

Keywords: cisplatin; carboplatin; oxaliplatin; platinum-based drugs; cancer treatment; vitamin B

Citation: Szefler, B.; Czeleń, P. Will the Interactions of Some Platinum (II)-Based Drugs with B-Vitamins Reduce Their Therapeutic Effect in Cancer Patients? Comparison of Chemotherapeutic Agents such as Cisplatin, Carboplatin and Oxaliplatin—A Review. *Int. J. Mol. Sci.* **2023**, *24*, 1548. https://doi.org/10.3390/ijms24021548

Academic Editors: Geoffrey Brown, Enikö Kallay and Andrzej Kutner

Received: 13 December 2022
Revised: 4 January 2023
Accepted: 10 January 2023
Published: 12 January 2023

1. Introduction

Cisplatin, oxaliplatin and carboplatin belong to the same group of anticancer agents, Pt (II) drugs, which contain a heavy metal, i.e., platinum, and have anticancer activity (Figure 1) [1–6]. Cisplatin (cis-diamminedichloroplatinum) and other drugs under consideration are metallic (platinum) coordination compounds with a square planar geometry (Figure 1). Carboplatin (Figure 1b) is a 1,1-cyclobutyldicarboxylat, while oxaliplatin (Figure 1c) is a complex with 1,2-diaminocyclohexane and an oxalate group (DACH) [1–6].

Cisplatin is a chemotherapeutic agent that has activity against a wide range of cancers [7,8], such as lung, ovarian, testicular, bladder, head and neck cancers. It is a drug that has shown high anticancer efficacy against various types of cancer, including carcinomas, lymphomas, germ cell tumors and sarcomas. Carboplatin found its use especially in the treatment of cancer of the ovary, testis, head, neck and small cell lung cancer [9]. Compared to cisplatin, carboplatin is less toxic but therefore less therapeutic and must be used at several times higher doses. It can be used as a monotherapeutic or in combination therapy with another agent [10–13]. Oxaliplatin shows anticancer equivalence to cisplatin in the treatment of esophageal and gastric cancer [14]. It is an alkylating and a cytostatic drug used in chemotherapy of malignant tumors mainly of colorectal cancer [14–23]. Oxaliplatin is used against a broad spectrum of tumors, including some cisplatin- and carboplatin-resistant cell lines [24,25].

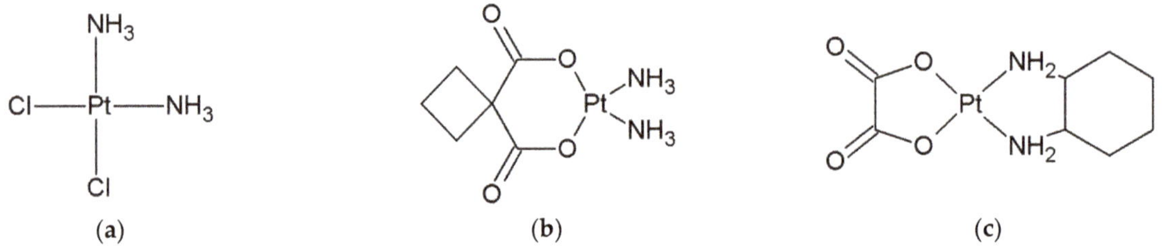

Figure 1. The structure of platinum (II)-based drugs (**a**) cisplatin, (**b**) carboplatin and (**c**) oxaliplatin.

Cisplatin, oxaliplatin and carboplatin have a similar mechanism of action [26–35] and work by interacting with the cell's genetic material, i.e., DNA nucleic acid (deoxyribonucleic acid).

These three Pt (II) drugs react with DNA by having the ability to crosslink with the canonical purine bases, primarily with guanine or adenine but in the case of oxaliplatin also with cytosine, and form crosslinks both within the molecule and between the DNA molecules (Figure 2) [22,31,36]. The adducts formed by these compounds can be mono adducts or intra- and interchain di adducts. The formation of incorrect bonds causes structural changes in the DNA strand, causing its breakage and thus damage to the DNA strand, disrupts DNA synthesis, prevents cell division, and then induces apoptosis in cancer cells [9,31,37–40].

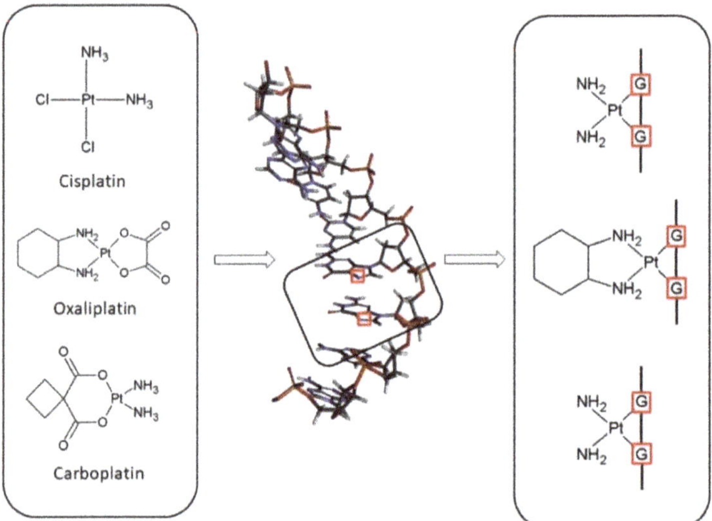

Figure 2. Scheme of interactions of platinum (II)-based drugs (cisplatin, carboplatin and oxaliplatin) with nucleobases from DNA.

The three drugs pass through the cell membrane via passive diffusion or active transport via copper transport CTR1 [41–44]. After entering the cell, the drug in the cell nucleus interacts with the nitrogen atom N_7 of the nucleobase, i.e., guanine or adenine, to form a monoadduct and then a di adduct (Figure 3) [45,46].

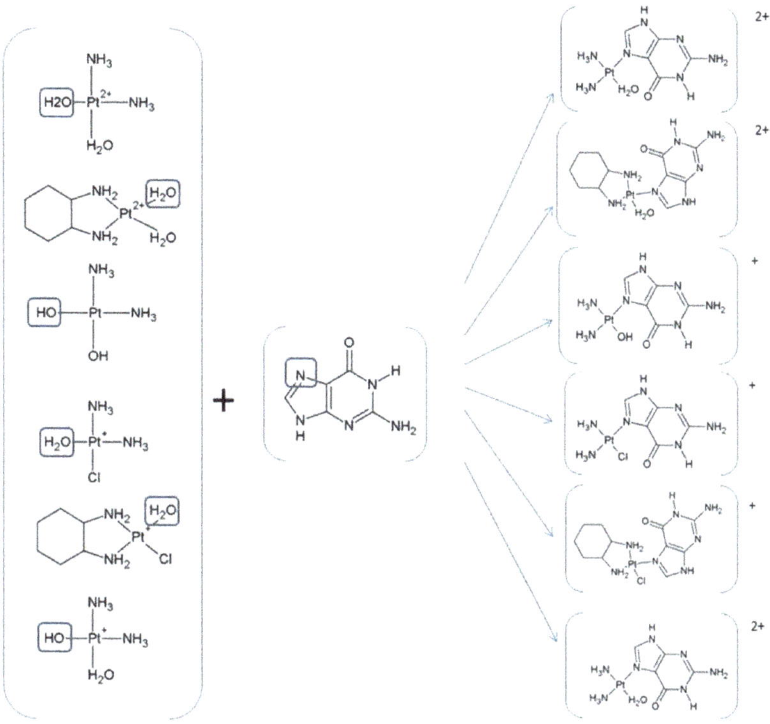

Figure 3. Scheme of complexes of guanine with mono-(di-)aqua and mono-(di-)hydroxy derivatives of three platinum (II)-based drugs.

Pt (II) drugs are administered to the patient in an inactive therapeutic form and in order to interact with the nucleobases in DNA. They must undergo a series of reactions (Figures 2–4).

Cisplatin and oxaliplatin have a similar mechanism of action where for the activation of the relatively inert platinum (II) complexes, the hydrolysis of non-active drugs to diaqua complexes via the monoaqua form is necessary [47–52] (Figures 2–4).

Inside the tumor cell, the molecule of cisplatin (oxaliplatin) undergoes hydrolysis wherein the chlorine ligand is replaced by a water molecule and forms monoaqua derivatives (cis~Pt-monoaqua) (Figures 3 and 4). This process occurs because of the low concentration of chloride ions (~3–20 mM) inside the cell. In the next step, monoaqua derivatives (cis~Pt-monoaqua, Pt-DACH-monoaqua) spontaneously transform to diaqua forms (cis~Pt-diaqua, Pt-DACH-diaqua) (Figure 3). The second step of hydrolysis occurs much more slowly ($K = 2.75 \cdot 10^{-5} s^{-1}$) compared to the first ($K = 5.18 \cdot 10^{-5} s^{-1}$) [45,53] in the case of cisplatin [47].

For the decomposition of carboplatin in water, two steps of the mechanism are needed. In the first one, the ring of 1,1-cyclobutanedicarboxylate must be opened where the malonato ligand is lost. The bond breakage between the platinum and the oxygen atoms in the carboplatin leads to the formation of a molecule with a total charge of one (Figures 3 and 4) [54]. Because of the presence of a water molecule, the monohydroxy derivatives of carboplatin $[(NH_3)_2Pt(OH)H_2O]^+$ undergo further hydrolysis to dihydroxy derivatives $(NH_3)_2Pt(OH)_2$, which via dihydroxylation form $[Pt(NH_3)_2(OH)]^+$ (Figures 3 and 4).

Monoaqua and diaqua derivatives of Pt (II) drugs such as cisplatin, carboplatin and oxaliplatin can easily generate the formation of adducts of platinum by interacting with nucleophilic molecules within the cell, such as DNA, RNA, and proteins [17,18,55].

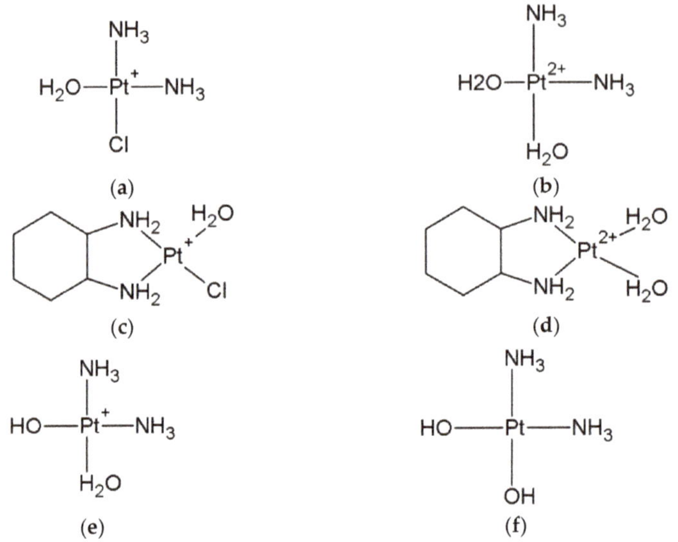

Figure 4. The products of hydrolysis of platinum (II)-based drugs as monoaqua (**a**) cis~Pt-Chloroaqua, (**c**) Pt-DACH-Chloroaqua, diaqua (**b**) cis~Pt-diaqua, (**d**) Pt-DACH-diaqua, monohydroxy (**e**) $Pt(NH_3)_2H_2O(OH)^+$ and dihydroxy derivatives (**f**) $Pt(NH_3)_2(OH)_2$.

Motivation of Study

During the study, the reactivity of hydrolysis products such as monoaqua and diaqua derivatives of three platinum (II)-based drugs, cisplatin, carboplatin and oxaliplatin, was analyzed [47–52]. Such products as [(cis~$[Pt(NH_3)_2Cl(H_2O)]^+$) and (cis~$[Pt(NH_3)_2(H_2O)_2]^{2+}$)] in the case of cisplatin, $[Pt(H_2O)ClDACH]^+$ and $[Pt(H_2O)_2DACH]^{2+}$ in the case of oxaliplatin and $Pt(NH_3)_2H_2O(OH)^+$ plus $Pt(NH_3)_2(OH)_2$ in the case of carboplatin can interact not only with nucleobases (Figures 3 and 4), which is the basis of their therapeutic action, but also with all compounds which have aromatic rings in their structures with lone-pair orbitals analogous to N_7 in purine (Figures 5 and 6). The group of such compounds includes vitamins from B group (Figure 5), namely, vitamins B1, B2, B3 and B6, with their alternative names thiamine, riboflavin, niacin and pyridoxal phosphate [1,56–63], respectively.

B vitamins not only occur naturally in some food products, but mainly their production takes place through synthesis. These B vitamins are contained in many elements of the daily diet, first of all, in vegetables such as carrot, beetroot and tomato, and in large significant doses in their juices and purees [1,56–63]. B vitamins are supplemented in many disorders of the human body. Vitamin B1 (thiamine) is a heterocyclic chemical compound composed of thiazole and pyrimidine rings connected by a methylene bridge. Thiamine participates in glucose metabolism in the human body [1,56–63]. The substance is used in states of vitamin B1 deficiency. It is used in the treatment of beriberi disease, Wernicke's encephalopathy, neuropathies, in the course of diabetes or alcoholism and pain syndromes in rheumatology and neurology. Prophylactically, thiamine is used in states of increased demand for this vitamin, i.e., hyperthyroidism, improper diet, chronic infections and fever, alcoholism, excessive physical effort, persistent diarrhea, general weakness and mental and physical exhaustion. Vitamin B2 (riboflavin) is an organic chemical compound, a combination of ribitol and flavin [1,56–63]. In the human body it acts as a vitamin, the deficiency of which can cause disorders in the functioning of the nervous system and inflammation of the mucous membranes. Vitamin B2 (Riboflavin) occurs in the form of riboflavin-5′-phosphate, which is gradually broken down in the body to riboflavin after ingestion. Vitamin B2-riboflavin-5′-phosphate is sought after by people oriented to conditions of the skin, eyesight and mucous membranes, for support for proper energy metabolism, support for iron absorption, support for the nervous system and mental condition and support in reducing tiredness and fatigue. Niacin, also known as nicotinic acid, is an organic compound and a form of vitamin B3, an essential human nutrient [1,56–63]. It can be produced by plants and animals from the amino acid tryptophan. It contributes to the maintenance of proper energy metabolism, helps in the proper functioning of the nervous system and maintains normal psychological functions. The role of niacin in the human body is wide. Namely, it regulates the burning of fats, proteins and carbohydrates, is necessary for the proper functioning of the brain and peripheral nervous system, participates in the synthesis of neurotransmitters, has antidepressant properties, lowers the level of cholesterol and triglycerides in the blood, dilates blood vessels, lowers blood pressure, regulates the work of the digestive tract (its motor and secretory functions), regulates the function of the liver and pancreas, participates in the synthesis of sex hormones, cortisol, thyroxine and insulin, participates in the formation of erythrocytes, participates in the detoxification processes of the body, improves the appearance of the skin, hair and nails, prevents pellagra and accelerates wound healing. Pyridoxal phosphate (vitamin B6), acts as a coenzyme or prosthetic group and is necessary for the operation of many enzymes such as enzymes involved in the biosynthesis of amino acids and amino acid-derived metabolites, but also enzymes found in the biosynthetic pathways of amino sugars and in the synthesis or catabolism of neurotransmitters [1,56–63]. Vitamin B6 can also inhibit DNA polymerases and several steroid receptors. Inadequate levels of pyridoxal phosphate in the brain can cause neurological dysfunction, particularly epilepsy.

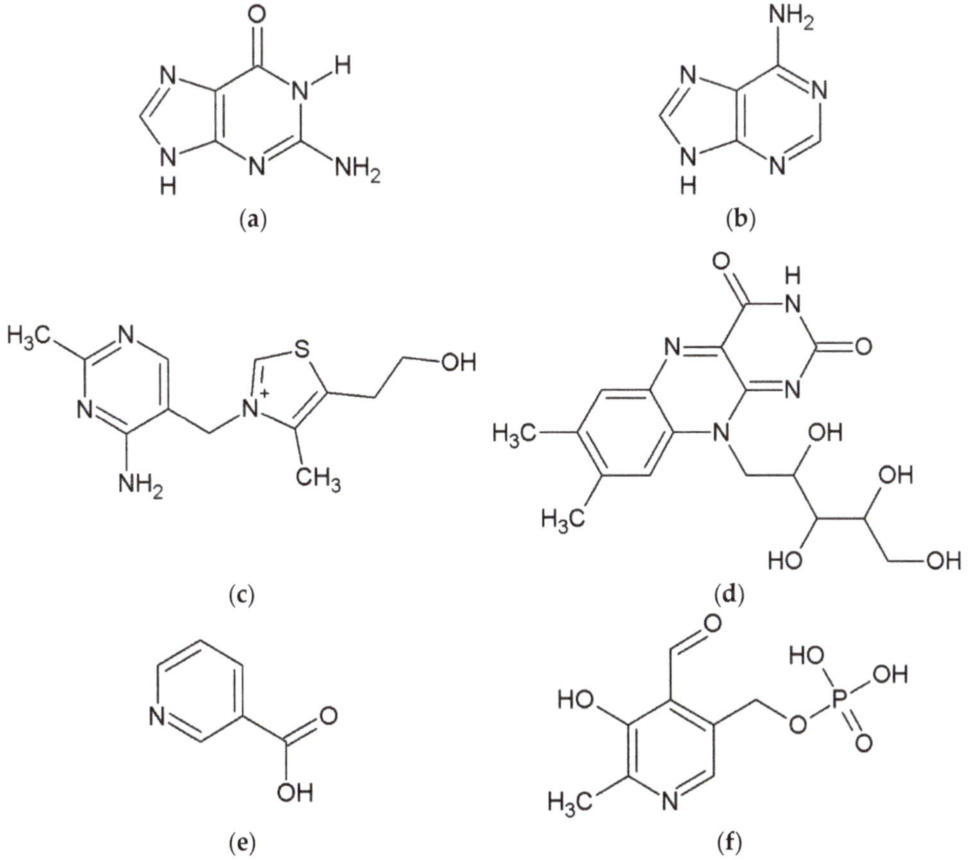

Figure 5. The structure of compounds used in reactions with platinum (II)-based drugs (**a**) guanine, (**b**) adenine (**c**) thiamine, (**d**) riboflavin, (**e**) niacin and (**f**) pyridoxal phosphate.

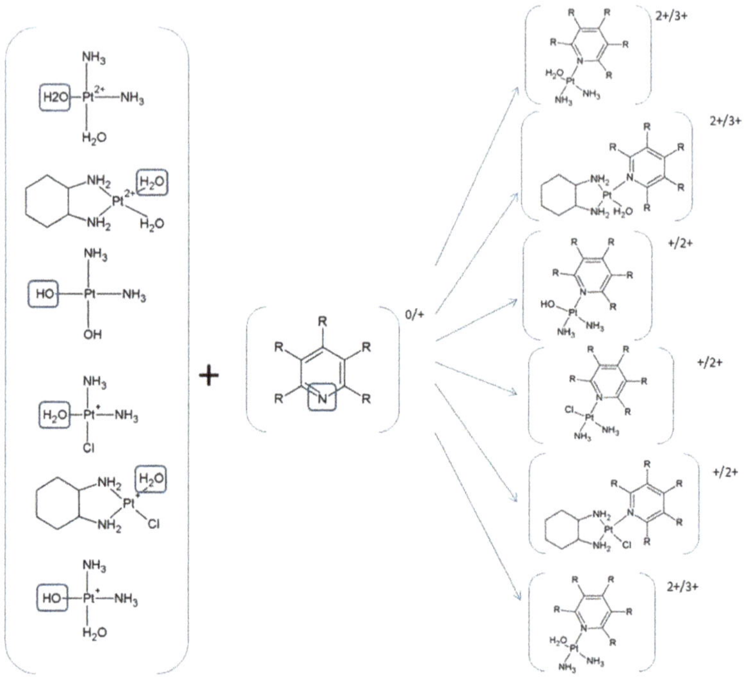

Figure 6. Scheme of complexes of B vitamins with mono-(di-)aqua and mono-(di-)hydroxy derivatives of three platinum (II)-based drugs.

Thus, B vitamins can be a competitive element for nucleobases of DNA, with which these drugs should ultimately bind. The schemes of interactions of the monoaqua and diaqua derivatives of three platinum (II)-based drugs, cisplatin, carboplatin and oxaliplatin, with nucleobases (Figures 2 and 3) and vitamins from B group are given below (Figure 6).

2. Results

In their structure, the B vitamins have a purine and/or pyrimidine rings and thus they have aromatic ring/rings with lone-pair orbitals analogous to N_7 in purine. Thiamine (vitamin B1) has the second addition center for interaction such as the nitrogen atom N_1. The affinities of all vitamins to cis-platinum were analyzed for the first time [49–52]. The estimated values of Gibbs free energy (ΔG_r) of reaction for mono- and diaqua platinum complexes of the studied vitamins were compared with the results for guanine in this review (Figures 7–10, Tables 1–4).

The energetic characteristics of the cis-Pt-Chloroaqua and Pt-DACH-Chloroaqua reactions with B vitamins and nucleobases obtained at the B3LYP/6-31G(d,p)/LANL2DZ level of theory are similar (Table 1, Figure 7) in water solution. The order of affinities of the studied vitamins to the drugs cisplatin and oxaliplatin are the same, from B3 vitamin via B1(N1 → N7) and B6 to B2 vitamin (Table 1, Figure 7). In the case of oxaliplatin, the order is slightly changed by vitamin B6. Among all B vitamins, B3 (niacin) most easily forms complexes with the hydrolysis products of the studied drugs, where ΔG_r values equal to -24.90 kcal/mol and -22.32 kcal/mol for cis~[Pt(NH$_3$)$_2$Cl]$^+$ and for [PtClDACH]$^+$, respectively (Table 1, Figure 7). Vitamin B2 (riboflavin) has the worst ability to create complexes with cis-Pt-Chloroaqua and Pt-DACH-Chloroaqua with values of Gibbs free energy equal to -3.75 kcal/mol and -1.46 kcal/mol (Table 1, Figure 7), respectively. Guanine has the highest affinity for both studied drugs, which was expected (Table 1, Figure 7); ΔG_r equals -28.17 kcal/mol in the case of cisplatin and -33.46 kcal/mol in the case of oxaliplatin.

Quite good affinity is shown by guanine also with the product of hydrolysis of carboplatin with a value of Gibbs free energy equal to -30.89 kcal/mol.

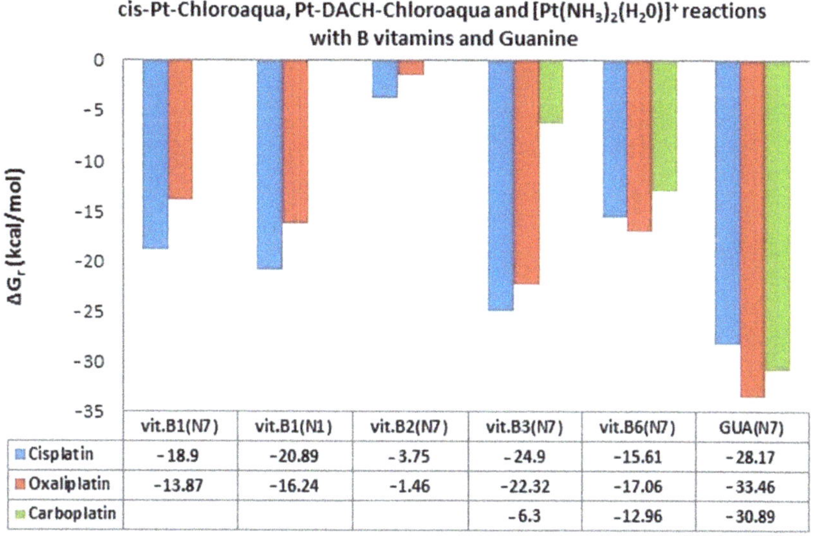

	vit.B1(N7)	vit.B1(N1)	vit.B2(N7)	vit.B3(N7)	vit.B6(N7)	GUA(N7)
Cisplatin	−18.9	−20.89	−3.75	−24.9	−15.61	−28.17
Oxaliplatin	−13.87	−16.24	−1.46	−22.32	−17.06	−33.46
Carboplatin				−6.3	−12.96	−30.89

Figure 7. Gibbs free energy (ΔG_r) values in kcal/mol of complexes of monoaqua derivatives of cisplatin, oxaliplatin and monohydroxy derivatives of carboplatin with B vitamins and guanine. Only negative values of ΔG_r (kcal/mol) are included in the figure at B3LYP/6-31G(d,p)/LANL2DZ level of theory.

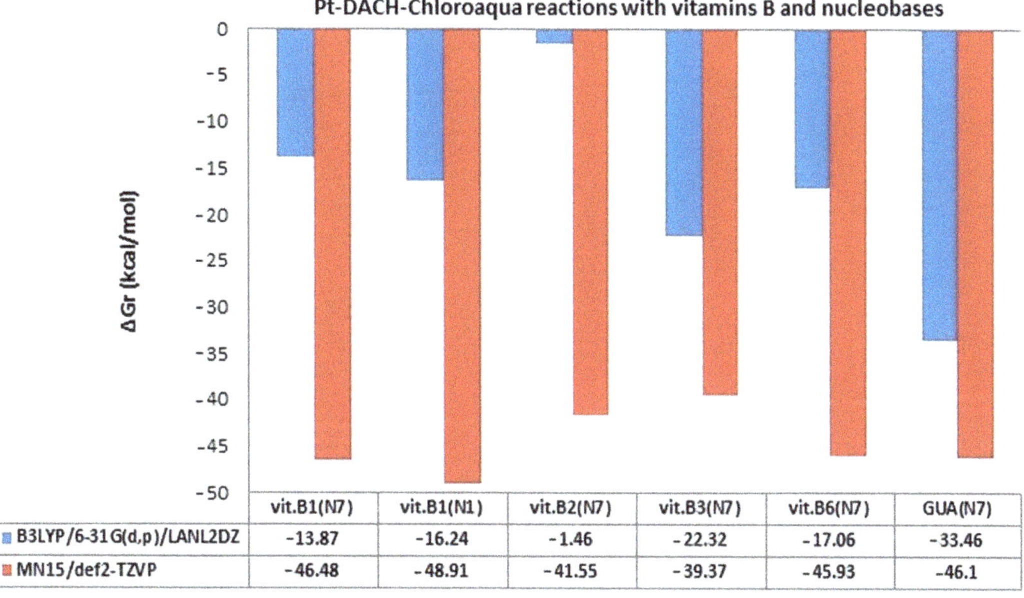

	vit.B1(N7)	vit.B1(N1)	vit.B2(N7)	vit.B3(N7)	vit.B6(N7)	GUA(N7)
B3LYP/6-31G(d,p)/LANL2DZ	−13.87	−16.24	−1.46	−22.32	−17.06	−33.46
MN15/def2-TZVP	−46.48	−48.91	−41.55	−39.37	−45.93	−46.1

Figure 8. Gibbs free energy (ΔG_r) values in kcal/mol of complexes of monoaqua derivatives of oxaliplatin (Pt-DACH-Chloroaqua) with B vitamins and guanine at B3LYP/6-31G(d,p)/LANL2DZ and MN15/def2-TZVP level of theory [52].

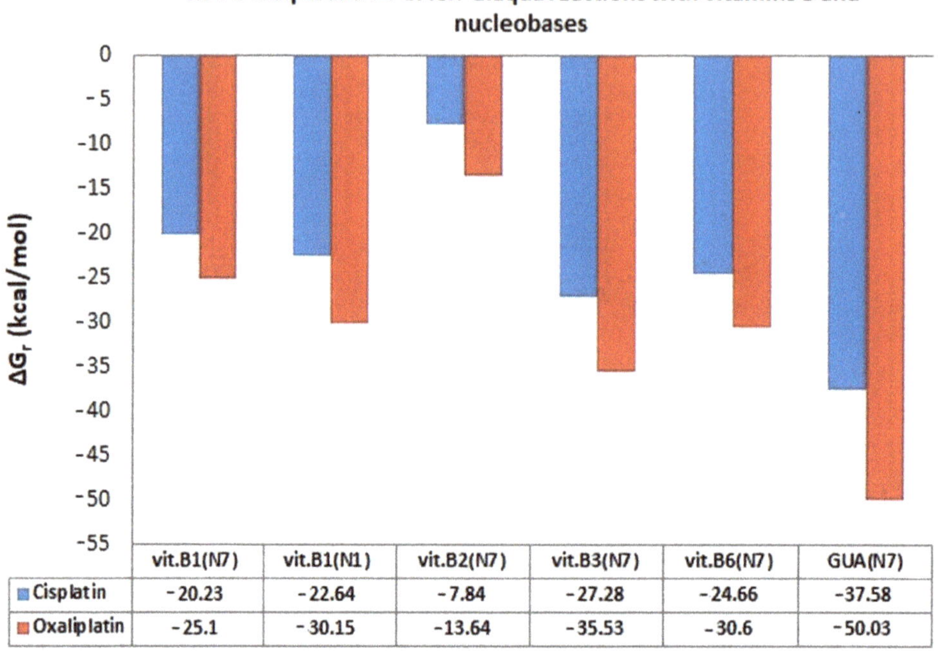

	vit.B1(N7)	vit.B1(N1)	vit.B2(N7)	vit.B3(N7)	vit.B6(N7)	GUA(N7)
Cisplatin	−20.23	−22.64	−7.84	−27.28	−24.66	−37.58
Oxaliplatin	−25.1	−30.15	−13.64	−35.53	−30.6	−50.03

Figure 9. Comparison of Gibbs free energy (ΔG_r) values in kcal/mol of complexes of diaqua derivatives of cisplatin and oxaliplatin (cis-Pt-diaqua and Pt-DACH-diaqua) at B3LYP/6-31G(d,p)/LANL2DZ level of theory [49,52].

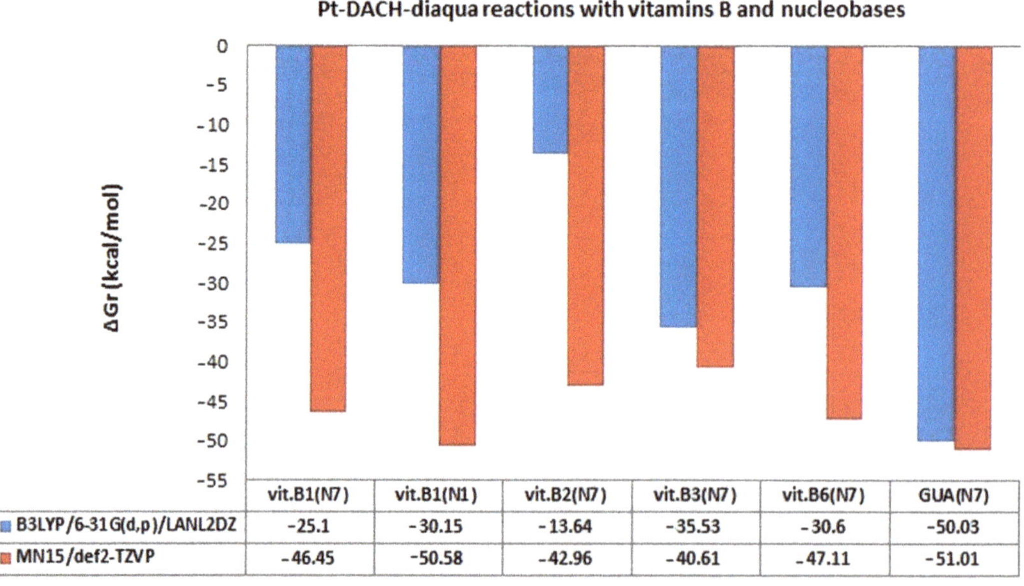

	vit.B1(N7)	vit.B1(N1)	vit.B2(N7)	vit.B3(N7)	vit.B6(N7)	GUA(N7)
B3LYP/6-31G(d,p)/LANL2DZ	−25.1	−30.15	−13.64	−35.53	−30.6	−50.03
MN15/def2-TZVP	−46.45	−50.58	−42.96	−40.61	−47.11	−51.01

Figure 10. Gibbs free energy (ΔG_r) values in kcal/mol of complexes of diaqua derivatives of oxaliplatin (Pt-DACH-diaqua) with B vitamins and guanine at B3LYP/6-31G(d,p)/LANL2DZ and MN15/def2-TZVP level of theory [52].

Table 1. The energetic characteristics of cis-Pt-Chloroaqua, Pt-DACH-Chloroaqua and $Pt(NH_3)_2H_2O(OH)^+$ in the case of cisplatin, oxaliplatin and carboplatin, respectively reactions with vitamins from B group and nucleobase as guanine obtained at the B3LYP/6-31G(d,p)/LANL2DZ level of theory in water solution. All energies are given in kcal/mol. Symbol Pt* stands for $Pt(NH_3)_2Cl$ of cisplatin, Pt** stands for PtClDACH of oxaliplatin and Pt*** stands for $Pt(NH_3)_2(H_2O)$ of carboplatin. Red bold text refers to reactions with oxaliplatin and blue to reactions with carboplatin [49–52].

ΔG_r (kcal/mol)			Reaction
Cisplatin	Oxaliplatin	Carboplatin	
−18.90	−13.87	149.53	$cis[Pt*/**/***]^{+/+/2+} + [B1_{(N7)}]^+ \rightarrow cis[Pt/**/***-B1_{(N7)}]^{2+/2+/3+}$
−20.89	−16.24	144.60	$cis[Pt*/**/***]^{+/+/2+} + [B1_{(N1)}]^+ \rightarrow cis[Pt/**/***-B1_{(N1)}]^{2+/2+/3+}$
−3.75	−1.46	233.04	$cis[Pt*/**/***]^{+/+/2+} + B2_{(N7)} \rightarrow cis[Pt/**/***-B2_{(N7)}]^{+/+/2+}$
−24.90	−22.32	−6.3	$cis[Pt*/**/***]^{+/+/2+} + B3_{(N7)} \rightarrow cis[Pt/**/***-B3_{(N7)}]^{+/+/2+}$
−15.61	−17.06	−12.96	$cis[Pt*/**/***]^{+/+/2+} + B6_{(N7)} \rightarrow cis[Pt/**/***-B6_{(N7)}]^{+/+/2+}$
−28.17	−33.46	−30.89	$cis[Pt*/**/***]^{+/+/2+} + GUA_{(N7)} \rightarrow cis[Pt/**/***-GUA_{(N7)})]^{+/+/2+}$

Table 2. The energetic characteristics of the Pt-DACH-Chloroaqua reactions with vitamin B and nucleobases obtained at the MN15/def2-TZVP level of theory. All energies are given in kcal/mol. Symbol Pt* stands for PtClDACH [52].

Number of Reaction	ΔG_r	Reaction
1	−46.48	$[Pt*]^+ + [B1_{(N7)}]^+ \rightarrow [Pt*-B1_{(N7)}]^{2+}$
2	−48.91	$[Pt*]^+ + [B1_{(N1)}]^+ \rightarrow [Pt*-B1_{(N1)}]^{2+}$
3	−41.55	$[Pt*]^+ + B2 \rightarrow [Pt*-B2]^+$
4	−39.37	$[Pt*]^+ + B3 \rightarrow [Pt*-B3]^+$
5	−45.93	$[Pt*]^+ + B6 \rightarrow [Pt*-B6]^+$
6	−46.10	$[Pt*]^+ + GUA \rightarrow [Pt*-GUA]^+$

Table 3. The energetic characteristics of the cis-Pt-diaqua in case of Cisplatin, the Pt-DACH-diaqua in case of Oxaliplatin and $[Pt(NH_3)_2(OH)]^+$ reactions with vitamins B and nucleobases obtained at the B3LYP/6-31G(d,p)/LANL2DZ level of theory in water solution. All energies are given in kcal/mol. Symbol Pt* stands for $Pt(NH_3)_2(H_2O)$ of Cisplatin, Pt** for PtH_2ODACH of Oxaliplatin and Pt*** for $Pt(NH_3)_2(OH)$ of Carboplatin. Red bold text refers to reactions with products of Oxaliplatin, blue bold text refers to reactions with products of Carboplatin [49,52].

ΔG_r (kcal/mol)			Reaction
Cisplatin	Oxaliplatin	Carboplatin	
−20.23	−25.10	149.53	$cis[Pt*/**/***]^{2+/2+/+} + [B1_{(N7)}]^+ \rightarrow cis[Pt/**-B1_{(N7)}]^{3+/3+/2+}$
−22.64	−30.15	144.60	$cis[Pt*/**/***]^{2+/2+/+} + [B1_{(N1)}]^+ \rightarrow cis[Pt/**-B1_{(N1)}]^{3+/3+/2+}$
−7.84	−13.64	233.04	$cis[Pt*/**/***]^{2+/2+/+} + B2_{(N7)} \rightarrow cis[Pt/**-B2_{(N7)}]^{2+/2+/+}$
−27.28	−35.53	−6.3	$cis[Pt*/**/***]^{2+/2+/+} + B3_{(N7)} \rightarrow cis[Pt/**-B3_{(N7)}]^{2+/2+/+}$
−24.66	−30.60	−12.96	$cis[Pt*/**/***]^{2+/2+/+} + B6_{(N7)} \rightarrow cis[Pt/**-B6_{(N7)}]^{2+/2+/+}$
−37.58	−50.03	−30.89	$cis[Pt*/**/***]^{2+/2+/+} + GUA_{(N7)} \rightarrow cis[Pt/**-GUA_{(N7)}]^{2+/2+/+}$

Relative affinity to carboplatin is shown by B6 vitamin and then B3 vitamin with ΔG_r values equal to −12.96 kcal/mol and −6.3 kcal/mol, respectively. The other tested vitamins from B group have no affinity for this drug and their Gibbs free energy values are positive (Table 1, Figure 7).

As seen in Figure 7, most of the B vitamins interact less with chloroaqua (monoaqua) products of oxaliplatin compared to cisplatin. Guanine forms bonds with chloroaqua products of cisplatin, oxaliplatin and carboplatin more easily compared to B vitamins. Products of hydrolysis of carboplatin have the weakest interaction with the tested vitamins.

The change of the calculation level to a higher level, such as MN15/def2-TZVP, in the case of monoaqua products of oxaliplatin causes the change of the order of affinity for the studied vitamins as follows: vit. $B1_{(N1)} \rightarrow$ vit. $B1_{(N7)} \rightarrow$ vit. B6$\rightarrow$ vit. B2 $\rightarrow$ vit. B3 (Table 2, Figure 8). The values of Gibbs free energy of $B1_{(N1)}$ vitamin and guanine are similar, -46.48 kcal/mol and -46.10 kcal/mol, respectively.

Table 4. The energetic characteristics of the Pt-DACH-diaqua reactions with vitamin B and nucleobases obtained at the MN15/def2-TZVP level of theory. All energies are given in kcal/mol. Pt** symbol is PtH$_2$ODACH [52].

Number of Reaction	ΔG_r	Reaction
1	-46.45	$[Pt^{**}]^{2+} + [B1_{(N7)}]^+ \rightarrow [Pt^{**}\text{-}B1_{(N7)}]^{3+}$
2	-50.58	$[Pt^{**}]^{2+} + [B1_{(N1)}]^+ \rightarrow [Pt^{**}\text{-}B1_{(N1)}]^{3+}$
3	-42.96	$[Pt^{**}]^{2+} + B2 \rightarrow [Pt^{**}\text{-}B2]^{2+}$
4	-40.61	$[Pt^{**}]^{2+} + B3 \rightarrow [Pt^{**}\text{-}B3]^{2+}$
5	-47.11	$[Pt^{**}]^{2+} + B6 \rightarrow [Pt^{**}\text{-}B6]^{2+}$
6	-51.01	$[Pt^{**}]^{2+} + GUA \rightarrow [Pt^{**}\text{-}GUA]^{2+}$

Increasing the calculation level of calculate increases the values of Gibbs free energy (Table 2, Figure 8).

Unexpectedly, guanine does not have the greatest affinity for the drug (Table 2, Figure 8) and takes the value of Gibbs free energy equal to -46.10 kcal/mol. Vitamin B1 shows the greatest affinity to the drug, with particular participation in the reaction coming from the N_1 nitrogen atom in the aromatic ring of the structure. Figure 8 shows that increasing the level of calculation from B3LYP/6-31G(d,p)/LANL2DZ to MN15/def2-TZVP significantly increases the values of ΔG_r of all B vitamins and guanine even two or three times (Table 2, Figure 8). The most spectacular affinity increase was observed for B2 vitamin, approximately a thirty-fold increase in the value of Gibbs Free Energy from -1.46 kcal/mol to -41.55 kcal/mol and at about 96.5%. Even so, in the case of other B vitamins, equally large increases were observed, i.e., over 40% for B3 vitamin, 63% for vitamin B6, 67% for $B1_{(N1)}$ vitamin and 70% for $B1_{(N7)}$ vitamin. This level of calculation clearly changes the affinity of B vitamins to the drug. It can be seen that vitamin B2 has the highest affinity for Oxaliplatin (-41.55 kcal/mol) compared to the B3LYP level of calculation (-1.46 kcal/mol), and its value does not significantly differ from the Gibbs Free Energy value for other vitamins from B group and Guanine.

Vitamin B3 (Niacin) easily forms complexes with cis-Pt-diaqua and Pt-DACH-diaqua, ($[cis\sim Pt(NH_3)_2(H_2O)_2]^{2+}$ and $[Pt(H2O)_2DACH]^{2+}$), where delta Gibbs Free Energies are equal -27.28 kcal/mol and -35.53 kcal/mol in water solution, respectively. The diaqua forms, compared to the chloro- forms (Table 3, Figure 9) have a similar affinity for B vitamins and Guanine (GUA $\rightarrow$ vit. B3 $\rightarrow$ vit. B6 $\rightarrow$vit. $B1_{(N1)} \rightarrow$ vit. $B1_{(N7)} \rightarrow$ vit. B2) at the B3LYP/6-31G(d,p)/LANL2DZ level of theory in water solution. The diaqua forms of Cisplatin and Oxaliplatin have the greatest affinity for Guanine (-37.58; -50.03) compared to B vitamins which show a stronger drug effect. Oxaliplatin hydrolysis products compared to Cisplatin show a higher affinity for both kinds of studied structures, namely nucleobases and B vitamins (Table 3, Figure 9). Vitamin B2 forms the weakest bonds with the diaqua products of Cisplatin and Oxaliplatin.

Diaqua forms of oxaliplatin show a higher affinity to B vitamins and guanine compared to cisplatin (Table 3, Figure 9). The highest increase in affinity was observed for guanine, more than 12 times, and next for vitamin B3, more than eight times. However, the order of affinity of both drugs to the tested vitamins and bases is the same.

As for Pt-Chloroaqua derivatives of the drugs, also in this case, increasing the calculation level causes the increase in the values of Gibbs free energy (Table 4, Figures 9 and 10). In the case of diaqua forms of oxaliplatin, the order of affinity for studied vitamins and nucleobases is the same, GUA $\rightarrow$ vit. $B1_{(N1)} \rightarrow$ vit. B6$\rightarrow$vit. $B1_{(N7)} \rightarrow$ vit. B2 $\rightarrow$ vit. B3, as for Pt-Chloroaqua derivatives. However, here the differences in values of ΔG_r between vitamin $B1_{(N7)}$ and guanine are larger by about 0.43 kcal/mol.

The transition from a lower level of calculation to a higher one, that is from B3LYP/6-31G(d,p)/LANL2DZ to MN15/def2-TZVP, causes the greatest increase of affinities for B2 vitamin and B1 vitamin (Table 3, Figure 10) in the case of diaqua derivatives. For vitamin B3 there was a three-fold increase (68%) and for vitamins $B1_{(N7)}$ and $B1_{(N1)}$ an almost two-fold increase (46%; 40%).

The possibility of drug interaction with macromolecules may be indicated by such a physicochemical parameter as a change in the value of the molar volume of the formed vitamin B (nucleobases)–drug complexes, which will be directly related to conformational changes of the substrates of the complexation reaction. Such studies at the calculation level MN15/def2-TZVP were carried out for the effects of oxaliplatin and vitamins from B group and such nucleobases as guanine. The values of molar volume (cm^3/mol) of studied complexes PtClDACH-Gua and PtH$_2$ODACH-Gua are 198.370 cm^3/mol and 210.671 cm^3/mol, respectively.

In order to carry out the topological and electronic studies of the structural changes after the replacement of a chlorine atom with a water molecule, AIM studies were carried out (Figure 11, Table 5). The BCPs were found for non-covalent and covalent interactions (Figure 11).

Figure 11. Graphic representation of PtClDACH-Gua (C1) and PtH$_2$ODACH-Gua (C2) complexes [52].

Table 5. BCPs (bond critical points) and potential energy density calculated at the MN15/def2-TZVP level of theory [52].

BCP	PtClDACH-Gua			BCP	PtH$_2$ODACH-Gua		
	$\rho_{BCP}(e \times a_0^{-3})$	$\nabla^2\rho_{BCP}(e \times a_0^{-5})$	V (r)		$\rho_{BCP}(e \times a_0^{-3})$	$\nabla^2\rho_{BCP}(e \times a_0^{-5})$	V (r)
Cl-Pt	1.02×10^{-1}	1.97×10^{-1}	-1.27×10^{-1}	H$_2$O-Pt	9.54×10^{-2}	4.69×10^{-1}	-1.65×10^{-1}
N-Pt	1.25×10^{-1}	4.53×10^{-1}	-2.06×10^{-1}	N-Pt	1.28×10^{-1}	4.31×10^{-1}	-2.07×10^{-1}
H$_2$N-Pt	1.25×10^{-1}	4.12×10^{-1}	-1.96×10^{-1}	H$_2$N-Pt	1.39×10^{-1}	3.77×10^{-1}	-2.12×10^{-1}
H$_2$N-Pt	1.26×10^{-1}	4.15×10^{-1}	-1.99×10^{-1}	H$_2$N-Pt	1.27×10^{-1}	4.04×10^{-1}	-1.99×10^{-1}
O $\dots$ HN	1.84×10^{-2}	7.27×10^{-2}	-1.23×10^{-2}	O $\dots$ HN	1.90×10^{-2}	7.36×10^{-2}	-1.27×10^{-2}

An analysis of the interactions of the Pt atom with its immediate surroundings was carried out. In the PtClDACH-Gua complex BCP intramolecular interactions between Pt and Cl atoms from oxaliplatin and between O atom from guanine and H atom from the NH$_2$ group of oxaliplatin moiety were found. In the case of PtH$_2$ODACH-Gua, the interactions between the Pt atom of the drug with the N atom and an O atom from the water molecule and two N atoms from guanine were found. There is a BCP between an oxygen atom from guanine and a hydrogen atom from the NH$_2$ group of the oxaliplatin moiety. This part of the study showed the network of interactions stabilizing the complexes. To describe the strength of the interactions, the potential electronic energy density was calculated. The BCP values were found between the Pt atom of the drug and the studied reagents. The electron density and its Laplacian values were changed with the appearance of a Cl atom or a H$_2$O molecule in the vicinity of the Pt atom of oxaliplatin.

To study the spectroscopic properties of drug–vitamin B or drug–guanine complexes, the TD-DFT (time-dependent density functional theory) was performed (Figure 12). The results were compared with experimental data (Section 3).

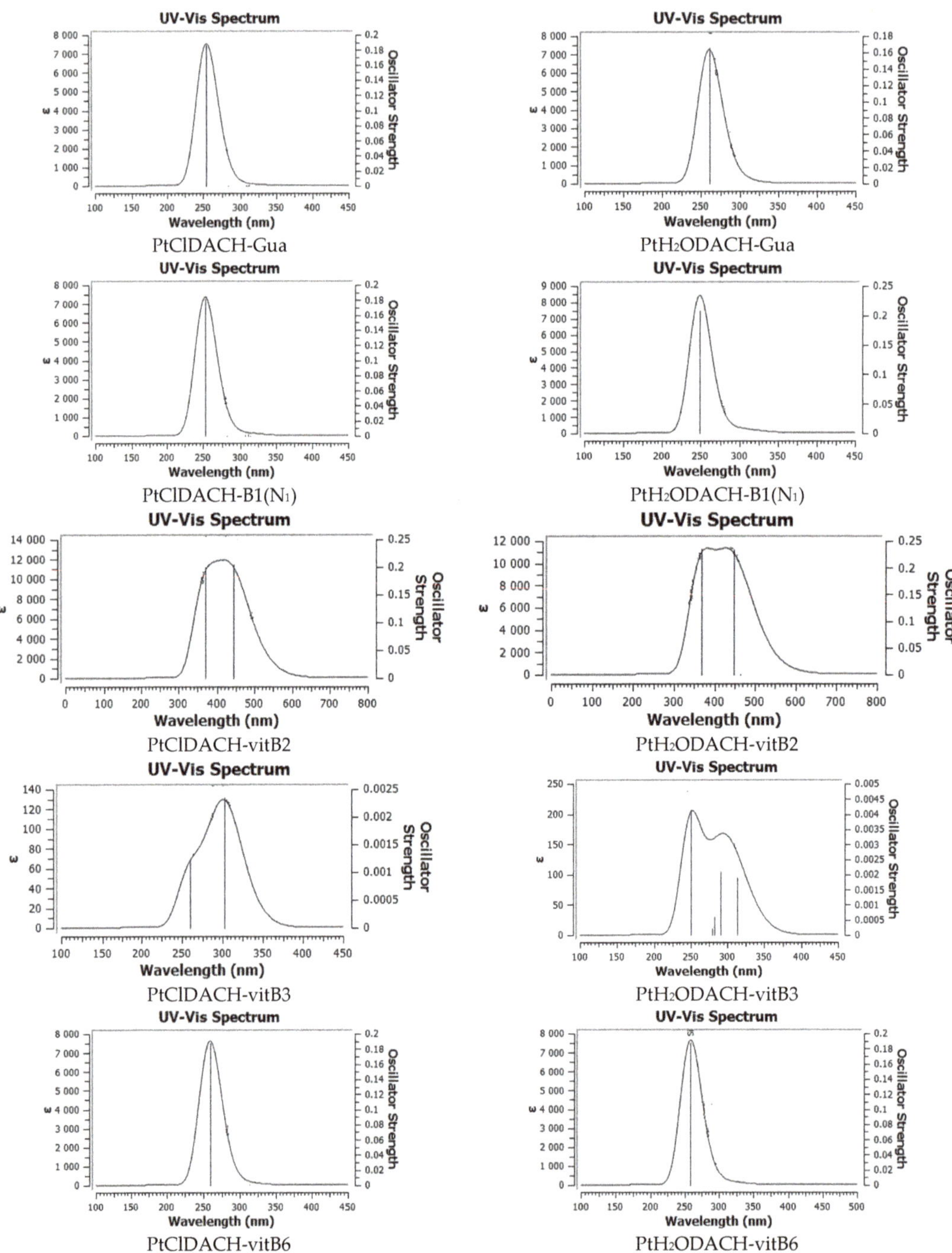

Figure 12. UV-Vis spectra calculated at MN15/def2-TZVP level of theory with PCM model and water as a solvent for selected oxaliplatin complexes with guanine and B vitamins [52].

In Figure 12, the excitation energies are given for singlet-singlet transitions (eV; nm) and the oscillator strength. Comparison with experimental values is described later in the article (Section 3).

3. Experimental Analysis

UV-Vis spectroscopic techniques were used for physico-chemical characterization of interactions of B6 vitamin (pyridoxine hydrochloride) and carboplatin (Figure 13). The study was performed in the wavelength range from 190 nm to 500 nm.

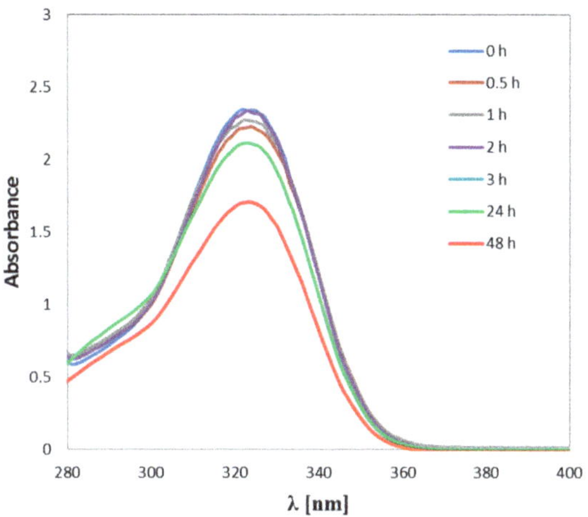

Figure 13. UV-Vis absorbance spectrum of mixture of carboplatin and pyridoxal phosphate (vit. B6) in an incubation buffer (1 mmol/L phosphate buffer, 4 mmol/L sodium-chloride, pH 7.4) [50].

Carboplatin and vitamin B6 with molar concentrations equal to $7.3 \times 10{-}4$ mol/L and $14.6 \times 10{-}3$ mol/L were prepared in the following incubation buffer: 1 mmol/L phosphate buffer and 4 mmol/L sodium chloride with pH equal to 7.4. Maximum absorbance of vitamin B6 was obtained at 323 nm wavelength and was almost equal to 3.0, which corresponds to 0.721 mmol/L concentration of vitamin B6 in the incubation buffer (Figure 13). Vitamin B6 was incubated with carboplatin in a 2:1 ratio at 37 °C. The absorbance of the mixture was measured at appropriate time intervals of incubation, namely, 0 h, 0.5 h, 1 h, 2 h, 4 h, 24 h and 48 h. The addition of the solution of carboplatin to the solution of vitamin B6 in the buffer causes a decrease of maximum absorbance for the studied vitamin by about 20% relative to the value of absorbance equal to 2.3 (time 0 h). It reflects a decrease in vitamin B6 concentration to a value of 0.624 mmol/L. Such a decrease of absorbance of the studied vitamin is caused by the formation of a drug–vitamin complex. However, a significant decrease by about 30% in maximum absorbance from the baseline of the complex (time 0 h) was observed after 48 h (1.75 absorbance) and a decrease by about 40% in the maximum absorbance of vitamin B6.

In the same way and under the same conditions, the absorbance maxima of all tested B vitamins and guanine with oxaliplatin (synonyms: [SP-4-2-(1R-trans)]-(1,2-Cyclohexanediamine-N,N′)[ethanedioata(2-)-O,O]platinum; oxaliplatinum; (SP-4-2)-[(1R,2R)-Cyclohexane-1,2-diamine-κN,κN′]-[ethanedioato(2-)-κO1,κO2]platinum) were determined. The initial concentrations of substrates in the reaction of complexation with oxaliplatin were 0.503 mmol/L for guanine, 0.178 mmol/L for vitamin B1 (thiamine), 0.107 mmol/L for vitamin B2 (riboflavin), 0.712 mmol/L for vitamin B3 (niacin) and 0.365 mmol/L for vitamin B6 (pyridoxal phosphate). The starting concentration for oxaliplatin was 0.252 mmol/L.

Both guanine and all tested B vitamins in complexes with oxaliplatin show more than one maximum absorbance (Figure 14). Guanine has two maxima, at 218 nm and 271 nm. However, in the first case there are no changes observed in the height of the absorbance maximum after the formation of the base–drug complex in the subsequent time intervals tested. Guanine–oxaliplatin complex shows a 0.188 high absorbance maximum at 271 nm (0 h), which went down to 0.153 and 0.124 after 24 h and 48 h, respectively. However, with the passage of time, the absorbance maximum is shifted towards a shorter wavelength, namely, 242 nm. Similarly, as the nucleobase, vitamin B1 has two absorbance maxima of other wavelengths at 231 nm and 266 nm. Here, a decrease of absorbance maximum is observed from 1.314 (0 h) by 1.200 (24 h) to 1.119 (48 h) in the first case and from 0.967 (0 h) by 0.903 (24 h) to 0.897 (48 h) in the second case (Figure 14). Vitamin B2 has maxima at 224 nm and 265 nm, which decrease accordingly from 1.875 to 1.780 and from 1.818 to 1.710 after 48h, respectively. After 24 h, the maximum absorbance of complex vitamin B2–oxaliplatin gained heights of 1.790 at 224 nm and 1.770 at 265 nm. Vitamin B3 has a maximum at 218 nm and 266 nm at baseline. In first place, the absorbances were reduced from 1.548 to 1.344 after 24 h and to 1.277 after 48 h. At 266 nm no decrease was observed of the absorbance equal to 0.918. Vitamin B6 shows also two absorbance maxima at 222 nm and 323 nm (Figure 14). In the first case, a decrease was observed of maximum absorbance from 2.254 (0 h) by 1.857 (24 h) to 1.784 (48 h), while in the second case, an absorbance drop from 1.142 (0 h) by 1.070 (24 h) to 1.035 (48 h) was observed.

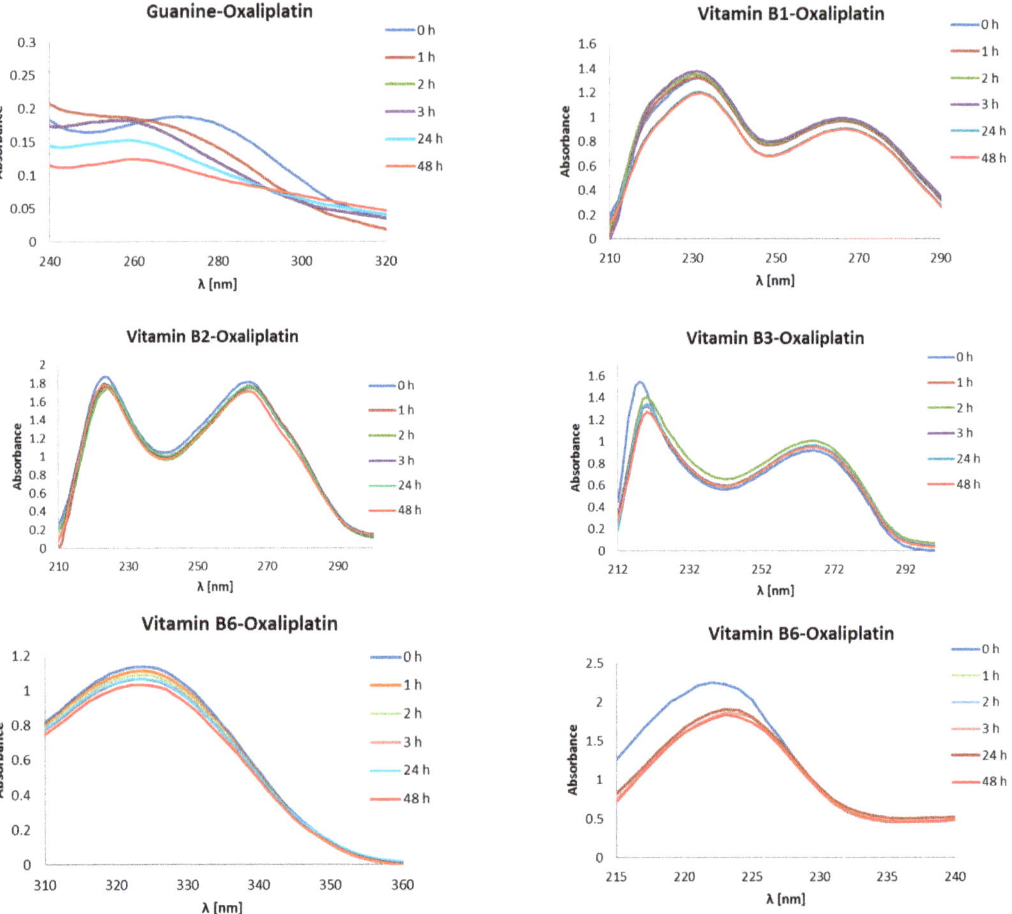

Figure 14. UV-Vis absorbance spectrum of mixture of oxaliplatin with B vitamins and guanine in an incubation buffer (1 mmol/L phosphate buffer, 4 mmol/L sodium-chloride, pH 7.4) [52].

After 168 h of incubation in the buffer, the vitamin–oxaliplatin and base–oxaliplatin complexes did not decrease in their concentration in relation to the concentrations of the tested complexes after 48 h of incubation. Thus, after 7 days of incubation, the height of the absorbance peaks was comparable to the height after 48 h. The exception is vitamin B2, where there was a decrease in the maximum height by about 10% compared to the value tested after 48 h (Figure 15).

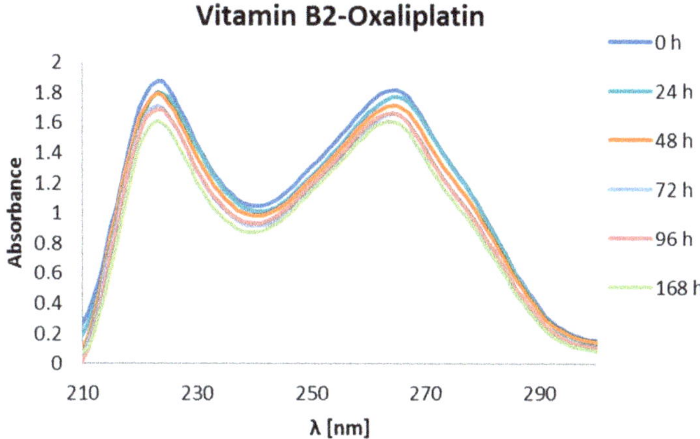

Figure 15. UV-Vis absorbance spectrum of mixture of oxaliplatin and riboflavin (vit. B2) in an incubation buffer (1 mmol/L phosphate buffer, 4 mmol/L sodium-chloride, pH 7.4) [52].

Experimental results reflect the results of ab initio calculations at the B3LYP/6-31G(d,p)/LANL2DZ level of calculation (GUA → vit. B3 → vit. B1 → vit. B6 → vit. B2).

In Figure 16 the experimental and the theoretical values of excitation energies are compared.

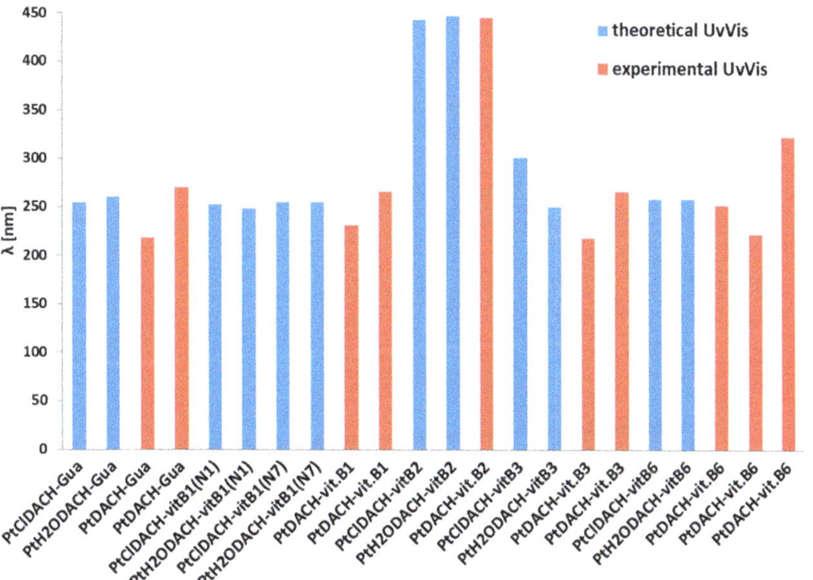

Figure 16. The experimental and the theoretical values of excitation energies (nm) of monoaqua and diaqua derivatives of oxaliplatin with B vitamin and guanine complexes [52].

In the case of the complex with Guanine, the computed values are 254.22 nm and 260.50 nm for PtClDACH-Gua and for PtH$_2$ODACH-Gua, respectively, while the experimental spectra showed close values of the absorbance maximum at 270 nm and at 218 nm (Figure 16). The PtClDACH-vit.B1(N$_1$) and PtClDACH-vit.B1(N$_7$) complexes have the maximum absorbance located at 252.64 nm and 255.04 nm, while PtH$_2$ODACH-vit.B1(N$_1$) and PtH$_2$ODACH-vit.B1(N$_7$) have it at 248.69 nm and 255.04 nm, respectively. The experimental data showed maxima of absorbance at 231 nm and 266 nm. The complexes with vitamin B2 showed this at 443.05 nm for PtClDACH-vit.B2 and 447.71 nm for PtH$_2$ODACH-vit.B2, which is in a good agreement with experimental data locating the maximum absorbance bands at 445 nm (as well as at 221 nm, 266 nm and 372 nm). The calculated results of vitamin B3 showed the most intensive singlet-singlet transitions at 250.47 nm for PtH$_2$ODACH-vit.B3 and at 301.58 nm for PtClDACH-vit.B3, while in the experimental spectrum the absorption maximum was found quite close at 266 nm and further at 218 nm. It was found that the absorbance maxima of computed absorption spectra for vitamin B6 occurred at 258.68 nm and at 258.55 nm for PtClDACH-vit.B6 and for PtH$_2$ODACH-vit.B6, respectively. The experimental spectra showed a close absorbance maximum at 252 nm; however, maxima at 222 nm and 323 nm were found. The computed UV-Vis spectra are shown in Figure 12, while the theoretical and experimental excitation energy for the complex of drugs and the studied structures are shown in Figure 16. Very good agreement has been proven between computed and experimental UV-Vis spectra. Theoretical spectra were calculated in water solution, at the MN15/def2-TZVP level of theory with the application of the polarizable continuum model (PCM).

4. Conclusions

Theoretical and experimental studies were performed in order to describe the interactions of selected Pt (II) derivative drugs with nucleobases and other compounds of analogous structure, such as vitamins from B group: thiamine (vitamin B1), riboflavin (vitamin B2), niacin (vitamin B3) and pyridoxal phosphate (vitamin B6). cisplatin, oxaliplatin and carboplatin were tested. Ab initio studies on complexes of nucleobases/B vitamins with drugs such as cisplatin, oxaliplatin and carboplatin were based on DFT and TD-DFT methods. Theoretical research was supported by UV-Vis spectrophotometric experimental studies. The performed research on the affinity of nucleobases and their competitive B vitamins to drugs explains not only their mechanism of action at the molecular level, but also explains their competitiveness and impact on anticancer therapy. During the study, the reactivity of hydrolysis products of studied drugs were analyzed such as mono- and diaqua derivatives, (cis~[Pt(NH$_3$)$_2$Cl(H$_2$O)]$^+$) and (cis~[Pt(NH$_3$)$_2$(H$_2$O)$_2$]$^{2+}$) in the case of cisplatin, [PtH$_2$OClDACH]$^+$ and [Pt(H$_2$O)$_2$DACH]$^{2+}$ in the case of oxaliplatin and Pt(NH$_3$)$_2$H$_2$O(OH)$^+$ plus Pt(NH$_3$)$_2$(OH)$_2$ in the case of carboplatin, which can interact not only with nucleobases, which is the basis of their therapeutic action, but also with all compounds which have aromatic rings in their structures. The research was carried out on two levels of calculation, namely, at the B3LYP/6-31G(d,p) and at the MN15/def2-TZVP levels of theory and their results were also compared with each other. The order of affinity of cisplatin and oxaliplatin for both monoaqua and diaqua derivatives of nucleobases and B vitamins is GUA → vit. B3 → vit. B6 → vit. B1$_{(N1)}$ → vit. B1$_{(N7)}$ → vit. B2, obtained at the B3LYP/6-31G(d,p) level of theory. However, in the case of cis~monoaqua derivatives, vitamin B6 alternates with vitamin B1 as GUA → vit. B3 → vit. B1$_{(N1)}$ → vit. B1$_{(N7)}$ → vit. B6 → vit. B2. With the changing of the calculation level to a higher one, the MN15/def2-TZVP level of theory, the affinity order changes to vit. B1$_{(N1,N7)}$ → GUA → vit. B6→ vit. B2 → vit. B3, as seen in the example of oxaliplatin. Carboplatin shows weak affinity to the studied structures, except for guaniane and vitamin B6, in both mono- and diaqua derivatives. Vitamin B3 weakly interacts with the products of carboplatin hydrolysis. Other vitamins do not interact with this drug. Theoretical studies confirm clinical observations and indicate high therapeutic effectiveness of oxaliplatin and then cisplatin in anticancer treatment, as well as confirm the competitiveness of B vitamins in relation to nucleobases.

The presence of intramolecular non-covalent interactions stabilizing the conformation of the drug–nucleobase/B vitamin complexes was confirmed via the application of the AIM theory. In addition, on the basis of the calculated and experimental spectra, the agreement between theoretical and experimental studies was proved. A decrease of concentration in the solutions of nucleobases or vitamins from B group was observed after their complexation with the tested Pt (II) drugs. However, it should be emphasized with certainty that B vitamins form weaker complexes with the products of hydrolysis of chemotherapeutics in relation to nucleobases.

5. Future Directions

The research on the stability of Pt (II) derivatives–B vitamins complexes and their physicochemical, thermodynamic and spectroscopic properties will be the main source of knowledge for a better understanding of the reactivity of these drugs with physiological target molecules such as nucleobases. This will open up further research on the behavior of other Pt (II) derivatives in this respect and will open up further studies for new potential Pt (II) drugs and drug delivery agents. In effect, all involved surveys will lead to the final results, that is, obtaining new alternative Pt (II) drugs and introducing new potential therapies used in patients with various types of cancer by using nanostructures as nanocarriers during targeted drug delivery, and will be completed with the synthesis of new potential medicines. The described results of potential interactions of Pt (II) anticancer drugs with compounds other than the ones in the target sites will contribute to the dissemination of knowledge about the use of an appropriate diet during chemotherapy.

Author Contributions: Conceptualization, B.S.; methodology, B.S.; software, B.S.; validation, B.S.; formal analysis, B.S.; investigation, B.S.; resources, B.S.; data curation, B.S.; writing—original draft preparation, B.S.; writing—review and editing, B.S. and P.C.; visualization, B.S. and P.C.; supervision, B.S. and P.C.; project administration, B.S.; funding acquisition, B.S. All authors have read and agreed to the published version of the manuscript.

Funding: This work was financed by Collegium Medicum, Nicolaus Copernicus University in Bydgoszcz.

Institutional Review Board Statement: Not applicable.

Informed Consent Statement: Not applicable.

Acknowledgments: This research was supported by PL-Grid Infrastructure (http://www.plgrid.pl/en). The authors gratefully acknowledge the generous grants of CPU time, facilities and support provided by the Wrocław Centre for Networking and Supercomputing (WCSS), the Poznań Supercomputing and Networking Center (PSNC), the Academic Computing Centre Cyfronet-Kraków (Prometheus supercomputer, part of the PL-Grid infrastructure) and the Centre of Informatics Tricity Academic Supercomputer and networK (CI TASK).

Conflicts of Interest: The authors declare no conflict of interest.

References

1. World Cancer Research Fund; American Institute for Cancer. *Diet, Nutrition, Physical Activity and Breast Cancer*; Continuous Update Project, Expert Report 2018; World Cancer Research Fund International: London, UK, 2018.
2. Wiseman, L.R.; Adkins, J.C.; Plosker, G.L.; Goa, K.L. Oxaliplatin: A review of its use in the management of metastatic colorectal cancer. *Drugs Aging* **1999**, *14*, 459–475. [CrossRef] [PubMed]
3. Rosenberg, B.; Vancamo, L.; Trosko, J.E.; Mansour, V.H. Platinum Compounds: A New Class of Potent Antitumour Agents. *Nature* **1969**, *222*, 385–386. [CrossRef] [PubMed]
4. Rosenberg, B.; Van Camp, L.; Krigas, T. Inhibition of cell division in Escherichia coli by electrolysis products from a platinum electrode. *Nature* **1965**, *205*, 698–699. [CrossRef] [PubMed]
5. Rixe, O.; Ortuzar, W.; Alvarez, M.; Parker, R.; Reed, E.; Paull, K.; Fojo, T. Oxaliplatin, tetraplatin, cisplatin, and carboplatin: Spectrum of activity in drug-resistant cell lines and in the cell lines of the national cancer institute's anticancer drug screen panel. *Biochem. Pharmacol.* **1996**, *52*, 1855–1865. [CrossRef]
6. Ho, G.Y.; Woodward, N.; Coward, J.I.G. Cisplatin versus carboplatin: Comparative review of therapeutic management in solid malignancies. *Crit. Rev. Oncol. Hematol.* **2016**, *102*, 37–46. [CrossRef]

7.   Frezza, M.; Hindo, S.; Chen, D.; Davenport, A.; Schmitt, S.; Tomco, D.; Dou, Q.P. Novel metals and metal complexes as platforms for cancer therapy. *Curr. Pharm. Des.* **2010**, *16*, 1813–1825. [CrossRef]
8.   Desoize, B.; Madoulet, C. Particular aspects of platinum compounds used at present in cancer treatment. *Crit. Rev. Oncol. Hematol.* **2002**, *42*, 317–325. [CrossRef]
9.   Fuertes, M.A.; Alonso, C.; Pérez, J.M. Biochemical modulation of cisplatin mechanisms of action: Enhancement of antitumor activity and circumvention of drug resistance. *Chem. Rev.* **2003**, *103*, 645–662. [CrossRef]
10.   Van Zyl, B.; Tang, D.; Bowden, N.A. Biomarkers of platinum resistance in ovarian cancer: What can we use to improve treatment. *Endocr. Relat. Cancer* **2018**, *25*, R303–R318. [CrossRef]
11.   Abotaleb, M.; Kubatka, P.; Caprnda, M.; Varghese, E.; Zolakova, B.; Zubor, P.; Opatrilova, R.; Kruzliak, P.; Stefanicka, P.; Büsselberg, D. Chemotherapeutic agents for the treatment of metastatic breast cancer: An update. *Biomed. Pharmacother.* **2018**, *101*, 458–477. [CrossRef]
12.   Rugo, H.S.; Olopade, O.I.; DeMichele, A.; Yau, C.; van 't Veer, L.J.; Buxton, M.B.; Hogarth, M.; Hylton, N.M.; Paoloni, M.; Perlmutter, J.; et al. Adaptive Randomization of Veliparib–Carboplatin Treatment in Breast Cancer. *N. Engl. J. Med.* **2016**, *375*, 23–34. [CrossRef] [PubMed]
13.   Mikuła-Pietrasik, J.; Witucka, A.; Pakuła, M.; Uruski, P.; Begier-Krasińska, B.; Niklas, A.; Tykarski, A.; Książek, K. Comprehensive review on how platinum- and taxane-based chemotherapy of ovarian cancer affects biology of normal cells. *Cell. Mol. Life Sci.* **2019**, *76*, 681–697. [CrossRef] [PubMed]
14.   Gebremedhn, E.G.; Shortland, P.J.; Mahns, D.A. The incidence of acute oxaliplatin-induced neuropathy and its impact on treatment in the first cycle: A systematic review. *BMC Cancer* **2018**, *18*, 410. [CrossRef] [PubMed]
15.   Chaney, S.G.; Campbell, S.L.; Bassett, E.; Wu, Y. Recognition and processing of cisplatin- and oxaliplatin-DNA adducts. *Crit. Rev. Oncol. Hematol.* **2005**, *53*, 3–11. [CrossRef] [PubMed]
16.   Di Francesco, A.M.; Ruggiero, A.; Riccardi, R. Cellular and molecular aspects of drugs of the future: Oxaliplatin. *Cell. Mol. Life Sci.* **2002**, *59*, 1914–1927. [CrossRef]
17.   Martinez-Balibrea, E.; Martínez-Cardús, A.; Ginés, A.; Ruiz de Porras, V.; Moutinho, C.; Layos, L.; Manzano, J.L.; Bugés, C.; Bystrup, S.; Esteller, M.; et al. Tumor-Related Molecular Mechanisms of Oxaliplatin Resistance. *Mol. Cancer Ther.* **2015**, *14*, 1767–1776. [CrossRef]
18.   Martin, L.P.; Hamilton, T.C.; Schilder, R.J. Platinum resistance: The role of DNA repair pathways. *Clin. Cancer Res.* **2008**, *14*, 1291–1295. [CrossRef]
19.   Meyerhardt, J.A.; Mayer, R.J. Systemic therapy for colorectal cancer. *N. Engl. J. Med.* **2005**, *352*, 476–487. [CrossRef]
20.   Seetharam, R.N. Oxaliplatin: Preclinical perspectives on the mechanisms of action, response and resistance. *Ecancermedicalscience* **2009**, *3*, 153. [CrossRef]
21.   Sarmah, A.; Saha, S.; Bagaria, P.; Kinkar Roy, R. On the complementarity of comprehensive decomposition analysis of stabilization energy (CDASE) Scheme and supermolecular approach. *Chem. Phys.* **2012**, *394*, 29–35. [CrossRef]
22.   Sarmah, A.; Roy, R.K. Understanding the preferential binding interaction of aqua-cisplatins with nucleobase guanine over adenine: A density functional reactivity theory based approach. *RSC Adv.* **2013**, *3*, 2822. [CrossRef]
23.   Wang, D.; Lippard, S.J. Cellular processing of platinum anticancer drugs. *Nat. Rev. Drug Discov.* **2005**, *4*, 307–320. [CrossRef] [PubMed]
24.   Göschl, S.; Schreiber-Brynzak, E.; Pichler, V.; Cseh, K.; Heffeter, P.; Jungwirth, U.; Jakupec, M.A.; Berger, W.; Keppler, B.K. Comparative studies of oxaliplatin-based platinum(iv) complexes in different in vitro and in vivo tumor models. *Metallomics* **2017**, *9*, 309–322. [CrossRef] [PubMed]
25.   Mantri, Y.; Lippard, S.J.; Baik, M.-H. Bifunctional binding of cisplatin to DNA: Why does cisplatin form 1,2-intrastrand cross-links with ag but not with GA? *J. Am. Chem. Soc.* **2007**, *129*, 5023–5030. [CrossRef] [PubMed]
26.   Spiegel, K.; Rothlisberger, U.; Carloni, P. Cisplatin Binding to DNA Oligomers from Hybrid Car-Parrinello/Molecular Dynamics Simulations. *J. Phys. Chem. B* **2004**, *108*, 2699–2707. [CrossRef]
27.   Kelland, L.R.; Farrell, N.P. *Platinum-Based Drugs in Cancer Therapy*; Humana Press: Totowa, NJ, USA, 2000; Volume 7, ISBN 1-59259-012-8.
28.   Schewe, G. *Platinum-Based Drugs in Cancer Therapy*; Humana Press: Totowa, NJ, USA, 2010; ISBN 9781617370915.
29.   Riddell, I.A.; Lippard, S.J. Cisplatin and oxaliplatin: Our current understanding of their actions. *Met. Ions Life Sci.* **2018**, *18*, 1–42.
30.   Riddell, I.A.; Lippard, S.J.; Brabec, V.; Kasparkova, J.; Menon, V.; Farrell, N.P.; Taylor, K.M. *Metallo-Drugs: Development and Action of Anticancer Agents*; Walter de Gruyter GmbH & Co KG: Berlin, Germany, 2018; Volume 18.
31.   Dasari, S.; Tchounwou, P.B. Cisplatin in cancer therapy: Molecular mechanisms of action. *Eur. J. Pharmacol.* **2014**, *740*, 364–378. [CrossRef]
32.   de Sousa, G.F.; Wlodarczyk, S.R.; Monteiro, G. Carboplatin: Molecular mechanisms of action associated with chemoresistance. *Braz. J. Pharm. Sci.* **2014**, *50*, 693–702. [CrossRef]
33.   Alcindor, T.; Beauger, N. Oxaliplatin: A review in the era of molecularly targeted therapy. *Curr. Oncol.* **2011**, *18*, 18–25. [CrossRef]
34.   Woynarowski, J.M.; Faivre, S.; Herzig, M.C.S.; Arnett, B.; Chapman, W.G.; Trevino, A.V.; Raymond, E.; Chaney, S.G.; Vaisman, A.; Varchenko, M.; et al. Oxaliplatin-induced damage of cellular DNA. *Mol. Pharmacol.* **2000**, *58*, 920–927. [CrossRef]
35.   Pasetto, L.M.; D'Andrea, M.R.; Rossi, E.; Monfardini, S. Oxaliplatin-related neurotoxicity: How and why? *Crit. Rev. Oncol. Hematol.* **2006**, *59*, 159–168. [CrossRef] [PubMed]

36. Hah, S.S.; Stivers, K.M.; De Vere White, R.W.; Henderson, P.T. Kinetics of carboplatin-DNA binding in genomic DNA and bladder cancer cells as determined by accelerator mass spectrometry. *Chem. Res. Toxicol.* **2006**, *19*, 622–626. [CrossRef] [PubMed]
37. Brabec, V.; Kasparkova, J. Modifications of DNA by platinum complexes. *Drug Resist. Updat.* **2005**, *8*, 131–146. [CrossRef]
38. Fuertes, M.A.; Castilla, J.; Alonso, C.; Pérez, J.M. Novel concepts in the development of platinum antitumor drugs. *Curr. Med. Chem. Anticancer. Agents* **2002**, *2*, 539–551. [CrossRef] [PubMed]
39. Fischer, J.; Robin Ganellin, C. *Analogue-Based Drug Discovery*; John Wiley & Sons: Hoboken, NJ, USA, 2006; ISBN 3527312579.
40. Hodgkinson, E.; Neville-Webbe, H.L.; Coleman, R.E. Magnesium depletion in patients receiving cisplatin-based chemotherapy. *Clin. Oncol. R. Coll. Radiol.* **2006**, *18*, 710–718. [CrossRef] [PubMed]
41. Larson, C.A.; Blair, B.G.; Safaei, R.; Howell, S.B. The role of the mammalian copper transporter 1 in the cellular accumulation of platinum-based drugs. *Mol. Pharmacol.* **2009**, *75*, 324–330. [CrossRef]
42. Ishida, S.; Lee, J.; Thiele, D.J.; Herskowitz, I. Uptake of the anticancer drug cisplatin mediated by the copper transporter Ctr1 in yeast and mammals. *Proc. Natl. Acad. Sci. USA* **2002**, *99*, 14298–14302. [CrossRef]
43. Howell, S.B.; Safaei, R.; Larson, C.A.; Sailor, M.J. Copper transporters and the cellular pharmacology of the platinum-containing cancer drugs. *Mol. Pharmacol.* **2010**, *77*, 887–894. [CrossRef]
44. Holzer, A.K.; Manorek, G.H.; Howell, S.B. Contribution of the major copper influx transporter CTR1 to the cellular accumulation of cisplatin, carboplatin, and oxaliplatin. *Mol. Pharmacol.* **2006**, *70*, 1390–1394. [CrossRef]
45. Johnstone, T.C.; Suntharalingam, K.; Lippard, S.J. The next generation of platinum drugs: Targeted Pt (II) agents, nanoparticle delivery, and Pt (IV) prodrugs. *Chem. Rev.* **2016**, *116*, 3436–3486. [CrossRef]
46. Han, C.H.; Khwaounjoo, P.; Hill, A.G.; Miskelly, G.M.; McKeage, M.J. Predicting effects on oxaliplatin clearance: In vitro, kinetic and clinical studies of calcium- and magnesium-mediated oxaliplatin degradation. *Sci. Rep.* **2017**, *7*, 4073. [CrossRef] [PubMed]
47. Baik, M.-H.; Friesner, R.A.; Lippard, S.J. Theoretical Study of Cisplatin Binding to Purine Bases: Why Does Cisplatin Prefer Guanine over Adenine? *J. Am. Chem. Soc.* **2003**, *125*, 14082–14092. [CrossRef] [PubMed]
48. Szefler, B.; Czeleń, P. Docking of cisplatin on fullerene derivatives and some cube rhombellane functionalized homeomorphs. *Symmetry* **2019**, *11*, 874. [CrossRef]
49. Szefler, B.; Czeleń, P.; Szczepanik, A.; Cysewski, P. Does the affinity of cisplatin to B-vitamins impair the therapeutic effect in the case of patients with lung cancer-consuming carrot or beet juice? *Anticancer Agents Med. Chem.* **2019**, *19*, 1775–1783. [CrossRef]
50. Szefler, B.; Czeleń, P.; Krawczyk, P. The Affinity of Carboplatin to B-Vitamins and Nucleobases. *Int. J. Mol. Sci.* **2021**, *22*, 3634. [CrossRef]
51. Szefler, B.; Czeleń, P.; Kruszewski, S.; Siomek-Górecka, A.; Krawczyk, P. The assessment of physicochemical properties of Cisplatin complexes with purines and vitamins B group. *J. Mol. Graph. Model.* **2022**, *113*, 108144. [CrossRef]
52. Szefler, B.; Czeleń, P.; Wojtkowiak, K.; Jezierska, A. Affinities to Oxaliplatin: Vitamins from B Group vs. Nucleobases. *Int. J. Mol. Sci.* **2022**, *23*, 10567. [CrossRef]
53. Farrell, N.P. Preclinical perspectives on the use of platinum compounds in cancer chemotherapy. *Semin. Oncol.* **2004**, *31*, 1–9. [CrossRef]
54. Shabalin, I.; Dauter, Z.; Jaskolski, M.; Minor, W.; Wlodawer, A. Crystallography and chemistry should always go together: A cautionary tale of protein complexes with cisplatin and carboplatin. *Acta Crystallogr. Sect. D Biol. Crystallogr.* **2015**, *71*, 1965–1979. [CrossRef]
55. Ahmad, S. Platinum–DNA interactions and subsequent cellular processes controlling sensitivity to anticancer platinum complexes. *Chem. Biodivers.* **2010**, *7*, 543–566. [CrossRef]
56. Stipanuk, M.H.; Caudill, M. *Biochemical, Physiological, and Molecular Aspects of Human Nutrition*; Saunders: St. Louis, MO, USA, 2013; ISBN 9781455746293.
57. Stargrove, M.B.; Treasure, J.; McKee, D.L. *Herb, Nutrient, and Drug Interactions: Clinical Implications and Therapeutic Strategies*; Mosby Elsevier: Amsterdam, The Netherlands, 2008; ISBN 9780323029643.
58. Polskie Towarzystwo Farmaceutyczne. *Farmakopea Polska X*; Urząd Rejestracji Produktów Leczniczych, Wyrobów Medycznych i Produktów Biobójczych: Warszawa, Poland, 2014; ISBN 978-83-63724-47-4.
59. Ghosal, A.; Said, H.M. Mechanism and regulation of vitamin B2 (riboflavin) uptake by mouse and human pancreatic β-cells/islets: Physiological and molecular aspects. *Am. J. Physiol. Liver Physiol.* **2012**, *303*, G1052–G1058. [CrossRef] [PubMed]
60. Kennedy, D. B Vitamins and the Brain: Mechanisms, Dose and Efficacy—A Review. *Nutrients* **2016**, *8*, 68. [CrossRef] [PubMed]
61. Zielińska-Dawidziak, M.; Grajek, K.; Olejnik, A.; Czaczyk, K.; Grajek, W. Transport of high concentration of thiamin, riboflavin and pyridoxine across intestinal epithelial cells Caco-2. *J. Nutr. Sci. Vitaminol.* **2008**, *54*, 423–429. [CrossRef]
62. White, E.; Patterson, R.E.; Kristal, A.R.; Thornquist, M.; King, I.; Shattuck, A.L.; Evans, I.; Satia-Abouta, J.; Littman, A.J.; Potter, J.D. VITamins and Lifestyle Cohort Study: Study Design and Characteristics of Supplement Users. *Am. J. Epidemiol.* **2004**, *159*, 83–93. [CrossRef] [PubMed]
63. Winkler, C.; Wirleitner, B.; Schroecksnadel, K.; Schennach, H.; Fuchs, D. Beer down-regulates activated peripheral blood mononuclear cells in vitro. *Int. Immunopharmacol.* **2006**, *6*, 390–395. [CrossRef] [PubMed]

International Journal of
Molecular Sciences

Article

Designing and Synthesis of New Isatin Derivatives as Potential CDK2 Inhibitors

Przemysław Czeleń [1,*], Agnieszka Skotnicka [2] and Beata Szefler [1]

[1] Department of Physical Chemistry, Faculty of Pharmacy, Collegium Medicum, Nicolaus Copernicus University, Kurpinskiego 5, 85-096 Bydgoszcz, Poland; beatas@cm.umk.pl

[2] Faculty of Chemical Technology and Engineering, Bydgoszcz University of Science and Technology, Seminaryjna 3, 85-326 Bydgoszcz, Poland; askot@pbs.edu.pl

* Correspondence: author: przemekcz@cm.umk.pl

Abstract: Tumors are still one of the main causes of death; therefore, the search for new therapeutic agents that will enable the implementation of effective treatment is a significant challenge for modern pharmacy. One of the important factors contributing to the development of neoplastic diseases is the overexpression of enzymes responsible for the regulation of cell division processes such as cyclin-dependent kinases. Numerous studies and examples of already-developed drugs confirm that isatin is a convenient basis for the development of new groups of inhibitors for this class of enzyme. Therefore, in this work, a new group of potential inhibitors of the CDK2 enzyme, utilizing isatin derivatives and substituted benzoylhydrazines, has been designed based on the application of computational chemistry methods, such as docking and molecular dynamics, and their inhibiting ability was assessed. In the cases of the selected compounds, a synthesis method was developed, and the selected physicochemical properties of the newly synthesized derivatives were estimated. As part of the completed project, new compounds are developed which are potential inhibitors of the CDK2 enzyme.

Keywords: isatin; CDK2; competitive inhibition; molecular dynamics; synthesis; spectroscopic properties

Citation: Czeleń, P.; Skotnicka, A.; Szefler, B. Designing and Synthesis of New Isatin Derivatives as Potential CDK2 Inhibitors. *Int. J. Mol. Sci.* **2022**, 23, 8046. https://doi.org/10.3390/ijms23148046

Academic Editors: Andrzej Kutner, Geoffrey Brown and Enikö Kallay

Received: 28 June 2022
Accepted: 19 July 2022
Published: 21 July 2022

Publisher's Note: MDPI stays neutral with regard to jurisdictional claims in published maps and institutional affiliations.

1. Introduction

Isatin (1H-indol-2,3-dione) is a chemical compound which, as natural alkaloid, may be extracted from plants of the Isatis genus, occurring all over the world at various latitudes [1–3]. The significant chemical and biological activities of isatin meant that this compound and its numerous derivatives, obtained from natural products, have widely been used in medicine for many centuries [4,5]. In terms of the pharmaceutical activity of such compounds, they have anti-inflammatory, antiviral, antibacterial, antifungal, anticonvulsant, anxiogenic and anticancer properties [6–10]. The large spectra of pharmacological activities of isatin derivatives are associated with the significant chemical variety of these compounds, within which one can distinguish hydrazones, thiosemicarbazones, oximes, spiro-oxindoles, imines and many more. The common element of this broad group of compounds is the oxindole system, which is the core repeatedly used in the development of new groups of competitive inhibitors [11–21]. The anticancer activity of new drugs is mainly related to the competitive inhibition of biological targets, the overexpression of which drives the accelerated growth of cancer cells. The group of enzymes associated with the development of cancers includes the set of SER/THR kinases classified as cyclin-dependent kinases (CDKs), especially CDK2. This enzyme plays a crucial role in the regulation of the cell cycle. The CDK2 and other cyclin-dependent kinases are also crucially important in the regulation of the other enzymes involved in transcription and replication processes [22–24]. A significant group of CDK2 inhibitors is isatin derivatives and its analogs, which contain a characteristic oxindole core in their structure. Numerous studies show that this type of compound has a notable affinity for the active site of this enzyme [12,16,25]. The location of the donors and acceptors of hydrogen bonds in the oxindol core ensures the possibility of creating stable interactions in the hinge region with aminoacids such as LEU83 and GLU81 [16,25]. The presence of additional hydrogen bond acceptors in oxindolic systems may provide an opportunity to create interactions with LYS33 and ASP145 [15,26,27]. Based on these guidelines, this work will develop a new group of isatin derivatives based on the reaction of isatin derivatives with the set of substituted benzoylhydrazines. The schematic representation of the isatin-based benzoylhydrazine structure is presented in Figure 1. During the creation of new potential inhibitors, the impact of the presence of various substituents (positions R2–R7) on their binding activity with the CDK2 active site will be evaluated.

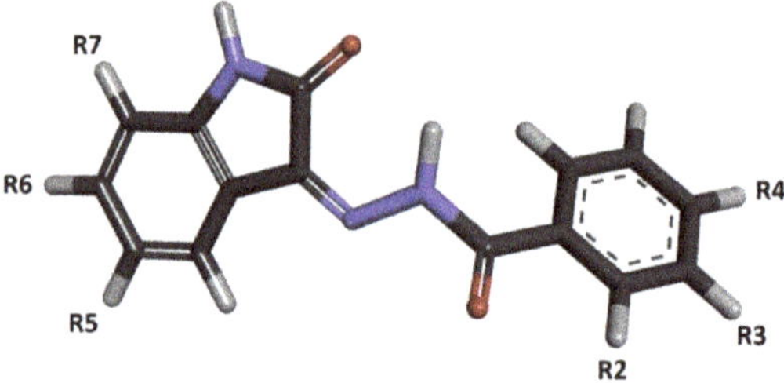

Figure 1. The graphic representation of isatin derivatives. Symbols R2–R7 represent places of addition of new chemical residues modifying the chemical properties of the native compound.

2. Results and Discussion

2.1. Design and Computational Analysis of Binding Activity

The isatin derivatives were created in two stages. Firstly, the modification of the isatin molecule was chosen. Based on the commercially available isatine derivatives containing substituents localized in positions 5, 6 and 7, model structures of inhibitors obtained by reaction with non-substituted benzoylhydrazide were created. The binding affinity of such molecules towards cyclin dependent kinase was evaluated using docking methods.

The outcome of these calculations is presented in Table 1. Considering all of the collected data, it can be observed that the least effective are modifications located in position 7, while the most favorable effects involving an increase in affinity are observed in the case of modifications located in positions 5 and 6. Among all of the considered structures, the highest value of affinity was recorded for the 5-nitroisatin derivative. The structure of the complex created by this molecule with the CDK2 active site is presented in Figure 2.

Table 1. The values of binding affinity (ΔG) of substituted isatin derivatives towards CDK2 active site. The reference value obtained for non-substituted isatin derivative is −8.9 (kcal/mol).

Name of Substituent	ΔG Binding Affinity (kcal/mol)		
	R5	R6	R7
- I	−9.20	−8.80	−8.00
- Br	−8.80	−9.10	−8.10
- CH$_3$	−9.40	−9.30	−8.60
- Cl	−8.80	−9.20	−8.30
- F	−9.20	−9.30	−8.60
- NO$_2$	−9.50	—	−8.30
- OCH$_3$	−9.10	−8.90	−8.20

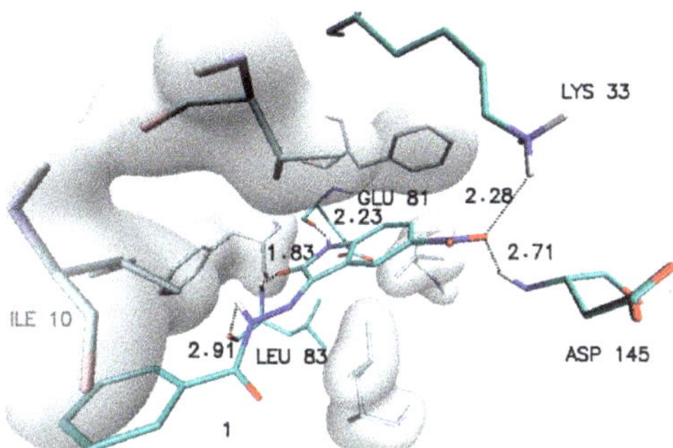

Figure 2. The graphic representation of interactions involved in stabilization of CDK2 complex with *N'*-[5-nitro-2-oxo-1,2-dihydro-3*H*-indol-3-ylidene]benzohydrazide (**1**).

The stability of such a system is maintained by a network of interactions including hydrogen bonds and hydrophobic interactions. Besides the hydrogen bonds created by atoms from a molecular core with GLU81 and LEU83, the activity of oxygen atoms from the nitro group was also observed, which participates in interactions with LYS33 and ASP145. Based on the values of binding affinity and the presence of additional binding factors, related with the existence of two additional hydrogen bond acceptors, 5-nitroisatin was chosen for the next stage of research. Based on commercially available substituted benzoylhydrazide derivatives, 30 new potential CDK2 inhibitors were created; the scheme of the foreseen modifications of the native molecule are presented in Figure 3. In the nomenclature adopted in this work, a native molecule based on unmodified benzoylhydrazide was marked with the number "1", while in the case of subsequent derivatives, the designation consists of a number symbolizing the place of substitution and the letter assigned for a specific chemical group. The binding capabilities towards the CDK2 active site of each newly created molecule were evaluated using docking methods. Table 2 presents the values of binding affinity and the inhibition constants obtained for all 5-nitroisatin-based benzoylhydrazines.

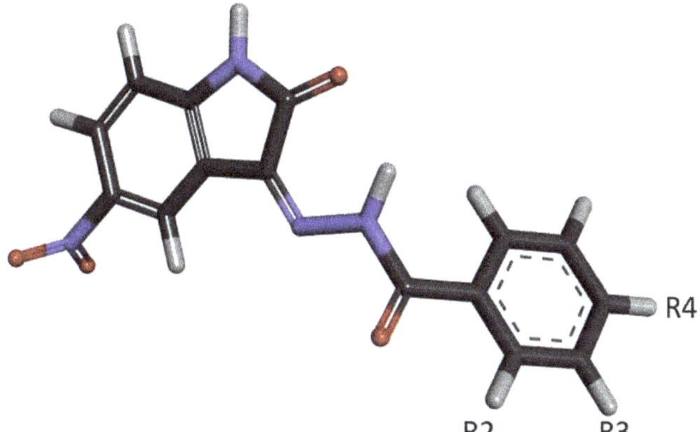

a -CH₃; **b** -CF₃; **c** -F; **d** -Br; **e** -Cl; **f** -NH₂; **g** -N(CH₃)₂; **h** -NO₂; **i** -OH; **j** -OCH₃;

Figure 3. The graphic representation of 5-nitroisatin-based benzoylhydrazines. The markings R2, R3 and R4 represent the places of chemical group substitution. The derivative markings consist of a number symbolizing the place of substitution and the letter assigned for a specific chemical group.

Table 2. The values of biding affinity and inhibition constants (IC) of 5-nitroisatin-based benzoylhydrazines towards CDK2 active site.

Chemical Group	Name	Binding Affinity (kcal/mol)	IC [nM]	Name	Binding Affinity (kcal/mol)	IC [nM]	Name	Binding Affinity (kcal/mol)	IC [nM]
-CH$_3$	2a	−9.60	91.9	3a	−10.00	46.8	4a	−9.57	96.6
-CF$_3$	2b	−10.00	46.8	3b	−10.00	46.8	4b	−9.57	96.6
-F	2c	−9.70	77.6	3c	−9.80	65.5	4c	−9.50	108.7
-Br	2d	−9.70	77.6	3d	−9.73	73.8	4d	−9.50	108.7
-Cl	2e	−9.70	77.6	3e	−9.80	65.5	4e	−9.50	108.7
-NH$_2$	2f	−9.70	77.6	3f	−9.70	77.6	4f	−9.40	128.7
-N(CH$_3$)$_2$	2g	−8.50	588.1	3g	−9.37	135.4	4g	−9.50	108.7
-NO$_2$	2h	−9.23	171.5	3h	−9.77	68.9	4h	−9.50	108.7
-OH	2i	−9.47	114.4	3i	−9.40	128.7	4i	−9.40	128.7
-OCH$_3$	2j	−9.27	160.3	3j	−9.47	114.4	4j	−9.37	135.4

The inhibition constant (IC) for ligand molecules was estimated based on the Van 't Hoff isotherm equation.

$$K_I = exp^{\left(\frac{\Delta G_b}{RT}\right)}$$

Comparing the values obtained for new molecules with the native one, it can be observed that the introduction of modifications in the aromatic system of the considered derivatives does not always increase their binding capacities and, in some cases, even lowers them. It is especially visible in the case of modifications located in the para position (R4), since none of the newly generated molecules with such modification demonstrated a noticeable increase in binding activity.

Different observations can be made in the case of modifications present in the ortho (R2) and meta (R3) positions. Among all considered substituents, the greatest influence on the binding capability was observed in the case of the trifluoromethyl group, the appearance of which caused the greatest increase in affinity value (−0.5 kcal/mol) and the most convenient value of the inhibition constant (46.8 nM). An analogous increase was also recorded for the methyl group located in the meta position (R3). In the case of the other considered substituents, the observed increases in affinity ranged from −0.2 to −0.3 kcal/mol. For all complexes formed by derivatives with an increase in affinity of at least −0.2 kcal/mol, structural stability was analyzed using molecular dynamics methods. The structures of the chosen complexes are presented in Figure 4 and Figure S1 (Supplementary Materials). Such systems are maintained by a network of numerous interactions of different types. The most important hydrogen bonds identified in all the analyzed systems are summarized in Table 3. Each considered molecule creates bindings with aminoacids such as LYS33, GLU81, LEU83 and ASP145. Based on the geometric hydrogen bond strength classification, interactions created with GLU81, LEU83 and LYS33 can be qualified as medium-strength hydrogen bonds, while interactions with ASP145 are classed weak bonds. In the case of molecules such as 2b, 2f and 3h, interactions created by atoms from chemical groups located in benzoylhydrazide part of the molecule were also observed.

The structural stability of considered complexes and the time evolution of such systems was evaluated based on an analysis of root mean square deviation (RMSD) values and structural descriptors, defining the interactions involved in complex stabilization such as distances and angles measured for hydrogen bond donors and acceptors identified for particular hydrogen bond. Figure S2 (Supplementary Materials) presents distributions of the RMSD values estimated for inhibitors and CDK2 protein. In each case, values were calculated with regard to the starting point, which was the geometry of the complex obtained during docking stage.

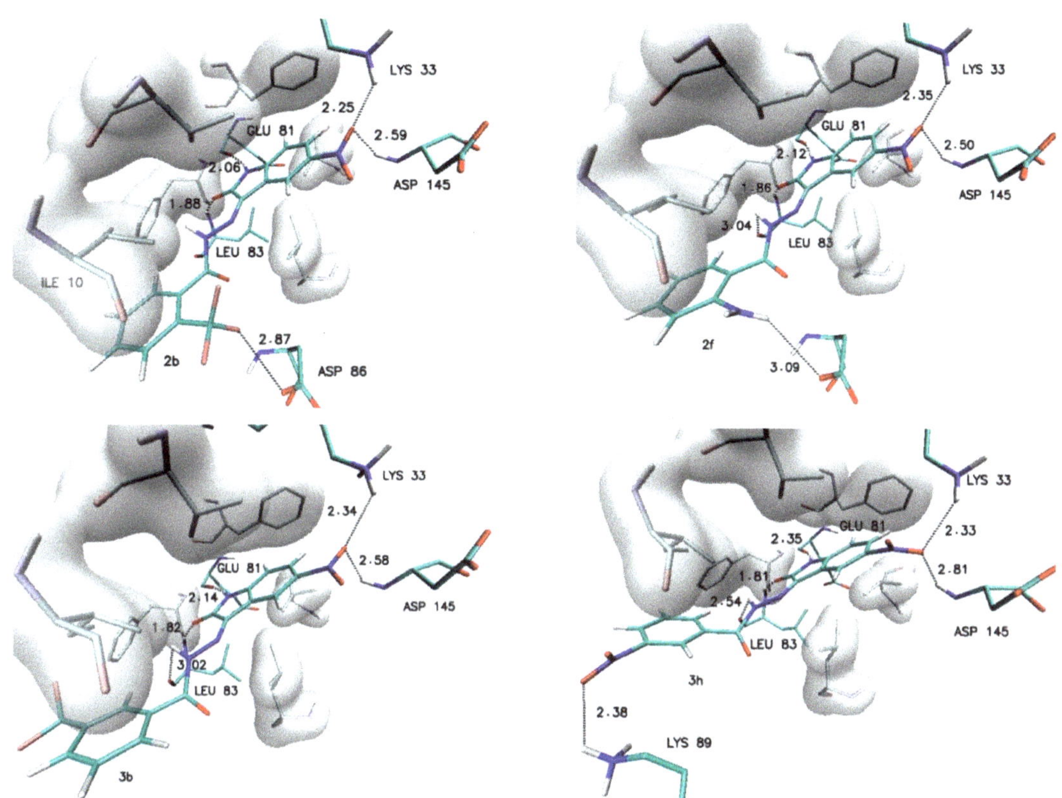

Figure 4. The graphic representation of interactions involved in stabilization of CDK2 complex with chosen 5-nitroisatin-based benzoylhydrazines.

Table 3. The values of hydrogen bond lengths [Å] involved in stabilization of 5-nitroisatin-based benzoylhydrazines complexes with the CDK2 active site.

Name	Hydrogen Bond Length [Å]						
	LYS33	GLU81	LEU83 H	LEU83 O	ASP145	ASP 86	LYS 89
1	2.28	2.23	1.83	2.91	2.71	—	—
2b	2.25	2.06	1.88	—	2.59	2.87	—
2c	2.36	2.05	1.83	—	2.53	—	—
2d	2.37	1.98	1.86	—	2.53	—	—
2e	2.34	2.03	1.84	—	2.55	—	—
2f	2.35	2.12	1.86	3.04	2.50	3.09	—
3a	2.28	2.19	1.81	2.95	2.64	—	—
3b	2.34	2.14	1.82	3.02	2.58	—	—
3c	2.31	2.23	1.83	2.89	2.66	—	—
3d	2.37	2.13	1.89	3.10	2.49	—	—
3e	2.26	2.20	1.84	2.93	2.64	—	—
3f	2.34	2.19	1.80	2.93	2.66	—	—
3h	2.33	2.35	1.81	2.54	2.81	—	2.38

An analysis of such distributions and averaged values is presented in Table 4; this shows that not all proposed derivatives have the ability to create stable complexes with the active place of the considered protein. In a few cases, such as molecules 2c, 2e, 3c and 3d, large fluctuations in RMSD values are observed, indicating the noticeable conformational instability of such molecules in the space of a CDK2 active site. The inability to maintain the original conformation obtained during docking or to adopt a new one ensuring a better fit to the active place may indicate insufficient matching, which is caused by the presence of a possible steric hindrance in the considered molecules. Completely different characteristics of the discussed properties are shown by molecules such as 1, 2b, 2d, 3b, 3e, 3f and 3h, for which uniform distributions and low standard deviation for the analyzed populations of RMSD values indicate the conformational stability of the analyzed systems.

Table 4. The averaged values of RMSD for ligands and CDK2 protein for all steps used during structural analysis.

Name	Ligand		CDK2	
	RMSD	**SD**	**RMSD**	**SD**
1	0.591	0.132	2.284	0.181
2b	0.698	0.146	2.745	0.419
2c	1.037	0.238	2.585	0.402
2d	0.525	0.101	2.316	0.209
2e	0.768	0.293	2.619	0.282
2f	0.689	0.177	2.265	0.245
3a	0.551	0.177	2.420	0.172
3b	0.708	0.155	2.618	0.164
3c	0.948	0.357	2.440	0.351
3d	0.603	0.204	2.504	0.400
3e	0.506	0.134	2.251	0.182
3f	0.475	0.085	2.431	0.263
3h	0.737	0.159	2.612	0.211

An important factor for assessing the stability of the complexes formed by potential inhibitors with the active site of a biological target is the evaluation of the durability of the interactions responsible for its maintenance. During the docking stage, the primary structures of the complexes were created, in which interactions such as hydrogen and halogen bonds were identified. Table 5 and Table S1 (Supplementary Materials) contain a list of the most important impacts that showed sufficient durability to exclude their incidental occurrence. Among all of the considered interactions, the most important and the most stable in each of the complexes are hydrogen bonds formed by the oxygen and the hydrogen atoms of the isatin core with LEU83 and GLU81. The presence of these hydrogen bonds was confirmed in 100% of conformers accumulated during molecular dynamics for each of the analyzed complexes. The observed differences between particular systems are visible in the distribution of the bond lengths dominant within the compared populations. In the case of bonds created with GLU81, over 90% of tested conformers creates at least medium-strength hydrogen bonds (distance $\leq$ 2 Å), the exceptions are only the complexes created by 2c and 3c molecules. The distances measured for hydrogen bonds created with LEU83 exhibit much larger diversity—in this case, there is at least a 70% share of medium-strength hydrogen bonds observed only for conformers from complexes created by 1, 2d, 2e, 2f, 3b, 3e and 3f molecules. The activity of oxygen atoms from the nitro group localized in the isatin core and the significance of hydrogen bonds created by them is diametrically different for complexes created by individual derivatives. Among the analyzed systems, one can distinguish complexes where the bonds with ASP145 almost disappeared and those created by LYS33 have been significantly reduced, as in the case of 2c and 3c complexes. A group of complexes is also noticeable where, for at least 80% of the conformers, stable bonds with at least one of the amino acids are observed, for example, complexes formed by the molecules 1, 2d, 2f, 3b, 3e, and 3h. The highest binding activity for the nitro group was recorded in the case of derivative 2b, in which stable hydrogen bonds were observed both for the ASP145 and LYS33 in over 90% of the analyzed conformers. Based on the structures of complexes obtained during docking for many of the derivatives considered, the possibility of creating a second hydrogen bond with LEU83 was expected. An analysis of the results of molecular dynamics simulation showed the existence of stable and numerous interactions of this type only for complexes created by 2b (85% of conformers) and 3f (98.9% of conformers) derivatives. An analysis of the interactions in the complex formed by the derivative 2f shows the presence of a halogen bond between the fluorine atoms of the 3-fluoromethyl group and the oxygen atom of ILE10. The distance between the donor and acceptor of this interaction in the scope of the entire simulation ranges from 2.6 to 3.2 Å. The presence of such an interaction contributes to the overall stabilizing effect of the analyzed complex.

Table 5. The cumulative analysis of the length of the interactions identified in CDK2 complexes with selected 5-nitroisatin-based benzoylhydrazines. The distances presented in the table represent the middle values of intervals with a width of 0.2 Å.

Interactions	Population %								
	Σ	1.6 Å	1.8 Å	2 Å	2.2 Å	2.4 Å	2.6 Å	2.8 Å	3 Å
1									
Ligand (H1) … (O) GLU81	100.0	2.6	54.6	37.1	5.3	0.5	0.0	0.0	0.0
Ligand (O1) … (HN) LEU83	100.0	0.6	28.2	46.4	19.2	4.6	0.9	0.1	0.0
Ligand (O) … (H) LYS33	82.8	0.1	0.1	0.1	0.1	0.1	0.1	0.1	0.1
Ligand (O) … (H) ASP145	58.8	0.0	0.6	2.9	6.6	9.1	11.9	14.3	13.5
Ligand (H5) … (O) LEU83	19.62	0.0	0.1	0.9	0.6	0.9	2.5	3.7	10.9
2b									
Ligand (H1) … (O) GLU81	100.0	4.1	49.5	37.1	7.8	1.3	0.2	0.0	0.0
Ligand (O2) … (HN) LEU83	99.9	0.6	22.6	40.9	21.9	9.3	3.4	1.2	0.1
Ligand (O) … (H) LYS33	91.5	0.0	5.1	13.4	18.1	17.6	16.4	11.4	9.5
Ligand (H5) … (O) LEU83	85.5	0.8	16.7	29.3	12.9	5.2	4.1	6.0	10.4
Ligand (F) … (O) ILE10	64.2	0.0	0.0	0.0	0.0	0.0	2.8	20.0	41.3
Ligand (O) … (H) ASP145	91.0	0.1	3.0	16.8	21.2	15.9	14.6	10.5	8.9
2d									
Ligand (H1) … (O) GLU81	100.0	2.1	49.1	40.2	7.4	1.1	0.1	0.0	0.0
Ligand (O1) … (HN) LEU83	100.0	0.2	25.1	46.3	20.1	6.8	1.6	0.1	0.0
Ligand (O) … (H) LYS33	81.2	0.2	8.6	17.4	19.7	15.2	8.5	6.9	4.7
Ligand (O) … (H) ASP145	33.3	0.0	0.0	0.6	1.8	3.1	7.0	9.1	11.6
2f									
Ligand (H1) … (O) GLU81	100.0	2.1	48.4	41.6	6.8	0.8	0.1	0.1	0.0
Ligand (O1) … (HN) LEU83	100.0	0.6	23.0	48.1	21.3	5.5	1.3	0.3	0.0
Ligand (O) … (H) LYS33	83.6	0.1	9.7	15.8	18.3	14.3	11.3	8.1	6.1
Ligand (O) … (H) ASP145	65.6	0.0	2.4	7.9	10.8	15.8	12.4	8.4	8.1
3b									
Ligand (H1) … (O) GLU81	100.0	3.3	54.2	35.9	6.1	0.4	0.1	0.0	0.0
Ligand (O1) … (HN) LEU83	100.0	0.1	24.1	46.7	21.9	5.6	1.3	0.3	0.1
Ligand (O) … (H) LYS33	79.3	0.1	5.4	12.2	16.3	14.3	12.0	10.8	8.3
Ligand (O) … (H) ASP145	66.3	0.0	0.3	4.4	11.2	15.9	12.9	9.9	11.5
Ligand (H5) … (O) LEU83	11.4	0.0	0.0	0.1	0.4	0.6	0.8	2.6	7.0

Table 5. *Cont.*

Interactions	Population %								
	Σ	1.6 Å	1.8 Å	2 Å	2.2 Å	2.4 Å	2.6 Å	2.8 Å	3 Å
3e									
Ligand (**H1**) … (**O**) GLU81	100.0	3.1	56.5	36.3	3.6	0.6	0.0	0.0	0.0
Ligand (**O1**) … (**HN**) LEU83	100.0	0.4	26.7	48.1	19.4	4.8	0.5	0.0	0.0
Ligand (**O**) … (**H**) LYS33	79.8	0.1	7.2	20.8	19.7	13.9	7.3	5.9	4.9
Ligand (**O**) … (**H**) ASP145	86.8	0.0	1.4	11.3	19.2	20.8	16.6	9.9	7.6
3f									
Ligand (**H3**) … (**O**) GLU81	100.0	1.9	49.3	39.4	8.4	0.9	0.1	0.0	0.0
Ligand (**O2**) … (**HN**) LEU83	100.0	0.1	12.1	39.1	31.1	14.4	2.6	0.8	0.0
Ligand (**O**) … (**H**) LYS33	71.3	0.1	3.9	10.4	14.8	12.6	11.5	10.8	7.1
Ligand (**O**) … (**H**) ASP145	55.5	0.0	0.8	4.9	7.9	9.1	8.6	11.8	12.5
Ligand (**H5**) … (**O**) LEU83	98.9	0.3	18.3	33.4	25.9	12.2	5.5	2.3	1.1
3h									
Ligand (**H1**) … (**O**) GLU81	100.0	1.9	48.0	42.4	7.1	0.6	0.1	0.0	0.0
Ligand (**O1**) … (**HN**) LEU83	100.0	0.5	22.3	47.9	21.8	5.8	1.6	0.1	0.1
Ligand (**O**) … (**H**) LYS33	84.4	0.2	12.8	19.9	16.4	13.5	10.8	5.9	4.8
Ligand (**O**) … (**H**) ASP145	67.6	0.0	2.6	4.4	8.2	10.6	11.1	14.6	16.2
Ligand (**H5**) … (**O**) LEU83	15.4	0.0	0.1	0.8	0.8	0.9	2.0	3.6	7.3

The stability of complexes was also evaluated using the molecular mechanics Poisson–Boltzmann surface area (MMPBSA) method, which helped to evaluate the values of binding affinity characterizing particular complexes. Based on the conformers from the last 60 ns of molecular dynamics simulation, enthalpic contributions to the binding affinity were estimated. The obtained values are presented in Figure 5; numeric labels represent accurate values of affinity, while the graphic markers correspond to a spread of values, taking into account the standard deviation. Based on the gathered values, the largest affinity toward the active site of CDK2 is shown by 2b, 3f and 3h derivatives. The energy characteristics of the systems corroborate the previous structural observations based on an analysis of the conformational properties of ligands and the stability of interactions involved in the maintenance of the complexes. Based on the summary characteristics of the analyzed systems, for the experimental phase, including the synthesis and characterization of molecular and spectroscopic properties, derivatives 1, 2b, 2d, 2f, 3b, 3e, 3f and 3h were selected.

2.2. Synthesis and NMR Data

Various synthetic procedures for 5-nitroisatin-based benzoylhydrazines can be found in the literature [28–30]. The simplest indicates that 5-nitroisatin (s1) is heated with substituted benzoylhydrazine (s2) in ethanol in the presence of acetic acid (Scheme 1). All the derivatives were analyzed using IR, NMR and also elemental analysis.

Structurally, the synthesized 5-nitroisatin-based benzoylhydrazines can exist in *cis*- and *trans*-isoforms. Although all of the derivatives demonstrated the presence of both isomers in the solution at room temperature, with the *cis*-isomer being the predominant one. It should be noted that in all cases, their ^{1}H NMR spectra showed a more intense signals for hydrogen atoms of *cis*-isomer. Moreover, the calculated energies showed that more stable are *cis*-isomers (Table 6).

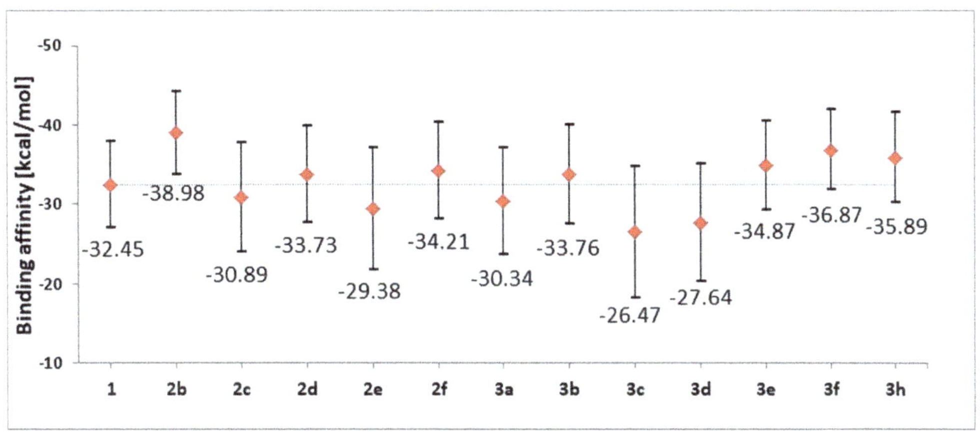

Figure 5. The values of binding enthalpy (kcal/mol) estimated for all complexes considered during molecular dynamics stage.

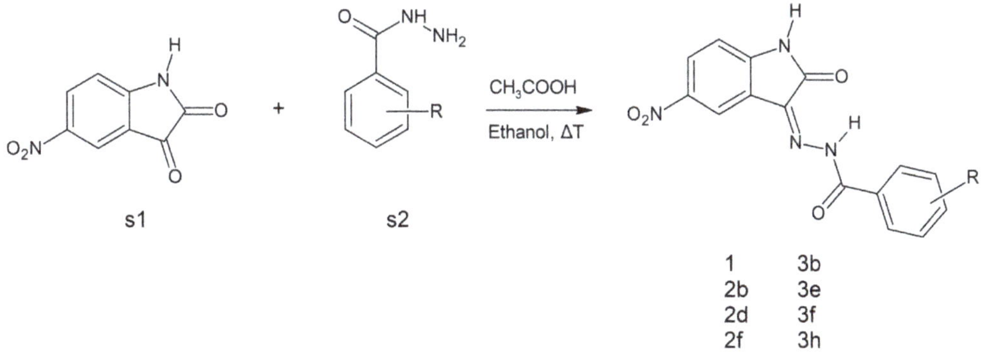

where R = H (1), 2-CF$_3$ (2b), 2-Br (2d), 2-NH$_2$ (2f), 3-CF$_3$ (3b), 3-Cl (3e), 3-NH$_2$ (3f), 3-NO$_2$ (3h)

Scheme 1. A schematic representation of the synthesis of the selected 5-nitroisatin-based benzoylhydrazines.

Table 6. Energies of *cis-* and *trans-*isomers of 5-nitroisatin-based benzoylhydrazines estimated in DMSO at the B3LYP/6-311 ++ G(d,p) level of theory. $\Delta E_{CIS\text{-}TRANS}$ represents the difference between *cis-* and *trans-*isomers ($\Delta E_{CIS\text{-}TRANS} = E_{CIS} - E_{TRANS}$).

Name	E_{CIS} [Hartree]	E_{TRANS} [Hartree]	$\Delta E_{CIS\text{-}TRANS}$ [Hartree]	$\Delta E_{CIS\text{-}TRANS}$ (kcal/mol)
1	−1097.75447	−1097.74754	−0.0069268	−4.347
2b	−1434.89555	−1434.88862	−0.0069287	−4.348
2d	−3671.28947	−3671.28262	−0.0068435	−4.294
2f	−1153.13766	−1153.13059	−0.0070714	−4.437
3b	−1434.90262	−1434.89550	−0.0071178	−4.466
3e	−1557.37551	−1557.36858	−0.0069289	−4.348
3f	−1153.13538	−1153.12863	−0.0067527	−4.237
3h	−1302.31909	−1302.31183	−0.0072546	−4.552

As shown in Figure 6, in the chemical shift structure of the *cis* form, there is a possibility of intramolecular hydrogen bonding between the hydrogen atom of the hydrazone section and the oxygen atom of the adjacent carbonyl group [31]. In addition, in derivatives substituted in position 2 (2b, 2d and 2f), the presence of an intramolecular hydrogen bond between the substituent and the carbonyl oxygen of benzoylhydrazine was observed. According to R.S. Hunoor et al. [32], the N–H group (hydrazine part) is sandwiched between carbonyl (C=O) and N=C azometine nitrogen and intramolecular hydrogen bonding with carbonyl oxygen of 5-nitroisatin. All these interactions make it difficult to unambiguously assign aromatic hydrogens and carbons to individual isomers of 2-substitued 5-nitroisatin-based benzoylhydrazines.

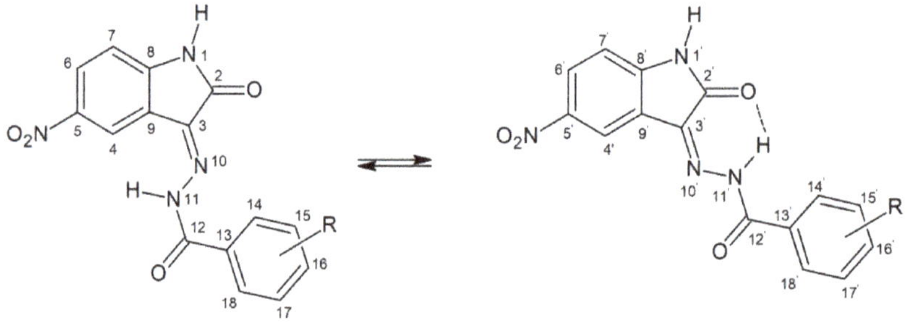

where R = H (1,1'), 2-CF$_3$ (2b, 2b'), 2-Br (2d, 2d'), 2-NH$_2$ (2f, 2f'), 3-CF$_3$ (3b, 3b'), 3-Cl (3e, 3e'), 3-NH$_2$ (3f, 3f '), 3-NO$_2$ (3h, 3h')

Figure 6. Possible geometry of the synthesized 5-nitroisatin-based benzoylhydrazines.

As previously mentioned, the compounds showed in Figure 6 were present in solution as *trans*- and *cis*-isomers. In all cases, their ^{1}H NMR spectra showed a signal with chemical shift in the range of 13.16–13.37 ppm, corresponding to the N–H of benzoylhydrazine involved in the relevant intramolecular hydrogen bond for *cis*-isomers. For the same proton the *trans*-isomer (without intramolecular hydrogen bond), the signal was observed in the range of 12.02–12.88 ppm. The ^{1}H-NMR signal of N–H of isatin can be seen at δ = 11.51–11.57 ppm and δ = 11.96–12.04 ppm in DMSO, which is a singlet, for *cis*- and *trans*-isomers, respectively (Figure 7). The first signal observed with the least chemical shift in aromatic protons relates to the H-7 and H-7$'$ of isatin, which appears as a doublet signal in the region of 6.85 to 7.18 ppm (J = 8.56–8.86 Hz). The next protons of isatin (H-4, H-4$'$ and H-6, H-6$'$) appeared in the region of 8.23 to 9.06 ppm. The aroyl (hydrazone) part of the target compounds showed appropriate signals in the range of 6.67–8.18 ppm, depending on the type of substituent. The carbon signals of the most characteristic carbon atoms, such as carbonyl carbon C-12/C-12$'$ and lactonyl carbon C-2/C-2$'$, were observed at ca. 167 and 164 ppm, respectively. The azomethine carbon C-3/C-3$'$ at ca. 137 ppm was not always recorded.

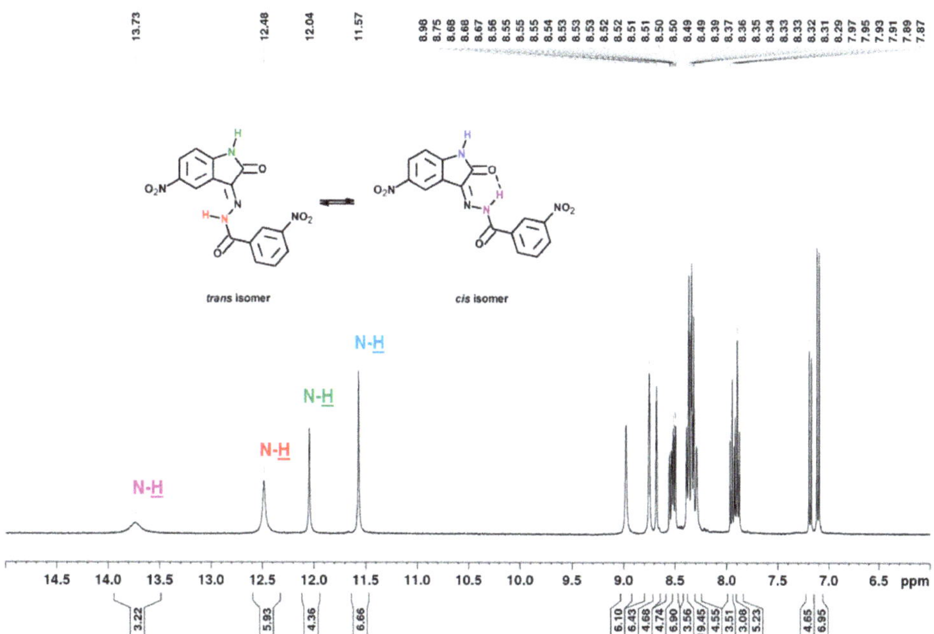

Figure 7. ^{1}H NMR spectrum (400 MHz) of 3-nitro-*N′*-[5-nitro-2-oxo-1,2-dihydro-3*H*-indol-3-ylidene]benzohydrazide (3h) in DMSO-d$_6$.

2.3. Infrared Spectral Studies

Vibrational spectroscopy is used extensively in the study of molecular conformations, the identification of functional groups and reaction kinetics, etc. Infrared spectral data for the synthesized 5-nitroisatin-based benzoylhydrazines are presented in Table 7. Sharp bands of medium intensity at 3167–3303 cm^{-1} in the analyzed compounds spectrums were due to stretching vibrations of the –NH group. The absorption band characteristic of ν(C=O) vibrations of isatin and benzoylhydrazine fragments appeared in the range 1697–1751 cm^{-1} and 1622–1698 cm^{-1}, respectively, and the absorption band at 1513–1528 cm^{-1} was assigned to azomethine ν(C=N) stretching.

Table 7. Characteristic band-strength values of the synthesized 5-nitroisatin-based benzoylhydrazines.

Compound	ν(–NH$_2$)		ν(N–H)	ν(C=O)		ν(C=N)	ν(N=O)
	ν$_{asym}$	ν$_{sym}$		Lactone	Hydrazide		
1	-	-	3167	1744	1625	1525	1340
2b	-	-	3232	1740	1690	1513	1315
2d	-	-	3303	1749	1698	1495	1339
2f	3490	3378	3197	1749	1623	1517	1334
3b	-	-	3197	1713	1685	1526	1341
3e	-	-	3189	1679	1676	1527	1340
3f	3481	3387	3204	1752	1622	1514	1339
3h	-	-	3197	1747	1684	1528	1341

2.4. Spectroscopic Properties

The spectral data of the synthesized compounds in different solvents are shown in Table 8. The study showed that new 5-nitroisatin-based benzoylhydrazines absorbed at around 320 nm and emitted (between 388 nm and 579 nm) over a wide spectral range with the extinction coefficients ranging from 1.92×10^4 M^{-1} cm^{-1} to 2.80×10^4 M^{-1} cm^{-1}.

The extinction coefficients did not achieve relatively high values. Additionally, no clear correlations between the solvent polarities and the molar extinction coefficients were found. The shape of the absorption spectra remained very similar to that of the parent compound (1) (R=H), and it was not significantly dependent on the electron-withdrawing or electron-releasing group in the different positions of the phenyl ring (Figure 8a). The position of the band of absorption did not depend on the polarity of the solvent (Figure 8b). The emission spectra were broad, with the single maximum of fluorescence that slightly shifted towards higher-wavelength values as the polarity of solvent increased.

Table 8. Spectroscopic properties of the synthesized 5-nitroisatin-based benzoylhydrazines in solvents of different polarity.

No.	Substituent	Solvent	λ_{ab} (nm)	λ_{fl} (nm)	ε ($\times 10^4$, $M^{-1} \cdot cm^{-1}$)	Stokes Shift (cm^{-1})
		DCM	320	522	a	12,092
1	H	MeOH	323	573	3.19	13,507
		MeCN	322	542	2.28	12,605
		DCM	315	517	a	12,403
2b	2-CF$_3$	MeOH	315	575	2.72	14,354
		MeCN	314	538	2.27	13,259
		DCM	317	523	a	12,425
2d	2-Br	MeOH	319	574	2.51	13,926
		MeCN	318	539	2.16	12,893
		DCM	320	437	a	8366
2f	2-NH$_2$	MeOH	320	417	1.92	13,926
		MeCN	318	382	1.98	5268
		DCM	319	522	a	12,190
3b	3-CF$_3$	MeOH	319	579	2.52	14,076
		MeCN	318	540	2.80	12,928
		DCM	319	530	a	12,480
3e	3-Cl	MeOH	325	556	1.32	12,783
		MeCN	320	514	1.18	11,794
		DCM	321	435	a	8164
3f	3-NH$_2$	MeOH	321	388	2.38	5379
		MeCN	320	527	2.27	12,274
		DCM	319	522	a	12,190
3h	3-NO$_2$	MeOH	321	577	2.10	13,821
		MeCN	320	547	1.88	12,968

a—compounds partially insoluble in DCM.

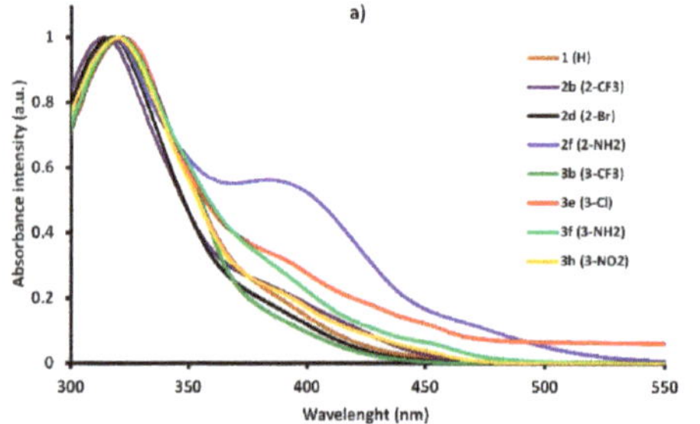

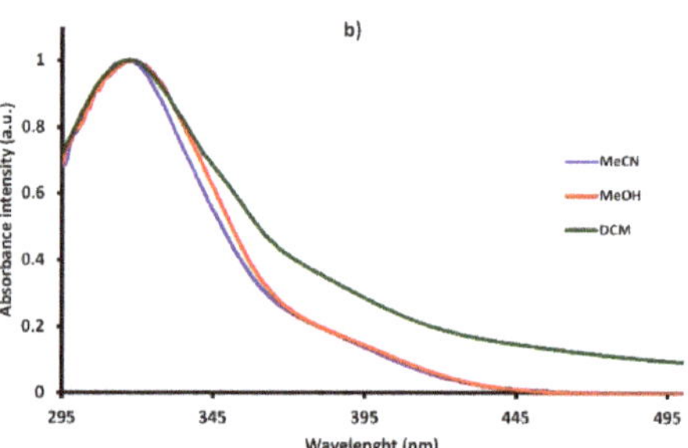

Figure 8. The normalized UV–Vis absorption of the synthesized 5-nitroisatin-based benzoylhydrazines (**a**) and the normalized UV–Vis absorption of 2-bromo-*N'*-[5-nitro-2-oxo-1,2-dihydro-3*H*-indol-3-ylidene]benzohydrazide (2d) in solvents of different polarity (**b**).

3. Materials and Methods

3.1. Computational Methods

3.1.1. The Docking Procedure

The structure of cyclin dependent kinase 2 (CDK2) was obtained from Brookhaven Protein Database PDB (CDK2—PDB ID 1E9H) [25]. The AutodockTools package [33] was used in all preliminary steps, including the identification and localization of the active site and the preparation of all structures of considered ligands and protein. In the case of each structure used during the molecular docking stage, all non-polar hydrogen atoms were removed. The localization of the center (x = 2.5, y = 35.0, z = 63.0) and the dimensions of the grid box have been adjusted to the size of the active site (16 × 14 × 22 Å). The docking procedure for all considered ligands was realized utilizing Autodock Vina package [34]. The value of exhaustiveness parameter was set at 20 for all realized simulations. In the case of each considered molecule, the docking procedure was repeated five times with a different value of the random seed.

3.1.2. The Molecular Dynamics Simulations

The structure of the CDK2 protein was described using ff14sb [35] force-field parameters, while in the case of the inhibitor molecules, gaff force field parameters were used. In the case of isatin derivatives, the values of charges were estimated based on the Merz–Kollmann scheme using the RESP procedure at the HF/6-31G* level [36]. The complexes of the CDK2 protein with considered inhibitors were neutralized and immersed in the periodic cube box of TIP3P water molecules. Each system was heated to 300 K in the preliminary stage; the control of the temperature was realized utilizing Langevin thermostat [37]. The 80-nanosecond molecular dynamics simulations were realized with an applied shake algorithm and periodic boundary conditions. The structural analysis of considered systems, including the stability of interactions involved in the maintenance of complexes, were realized utilizing a VMD package [38]. The definition of hydrogen bonds was based on the following boundary criteria: distance D (donor)–A (acceptor) < 3.5 Å, distance H–A < 3 Å and angle D–H–A > 90°. The affinity of isatin derivatives towards the CDK2 active site was evaluated using the molecular mechanics Poisson–Boltzmann surface area (MMPBSA) [39]. The molecular dynamics simulations were realized using the AMBER 14 package [40].

3.1.3. The Quantum Mechanics Calculations

The geometries of all the isatin derivatives were calculated using density functional theory (DFT) implemented in the Gaussian 09 package [41]. All calculations were realized using B3LYP functional at 6-311 + G (d,p) level of theory. The solvatation effects during calculations were included by applying a self-consistent reaction field (SCRF) approach [42] based on accurate numerical solutions of the Poisson–Boltzmann equation [43,44].

3.2. Materials

All reagents and solvents were purchased from Sigma-Aldrich (Poznań, Poland) and used without further purification. The highest (≥99%) purity of all used chemicals was required for spectroscopic studies.

3.3. Synthesis

The general procedure for the synthesis of the *N'*-[5-nitro-2-oxo-1,2-dihydro-3*H*-indol-3-ylidene]benzohydrazide.

Equimolar amounts of 5-nitroisatin (s1) (0.002 mol) and substituted benzolhydrazine (s2) (0.002 mol) were added to 96% ethanol (50 mL) containing 3 drops of glacial acetic acid. The mixture was heated under reflux for 5 h and then cooled to room temperature. The resulting solid was collected by filtration, washed with cold ethanol and recrystallized from ethanol to give the following compounds (1, 2b, 2d, 2f, 3b, 3e, 3f, and 3h).

The elemental analysis is as follows:

N′-[5-nitro-2-oxo-1,2-dihydro-3*H*-indol-3-ylidene]benzohydrazide (1) [30] Yellow solid, yield 85%, d.t. 338.4 °C, IR (ATR), cm^{-1}: 3167, 1744, 1625, 1525, 1340. *Trans*-isomer (1): ^{1}H NMR (DMSO-d$_6$ from TMS) δ (ppm): 12.22 (s, 1H), 11.56 (s, 1H), 8.89 (s, 1H), 8.34 (dd,1H), 7.96 (m, 2H), 7.68 (m, 6H), 7.10 (d, *J* = 8.76, 1H). ^{13}C NMR δ (ppm): 167.55, 165.44, 148.02, 142.31, 139.06, 132.96, 129.73, 129.18, 129.04, 128.06, 122.76, 115.78, 111.21. *Cis*-isomer (1′): ^{1}H NMR (DMSO-d$_6$ from TMS) δ (ppm): 13.74 (s, 1H), 12.03 (s, 1H), 8.31 (m, 2H), 7.92 (m, 2H), 7.68 (m, 6H), 7.18 (d, *J* = 8.24 Hz, 1H). ^{13}C NMR δ (ppm): 163.87, 149.86, 143.41, 133.63, 132.18, 121.08, 116.29, 112.03. C$_{15}$H$_{10}$N$_4$O$_4$, Calcd. C, 58.07, H, 3.25, N, 18.06. Found C, 58.19, H, 3.30, N, 21.48.

2-trifluoromethyl-*N′*-[5-nitro-2-oxo-1,2-dihydro-3*H*-indol-3-ylidene]benzohydrazide (2b). Yellow solid, yield 87%, m.p. 267.4 °C, d.t. 310.3 °C, IR (ATR), cm^{-1}: 3232, 1740, 1690, 1513, 1315. ^{1}H NMR (DMSO-d$_6$ from TMS) δ (ppm): 13.18 (s, 1H, 2b′), 12.88 (s, 1H, 2b), 11.96 (s, 1H, 2b′), 11.51 (s, 1H, 2b), 9.17 (bs, 1H), 8.85 (bs, 1H), 8.32 (m, 2H, 2b and 2b′), 7,87 (m, 8H, 2b and 2b′), 7.14 (d, *J* = 8.56 Hz, 1H, 2b′), 7.07 (d, *J* = 8.86 Hz, 1H, 2b). ^{13}C NMR δ (ppm): 165.14, 163.46, 150.03, 148.30, 143.31, 142.42, 134.20, 132.92, 131.32, 129.38, 128.26, 126.91, 125.54, 122.82, 122.46, 120.75, 115.38, 112.05, 111.21. C$_{16}$H$_9$F$_3$N$_4$O$_4$, Calcd. C, 50.80, H, 2.40, N, 14.81. Found C, 50.74, H, 2.34, N, 14,93.

2-bromo-*N′*-[5-nitro-2-oxo-1,2-dihydro-3*H*-indol-3-ylidene]benzohydrazide (2d) Yellow solid, yield 85%, m.p. 268.4 °C, d.t. 305.9 °C, IR (ATR), cm^{-1}: 3303, 1749, 1698, 1495, 1339. ^{1}H NMR (DMSO-d$_6$ from TMS) δ (ppm): 13.16 (s, 1H, 2d′), 12.43 (s, 1H, 2d), 11.97 (s, 1H, 2d′), 11.52 (s, 1H, 2d), 9.06 (s, 1H, 2d), 8.34 (dd, 2H, 2d and 2d′), 7,79 (m, 2H, 2d and 2d′), 7.55 (m, 6H, 2d and 2d′), 7.15 (d, *J* = 8.72 Hz, 1H, 2d′), 7.07 (d, *J* = 8.72 Hz, 1H, 2d). ^{13}C NMR δ (ppm): 165.20, 149.96, 148.24, 143.32, 142.43, 137.26, 135.64, 133.20, 132.24, 130.05, 129.31, 128.23, 122.49, 120.85, 119.68, 115.43, 112.06, 111.20. C$_{15}$H$_9$BrN$_4$O$_4$, Calcd. C, 46.29, H, 2.33, N, 14.40. Found C, 46.00, H, 2.43, N, 14.59.

2-Amino-*N′*-[5-nitro-2-oxo-1,2-dihydro-3*H*-indol-3-ylidene]benzohydrazide (2f) Dark orange solid, yield 91%, d.t. 320.1 °C, IR (ATR), cm^{-1}: 3490, 3378, 3197, 1749, 1623, 1517, 1334. *Trans*-isomer (2f): ^{1}H NMR (DMSO-d$_6$ from TMS) δ (ppm): 11.52 (s, 1H), 8.83 (d, *J* = 2.16 Hz, 1H), 8.32 (m, 1H), 7.67 (m, 1H), 7.31 (m, 2H), 7.09 (d, *J* = 8.67 Hz, 1H), 6.87 (m, 2H), 6.67 (m, 2H). ^{13}C NMR δ (ppm): 167.92, 165.56, 150.91, 149.56, 142.25, 138.28, 133.83, 130.01, 128.88, 122.36, 117.84, 115.92, 113.01, 111.11. *Cis*-isomer (2f′): ^{1}H NMR (DMSO-d$_6$ from TMS) δ (ppm): 13.64 (s, 1H), 11.97 (s, 1H), 8.30 (s, 1H), 8.27 (m, 1H), 7.45 (dd, 1H), 7.31 (m, 2H), 7.15 (d, *J* = 8.67 Hz, 1H), 6.87 (m, 2H), 6.67 (m, 2H). ^{13}C NMR δ (ppm):165.21, 163.86, 151.90, 147.66, 143.33, 135.69, 134.37, 127.65, 127.60, 117.42, 115.66, 111.90, 111.60. C$_{15}$H$_{11}$N$_5$O$_4$, Calcd. C, 55.39, H, 3.41, N, 21.53. Found C, 55.20, H, 3.51, N, 21.62.

3-trifluoromethyl-*N′*-[5-nitro-2-oxo-1,2-dihydro-3*H*-indol-3-ylidene]benzohydrazide (3b) Yellow solid, yield 86%, m.p. 324.5 °C, d.t. 328.7 °C, IR (ATR), cm^{-1}: 3197, 1713, 1685, 1526, 1341. *Cis*-isomer (3b′): ^{1}H NMR (DMSO-d$_6$ from TMS) δ (ppm): 13.68 (s, 1H), 12.01 (s, 1H), 8.30 (dd, 1H), 8.23 (s, 1H), 8.18 (s, 1H), 8.17 (d, *J* = 8.12 Hz, 1H), 8.09 (d, *J* = 7.80 Hz, 1H), 7.89 (t, 1H), 7.15 (d, *J* = 8.68 Hz, 1H). ^{13}C NMR δ (ppm): 163.67, 148.12, 143.40, 137.28, 133.25, 132.10, 131.00, 130.37, 129.72, 128.22, 125.52, 124.99, 122.811, 120.84, 116.32, 112.09. Only the *cis*-isomer was observed. C$_{16}$H$_9$F$_3$N$_4$O$_4$, Calcd. C, 50.80, H, 2.40, N, 14.81. Found C, 50.92, H, 2.36, N, 14,73.

3-chloro-*N′*-[5-nitro-2-oxo-1,2-dihydro-3*H*-indol-3-ylidene]benzohydrazide (3e). Yellow solid, yield 85%, d.t. 358.5 °C, IR (ATR), cm^{-1}: 3189, 1679, 1676, 1527, 1340. *Trans*-isomer (3e): ^{1}H NMR (DMSO-d$_6$ from TMS) δ (ppm): 12.29 (s, 1H), 11.57 (s, 1H), 8.92 (s, 1H), 8.35 (dd, 1H), 8.00 (s, 1H), 7.91 (d, *J* = 7.8 Hz, 1H), 7.76 (m, 1H), 7.64 (t, 1H), 7.11 (d, *J* = 8.75, 1H). ^{13}C NMR δ (ppm): 166.49, 165.34, 150.00, 142.33, 135.38, 133.71, 132.62, 130.99, 129.31, 129.00, 127.95, 122.90, 115.68, 111.25. *Cis*-isomer (3e′): ^{1}H NMR (DMSO-d$_6$ from TMS) δ (ppm): 13.67 (s, 1H), 12.04 (s, 1H), 8.31 (m, 1H), 8.30 (s, 1H), 7.93 (m, 1H), 7.86 (d, *J* = 7.84 Hz, 1H), 7.80 (m, 1H), 7.69 (t, 1H), 7.18 (d, *J* = 8.6 Hz, 1H). ^{13}C NMR δ (ppm): 148.14, 134.39, 133.33, 131.70, 129.31, 128.24, 116.36, 112.10. C$_{15}$H$_9$ClN$_4$O$_4$, Calcd. C, 52.26, H, 2.63, N, 16.25. Found C, 52.16, H, 2.72, N, 16.30.

3-Amino-N'-[5-nitro-2-oxo-1,2-dihydro-3H-indol-3-ylidene]benzohydrazide (3f) Yellow solid, yield 92%, d.t. 327.4 °C, IR (ATR), cm^{-1}: 3481, 3387, 3204, 1752, 1622, 1514, 1339. *Trans*-isomer (3f): ^{1}H NMR (DMSO-d$_6$ from TMS) δ (ppm): 12.02 (s, 1H), 11.53 (s, 1H), 8.80 (d, J = 1.96 Hz, 1H), 8.33 (dd, 1H), 7.24 (m, 1H), 7.18 (s, 1H), 7.09 (m, 2H), 6.85 (m, 1H), 5.48 (bs, 4H). ^{13}C NMR δ (ppm): 163.86, 150.04, 149.54, 143.37, 139.06, 134.02, 129.51, 129.06, 122.64, 118.26, 115.89, 114.07, 111.76. *Cis*-isomer (3f'): ^{1}H NMR (DMSO-d$_6$ from TMS) δ (ppm): 13.65 (s, 1H), 11.99 (s, 1H), 8.30 (m, 2H), 7.24 (m, 1H), 7.21 (m, 3H), 6.98 (d, J = 8.24 Hz, 1H), 5.48 (bs, 4H). ^{13}C NMR δ (ppm): 167.54, 164.07, 149.77, 147.51, 142.27, 136.55, 132.83, 130.12, 127.87, 121.19, 118.78, 116.17, 114.37, 112.99, 111.96. $C_{15}H_{11}N_5O_4$, Calcd. C, 55.39, H, 3.41, N, 21.53. Found C, 55.43, H, 3.40, N, 21.50.

3-nitro-N'-[5-nitro-2-oxo-1,2-dihydro-3H-indol-3-ylidene]benzohydrazide (3h). Yellow solid, yield 92%, m.p. 275.2 °C, d.t. 345.5 °C, IR (ATR), cm^{-1}: 3197, 1747, 1684, 1528, 1341. *Trans*-isomer (3h): ^{1}H NMR (DMSO-d$_6$ from TMS) δ (ppm): 12.48 (s, 1H), 11.57 (s, 1H), 8.98 (s, 1H), 8.75 (s, 1H), 8.51 (m, 1H), 8.37 (m, 2H), 7.89 (t, 1H), 7.11 (d, J = 8.72 Hz, 1H). ^{13}C NMR δ (ppm): 165.24, 150.06, 148.05, 142.37, 135.73, 134.81, 134.02, 130.65, 129.37, 127.17, 122.85, 124.13, 115.61, 111.27. *Cis*-isomer (3h'): ^{1}H NMR (DMSO-d$_6$ from TMS) δ (ppm): 13.73 (s, 1H), 12.04 (s, 1H), 8.68 (s, 1H), 8.54 (m, 1H), 8.33 (m, 2H), 8.29 (s, 1H), 7.18 (d, J = 8.72 Hz, 1H). ^{13}C NMR δ (ppm): 163.65, 148.45, 148.20, 143.41, 134.81, 134.20, 133.63, 131.44, 129.37, 128.30, 123.12, 120.80, 116.42, 112.10. $C_{15}H_9N_5O_6$, Calcd. C, 50.71, H, 2.55, N, 19.71. Found C, 50.86, H, 2.52, N, 19,59.

3.4. Experimental Measurements

3.4.1. NMR Measurements

The ^{1}H NMR spectra were recorded using an Ascend III spectrometer operating at 400 MHz, Bruker. Chloroform was used as solvent and tetramethylsilane (TMS) as internal standard. Chemical shifts (δ) are reported in ppm relative to TMS and coupling constants (J) in Hz.

3.4.2. Elemental Analysis Measurements

The elemental analysis was conducted with a Vario MACRO 11.45–0000, Elemental Analyser System GmbH, operating with the VARIOEL software (version 5.14.4.22).

3.4.3. UV–VIS Measurements

The absorption and emission spectra were measured at room temperature in quartz cuvette (1 cm) using an Agilent Technology UV–Vis Cary 60 Spectrophotometer and a Hitachi F-7000 Spectrofluorometer, respectively.

3.4.4. FTIR Measurements

The infrared spectra were recorded using reflectance spectroscopy measurements realized using PerkinElmer's FTIR Spectrum Two, spectrophotometer equipped with diamond ATR in the range of 4000–400 cm^{-1}.

3.4.5. The Calorimetric Measurements

The thermal stability of 5-nitroisatin derivatives, considering the determination of the melting point (m.p.) and thermal decomposition temperature (d.t.), was realized using the DSC 6000 from PerkinElmer, Waltham, MA, USA. The calorimetric measurements were conducted with a heating rate of 10 K/min and 20 mL/min nitrogen flow to provide an inert atmosphere. The initial calibration the calorimeter was realized with indium and zinc standards; during measurements, standard aluminum pans were used.

4. Conclusions

The conducted research allowed for the development of a new group of 5-nitroisatin-based benzoylhydrazines, which show a significant affinity for the active site of CDK2. Based on the conducted analyses, it was possible to select groups of chemical substituents—

the substitution of which, both in the isatin core and in the aromatic benzoylhydrazine ring, contributed to the improvement of the binding capacity of the considered derivatives. The conducted research allowed for the accurate characterization of the interactions responsible for maintaining the complexes of the considered inhibitors with the CDK2 enzyme. The most promising potential inhibitors, with the highest affinity values and forming the most stable complexes with the biological target, were directed to the experimental phase. A procedure for the synthesis of 5-nitroisatin derivatives was developed, and eight new compounds were obtained with a good yield. The structures of the newly synthesized compounds were confirmed using IR and NMR spectroscopy. It needs to be highlighted that the synthesized 5-nitroisatin-based benzoylhydrazines were present in the solution as *trans*- and *cis*-isomers. The absorption and fluorescence maxima were determined. It was found that the position of the bands of absorption and emission did not depend on the polarity of the solvent.

Based on the collected data—considering the binding capacity, the structural stability of the complexes and the molecular properties of the studied group of compounds—it can be concluded that the newly developed isatin derivatives can be used in the development of new anti-cancer therapies.

Supplementary Materials: The following supporting information can be downloaded at: https://www.mdpi.com/article/10.3390/ijms23148046/s1.

Author Contributions: Conceptualization, P.C.; methodology, P.C. and A.S.; software, P.C.; docking and molecular dynamics P.C.; ab initio calculations P.C. and B.S.; validation, P.C. and A.S.; formal analysis, P.C. and A.S.; investigation, P.C.; resources, P.C.; synthesis A.S. and P.C.; UV-VIS, NMR measurements A.S.; IR and DSC measurements P.C., A.S. and B.S.; data curation, P.C. and A.S.; writing—original draft preparation, P.C. and A.S.; review and editing P.C., A.S. and B.S.; visualization, P.C. and A.S.; supervision, P.C.; project administration, P.C.; funding acquisition, P.C. All authors have read and agreed to the published version of the manuscript.

Funding: This research received no external funding.

Institutional Review Board Statement: Not applicable.

Informed Consent Statement: Not applicable.

Data Availability Statement: Data is contained within the article or supplementary material.

Acknowledgments: This research was supported by PL-Grid Infrastructure (http://www.plgrid.pl/en).

Conflicts of Interest: The authors declare that there are no conflict of interest.

References

1. Zhou, J.; Qu, F. Analysis of the Extracts of Isatis tinctoria by New Analytical Approaches of HPLC, MS and NMR. *Afr. J. Tradit. Complement. Altern. Med.* **2011**, *8*, 33–45. [CrossRef]
2. Bergman, J.; Lindström, J.-O.; Tilstam, U. The structure and properties of some indolic constituents in Couroupita guianensis aubl. *Tetrahedron* **1985**, *41*, 2879–2881. [CrossRef]
3. Bayly, M.J.; Duretto, M.F.; Holmes, G.D.; Forster, P.I.; Cantrill, D.J.; Ladiges, P.Y.; Bayly, M.J.; Duretto, M.F.; Holmes, G.D.; Forster, P.I.; et al. Transfer of the New Caledonian genus Boronella to Boronia (Rutaceae) based on analyses of cpDNA and nrDNA. *Aust. Syst. Bot.* **2015**, *28*, 111. [CrossRef]
4. Xiao, Z.; Hao, Y.; Liu, B.; Qian, L. Indirubin and Meisoindigo in the Treatment of Chronic Myelogenous Leukemia in China. *Leuk. Lymphoma* **2002**, *43*, 1763–1768. [CrossRef]
5. Hoessel, R.; Leclerc, S.; Endicott, J.A.; Nobel, M.E.M.; Lawrie, A.; Tunnah, P.; Leost, M.; Damiens, E.; Marie, D.; Marko, D.; et al. Indirubin, the active constituent of a Chinese antileukaemia medicine, inhibits cyclin-dependent kinases. *Nat. Cell Biol.* **1999**, *1*, 60–67. [CrossRef]
6. Cheke, R.S.; Firke, S.D.; Patil, R.R.; Bari, S.B. ISATIN: New Hope Against Convulsion. *Cent. Nerv. Syst. Agents Med. Chem.* **2018**, *18*, 76–101. [CrossRef]
7. Bharathi Dileepan, A.G.; Daniel Prakash, T.; Ganesh Kumar, A.; Shameela Rajam, P.; Violet Dhayabaran, V.; Rajaram, R. Isatin based macrocyclic Schiff base ligands as novel candidates for antimicrobial and antioxidant drug design: In vitro DNA binding and biological studies. *J. Photochem. Photobiol. B Biol.* **2018**, *183*, 191–200. [CrossRef]
8. Guo, H. Isatin derivatives and their anti-bacterial activities. *Eur. J. Med. Chem.* **2019**, *164*, 678–688. [CrossRef]

9. Zhang, M.-Z.; Chen, Q.; Yang, G.-F. A review on recent developments of indole-containing antiviral agents. *Eur. J. Med. Chem.* **2015**, *89*, 421–441. [CrossRef]
10. Prakash, C.R.; Theivendren, P.; Raja, S.; Prakash, C.R.; Theivendren, P.; Raja, S. Indolin-2-Ones in Clinical Trials as Potential Kinase Inhibitors: A Review. *Pharmacol. Pharm.* **2012**, *3*, 62–71. [CrossRef]
11. Nikoulina, S.E.; Ciaraldi, T.P.; Mudaliar, S.; Carter, L.; Johnson, K.; Henry, R.R. Inhibition of glycogen synthase kinase 3 improves insulin action and glucose metabolism in human skeletal muscle. *Diabetes* **2002**, *51*, 2190–2198. [CrossRef]
12. Kim, S.-A.; Kwon, S.-M.; Kim, J.-A.; Kang, K.W.; Yoon, J.-H.; Ahn, S.-G. 5′-Nitro-indirubinoxime, an indirubin derivative, suppresses metastatic ability of human head and neck cancer cells through the inhibition of Integrin β1/FAK/Akt signaling. *Cancer Lett.* **2011**, *306*, 197–204. [CrossRef]
13. Martin, L.; Magnaudeix, A.; Wilson, C.M.; Yardin, C.; Terro, F. The new indirubin derivative inhibitors of glycogen synthase kinase-3, 6-BIDECO and 6-BIMYEO, prevent tau phosphorylation and apoptosis induced by the inhibition of protein phosphatase-2A by okadaic acid in cultured neurons. *J. Neurosci. Res.* **2011**, *89*, 1802–1811. [CrossRef]
14. Avila, J.; Hernández, F. GSK-3 inhibitors for Alzheimer's disease. *Expert Rev. Neurother.* **2007**, *7*, 1527–1533. [CrossRef]
15. Czeleń, P. Investigation of the Inhibition Potential of New Oxindole Derivatives and Assessment of Their Usefulness for Targeted Therapy. *Symmetry* **2019**, *11*, 974. [CrossRef]
16. Bramson, H.N.; Holmes, W.D.; Hunter, R.N.; Lackey, K.E.; Lovejoy, B.; Luzzio, M.J.; Montana, V.; Rocque, W.J.; Rusnak, D.; Shewchuk, L.; et al. Oxindole-based inhibitors of cyclin-dependent kinase 2 (CDK2): Design, synthesis, enzymatic activities, and X-ray crystallographic analysis. *J. Med. Chem.* **2001**, *44*, 4339–4358. [CrossRef]
17. Czeleń, P.; Szefler, B. The oxindole derivatives, new promising gsk-3β inhibitors as one of the potential treatments for alzheimer's disease—A molecular dynamics approach. *Biology* **2021**, *10*, 332. [CrossRef]
18. Shin, E.-K.; Kim, J.-K. Indirubin derivative E804 inhibits angiogenesis. *BMC Cancer* **2012**, *12*, 164. [CrossRef]
19. Lee, M.-Y.; Liu, Y.-W.; Chen, M.-H.; Ww, J.-Y.; Ho, H.-Y.; Wang, Q.-F.; Chuang, J.-J. Indirubin-3′-monoxime promotes autophagic and apoptotic death in JM1 human acute lymphoblastic leukemia cells and K562 human chronic myelogenous leukemia cells. *Oncol. Rep.* **2013**, *29*, 2072–2078. [CrossRef]
20. Lozinskaya, N.A.; Babkov, D.A.; Zaryanova, E.V.; Bezsonova, E.N.; Efremov, A.M.; Tsymlyakov, M.D.; Anikina, L.V.; Zakharyascheva, O.Y.; Borisov, A.V.; Perfilova, V.N.; et al. Synthesis and biological evaluation of 3-substituted 2-oxindole derivatives as new glycogen synthase kinase 3β inhibitors. *Bioorg. Med. Chem.* **2019**, *27*, 1804–1817. [CrossRef]
21. de Paiva, R.E.F.; Vieira, E.G.; da Silva, D.R.; Wegermann, C.A.; Ferreira, A.M.C. Anticancer Compounds Based on Isatin-Derivatives: Strategies to Ameliorate Selectivity and Efficiency. *Front. Mol. Biosci.* **2021**, *7*, 511.
22. Besson, A.; Dowdy, S.F.; Roberts, J.M. CDK Inhibitors: Cell Cycle Regulators and Beyond. *Dev. Cell* **2008**, *14*, 159–169. [CrossRef]
23. Malumbres, M.; Barbacid, M. Cell cycle, CDKs and cancer: A changing paradigm. *Nat. Rev. Cancer* **2009**, *9*, 153–166. [CrossRef]
24. Morgan, D.O. CYCLIN-DEPENDENT KINASES: Engines, Clocks, and Microprocessors. *Annu. Rev. Cell Dev. Biol.* **1997**, *13*, 261–291. [CrossRef]
25. Davies, T.G.; Tunnah, P.; Meijer, L.; Marko, D.; Eisenbrand, G.; Endicott, J.A.; Noble, M.E. Inhibitor binding to active and inactive CDK2: The crystal structure of CDK2-cyclin A/indirubin-5-sulphonate. *Structure* **2001**, *9*, 389–397. [CrossRef]
26. Czeleń, P.; Szefler, B. Molecular dynamics study of the inhibitory effects of ChEMBL474807 on the enzymes GSK-3β and CDK-2. *J. Mol. Model.* **2015**, *21*, 74. [CrossRef]
27. Czeleń, P. Molecular dynamics study on inhibition mechanism of CDK-2 and GSK-3β by CHEMBL272026 molecule. *Struct. Chem.* **2016**, *27*, 1807–1818. [CrossRef]
28. Katiyar, A.; Hegde, M.; Kumar, S.; Gopalakrishnan, V.; Bhatelia, K.D.; Ananthaswamy, K.; Ramareddy, S.A.; de Clercq, E.; Choudhary, B.; Schols, D.; et al. Synthesis and evaluation of the biological activity of N′-[2-oxo-1,2 dihydro-3H-indol-3-ylidene] benzohydrazides as potential anticancer agents. *RSC Adv.* **2015**, *5*, 45492–45501. [CrossRef]
29. Debnath, K.; Pathak, S.; Pramanik, A. Facile synthesis of ninhydrin and isatin based hydrazones in water using PEG-OSO3H as a highly efficient and homogeneous polymeric acid-surfactant combined catalyst. *Tetrahedron Lett.* **2013**, *54*, 4110–4115. [CrossRef]
30. Abo-Ashour, M.F.; Eldehna, W.M.; Nocentini, A.; Ibrahim, H.S.; Bua, S.; Abou-Seri, S.M.; Supuran, C.T. Novel hydrazido benzenesulfonamides-isatin conjugates: Synthesis, carbonic anhydrase inhibitory activity and molecular modeling studies. *Eur. J. Med. Chem.* **2018**, *157*, 28–36. [CrossRef]
31. Emami, S.; Valipour, M.; Komishani, F.K.; Sadati-Ashrafi, F.; Rasoulian, M.; Ghasemian, M.; Tajbakhsh, M.; Masihi, P.H.; Shakiba, A.; Irannejad, H.; et al. Synthesis, in silico, in vitro and in vivo evaluations of isatin aroylhydrazones as highly potent anticonvulsant agents. *Bioorg. Chem.* **2021**, *112*, 104943. [CrossRef]
32. Hunoor, R.S.; Patil, B.R.; Badiger, D.S.; Chandrashekhar, V.M.; Muchchandi, I.S.; Gudasi, K.B. Co(II), Ni(II), Cu(II) and Zn(II) complexes of isatinyl-2-aminobenzoylhydrazone: Synthesis, characterization and anticancer activity. *Appl. Organomet. Chem.* **2015**, *29*, 101–108. [CrossRef]
33. Bartashevich, E.V.; Potemkin, V.A.; Grishina, M.A.; Belik, A.V. A Method for Multiconformational Modeling of the Three-Dimensional Shape of a Molecule. *J. Struct. Chem.* **2002**, *43*, 1033–1039. [CrossRef]
34. Trott, O.; Olson, A.J. AutoDock Vina: Improving the speed and accuracy of docking with a new scoring function, efficient optimization, and multithreading. *J. Comput. Chem.* **2010**, *31*, 455–461. [CrossRef]
35. Maier, J.A.; Martinez, C.; Kasavajhala, K.; Wickstrom, L.; Hauser, K.E.; Simmerling, C. ff14SB: Improving the Accuracy of Protein Side Chain and Backbone Parameters from ff99SB. *J. Chem. Theory Comput.* **2015**, *11*, 3696–3713. [CrossRef]

36. Bayly, C.I.; Cieplak, P.; Cornell, W.; Kollman, P.A. A well-behaved electrostatic potential based method using charge restraints for deriving atomic charges: The RESP model. *J. Phys. Chem.* **1993**, *97*, 10269–10280. [CrossRef]
37. Adelman, S.A. Generalized Langevin equation approach for atom/solid-surface scattering: General formulation for classical scattering off harmonic solids. *J. Chem. Phys.* **1976**, *64*, 2375. [CrossRef]
38. Humphrey, W.; Dalke, A.; Schulten, K. VMD: Visual molecular dynamics. *J. Mol. Graph.* **1996**, *14*, 33–38. [CrossRef]
39. Miller, B.R.; McGee, T.D.; Swails, J.M.; Homeyer, N.; Gohlke, H.; Roitberg, A.E. *MMPBSA.py*: An Efficient Program for End-State Free Energy Calculations. *J. Chem. Theory Comput.* **2012**, *8*, 3314–3321. [CrossRef]
40. Case, D.A.; Babin, V.; Berryman, J.T.; Betz, R.M.; Cai, Q.; Cerutti, D.S.; Cheatham, T.E.; Darden, T.A., III; Duke, R.E.; Gohlke, H.; et al. AMBER 14 2014. Available online: https://www.researchgate.net/publication/270588529_Amber_2014 (accessed on 27 June 2022).
41. Frisch, M.J.; Trucks, G.W.; Schlegel, H.B.; Scuseria, G.E.; Robb, M.A.; Cheeseman, J.R.; Scalmani, G.; Barone, V.; Petersson, G.A.; Nakatsuji, H.; et al. *Gaussian 09, Revision A.02*; Carnegie Mellon University: Pittsburgh, PA, USA, 2016.
42. Cramer, C.J.; Truhlar, D.G. Implicit Solvation Models: Equilibria, Structure, Spectra, and Dynamics. *Chem. Rev.* **1999**, *99*, 2161–2200. [CrossRef]
43. Barone, V.; Cossi, M.; Tomasi, J. A new definition of cavities for the computation of solvation free energies by the polarizable continuum model. *J. Chem. Phys.* **1997**, *107*, 3210–3221. [CrossRef]
44. Marten, B.; Kim, K.; Cortis, C.; Friesner, R.A.; Murphy, R.B.; Ringnalda, M.N.; Sitkoff, D.; Honig, B. New model for calculation of solvation free energies: Correction of self-consistent reaction field continuum dielectric theory for short-range hydrogen-bonding effects. *J. Phys. Chem.* **1996**, *100*, 11775–11788. [CrossRef]

International Journal of
Molecular Sciences

Article

In Silico Prediction of the Metabolic Resistance of Vitamin D Analogs against CYP3A4 Metabolizing Enzyme

Teresa Żołek [1,*], Kaori Yasuda [2], Geoffrey Brown [3], Toshiyuki Sakaki [2] and Andrzej Kutner [4]

[1] Department of Organic Chemistry, Faculty of Pharmacy, Medical University of Warsaw, 1 Banacha, 02-097 Warsaw, Poland

[2] Department of Pharmaceutical Engineering, Toyama Prefectural University, Toyama 939-0398, Japan; kyasuda@pu-tayama.ac.jp (K.Y.); tsakaki@pu-toyama.ac.jp (T.S.)

[3] School of Biomedical Sciences, Institute of Clinical Sciences, College of Medical and Dental Sciences, University of Birmingham, Edgbaston, Birmingham B15 2TT, UK; g.brown@bham.ac.uk

[4] Department of Bioanalysis and Drug Analysis, Faculty of Pharmacy, Medical University of Warsaw, 1 Banacha, 02-097 Warsaw, Poland; andrzej.kutner@wum.edu.pl

* Correspondence: tzolek@wum.edu.pl; Tel.: +48-2-2572-0643

Abstract: The microsomal cytochrome P450 3A4 (CYP3A4) and mitochondrial cytochrome P450 24A1 (CYP24A1) hydroxylating enzymes both metabolize vitamin D and its analogs. The three-dimensional (3D) structure of the full-length native human CYP3A4 has been solved, but the respective structure of the main vitamin D hydroxylating CYP24A1 enzyme is unknown. The structures of recombinant CYP24A1 enzymes have been solved; however, from studies of the vitamin D receptor, the use of a truncated protein for docking studies of ligands led to incorrect results. As the structure of the native CYP3A4 protein is known, we performed rigid docking supported by molecular dynamic simulation using CYP3A4 to predict the metabolic conversion of analogs of 1,25-dihydroxyvitamin D2 (1,25D2). This is highly important to the design of novel vitamin D-based drug candidates of reasonable metabolic stability as CYP3A4 metabolizes ca. 50% of the drug substances. The use of the 3D structure data of human CYP3A4 has allowed us to explain the substantial differences in the metabolic conversion of the side-chain geometric analogs of 1,25D2. The calculated free enthalpy of the binding of an analog of 1,25D2 to CYP3A4 agreed with the experimentally observed conversion of the analog by CYP24A1. The metabolic conversion of an analog of 1,25D2 to the main vitamin D hydroxylating enzyme CYP24A1, of unknown 3D structure, can be explained by the binding strength of the analog to the known 3D structure of the CYP3A4 enzyme.

Keywords: vitamin D analogs; cytochrome P450 3A4; cytochrome P450 24A1; in silico prediction; molecular dynamics

Citation: Żołek, T.; Yasuda, K.; Brown, G.; Sakaki, T.; Kutner, A. In Silico Prediction of the Metabolic Resistance of Vitamin D Analogs against CYP3A4 Metabolizing Enzyme. *Int. J. Mol. Sci.* **2022**, *23*, 7845. https://doi.org/10.3390/ijms23147845

Academic Editors: Patrick M. Dansette and Paulino Gómez-Puertas

Received: 20 May 2022
Accepted: 14 July 2022
Published: 16 July 2022

Publisher's Note: MDPI stays neutral with regard to jurisdictional claims in published maps and institutional affiliations.

1. Introduction

1,25-dihydroxycholecalciferol (1,25-dihydroxyvitamin D_3, calcitriol, 1,25D3) and 1,25-dihydroxyergocalciferol (1,25-dihydroxyvitamin D_2, 1,25D2) (Figure 1) are no longer considered just as "vital amines" [1,2] that maintain the calcium and phosphate homeostasis [3], they are also pleiotropic hormones [4] that regulate key physiological processes [5,6]. Several active vitamin D metabolites, their precursors, and synthetic analogs are used as drug substances in the treatment of bone diseases (type I rickets, osteomalacia, hypoparathyroidism, pseudohypoparathyroidism, renal osteodystrophy, and osteoporosis), and hyperproliferative skin diseases such as psoriasis [7,8]. Efforts are now focused on fortifying standard anticancer chemotherapy by the addition of synthetic vitamin D analogs to improve efficacy and delay cancer recurrence [9]. The active forms of vitamin D_3 (1,25D3) and D_2 (1,25D2), jointly denoted as 1,25D, cannot be used for this purpose because they are highly calcaemic. They also have a low resistance to vitamin D metabolizing enzymes [10]. The current search is for vitamin D analogs of therapeutic potential against leukemia [7,11,12] and solid tumors, including breast [13], lung [14], prostate [15], colorectal [16,17], skin [18], and ovarian cancer [19]. We have developed analogs of 1,25D that exceed the potency of the parent 1,25D against cancer cells in vitro and in vivo, and that are less calcaemic and more enzymatically stable [10].

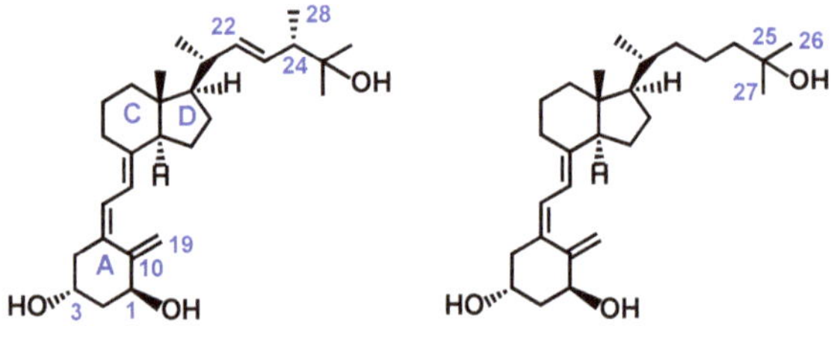

Figure 1. Structural formulas of 1α,25-dihydroxyergocalciferol (1,25-dihydroxyvitamin D_2, 1,25D2) and 1α,25-dihydroxycholecalciferol (1,25-dihydroxyvitamin D_3, 1,25D3).

The rational design of vitamin D analogs is hampered by the lack of the 3D structure of the native full-length human vitamin D receptor (*h*VDR); the use of an engineered and truncated *h*VDRΔ [20] might result in a misleading outcome. For example, an early and controversial finding is that the *h*VDRΔ structure has an identical conformation, but that there is overlapping of all of the hydroxyls for 1,25D3 and two other analogs of very different structures when 1,25D3 and the analogs are bound to the ligand-binding pocket [21]. This intriguing finding has never been explained. Therefore, the use of the native integral vitamin D protein is recommended for the reliable design of analogs [22]. There is also the need to predict the metabolic resistance of potential anticancer analogs of 1,25D; this is important to a drug candidate prior to initiating a multistep synthesis followed by broad biological screening.

Cytochrome P450 hydroxylates vitamin D and its analogs. It also metabolizes several xenobiotics, including various drug substances. For instance, we postulated that our vitamin D analogs and the drug substance imatinib compete for the active site of the same cytochrome P450 enzyme [23]. This competition delays the deactivating metabolism of a vitamin D analog and prolongs the synergistic anticancer activity of both agents. Our concept has been confirmed recently [24] by studies that have shown that there is a pharmacodynamic synergy between ginsengoside Rh2 and 1,25D3 regarding the growth inhibition and apoptosis of human prostate cancer cells. For human hepatic microsomes in vitro, this led to inhibition of cytochrome P450 3A4-mediated metabolism and inactivation of 1,25D3. Of the several cytochrome P450 hydroxylating enzymes that are involved in metabolizing 1,25D and analogs, cytochrome P450 24A1 (CYP24A1) is primarily considered. The expression of CYP24A1 mRNA is used as a measure of the transactivating activity of a vitamin D analog. We have already determined the metabolic conversion of 1,25D analogs to CYP24A1 using the membrane fraction from recombinant *Escherichia coli* cells that expressed *h*CYP24A1 [25,26]. The crystal structure of rat recombinant CYP24A1 (Δ2-32, S57D mutant) has been reported (SSRL BL9-2 and BL12-2) [27,28], but the structure of the native full-length human CYP24A1 is still lacking. Therefore, taking-into-account the uncertain results from the modeling of analog binding to the pocket of the recombinant VDR, we opted for the modeling of metabolic resistance 1,25D analogs to the native cytochrome P450 3A4 (CYP3A4). The crystallographic 3D structure of human CYP3A4, as solved [29], conforms with the protein fold that is typical for the cytochrome P450 superfamily. Our approach is additionally supported by the very recent finding that the docking of peptidomimetic ligands to the cysteine-like protease of SARS-CoV-1 3CLpro (of the known 3D structure) facilitated the design of potent inhibitors with antiviral potency against SARS-CoV-2 3CLpro (of unknown 3D structure) [30]. Therefore, we postulated that by using the 3D structure of the native human CYP3A4 we would be able to predict the resistance of our new vitamin D analogs to CYP24A1 metabolism.

To this end, we used, for the first time, the 3D crystallographic data of cytochrome CYP3A4 (RCSD PDB: 2V0M.pdb) [31,32] for our theoretical simulations of vitamin D analogs. CYP3A4 is a hepatic microsomal 24-hydroxylase of 1α-hydroxyvitamin D_2 and its analogs [33], and is also a vitamin-25 hydroxylase. We have previously shown that the inducible enzyme CYP3A4 is also a source of oxidative metabolism of 1,25D3 in the human liver and small intestine [33,34].

CYP3A4 is multifunctional and also metabolizes endobiotics and xenobiotics [35]. This enzyme is responsible for the deactivating hydroxylation of ca. 50% of drug substances [29,36]. Therefore, we used CYP3A4 in our study because we were interested in the metabolic stability of our vitamin D-based potential drug candidates. Additionally, both CYP24A1 and CYP3A4 show the same monooxygenase activity and catalyze the side-chain oxidation of the hormonal form of vitamin D to the same side-chain hydroxylated metabolites [37]. In this study, we provide unprecedented evidence that the metabolic conversion of analogs of 1,25D2 by CYP24A1 can be correlated with the free enthalpy of binding of the analogs to CYP3A4. By using the 3D structural data of CYP3A4 as a starting parameter, we have been able to explain the differences in the metabolic resistance of the side-chain geometric analogs of 1,25D. We have also been able to predict the metabolic resistance of structurally related analogs of 1,25D2 with a branched side-chain, and in turn, direct the design and synthesis of vitamin D analogs toward more promising drug candidates.

The interactions of 1,25D analogs with CYP3A4 were elucidated by computational approaches which included molecular docking, molecular dynamics simulations, binding free enthalpy calculations, and density functional theory (DFT) calculations. The relatively large active site of CYP3A4 shows large flexibility, allowing for side-chain fluctuations and the presence of various functional groups. There is also a large malleability that allows the adoption of a different shape depending on the ligand structure; making CYP3A4 a challenging protein to simulate [38]. Based on our experience of simulating molecular mechanisms for similar systems [39,40], simulations of CYP3A4 complexed with vitamin D analogs were carefully compared in terms of their structural and energetic properties. Their molecular structures were used to explore the possible binding modes with the amino acids of the active site of CYP3A4, as well as to show their different capacities for interaction with heme groups, which has biological consequences. Our calculations were compared with the previously determined metabolic resistance of analogs of 1,25D2 against the CYP24A1 hydroxylating enzyme. In order to check the extent of the correlation observed for the series of structurally related 1,25D2 analogs, preliminary studies were performed using two analogs of 1,25D3 that have very different structures.

2. Results and Discussion

2.1. Potential Binding Site of 1,25D Analogs in CYP3A4

Molecular docking studies were carried out to predict the binding site and the orientation at the active site of CYP3A4 for ten analogs of 1,25D (eight analogs of 1,25D2 and two analogs of 1,25D3). The orientation and residue bonding of ketoconazole (as a reference inhibitor for the testing of the binding affinities to the CYP3A4) were inferred from the X-ray structure of human CYP3A4 (PDB entry 2V0M). This served as a model for the simulations of the interactions of vitamin D analogs with the enzyme. We supposed that the size of the molecular structure of vitamin D analogs was comparable to that of ketoconazole, and assumed that the docking of 1,25D analogs into the activity domain did not significantly perturb the crystal structure of the 2V0M. To validate the suitability of the selected docking model, we first re-docked ketoconazole to ensure that its bonding to the CYP3A4 binding site was consistent with the original structure of 2V0M. According to the ketoconazole-docking simulation, the complex showed that eight significant amino acids are involved in the interactions: Leu-210, Phe-241, Ile-301, Ala-305, Ala-370, Arg-372, Gly-481, and Leu-482, as well as the heme group, and are indispensable for the activity of the classic inhibitors. Next, all of our analogs were docked to the CYP3A4 pocket. The analogs effectively filled the active site cavity of CYP3A4. The position of their 25-hydroxyl was located close to the heme ring, indicating that it might be the potential metabolic site (see Figure 2).

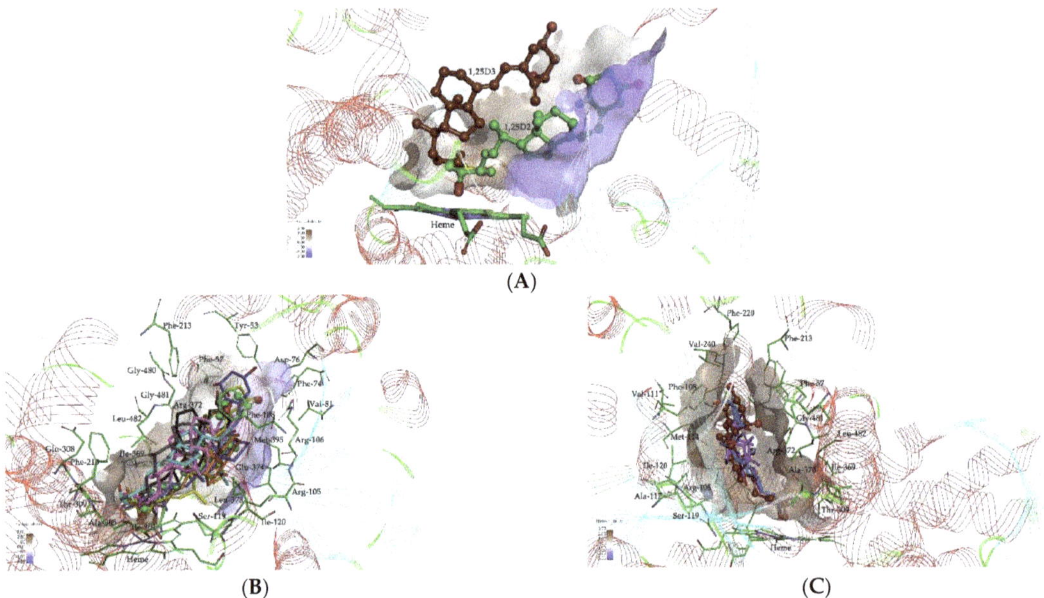

Figure 2. Close-up view of an active site of the CYP3A4 structure. (**A**) View of the binding of 1,25D2 (green) and 1,25D3 (brown) in the active site CYP3A4. (**B,C**) View of 1,25D2 and 1,25D3 analogs in the pocket of CYP3A4. Predicted superposition of compounds. (**B**) 1,25D2 (C atoms shown as green), PRI-5106 (C atoms shown as dark green), PRI-5105 (C atoms shown as magenta), PRI-1916 (C atoms shown as cyan), PRI-1917 (C atoms shown as pink), PRI-1906 (C atoms shown as blue), PRI-1907 (C atoms shown as grey), PRI-5201 (C atoms shown as yellow), and PRI-5202 (C atoms shown as orange). (**C**) PRI-1901 (C atoms shown as purple) and PRI-2205 (C atoms shown as blue). Surface hydrophobicity is depicted by the shaded colors: negative values (blue) correspond to hydrophilic residues, whereas positive values (brown) correspond to hydrophobic residues.

As shown in Figure 2B,C, the inner wall of the pocket of CYP3A4 is formed by hydrophilic side chains (Asp-76, Arg-105, Arg-106, Ser-119, Thr-309, Arg-372, and Glu-374) and a hydrophobic region (formed by Phe-57, Phe-108, Ile-120, Leu-211, Ile-301, Phe-304, Ala-305, Ile-369, Ala-370, Leu-373, Gly-481, Leu-482, and Hem-499). The predicted location of 1,25D analogs in the CYP3A4 pocket is different for the 1,25D2 and 1,25D3 analogs (see Figure 2A). The 1,25D2 analogs interacted with the CYP3A4 pocket using hydrogen bonds and hydrophobic and van der Waals interactions, while the interaction of 1,25D3 and its analogs were mainly hydrophobic in nature. To examine the stability of the 1,25D analogs bound at the active site of CYP3A4, MD simulations were performed on the CYP3A4-liganded analog complex.

2.2. Importance of Side-Chain Geometry of 1,25D2 Analogs in the CYP3A4 Binding Site

The structural formulas of PRI-1906 and PRI-1916, and PRI-1907 and PRI-1917 (Figure 3) are quite similar, but the extent to which they bind to CYP24A1 is different. Experimental data indicated that the binding to CYP24A1 of the C-25 dimethyl analog PRI-1916 of (24Z) geometry was much lower than that of the (24E)-25-dimethyl analog PRI-1906. Similarly, the binding of (24Z)-25-diethyl analog PRI-1917 was lower than that of the respective (24E)-25-diethyl analog PRI-1907 [33].

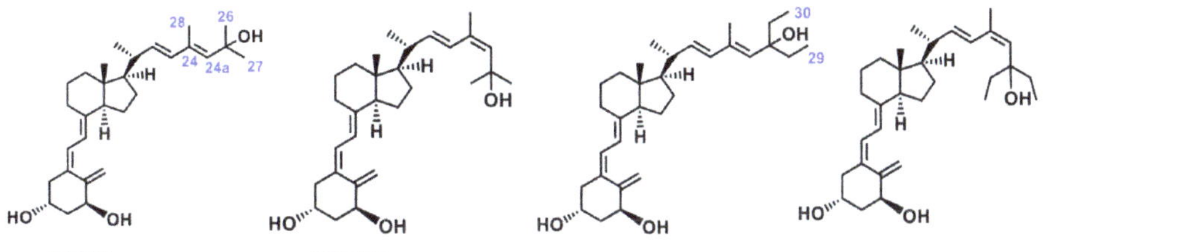

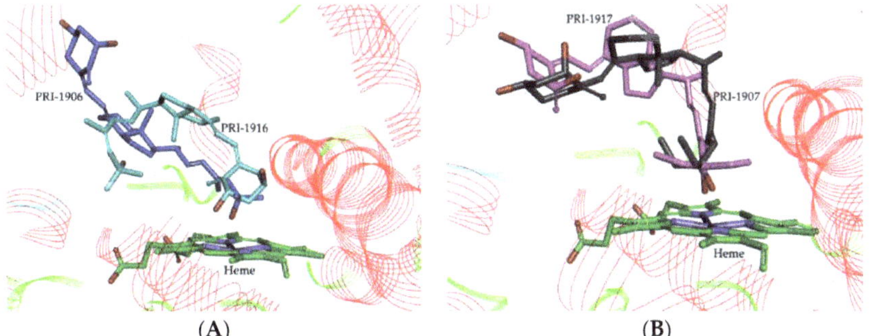

Figure 3. Structural formulas of the analogs of 1α,25-dihydroxyvitamin D$_2$ (PRI-1906, PRI-1907, PRI-1916, and PRI-1917).

Theoretical calculations showed that the geometry of the side-chain of the series of four analogs of 1,25D2 also followed the same trend as the predicted stereoselective activity of cis/trans isomers in the human CYP3A4 model. The MD resulting orientations of analogs, presented in Figure 4, define the geometric preferences of the four compounds on the active site of CYP3A4.

Figure 4. MD simulated conformational differences between 1,25D2 analogs located above the heme ring. (**A**) PRI-1906 (C atoms shown as blue) and PRI-1916 (C atoms shown as cyan); (**B**) PRI-1907 (C atoms shown as grey) and PRI-1917 (C atoms shown as pink).

Analogs PRI-1906 and PRI-1916 bind to the active site in different ways and with different affinities. Due to the side-chain geometry, PRI-1916 was predicted to adopt the U-shaped conformation in the CYP3A4 pocket, while (24E)-25-dimethyl PRI-1906 adopted an extended conformation. The most noticeable difference was that the U-shaped side-chain A-ring of PRI-1916 was rotated by 180° to the heme group, compared with the extended side-chain of the A-ring of PRI-1906 (Figure 4A). This could explain the lower binding of PRI-1916 for CYP3A4. PRI-1916 was not able to form a significant portion of the hydrophobic interactions as for PRI-1906 and with residues Phe-57, Ile-120, Ala-305, Ile-369, Leu-482, and Arg-372. On the other hand, the A-ring 3-hydroxyl of PRI-1906 engaged in a strong hydrogen bond at the binding site with the Asp-76 residue (see Figure 5 and Figure S1 in Supplementary Materials). For PRI-1907/CYP3A4 and PRI-1917/CYP3A4 complexes, the position of their 25-hydroxyl was almost the same in the active site of the receptor (Figure 4B), but the presence of the (24E)-25-diethyls of PRI-1907 affected the parallel arrangement of A- and CD-rings to the heme group.

The parallel arrangement of the A- and CD-rings to the heme group is conducive to the formation of very strong hydrogen bonds by the A-ring hydroxyls with the amino acid residues Ala-370, Arg-372, Glu-374, and Gly-481 in CYP3A4 (see Figure 5 and Figure S1 in Supplementary Materials). Furthermore, the 1- and 3-hydroxyl of PRI-1907 formed hydrogen bonds with water molecules. In contrast, the docked analog PRI-1917, with (24Z)-25-diethyls in the side-chain, showed the perpendicular arrangement of A- and CD-rings to the heme group that caused only the 1-hydroxyl to be involved in a hydrogen bond with amino acid residue Gly-481. This could lead to a lower affinity of PRI-1917 for the CYP3A4 cavity compared to that of PRI-1907 (see Figure 5 and Figure S1 in Supplementary Materials).

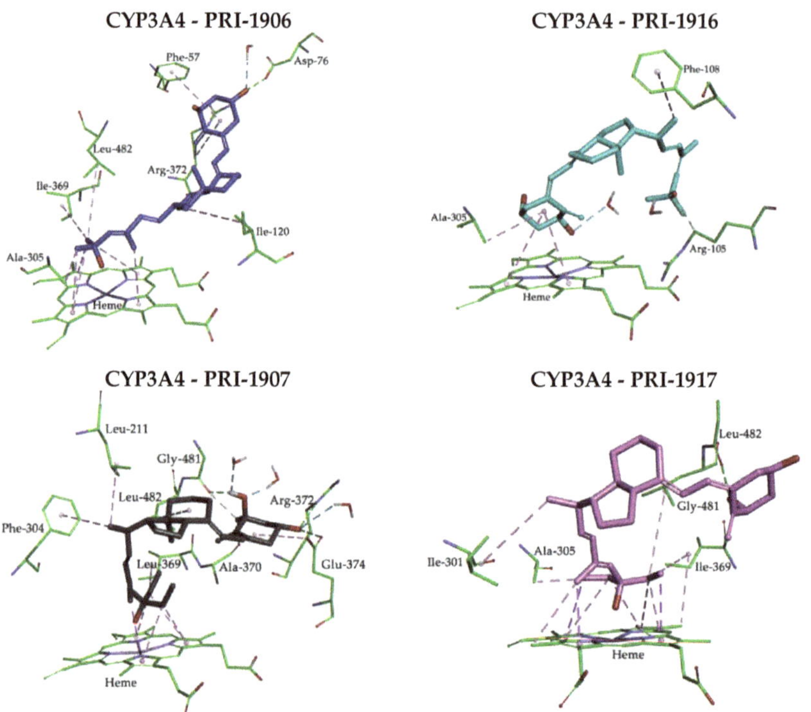

Figure 5. Molecular interactions between 1,25D2 analogs (PRI-1906, PRI-1907, PRI-1916, and PRI-1917) and CYP3A4 resulting from MD simulations. Hydrophobic interactions are shown as magenta dashed lines and hydrogen bonds as green.

2.3. Molecular Interactions between the CYP3A4 Binding Site and 1,25D2 Analogs

MD simulations of the CYP3A4-vitamin D analog complexes were performed to further investigate the binding of ligands and the effect of their structures on the intermolecular interactions. The MD resulting orientations of analogs, presented in Figure 5, Figure 6, Figure 7 and Figure S2 in Supplementary Materials, define the key interactions in the active site of the human CYP3A4 model. The binding free enthalpy values, calculated using the MM-PBSA method, are shown in Table 1, together with the experimental metabolic conversions by CYP24A1. A correlation was observed between the experimental metabolic conversion by CYP24A1 and the theoretically estimated binding free enthalpy of the CYP3A4-ligand complexes (ΔG_{bind}) (see Figure S3 in Supplementary Materials). The correlation coefficient, equal to $R^2 = 0.853$, confirmed that the calculated enthalpy of binding to the vitamin D hydroxylating enzyme CYP3A4, with the crystal structure solved, shows good agreement with the experimental metabolic conversion by the main vitamin D hydroxylating enzyme CYP24A1. In other words, the experimental metabolic conversion of an analog by CYP24A1 advantageously decreases with the decrease in the calculated free enthalpy of the binding of an analog of 1,25D2 to CYP3A4.

The lowest value for the ΔG_{bind} enthalpy was obtained for PRI-1907. This is most likely a consequence of a good fit, as mentioned above, and the formation of very strong hydrogen bonds with amino acids that have an important role in the CYP3A4 active site. The analog of PRI-5202, which lacks a 19-methylene, showed a similarly folded conformation in the pocket of CYP3A4 but its location was rather different (see Figure S2A in Supplementary Materials). Apart from the typical interactions, the side-chain of PRI-5202 allowed for favorable interactions of the 28-methyl with Ile-301 and 26, 27-methylene with amino acid residues Ala-305, Ile-369, and Leu-482. The A-ring 3-hydroxyl was involved in additional hydrogen bonds with the Phe-57 residue (Figures 6 and S1 in Supplementary Materials). Importantly, PRI-5202 is also highly solvated by water molecules that are present inside the pocket of CYP3A4.

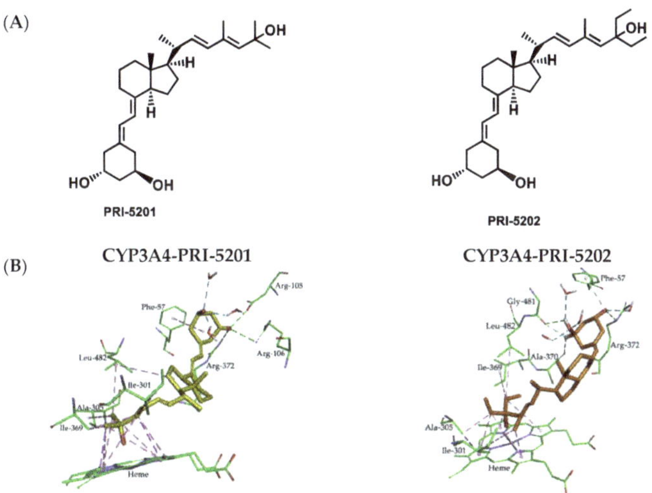

Figure 6. (**A**) Structural formulas of the analogs of 1α,25-dihydroxyvitamin D$_2$ (PRI-5201 and PRI-5202). (**B**) Molecular interactions between 1,25D2 analogs (PRI-5201 and PRI-5202) and CYP3A4 resulting from MD simulations. Hydrophobic interactions are shown as magenta dashed lines and hydrogen bonds as green.

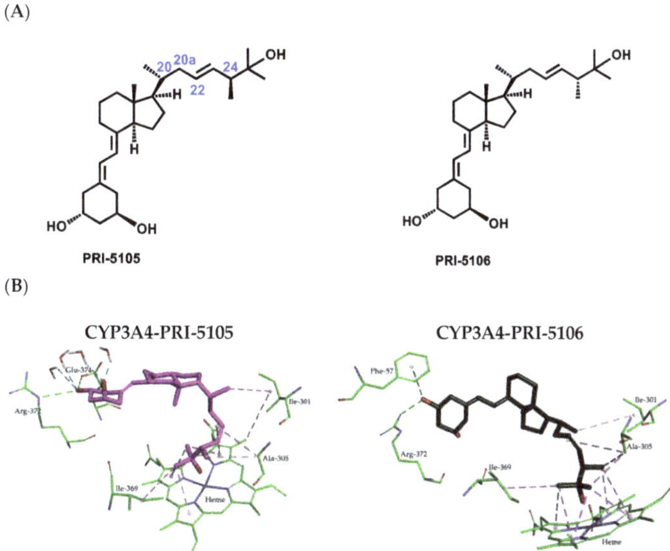

Figure 7. (**A**) Structural formulas the analogs of 1α,25-dihydroxyvitamin D$_2$ (PRI-5105 and PRI-5106). (**B**) Molecular interactions between 1,25D2 analogs (PRI-5105 and PRI-5106) and CYP3A4 resulting from MD simulations. Hydrophobic interactions are shown as magenta dashed lines and hydrogen bonds as green.

Table 1. The experimental metabolic conversion of 1,25D2 and its analogs by CYP24A1 [10,16,41] and the calculated free enthalpy of binding to CYP3A4.

Vitamin D Metabolite and Its Analogs	Metabolic Conversion by CYP24A1 (%) $\pm$ SD	Free Enthalpy of Binding to CYP3A4 * ΔG_{bind} (kcal/mol)
1,25D2	34 ± 5	-24
PRI-5106	26 ± 3	-37
PRI-5105	25 ± 2	-27
PRI-1916	13 ± 4	-52
PRI-1917	10 ± 1	-65
PRI-1906	1.9 ± 0.7	-76
PRI-1907	0.9 ± 0.1	-120
PRI-5201	1.9 ± 0.1	-91
PRI-5202	2.0 ± 0.2	-100

* Rounded to the absolute value.

The relatively high values of enthalpy obtained for both PRI-5105 and PRI-5106 result from the presence of an additional 28-methyl at C-24, irrespective of the (24*S*)-or (24*R*)-absolute configuration. PRI-5105 and its 24-diastereomer PRI-5106 were predicted to be accommodated in the CYP3A4 pocket in twisted conformations (see Figure S2B in Supplementary Materials). These conformations are energetically unfavorable and decrease the hydrophobic interactions with the surrounding amino acids (see Figure 7). The molecular flexibility conferred by one carbon extension (C-20a) to the side-chain of PRI-5105 and PRI-5106 results in a decreased metabolic resistance, emphasizing the importance of side-chain rigidity and the correct position in the active site. Notably, these analogs showed relatively low resistance to CYP24A1-dependent metabolic conversion (Table 1). The MD simulation showed that the location within the cavity of analogs PRI-5105 and PRI-5106 is largely the same, and the molecules are surrounded purely by the interior pocket residues of Phe-57, Ile-301, Ala-305, Ile-369, Arg-372, and Glu-374 which create hydrophobic interactions and hydrogen bonds.

Compared to PRI-1906, PRI-5201 has the 19-*nor* modification and the PRI-5201/CYP3A4 complex showed an extended conformation, as seen for the PRI-1906/CYP3A4 complex, but the position of 1-and 3-hydroxyls at the A-ring of PRI-5201 formed a very strong hydrogen bond with Asp-76, Arg-106, Arg-372, and three water molecules (see Figure 6 and Figure S1 in Supplementary Materials). This is likely responsible for the higher affinity of PRI-5201, versus that of PRI-1906.

2.4. Metabolic Conversion of 1,25D3 Analogs PRI-1901 and PRI-2205 to hCYP24A1-Mediated Degradation

The membrane fraction prepared from the recombinant *Escherichia coli* cells that expressed *h*CYP24A1 [25,26] was used to examine the metabolism of 1,25Ds, as previously done for the 1,25D2 analogs. The HPLC metabolic profiles of PRI-1901 and PRI-2205 are shown in Figure 8.

The metabolic conversion of PRI-2205 was $44 \pm 6\%$ and the same as that observed for the parent 1,25D3 [10]. This suggests that modifications such as introducing an isolated double bond in the regular vitamin D_3 side-chain at C-22 as well as the formal transfer of the 19-methylene from C-10 to C-4 position in the A-ring do not improve the metabolic resistance of analogs. Quite unexpectedly, the extension of the side-chain by two carbon units in PRI-1901 increased the metabolic conversion up to $58 \pm 9\%$.

2.5. Molecular Interactions between CYP3A4 Binding Site and 1,25D3 Analogs

Based on the experimental data obtained for the metabolic conversion of 1,25D3 analogs by the CYP24A1 enzyme, we examined the nature of the mode of action that can be created between analogs and the CYP3A4 enzyme. This was investigated for the PRI-1901 and PRI-2205 analogs of 1,25D3 by using the human CYP3A4 model (see Figure 9). The calculated binding free enthalpy values for PRI-1901 and PRI-2205 were -62 kcal/mol and −64 kcal/mol, respectively (compared to −29 kcal/mol for 1,25D3). The interactions inside the complexes of PRI-1901 and PRI-2205 are shown in Figure 9 and Figure S4 in the Supplementary Materials.

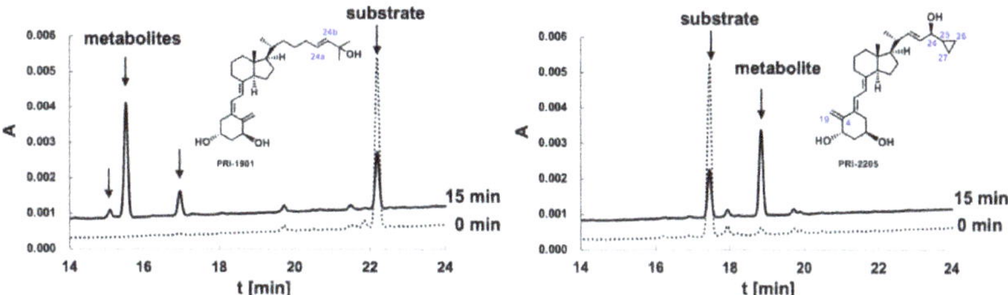

Figure 8. HPLC profiles of 1,25D3 analogs (PRI-1901 and PRI-2205) and their metabolites, before and after incubation with *h*CYP24A1. UV absorbance was recorded at 254 nm. The upper chromatograms represent profiles of the reaction mixture following incubation with *h*CYP24A1 for 15 min. The lower chromatograms were obtained at the starting point.

(A)

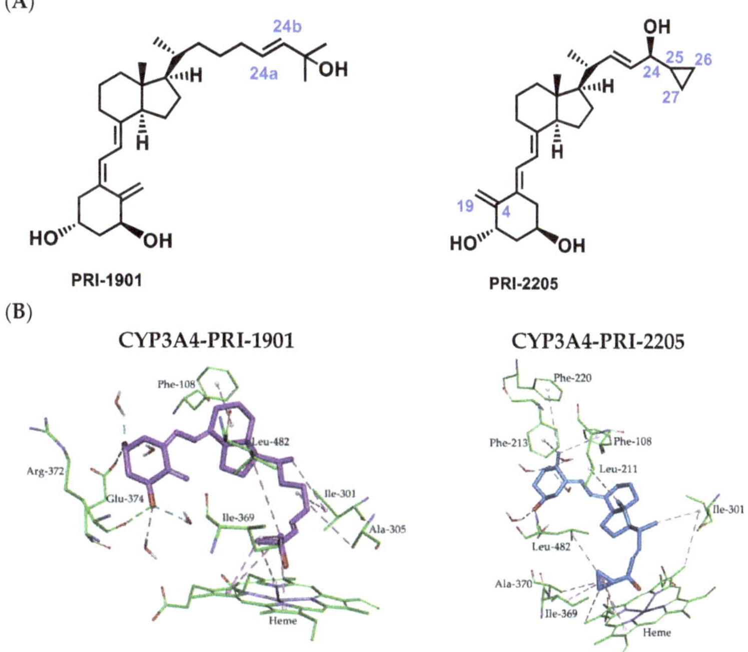

Figure 9. **(A)** Structural formulas of the analogs of 1α,25-dihydroxyvitamin D$_3$ (PRI-1901 and PRI-2205). **(B)** Molecular interactions between 1,25D3 analogs (PRI-1901 and PRI-2205) and CYP3A4 resulting from MD simulations. Hydrophobic interactions are shown as magenta dashed lines and hydrogen bonds as green.

Analog PRI-1901, which has a (24aE) double bond that is located in the vicinity of the C-25 center in the side-chain, is more flexible and adopts a U-shaped conformation. This does not allow for the correct disposition in the active site. The binding model for the PRI-1901 and CYP3A4 protein revealed that the 1,25D3 analog forms two hydrogen bonds with the two residues of the active site of CYP3A4 (Arg-372 and Glu-374) and that there are hydrophobic interactions with the amino acid residues Phe-108, Ile-301, Ala-305, Ile-369, and Leu-482. On the other hand, the higher affinity of PRI-2205 can be attributed to its extended conformation and may also be due to the presence of the cyclopropyl ring on the C-25 carbon atom at the side-chain which forms hydrophobic interactions with the amino acid residues Ile-369, Ala-370, and Leu-482 as well as with the heme ring. It can be seen (Figure 9) that the 19-methylene at the C-4 position makes strong π-alkyl interactions with the key residues of the active site of CYP3A4, such as the aromatic ring of Phe-108, Phe-213, and Phe-220. Furthermore, the positions of the A-ring hydroxyls of PRI-2205 allow the formation of a network of hydrogen bonds with water molecules which stabilize the binding of the PRI-2205 in the active site of CYP3A4.

3. Materials and Methods

3.1. Theoretical Calculations

3.1.1. Dataset Preparation

In this study, we made use of our previously published and new experimental data for the metabolic resistance of analogs of 1,25D2 (PRI-5106, PRI-5105, PRI-1916, PRI-1917, PRI-1906, PRI-1907, PRI-5201, and PRI-5202) and 1,25D3 (PRI-1901 and PRI-2205) [16,41,42]. The starting conformations of the analogs were constructed based on the solid-state diffraction data of structurally related compounds to eliminate any subjectivity in generating the three-dimensional structure [43]. The molecular structures of the 1,25D analogs were optimized using the density functional theory (DFT) at the B3LYP/6-311$^+$G(d,p) level implemented in the Gaussian 16 program [44]. The electrostatic potential (ESP) of the atomic partial charges of the atoms was computed using the Breneman model [45] which reproduces the molecular electrostatic potential. The crystal structure of human CYP3A4 (PDB entry 2V0M) was obtained from the RCSB Protein Data Bank [32]. The PDB file presents a co-crystal with a ketoconazole ligand in the active site. Ligand (except for the heme group), water molecules, and inorganic ions were removed before making the calculations, and hydrogen atoms were added to reflect the physiological pH. Prior to the analysis, the iron atom in the heme group was constrained to preserve its bonding to the nitrogen atoms of the heme after the application of the CHARMm force field.

3.1.2. Molecular Docking and Dynamic Simulation

Docking and molecular dynamic (MD) simulations were performed using the Discovery Studio 2020 with a visual interface BIOVIA [46]. To identify the starting structures for the subsequent computations of the binding affinity of 1,25D analogs for CYP3A4; a rigid docking procedure was performed using the CDOCKER protocol of Discovery Studio 2020. The active site was defined with a radius of 15 Å around the analog present in the 2V0M crystal. The analogs were allowed to interact with the residues within the binding site spheres to generate ten conformations. The best poses predicted by CDOCKER were used as the starting points in the MD simulation.

MD simulations were run using the CHARMm force field [47] implemented in the module of Discovery Studio 2020. The molecular parameters and atomic charges for the protein were taken from a set of CHARMm force field parameters. Each model of the CYP3A4-analog complex was inserted into a cubic box of water molecules (TIP3P models) [48] extending up to 10 Å from any solute atom. Counter-ions (Na$^+$, Cl$^-$) were added randomly to each complex at a concentration of ~0.15 M, close to physiological conditions, by using the Solvation Module of Discovery Studio 2020. All energy minimization and MD simulations were performed using the Particle Mesh Ewald (PME) algorithm [49] for the correct treatment of electrostatic interactions [50] and periodic boundary conditions. Prior to MD simulations, all systems were minimized based on the steepest descent method with 3000 steps followed by 3000 conjugate gradient energy-minimization steps (until the RMS gradient of the structure was below 0.01 kcal/mol·Å) with an applied restraint potential of 10 kcal/mol·Å^2 for protein. The conjugate gradient algorithm without restraint was further carried out with an additional full minimization of 1000 steps. A gradual heating MD simulation from 50 to 300 K was executed for 50 ps. Following heating, an equilibration estimating 100 ps of each system was conducted (the operating temperature was kept constant at 300 K). In the stages of heating and equilibration, the protein was fixed with a force constant of 5 kcal mol^{-1} Å^{-2}. Using two phases, the equilibrated system was taken as the starting structure for production runs. In the first phase, NVT was performed at 300 K for 10 ns in the presence of a weak harmonic restraint on the solute, and in the second phase, NPT at 300 K and 1 bar for 10 ns. During the 20 ns simulations, snapshots were stored every 1 ps and used for the analysis. All simulations were run with SHAKE on hydrogen atoms with a 2 fs time step and Langevin thermostat for temperature control [51]. The integration time step was set to 1 fs. The cut-off distance for non-bonded terms was set to 10 Å.

3.1.3. Binding Free Enthalpy Calculation

Even though docking combined with MD simulations provides a clear image of the shape complementarity between the analog and the protein, there is the need for additional and essential information regarding the free enthalpy of binding. This assesses the affinity of an analog to its target. To compare the stability of each binding mode while considering solvent effects, we calculated the binding free enthalpy (ΔG_{bind}) for each mode by using the MM/PBSA method in Discovery Studio 2020. The ΔG_{bind} of 1,25D analogs to CYP3A4 was calculated using Equation (1):

$$\Delta G_{bind} = G_{CYP3A4\text{-}analog} - G_{CYP3A4} - G_{analog} \tag{1}$$

where $G_{CYP3A4\text{-}analog}$ is the free enthalpy of complex, G_{CYP3A4} is the free enthalpy of CYP3A4, and G_{analog} is the free enthalpy of tested analogs. Binding free enthalpy was calculated based on the average structures obtained from the 20 ns of MD trajectories.

3.2. Metabolic Conversion of 1,25D3 Analogs PRI-1901 and PRI-2205 to CYP24A1-Mediated Degradation

CYP24A1-mediated degradation of the two analogs of 1,25D3 (PRI-1901 and PRI-2205) was determined using recombinant human CYP24A1 (hCYP24A1) as described previously [25,26]. The reaction mixture containing 2.0 µM bovine adrenodoxin, 0.2 µM bovine adrenodoxin reductase, 20 nM CYP24A1, 5 µM 1,25D3 analog, 1 mM NADPH, 100 mM Tris-HCl (pH 7.4), and 1 mM EDTA in total volume of 100 mL, was incubated at 37 °C for 15 min. The reaction was quenched by the addition of four volumes of chloroform/methanol (3:1) and vigorous shaking. The organic phase was collected and dried down under reduced pressure. The residue was dissolved in acetonitrile and centrifuged at 20,000× g for 15 min. The supernatant (applied volume 40 µL) was submitted to HPLC Capcell-Pak C18 UG120 4.6 mm × 250 mm column (Phenomenex, Tokyo, Japan) at a flow rate of 1 mL/ according to previously described conditions [25,26]. Samples were eluted by linear gradients of water-acetonitrile 20–100% solution (0–25 min) followed by 100% acetonitrile (25–40 min). The eluted metabolites of 1,25D3 analogs were detected by UV absorbance at 254 nm. The amount of a 1,25D3 analog in the eluates was calculated from the peak area. The percent of metabolic conversion of the analog was calculated as a ratio of the peak area of metabolites to the sum of the peak areas of the remaining 1,25D3 analog and the metabolite or metabolites (assumed as 100%).

4. Conclusions

This work has shown for the first time how the mode of action of 1,25D analogs at the active site of CYP3A4 can be predicted using simulation models. The prediction is of high importance for vitamin D-based drug candidates as CYP3A4 deactivates ca. 50% of drug substances. The results of the computational analysis have revealed that the molecular interactions with the active site of the CYP3A4 enzyme were very different for the 25-dimethyl analogs PRI-1906 and PRI-1916, as well as for the 25-diethyl analogs PRI-1907 and PRI-1917. For PRI-1906 and PRI-1916, there was a head (A-ring) and tail (side-chain) inversion; while for PRI-1907 and PRI-1917 there was a completely different arrangement of the CD-ring system. Differences of the same kind might occur regarding the binding of the above analogs to the CYP24A1 enzyme. The differences in molecular interactions explain why pairs of analogs differing only in the (24E)-and (24Z)-geometry of the side-chain showed substantial differences in their metabolic resistance. The enzymatic conversion of (24*E*)-analogs PRI-1906 and PRI-1907 by CYP24A1 was relatively very low, while the conversion of (24*Z*)-analogs PRI-1916 and PRI-1917 was six times and ten-times higher, respectively. As we anticipated, the calculated free enthalpy of binding of a vitamin D metabolite or analog to CYP3A4 showed agreement with the experimental conversion of an analog by CYP24A1. The very low values obtained for the enthalpy for the analogs PRI-1906, PRI-1907, PRI-5201, and PRI-5202 correspond to the lowest values of metabolic conversion. Similarly, high values for the enthalpy of the analogs PRI-1916 and PRI-1917 correspond with their high metabolic conversion. Finally, the relatively high values of enthalpy for both 1,25D2 and 1,25D3 correspond well with their very high metabolic conversion. Quite unexpectedly, the predicted location of 1,25D analogs in the CYP3A4 pocket is different for 1,25D2 and 1,25D3 analogs. The 1,25D2 analogs interact with the CYP3A4 pocket using hydrogen bonds, and hydrophobic and van der Waals interactions, while the interaction of 1,25D3 and its analogs are mainly hydrophobic in nature. Our metabolic studies of the two series of analogs of 1,25D2 and 1,25D3 have revealed that the position at C-10 or C-4 and the absence of 19-methylene in the A-ring, as well as the modifications that retained the regular size of the side-chain, did not influence the metabolic resistance of the analog. Regarding the structural modifications of vitamin D, a combination of both an extension by one carbon unit and introducing a conjugated system of double bonds in the side-chain results in the most pronounced increase in the resistance of an analog to *h*CYP24A1-mediated degradation.

In summary, we have shown that the metabolic resistance of a vitamin D analog to the CYP24A1 degrading enzyme, of unknown 3D structure, can be estimated by the binding strengths of the analogs to the CYP3A4 enzyme, which is of known 3D structure. The metabolic conversion of vitamin D analogs by CYP24A1 can be related to the free enthalpy of binding to CYP3A4 only for analogs of 1,25D2 that have very similar structures. Accordingly, examination of the correlation for analogs of 1,25D3 requires an extensive series of analogs that have a similar structure.

Supplementary Materials: The following supporting information can be downloaded at: https://www.mdpi.com/article/10.3390/ijms23147845/s1.

Author Contributions: Conceptualization, supervision, A.K.; methodology, formal analysis, investigation, visualization, T.Ż.; formal analysis, investigation, K.Y.; writing-original draft, T.Ż. and A.K.; writing—review and editing, T.Ż., A.K., G.B., T.S. and K.Y.; funding acquisition, G.B. All authors have read and agreed to the published version of the manuscript.

Funding: The research leading to these results has received funding from the People Program (Marie Curie Actions) of the European Union's Seventh Framework Program FP7/2007-2013 under Research Executive Agency grant agreement No. 315902. G.B. and A.K. are partners within the Marie Curie Initial Training Network DECIDE (decision making within cells and differentiation entities therapies. This research was co-funded from the internal sources of the Medical University of Warsaw, Poland.

Institutional Review Board Statement: Not applicable.

Informed Consent Statement: Not applicable.

Data Availability Statement: All relevant data are provided in the article.

Conflicts of Interest: The authors declare no conflict of interest.

References

1. Mozolowski, W. Jędrzej Sniadecki (1768–1838) on the Cure of Rickets. *Nature* **1939**, *3612*, 121. [CrossRef]
2. Funk, C. The etiology of the deficiency diseases. *J. State Med.* **1912**, *20*, 341–368.
3. Bikle, D.D. Vitamin D: Newer Concepts of Its Metabolism and Function at the Basic and Clinical Level. *J. Endocr. Soc.* **2020**, *4*, bvz038. [CrossRef] [PubMed]
4. Cancela, L.; Theofan, G.; Norman, A.W. *The Pleiotropic Vitamin D Hormone*; Cooke, B.A., van der Molen, H.J., Eds.; New Comprehensive Biochemistry; Elsevier: Amsterdam, The Netherlands, 1988; Volume 18, Part A.
5. Carlberg, C.; Muñoz, A. An update on vitamin D signaling and cancer. *Semin. Cancer Biol.* **2020**, *79*, 217–230. [CrossRef]

6. Pike, J.W.; Meyer, M.B. New Approaches to Assess Mechanisms of Action of Selective Vitamin D Analogues. *Int. J. Mol. Sci.* **2021**, *22*, 12352. [CrossRef] [PubMed]

7. Kutner, A.; Brown, G. Vitamins D: Relationship between Structure and Biological Activity. *Int. J. Mol. Sci.* **2018**, *19*, 2119. [CrossRef]

8. Bouillon, R.; Marcocci, C.; Carmeliet, G.; Bikle, D.; White, J.H.; Dawson-Hughes, B.; Lips, P.; Munns, C.F.; Lazaretti-Castro, M.; Giustina, A.; et al. Skeletal and Extraskeletal Actions of Vitamin D: Current Evidence and Outstanding Questions. *Endocr. Rev.* **2018**, *40*, 1109–1151. [CrossRef]

9. Milczarek, M.; Rossowska, J.; Klopotowska, D.; Stachowicz, M.; Kutner, A.; Wietrzyk, J. Tacalcitol increases the sensitivity of colorectal cancer cells to 5-fluorouracil by downregulating the thymidylate synthase. *J. Steroid Biochem. Mol. Biol.* **2019**, *190*, 139–151. [CrossRef]

10. Nachliely, M.; Trachtenberg, A.; Khalfin, B.; Nalbandyan, K.; Cohen-Lahav, M.; Yasuda, K.; Sakaki, T.; Kutner, A.; Danilenko, M. Dimethyl fumarate and vitamin D derivatives cooperatively enhance VDR and Nrf2 signaling in differentiating AML cells in vitro and inhibit leukemia progression in a xenograft mouse model. *J. Steroid Biochem. Mol. Biol.* **2019**, *188*, 8–16. [CrossRef]

11. Trynda, J.; Turlej, E.; Milczarek, M.; Pietraszek, A.; Chodyński, M.; Kutner, A.; Wietrzyk, J. Antiproliferative Activity and in Vivo Toxicity of Double-Point Modified Analogs of 1,25-Dihydroxyergocalciferol. *Int. J. Mol. Sci.* **2015**, *16*, 24873–24894. [CrossRef]

12. Corcoran, A.; Nadkarni, S.; Yasuda, K.; Sakaki, T.; Brown, G.; Kutner, A.; Marcinkowska, E. Biological Evaluation of Double Point Modified Analogues of 1,25-Dihydroxyvitamin D2 as Potential Anti-Leukemic Agents. *Int. J. Mol. Sci.* **2016**, *17*, 91. [CrossRef] [PubMed]

13. Filip-Psurska, B.; Psurski, M.; Anisiewicz, A.; Libako, P.; Zbrojewicz, E.; Maciejewska, M.; Chodyński, M.; Kutner, A.; Wietrzyk, J. Vitamin D Compounds PRI-2191 and PRI-2205 Enhance Anastrozole Activity in Human Breast Cancer Models. *Int. J. Mol. Sci.* **2021**, *22*, 2781. [CrossRef]

14. Maj, E.; Maj, B.; Bobak, K.; Gos, M.; Chodyński, M.; Kutner, A.; Wietrzyk, J. Differential Response of Lung Cancer Cells, with Various Driver Mutations, to Plant Polyphenol Resveratrol and Vitamin D Active Metabolite PRI-2191. *Int. J. Mol. Sci.* **2021**, *22*, 2354. [CrossRef] [PubMed]

15. Baurska, H.; Klopot, A.; Kielbinski, M.; Chrobak, A.; Wijas, E.; Kutner, A.; Marcinkowska, E. Structure–function analysis of vitamin D2 analogs as potential inducers of leukemia differentiation and inhibitors of prostate cancer proliferation. *J. Steroid Biochem. Mol. Biol.* **2011**, *126*, 46–54. [CrossRef] [PubMed]

16. Milczarek, M.; Chodyński, M.; Pietraszek, A.; Stachowicz-Suhs, M.; Yasuda, K.; Sakaki, T.; Wietrzyk, J.; Kutner, A. Synthesis, CYP24A1-Dependent Metabolism and Antiproliferative Potential against Colorectal Cancer Cells of 1,25-Dihydroxyvitamin D2 Derivatives Modified at the Side Chain and the A-Ring. *Int. J. Mol. Sci.* **2020**, *21*, 642. [CrossRef]

17. Kotlarz, A.; Przybyszewska, M.; Swoboda, P.; Miłoszewska, J.; Grygorowicz, M.A.; Kutner, A.; Markowicz, S. Differential interference of vitamin D analogs PRI-1906, PRI-2191, and PRI-2205 with the renewal of human colon cancer cells refractory to treatment with 5-fluorouracil. *Tumor Biol.* **2015**, *37*, 4699–4709. [CrossRef] [PubMed]

18. Piotrowska, A.; Wierzbicka, J.; Kwiatkowska, K.; Chodyński, M.; Kutner, A.; Żmijewski, M.A. Antiproliferative activity of side-chain truncated vitamin D analogs (PRI-1203 and PRI-1204) against human malignant melanoma cell lines. *Eur. J. Pharmacol.* **2020**, *881*, 173170. [CrossRef] [PubMed]

19. Piątek, K.; Kutner, D.A.; Castillo-Tong, D.C.; Manhardt, T.; Kupper, N.; Nowak, U.; Chodyński, M.; Marcinkowska, E.; Kallay, E.; Schepelmann., M. Vitamin D Analogs Regulate the Vitamin D System and Proliferation in Ovarian Cancer Cells. *Int. J. Mol. Sci.* **2022**, *23*, 172. [CrossRef] [PubMed]

20. Rochel, N.; Wurtz, J.; Mitschler, A.; Klaholz, B.; Moras, D. The Crystal Structure of the Nuclear Receptor for Vitamin D Bound to Its Natural Ligand. *Mol. Cell* **2000**, *5*, 173–179. [CrossRef]

21. Tocchini-Valentini, G.; Rochel, N.; Wurtz, J.M.; Mitschler, A.; Moras, D. Crystal structures of the vitamin D receptor complexed to superagonist 20-epi ligands. *Proc. Natl. Acad. Sci. USA* **2001**, *98*, 5491–5496. [CrossRef]

22. Brélivet, Y.; Rochel, N.; Moras, D. Structural analysis of nuclear receptors: From isolated domains to integral proteins. *Mol. Cell. Endocrinol.* **2012**, *348*, 466–473. [CrossRef]

23. Kotlarz, A.; Przybyszewska, M.; Swoboda, P.; Neska, J.; Miłoszewska, J.; Grygorowicz, M.A.; Kutner, A.; Markowicz, S. Imatinib inhibits the regrowth of human colon cancer cells after treatment with 5-FU and cooperates with vitamin D analogue PRI-2191 in the downregulation of expression of stemness-related genes in 5-FU refractory cells. *J. Steroid Biochem. Mol. Biol.* **2019**, *189*, 48–62. [CrossRef] [PubMed]

24. Ben-Eltriki, M.; Deb, S.; Guns, E.S.T. 1α,25-Dihydroxyvitamin D3 synergistically enhances anticancer effects of ginsenoside Rh2 in human prostate cancer cells. *J. Steroid Biochem. Mol. Biol.* **2021**, *209*, 105828. [CrossRef] [PubMed]

25. Kawagoe, F.; Mototani, S.; Yasuda, K.; Mano, H.; Sakaki, T.; Kittaka, A. Stereoselective Synthesis of 24-Fluoro-25-Hydroxyvitamin D3 Analogues and Their Stability to hCYP24A1-Dependent Catabolism. *Int. J. Mol. Sci.* **2021**, *22*, 11863. [CrossRef]

26. Urushino, N.; Yasuda, K.; Ikushiro, S.; Kamakura, M.; Ohta, M.; Sakaki, T. Metabolism of 1α,25-dihydroxyvitamin D2 by human CYP24A1. *Biochem. Biophys. Res. Commun.* **2009**, *384*, 144–148. [CrossRef]

27. Annalora, A.J.; Goodin, D.B.; Hong, W.-X.; Zhang, Q.; Johnson, E.F.; Stout, C.D. Crystal Structure of CYP24A1, a Mitochondrial Cytochrome P450 Involved in Vitamin D Metabolism. *J. Mol. Biol.* **2010**, *396*, 441–451. [CrossRef] [PubMed]

28. Soltis, S.M.; Cohen, A.E.; Deacon, A.; Eriksson, T.; González, A.; McPhillips, S.; Chui, H.; Dunten, P.; Hollenbeck, M.; Mathews, I.; et al. New paradigm for macromolecular crystallography experiments at SSRL: Automated crystal screening and remote data collection. *Acta Crystallogr. Sect. D Biol. Crystallogr.* **2008**, *64*, 1210–1221. [CrossRef]

29. Williams, P.A.; Cosme, J.; Vinković, D.M.; Ward, A.; Angove, H.C.; Day, P.J.; Vonrhein, C.; Tickle, I.J.; Jhoti, H. Crystal Structures of Human Cytochrome P450 3A4 Bound to Metyrapone and Progesterone. *Science* **2004**, *305*, 683–686. [CrossRef]

30. Hoffman, R.L.; Kania, R.S.; Brothers, M.A.; Davies, J.F.; Ferre, R.A.; Gajiwala, K.S.; He, M.; Hogan, R.J.; Kozminski, K.; Li, L.Y.; et al. Discovery of Ketone-Based Covalent Inhibitors of Coronavirus 3CL Proteases for the Potential Therapeutic Treatment of COVID-19. *J. Med. Chem.* **2020**, *63*, 12725–12747. [CrossRef]

31. Yano, J.K.; Wester, M.R.; Schoch, G.A.; Griffin, K.J.; Stout, C.D.; Johnson, E.F. The Structure of Human Microsomal Cytochrome P450 3A4 Determined by X-ray Crystallography to 2.05-Å Resolution. *J. Biol. Chem.* **2004**, *279*, 38091–38094. [CrossRef]

32. Ekroos, M.; Sjögren, T. Structural basis for ligand promiscuity in cytochrome P450 3A4. *Proc. Natl. Acad. Sci. USA* **2006**, *103*, 13682–13687. [CrossRef]

33. Gupta, R.P.; He, Y.A.; Patrick, K.S.; Halpert, J.R.; Bell, N.H. CYP3A4 Is a Vitamin D-24- and 25-Hydroxylase: Analysis of Structure Function by Site-Directed Mutagenesis. *J. Clin. Endocrinol. Metab.* **2005**, *90*, 1210–1219. [CrossRef] [PubMed]

34. Xu, Y.; Hashizume, T.; Shuhart, M.C.; Davis, C.L.; Nelson, W.L.; Sakaki, T.; Kalhorn, T.F.; Watkins, P.B.; Schuetz, E.G.; Thummel, K.E. Intestinal and Hepatic CYP3A4 Catalyze Hydroxylation of 1α,25-Dihydroxyvitamin D$_3$: Implications for Drug-Induced Osteomalacia. *Mol. Pharmacol.* **2005**, *69*, 56–65. [CrossRef] [PubMed]

35. Pavek, P.; Dusek, J.; Smutny, T.; Lochman, L.; Kucera, R.; Skoda, J.; Smutna, L.; Kamaraj, R.; Soucek, P.; Vrzal, R.; et al. Gene Expression Profiling of 1. *Mol. Nutr. Food Res.* **2022**, *66*, 2200070. [CrossRef]

36. Wang, Z.; Schuetz, E.G.; Xu, Y.; Thummel, K.E. Interplay between vitamin D and the drug metabolizing enzyme CYP3A4. *J. Steroid. Biochem. Mol. Biol.* **2012**, *136*, 54–58. [CrossRef] [PubMed]

37. Gomaa, M.S.; Simons, C.; Brancale, A. Homology model of 1α,25-dihydroxyvitamin D3 24-hydroxylase cytochrome P450 24A1 (CYP24A1): Active site architecture and ligand binding. *J. Steroid Biochem. Mol. Biol.* **2007**, *104*, 53–60. [CrossRef]

38. Stjernschantz, E.; Vermeulen, N.P.; Oostenbrink, C. Computational prediction of drug binding and rationalisation of selectivity towards cytochromes P450. *Expert Opin. Drug Metab. Toxicol.* **2008**, *4*, 513–527. [CrossRef]

39. Żołek, T.; Dömötör, O.; Ostrowska, K.; Enyedy, É.A.; Maciejewska, D. Evaluation of blood-brain barrier penetration and examination of binding to human serum albumin of 7-O-arylpiperazinylcoumarins as potential antipsychotic agents. *Bioorg. Chem.* **2019**, *84*, 211–225. [CrossRef]

40. Żołek, T.; Dömötör, O.; Rezler, M.; Enyedy, A.; Maciejewska, D. Deposition of pentamidine analogues in the human body—Spectroscopic and computational approaches. *Eur. J. Pharm. Sci.* **2021**, *161*, 105779. [CrossRef]

41. Bolla, N.R.; Corcoran, A.; Yasuda, K.; Chodyński, M.; Krajewski, K.; Cmoch, P.; Marcinkowska, E.; Brown, G.; Sakaki, T.; Kutner, A. Synthesis and evaluation of geometric analogs of 1α,25-dihydroxyvitamin D2 as potential therapeutics. *J. Steroid. Biochem. Mol. Biol.* **2015**, *164*, 50–55. [CrossRef]

42. Pietraszek, A.; Malińska, M.; Chodyński, M.; Krupa, M.; Krajewski, K.; Cmoch, P.; Woźniak, K.; Kutner, A. Synthesis and crystallographic study of 1,25-dihydroxyergocalciferol analogs. *Steroids* **2013**, *78*, 1003–1014. [CrossRef]

43. Suwinska, K.; Kutner, A. Crystal and molecular structure of 1,25-dihydroxycholecalciferol. *Acta Crystallogr. Sect. B Struct. Sci.* **1996**, *52*, 550–554. [CrossRef]

44. Frisch, M.J.; Trucks, G.W.; Schlegel, H.B.; Scuseria, G.E.; Robb, M.A.; Cheeseman, J.R.; Scalmani, G.; Barone, V.; Petersson, G.A.; Nakatsuji, H.; et al. Gaussian, Inc., Wallingford CT. 2016. Available online: https://gaussian.com/citation/ (accessed on 13 July 2022).

45. Breneman, C.M.; Wiberg, K.B. Determining atom-centered monopoles from molecular electrostatic potentials. The need for high sampling density in formamide conformational analysis. *J. Comput. Chem.* **1990**, *11*, 361–373. [CrossRef]

46. BIOVIA, Dassault Systèmes, Discovery Studio Modeling Environment, San Diego: Dassault Systèmes. 2021. Available online: https://www.3ds.com/products-services/biovia/products/molecular-modeling-simulation/biovia-discovery-studio/ (accessed on 13 July 2022).

47. Brooks, B.R.; Bruccoleri, R.E.; Olafson, B.D.; States, D.J.; Swaminathan, S.; Karplus, M. CHARMM: A program for macromolecular energy, minimization, and dynamics calculations. *J. Comput. Chem.* **1983**, *4*, 187–217. [CrossRef]

48. Jorgensen, W.L.; Chandrasekhar, J.; Madura, J.D.; Impey, R.W.; Klein, M.L. Comparison of simple potential functions for simulating liquid water. *J. Chem. Phys.* **1983**, *79*, 926–935. [CrossRef]

49. Essmann, U.; Perera, L.; Berkowitz, M.L.; Darden, T.; Lee, H.; Pedersen, L.G. A smooth particle mesh Ewald method. *J. Chem. Phys.* **1995**, *103*, 8577–8593. [CrossRef]

50. Sagui, C.; Darden, T.A. MOLECULAR DYNAMICS SIMULATIONS OF BIOMOLECULES: Long-Range Electrostatic Effects. *Annu. Rev. Biophys. Biomol. Struct.* **1999**, *28*, 155–179. [CrossRef]

51. Ryckaert, J.-P.; Ciccotti, G.; Berendsen, H.J.C. Numerical integration of the cartesian equations of motion of a system with constraints: Molecular dynamics of n-alkanes. *J. Comput. Phys.* **1977**, *23*, 327–341. [CrossRef]

International Journal of
Molecular Sciences

Article

The Effects of Genistein at Different Concentrations on *MCF-7* Breast Cancer Cells and *BJ* Dermal Fibroblasts

Magda Aleksandra Pawlicka *, Szymon Zmorzyński, Sylwia Popek-Marciniec and Agata Anna Filip

Department of Cancer Genetics with Cytogenetic Laboratory, Medical University of Lublin, Radziwiłłowska 11 Street, 20-080 Lublin, Poland
* Correspondence: magda.pawlicka1@gmail.com; Tel.: +48-509-647-783

Abstract: This study aimed to evaluate the safety and potential use of soy isoflavones in the treatment of skin problems, difficult-to-heal wounds and postoperative scars in women after the oncological treatment of breast cancer. The effects of different concentrations of genistein as a representative of soy isoflavonoids on *MCF-7* tumor cells and *BJ* skin fibroblasts cultured in vitro were assessed. Genistein affects both healthy dermal *BJ* fibroblasts and cancerous *MCF-7* cells. The effect of the tested isoflavonoid is closely related to its concentration. High concentrations of genistein destroy *MCF-7* cancer cells, regardless of the exposure time, with a much greater effect on reducing cancer cell numbers at longer times (48 h). Lower concentrations of genistein (10 and 20 μM) increase the abundance of dermal fibroblasts. However, higher concentrations of genistein (50 μM and higher) are detrimental to fibroblasts at longer exposure times (48 h). Our studies indicate that although genistein shows high potential for use in the treatment of skin problems, wounds and surgical scars in women during and after breast cancer treatment, it is not completely safe. Introducing isoflavonoids to treatment requires further research into their mechanisms of action at the molecular level, taking into account genetic and immunological aspects. It is also necessary to conduct research in in vivo models, which will allow for eliminating adverse side effects of therapy.

Keywords: genistein; breast cancer; *MCF-7* cells; skin; fibroblasts

Citation: Pawlicka, M.A.; Zmorzyński, S.; Popek-Marciniec, S.; Filip, A.A. The Effects of Genistein at Different Concentrations on *MCF-7* Breast Cancer Cells and *BJ* Dermal Fibroblasts. *Int. J. Mol. Sci.* **2022**, 23, 12360. https://doi.org/10.3390/ijms232012360

Academic Editors: Geoffrey Brown, Enikö Kallay and Andrzej Kutner

Received: 31 August 2022
Accepted: 13 October 2022
Published: 15 October 2022

Publisher's Note: MDPI stays neutral with regard to jurisdictional claims in published maps and institutional affiliations.

1. Introduction

Breast cancer is a malignant tumor of the mammary gland originating from epithelial tissue and is also the most common malignant tumor in women [1]. In 2020, female breast cancer surpassed lung cancer as the most commonly diagnosed cancer, with an estimated 2.3 million new cases (11.7% of total cancers) [2]. Depending on the type of cancer, stage of the disease and patient's condition, different treatment methods are applied [3,4]. The most used method is local or radical surgery supplemented with radiotherapy and/or chemotherapy [5]. There are also trials of novel treatments with fewer side effects, e.g., photodynamic therapy [6]. Although these methods show great promise, at this point they are rarely used due to the high cost of therapy. Modern medicine allows for breast reconstructive surgery, which can be performed during subtraction surgery or at another time [7]. However, regardless of the decision made by the surgeon and the patient, surgery is usually followed by an extensive scar. A complementary treatment used to eliminate the remaining cancer cells can also adversely affect postoperative wound healing and proper scar formation, which is not only an aesthetic defect but can also impede mobility and be painful. Therefore, it is essential to implement as soon as possible a therapy supporting these two processes [8]. Several preparations are available for the treatment of scars, but few of them have been sufficiently tested for safety in oncology patients. Many active substances can potentially stimulate ongoing tumorigenesis in tissues, but there is a lack of scar treatment products that would be safe for oncological patients but effective at the same time.

Soy isoflavonoids are classified as phytoestrogens, plant equivalents of female hormones [9,10]. Soy phytoestrogens, especially genistein and daidzein, in their structure and action resemble the female hormone estradiol [11]. Therefore, applied topically, they could bind to estrogen receptors in the skin and thus show a direct effect on the changes occurring in its cells [12,13]. The action of phytoestrogens on the skin and wound healing has been supported by numerous studies [14–32]. These compounds primarily affect the production of collagen, elastin and hyaluronic acid [23]. They also show strong soothing and anti-inflammatory properties. As flavonoid substances are characterized by strong anti-free solid radical properties, they also inhibit the activity of enzymes from the metalloproteinase class (MMP-1 and MMP-3), which are activated under the influence of UV radiation. Metalloproteinase MMP-1 breaks down collagen fibers, whereas MMP-3 breaks down the proteins responsible for proper communication between the epidermis and the dermis [33].

The aim of this study was to evaluate the safety and potential use of soy isoflavones in the treatment of skin problems, difficult-to-heal wounds and postoperative scars in women after oncological treatment of breast cancer. Therefore, a study was conducted to analyze the effects of different concentrations of genistein, as a representative of soy isoflavonoids, on *MCF-7* cancer cells and *BJ* dermal fibroblasts.

2. Results

Figures 1 and 2 show the cytotoxic effects of genistein at different concentrations after 24 h and 48 h on tumor cells *MCF-7* and non-tumor cells *BJ* in MTT test. After 24 h of treatment, genistein was cytotoxic for *MCF-7* cells in concentrations over 80 µM and for *BJ* cells at its highest concentration (200 µM). It could also be observed that genistein at lower concentrations stimulated fibroblast growth. Treatment with genistein 50, 80, 100, 150 and 200 µM for 48 h, induced cytotoxicity in both non-tumor *BJ* and tumor *MCF-7* cells.

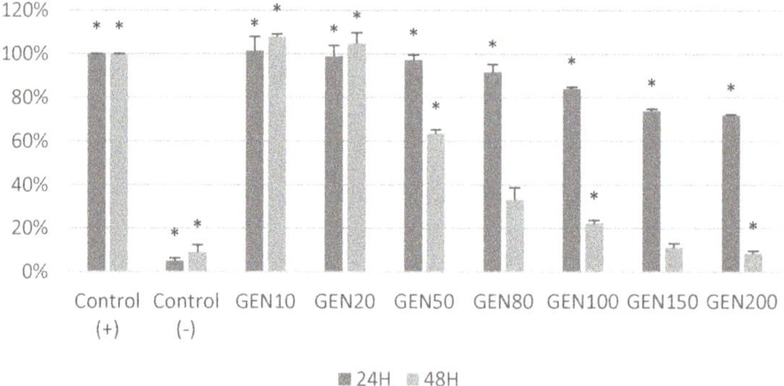

Figure 1. Cytotoxic effect of genistein at different concentrations after 24 h and 48 h on tumor *MCF-7* cells in MTT test. Results are the mean ± SD of n = 5. * Statistically significant ($p < 0.05$; ANOVA followed by Dunnett's test compared with control group). Cells cultured with 0.1% DMSO were considered Control (+). Cells cultured with 15% DMSO were considered as Control (−).

Figures 3 and 4 show the influence of different concentrations of genistein after 24 h or 48 h on apoptosis of tumor *MCF-7* cells and non-tumor *BJ* cells. After 24 h, among *MCF-7* tumor line, apoptotic cells are observed at concentration as low as 20 µM. At genistein concentrations of 50 µM and higher, a reduction in the number of viable cells is observed. Among the cells of the non-tumor *BJ* line, apoptotic cells are observed only at a concentration of 150 µM. At lower concentrations of genistein, an increased number of viable fibroblasts can be observed. After 48 h incubation of *MCF-7* cells with genistein, apoptosis and a significant decrease in the number of viable cells is observed from a concentration of 20 µM. For this exposure time, the *BJ* fibroblast population is reduced already at genistein concentrations of 50 µM.

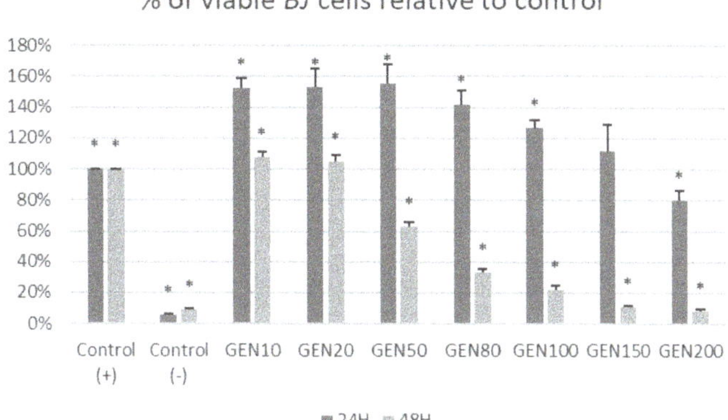

Figure 2. Cytotoxic effect of genistein at different concentrations after 24 h and 48 h on non-tumor *BJ* cells in MTT test. Results are the mean ± SD of n = 5. * Statistically significant ($p < 0.05$; ANOVA followed by Dunnett's test compared with control group). Cells cultured with 0.1% DMSO were considered Control (+). Cells cultured with 15% DMSO were considered as Control (−).

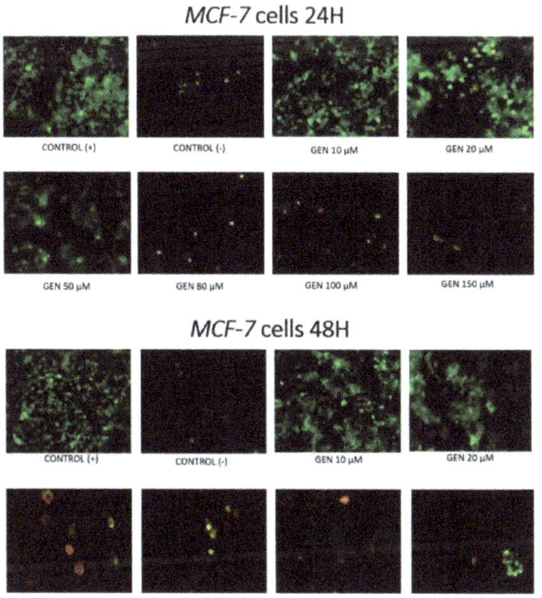

Figure 3. Induction of apoptosis by genistein at different concentrations in tumor *MCF-7* cell lines in fluorescent microscope (magnification 400×). Cells were stained with annexin-Cy3.18 (AnnCy3) and 6-Carboxyfluorescein diacetate (6-CFDA). AnnCy3 binds to phosphatidylserine present in the outer leaflet of plasma membrane of cells starting the apoptotic process. The binding is observed as red fluorescence. 6-CFDA is used to measure viability. When this non-fluorescent compound enters living cells, esterases hydrolyze it, producing fluorescent compound 6-carboxyfluorescein (6-CF). This appears as green fluorescence. Cells cultured with 0.1% DMSO were considered as Control (+). Cells cultured with 15% DMSO were considered as Control (−).

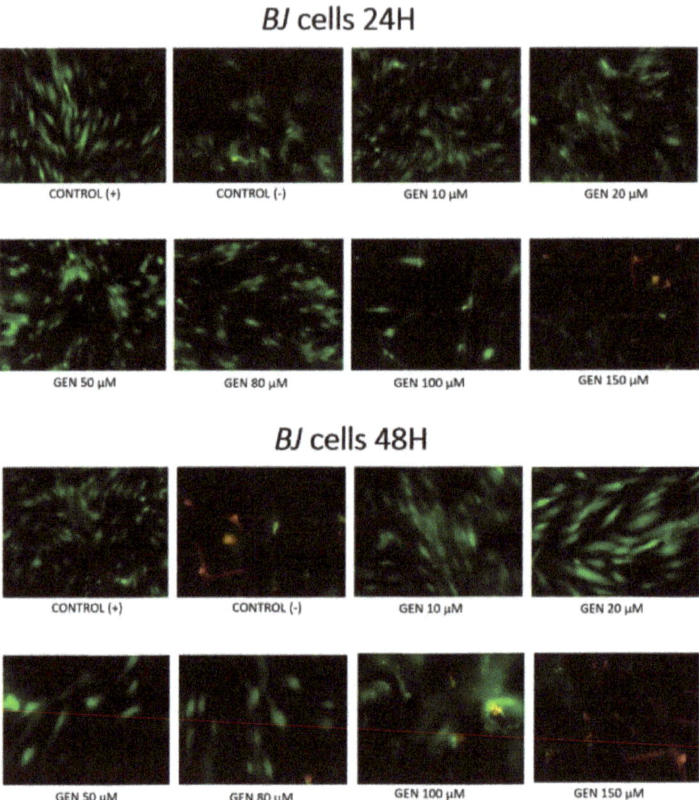

Figure 4. Induction of apoptosis by genistein at different concentrations in non-tumor *BJ* cell lines in fluorescent microscope (magnification 400×). Cells were stained with annexin-Cy3.18 (AnnCy3) and 6-Carboxyfluorescein diacetate (6-CFDA). AnnCy3 binds to phosphatidylserine present in the outer leaflet of plasma membrane of cells starting the apoptotic process. The binding is observed as red fluorescence. 6-CFDA is used to measure viability. When this non-fluorescent compound enters living cells, esterases hydrolyze it, producing fluorescent compound 6-carboxyfluorescein (6-CF). This appears as green fluorescence. Cells cultured with 0.1% DMSO were considered Control (+). Cells cultured with 15% DMSO were considered Control (−).

Figures 5 and 6 show the impact of different concentrations of genistein on viability of tumor *MCF-7* cells and non-tumor *BJ* cells. A reduction in the number of *MCF-7* tumor cells above the genistein 20 µM concentration was observed at both exposure times. In the fibroblast population, at 24 h genistein exposure, a decrease in cell number was observed only at a concentration of 150 and 200 µM. When *BJ* line cells were incubated with genistein for 48 h, a significant decrease in population size was observed as low as 50 µM concentration.

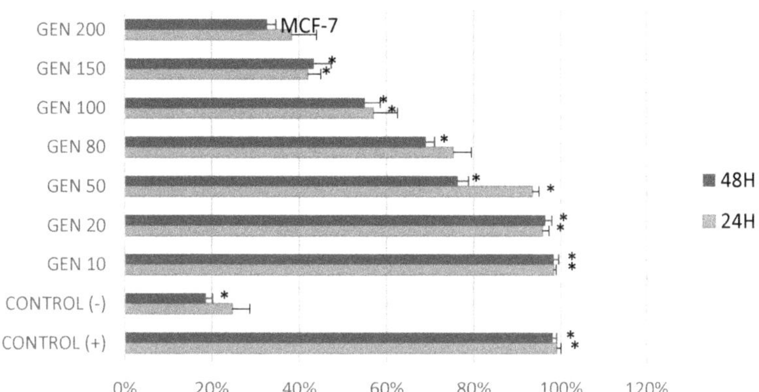

Figure 5. Percentage of viable MCF7 cells [%] in trypan blue exclusion assay after treatment with genistein at different concentration [μM]. Results are the mean ± SD of n = 3. * Statistically significant ($p < 0.05$; ANOVA followed by Dunnett's test compared with control group). Cells cultured with 0.1% DMSO were considered Control (+). Cells cultured with 15% DMSO were considered Control (−).

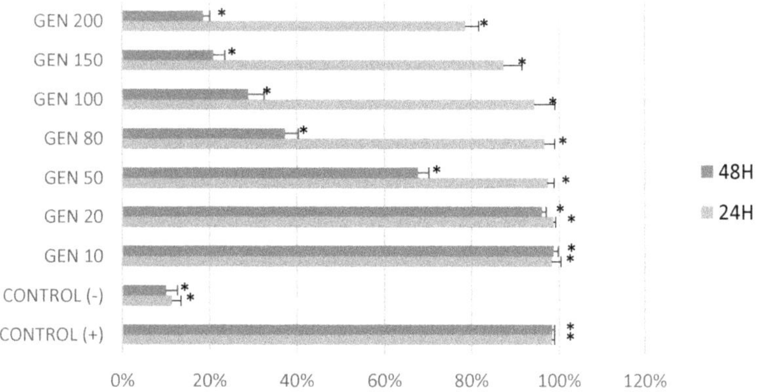

Figure 6. Percentage of viable *BJ* cells [%] in trypan blue exclusion assay after treatment with genistein at different concentration [μM]. Results are the mean ± SD of n = 3. * Statistically significant ($p < 0.05$; ANOVA followed by Dunnett's test compared with control group). Cells cultured with 0.1% DMSO were considered Control (+). Cells cultured with 15% DMSO were considered Control (−).

Gene expression analysis showed that all genistein concentrations resulted in change of all gene expression (Figure 7). Genistein at a concentration of 20 μM also strongly stimulated the expression of the *MKI67* in the *MCF-7* line and mildly stimulated its expression in fibroblasts. The highest concentration of genistein negatively affected the expression of the *MKI67* in both tumor cells and dermal fibroblasts. Genistein at concentrations of 10 and 20 μM increased *BIRC5* gene expression in the *BJ* line, while at the highest concentration, it strongly decreased it. The lowest concentration of phytoestrogen also increased the expression of the *BIRC5* gene in the tumor line, while concentrations of 20 and 50 μM decreased it. The lowest concentration of genistein strongly stimulated *AKT1* expression in the *BJ* line, while the other concentrations resulted in decreased expression. All concentrations of genistein decreased *AKT1* expression, with a decreasing expression in proportion to increasing concentrations. It was observed that genistein increased *BCL2* expression in fibroblasts at all concentrations, with the strongest at a concentration of 10 µM. The lowest concentration of the phytoestrogen gently increased the expression of the analyzed gene, while the other concentrations caused a decrease.

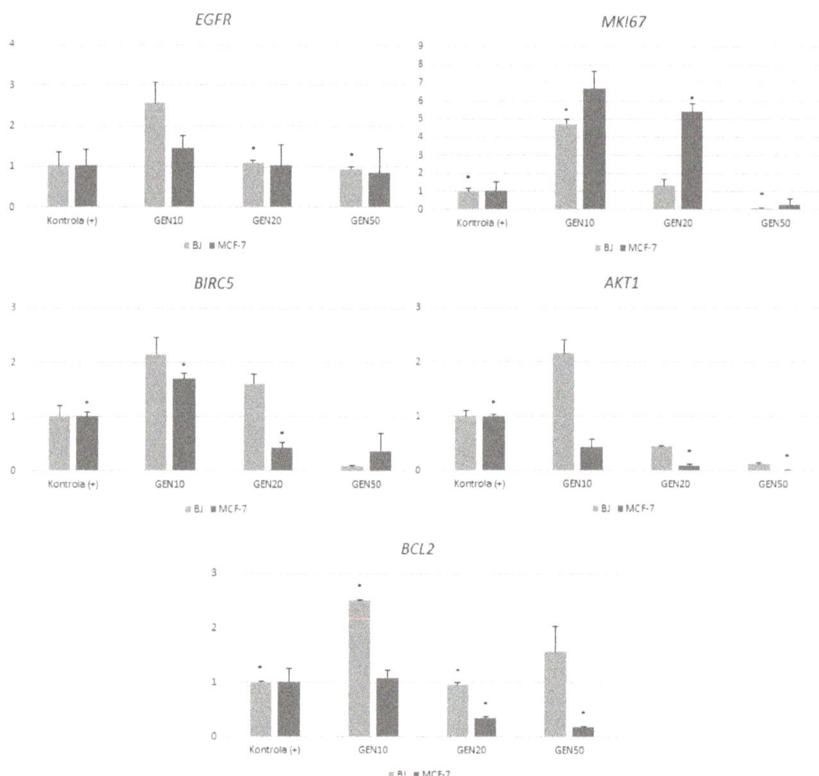

Figure 7. Analysis of gene expression in *MCF-7* and *BJ* cells treated with genistein at different concentrations 10, 20 and 50 µM) after 24 h Results are the mean value SD of n = 2. * Statistically significant ($p < 0.05$ ANOVA Friedman along with Dunn's multiple comparisons test compared with control group). Cells cultured with 0,1% DMSO were considered as Control (+). Cells cultured with 15% DMSO were considered as Control (−).

3. Discussion

The tests performed clearly indicate that genistein affects both healthy dermal *BJ* fibroblasts and cancerous *MCF-7* cells. The effect of the tested isoflavonoid is closely related to its concentration. The authors' study indicates that high concentrations of genistein destroy *MCF-7* cancer cells, regardless of the exposure time, with a much greater effect reduces cancer cell numbers at longer times (48 h). Lower concentrations of genistein (10 and 20 µM) increase the abundance of dermal fibroblasts. However, higher concentrations of genistein (50 µM and higher) prove to be detrimental to fibroblasts at longer exposure times (48 h).

The results of the *MCF-7* cell assay are in line with those of Kabała et al., who used two cancer lines, *MCF-7* and *MDA-MV-231* in their research. The conducted study showed a high dose-dependent effect in cell viability assays. The IC50 value was 47.5 µM for genistein. Isoflavonoid-induced apoptosis was dose- and time-dependent for both cell lines, with isoflavonoids being more active on *MCF-7* cells [34].

Choi also conducted similar studies et al. When *MCF-7* cells were treated with 50, 100, 150 and 200 µM of genistein for 24, 48 or 72 h, cell growth was significantly decreased in a concentration-dependent manner. After 48 h of treatment with 50 µM genistein,

genistein-induced apoptosis features were observed in *MCF-7*. This observation is based on the finding that the expression of B-cell lymphoma protein 2 (Bcl-2) was decreased, whereas the expression of Bcl-2-related X protein (Bax) was induced by genistein. *BCL2* gene family includes genes associated with regulation of apoptosis; BCL2 protein itself acts as antiapoptotic factor, while BAX is proapoptotic one [34]. The results of their study suggested that the induction of apoptosis by genistein was associated with the modulation of estrogen receptor-α (ERα) and cyclin D1 expression, which resulted in the inhibition of *MCF-7* cells proliferation [35].

An important problem in the use of isoflavonoids in cancer therapy turns out to be the maximum physiological concentrations of the tested compounds. This problem was addressed by Tsuboy et al., who found that the maximum concentration of genistein that can be achieved in tissues is 30 μM [36]. Tsuboy et al. also determined that only supraphysiological levels of Gen (50 and 100 μM) were cytotoxic to these cell lines; concentrations of 10 and 25 μM did not induce apoptosis or significant changes in the expression of the genes. Positive results were found only in cell cycle analysis: G0/G1. Therefore, despite the lack of induction of apoptosis, genistein at physiologically relevant serum concentrations still exerts chemopreventive effects through cell cycle modulation [36].

The use of soy isoflavonoids for skin problems after cancer treatment therefore requires studies on lower concentrations of genistein. Uifălean at al. conducted such a study and showed that at relatively low concentrations (1.56–13.06 μM), genistein stimulated the cell growth of *MCF-7*, in contrast with control, while higher concentrations of this compound had an inhibitory effect. The inhibitory effects of genistein are due to various molecular mechanisms: high doses of isoflavones (typically > 20 μM Gen) have been shown to stimulate apoptosis, inhibit cell proliferation and survival, and have antioxidant and angiogenesis inhibitory effects in breast cancer cells. In the present authors' study, the MTT assay indicated mild stimulation of tumor cell proliferation at a genistein concentration of 10 μM [37]. Therefore, the safest achievable systemic concentration of genistein that will have a beneficial effect on wound healing and scar formation after cancer treatment while at the same time not stimulating cancer cells may be 20 μM.

There are several papers in which authors have analyzed the molecular mechanisms and basic signaling pathways involved in both the stimulatory and inhibitory effects of genistein [37–42]. Lavigne et al. analyzed gene expression patterns in *MCF-7* line cells treated with physiological (1 and 5 μM) and pharmacological (25 μM) concentrations of genistein. Studies using oligonucleotide microarrays showed that genistein alters the expression of genes belonging to a broad spectrum of pathways, including the estrogen pathway and the p53 pathway. *TP53* gene encodes proteins that are important for DNA damage repair and apoptosis. At doses of 1 and 5 μM, genistein induced an expression pattern suggestive of increased mitogenic activity, confirming the proliferative response to genistein observed in the *MCF-7* cell cultures we conducted, while at a dose of 25 μM, genistein induced a pattern that likely contributes to increased apoptosis, decreased proliferation and reduced total cell number [38], which is also consistent with the cell culture results I obtained.

In their study, Sreenivasa et al. showed that genistein inhibits the growth of *MCF-7* cell line in a dose-dependent manner. The genistein-induced growth inhibition is accompanied by a reduction in the number of mitotic cells and overexpression of the cyclin-dependent kinase inhibitor p21WAF1, leading to cell cycle arrest. In addition, telomere length was significantly reduced in genistein-treated cancer cells. The analysis of a number of apoptosis-related genes showed the inhibition of Akt activity without affecting the level of Akt protein expression, as well as decreased expression of the pro-apoptotic gene *BAD*. The expression of one of the genes belonging to this group, *AKT1*, was also analyzed in our study. AKT1 is one of 3 closely related serine/threonine-protein kinases (AKT1, AKT2 and AKT3) called the AKT kinase, which regulate many processes including metabolism, proliferation, cell survival, growth, and angiogenesis. This is mediated through serine and/or threonine phosphorylation of a range of downstream substrates. The results of our study suggest a reduction in its expression at all tested genistein concentrations. Based on our own and other authors' results, we can conclude that the inhibition of cell division by genistein is mediated in part by a decrease in telomere length, a decrease in mitotic divisions and inhibition of Akt activation, leading to the induction of apoptosis [39].

The antiproliferative effect of genistein on *MCF-7* cells and the molecular mechanisms involved in this effect were also analyzed by Prietsch et al. Their study showed that genistein induced phosphatidylserine externalization and LC3A/B immunopositivity in *MCF-7* cells, indicating apoptosis and cell death by autophagy. Genistein increased the proapoptotic BAX/Bcl-2 ratio threefold and induced a 20-fold reduction in anti-apoptotic survivin (BIRC-5). The reduced expression of the *BIRC-5* gene was also consistent with our analysis [40].

Chen et al. found that genistein inhibited *MCF-7* cell proliferation and induced cell apoptosis through the inactivation of IGF-1R and p-Akt and decreased the Bcl-2/Bax protein ratio. These results suggest that genistein inhibited cell proliferation by inactivating the IGF-1R-PI3 K/Akt pathway and reducing mRNA and protein expression of Bcl-2/Bax [41,42].

4. Materials and Methods

The *MCF-7* human breast cancer cells, media, trypsin and DMSO were purchased from the ATCC (Manassas, VA, USA), respectively, for use in the present study. *BJ* human skin fibroblasts were a gift from the Department of Synthesis and Chemical Technology of Medicinal Products in Medical University of Lublin. All plasticware used for cell culture were purchased from Sarsted (Nümbrecht, Germany). Insulin, Genistein, Trypan Blue Solution and Annexin V-Cy3TM Apoptosis Detection Kit were purchased from Sigma-Aldrich Co. (St. Louis, MO, USA). The Antibiotic-Antimycotic was purchased from Thermo-Fisher (Waltham, MA, USA). The Fetal Bovine Serum (FBS) was purchased form Biomed. The Cell Proliferation Kit (MTT) was purchased from Roche Diagnostics GmbH (Hong Kong, China), respectively. Genistein was dissolved in DMSO (final concentration of 0.1% in medium).

The *MCF-7* human breast cancer cells and *BJ* human skin fibroblasts were maintained in The Department of Cancer Genetics of Medical University in Lublin. The *MCF-7* cells were cultured in EMEM (Eagle's medium), supplemented with 10% FBS, human recombinant insulin (0.01 mg/mL) and antibiotics (50 U/mL penicillin and 50 µg/mL streptomycin). The *BJ* cells were cultured in EMEM (Eagle's medium), supplemented with 10% FBS and antibiotics (10,000 U/mL of penicillin, 10,000 µg/mL of streptomycin, and 25 µg/mL of Gibco Amphotericin B). Both of cultures were maintained at 37 °C in a humidified atmosphere containing 5% CO_2.

Cytotoxicity was examined using MTT assay. Cells were plated at 2.5–5×10^5 cells/well in a 96-well tissue culture plate and incubated for 24 h following which they were exposed to genistein solutions at concentrations of 10, 20, 50, 80, 100, 150 and 200 µM and DMSO solutions at concentration of 10% and 15%. Following incubation for 24 and 48 h, the plated cells were incubated with MTT (final concentration 0.5 mg/mL) for 4 h at 37 °C. Then the Solubilization solution was added into each well (100 µL). The plates were incubated overnight at 37 °C, 5% CO_2, so that the complete dissolution of formazan was achieved. The absorbance of MTT formazan was determined at 570 nm using a microplate ELISA reader.

The apoptosis was detected using Annexin V-Cy3TM Apoptosis Detection Kit. Cells were plated at 0.5–1.0×10^6 cells/ml on Petri dishes and incubated for 24 h following witch they were exposed to genistein solutions at concentrations of 10, 20, 50, 80 and 100 µM and DMSO solution at concentration of 10%. Following incubation for 24 h, cells were washed three times using binding buffer solution. Then, double label staining solution (AnnCy3 and 6-CFDA) was added to Petri dishes covered in aluminum foil. Cells were incubated for 10 min at room temperature and washed five times with binding buffer to remove excess label from the cells. Following that, the binding buffer was added to the cells and the dishes were covered with a coverslip and observed under a florescence microscope, using the correct filter and light source, and then photographed.

To quantify the percentage of cells that underwent apoptosis, a trypan blue assay was additionally performed. Cells were plated at 0.5–1.0 × 10^6 cells/ml on Petri dishes and incubated for 24 h after which they were exposed to genistein solutions at concentrations of 10, 20, 50, 80 and 100 µM and DMSO solution at a of 10%. Following incubation for 24 h, the cells were detached from the plates with trypsin solution. An equal amount of trypan blue was added to the cell pellet and incubated for 2 min at room temperature. The suspension was then applied to a primary slide, covered with a coverslip and observed under the microscope. The number of dead stained cells per 100 total cells was then calculated.

Expression of *EGFR, BCL2, MKI7, BIRC5* and *AKT1* genes was measured using SYBR Green PCR Mastermix (Thermo Scientific), in duplicate in 96-well plates using the 7500 Real-Time PCR System from Applied Biosystems (Foster City, CA, USA). The results were then analyzed using the 7500 System software (Applied Biosystems). Each reaction was normalized to *GAPDH* reporter gene expression, and relative expression was calculated using the ΔΔCt method [43]. The sequences of the primers used are listed in Table 1.

Table 1. Sequences of primers.

Gene	Forward Sequence	Reverse Sequence	Product Size (bp)
BCl2	ATCGCCCTGTGGATGACTGAGT	GCCAGGAGAAATCAAACAGAGGC	140
MKI67	GAAAGAGTGGCAACCTGCCTTC	GCACCAAGTTTTACTACATCTGCC	151
EGFR	AACACCCTGGTCTGGAAGTACG	TCGTTGGACAGCCTTCAAGACC	106
AKT1	TGGACTACCTGCACTCGGAGAA	GTGCCGCAAAAGGTCTTCATGG	154
BIRC5	CCACTGAGAACGAGCCAGACTT	GTATTACAGGCGTAAGCCACCG	115

5. Conclusions

Our studies indicate that despite the high potential of genistein for use in the treatment of skin problems, wounds, and surgical scars in women during and after breast cancer treatment, it cannot be recognized as totally safe. Soy isoflavonoids are substances with very high biological activity, so their introduction into treatment requires further studies on their mechanisms of action at the molecular level considering genetic and immunological aspects. It is also necessary to conduct studies in in vivo models. This will allow for the elimination of possible side effects due to adverse effects on the body or therapy.

Author Contributions: Conceptualization, M.A.P.; Data curation, M.A.P.; Formal analysis, M.A.P.; Funding acquisition, A.A.F.; Investigation, M.A.P.; Methodology, A.A.F., S.Z. and S.P.-M.; Project administration, M.A.P.; Resources, S.Z., S.P.-M. and A.A.F.; Software, S.Z. and S.P.-M.; Supervision, A.A.F.; Visualisation, M.A.P.; Writing—original draft, M.A.P.; Writing—review & editing, A.A.F. All authors have read and agreed to the published version of the manuscript.

Funding: This research received no external funding.

Institutional Review Board Statement: Not applicable.

Informed Consent Statement: Not applicable.

Data Availability Statement: All data supporting and reported results available at e-mail address magda.pawlicka1@gmail.com.

Conflicts of Interest: The authors declare no conflict of interest.

References

1.   Makki, J. Diversity of Breast Carcinoma: Histological Subtypes and Clinical Relevance. *Clin. Med. Insights Pathol.* **2015**, *8*, 23–31. [CrossRef] [PubMed]

2.   World Health Organization (WHO). Global Health Estimates 2020: Deaths by Cause, Age, Sex, by Country and by Region, 2000–2019. 2020. Available online: https://www.who.int/data/gho/data/themes/mortality-and-global-health-estimates/ghe-leading-causes-of-death (accessed on 16 October 2021).

3.   Krzemieniecki, K.; Komorowski, K.; Wysocki. *Interna Szczeklika. Podręcznik Chorób Wewnętrznych*; Medycyna praktyczna: Cracow, Poland, 2013.

4.   Krzakowski, M.; Potemski, P.; Warzocha, K. *Onkologia kliniczna tom 2*; Via Medica: Gdansk, Poland, 2015.

5.   McDonald, E.; Clark, A.; Tchou, J.; Zhang, P.; Freedman, G. Clinical Diagnosis and Management of Breast Cancer. *J. Nucl. Med. Febr.* **2016**, *57*, 9S–16S. [CrossRef] [PubMed]

6.   Gamal, H.; Tawfik, W.; Fahmy, H.M.; El-Sayyad, H.H. Breakthroughs of using Photodynamic Therapy and Gold Nanoparticles in Cancer Treatment. In Proceedings of the 021 IEEE International Conference on Nanoelectronics, Nanophotonics, Nanomaterials, Nanobioscience & Nanotechnology (5NANO), Kottayam, Kerala, India, 29–30 April 2021; pp. 1–4. [CrossRef]

7.   Somogyi, R. Breast reconstruction. *Can. Fam. Physician* **2018**, *64*, 424–432.

8.   Gass, J.; Mitchell, S.; Hanna, M. How do breast cancer surgery scars impact survivorship? Findings from a nationwide survey in the United States. *BMC Cancer* **2019**, *19*, 342. [CrossRef] [PubMed]

9.   Jabłońska, B.; Brańka, J.; Lampe, P. Od Halsteda do leczenia oszczędzającego, czyli krótka historia chirurgii raka gruczołu piersiowego. *Postępy Nauk. Med.* **2011**, *1*, 29–32.

10. Křížová, L.; Dadáková, K.; Kašparovská, J.; Kašparovský, T. Isoflavoes. *Molecules* **2019**, *24*, 1076. [CrossRef] [PubMed]

11. Kim, Y.M.; Huh, J.S.; Lim, Y.; Cho, M. Soy Isoflavone Glycitin (4′-Hydroxy-6-Methoxyisoflavone-7-D-Glucoside) Promotes Human Dermal Fibroblast Cell Proliferation and Migration via TGF-β Signaling. *Phytother. Res.* **2015**, *29*, 757–769. [CrossRef] [PubMed]

12. Czerpak, R.; Pietryczuk, A.; Jabłońska-Trypuć, A.; Obrębska, K. Aktywność biologiczna izoflawonoidów i ich znaczenie terapeutyczne i kosmetyczne. *Borgis. Postępy Fitoter.* **2009**, *2*, 113–121.

13. Irrera, N.; Pizzino, G.; D'Anna, R.; Vaccaro, M.; Arcoraci, V.; Squadrito, F.; Altavilla, D.; Bittol, A. Dietary Management of Skin Health: The Role of Genistein. *Nutrients* **2017**, *9*, 622. [CrossRef] [PubMed]

14. Chin, G.S.; Liu, W.; Steinbrech, D.; Hsu, M.; Levinson, H.; Longaker, M.T. Cellular signaling by tyrosine phosphorylation in keloid and normal human dermal fibroblasts. *Plast. Reconstr. Surg.* **2000**, *106*, 1532–1540. [CrossRef]

15. Cao, C.; Li, S.; Dai, X.; Chen, Y.; Feng, Z.; Zhao, Y.; Wu, J. Genistein inhibits proliferation and functions of hypertrophic scar fibroblasts. *Burns* **2009**, *35*, 89–97. [CrossRef] [PubMed]

16. Jurzak, M.; Adamczyk, K.; Antończak, P.; Garncarczyk, A.; Kuśmierz, D.; Latocha, M. Evaluation of genistein ability to modulate CTGF mRNA/protein expression, genes expression of TGFβ isoforms and expression of selected genes regulating cell cycle in keloid fibroblasts in vitro. *Acta Pol. Pharm.* **2014**, *71*, 972–986. [PubMed]

17. Jurzak, M.; Adamczyk, K. Influence of genistein on c-Jun, c-Fos and Fos-B of AP-1 subunits expression in skin keratinocytes, fibroblasts and keloid fibroblasts cultured in vitro. *Acta Pol. Pharm.* **2013**, *70*, 205–213. [PubMed]

18. Sienkiewicz, P.; Surazyński, A.; Pałka, J.; Miltyk, W. Nutritional concentration of genistein protects human dermal fibroblasts from oxidative stress-induced collagen biosynthesis inhibition through IGF-I receptor-mediated signaling. *Acta Pol. Pharm.* **2008**, *65*, 203–211. [PubMed]

19. Park, E.; Lee, S.M.; Jung, I.K.; Lim, Y.; Kim, J.H. Effects of genistein on early-stage cutaneous wound healing. *Biochem. Biophys. Res. Commun.* **2011**, *410*, 514–519. [CrossRef] [PubMed]

20. Emmerson, E.; Campbell, L.; Ashcroft, G.S.; Hardman, M.J. The phytoestrogen genistein promotes wound healing by multiple independent mechanisms. *Mol. Cell. Endocrinol.* **2010**, *321*, 184–193. [CrossRef] [PubMed]

21. Marini, H.; Polito, F.; Altavilla, D.; Irrera, N.; Minutoli, L.; Calò, M.; Adamo, E.B.; Vaccaro, M.; Squadrito, F.; Bitto, A. Genistein aglycone improves skin repair in an incisional model of wound healing: A comparison with raloxifene and oestradiol in ovariectomized rats. *Br. J. Pharmacol.* **2010**, *160*, 1185–1194. [CrossRef] [PubMed]

22. Polito, F.; Marini, H.; Bitto, A.; Irrera, N.; Vaccaro, M.; Adamo, E.B.; Micali, A.; Squadrito, F.; Minutoli, L.; Altavilla, D. Genistein aglycone, a soy-derived isoflavone, improves skin changes induced by ovariectomy in rats. *Br. J. Pharmacol.* **2012**, *165*, 994–1005. [CrossRef]

23. Kloska, A.; Jakóbkiewicz-Banecka, J.; Narajczyk, M.; Banecka-Majkutewicz, Z.; Węgrzyn, G. Effects of flavonoids on glycosamino-glycan synthesis: Implications for substrate reduction therapy in Sanfilippo disease and other mucopolysaccharidoses. *Metab. Brain Dis.* **2011**, *26*, 1–8. [CrossRef]

24. Isoherranen, K.; Punnonen, K.; Jansen, C.; Uotila, P. Ultraviolet irradiation induces cyclooxygenase-2 expression in keratinocytes. *Br. J. Dermatol.* **1999**, *140*, 1017–1022. [CrossRef]

25. Iovine, B.; Iannella, M.L.; Gasparri, F.; Monfrecola, G.; Bevilacqua, M.A. Synergic Effect of Genistein and Daidzein on UVB-Induced DNA Damage: An Effective Photoprotective Combination. *J. Biomed. Biotechnol.* **2011**, *2011*, 692846. [CrossRef]

26. Wang, Y.N.; Wu, W.; Chen, H.C.; Fang, H. Genistein protects against UVB-induced senescence-like characteristics in human dermal fibroblast by p66Shc down-regulation. *J. Dermatol. Sci.* **2010**, *58*, 19–27. [CrossRef]

27. Kessel, B. Alternatives to estrogen for menopausal women. *Soc. Exp. Biol. Med.* **1998**, *217*, 38–44. [CrossRef] [PubMed]

28. Izumi, T.; Saito, M.; Obata, A.; Arii, M.; Yamaguchi, H.; Matsuyama, A. Oral intake of soy isoflavone aglycone improves the aged skin of adult women. *J. Nutr. Sci. Vitaminol. Tokyo* **2007**, *53*, 57–62. [CrossRef] [PubMed]

29. Silva, L.A.; Ferraz Carbonel, A.A.; de Moraes, A.R.B.; Simões, R.S.; Sasso, G.R.D.S.; Goes, L.; Nunes, W.; Simões, M.J.; Patriarca, M.T. Collagen concentration on the facial skin of postmenopausal women after topical treatment with estradiol and genistein: A randomized double-blind controlled trial. *Gynecol. Endocrinol.* **2017**, *33*, 845–848. [CrossRef]

30. Patriarca, M.T.; Barbosa de Moraes, A.R.; Nader, H.B.; Petri, V.; Martins, J.R.; Gomes, R.C.; Soares, J.M., Jr. Hyaluronic acid concentration in postmenopausal facial skin after topical estradiol and genistein treatment: A double-blind, randomized clinical trial of efficacy. *Menopause* **2013**, *20*, 336–341. [CrossRef] [PubMed]

31. Moraes, A.B.; Haidar, M.A.; Soares Júnior, J.M.; Simões, M.J.; Baracat, E.C.; Patriarca, M.T. The effects of topical isoflavones on postmenopausal skin: Double-blind and randomized clinical trial of efficacy. *Eur. J. Obstet. Gynecol. Reprod. Biol.* **2009**, *146*, 188–192. [CrossRef] [PubMed]

32. Wei, H.; Saladi, R.; Lu, Y.; Wang, Y.; Palep, S.R.; Moore, J.; Phelps, R.; Shyong, E.; Lebwohl, M.G. Isoflavone genistein: Photoprotection and clinical implications in dermatology. *J. Nutr.* **2003**, *133* (Suppl. 1), 3811S–3819S. [CrossRef]

33. Sharifi-Rad, J.; Quispe, C.; Imran, M.; Rauf, A.; Nadeem, M.; Gondal, T.A.; Ahmad, B.; Atif, M.; Mubarak, M.S.; Sytar, O. Genistein: An Integrative Overview of Its Mode of Action, Pharmacological Properties, and Health Benefits. *Oxidative Med. Cell. Longev.* **2021**, *2021*, 3268136. [CrossRef]

34. Kabała-Dzik, A.; Rzepecka-Stojko, A.; Kubina, R.; Iriti, M.; Wojtyczka, R.D.; Buszman, E.; Stojko, J. Flavonoids, bioactive components of propolis, exhibit cytotoxic activity and induce cell cycle arrest and apoptosis in human breast cancer cells MDA-MB-231 and *MCF-7*—A comparative study. *Cell. Mol. Biol.* **2018**, *64*, 1. [CrossRef]

35. Choi, E.J.; Jung, J.Y.; Kim, G.H. Genistein inhibits the proliferation and differentiation of *MCF-7* and 3T3-L1 cells via the regulation of ERα expression and induction of apoptosis. *Exp. Ther. Med.* **2014**, *8*, 454–458. [CrossRef] [PubMed]

36. Tsuboy, M.S.; Marcarini, J.C.; de Souza, A.O.; de Paula, N.A.; Dorta, D.J.; Mantovani, M.S.; Ribeirol, L.R. Genistein at Maximal Physiologic Serum Levels Induces G0/G1 Arrest in *MCF-7* and HB4a Cells, But Not Apoptosis. *J. Med. Food.* **2014**, *17*, 218–225. [CrossRef] [PubMed]

37. Uifălean, A.; Schneider, S.; Gierok, P.; Ionescu, C.; Iuga, C.A.; Lalk, M. The Impact of Soy Isoflavones on *MCF-7* and MDA-MB-231 Breast Cancer Cells Using a Global Metabolomic Approach. *Int. J. Mol. Sci.* **2016**, *17*, 1443. [CrossRef] [PubMed]

38. Lavigne, J.A.; Takahashi, Y.; Chandramouli, G.V.R.; Liu, H.; Perkins, S.N.; Hursting, S.D.; Wang, T.T.Y. Concentration-dependent effects of genistein on global gene expression in *MCF-7* breast cancer cells: An oligo microarray study. *Breast. Cancer Res. Treat.* **2008**, *110*, 85–98. [CrossRef] [PubMed]

39. Chinni, S.R.; Alhasan, S.A.; Multani, A.S.; Pathak, S.; Sarkar, F.H. Pleotropic effects of genistein on *MCF-7* breast cancer cells. *Int. J. Mol. Med.* **2003**, *12*, 29–34. [CrossRef]

40. Prietsch, R.F.; Monte, L.G.; da Silva, F.A.; Beira, F.T.; del Pino, F.A.B.; Campos, V.F.; Collares, T.; Pinto, L.S.; Spanevello, R.M.; Gamaro, G.D.; et al. Genistein induces apoptosis and autophagy in human breast *MCF-7* cells by modulating the expression of proapoptotic factors and oxidative stress enzymes. *Mol. Cell Biochem.* **2014**, *390*, 235–242. [CrossRef]

41. Chen, J.; Lin, C.; Yong, W.; Ye, Y.; Huang, Z. Calycosin and genistein induce apoptosis by inactivation of HOTAIR/p-Akt signaling pathway in human breast cancer *MCF-7* cells. *Cell Physiol. Biochem.* **2015**, *35*, 722–728. [CrossRef]

42. Chen, J.; Duan, Y.; Zhang, X.; Ye, Y.; Ge, B.; Chen, J. Genistein induces apoptosis by the inactivation of the IGF-1R/p-Akt signaling pathway in *MCF-7* human breast cancer cells. *Food Funct.* **2015**, *6*, 995–1000. [CrossRef]

43. Livak, K.J.; Schmittgen, T.D. Analysis of relative gene expression data using real-time quantitative PCR and the 2(-Delta Delta C(T)) Method. *Methods* **2001**, *25*, 402–408. [CrossRef]

International Journal of
Molecular Sciences

Article

Thiogenistein—Antioxidant Chemistry, Antitumor Activity, and Structure Elucidation of New Oxidation Products

Elżbieta U. Stolarczyk [1], Weronika Strzempek [2,3], Marta Łaszcz [1,4], Andrzej Leś [5], Elżbieta Menaszek [3] and Krzysztof Stolarczyk [5,*]

[1] Analytical Department, Łukasiewicz Research Network–Industrial Chemistry Institute, 8 Rydygiera Street, 01-793 Warsaw, Poland; elzbieta.stolarczyk@ichp.pl (E.U.S.); m.laszcz@nil.gov.pl (M.Ł.)
[2] Faculty of Chemistry, Jagiellonian University, 2 Gronostajowa Str., 30-387 Krakow, Poland; weronika.strzempek@doctoral.uj.edu.pl
[3] Faculty of Pharmacy, Collegium Medicum, Jagiellonian University, 9 Medyczna Str., 30-068 Krakow, Poland; elzbieta.menaszek@uj.edu.pl
[4] Department of Falsified Medicines and Medical Devices, National Medicines Institute, Chełmska 30/34, 00-725 Warsaw, Poland
[5] Faculty of Chemistry, University of Warsaw, 1 Pasteura Street, 02-093 Warsaw, Poland; ales@chem.uw.edu.pl
* Correspondence: kstolar@chem.uw.edu.pl; Tel.: +48-22-55-26-351

Abstract: Isoflavonoids such as genistein (GE) are well known antioxidants. The predictive biological activity of structurally new compounds such as thiogenistein (TGE)–a new analogue of GE–becomes an interesting way to design new drug candidates with promising properties. Two oxidation strategies were used to characterize TGE oxidation products: the first in solution and the second on the 2D surface of the Au electrode as a self-assembling TGE monolayer. The structure elucidation of products generated by different oxidation strategies was performed. The electrospray ionization mass spectrometry (ESI-MS) was used for identifying the product of electrochemical and hydrogen peroxide oxidation in the solution. Fourier transform infrared spectroscopy (FT-IR) with the ATR mode was used to identify a product after hydrogen peroxide treatment of TGE on the 2D surface. The density functional theory was used to support the experimental results for the estimation of antioxidant activity of TGE as well as for the molecular modeling of oxidation products. The biological studies were performed simultaneously to assess the suitability of TGE for antioxidant and antitumor properties. It was found that TGE was characterized by a high cytotoxic activity toward human breast cancer cells. The research was also carried out on mice macrophages, disclosing that TGE neutralized the production of the LPS-induced reactive oxygen species (ROS) and exhibits ABTS (2,2′-azino-bis-3-(ethylbenzothiazoline-6-sulphonic acid) radical scavenging ability. In the presented study, we identified the main oxidation products of TGE generated under different environmental conditions. The electroactive centers of TGE were identified and its oxidation mechanisms were proposed. TGE redox properties can be related to its various pharmacological activities. Our new thiolated analogue of genistein neutralizes the LPS-induced ROS production better than GE. Additionally, TGE shows a high cytotoxic activity against human breast cancer cells. The viability of MCF-7 (estrogen-positive cells) drops two times after a 72-h incubation with 12.5 µM TGE (viability 53.86%) compared to genistein (viability 94.46%).

Keywords: antioxidant; antitumor; biologically active compounds; electrochemistry; electrospray ionization; structure elucidation; identification; molecular modeling; oxidation mechanisms; spectroscopic data

Citation: Stolarczyk, E.U.; Strzempek, W.; Łaszcz, M.; Leś, A.; Menaszek, E.; Stolarczyk, K. Thiogenistein—Antioxidant Chemistry, Antitumor Activity, and Structure Elucidation of New Oxidation Products. *Int. J. Mol. Sci.* **2022**, *23*, 7816. https://doi.org/10.3390/ijms23147816

Academic Editor: Jae Youl Cho

Received: 23 June 2022
Accepted: 11 July 2022
Published: 15 July 2022

Publisher's Note: MDPI stays neutral with regard to jurisdictional claims in published maps and institutional affiliations.

1. Introduction

Reactive oxygen species (ROS) and reactive nitrogen species (RNS) are the main sources of oxidative stress in biological systems. ROS and RNS can react with proteins, lipids, and nucleic acids, giving rise to damage at various sites within the cell [1]. The cell contains various enzymes and antioxidants (AOXs) to provide protection and avoid damage. AOXs are compounds that significantly delay or inhibit oxidation of the oxidisable substrate at relatively low concentrations [2]. Flavonoids and isoflavonoids such as genistein (GE) are well-known antioxidants and can help protect cells against reducing carcinogenesis. Flavonoids exhibit a broad spectrum of biological activity, including antioxidant, antitumor, antibacterial, antiviral, anti-inflammatory, anti-allergenic, and vasodilatory actions [3–8]. Additionally, they play an important role in the prevention and treatment of hormone-dependent diseases. Genistein is thought to act as an anticancer agent in a large part through its ability to scavenge oxidants involved in carcinogenesis. However, the clinical and therapeutic use of genistein still suffers from many problems related mainly to low lipid and water solubility [6] and poor bioavailability [7]. Therefore, a lot of studies have focused on obtaining genistein derivatives (mainly modified by glycosylation, alkylation, esterification, and hydroxylation) and their drug delivery systems that achieved the required pharmacological activity but also showed fewer side effects and therapeutic limitations. Thiolated genistein (TGE), described by Sidoryk and co-workers [8], is composed of the genistein residue bound at the 7-OH site to the hydroxyethyl linker and thioglycolic acid residue. Such a new construction of genistein has several interesting properties. The predictive biological activity of structurally new compounds such as thiogenistein–a new analogue of genistein–becomes an interesting way of designing new drug candidates with promising properties.

The antioxidant activity of flavonoids attracted attention because they cannot only scavenge free radicals, but also reduce free radical formation [9]. The antioxidant activity of flavonoids is primarily exerted by phenolic hydroxyl groups. Interestingly, in certain flavonoids, bond dissociation of the C−H bonds located at C3 of the C-ring competes with the O-H bond, suggesting that dissociation of the C-H bond can be involved in antioxidant activity [9]. The chemical structure plays a fundamental role in the antioxidant activity of substances. It is known that the existence of a certain hydroxylation pattern, particularly, in the B-ring of the flavonoid structure and/or a C2=C3 double bond in conjugation with a C4-carbonyl group, O3−H groups, and methoxyl groups, increases antioxidant activities [10]. The computed here molecular structure of TGE exhibits several properties supposed to shed more light on the molecular basis of this new compound's antioxidant activity. They possess the ability for effective radical scavenging, an important (substituted) phenolic ring (B), it has a C-2,3 double bond in conjugation with a 4-oxo function in the C-ring. A rather minor influence of the C-ring on electron delocalization from the B-ring is expected. In the TGE molecule, the C and B rings are twisted relative to each other (about 40 degrees). Moreover, the length of the C3-C1' bond between them is closer to be single (0.149 nm vs. 0.151 nm for single C-C) than double (0.140 nm, as in the B ring). Some of these properties can be recognized in flavonoids (e.g., apigenin) and isoflavonoids (e.g., genistein and daidzein). It is very challenging to correlate the molecular characteristics to antioxidant activity, e.g., [11,12]. Previously reported studies on polyphenolic flavonoid structures have shown that their radical scavenging activity is related to the presence of phenolic hydroxyl groups, through their H-donating properties acting via the hydrogen atom transfer (HAT) or the single electron transfer followed by proton transfer (SETPT) mechanisms [13,14]. It is known that compounds that possess strong radical scavenging abilities are oxidized at lower potentials [15,16]. However, the flaw in this way of thinking is the possible autoxidation of strong radical-scavenging compounds, so their pro-oxidant properties. A balance is needed between antioxidant and pro-oxidant activities. Specific products are formed during these antioxidant reactions. The knowledge of the oxidation mechanisms is crucial to ascertain the influence of the redox behavior on the pharmacological, nutritional, and chemical properties of our new thiolated analogue of genistein and for predicting the mechanism of metabolism.

The aim of the current study was to apply different strategies for the generation of oxidation products of TGE in solution and on 2D surfaces and the subsequent identification of the unknown products. For the 2D oxidation study, TGE was chemically attached to the gold surfaces as the self-assembled monolayers (SAMs). TGE in SAMs has limited conformational space, it is more susceptible for intermolecular interactions, as well as for the oxidation agent attacking various TGE molecular fragments. A multitude of experimental approaches is needed for the characterization of SAMs after oxidation. Attenuated total reflectance spectroscopy (ATR) is a variant of infrared spectroscopy (IR) and is frequently employed to investigate SAMs [17,18]. The multi-tool analytical approach was based on electrochemistry (EC) and mass spectrometry with electrospray ionization (ESI-MS) for the TGE solution oxidation study. The EC was coupled directly to ESI-MS. TGE oxidation products obtained by using electrochemical potentials and reactions with hydrogen peroxide were identified. The use of electrochemical potentials and H_2O_2 will model the action of oxidative agents on TGE in various environments. This helps to predict a possible pattern of TGE metabolism. Both physicochemical methods are precisely controlled and have great potential as fast alternatives to in vitro assays. The electroactive centers of TGE are identified and their oxidation mechanisms are determined. It is worth mentioning here that electrochemical techniques have also been applied for showing similarities between electrochemical and biochemical reactions with the use of the structure–activity relationship (SAR) for flavonoids [16,19] possessing redox activity. The molecular modeling and the quantum mechanical density functional calculations are also performed on model systems to support a discussion.

The anticancer and antioxidative activity of the new analogue is evaluated based on cell culture, the radical scavenging activity (ABTS assay), and the generated ROS level (DCFH-DA assay). The research is carried out on one human breast cancer (MCF-7 estrogen positive and MDA-MB-231 estrogen negative cells line) and also on LPS-induced mice macrophages (RAW 264.7 line).

To summarize, this work becomes a continuation of our former study on physicochemical, biological, and theoretical properties of a new compound, TGE [20]. Here, we estimated the oxidation with electrochemical and chemical (H_2O_2) methods by identifying the main oxidation products with MS/MS spectrometry and IR spectroscopy, antitumor and antioxidant properties by biological methods, and by theoretical modeling predicting antioxidant activity. It was tested whether the modification of the structure, by introducing an HS-CH_2-COO-CH_2-CH_2- group, would produce an analogue with a higher antitumor and antioxidant potential.

The study of the effect of TGE chemistry on its biological activity is an important aspect. We hope that understanding these effects will help to determine key molecular fragments involved in the subsequent biochemical processes in living organisms and extend information about their role in metabolic processes and pharmacological activity [21].

2. Results and Discussion

Genistein is one of the isoflavones with extensively studied antioxidant and antitumor potentials. It modulates various steps of the cell cycle, angiogenesis, apoptosis, and metastasis in different types of cancers [22,23]. Therefore, the number of studies on the synthesis and biological evaluation of new genistein derivatives is increasing every year. In our research, we focused on a new thioderivative of genistein and its biological and antioxidant properties.

2.1. ABTS Radical Scavenging Assay

Radical scavenging activity (RSA%) measured by ABTS correlates with the electron or hydrogen donating ability of antioxidants, and it is the mainly used method to evaluate the antioxidant activity of studied compounds. The inhibitory effects of TGE and GE on ABTS radicals are shown in Figure 1.

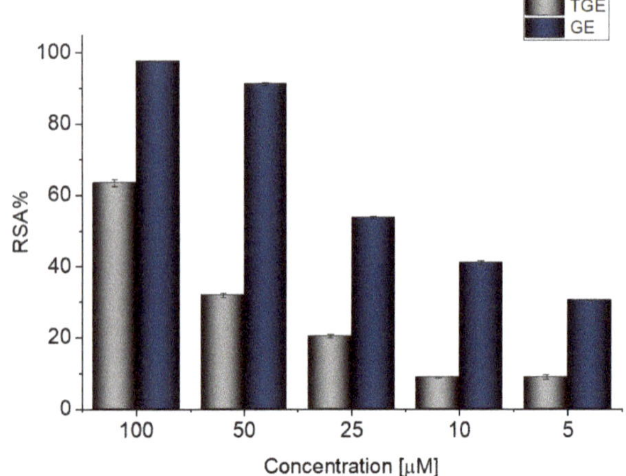

Compound	TGE	GE
IC$_{50}$ (µM)	76.93 ± 1.04	19.93 ± 0.12

Figure 1. Determination of the antioxidant activity of GE and TGE by ABTS radical scavenging assay. The data shown were obtained after 30 min of incubation. The value of IC$_{50}$ (half-maximal inhibitory concentration) in the figure inset is represented as mean ± SD (n = 3).

The previously described studies of the flavonoids polyphenolic structures have shown that hydroxyl groups and their structural arrangements induce antioxidant properties through the H-donating transfer. After 30 min of incubation, both TGE and GE possessed inhibitory effects on the ABTS radical in a dose-dependent manner. However, the test performed for TGE (4′,5-hydroxyisoflavone) versus GE (4′,5,7-dihydroxyisoflavone) confirmed that conversion of the 7-hydroxy structure in the A ring of GE to the thiolated residue in the A ring of TGE reduced the value of RSA% TGE to 64.97%, compared to 97.35% GE. The IC$_{50}$ values of TGE and GE were 76.93 (±1.04) and 19.93 (±0.12) µM, respectively.

2.2. H$_2$DCF-DA ROS Detection Assay

The inhibition activity of LPS-induced ROS production by TGE at different concentrations in macrophage cells was determined by using the DCFH-DA assay. The LPS-treated (10 µg/mL) macrophages showed an increase in reactive oxygen species production of approximately 60% more ROS compared to non-stimulated cells (Figure 2). The addition of TGE and GE solutions in the tested range significantly reduced the LPS-induced ROS production in RAW 264.7 cells. In the case of cells treated with the two highest concentrations of TGE, 200 and 100 µM observed an approximate 58.29% (±0.40%) and 54.82% (±0.93%) reduction, respectively. The level of ROS was similar to those generated by non-stimulated macrophages. During the concentration of 6.25 µM, the effect for TGE and GE is similar. Compared to the LPS-induced cells, the ROS level is decreased by 34.2% (±2.58%) and 36.94% (±1.74%) for TGE and GE, respectively.

The presence of the -SH group in the new substituent may be responsible for the correct thiol–disulfide balance and the associated oxidation–reduction potential of cells. A similar mechanism is observed for glutathione, which is a natural antioxidant [20,24].

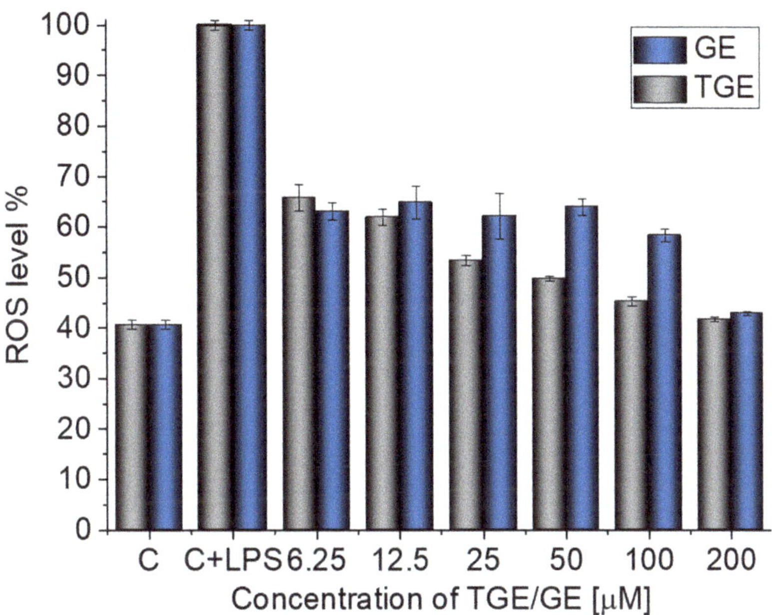

Figure 2. Effective inhibition of LPS-induced ROS production by TGE and GE solutions in macrophage cells after 24 h of incubation (C+LPS means cells stimulated by LPS and C means non-stimulated cells/control group).

2.3. In Vitro Study–Cell Viability, Cytotoxicity

Studies on synthetic genistein derivatives, including combinations with glycosides, for example, prove the effectiveness of the anti-tumor activity in in vitro systems. The potency of the anti-tumor activity of genistein glycosides varies depending on the sugar groups attached. For example, the addition of acetylated sugar hydroxyls to genistein resulted in a greater selectivity for cancer cells. Additionally, it was confirmed that the potency of the antitumor activity of genistein and its derivatives varies in different tumor types, depending on their selectivity for the target molecules. Our previous studies have shown that the substitution of the -OH group at position seven of the A ring by an ethyl linker and a thioglycolic acid residue allows us to obtain a derivative showing a higher cytotoxic activity against human prostate cancer DU145 cells, but also a lower toxicity against normal prostate epithelial cells (PNT2) compared to its precursor–genistein [20]. Based on previous promising results, our research has been extended to include two breast cancer lines: the estrogen-positive MCF-7 and the estrogen-negative MDA-MB-231.

Antitumor Activity–Breast Cancer

Breast cancer belongs to the group of heterogeneous diseases with many clinical, molecular, and histopathological forms, which makes obtaining effective chemotherapy problematic. Seventy percent of them are associated with the expression of the estrogen-α (ERα) receptor. Therefore, as a model to determine the anti-tumor activity of TGE, two breast cancer lines, MCF-7, an estrogen-positive breast cancer, and MDA-MB-231, an estrogen-negative breast cancer, are used.

For both lines, similar relationships were obtained, as in the case of prostate cancer [20]. After 72 h (Figure 3c), the viability drops to 8.76% (±0.20%) for TGE. For the sake of comparison, the administration of genistein reduced this parameter only to 31.42% (±0.73%). The use of the lower concentration of 12.5 μM TGE is sufficient to reduce the viability of the cells to 53.86% (±2.79%). When the same concentration of GE is administered, there is no effect visible in the reduction of viability. This indicates that TGE is more effective in inhibiting the growth of these ERα-positive breast cancer cells at much lower concentrations than genistein.

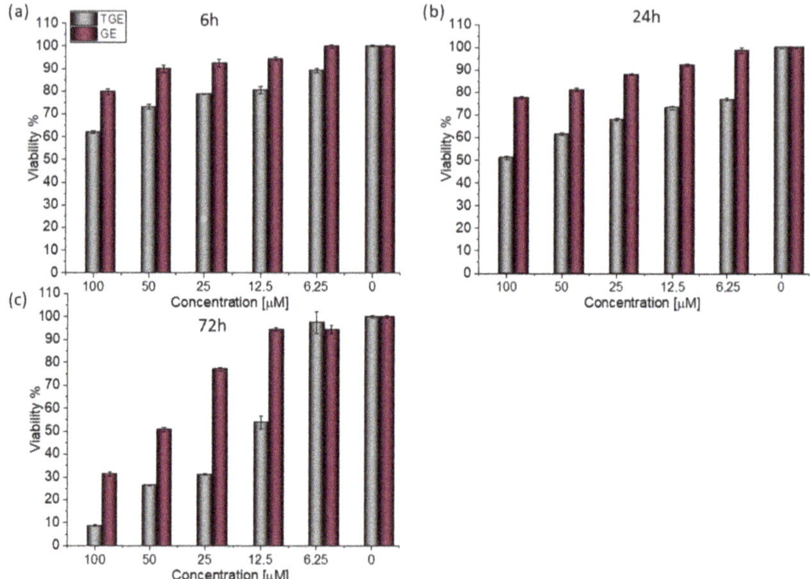

Figure 3. The viability of MCF-7 cells treated by different concentrations of GE and TGE after (**a**) 6 h, (**b**) 24 h, and (**c**) 72 h (determined by PrestoBlue™ test). Presented data are representative of two independent experiments and are expressed as the mean ± SD. The error bars represent the ±SD.

Based on the value of the relative fluorescence intensity (RFU) recorded after 6, 24, and 72 h of incubation of MCF-7 cells with 100 μM TGE, not only an arrest of proliferation was observed, but also a reduction in the number of living cells, which may indicate their death (Figure 4). This effect was also observed for the highest (100 μM) concentration of TGE, while it was not visible for pure GE. Note that genistein is one of the major soy isoflavones and exhibits a relative nontoxicity. However, the structural similarity of genistein with 17β-estradiol, means that GE can bind to Erα+ receptors and induce both agonistic and antagonistic effects on MCF-7 cells proliferation. Thus, at a lower dose, GE may stimulate ERα+ cell growth and entry into the cell cycle [25,26]. In the present study, following treatment for three points of time, TGE significantly inhibited the proliferation of MCF-7 cells in a concentration-dependent manner, may offset effects related to a biphasic effect, and could be beneficial for breast cancer even in lower concentrations.

In the case of the MDA-MB-231 breast cancer line (estrogen-negative neoplasm), antitumor activity is also observed after the administration of TGE (Figure 5).

After 72 h of incubation with 100 μM solutions, cell viability decreased to 9.99% (±0.05%). For comparison, the viability of cells incubated with the two highest concentrations of GE was 19.12% (±0.21%) and 67.68% (±0.26%) (Figure 6). Based on the available literature, this effect may be associated with the induction of MDA-MB-231 cells apoptosis by genistein and its derivative and inhibition of proliferation by arresting the cell cycle in the phase G2/M [27]. Additionally, the research presented by H. Pan and coworkers reveals that GE could inhibit the activity of NF-κB via the Notch-1, which may be an explanation mechanism of the downregulation of proliferation. Moreover, it was confirmed that GE downregulated the expression of cyclin B1, Bcl-2, and Bcl-xL in MDA-MB-231 cells, possibly mediated by NF-κB activation via the Notch-1 signaling pathway.

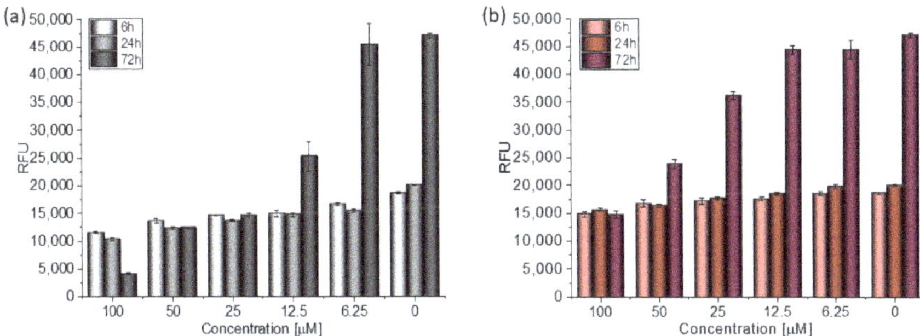

Figure 4. The proliferation rate of MCF-7 cells correlated with relative fluorescence units (RFU) obtained from all cells in the samples with PrestoBlue™ test after 6 h, 24 h, and 72 h of incubation with (**a**) TGE and (**b**) GE. Untreated cells were used as references. The obtained results are proportional to the number of cells. The data are representative of two independent experiments and are expressed as the mean ± SD. The error bars represent the ±SD.

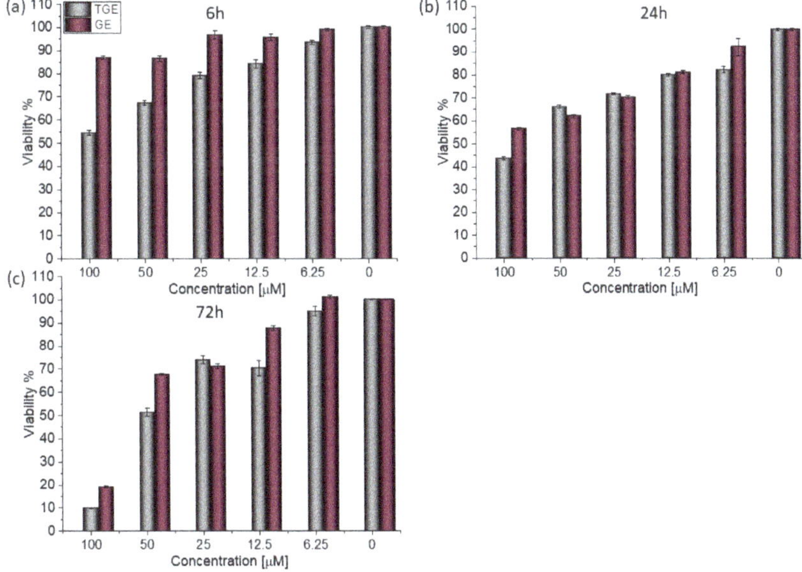

Figure 5. The viability of MDA-MB-231 cells treated by different concentrations of GE and TGE after (**a**) 6 h, (**b**) 24 h, and (**c**) 72 h (determined by PrestoBlue™ test). Presented data are representative of two independent experiments and are expressed as the mean ± SD. The error bars represent the ±SD.

In vitro studies would suggest that changing the hydroxyl for an ether containing an –SH group on the C7 carbon in the A ring may increase the cytotoxic properties of TGE. This effect can be attributed to the presence of a highly reactive -SH group [20,28]. Marik and coworkers also proved that the substitution of the -OH group at position seven of the A ring by aliphatic chains and heterocyclic 1,2,3-triazole moieties enhances the cytotoxic properties. Additionally, modifying the structure of genistein by incorporating structural features with bulky and flexible lipophilic substitutions on the genistein scaffold can increase their binding affinities for estrogen α-receptors. These results indicate that the 7-O-substitutes of genistein are an effective approach to obtaining compounds with an improved antiproliferative activity [29].

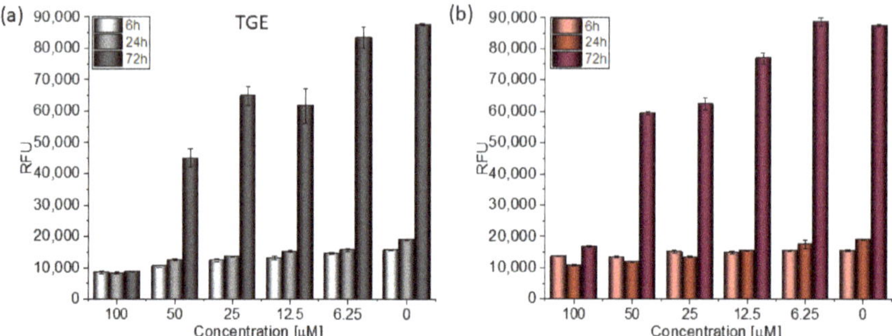

Figure 6. The proliferation rate of MDA-MB-231 cells correlated with relative fluorescence units (RFU) obtained from all cells in the samples with PrestoBlue™ test after 6 h, 24 h, and 72 h of incubation with (**a**) TGE and (**b**) GE. Untreated cells were used as references. The obtained results are proportional to the number of cells. The data are representative of two independent experiments and are expressed as the mean ± SD. The error bars represent the ±SD.

To sum up, both TGE and GE have the ability to act against neoplastic activity towards estrogen-negative and estrogen-positive cells, which makes them potential substances in the fight against breast cancer. The obtained results constitute the basis for extending the research to include the characteristics of the action of the new derivative both on individual phases of the cell cycle, as well as the induction of biochemical markers and targeted receptor activity.

2.4. Antioxidant Chemistry–Oxidation of TGE by Potential and With Hydrogen Peroxide

The ROXY™ electrochemical system is a technique that can be applied for mimicking drug metabolism [30,31] or for interaction studies [32] as well as other electrochemical cells coupled on-line to the electrospray ionisation (ESI)-MS for study of biological redox reactions [33]. In the work of Sagandykova and co-workers [34], the electrochemical unit coupled to ESI-MS was used for fragmentation activity relationships of selected flavonoids. Selected compounds were subjected to electrochemical degradation and the differences in formed products were analysed for correlation with their structures and antioxidant activity.

The TGE itself as a template was used for the interpretation of the unknown structures of products created during the oxidation of TGE. The product ion analysis was obtained and specific product ions and neutral losses were assigned to the substructures of the molecule. To the best of our knowledge, the MS fragmentation of TGE has not been discussed in the literature. The EPI spectrum of pseudomolecular ions displayed a lot of product ions, which are presented in Table 1. The structures of characteristic fragments, which are proposed, are presented in Figure S1. There are many strategies to identify unknown structures. The one used in this article is based on the assumption that much of the parent compound structure will be retained in the new products or decomposition products. The product ion mass spectrum of the parent compound and fragmentation pattern of the parent compound are used as the templates for the identification of the unknown structure. An orthogonal approach was used to identify new compounds. On the one hand, identification is based on the fragmentation spectrum and on the other hand, consideration of the mechanisms (logical oxidation pathways) that generate the formation of a particular structure.

The formation of the TGE electrochemical oxidation product was monitored by ESI-MS in negative polarization and the results are summarised in Table 1. The formation of the TGE degradation product after treatment with hydrogen peroxide is summarised in Table 2. Very interesting results were obtained using two oxidizing agents simultaneously: hydrogen peroxide and potential, which are summarized in Table 3. TGE without being subjected to oxidation was analysed as a controls sample and is subject only to possible transformation in ESI source, which is an electrochemical cell by itself but with constant potential [35]; in this case, 4.5 kV. This data may be also indicative of electrochemical reactions and provide insights into structure–activity studies.

Table 1. Summary of proposed structures of main oxidation products of TGE formed after oxidation with potential.

Product Mass, Characteristic Ions *m/z*, (rel. int., % for EPI)	**No. Oxidation Product/Intensity**	**Proposed Metabolic Reaction-Suggested Reactions**
TGE $[M\text{-}H]^-$ = 387 Da; DP (−80), CE (−40) 0V, *m/z* (EPI−) 387: 369(1); 343(1); 327(1); 313(1); 299(1); 295(1); 286(1); 269(100); 268(13); 255(1); 241(1); 225(1); 224(1); 211(1); 201(1); 199(1); 196(1); 181(1); 157(1) 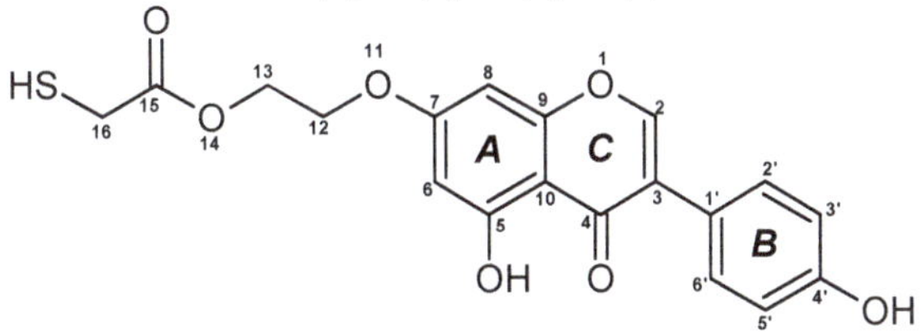 Electrode: BDD; pulse 5: pulse of 1 s/+2.5 V + pulse of 0.5 s/−0.3 V ***m/z*: EPI−**		
402: 384(4); 374(8); 368(3); 358(11); 342(3); 339(5); 329(9); 326(11); 317(7); 312(9); 303(2); 294(5); 284(20); 283(7); 268(22); 255(7); 241(4); 195(9); 177(10) M = 402 Da: 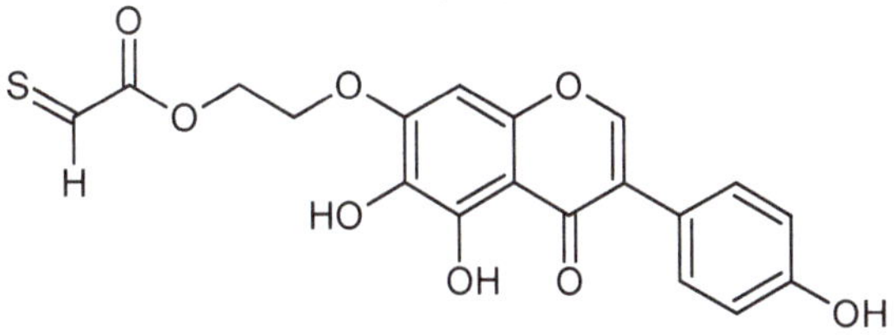 15-thioformyl-6-hydroxy-TGE (1)	1/Major	O gain (+16) + 2H loss (−2)
M = 404 Da: 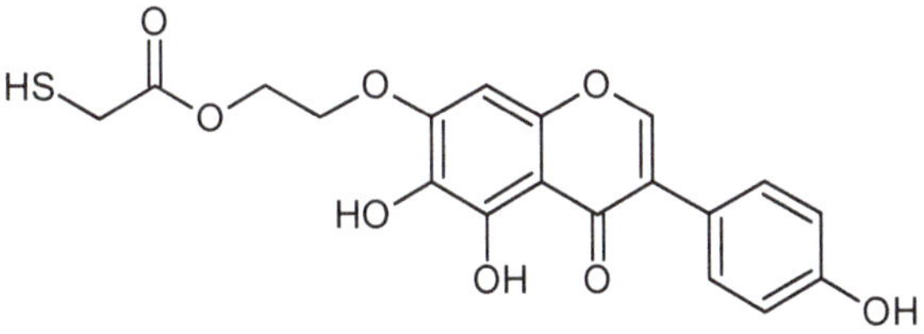 6-hydroxy-TGE (2)	2/Minor	O gain (+16)

Table 1. *Cont.*

Product Mass, Characteristic Ions *m/z*, (rel. int., % for EPI)	No. Oxidation Product/Intensity	Proposed Metabolic Reaction-Suggested Reactions
419: 401(1); 391(57); 385(29); 375(57); 371(14); 357(14); 351 (43); 347(1); 329(43); 327(29); 315(29); 301(29); 299(29); 295 (29); 285(28); 283(42); 271(14); 269(86); 257(14); 255(28); 227(14); 195(43); 177(14); 165(57); 151(14) M = 420 Da: 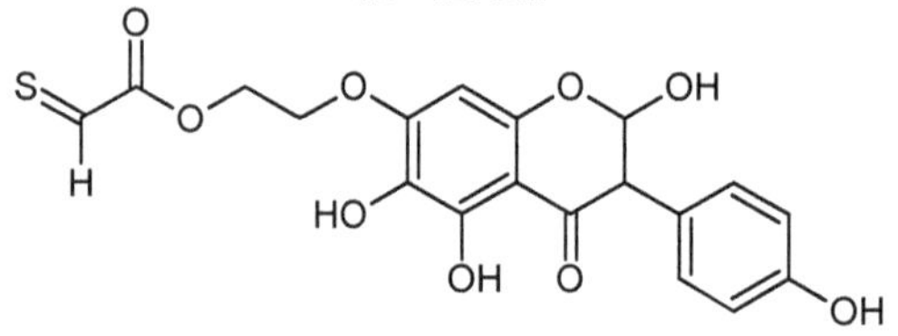 15-thioformyl-2,3-dihydro-2,6-dihydroxy-TGE (3)	3/Medium	O gain (+16) + H$_2$O gain (+18) + 2H loss (−2)
M = 422 Da: 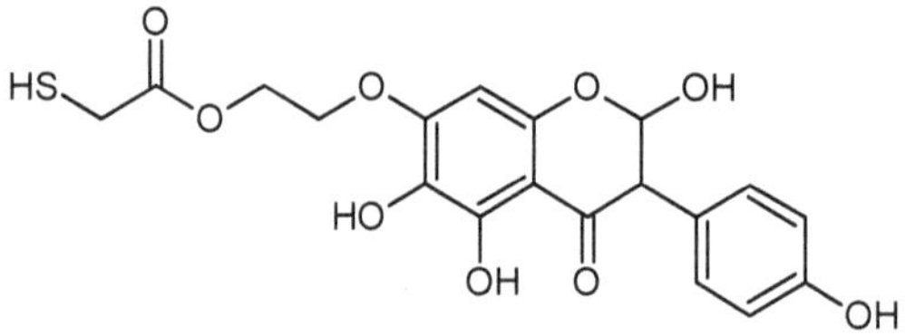 2,3-dihydro-2,6-dihydroxy-TGE (4)	4/Minor	O gain (+16) + H$_2$O gain (+18)
423: 405(6); 395(25); 359(1); 339(19); 331(6); 307(1); 283 (6); 269(6); 257(100); 255(1); 229(13); 212(6); 185(13); 176(12); 120(13) M = 424 Da: 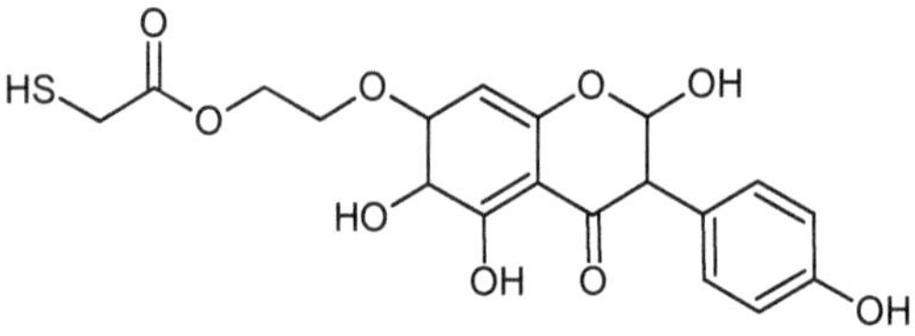 2,3,6,7-tetrahydro-2,6-dihydroxy-TGE (5)	5/Minor	2H$_2$O gain (+36)
433: 415(9); 405(13); 399(4); 387(9); 369(9); 315(9); 287(9); 269(100); 243(4); 213(9); 193(17); 177(4); 139(4) M = 434 Da: 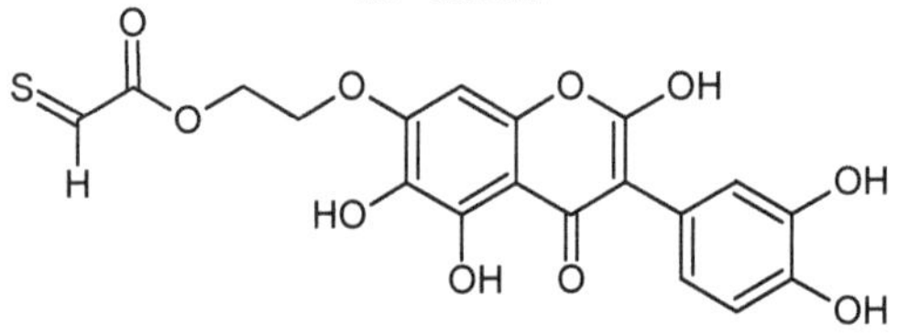 15-thioformyl-2,6,3′-trihydroxy-TGE (6) or 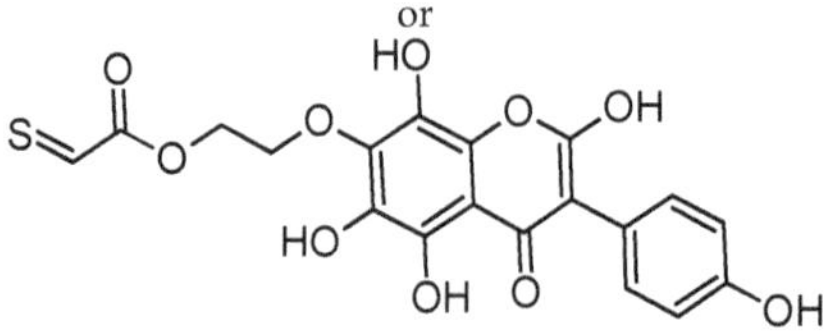	6/Medium	3O gain (+48) + 2H loss (−2)

Table 1. *Cont.*

Product Mass, Characteristic Ions *m/z*, (rel. int., % for EPI)	No. Oxidation Product/Intensity	Proposed Metabolic Reaction-Suggested Reactions
435: 417(7); 389(7); 372(1); 363(2); 345(5); 301(10); 286(6); 280(1); 271(17); 269(100); 256 (3); 255(4); 241(2); 227(3); 212(7); 194(7); 176(1); 150(48) M = 436 Da: 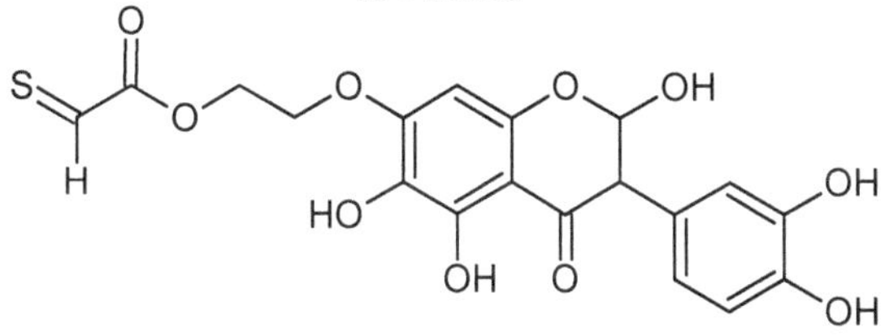 15-thioformyl-2,3-dihydro-2,6,3′-trihydroxy-TGE (7) or 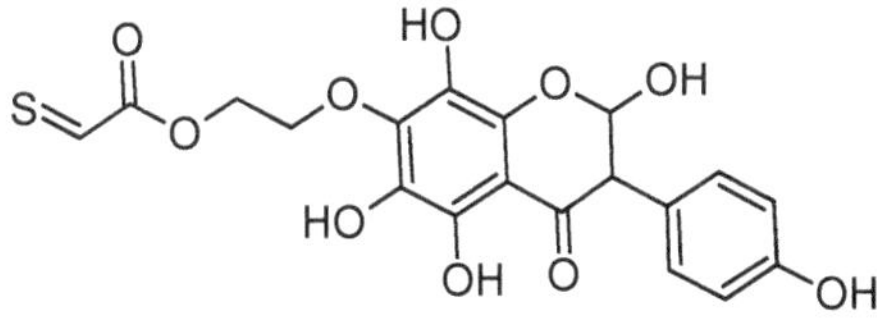 or 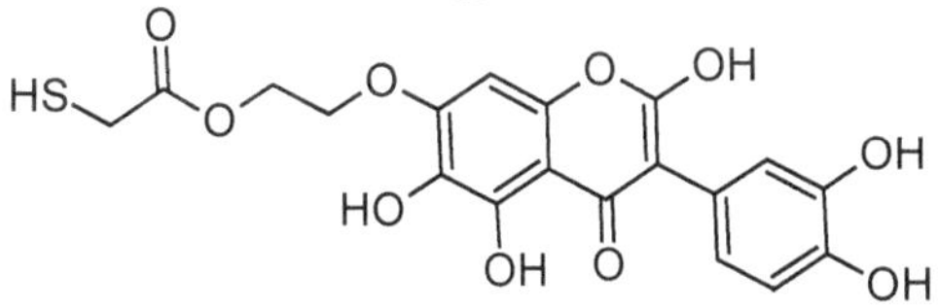 or 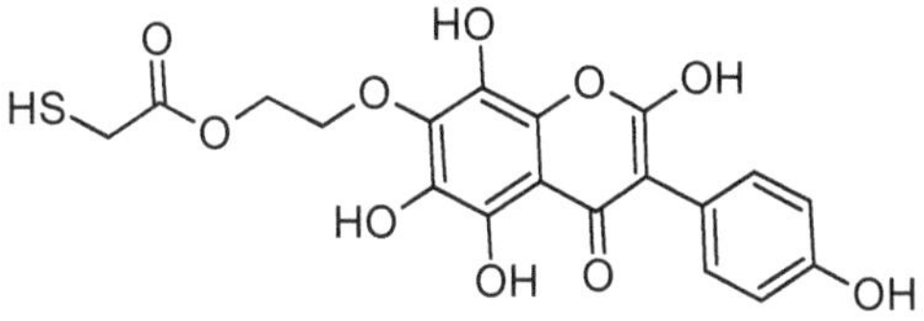	7/Medium	2H loss (−2) + 2O gain (+32) + H_2O gain (+18) or 3O gain (+48)

In the TGE mass spectrum recorded for the reference/control sample, the ion at *m/z* 773 is presented. This is a dimer of the TGE molecule formed by bridging as a disulfide. The compound of M = 773 Da was decomposed when two oxidizing agents were applied simultaneously: hydrogen peroxide and potential. In addition, ions at *m/z* 355 and 339 are present in the reference spectrum of TGE. They are probably the result of uncontrolled fragmentation in the ion source and do not appear in the TGE fragmentation spectrum. The ions at *m/z* 355 and 339 are formed by the loss of two oxygen molecules and three oxygen molecules, respectively. The observed ion at m/z 369 is formed by the loss of a water molecule and is a characteristic signal in the TGE fragmentation spectrum. Furthermore, these are ions whose intensity increases during potential oxidation. Additionally, TGE adducts/polymerization with oxidation products are observed, exemplified by *m/z* ions 836, 819, 811, 809, 795, 773, 749, and 673 with low intensities and very complicated mass spectra. Therefore, these spectra were not analyzed.

Table 2. Summary of proposed structures of main products of TGE formed after oxidation with 0.34 M H_2O_2, 0 V.

Product Mass, Characteristic Ions, *m/z*, (rel. int., % for EPI)	No. Oxidation Product/Intensity	Proposed Metabolic Reaction–Suggested Reactions
TGE [M-H]$^-$ = 387 Da *m/z*: EPI–		
435: 349(1); 286(1); 269(100); 241(1); 224(1); 213(1); 201(1); 181(1); 165(1); 157 (1); 121(2) M = 436 Da: 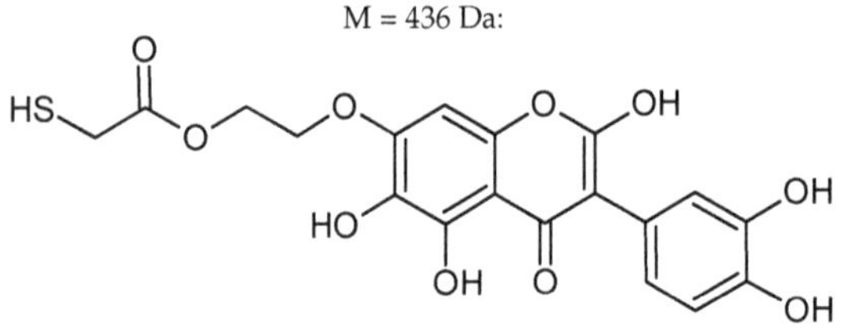 2,6,3′-trihydroxy-TGE (8)	8/Major	3O gain (+48)
M = 438 Da: 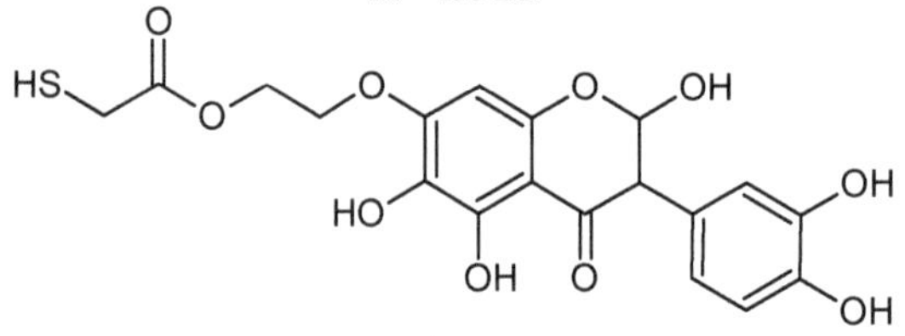 2,3-dihydro-2,6,3′-trihydroxy-TGE (9)	9/Minor	2O gain (+32) + H_2O gain (+18)
451: 423(2); 285(100); 283(2); 269(3); 257(5); 229(2); 217(3); 213(1); 121(2) M = 452 Da: 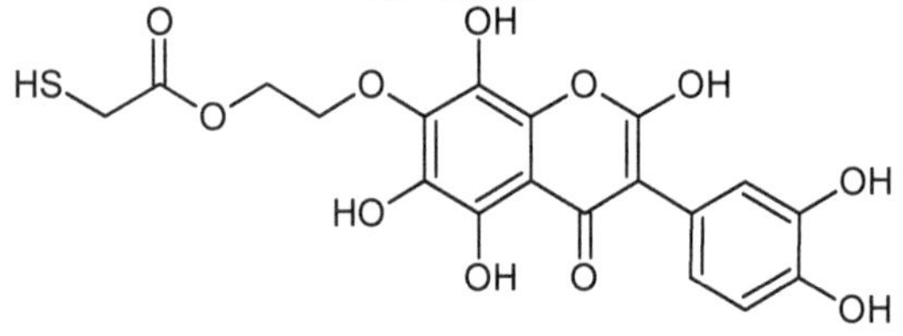 2,6,8,3′-tetrahydroxy-TGE (10)	10/Minor	4O gain (+64)
453: 438(67); 435(17); 425(50); 409(17); 407(33); 395(17); 364(100); 338(17); 309(33); 295(17); 287(67); 285(100); 269(17); 241(17); 229(17); 123(33) M = 454 Da: 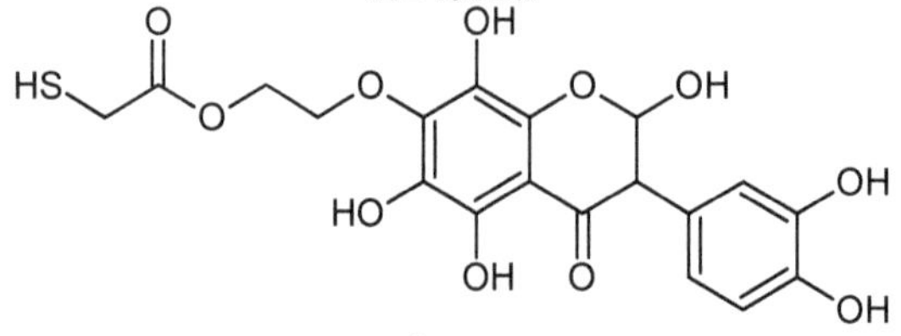 2,3-dihydro-2,6,8,3′-tetrahydroxy-TGE (11)	11/Medium	3O gain (+48) + H_2O gain (+18)

Table 2. *Cont.*

Product Mass, Characteristic Ions, *m/z*, (rel. int., % for EPI)	No. Oxidation Product/Intensity	Proposed Metabolic Reaction–Suggested Reactions
467: 450(2); 449 (1); 434(61); 433(1); 421(2); 405(1); 370(4); 369(1); 328(13); 317(2); 301(10); 284(20); 283(17); 268(100); 267(7); 255(6); 239(4); 211(2); 193(6); 151(3); 121(3) M = 468 Da: 2, 6, 8,3′,5′-pentahydroxy-TGE (12)	12/Minor	5O gain (+80)
469: 454(5); 441(7); 437(30); 425(6); 423(20); 411(5); 409(100); 397(4); 383(8); 380(4); 355(2); 353(20); 339(5); 325(13); 317(3); 311(7); 307(11); 303(7); 295(2); 285(28); 283(4); 269(8); 257(3); 241(4); 225(1); 199(2); 185(2); 171(6); 151(4); 147(2); 121(2) M = 470 Da: 2,3-dihydro-2,6,8,3′,5′-pentahydroxy-TGE (13)	13/Medium	4O gain (+64) + H_2O gain (+18)
485: 467(1); 451(3); 439(2); 437(30); 435(11); 409(100); 407(3); 393(6); 375(3); 350(3); 319(2); 301(5); 285(5); 269(5); 257(2); 243(1); 227(2); 199(2); 183(1); 151(2); 143(1); 121(2) M = 486 Da: 2,3-dihydro-2,6,8,3′,5′,6′-hexahydroxy-TGE (14)	14/Medium	5O gain (+80) + H_2O gain (+18)

Table 2. *Cont.*

Product Mass, Characteristic Ions, *m/z*, (rel. int., % for EPI)	No. Oxidation Product/Intensity	Proposed Metabolic Reaction–Suggested Reactions
501: 483(1); 473(1); 453(11); 437(7); 425(10); 409(100); 407(5); 393(5); 379(5); 364(8); 352(3); 336(3); 309(3); 291(2); 282(5); 268(2); 155(1); 139(1); 123(1) M = 502 Da: 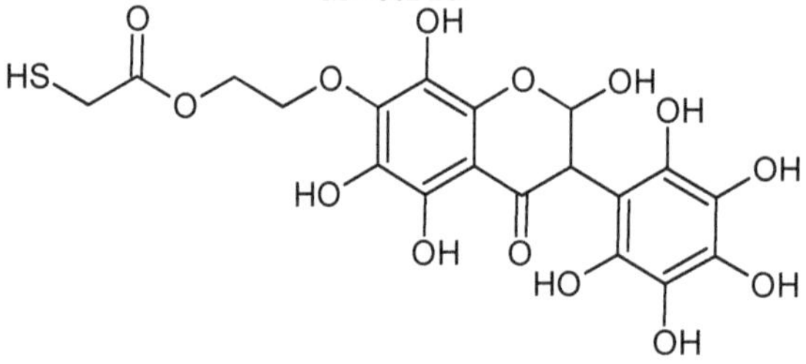 2,3-dihydro-2,6,8,2′,3′,5′,6′-heptahydroxy-TGE (15)	15/Minor	6O gain (+96) + H_2O gain (+18)
520: 502(3); 476(6); 474(98); 457(58); 448(10); 435(83); 430(29); 429(6); 406(5); 404(8); 390(5); 372(3); 358(6); 353(6); 338(5); 326(6); 306(3); 283(4); 269(100); 197(5); 177(8); 121(10) M = 520 Da: 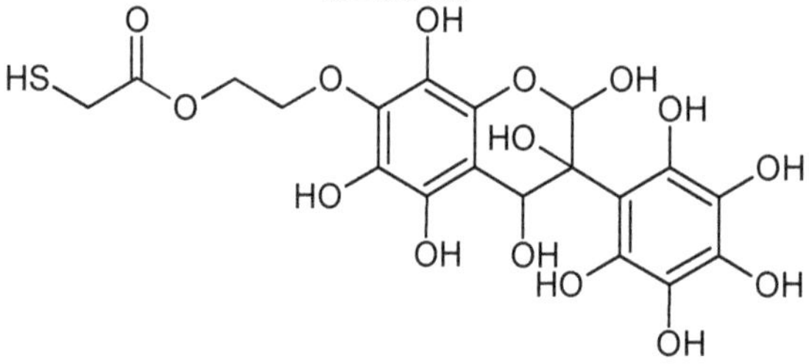 2-hydro-2,3,4,6,8,2′,3′,5′,6′-nonahydroxy-TGE (16)	16/Minor	6O gain (+96) + $2H_2O$ gain (+36)
536: 518(1); 502(1); 490(3); 473(11); 455(1); 454(1); 435(86); 429(1); 402(1); 384(1); 382(1); 370(3); 353(3); 338(1); 308(1); 285(2); 269(100); 214(2); 165(1); 121(2) M = 536 Da: 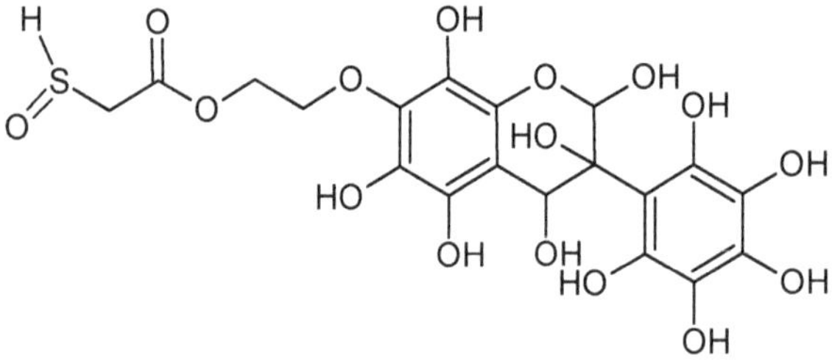 16-sulfenic-2-hydro-2,3,4,6,8,2′,3′,5′,6′-nonahydroxy-TGE (17)	17/Minor	7O gain (+112) + $2H_2O$ gain (+36)

Table 3. Summary of proposed structures of main oxidations products of TGE formed after 0.34 M H_2O_2 and potential.

Product Mass, Characteristic Ions *m/z*, (rel. int., % for EPI)	No. Oxidation Product/Intensity	Proposed Metabolic Reaction-Suggested Reactions
TGE [M-H]$^-$ = 387 Da *m/z*: EPI—		
423: 395(7); 339(7); 307(27); 287(3), 269(3); 257(100); 229(10); 213(1); 185(10); 177(3); 159(1); 143(1); 135(1); 121(10) M = 424 Da: 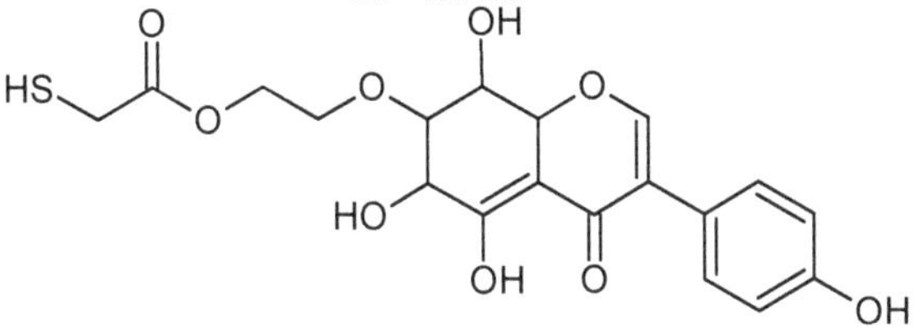 6,7,8,9-tetrahydro-6,8-dihydroxy-TGE (18)	18/Medium	$2H_2O$ gain (+36)
435: 349(1); 286(1); 269(100); 241(1); 224(1); 213(1); 201(1); 181(1); 165(1); 157 (1); 121(2) M = 436 Da: 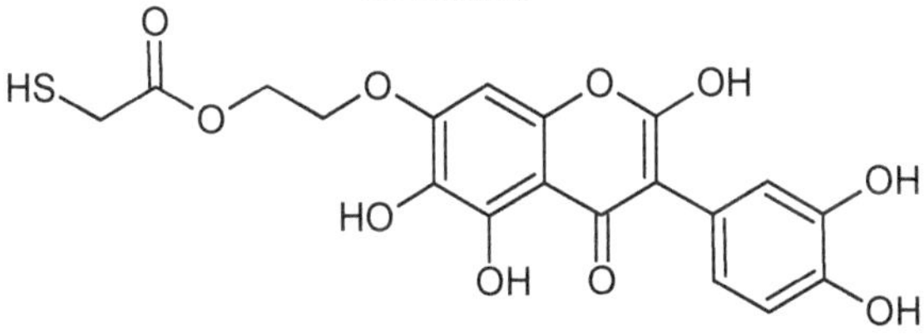 2,6,3′-trihydroxy-TGE (8)	8/Major	3O gain (+48)
M = 438 Da: 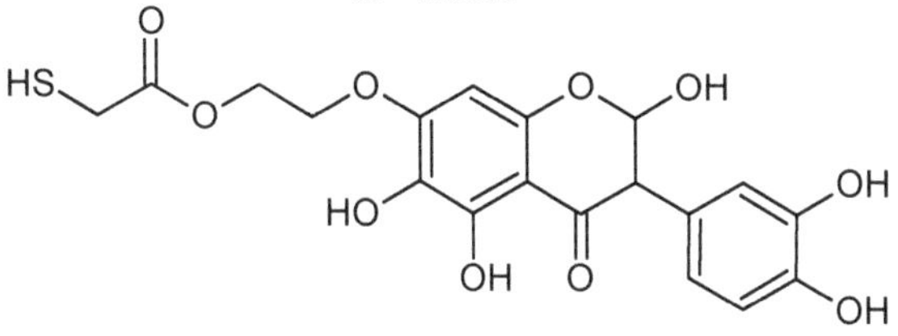 2,3-dihydro-2,6,3′-trihydroxy-TGE (9)	9/Minor	2O gain (+32) + H_2O gain (+18)
451: 423(2); 285(100); 283(2); 269(3); 257(5); 229(2); 217(3); 213(1); 121(2) M = 452 Da: 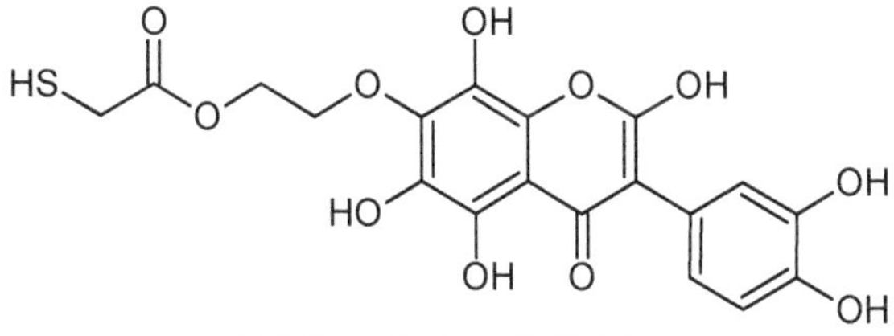 2,6,8,3′-tetrahydroxy-TGE (10)	10/Medium	4O gain (+64)

Table 3. *Cont.*

Product Mass, Characteristic Ions *m/z*, (rel. int., % for EPI)	No. Oxidation Product/Intensity	Proposed Metabolic Reaction-Suggested Reactions
453: 438(67); 435(17); 425(50); 409(17); 407(33); 395(17); 364(100); 338(17); 309(33); 295(17); 287(67); 285(100); 269(17); 241(17); 229(17); 123(33) M = 454 Da: 2,3-dihydro-2,6,8,3′-tetrahydroxy-TGE (11)	11/Medium	3O gain (+48) + H_2O gain (+18)
467: 450(2); 434(61); 421(2); 370(4); 328(13); 317(2); 301(10); 284(20); 268(100); 255(6); 239(4); 211(2); 193(6); 151(3); 121(3) M = 468 Da: 2,6,8,3′,5′-pentahydroxy-TGE (12)	12/Medium	5O gain (+80)
469: 454(5); 441(7); 437(30); 425(6); 423(20); 411(5); 409(100); 397(4); 383(8); 380(4); 355(2); 353(20); 339(5); 325(13); 317(3); 311(7); 307(11); 303(7); 295(2); 285(28); 283(4); 269(8); 257(3); 241(4); 225(1); 199(2); 185(2); 171(6); 151(4); 147(2); 121(2) M = 470 Da: 2,3-dihydro-2,6,8,3′,5′-pentahydroxy-TGE (13)	13/Minor	4O gain (+64) + H_2O gain (+18)
473: 457(1); 409(1); 399(13); 351(13); 323(1); 313(25); 307(100); 306(12); 305(1); 295(25); 284(1); 281(13); 269(63); 239(1); 223(13); 181(1); 177(88); 165(1); 139(1); 121(50) M = 474 Da: 2,3,6,7,8,9-hexahydro-2,6,8,3′,5′-pentahydroxy-TGE (19)	19/Medium	2O gain (+32) + 3 H_2O gain (+54)

Table 3. Cont.

Product Mass, Characteristic Ions *m/z*, (rel. int., % for EPI)	No. Oxidation Product/Intensity	Proposed Metabolic Reaction-Suggested Reactions
485: 467(1); 451(3); 439(2); 437(30); 435(11); 409(100); 407(3); 393(6); 375(3); 350(3); 319(2); 301(5); 285(5); 269(5); 257(2); 243(1); 227(2); 199(2); 183(1); 151(2); 143(1); 121(2) M = 486 Da: 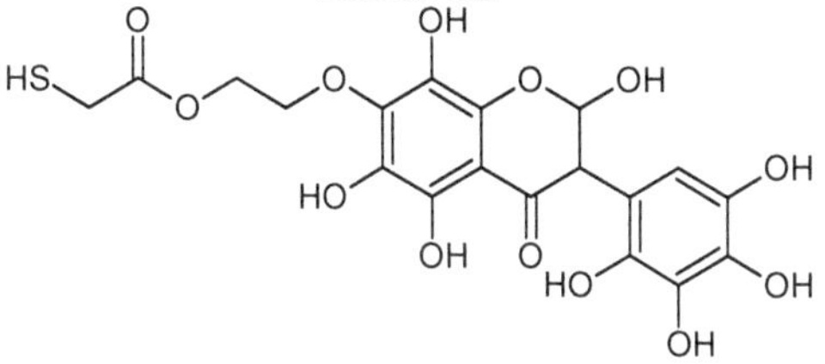 2,3-dihydro-2,6,8,3′,5′,6′-hexahydroxy-TGE (14)	14/Minor	5O gain (+80)+ H_2O gain (+18)
501: 483(2); 473(1); 453(3); 437(9); 425(8); 409(100); 407(9); 393(5); 379(5); 364(11); 352(3); 336(1); 309(3); 291(7); 282(5); 268(1); 155(1); 139(1); 123(1) M = 502 Da: 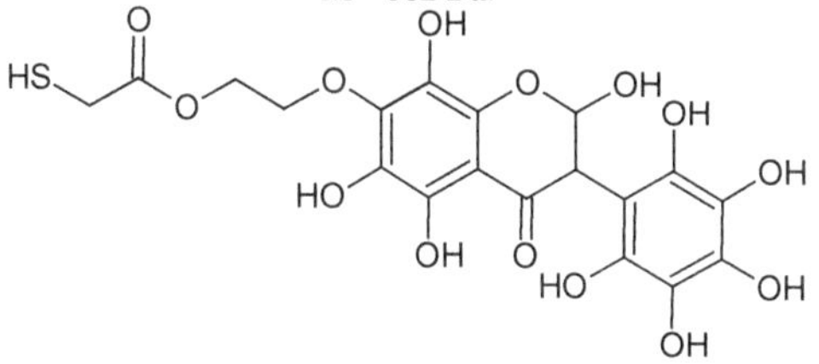 2-hydro-2,6,8,2′,3′,5′,6′-heptahydroxy-TGE (15)	15/Minor	6O gain (+96) + H_2O gain (+18)
520: 502(6); 476(2); 474(30); 457(16); 448(7); 435(13); 430(12); 429(7); 404(5); 390(1); 358(3); 353(7); 338(4); 326(9); 306(4); 283(3); 269(100); 197(6); 177(5); 121(12) M = 520 Da: 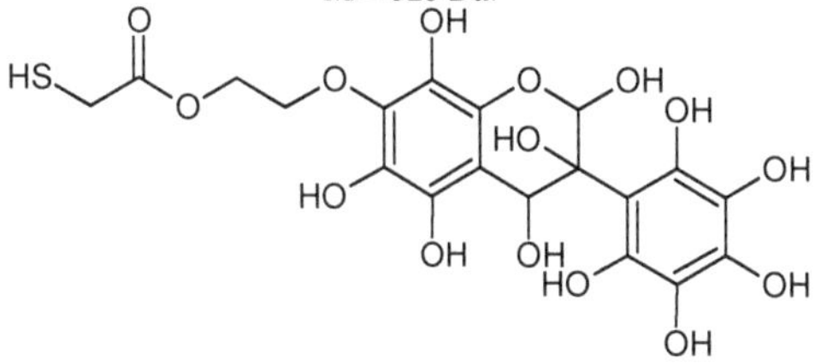 2,3-dihydro-2,3,4,6,8,2′,3′,5′,6′-nonahydroxy-TGE (16)	16/Minor	6O gain (+96) + $2H_2O$ gain (+36)
536: 518(2); 502(2); 490(2); 473(4); 455(1); 435(12); 429(1); 402(1); 370(3); 353(4); 338(1); 308(1); 285(2); 269(100); 214(1); 165(1); 121(3) M = 536 Da: 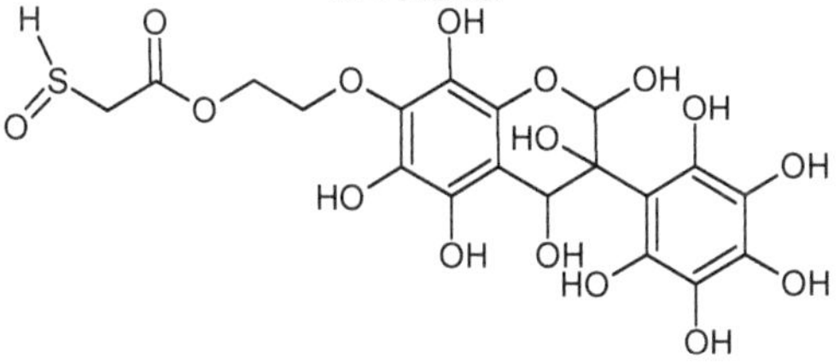 16-sulfenic-2-hydro-2,3,4,6,8,2′,3′,5′,6′-nonahydroxy-TGE (17)	17/Minor	7O gain (+112) + $2H_2O$ gain (+36)

In addition to the described background, there are analytically significant signals in the spectra. Based on the fragmentation spectra, structures were proposed for **19** new compounds–oxidation products–in three different environments. Compounds **1–7** are characteristic of potential oxidation (Table 1). Compounds **8–17** are formed by oxidation with hydrogen peroxide (Table 2). The combination of the two oxidizing agents' potential and hydrogen peroxide generates the formation of compounds **8–17**, the same as in the oxidation with hydrogen peroxide, and in addition, two new compounds **18** and **19** appear (Table 3). The proposed fragmentation pathway of new compounds and structures of characteristic fragments are presented in Figures S2–S15. Based on the proposed TGE potential oxidation products, four oxidation mechanisms are proposed for oxidation with potential (Figure 7a), two different TGE oxidation pathways for oxidation with hydrogen peroxide (Figure 8a) and the same oxidation pathways 1 and 2 as for hydrogen peroxide, and two new pathways 3 and 4 for the two oxidizing factors potential and hydrogen peroxide occurring simultaneously (Figure 9a). The structures for the major compounds according to the proposed oxidation mechanisms are shown in Figures 7, 8 and 9b.

The compounds of M = 404, 406, 418, 420, 440, 442, 456, and 458 Da occur in the proposed oxidation pathways but are not observed in the spectra. These compounds are unstable and are rapidly converted to further structures, so they are seen with low intensities or not observed at all. It is likely that in an active environment these compounds are unstable and undergo further oxidation. However, the formation of compounds that differ by 2Da and that can be combined into characteristic pairs, i.e., 404/406, 418/420, 420/422, 422/424, 434/436, 436/438, 452/454, and 468/470 related to oxidation with the attachment of a water molecule to TGE structure or oxidation alone, was observed. Compounds of M = 402 Da, M = 486 Da, 502 Da, 520 Da, 536 Da, and 474 Da also occupy an important place in the oxidation tracts.

The most intense ion in the MS spectrum of TGE, after potential oxidation, is the ion at *m/z* 402 (compound **1**). The EPI spectrum of this pseudomolecular ion displayed a lot of products ions, which are presented in Table 1. The structures of characteristic fragments, which are proposed, are presented in Figure S2. Based on these data, the structure of the compound **1** has been assigned as 15-thioformyl-6-hydroxy-TGE. During the formation of this compound, dehydrogenation occurs first in the aliphatic side chain (the mildest form of oxidation) and then oxidation in the A ring of TGE. This compound contains an additional -OH group in the A ring in the ortho position. There are ions *m/z* 384, 268 in the fragmentation spectrum that indicate this order. Fragmentation proceeds by elimination of the water molecule from the C ring, therefore two -OH groups are adjacent to each other (the ion at *m/z* 384). Additionally, the ion confirming such an arrangement of -OH groups is fragmented at *m/z* 177. From the fragment at *m/z* 384, the neutral molecule of M = 116 Da (S=CHCOOCHCH$_2$) is eliminated, generating a fragment at *m/z* 268, while from the fragment at *m/z* 374, the neutral molecule of M = 132 Da (S=CHCOOCH$_2$CHO) is eliminated, generating a fragment at *m/z* 241. These eliminations indicate the occurrence of dehydrogenation at sulfur. It was initially assumed that this compound is formed by oxidation in the C ring; however, a correlation was observed between the compound of M = 402 Da and the compound of M = 420 Da, where the compound of M = 420 Da (compound **3**) is formed by the attachment of a water molecule. Therefore, the compound of M = 402 Da must have a double bond in the C-ring to which a molecular water attaches to form the compound of M = 420 Da. The EPI spectrum of pseudomolecular ions at *m/z* 419 displayed a lot of product ions, which are presented in Table 1. The structures of characteristic fragments, which are proposed, are presented in Figure S3. Based on these data, the structure of compound **3** has been assigned as 15-thioformyl-2,3-dihydro-2,6-dihydroxy-TGE. A mass of 424 Da is formed during both oxidation with potential and oxidation with hydrogen peroxide with potential, but these are two different compounds due to their different fragmentation spectra, the interpretations of which are shown in Figures S4 and S5, respectively. In the fragmentation spectrum of the compound formed during the action of hydrogen peroxide with potential, there is a characteristic fragment *m/z* 339 formed by loss of the neutral moiety of M = 56 Da (O=C=C=O) from the ion at *m/z* 395. Therefore, compound **18** is formed by the attachment of two water molecules to the A ring of TGE. Based on these data, the structure of the compound **18** has been assigned as 6,7,8,9-tetrahydro-6,8-dihydroxy-TGE. In contrast, compound **5** is formed by the attachment of two water molecules, with the first oxidation going to A ring and the second to C ring. Based on these data, the structure of the compound **5** has been assigned as 2,3,6,7-tetrahydro-2,6-dihydroxy-TGE. During oxidation by potential, a compound of M = 436 Da (compound **7**) is formed from compound of 420 Da by oxidation in the ortho position. In this case, the fragmentation spectrum is not clear, and it is not possible to indicate whether this oxidation involves the A or B ring, but it is certainly an oxidation in the ortho position. The rationale behind this interpretation is the proposed structure for the fragment at *m/z* 176. However, it was observed that the neutral molecule S=C=C=O falls off, which confirms the assumption of oxidation by dehydrogenation in the aliphatic part of the TGE. The proposed fragmentation pathway of compound **7** and structures of characteristic fragments, which confirmed the proposed structures, is presented in Figure S6. The product of mass at M = 436 Da (compound **8**) is also formed when hydrogen peroxide and potential with hydrogen peroxide are applied, but it is a different compound due to the different fragmentation spectra. The proposed fragmentation pathway of impurity eight and structures of characteristic fragments, which confirmed the proposed structures, is presented in Figure S7. These compounds are formed by two different oxidation mechanisms. The fragmentation spectrum of compound **8** shows a characteristic fragment at *m/z* 349, which confirms the occurrence of two -OH groups in the B-ring by eliminating the neutral fragment of M = 86 Da. In addition, this elimination shows the occurrence of the -OH group in the C-2 position of TGE. Based on these data, the structure of compound **8** has been assigned as 2,6,3′-trihydroxy-TGE.

Further oxidation of compound **8** with hydrogen peroxide proceeds to carbon C- 8 in the A ring of TGE, generating a compound of mass 452 Da (compound **10**). The proposed fragmentation pathway of compound **10** and the structures of characteristic fragments that confirmed the proposed structure are shown in Figure S8. Based on these data, the structure of compound **10** has been assigned as 2,6,8,3′-tetrahydroxy-TGE. The next oxidation in this sequence proceeds to carbon 5′ in the B ring of TGE, generating the compound of mass 468 Da (compound **12**). The proposed fragmentation pathway of compound **12** and the structures of the characteristic fragments that confirmed the proposed structure are shown in Figure S9. Based on these data, the structure of compound **12** has been assigned as 2,6,8,3′,5′-pentahydroxy-TGE. Compound **12** completes the first oxidation sequence with hydrogen peroxide. In the proposed second sequence of TGE oxidation with hydrogen peroxide, an interesting structure is the compound of M = 520 Da (compound **16**), for which the fragmentation spectrum is characterized by strong signals for ions at *m/z* 474, 457, 435, 430, and 269 (Figure S10). In addition to the saturation of all carbons of A, C, and B rings, additionally in the C ring, an enol is formed from ketone because there is one hydrogen atom in the neighborhood. The interpretation of the fragmentation spectrum for the compound with M = 520 Da shows that the B-ring (saturated with -OH groups) with all -OH groups are very stable. It passes unchanged through all fragments except the ions at *m/z* 435, 406, 390, 178, or *m/z* 122. These are the fragments with the "stripped" B ring. However, it is the *m/z* 435 ion that is the precursor to the *m/z* 390 and 406 ions. The *m/z* 435 ion comes directly from the *m/z* 502 ion because there are no intermediate fragments in the fragmentation spectrum. Based on these data, the structure of the compound **16** has been assigned as 2-hydro-2,3,4,6,8,2′,3′,5′,6′-nonahydroxy-TGE. When it would seem that the oxidation of the TGE molecule with hydrogen peroxide would end with the compound of M = 520 Da, due to the saturation of all carbons, a compound of M = 536 Da appears, as compound **17**. The interpretation of the fragmentation spectrum for compound of M = 536 Da (Figure S11) confirms the presence of oxygen at the sulfur due to several characteristic fragments, e.g., at *m/z* 474, which is formed by elimination of the -OSCH$_2$ group, or the *m/z* 454 ion, which is formed by loss of the neutral molecule O=SH-CH$_3$.

Based on these data, the structure of the compound **17** has been assigned as 16-sulfenic-2-hydro-2,3,4,6,8,2′,3′,5′,6′-nonahydroxy-TGE.

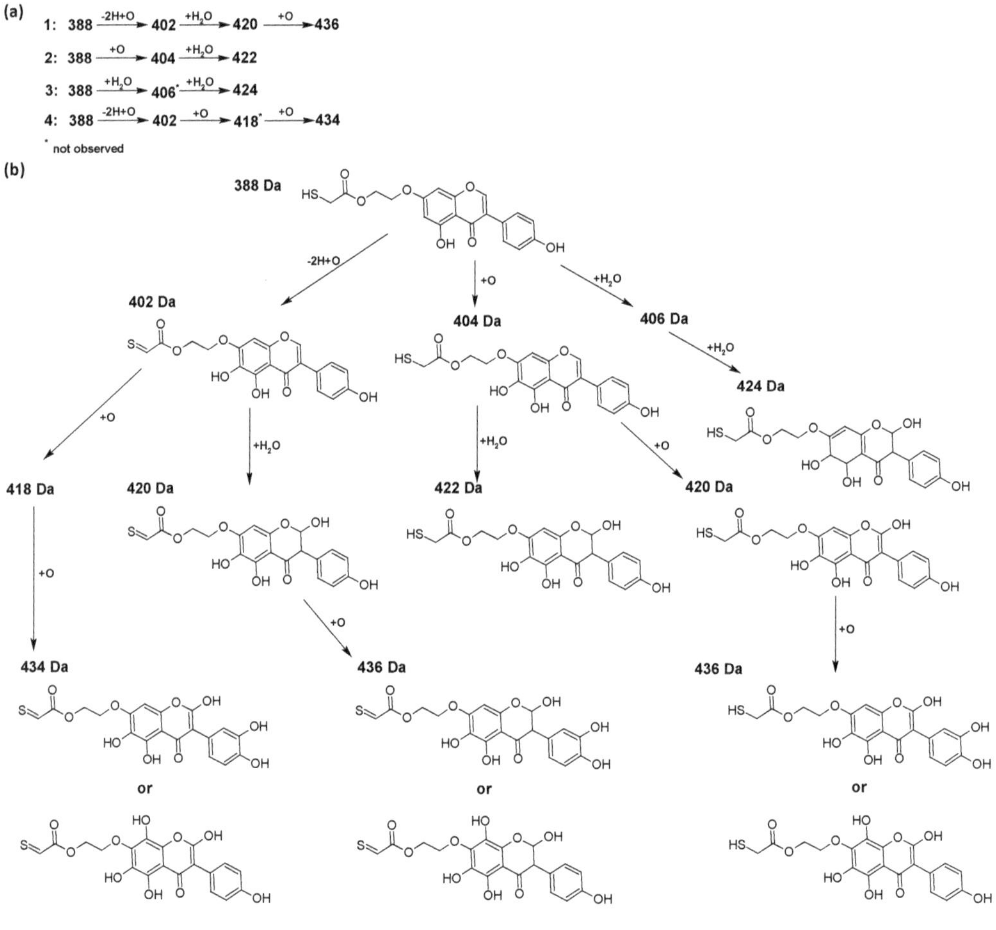

(a)

1: $388 \xrightarrow{-2H+O} 402 \xrightarrow{+H_2O} 420 \xrightarrow{+O} 436$

2: $388 \xrightarrow{+O} 404 \xrightarrow{+H_2O} 422$

3: $388 \xrightarrow{+H_2O} 406^{*} \xrightarrow{+H_2O} 424$

4: $388 \xrightarrow{-2H+O} 402 \xrightarrow{+O} 418^{*} \xrightarrow{+O} 434$

* not observed

(b)

Figure 7. The proposed oxidation pathways of TGE (**a**) and structures of main reaction products (**b**) formed by the oxidation with potential.

During the TGE oxidation with hydrogen peroxide and potential simultaneously, an additional oxidation sequence (No. 3, 4) is observed that is not present during the other oxidizing agents. This is a sequence containing a different compound of M = 424 Da (compound **18**) than in the case of oxidation by potential. It is formed by the oxidation of the A ring at positions C-6 and C-8 of TGE as a molecular water addition. The proposed a fragmentation pathway of compound **18** and structures of characteristic fragments, which confirmed the proposed structures, is presented in Figure S5. Based on these data, the structure of compound **18** has been assigned as 6,7,8,9-tetrahydro-6, 8-dihydroxy-TGE. Compound **19** with M = 474 Da completes this oxidation sequence. The suggested fragmentation pathway of compound **19** and the structures of the characteristic fragments that confirmed the proposed structure are shown in Figure S12. Based on these data, the structure of compound **19** has been assigned as 2,3,6,7,8,9-hexahydro-2,6,8,3′,5′-pentahydroxy-TGE. Only in this case, there is an attachment of two water molecules. Oxidation sequence No. 3 and 4 (Figure 9a) are characteristic of the action of hydrogen peroxide and potential simultaneously. Thus, oxidation of the compound alone does not occur. The potential favors the attachment of water molecules.

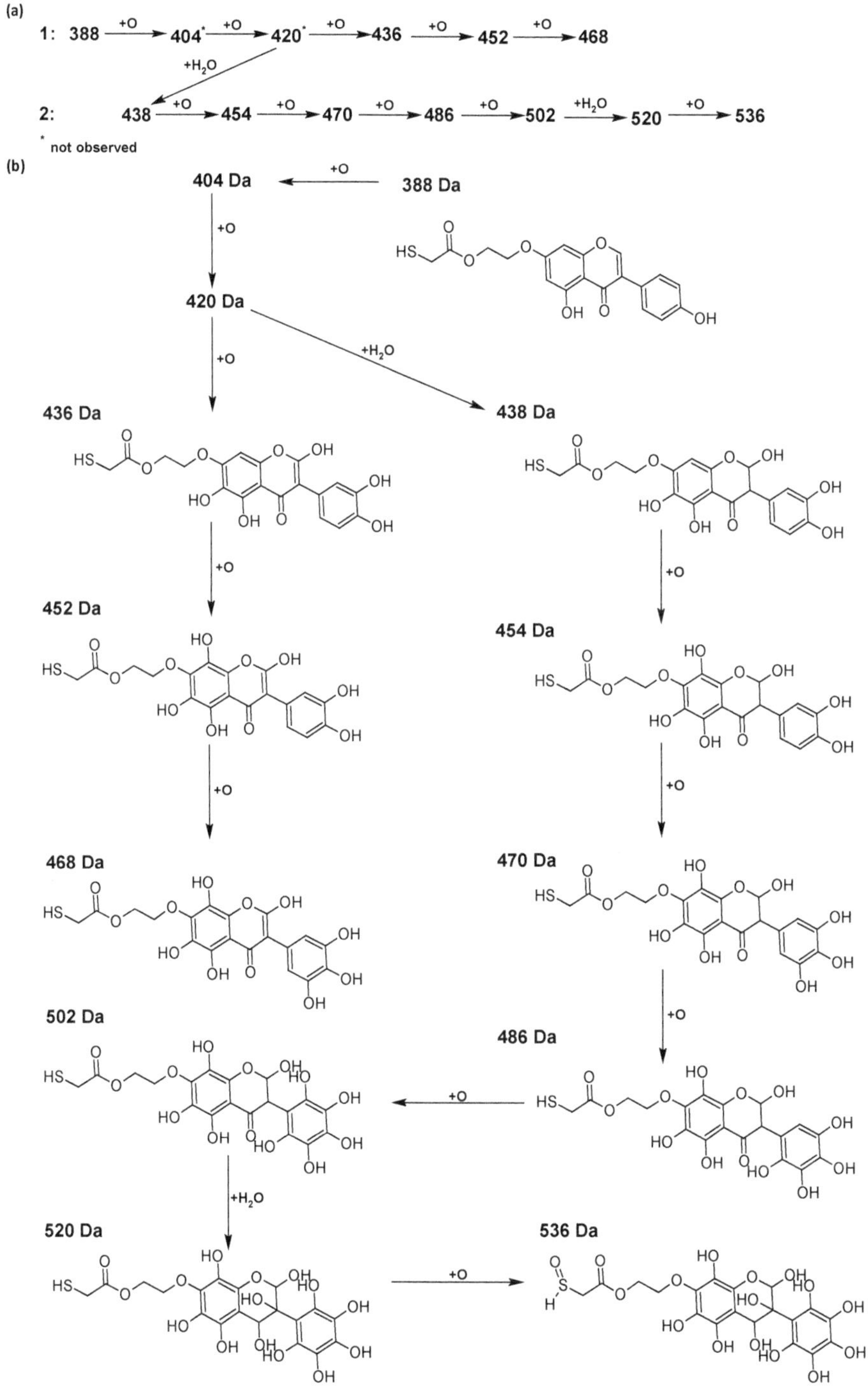

Figure 8. The proposed oxidation pathways of TGE (**a**) and structures of main reaction products (**b**) formed by the oxidation with hydrogen peroxide.

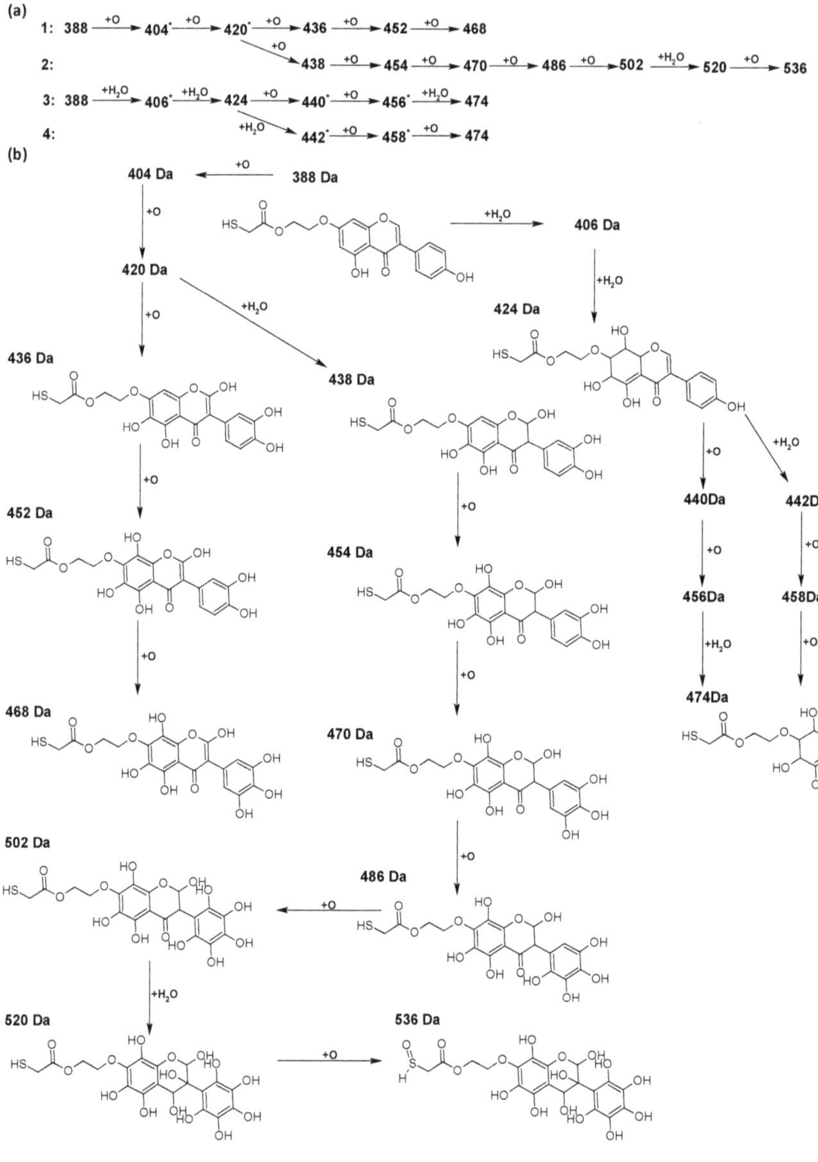

Figure 9. The proposed oxidation pathways of TGE (**a**) and structures of main reaction products (**b**) formed by the oxidation with potential and hydrogen peroxide. * not observed.

In summary, three forms of TGE oxidation were observed: the oxidation by water molecule attachment, the oxidation by oxygen attachment, or the dehydrogenation as a mild form of oxidation. The first attack of oxygen probably occurs in the C ring of TGE in the case of oxidation with potential. In addition, the formation of compounds resulting from oxidation on rings with dehydrogenation in the side chain is observed during potential oxidation. Compound **5** is formed by the attachment of two water molecules to the A and C rings. Compounds **3**, **4**, **9**, and **16** are formed by the attachment of a water molecule to the C ring. Compound **18** is formed by the attachment of two water molecules to the A ring. Compound **19** is formed by attaching two water molecules to the A ring and one to the C ring after prior oxidation of the B ring of TGE. The substitution in the A ring is due to an inductive effect. The electrodonor group -OH causes a partial positive charge to appear on the adjacent C atom. The elimination of the neutral moiety -CH$_2$O is observed in the fragmentation spectrum. This is a characteristic elimination in fragmentation spectra of compounds that are formed by the oxidation by attachment of a water molecule. It has been observed that the experiment with potential favors the attachment of a water molecule. In this case, a higher number of water molecules are observed in the A ring. In the experiments without the action with potential, there are the water molecules attached mainly to the double bond in the C ring of TGE.

Our experiments are important for predicting the physicochemical properties of the new compound TGE. Electrochemistry, which helps determine metabolism patterns, is in turn considered to have a great potential as a fast alternative to in vitro assays [36]. The prooxidation activity of flavonoids can be directly related to the number of hydroxyl groups in the molecule. Mono- and dihydroxyl flavonoids usually do not show such activity, while in the case of polyhydroxyl compounds (especially when OH groups are attached to the B ring), a significant increase in the production of reactive oxygen species (ROS) is observed. From the present work, it appears that the TGE antioxidant can undergo modifications of the A, C, and B rings upon oxidation. This is a much more extensive structural variation than formerly was thought, that changes occur mainly in the B ring. What is particularly interesting about the results of our product studies with TGE is the variety of antioxidant chemistries displayed by the A ring. TGE produced products that are not typical of well-documented genistein antioxidant reactions: the B ring hydroxylation and formation of radical addition products. We do not observe structures derived from the two significant metabolites of genistein identified by Kullig et al. [37], i.e., 5,7,8,4′-tetrahydroxyisoflavone and 5,7,3′,4′-tetrahydroxyisoflavone, on oxidative metabolism in human liver microsomes. However, it has been reported that GE can be metabolized by the cytochrome P450s to hydroxylated metabolites (6-, 8-, and 3′-hydroxyGEN) [38] and by the gut microbiota to dihydrogenistein (DGEN), 5-hydroxy-equol (5-OH-equol), and 6′-hydroxy-O-desmethylangolensin (6′-OH-DMA) [39,40]. We observe that the oxidation by potential proceeds much further (more -OH groups) than was found by Kullig et al. [37] for genistein metabolites in microsomes. In addition, oxidation products are observed that result from dehydrogenation at the -SH group, which are new compounds. The characterization of these product structures provides potential markers of TGE antioxidant reactions. Sensitive assays for specific TGE oxidation products may thus be useful markers for antioxidant reactions of the isoflavone in biological systems.

2.5. Description of Qualitative Changes Observed in the IR-ATR Spectra during the Oxidation of the TGE Monolayer on the Au Electrode

The free radical peroxidation of unsaturated lipids in biomembranes disrupts the various important structural functions associated with this natural protective barrier for cells; as a result of this oxidation, various in vivo pathologies can occur [41]. It is known that genistein and its metabolites possess an ability to inhibit lipid peroxidation in biological membranes. It is supposed that molecules are localized near the lipid–water interference of the membrane or/and they are incorporated into its hydrophobic core, which causes an increase in membrane rigidity and stability to the diffusion of free radicals [42]. The aim of our preliminary studies of TGE oxidation on the surface of the Au 2D electrode is to mimic the oxidation of a new GE derivative on the surface of a biomembrane.

Vibrational Studies on Oxidative Monolayers–Simulation of Active Substance Immobilization on the Surface of Biological Membranes

At the beginning, it should be emphasized that the following assumption was made regarding the interpretation of IR spectra, that in a tightly packed monolayer, limited by the surface of the electrode, more oxidation products of TGE by H$_2$O$_2$ are possible due to the differences in the inner and outer environment of the molecule.

Figure 10a (3600–2500 cm^{-1}), 10b (1800–1420 cm^{-1}), and 10c (1420–800 cm^{-1}) show the comparison of IR-ATR spectra of TGE on the Au electrode before and after oxidation by

H_2O_2 (spectra: I–1 mM 5 min, II–1 mM 15 h, and III–2 mM 24h). The presence of hydroxyl groups in the TGE molecule before and after oxidation is proven in the IR spectrum by the presence of the broad band between 3600 and 3200 cm^{-1}.The presence of a broad band may suggest the absence of hydrogen bonding with hydroxyl groups in the TGE molecule before and after oxidation. At the same time, it should be taken into account that during oxidation more OH groups are added, but in a tightly packed monolayer the outer molecules are much more susceptible to oxidation (which means more hydroxyl groups) than the inner ones because the whole molecule is exposed to H_2O_2; for the inner molecules only the B ring will be the most susceptible to oxidation.

Bands from C-H stretching vibrations from the methylene group of the TGE molecule are observed at 2959, 2927, and 2855 cm^{-1}. It can be observed that relative intensities, as well as wavenumbers of these bands, remain unchanged during oxidation.

The band from the carbonyl stretching vibration from the chain [20] has the same value of 1738 cm^{-1} before and after oxidation. However, the band from the carbonyl stretching vibration assigned to the ring [20] is shifted towards higher wavenumbers from 1659 to 1667 cm^{-1}, which indicates a weakening or breaking of the intramolecular C=O–O-H hydrogen bond. This can be caused by the oxidation of the –OH group in the A ring into a C=O group or by the presence of additional –OH groups as it can be presented in the compounds of M = 436, 452 and 468 Da, as can see in Figure 8.

In IR spectra of TGE, both before and after oxidation, bands from 1615 to 1518 cm^{-1} mainly originate from C=C vibrations. However, their values are slightly shifted towards higher wavenumbers after oxidation, which can indicate changes in aromatic rings during oxidation.

In the range of deformation, vibrations of hydroxyl groups 1400–1300 cm^{-1} following changes that occurred during oxidation are observed: the band at 1362 cm^{-1} disappears and the band at 1316 cm^{-1} rises.

In IR spectra of TGE before oxidation, the doublet at 1179 cm^{-1} and 1158 cm^{-1} originate from C-O-C asymmetric stretching and C-OH stretching vibrations. During oxidation, the band at 1158 cm^{-1} shifts into 1167 cm^{-1} and the new band is formed at 1110 cm^{-1}, which probably originates from stretching C-O vibrations and can prove the presence of new -OH groups after oxidation.

In the TGE spectrum before oxidation, the main band from C-H deformation out-of-plane vibrations from aromatic rings is at 841 cm^{-1}. During oxidation, bands at 869 cm^{-1} and 850 cm^{-1} are observed that can prove that the oxidation of TGE influences the rings of the molecule.

The comparison of IR spectra does not indicate the polymerization of TGE molecules during oxidation by H_2O_2, which is in agreement with the results obtained by Xu and co-workers for the caffeic acid monomer without the presence of horseradish peroxidase [43]. On the other hand, polymerization by Ar-Ar (B rings) is possible but difficult to prove. Likewise, polymerization via Ar-O-Ar (B-rings) is also possible, but the expected aryl ether band is in the range of 1270–1230 cm^{-1} and is found in the spectrum of the initial sample.

From the comparison of the theoretical IR spectra (obtained by the B3LYP calculations) of TGE vs. M436, TGE vs. M452, and TGE vs. M468 (Figures S16–S18), one can see that introducing the OH-groups to the skeleton of the TGE molecule results in the significant modification of the theoretical TGE spectrum in the region of 1300–900 cm^{-1}. This is actually the region in the experimental spectrum presented above in Figure 10c as spectrum III (2 mM 20 min), corresponding to the intensive oxidation by H_2O_2. However, in this experimental spectrum, only the bands at: 1110, 869, and 850 cm^{-1} have an increased intensity. This qualitative comparison supports our belief that in the 2D area the molecules with additional OH groups in the B ring should dominate.

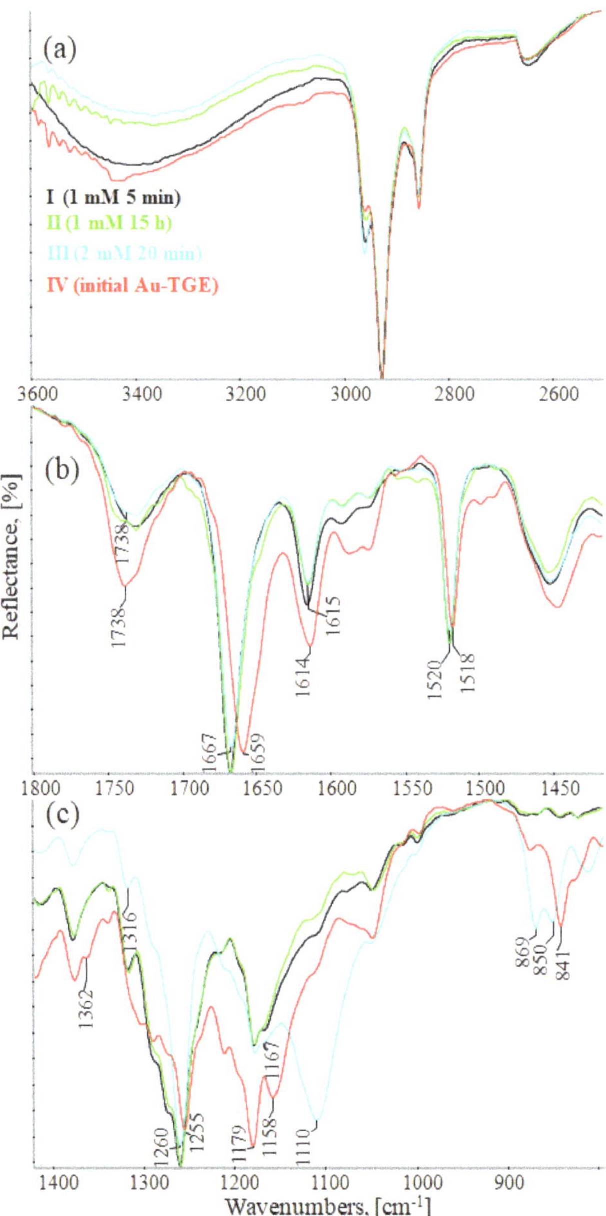

Figure 10. IR-ATR spectra of Au-TGE before and after oxidation with H_2O_2, (**a**) from 3600 to 2500 cm^{-1}; (**b**) from 1800 to 1425 cm^{-1}; (**c**) from 1400 to 800 cm^{-1}.

2.6. Molecular Modeling and the Quantum Mechanical Density Functional Calculations

A theoretical estimation of the antioxidant activity of TGE was also studied. There exists an extensive discussion in the literature on how to define the antioxidant power of selected species, starting from colorimetric determination of the total oxidant status (TAS) and the total antioxidant capacity (TAC) [44–46]. Here, we apply a theoretical method based on the density functional theory allowing the comparison of various antioxidants using two parameters, i.e., the bond dissociation energy (BDE) and the Gibbs free energy of the antioxidant reaction with H_2O_2 [47,48]. As long as many of the antioxidants were investigated experimentally, their antioxidant effectiveness was evaluated, and the above-mentioned parameters were calculated, there is a possibility to evaluate the antioxidant power of new analogues theoretically by a simple comparison. This method allows us to locate TGE within the series or the scale of known antioxidants.

One of the best reliable thermodynamic parameters to describe the hydrogen atom transfer (HAT) mechanism of antioxidant activity is the bond dissociation enthalpy (BDE). A corresponding model reaction is given below:

$$\text{Ge-OH} + \text{R}^{\bullet} \rightarrow \text{Ge-O}^{\bullet} + \text{R-H} \tag{1}$$

where Ge-OH denotes, for example, the genistein with the OH group at the $C4'$ position. R is the radical whose unpaired electron is abstracted by the antioxidant (here the genistein molecule). In this way, the free radical ($\text{R}^{\bullet}$) is deactivated by the antioxidant. Such a picture oversimplifies a more complex mechanism because apart from electron transfer the proton is transferred as well.

The bond dissociation enthalpy is defined as follows:

$$\text{BDE} = H(\text{Ge-O}^{\bullet}) + H(\text{H}^{\bullet}) - H(\text{Ge-OH}) \tag{2}$$

where H denotes the enthalpy.

According to this mechanism, the hydroxyl group of the antioxidant compound (Ge-OH) releases the hydrogen atom ($\text{H}^{\bullet}$), being transformed to the respective free radical ($\text{Ge-O}^{\bullet}$). The higher antioxidant activity corresponds to the weaker O-H bond. In Table 4, the bond dissociation enthalpy (BDE) was estimated with the density functional theory (DFT) and one of its implementations, i.e., the B3LYP/6-311++G(d,p) method for genistein and thiogenistein. For comparison, analogical calculations were carried out for phenol, Trolox, and curcumin. Trolox is a popular antioxidant compound frequently used in the laboratory studies. The BDE was calculated for isolated (non-interacting with environment) molecules.

Table 4. The O-H bond dissociation enthalpy (BDE), in kcal/mol as predicted with the B3LYP/6-311++G(d,p) DFT method.

Compound	BDE
Phenol	82.90
Genistein	81.56
Thiogenistein	81.82
Trolox	73.91
Curcumin (*)	75.56

(*) this calculation was carried out with the B3LYP/6-31G(d,p) method.

A series of the BDE values shows that the lowest BDE corresponds to Trolox while the highest BDE can be ascribed to phenol. The BDE values of genistein and thiogenistein are comparable and smaller than the BDE of phenol and curcumin. It suggests that the antioxidant activity of genistein and thiogenistein is governed by the OH-substituted core of these compounds. One can mention that the phenolic residue appears in a plethora of antioxidants [49].

The antioxidant activity of TGE and the related species was estimated also for the reaction with the $\text{ABTS}^{\bullet+}$ radical cation (Equation (3)). This is a one electron reduction reaction of the $\text{ABTS}^{\bullet+}$ radical cation. A corresponding reaction is given below:

$$\text{Antioxidant} + \text{ABTS}^{\bullet+} \rightarrow \text{Antioxidant}^{\bullet+} + \text{ABTS} \tag{3}$$

With the use of the B3LYP/6-31G(d,p) method, the Gibbs free energy output of the Equation (3) reaction was estimated. The corresponding values are presented in the Table 5.

Table 5. The Gibbs free energy difference for the Equation (3) reaction according to the B3LYP/6-31G(d,p) density functional calculations.

Antioxidant	ΔG, kcal/mol
Genistein (GE)	16.02
Genistein thiolated at O-7 (TGE)	15.37
Genistein thiolated at O-4'	14.88
Genistein thiolated at O-5	14.99
Genistein thiolated at O-7 by a modified thiol residue with the extra -CH_2- group (M26P)	14.49
Trolox	11.44

The ΔG results presented in Table 5 may reflect real reactions only qualitatively assuming that the values of Trolox are the reference point and that the ΔG values above 11.44 kcal/mol denote a poorer antioxidant activity than is exhibited by Trolox. From Table 5, one can conclude that the antioxidant activity of the thiolated genistein, i.e., TGE and M26P (genistein substituted at O-7 with a modified thiol residue by the extra -CH_2- group), seems to be rather poorer than Trolox.

A slight prevalence of the antioxidant activity of TGE over GE predicted with the B3LYP/6-31G(d,p) method (15.37–16.02 = −0.65 kcal/mol) was verified with the B3LYP/6-311++G(d,p) methods to be 1.22 kcal/mol in favour of TGE. This result ensures us that TGE is a better antioxidant than GE.

Moreover, the genistein substituted with the thiol residues at the B ring at O-4' as well as in the A ring at O-5 seems to be a slightly better antioxidant than the genistein substituted at the A ring at O-7 (TGE). This is a promising line for further studies on new antioxidants.

The predictive activity of new derivatives of naturally available compounds based on structural changes is important for considering such molecules as drug candidates and for creating new structures with promising properties. Flavonoids are well-known antioxidants, and their redox properties can be related to their pharmacological activity. The current state of knowledge on the biological activity of flavonoid compounds indicates unequivocally that their positive effect on the human body results mainly from antioxidant properties. The antioxidant activity of individual flavonoids depends on the number of hydroxyl groups and their position. Para and ortho positions enhance these properties. However, observations in in vitro studies are not always confirmed in in vivo studies, particularly with regard to the number of -OH groups [50,51]. The antioxidant activity of flavonoids is possible through different mechanisms of action, among others; also, indirectly, flavonoids can chelate transition metal ions (copper, iron), which prevents the formation of reactive hydroxyl radicals in cells. The presence of "soft" donor atoms, such as nitrogen or sulfur (thiogenistein), can determine the formation of active complexes with iron ions, which are characterized by a strong anticancer activity. In contrast, compounds contained in their structure, "hard" donor atoms such as oxygen, lead to the formation of redox-inactive complexes by blocking the coordination sphere [52,53].

3. Materials and Methods

3.1. Materials

The reagents used in the study were obtained from POCh (Gliwice, Poland) and Sigma-Aldrich (Saint Louis, MO, USA). They were of the highest purity and used without prior purification. All solutions were prepared with deoxidized water, distilled, and cleaned in a "Milli-Q" filter apparatus (Millipore Corporation, Bedford, MA, USA). Its final resistance was 18.2 MΩ/cm.

TGE was synthesized in Łukasiewicz Research Network–Industrial Chemistry Institute (Łukasiewicz-ICHP), Warsaw, Poland. Ammonium formate buffer was obtained from Sigma-Aldrich (99%, LC–MS grade, Saint Louis, MO, USA). The pH of the solutions used in the experiments was determined using a commercially available Mettler Toledo (Greifensee, Switzerland) pH meter.

The monolayers of the TGE compound on the gold electrodes were prepared in the self-assembly process. First, the monolayers of the TGE compound were prepared on a gold surface by immersing the purified gold electrodes in ethanolic solutions containing 1 mM TGE. After their removal, the electrodes were rinsed thoroughly with ethanol, water, and ethanol to wash off the physically adsorbed molecules and left to dry in the air.

3.2. Antioxidant Study

3.2.1. ABTS Radical Scavenging Assay

The ABTS radical scavenging activity was measured based on the following slightly modified method proposed by Re and co-workers [54]. Briefly, the ABTS-mixture solution was prepared by mixing 2.45 mM potassium peroxodisulfate (the final concentration) and 7 mM ABTS water solutions. For the formation of free radicals ABTS$^{\cdot+}$, the final solution was left in the dark at room temperature for 16 h. The obtained radical was stable in this form for about 72 h and stored in the dark at room temperature. For the study of thiogenistein scavenging activity, ABTS$^{\cdot+}$ solution was diluted with ethanol to an absorbance at 734 nm reached 1.3 ($\pm$0.03). Stock solutions of genistein and thiogenistein in DMSO were diluted such that, after the introduction of each dilution into the 1 mL of ABTS$^{\cdot+}$ solution, they reached a final concentration of 100 μM, 50 μM, 25 μM, and 10 μM (<1% DMSO). The absorbance of the resulting solutions was measured at 734 nm in a microplate reader POLARstar Omega (BMG Labtech, Ortenberg, Germany) after 5, 10, 15, 30, and 45 min after dark incubation. Genistein was used as a reference standard. The percent sweep (RSA %) was calculated according to the equation below:

$$\mathrm{RSA\%} = \frac{\left(\mathrm{ABTS}^{\bullet+}\,\mathrm{absorbance} - \mathrm{sample\ absorbance}\right)\cdot 100}{\mathrm{ABTS}^{\bullet+}\,\mathrm{absorbance}}\,[\%]$$

The results were expressed as mean $\pm$ standard deviation (SD) from 3 samples for each experimental group.

3.2.2. H$_2$DCF-DA ROS Detection Assay

The level of generated ROS was measured in the murine macrophage cells line (RAW 264.7, ATTC). Cells were grown in Dulbecco's Modified Eagle's Medium (DMEM, Sigma), with the addition of 10% (v/v) fetal bovine serum (FBS, Sigma). Cells were seeded at 6×10^4 cells per well in a 96-well plate and incubated for 24 h in a humidified atmosphere at 37 °C with 5% CO$_2$. Then, cells were treated with different concentrations of GE and TGE and with 10 μg/mL of LPS. Oxidative species produced by cells incubated with GE and TGE for 24 h were detected with the dichloro-dihydrofluorescein diacetate (DCFH-DA) assay. The DCFH-DA is internalized within the cells in reduced form. In the presence of ROS, the probe is oxidized into the fluorescent form, thus, the measured intensity is proportional to the intracellular oxidative stress. The RSA%, and ROS generation levels were evaluated following the manufacturer's protocols for absorbance, fluorescence, or luminescence measurement using a microplate reader POLARstar Omega (BMG Labtech, Ortenberg, Germany). The results were expressed as mean $\pm$ standard deviation (SD) from 6 samples for each experimental group.

3.3. In Vitro Study–Cell Viability, Cytotoxicity

Two models of neoplastic cells, breast (MCF-7 estrogen positive and MDA-MB-231 estrogen negative cells) were selected for the study. Breast cancer cells, MCF-7 (ATTC), were grown in the Eagle's Minimum Essential Medium (EMEM) with the addition of 10% (v/v) fetal bovine serum (FBS, ATTC) and 0.1 mg/mL human recombinant insulin (Gibco). MDA-MB-231 (ATTC) cells were maintained of complete growth medium Leibovitz's L-15 supplemented by 10% FBS (ATTC).

Genistein and thiogenistein in a DMSO stock solution were each diluted in the growth culture medium and added in quintuplicates to the wells in the final concentration (100, 50, 25, 12.5, and 6.25 µM). The maximum content of DMSO was <1%. Both cell lines were incubated with the addition of drugs for 6 h, 24 h, and 72 h after that, the cells were washed three times in Hank's Balanced Salt Solution (Gibco) and analysed by the viability (PrestoBlue™, Thermo Fisher, USA) and cytotoxicity (ToxiLight, and ToxiLight 100% Lysis Control, Lonza, USA) assays. The obtained results were compared statistically using the *t*-test.

The cell viability, toxicity, RSA%, and ROS generation level were evaluated following the manufacturer's protocols for absorbance, fluorescence, or luminescence measurement using a microplate reader POLARstar Omega (BMG Labtech, Ortenberg, Germany).

3.4. Electrochemical Measurements

TGE was injected to electrochemical treatment in 20 mM ammonium formate–acetonitrile (1:1) adjusted to pH 7.4 with ammonia solution using 1 mL syringe (Hamilton, Reno, NV, USA). The final concentration of TGE was 4 µg/mL.

Electrochemical degradation was carried out using the ROXY potentiostat (Antec Scientific, Zoeterwoude, the Netherlands) with a flow rate of 10 µL/min at room temperature and with the use of a BDD electrode in a potential range from 0 to 2500 mV. The system was composed of a palladium (Pd) counter electrode and a HyREF (Pd/H_2) reference electrode.

3.5. MS Spectrometry

The MS analysis was performed on an MS/MS mass spectrometer model 4000 Q TRAP (Applied Biosystems, Concord, Ontario, Canada). The quadrupole/linear ion trap is a hybrid system in which the final quadrupole can operate as a conventional mass filter or as a linear IT (ion trap) with an axial ion inject. The analysis was performed in the EMS and EPI (Enhanced Product Ion Scan) modes with an electrospray ionization source (ESI). The EPI was performed by fragmenting the pseudomolecular ions of TGE and its degradation compounds. Spray voltage was used in a negative ionization mode at −4500 V, curtain gas pressure was set at 20 psi, the capillary temperature was set at 200 °C, and declustering potential was set at −80 V. The collision energies applied were optimized for each compound individually.

3.6. IR Measurements

The infrared spectra were recorded on the Nicolet iS10 FT-IR spectrometer (Thermo Scientific, Waltham, MA, USA) using ATR sampling module, on the diamond crystal, in the range from 4000 to 650 cm^{-1}, with the spectral resolution of 4 cm^{-1}. For one spectrum 1000 scans were recorded. All data were analyzed using Omnic software. TGE monolayers were treated with hydrogen peroxide by immersing the electrodes in a solution of concentrations of 1 mM (5 min, 15 h) and 2 mM (20 min). After this, the electrodes were removed washed with ethanol and dried.

3.7. Quantum Mechanical Modeling

Quantum mechanical modeling was performed using the density functional B3LYP method with medium-size Gaussian basis sets, i.e., 6-31G(d), 6-31G(d,p), and 6-311++G(d,p) for the H, C, N, O, and S atoms depending on the system investigated. Molecular geometry was optimized with the Berny algorithm implemented in the Gaussian G16 program [55]. The minimum was confirmed with all positive harmonic frequencies. All the calculations were performed on the HPC cluster in the Interdisciplinary Centre for Mathematical and Computational Modelling at the University of Warsaw, Poland.

4. Conclusions

In the presented study, we identified the main oxidation products of TGE generated under different environmental conditions, which can in effect determine the input of a particular structural feature to the activity. Many experimental approaches were used for the characterization of TGE after oxidation. They were based on electrochemistry and mass spectrometry (ESI-MS). The electroactive centers of TGE were identified and its oxidation mechanisms were suggested. The main reaction products formed by the oxidation of TGE were identified. These product structures indicate reactivity of the thiogenistein C, A, and B rings in reactions with peroxyl radicals and with a potential. Three types of oxidations were observed: the oxidation by oxygen attachment, the oxidation by water attachment, and dehydrogenation as the mildest form of oxidation occurring during potentiation. Four main oxidation pathways were also observed during the potential oxidation. When hydrogen peroxide was used, the oxidation mechanism is bidirectional. When both oxidizing agents were used simultaneously, two new pathways were observed in addition. It was observed that where the oxidation occurs without potential (hydrogen peroxide, potential + hydrogen peroxide) the water molecule joins the double bond of the C ring first. Where oxidation with potential was used, more water molecules were observed to attach to the rings, including the A ring. It was observed that the A ring was more favored for oxidation due to the presence of an inductive effect. To summarize, our results show that the A ring is the critical active site in TGE that contributes to its *OH trapping efficacy, which is similar to the GE MGO trapping efficacy [56].

IR-ATR studies indicated that in the case of a tightly packed monolayer, the B ring was the most susceptible to oxidation, as it is the most protruding one. It should be taken into account that in the outer molecules bounding the 2D surface, the A ring may also undergo oxidation. For this reason, it is possible to obtain different TGE oxidation products in the monolayer.

The anticancer and antioxidative activity of our new thiolated analogue of genistein was evaluated based on cell culture and the ABTS$^{·+}$ radical cation. TGE has a high cytotoxic activity towards the human breast cancer cells (MCF-7 estrogen positive and MDA-MB-231 estrogen negative) and neutralizes the LPS-induced reactive oxygen species production better than GE, even though the thio-linker blocks one of its -OH groups. This can be interpreted by our results that the oxidation of TGE, on the one hand, takes place by passing the hydrogen atom to free oxygen radicals, and on the other hand, it can occur by attaching additional -OH groups to the frame of the molecule.

In this work, antioxidant activity was calculated by density functional theory studies for a possible correlation with the structures of TGE and products of electrochemical and oxidation conversion and investigating the mechanism of reactions of the TGE with the ABTS$^+$. The model system used for these studies is reasonably predictive of antioxidant chemistry in more complex biological systems.

The understanding of the biological activity of antioxidant compounds is a very important aspect of human life. We hope that understanding these procedures will help to determine key molecular fragments involved in the subsequent biochemical processes in living organisms and give information about their role in metabolic processes and pharmacological activity [57]. We plan to study whether reactions of TGE and peroxyl radicals in lipid bilayers in vitro and in biological membranes in vitro and in vivo yield products similar to those observed in this study in a homogeneous solution system.

Supplementary Materials: The following supporting information can be downloaded at: https://www.mdpi.com/article/10.3390/ijms23147816/s1.

Author Contributions: E.U.S.–idea, structure, and design of the paper; MS study; analyzed the data; wrote part of the manuscript; W.S.–in vitro study; wrote part of the paper; M.Ł.–ATR study; wrote part of the manuscript; A.L.–molecular modeling; a quantum mechanical density functional calculations; wrote part of the paper; E.M.–supervision of biological research; K.S.–idea for the paper, conceived and designed electrochemical experiments, performed the experiments; analyzed the data; wrote part of the paper. All authors have read and agreed to the published version of the manuscript.

Funding: This research received no external funding.

Institutional Review Board Statement: Not applicable.

Informed Consent Statement: Not applicable.

Data Availability Statement: Not applicable.

Acknowledgments: The study has been supported by the Polish Ministry of Science and Higher Education statutory grant no. 841343B. The calculations were performed at the Interdisciplinary Centre for Mathematical and Computational Modeling of the University of Warsaw (ICM UW, computer grant G18-6 and g85-962), which is kindly acknowledged for allocating facilities and computer time. The authors would like to thank Marta Zezula and Katarzyna Sidoryk for their support in this study.

Conflicts of Interest: The authors declare no conflict of interest.

References

1. Plattner, S.; Erb, R.; Chervet, J.P.; Oberacher, H. Studying the Reducing Potencies of Antioxidants with the Electrochemistry Inherently Present in Electrospray Ionization-Mass Spectrometry. *Anal. Bioanal. Chem.* **2014**, *406*, 213–224. [CrossRef] [PubMed]
2. Gutteridge, J.M. Free Radicals and Aging. *Rev. Clin. Gerontol.* **1994**, *4*, 279–288. [CrossRef]
3. Wang, T.Y.; Li, Q.; Bi, K.S. Bioactive Flavonoids in Medicinal Plants: Structure, Activity and Biological Fate. *Asian J. Pharm. Sci.* **2018**, *13*, 12–23. [CrossRef]
4. Harborne, J.B.; Williams, C.A. Advances in flavonoid Research since 1992. *Phytochemistry* **2000**, *55*, 481–504. [CrossRef]
5. Heim, K.E.; Tagliaferro, A.R.; Bobilya, D.J. Flavonoid Antioxidants: Chemistry, Metabolism and Structure-Activity Relationships. *J. Nutr. Biochem.* **2002**, *13*, 572–584. [CrossRef]
6. Waldmann, S.; Almukainzi, M.; Bou-Chacra, N.; Amidon, G.; Lee, B.; Feng, J.; Kanfer, I.; Zuo, J.; Wei, H.; Bolger, M.; et al. Provisional Biopharmaceutical Classification of Some Common Herbs Used in Western Medicine. *Mol. Pharm.* **2012**, *9*, 815–822. [CrossRef]
7. Sepehr, E.; Cooke, G.; Robertson, P.; Gilani, G.S. Bioavailability of Soy Isoflavones in Rats Part I: Application of Accurate Methodology for Studying the Effects of Gender and Source of Isoflavones. *Mol. Nutr. Food Res.* **2007**, *51*, 799–812. [CrossRef]
8. Sidoryk, K.; Michalak, O.; Kubiszewski, M.; Leś, A.; Cybulski, M.; Stolarczyk, E.U.; Doubsky, J. Synthesis of Thiol Derivatives of Biological Active Compounds for Nanotechnology Application. *Molecules* **2020**, *25*, 3470. [CrossRef]
9. Vo, Q.V.; Nam, P.C.; Thong, N.M.; Trung, N.T.; Phan, C.T.D.; Mechler, A. Antioxidant Motifs in Flavonoids: O−H versus C−H Bond Dissociation. *ACS Omega* **2019**, *4*, 8935–8942. [CrossRef]
10. Chen, Z.; Liu, Q.; Zhao, Z.; Bai, B.; Sun, Z.; Cai, L.; Fu, Y.; Ma, Y.; Wang, Q.; Xi, G. Effect of Hydroxyl on Antioxidant Properties of 2,3-Dihydro-3,5-Dihydroxy-6-Methyl-4: H -Pyran-4-One to Scavenge Free Radicals. *RSC Adv.* **2021**, *11*, 34456–34461. [CrossRef]
11. Han, R.M.; Tian, Y.X.; Liu, Y.; Chen, C.H.; Ai, X.C.; Zhang, J.P.; Skibsted, L.H. Comparison of Flavonoids and Isoflavonoids as Antioxidantsz. *J. Agric. Food Chem.* **2009**, *57*, 3780–3785. [CrossRef]
12. Castellano, G.; Torrens, F. Quantitative Structure-Antioxidant Activity Models of Isoflavonoids: A Theoretical Study. *Int. J. Mol. Sci.* **2015**, *16*, 12891–12906. [CrossRef] [PubMed]
13. Wright, J.; Johnson, E.; DiLabio, G. Predicting the Activity of Phenolic Antioxidants: Theoretical Method, Analysis of Substituent Effects, and Application to Major Families of Antioxidants. *J. Am. Chem. Soc.* **2001**, *123*, 1173–1183. [CrossRef]
14. Mahmoud, M.A.A.; Chedea, V.S.; Detsi, A.; Kefalas, P. Ascorbic Acid Modifies the Free Radical Scavenging Behaviour of Catechin: An Insight into the Mechanism. *Food Res. Int.* **2013**, *51*, 907–913. [CrossRef]
15. Proença, C.; Freitas, M.; Ribeiro, D. Inhibition of Protein Tyrosine Phosphatase 1B by Flavonoids: A Structure–Activity Relationship Study. *Food Chem. Toxicol.* **2018**, *111*, 474–481. [CrossRef] [PubMed]
16. Firuzia, O.; Lacanna, A.; Petrucci, R.; Marrosu, G.S.L. Evaluation of the Antioxidant Activity of Flavonoids by "Ferric Reducing Antioxidant Power" Assay and Cyclic Voltammetry. *Biochim. Biophys. Acta-Gen. Subj.* **2005**, *1721*, 174–184. [CrossRef] [PubMed]
17. Marshall, G.M.; Bensebaa, F.; Dubowski, J.J. Observation of Surface Enhanced IR Absorption Coefficient in Alkanethiol Based Self-Assembled Monolayers on GaAs(001). *J. Appl. Phys.* **2009**, *105*, 094310. [CrossRef]
18. Skoda, M.W.A.; Jacobs, R.M.J.; Willis, J.; Schreiber, F. Hydration of Oligo(Ethylene Glycol) Self-Assembled Monolayers Studied Using Polarization Modulation Infrared Spectroscopy. *Langmuir* **2007**, *23*, 970–974. [CrossRef]
19. Galato, D.; Ckless, K.; Susin, M.F.; Giacomelli, C.; Ribeiro-do-Valle, R.M.; Spinelli, A. Antioxidant Capacity of Phenolic and Related Compounds: Correlation among Electrochemical, Visible Spectroscopy Methods and Structure-Antioxidant Activity. *Redox Rep.* **2001**, *6*, 243–250. [CrossRef]
20. Stolarczyk, E.U.; Strzempek, W.; Łaszcz, M.; Leś, A.; Menaszek, E.; Sidoryk, K.; Stolarczyk, K. Anti-Cancer and Electrochemical Properties of Thiogenistein–New Biologically Active Compound. *Int. J. Mol. Sci.* **2021**, *22*, 8783. [CrossRef]
21. Gupta, V.K.; Jain, R.; Radhapyari, K.; Jadon, N.; Agarwal, S. Voltammetric Techniques for the Assay of Pharmaceuticals-A Review. *Anal. Biochem.* **2011**, *408*, 179–196. [CrossRef]
22. Stolarczyk, E.U.; Stolarczyk, K.; Łaszcz, M.; Kubiszewski, M.; Maruszak, W.; Olejarz, W.; Bryk, D. Synthesis and Characterization of Genistein Conjugated with Gold Nanoparticles and the Study of Their Cytotoxic Properties. *Eur. J. Pharm. Sci.* **2017**, *96*, 176–185. [CrossRef] [PubMed]

23. Tuli, H.S.; Tuorkey, M.J.; Thakral, F.; Sak, K.; Kumar, M.; Sharma, A.K.; Sharma, U.; Jain, A.; Aggarwal, V.; Bishayee, A. Molecular Mechanisms of Action of Genistein in Cancer: Recent Advances. *Front. Pharmacol.* **2019**, *10*, 1336. [CrossRef] [PubMed]

24. Grosso, R.; de-Paz, M.V. Thiolated-Polymer-Based Nanoparticles as an Avant-Garde Approach for Anticancer Therapies–Reviewing Thiomers from Chitosan and Hyaluronic Acid. *Pharmaceutics* **2021**, *13*, 854. [CrossRef] [PubMed]

25. Mueller, S.O.; Simon, S.; Chae, K.; Metzler, M.; Korach, K.S. Phytoestrogens and Their Human Metabolites Show Distinct Agonistic and Antagonistic Properties on Estrogen Receptor α (ERα) and ERβ in Human Cells. *Toxicol. Sci.* **2004**, *80*, 14–25. [CrossRef] [PubMed]

26. Bovee, T.F.H.; Schoonen, W.G.E.J.; Hamers, A.R.M.; Bento, M.J.; Peijnenburg, A.A.C.M. Screening of Synthetic and Plant-Derived Compounds for (Anti)Estrogenic and (Anti)Androgenic Activities. *Anal. Bioanal. Chem.* **2008**, *390*, 1111–1119. [CrossRef]

27. Pan, H.; Zhou, W.; He, W.; Liu, X.; Ding, Q.; Ling, L.; Zha, X.; Wang, S. Genistein Inhibits MDA-MB-231 Triple-Negative Breast Cancer Cell Growth by Inhibiting NF-KB Activity via the Notch-1 Pathway. *Int. J. Mol. Med.* **2012**, *30*, 337–343. [CrossRef]

28. Zhang, W.; Bao, L.; Zhang, X.; He, J.; Wei, G. Electropolymerization Treatment of Phenol Wastewater and the Reclamation of Phenol. *Water Environ. Res.* **2012**, *84*, 2028–2036. [CrossRef]

29. Marik, R.; Allu, M.; Anchoori, R.; Stearns, V.; Umbricht, C.B.; Khan, S. Potent Genistein Derivatives as Inhibitors of Estrogen Receptor Alpha-Positive Breast Cancer. *Cancer Biol. Ther.* **2011**, *11*, 883–892. [CrossRef]

30. Szultka-Młyńska, M.; Bajkacz, S.; Baranowska, I.; Buszewski, B. Structural Characterization of Electrochemically and in Vivo Generated Potential Metabolites of Selected Cardiovascular Drugs by EC-UHPLC/ESI-MS Using an Experimental Design Approach. *Talanta* **2018**, *176*, 262–276. [CrossRef]

31. Szultka-Młyńska, M.; Bajkacz, S.; Kaca, M.; Baranowska, I.; Buszewski, B. Electrochemical Simulation of Three Novel Cardiovascular Drugs Phase I Metabolism and Development of a New Method for Determination of Them by Liquid Chromatography Coupled with Tandem Mass Spectrometry. *J. Chromatogr. B Anal. Technol. Biomed. Life Sci.* **2018**, *1093–1094*, 100–112. [CrossRef] [PubMed]

32. Kallinich, C.; Schefer, S.; Rohn, S. Analysis of Protein-Phenolic Compound Modifications Using Electrochemistry Coupled to Mass Spectrometry. *Molecules* **2018**, *23*, 264. [CrossRef] [PubMed]

33. Deng, H.; Van Berkel, G.J. A Thin-Layer Electrochemical Flow Cell Coupled on-Line with Electrospray-Mass Spectrometry for the Study of Biological Redox Reactions. *Electroanalysis* **1999**, *11*, 857–865. [CrossRef]

34. Sagandykova, G.N.; Szultka-Młyńska, M.; Walczak-Skierska, J.; Pomastowski, P.P.; Buszewski, B. Combination of Electrochemical Unit and ESI-MS in Fragmentation of Flavonoids. *Phytochem. Anal.* **2021**, *32*, 601–620. [CrossRef]

35. Konermann, L.; Ahadi, E.; Rodriguez, A.D.; Vahidi, S. Unraveling the Mechanism of Electrospray Ionization. *Anal. Chem.* **2013**, *85*, 2–9. [CrossRef]

36. Potęga, A.; Żelaszczyk, D.; Mazerska, Z. Electrochemical Simulation of Metabolism for Antitumor-Active Imidazoacridinone C-1311 and in Silico Prediction of Drug Metabolic Reactions. *J. Pharm. Biomed. Anal.* **2019**, *169*, 269–278. [CrossRef]

37. Kulling, S.E.; Honig, D.M.; Metzler, M. Oxidative Metabolism of the Soy Isoflavones Daidzein and Genistein in Humans in Vitro and in Vivo. *J. Agric. Food Chem.* **2001**, *49*, 3024–3033. [CrossRef]

38. Roberts-Kirchhoff, E.S.; Crowley, J.R.; Hollenberg, P.F.; Kim, H. Metabolism of Genistein by Rat and Human Cytochrome P450s. *Chem. Res. Toxicol.* **1999**, *12*, 610–616. [CrossRef]

39. Maessen, D.E.; Stehouwer, C.D.; Schalkwijk, C.G. The Role of Methylglyoxal and the Glyoxalase System in Diabetes and Other Age-Related Diseases. *Clin. Sci.* **2015**, *128*, 839–861. [CrossRef]

40. Kwak, H.S.; Park, S.Y.; Kim, M.G.; Yim, C.H.; Yoon, H.K.; Han, K.O. Marked Individual Variation in Isoflavone Metabolism after a Soy Challenge Can Modulate the Skeletal Effect of Isoflavones in Premenopausal Women. *J. Korean Med. Sci.* **2009**, *24*, 867–873. [CrossRef]

41. Barclay, L.R.C. 1992 Syntex Award Lecture. Model Biomembranes: Quantitative Studies of Peroxidation, Antioxidant Action, Partitioning, and Oxidative Stress. *Can. J. Chem.* **1993**, *71*, 1–16. [CrossRef]

42. Arora, A.; Byrem, T.M.; Nair, M.G.; Strasburg, G.M. Modulation of Liposomal Membrane Fluidity by Flavonoids and Isoflavonoids. *Arch. Biochem. Biophys.* **2000**, *373*, 102–109. [CrossRef] [PubMed]

43. Xu, P.; Uyama, H.; Whitten, J.E.; Kobayashi, S.; Kaplan, D.L. Peroxidase-Catalyzed in Situ Polymerization of Surface Orientated Caffeic Acid. *J. Am. Chem. Soc.* **2005**, *127*, 11745–11753. [CrossRef]

44. Zhang, T.; Andrukhov, O.; Haririan, H.; Müller-Kern, M.; Liu, S.; Liu, Z.; Rausch-Fan, X. Total Antioxidant Capacity and Total Oxidant Status in Saliva of Periodontitis Patients in Relation to Bacterial Load. *Front. Cell. Infect. Microbiol.* **2016**, *5*, 97. [CrossRef] [PubMed]

45. Zhang, J.; Du, F.; Peng, B.; Lu, R.; Gao, H.; Zhou, Z. Structure, Electronic Properties, and Radical Scavenging Mechanisms of Daidzein, Genistein, Formononetin, and Biochanin A: A Density Functional Study. *J. Mol. Struct. THEOCHEM* **2010**, *955*, 1–6. [CrossRef]

46. Eskandari, M.; Rembiesa, J.; Startaitė, L.; Holefors, A.; Valančiūtė, A.; Faridbod, F.; Ganjali, M.R.; Engblom, J.; Ruzgas, T. Polyphenol-Hydrogen Peroxide Reactions in Skin: In Vitro Model Relevant to Study ROS Reactions at Inflammation. *Anal. Chim. Acta* **2019**, *1075*, 91–97. [CrossRef]

47. Di Meo, F.; Lemaur, V.; Cornil, J.; Lazzaroni, R.; Duroux, J.L.; Olivier, Y.; Trouillas, P. Free Radical Scavenging by Natural Polyphenols: Atom versus Electron Transfer. *J. Phys. Chem. A* **2013**, *117*, 2082–2092. [CrossRef]

48. Vásquez-Espinal, A.; Yañez, O.; Osorio, E.; Areche, C.; García-Beltrán, O.; Ruiz, L.M.; Cassels, B.K.; Tiznado, W. Structure-Antioxidant Activity Relationships in Boldine and Glaucine: A DFT Study. *New J. Chem.* **2021**, *45*, 590–596. [CrossRef]
49. Chiorcea-Paquim, A.M.; Enache, T.A.; De Souza Gil, E.; Oliveira-Brett, A.M. Natural Phenolic Antioxidants Electrochemistry: Towards a New Food Science Methodology. *Compr. Rev. Food Sci. Food Saf.* **2020**, *19*, 1680–1726. [CrossRef]
50. Rice-Evans, C.A.; Miller, N.J.; Paganga, G. Structure-Antioxidant Activity Relationships of Flavonoids and Phenolic Acids. *Free Radic. Biol. Med.* **1996**, *20*, 933–956. [CrossRef]
51. Majewska, M.; Czeczot, H. Flawonoidy w Profilaktyce i Terapii. *Farm. Pol.* **2009**, *65*, 369–377.
52. Kalinowski, D.S.; Yu, Y.; Sharpe, P.C.; Islam, Y.; Liao, T.; Lovejoy, D.B.; Kumar, N.; Bernhardt, P.V.; Richardson, D.R. Design, Synthesis, and Characterization of Novel Iron Chelators: Structure-Activity Relationships of the 2-Benzoylpyridine Thiosemicarbazone Series and Their 3-Nitrobenzoyl Analogues as Potent Antitumor Agents. *J. Med. Chem.* **2007**, *50*, 3716–3729. [CrossRef] [PubMed]
53. Richardson, D.R.; Sharpe, P.C.; Lovejoy, D.B.; Senaratne, D.; Kalinowski, D.S.; Islam, M.; Bernhardt, P.V. Dipyridyl Thiosemicarbazone Chelators with Potent and Selective Antitumor Activity Form Iron Complexes with Redox Activity. *J. Med. Chem.* **2006**, *49*, 651–6521. [CrossRef] [PubMed]
54. Re, R.; Pellegrini, N.; Proteggente, A.; Pannala, A.; Yang, M.; Rice-Evans, C. Antioxidant activity applying an improved abts radical cation decolorization assay roberta. *Free Radic. Biol. Med.* **1999**, *26*, 1231–1237. [CrossRef]
55. Frisch, M.J.; Truck, G.W.; Schlegel, H.B.; Scuseria, G.E.; Robb, M.A.; Cheeseman, J.R.; Scalmani, G.; Barone, V.; Petersson, G.A.; Nakatsuji, H.; et al. *Gaussian G16 Gaussian 16, Revis. A.03*; Gaussian, Inc.: Wallingford, CT, USA, 2016. Available online: http://www.gaussian.com (accessed on 1 December 2021).
56. Wang, P.; Chen, H.; Sang, S. Trapping Methylglyoxal by Genistein and Its Metabolites in Mice. *Chem. Res. Toxicol.* **2016**, *29*, 406–414. [CrossRef]
57. Santos-Sánchez, N.F.; Salas-Coronado, R.; Villanueva-Cañongo, C.; Hernández-Carlos, B. Antioxidant Compounds and Their Antioxidant Mechanism. In *Antioxidants*; Shalaby, E., Ed.; IntechOpen: London, UK, 2019. [CrossRef]

International Journal of
Molecular Sciences

Article

Discovery of New 3,3-Diethylazetidine-2,4-dione Based Thiazoles as Nanomolar Human Neutrophil Elastase Inhibitors with Broad-Spectrum Antiproliferative Activity

Beata Donarska [1], Marta Świtalska [2], Joanna Wietrzyk [2], Wojciech Płaziński [3,4], Magdalena Mizerska-Kowalska [5], Barbara Zdzisińska [5] and Krzysztof Z. Łączkowski [1,*]

[1] Department of Chemical Technology and Pharmaceuticals, Faculty of Pharmacy, Collegium Medicum, Nicolaus Copernicus University, Jurasza 2, 85-089 Bydgoszcz, Poland; 265000@stud.umk.pl
[2] Hirszfeld Institute of Immunology and Experimental Therapy, Polish Academy of Sciences, Rudolfa Weigla 12, 53-114 Wrocław, Poland; marta.switalska@hirszfeld.pl (M.Ś.); joanna.wietrzyk@hirszfeld.pl (J.W.)
[3] Jerzy Haber Institute of Catalysis and Surface Chemistry, Polish Academy of Sciences, Niezapominajek 8, 30-239 Cracow, Poland; wojtek_plazinski@tlen.pl
[4] Department of Biopharmacy, Medical University of Lublin, Chodzki 4a, 20-093 Lublin, Poland
[5] Department of Virology and Immunology, Faculty of Biology and Biotechnology, Maria Curie-Skłodowska University, Akademicka 19 Street, 20-033 Lublin, Poland; magdalena.mizerska-kowalska@mail.umcs.pl (M.M.-K.); basiaz@poczta.umcs.lublin.pl (B.Z.)
* Correspondence: krzysztof.laczkowski@cm.umk.pl

Citation: Donarska, B.; Świtalska, M.; Wietrzyk, J.; Płaziński, W.; Mizerska-Kowalska, M.; Zdzisińska, B.; Łączkowski, K.Z. Discovery of New 3,3-Diethylazetidine-2,4-dione Based Thiazoles as Nanomolar Human Neutrophil Elastase Inhibitors with Broad-Spectrum Antiproliferative Activity. *Int. J. Mol. Sci.* **2022**, 23, 7566. https://doi.org/10.3390/ijms23147566

Academic Editor: Sotiris K Hadjikakou

Received: 7 June 2022
Accepted: 6 July 2022
Published: 8 July 2022

Publisher's Note: MDPI stays neutral with regard to jurisdictional claims in published maps and institutional affiliations.

Abstract: A series of 3,3-diethylazetidine-2,4-dione based thiazoles **3a–3j** were designed and synthesized as new human neutrophil elastase (HNE) inhibitors in nanomolar range. The representative compounds **3c**, **3e**, and **3h** exhibit high HNE inhibitory activity with IC_{50} values of 35.02–44.59 nM, with mixed mechanism of action. Additionally, the most active compounds **3c** and **3e** demonstrate high stability under physiological conditions. The molecular docking study showed good correlation of the binding energies with the IC_{50} values, suggesting that the inhibition properties are largely dependent on the stage of ligand alignment in the binding cavity. The inhibition properties are correlated with the energy level of substrates of the reaction of ligand with Ser195. Moreover, most compounds showed high and broad-spectrum antiproliferative activity against human leukemia (MV4-11), human lung carcinoma (A549), human breast adenocarcinoma (MDA-MB-231), and urinary bladder carcinoma (UMUC-3), with IC_{50} values of 4.59–9.86 μM. Additionally, compounds **3c** and **3e** can induce cell cycle arrest at the G2/M phase and apoptosis via caspase-3 activation, leading to inhibition of A549 cell proliferation. These findings suggest that these new types of drugs could be used to treat cancer and other diseases in which immunoreactive HNE is produced.

Keywords: human neutrophil elastase; antiproliferative activity; thiazole; molecular docking

1. Introduction

Human neutrophil elastase (HNE, E.C. 3.4.21.37) is a 30 kDa serine protease of the chymotrypsin family stored in the azurophilic granules of neutrophils. This enzyme can hydrolyse extracellular matrix proteins such as elastin, fibronectin, proteoglycans, and collagens. Due to the fact that HNE is the main mediator of inflammation, it participates in the pathogenesis of various inflammatory diseases, including acute lung injury (ALI), acute respiratory distress syndrome (ARDS), chronic obstructive pulmonary disease (COPD), and rheumatoid arthritis and Wegener's granulomatosis [1–5]. Due to the possibility of tissue damage, HNE activity is regulated by specific endogenous inhibitors such as α1-antitrypsin, α2-macroglobulin, elafin and secretory leukocyte protease inhibitor [6–8]. Human neutrophil elastase has also been associated with the metastasis and progression of various types of cancer via its degradation of extracellular matrix components. A high immunoreactivity of HNE in patients with breast cancer and lung cancer is correlated with poor prognosis and a reduction in the survival rate [9–16]. HNE is responsible for cellular proliferation of lung cancer cells [17].

Therefore, human neutrophil elastase has become an important therapeutic target, which has resulted in the intensive search for small, effective HNE inhibitors as promising therapeutics. In recent years, many HNE inhibitors belonging to different classes of compounds have been designed (Figure 1), such as sivelestat sodium hydrate A (ONO-5046, Elaspol) [18], *N*-benzoylindazoles B [19], indoles [20], 1*H*-pyrrolo [2,3-b]pyridines C [21], isoxazol-5(2*H*)-ones D [22], heteroaryl oxime esters E [23], sulfonyl fluorides (SuFEx) F [24], and phthalimides [25]. A very interesting group of compounds that caught our attention are azetidine-2,4-diones, also known as 4-oxo-β-lactams [26]. These compounds showed the ability to serine acylation in HNE, and further research on the structure optimization showed that *N*-phenyl derivatives of 3,3-diethylazetidine-2,4-dione G (Figure 1) show much higher activity and selectivity towards HNE [27–32].

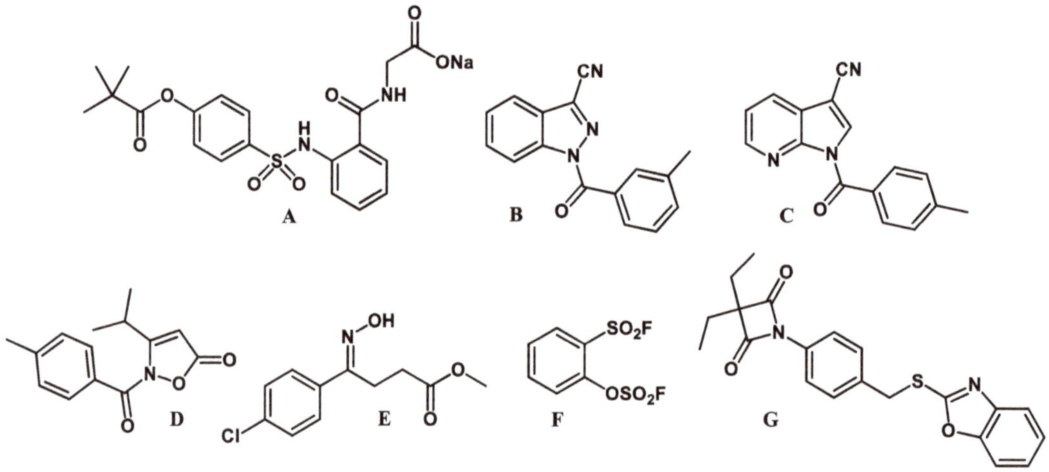

Figure 1. Structures of selected HNE inhibitors.

Due to the broad spectrum of activity, the thiazole scaffold is currently one of the most studied heterocyclic compounds [33,34]. We previously reported that compounds containing the thiazole ring show high antitumor activity and are characterized by low toxicity to healthy cells [35–37].

The aim of our project was two-fold. Firstly, we aimed at designing hybrid molecules containing the 3,3-diethylazetidine-2,4-dione and thiazole systems, which have the ability to inhibit human neutrophil elastase. Secondly, we investigated whether such derivatives exhibit anticancer activity. In addition, through the introduction of different substituents, we modified the electronic properties and geometry of the molecules. The inhibition of human neutrophil elastase was carried out by the spectrofluorimetric method using N-methoxysuccinyl-Ala-Ala-Pro-Val-7-amido-4-methylcoumarin (MeOSuc-AAPV-AMC) as substrate. The antiproliferative activity was tested against four human cancer cell lines, leukemia (MV4-11), lung (A549), breast (MDA-MB-231), urinary bladder (UMUC-3), and normal mouse fibroblast (BALB/3T3) cells using the 3-(4,5-dimethylthiazol-2-yl)-2,5-diphenyltetrazoliun bromide (MTT) or sulforhodamine B (SRB) assays. The molecular docking study of the inhibitors to the HNE binding site was used to analyze the influence of pharmacophore group on the activity of the studied compounds.

2. Results and Discussion

2.1. Chemical Synthesis

The synthetic pathway leading to the final thiazole derivatives **3a–3j** is reported in Scheme 1. In the first step 1-(4-(2-chloroacetyl)phenyl)-3,3-diethylazetidine-2,4-dione (**2**) was obtained starting from previously described 1-(4-aminophenyl)-2-chloroethanone (**1**) [38] treated with the appropriate 2,2-diethyl malonyl dichloride and triethylamine in dry dichloromethane. Next, using the Hantzsch reaction of precursor **2** with ten different thioureas, appropriate thiazole derivatives **3a–3j** were obtained with 52–99% yields. The structure of all compounds was confirmed on the basis of spectral data, including ^{1}H NMR (400 MHz), ^{13}C NMR (100 MHz), and ESI-HRMS analysis (see Supplementary material). The ^{1}H NMR spectrum of precursor 2 showed characteristic signals derived from two ethyl and a chloroacetyl groups at 0.98, 1.86 and 5.19 ppm, respectively. There is also a characteristic signal from the carbonyl group of the chloroacetyl fragment in the ^{13}C NMR spectrum at about 191 ppm. The final thiazoles **3a–3j** showed characteristic singlets at (7.30–7.51) ppm derived from thiazole-5*H* proton, whereas in the ^{13}C NMR spectrum we observe characteristic peaks at about 9.40 and 23.30 ppm, and at about 172 ppm derived from the 3,3-diethylazetidine-2,4-dione system. The high resolution mass spectrometry spectra fully support the proposed structures of the target compounds.

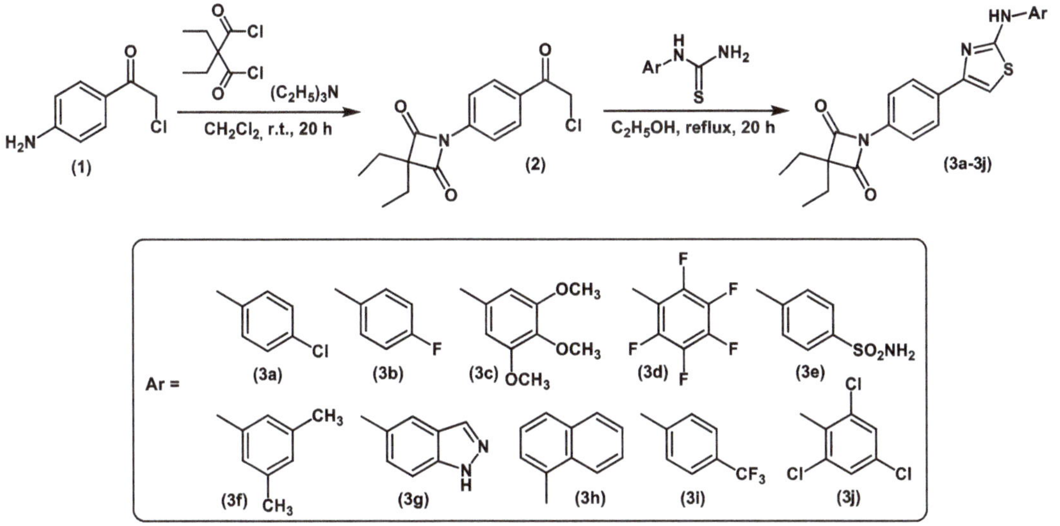

Scheme 1. Synthesis of the 3,3-diethylazetidine-2,4-dione based thiazoles **3a–3j**.

2.2. Elastase Inhibitory Activity and Kinetic Analysis

Inhibition of human neutrophil elastase was carried out by the spectrofluorimetric method using N-methoxysuccinyl-Ala-Ala-Pro-Val-7-amido-4-methyl-coumarin (MeOSuc-AAPV-AMC) as a substrate, and compared with sivelestat, which is a selective inhibitor of human neutrophil elastase. The results are summarized in Table 1. Curves for the determination of IC$_{50}$ values for inhibition of HNE activity of compounds **3a–3j** and sivelestat can be found in Supporting Information (see Figures S1–S11). All tested compounds **3a–3j** showed the ability to inhibit HNE in nanomolar concentration, with IC$_{50}$ values in the range of 35.02–312.19 nM. Among the tested compounds, the highest activity was demonstrated by thiazoles **3c**, **3e** and **3h** with IC$_{50}$ values 38.25, 35.02 and 44.59 nM, respectively. These compounds contain the trimethoxyphenyl, benzenesulfonamide and naphthyl groups. These values are only two times lower than the inhibition value for standard sivelestat, which was 18.78 nM. It can be seen that the HNE inhibitory activity increases in a series of compounds **3b**, **3i** and **3d** containing 4-fluorophenyl, 4-trifluoromethylphenyl and perfluorophenyl groups with IC_{50} values 176.28, 167.57, and 110.02 nM, respectively. This is probably due to an increased electron-acceptor or steric effect. The opposite effect is observed for compounds **3a** and **3j** containing the 4-chlorophenyl and 2,4,6-trichlorophenyl groups with IC_{50} 227.29 and 248.74 nM, respectively. Also, the same relationship can be seen for derivatives **3f** and **3j** containing the 3,5-dimethylphenyl and 2,4,6-trichlorophenyl groups with IC_{50} values of 243.96 and 248.74 nM, respectively.

Table 1. Human neutrophil elastase inhibitory activity of 3,3-diethylazetidine-2,4-dione based thiazoles **3a**–**3j** and kinetic analysis of their mechanism of action.

Thiazole Derivatives	$IC_{50} \pm SD$ [nM]	Dose [μM]	V_{max}	K_m	Inhibition Type	K_i [nM]	K_{is} [nM]
3a	227.29 ± 35.08	100.00 250.00	53.19 55.25	195.19 264.55	mixed	590.41	426.00
3b	176.28 ± 13.49	100.00 250.00	63.29 15.95	280.95 112.31	mixed	72.15	65.50
3c	38.25 ± 4.91	50.00 100.00	40.65 25.58	193.65 207.26	mixed	66.06	99.50
3d	110.02 ± 4.41	100.00 250.00	30.49 10.59	126.00 90.08	mixed	138.22	49.33
3e	35.02 ± 3.88	50.00 100.00	28.82 28.57	210.77 355.14	mixed	35.78	240.00
3f	243.96 ± 19.40	100.00 250.00	44.64 37.31	186.93 204.42	mixed	144.40	426.00
3g	312.19 ± 5.76	250.00 500.00	56.50 15.97	229.24 74.35	mixed	137.24	171.25
3h	44.59 ± 6.04	50.00 100.00	29.15 15.04	190.67 184.09	mixed	33.94	46.50
3i	167.57 ± 18.09	100.00 250.00	75.76 32.26	284.78 154.24	mixed	596.64	425.00
3j	248.74 ± 32.76	250.00 500.00	24.63 6.88	132.14 71.54	mixed	209.35	38.00
Sivelestat	18.78 ± 0.13	–	–	–	–	–	–

V_{max}—maximum velocity, K_m—inhibition constant, K_i—enzyme-inhibitor dissociation constants, K_{is}—enzyme-substrate-inhibitor dissociation constants.

Compound **3g** containing the indazole group shows the lowest inhibitory activity, IC_{50} 312.19 nM, and replacement of this group with a naphthyl group significantly increases the inhibitory activity.

Next, kinetic analysis of the mechanism of inhibition of human neutrophil elastase by thiazoles **3a**–**3j** was determined using double-reciprocal plots of Lineweaver-Burk plots and Dixon analysis. The results are shown in Table 1. In our research, we showed that all derivatives **3a**–**3j** showed a mixed mechanism of inhibition of elastase, indicating that the tested compounds can inhibit elastase in two different models.

The first model is one in which the inhibitor binds better to the free enzyme (competitive mechanism), while the second model is one in which the inhibitor binds better to the enzyme-substrate complex (non-competitive mechanism). Next the secondary plots of slope versus concentration of test compounds were plotted to determine the enzyme-inhibitor dissociation constants K_i, and similarly as was used to determine enzyme-substrate-inhibitor dissociation constants K_{is}, the secondary plots of intercept versus concentration of tested compounds were plotted. The K_i values for the compounds **3c**, **3e**, **3f**, **3g**, and **3h** were lower than the K_{is} values, suggesting that these inhibitors bind more strongly to the free enzyme than the enzyme-substrate complex (mixed type I inhibition). For the remaining compounds, the K_i values are higher than the K_{is} values, indicating that these compounds have a higher affinity for the enzyme-substrate complex (mixed type II inhibition). The plot for the most active inhibitor **3e** is presented in Figure 2.

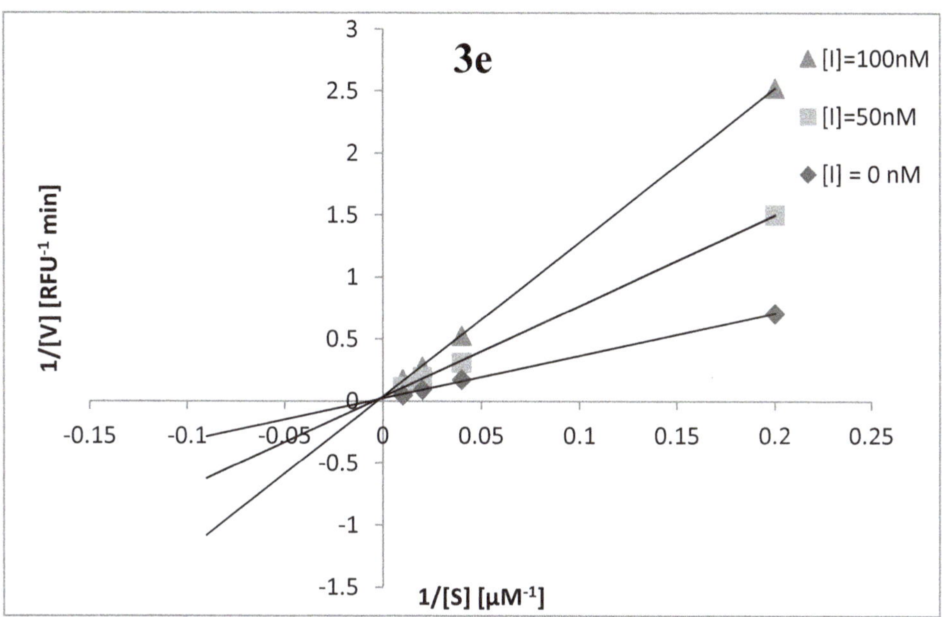

Figure 2. Lineweaver-Burk plots for inhibition of human neutrophil elastase in the presence of compound **3e**.

2.3. Molecular Docking Study

According to reference [32], the studied compounds are likely to be bound to protein by the formation of a covalent bond between serine and the azetidine-2,4-dione moiety, followed by the decomposition of the ligand molecule into two fragments. Thus, either standard (non-covalent) docking or the covalent docking involving the whole ligand molecule do not seem to be suitable for predicting the inhibition properties due to the following facts: (i) the low-energy ligand poses do not necessarily correspond to the conformations that facilitate the reaction with serine, thus, they may be chemically meaningless; (ii) after the ligand decomposition, the covalently bound ligand fragment is exactly the same for all considered compounds, thus, performing covalent docking would not produce any distinguishable results. To overcome these difficulties, we decided to carry out the standard docking that takes into account the complete molecular structure of all ligands but analyze only those poses that fulfil additional structural criteria. Namely, it was assumed that in order to enable the chemical reaction between the azetidine-2,4-dione moiety and serine (according to the structural data in the PDB database, covalent ligand bonding occurs always with contribution of Ser195) the distance between the serine hydroxyl oxygen atom and any of the two oxygens in the azetidine-2,4-dione group must be smaller than 0.3 nm. The poses that do not fulfil these cutoff-based criteria were discarded in further analysis. (Docking for sivelestat was carried out as well and an analogous distance-based cutoff was applied. According to reference [39], covalent binding occurs involving the carbonyl group; this was accounted for when defining the corresponding cutoff).

This procedure can be considered as closely related to the covalent docking of whole ligand molecules or to docking with energy constraints on the interatomic distance. Here, the role of either additional covalent bonds or the constraint-inherent energy contribution is played by 'logical constraint', i.e., filtering out the ligand poses that fall outside the cutoff criteria. The molecular docking study was used to analyze the influence of the pharmacophore group on the activity of the studied compounds. The binding energies found during docking simulations are given in Table 2 and graphically illustrated in Figure 3. Each of the final energy values was averaged over the set of poses exhibiting the same structural orientation in the binding cavity, as confirmed by separate RMSD calculations and visual inspection (Figure 3B). All of the obtained ligand-protein interaction energies display similar magnitude, varying in the range of ~ -7.3–-8.5 kcal/mol. The most favorable interactions are exhibited by compounds **3c**, **3e**, **3h** and sivelestat. These compounds displayed their high inhibition potencies in the experimental studies. Moreover, the binding energies calculated for all sets of compounds are fairly well correlated with the experimentally determined values of IC_{50} (Figure 3A). Thus, one can conclude that a reasonable agreement between the theoretical and experimental results exists, especially when considering the mechanism of ligand binding which can be tackled by a docking study only in an indirect manner. At the same time one can observe that the compounds exhibiting the highest IC_{50} values were not always correctly identified in the calculations; this is discussed later.

Table 2. The docking results with respect to the studied ligands. The calculations were averaged over the five structures available in the PDB database; the corresponding standard deviations are given as well.

Compound	Binding Energy $\pm$ SD [kcal/mol]
3a	-7.32 ± 0.24
3b	-7.81 ± 0.23
3c	-8.39 ± 0.54
3d	-8.15 ± 0.31
3e	-8.42 ± 0.63
3f	-7.82 ± 0.26
3g	-8.23 ± 0.22
3h	-8.53 ± 0.21
3i	-7.98 ± 0.14
3j	-7.76 ± 0.08
Sivelestat	-8.51 ± 0.22

The relatively good correlation of the binding energies with the IC_{50} values suggests that the inhibition properties are largely dependent on the stage of ligand alignment in the binding cavity. In other words, the inhibition properties are correlated with the energy level of substrates of the reaction of ligand with Ser195.

The very small scatter of ligand-protein interaction energies across the studied set of compounds (1.2 kcal/mol) can be explained in terms of their high structural similarity (they differ only by the type of one substituent) combined with extremely close locations and orientations exhibited by them when interacting with the binding cavity.

The results of the docking studies were also analyzed with respect to the mechanistic protein-ligand interaction pattern that is essential for interpretation of the obtained binding energy values. The summary given below relies on analysing the ligand-protein contacts that take place if the distance between any corresponding atom pair is smaller than the arbitrarily accepted value of 0.4 nm. The provided description can be related to all the studied compounds due to their nearly identical orientations in the binding cavity, which was partially imposed by cutoff criteria (Figure 3). The accuracy limits in this context result mainly from the molecular topology of the ligand, i.e., different chemical character of a single moiety). The graphical illustration of the docking results (on example of **3e**) is given in Figure 3C.

The network of interactions responsible for distinguishing between potencies of particular compounds is created by Trp172, Thr175, Phe215 and (partially) Arg217. Independently of the compound, its limiting, aromatic substituent always maintains contact with Phe215 and Trp172 via attractive π-π or CH-π interactions. The presence of interactions with Thr175 is dependent on the character of the substituent and concerns only those compounds which contain polar groups in this region of molecule. Interactions with Arg217 occurs mainly with thiazole ring (as in Figure 3C and according to description below), however, the Arg217 sidechain is the most flexible among all sidechains considered explicitly during docking and can exhibit interactions also with the limiting substituent when the latter is polar. Although corresponding docking poses are only of the secondary type, the difference in energy levels is rather small (<0.3 kcal/mol), which suggests the consideration of this contact in the discussion.

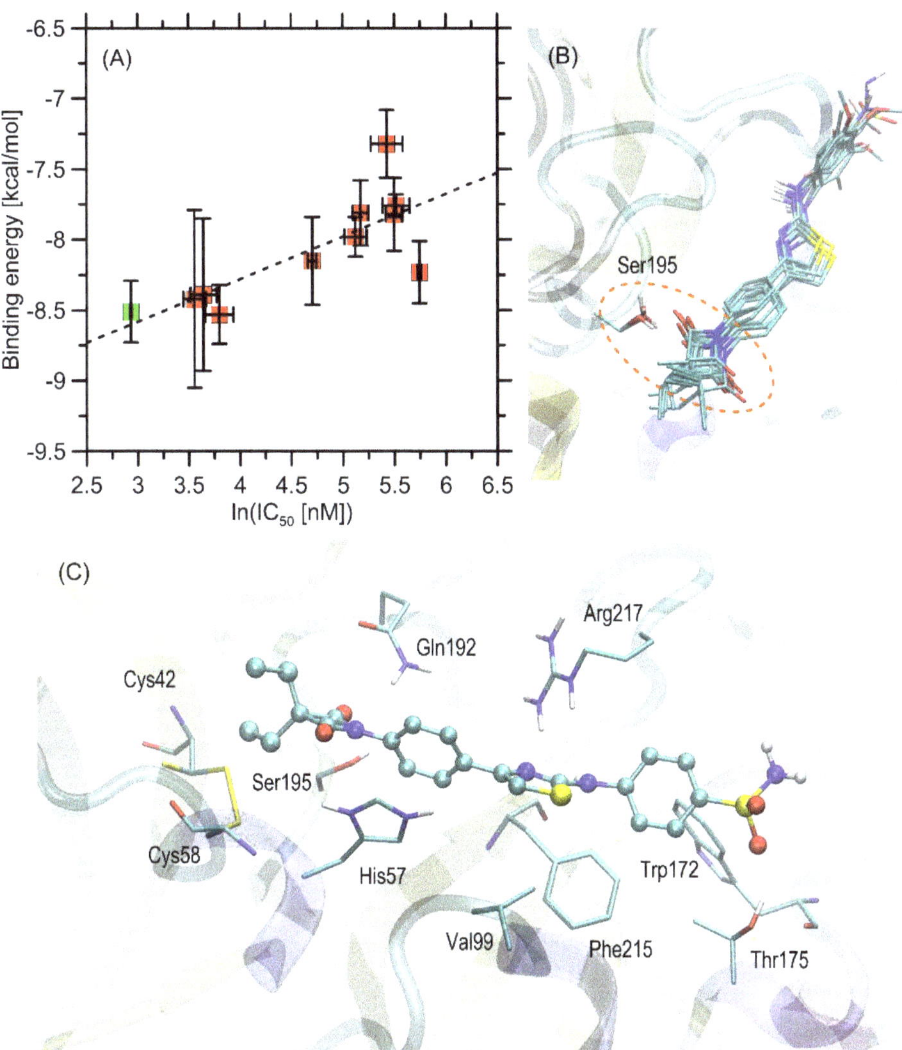

Figure 3. (**A**) The correlation between the calculated binding energies, averaged over the five elastase structures available in the PDB database, and the corresponding (experimental) values of IC_{50}. The horizontal bars represent the standard deviation values within the set of binding energies. Green symbol corresponds to sivelestat whereas red ones correspond to the remaining compounds. (**B**) The superposed positions of all studied ligands (stick representation) in the binding cavity of the 1bma structure of elastase. All of these positions were identified as the optimal ones during the standard (non-covalent) docking procedure with additional conditions imposing the maximal distance of oxygen atoms at the azetidine-2,4-dione group and Ser195 as equal to 0.3 nm. (**C**) The location of the **3e** ligand molecule bound to the 1bma structure of elastase and found in the docking procedure. The ligand molecule is shown in a ball-and-stick representation, whereas all the closest amino-acid residues are represented by sticks. The description of the interaction types is given in the text.

As the physical-chemical character of interactions formed by this crucial substituent and neighboring residues is not uniform and involves e.g., hydrogen bonding donation and acceptance, CH-π and π-π stacking, the detailed interpretation of the dependence of the moiety type on the compound potency is not straightforward. It seems that binding energies determined in the study result from the detailed balance of either favorable or non-favorable interactions of competitive character.

For instance, the compounds exhibiting the most favorable binding energies (**3c**, **3e** and **3h**) have a distinctly different character of their substituents and the corresponding favorable interactions with protein can be rationalized differently. Compounds **3c** and **3e** contain a relatively large number of hydrogen bonding acceptors and/or donors which enables attractive interactions with Arg217 and Thr175. Compound **3h** does not contain any polar group in the corresponding substituent, but this is compensated by the enhanced magnitude of the π-π interactions of the condensed aromatic ring with Trp172 and Phe215. Compound **3g**, also bearing a larger, aromatic (indazole) ring is not capable of creating equally attractive interactions due to different substitution positions and, subsequently, the altered relative orientation of the ring with respect to the Phe215 sidechain. Finally, the energy balance, in particular the intermolecular contacts, depends partially on the factors that cannot be explicitly considered during docking e.g., the presence of solvent or the mobility of the protein backbone. The latter factor is demonstrated by a high variability of the determined binding energies and their dependence on the geometry of the protein backbone around Val99, as observed for compounds **3c** and **3e**.

The remaining ligand-protein interactions concern the polar interactions of Arg217 with thiazole ring (hydrogen bonding) and CH-π stacking also involving the thiazole ring and Val99 sidechain. The phenyl ring of the ligand maintains contact with His57 (aromatic π-π interactions) and Gln192. The same Gln192 is located in the vicinity of the azetidine-2,4-dione group where it is able to form hydrogen bonding with the nitrogen atom, stabilizing the arrangement of the azetidine-2,4-dione moiety with respect to Ser195. Thus, this type of interaction is probably essential in the context of the formation of covalent bonding with Ser195.

The part of the ligand molecule that contains the azetidine-2,4-dione moiety is especially crucial for the occurrence of the reaction with Ser195. Due to imposed logical constraints, the attractive, hydrogen bonding-mediated interactions are always observed in the case of one of the oxygen atoms in the azetidine-2,4-dione group and neighboring Ser195 sidechain. Apart from this contact (and the presence of Gln192, mentioned above) no other specific, attractive interactions can be ascribed to this part of the ligand molecule. The vicinity of the two cysteines (Cys42 and Cys58) and the aliphatic substituents of the azetidine-2,4-dione group seems to be only an opportunistic consequence of other protein-ligand interactions discussed above.

Thus, to summarize, the docking results showed that the potency of compounds is correlated with the energy level of the ligand-protein complex (substrates for the subsequent ligand-protein reactions resulting in the formation of a covalent bond) but only when a certain arrangement of the ligand is achieved, enabling for occurring further stages of process. This arrangement relies on close contact of the azetidine-2,4-dione group and the Ser195 sidechain, maintained through hydrogen bonding and stabilized by interactions with Gln192. The order of potencies of compounds considered in the present study depends on the interactions with Trp172, Thr175, Phe215 and (partially) Arg217.

2.4. Antiproliferative Activity

3,3-Diethylazetidine-2,4-dione based thiazoles **3a–3j** were assessed for their antiproliferative activity against a panel of four cancer cell lines: human biphenotypic B myelomonocytic leukemia (MV4-11), lung carcinoma (A549), human breast adenocarcinoma (MDA-MB-231), urinary bladder carcinoma (UMUC-3), and normal mouse fibroblast (BALB/3T3) cells. Biological studies were carried out using the 3-(4,5-dimethylthiazol-2-yl)-2,5-diphenyltetrazolium bromide (MTT) or sulforhodamine B (SRB) assays. The results are summarized in Table 3.

Curves for the determination of IC$_{50}$ values of antiproliferative activity of compounds **3a–3j** can be found in Supporting Information (see Figures S12–S16). All derivatives **3a–3j** showed high activity against MV-4-11 cells, with IC$_{50}$ values between 4.59–7.15 µM. The antiproliferative activity of the tested compounds against human lung carcinoma (A549) cells is also very high, with IC$_{50}$ values in the range of 5.59–7.31 µM. Some decrease in activity for this type of cancer was observed for compounds **3d** and **3h**, with IC$_{50}$ values 15.28 and 9.20 µM, respectively. Our research also showed that compounds **3a–3c**, **3e** and **3g** have very high activity against human breast adenocarcinoma (MDA-MB-231) with IC$_{50}$ values between 6.19 and 9.86 µM. Compounds **3f**, **3h**, **3i** and **3j** showed good activity with IC$_{50}$ values in the range of 11.65–14.64 µM. In this type of cancer, compound **3d** showed the lowest activity with an IC$_{50}$ value of 72.09 µM. Studies with urinary bladder carcinoma (UMUC-3) cells showed that four compounds **3b**, **3c**, **3e**, an **3g** showed good activity against this type of cancer, with IC$_{50}$ values between 6.90 and 9.07 µM.

Table 3. Antiproliferative activity of thiazoles **3a–3j** against human biphenotypic B myelomonocytic leukemia (MV4-11), lung carcinoma (A549), human breast adenocarcinoma (MDA-MB-231), urinary bladder carcinoma (UMUC-3), and normal mouse fibroblast BALB/3T3 cells.

Thiazole Derivatives	IC$_{50}$ ± SD [µM]				
	MV-4-11	**A-549**	**MDA-MB-231**	**UMUC-3**	**Balb/3T3**
3a	7.15 ± 0.94	7.01 ± 0.47	9.86 ± 0.94	10.68 ± 1.41	10.87 ± 1.18
3b	5.92 ± 1.71	7.31 ± 0.46	9.34 ± 1.12	9.07 ± 0.49	11.49 ± 1.47
3c	4.59 ± 2.03	6.40 ± 0.08	6.38 ± 0.71	6.90 ± 0.21	7.42 ± 1.04
3d	5.40 ± 2.06	15.28 ± 4.98	72.09 ± 25.98	62.73 ± 9.98	46.52 ± 4.99
3e	6.23 ± 0.42	6.32 ± 0.89	6.42 ± 1.70	7.49 ± 0.64	6.57 ± 0.21
3f	5.89 ± 1.67	7.28 ± 1.07	12.48 ± 1.79	19.70 ± 5.49	23.83 ± 4.53
3g	5.38 ± 0.46	5.59 ± 0.16	6.19 ± 0.42	7.89 ± 0.23	6.03 ± 0.46
3h	6.64 ± 1.59	9.20 ± 1.36	14.64 ± 3.17	13.31 ± 1.81	22.44 ± 8.62
3i	5.47 ± 1.52	6.42 ± 0.72	11.65 ± 3.70	10.67 ± 3.05	10.13 ± 0.87
3j	5.11 ± 1.42	6.21 ± 0.73	14.30 ± 3.45	25.29 ± 9.94	34.48 ± 22.72
cisplatin	1.33 ± 0.17	4.09 ± 1.00	19.43 ± 6.80	1.73 ± 0.17	8.04 ± 3.16

IC$_{50}$—compound concentration leading to 50% inhibition of cell proliferation. IC values were calculated for each experiment separately (experiments were repeated 3–5 times) and data are presented as mean values ± SD, calculated using Prolab-3 system based on Cheburator 0.4 software [40].

The less potent of the series for this cancer cell line was compounds **3a**, **3h** and **3i**, with IC$_{50}$ values in the range of 10.67–13.31 µM. Compounds **3d**, **3f** and **3j** showed the lowest activity with IC$_{50}$ value in the range of 19.70–62.73 µM. Due to the lack of a standard drug that is structurally similar to our compounds, we decided to use cisplatin, which is active against all cancer cell lines tested. Studies have shown that our compounds, except **3d**, are more active against MDA-MB-231 than cisplatin. Also, most compounds show less toxicity than cisplatin.

Anticancer drugs are toxic to cancer cells, but also show some toxicity to healthy cells. Therefore, in the next step, we determined the selectivity index (SI) of the cytotoxic activity of the 3,3-diethylazetidine-2,4-dione based thiazoles by comparing the cytotoxic activity (IC$_{50}$) of compounds against the normal fibroblasts BALB/3T3 with the cytotoxic activity (IC$_{50}$) of cancer cell lines (Table 4). The selectivity indexes for thiazoles **3a–3j** against MV4-11 and A-549 cells were in general much higher than for MDA-MB-231 and UMUC-3 cells. The compounds **3f**, **3h** and **3j** showed the higher selectivity index against all tested cancer cell lines.

Table 4. The selectivity index (SI) of tested compounds. The SI index = IC_{50} for normal cell line (Balb/3T3)/IC_{50} for appropriate cancer cell line.

Thiazole Derivatives	Cell Lines/Calculated Selectivity Index SI			
	MV-4-11	A-549	MDA-MB-231	UMUC-3
3a	1.52	1.55	1.10	1.02
3b	1.94	1.57	1.23	1.27
3c	1.62	1.16	1.16	1.07
3d	8.61	3.04	0.64	0.74
3e	1.05	1.04	1.02	0.88
3f	4.04	3.27	1.90	1.21
3g	1.12	1.08	0.97	0.76
3h	3.38	2.44	1.53	1.69
3i	1.85	1.58	0.87	0.95
3j	6.75	5.55	2.41	1.36

A beneficial SI > 1.0 indicates a drug with efficacy against cancer cells greater than toxicity against normal cells.

2.5. Apoptosis and Cell Cycle Assays

To explore the antiproliferative activity of the tested compounds, the influence of the most active ones, i.e., **3c** and **3e**, on apoptosis induction and caspase-3 activation and the distribution of cell cycle phases in the A549 cells after 48-h incubation were analyzed by flow cytometry. The rate of cell apoptosis and necrosis was detected using Annexin V-FITC/PI double staining. As shown in Figure 4A,B, both compounds were found to induce apoptosis in the A549 cells in a concentration-dependent manner, whereas no induction of necrosis was observed. Compound **3e** turned out to be a stronger inducer of apoptosis than **3c**. The percentage of A549 cells undergoing apoptosis increased significantly from $6.05 \pm 0.17\%$ in the control to $8.79 \pm 0.50\%$, $14.45 \pm 1.02\%$, and $23.39 \pm 0.31\%$ after the incubation with 5, 10, or 20 μM of the **3e**, respectively (Figure 4B). In turn, compound **3c** induced apoptosis only at the concentrations of 10 and 20 μM, and the percentage of cells undergoing apoptosis increased from $6.05 \pm 0.17\%$ in the control to $10.14 \pm 0.76\%$ and $14.67 \pm 0.68\%$, respectively (Figure 4B). Moreover, compound **3e** at a concentration of 5 μM induced apoptosis to a similar level as cisplatin (CDDP) used at a concentration of 4 μM. To identify the mechanism of the **3c**- and **3e**-induced apoptosis in the A549 cells, the activation of effector caspase-3 was evaluated by flow cytometry. As shown in Figure 5A,B, the treatment of the cells with both compounds resulted in a concentration-dependent increase in the number of cells with active caspase 3. To further explore the antiproliferative activity of compounds **3c** and **3e**, the influence of both compounds on the distribution of cell cycle phases in the A549 cells was analyzed. The **3c** or **3e** treatment resulted in an increased number of A549 cells in the G2/M phase and the sub-G1 apoptotic subpopulation with a concomitant reduction of the cell number in the G0/G1 phase (Figure 6A–C). Compared to the control, compound **3c** used at a concentration of 5, 10, or 20 μM significantly elevated the G2/M fraction from $6.20 \pm 0.23\%$ to $9.69 \pm 0.77\%$, $15.68 \pm 5.31\%$, and $29.04 \pm 1.09\%$, respectively (Figure 6B). The exposure of the cells to 0, 5, 10, and 20 μM of compound **3e** resulted in the accumulation of cells in the G2/M phase from $6.20 \pm 0.23\%$ to $9.08 \pm 0.69\%$, $10.83 \pm 1.06\%$, and $35.78 \pm 0.46\%$, respectively (Figure 6C). The G2 checkpoint prevents cells from entering mitosis when DNA is damaged, providing an opportunity for repair and stopping the proliferation of damaged cells [41]. The cell cycle arrest at the G2/M phase indicates that the damage of intracellular DNA is difficult to repair [42]. Apoptosis is a crucial process involved in the regulation of tumor formation and in the treatment response; therefore, cancer therapy has been linked to activation of the apoptosis signal transduction pathway. Caspase-3, an executioner caspase, plays an important role in apoptosis and becomes a primary target for cancer treatment [43]. In conclusion, these results indicate that compounds **3c** and **3e** can induce cell cycle arrest at the G2/M phase and apoptosis via caspase-3 activation, leading to the inhibition of A549 cell proliferation.

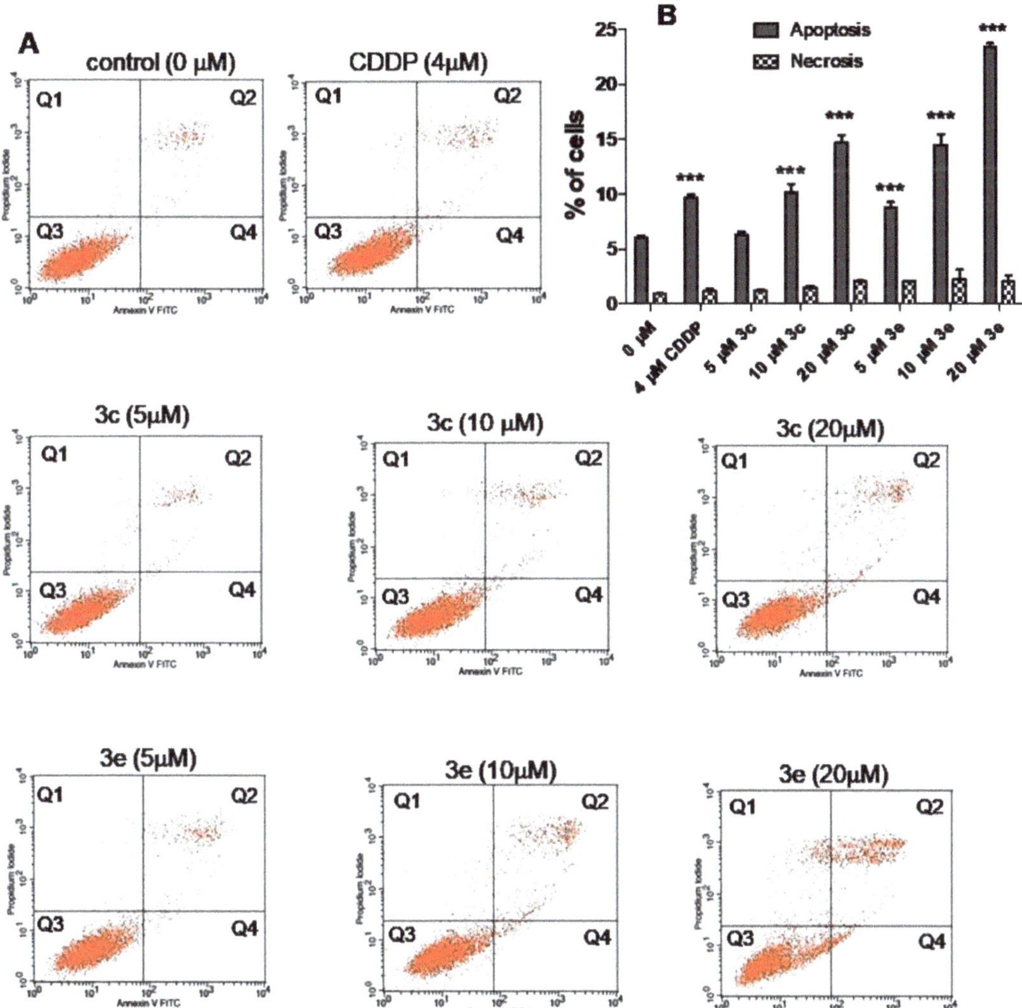

Figure 4. Effect of compounds **3c** and **3e** and cisplatin (CDDP) on apoptosis induction in the A549 cell line. After the 48-h exposure to the indicated concentrations of the compounds, the cells were stained with annexin (An) V-FITC)/propidium iodide (PI) and examined with flow cytometry. (**A**) The representative graphs obtained by flow cytometric analysis: dot plots indicate the percentage of An−/PI+ necrotic cells (Q1), An+ /PI+ late apoptotic cells (Q2), An−/PI− viable cells (Q3), and An+ /PI− early apoptotic cells (Q4). (**B**) Statistical analysis of the apoptotic (early + late apoptosis) and necrotic rate populations after the **3c**, **3e**, and cisplatin (CDDP) treatment. Data are presented as mean ± SD from three independent experiments; *** $p < 0.001$ in comparison to the control; one-way ANOVA followed by Dunnett's post hoc test.

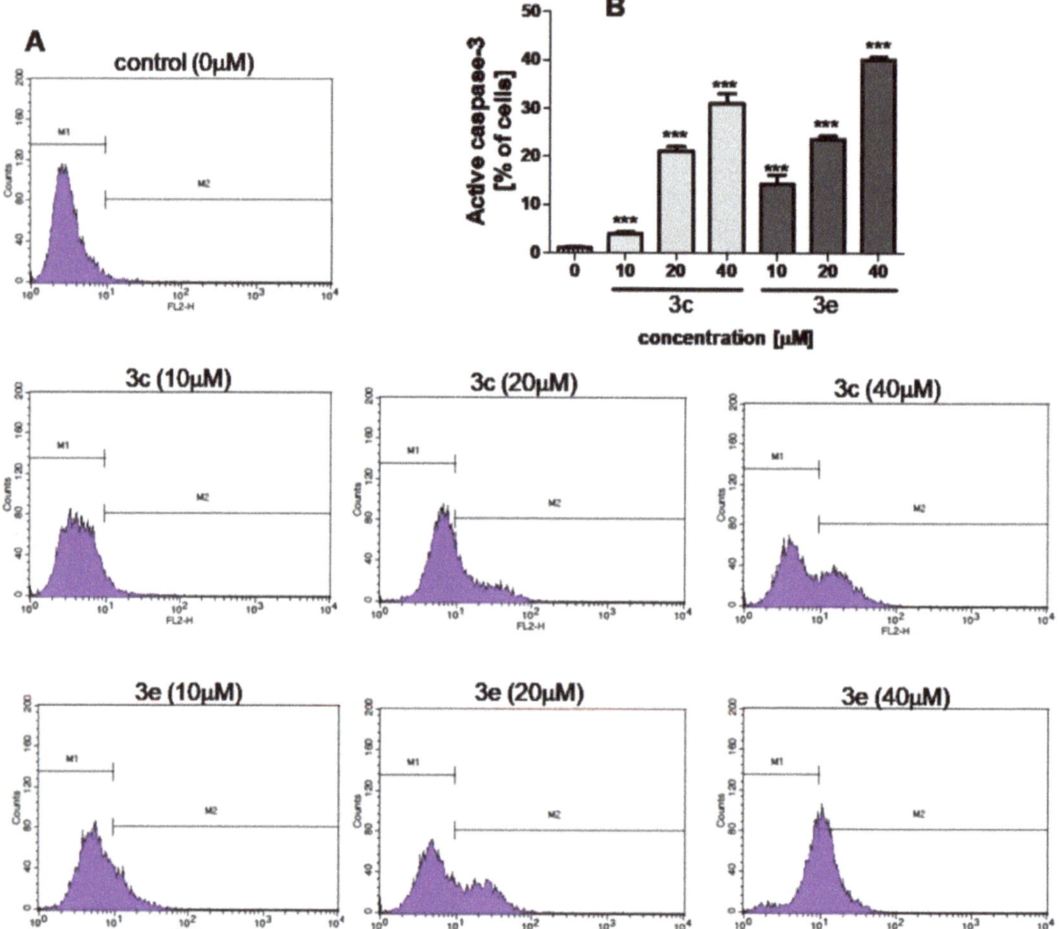

Figure 5. Flow cytometry analysis of active caspase-3 in A549 cells treated with compounds **3c** and **3e** for 48 h. (**A**) Representative histograms of A549 cell culture—M1 and M2 gates represent peaks for viable (caspase-3 negative cells) and apoptotic cell fractions (caspase-3 positive cells detected after cell staining with PE-conjugated anti-active caspase-3 antibodies), respectively. (**B**) Quantification of caspase-3 activity in A549 cell culture. The results are mean values ± SD of three independent experiments; statistical significance at $p < 0.001$ *** in comparison to the control; one-way ANOVA followed by Dunnett's post hoc test.

2.6. Compounds Stability

Testing the stability of new drugs is of great importance for obtaining a good therapeutic profile as well as the safety of their future use. Degradation impurities can reduce drug efficacy and generate undesirable side effects. Therefore, two of the most active 3,3-diethylazetidine-2,4-dione based thiazoles **3c** and **3e** were evaluated for chemical stability in an aqueous phosphate buffer at pH 7.3 imitating physiological conditions using spectrophotometric analysis (Table 5). An exemplary spectrum for compound **3e** is presented in Figures 7 and 8. The absorbance maxima of compound **3c** at 293 nm and at 262 and 322 nm for compound **3e** decreased over time, indicating spontaneous hydrolysis with a $t_{1/2}$ of 46.2 and 39.6 min, respectively. The tested compounds demonstrate high stability under physiological pH conditions.

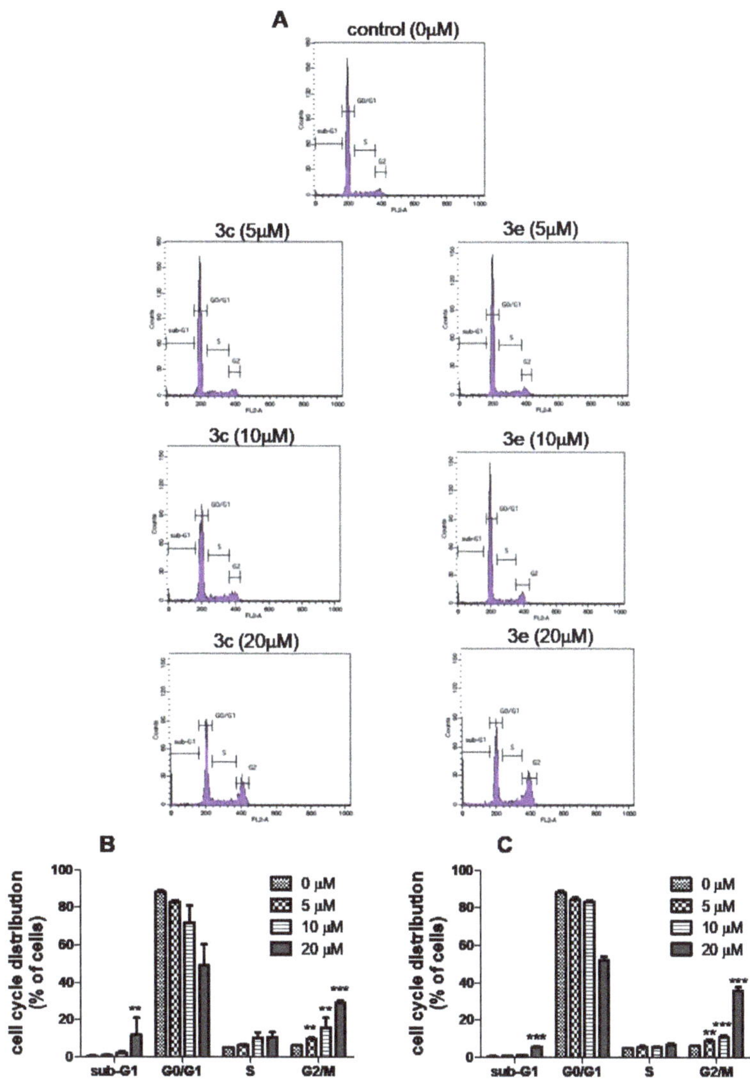

Figure 6. Effect of compounds **3c** and **3e** on cell cycle distribution in A549 cells. After the treatment with the indicated concentrations of compounds **3c** and **3e** for 48 h, the cells were stained with propidium iodide and analyzed by flow cytometry. (**A**) Representative DNA histograms of A549 cells treated with compounds **3c** and **3e**. (**B,C**) Statistical analysis of the percentages of A549 cells in the sub-G1, G0/G1, S, and G2/M phases after treatment with compounds **3c** and **3e**, respectively. Data are expressed as means ± SD from three independent experiments. ** $p < 0.01$ and *** $p < 0.001$ in comparison to the control; one-way ANOVA followed by Tukey's post hoc test.

Table 5. Half-life (t1/2) for the spontaneous hydrolysis of selected compounds **3c** and **3e**.

Thiazole Derivatives	λ [nm]	k [min^{-1}]	t$_{1/2}$ [min]
3c	293	0.015	46.2
3e	262, 322	0.0175	39.6

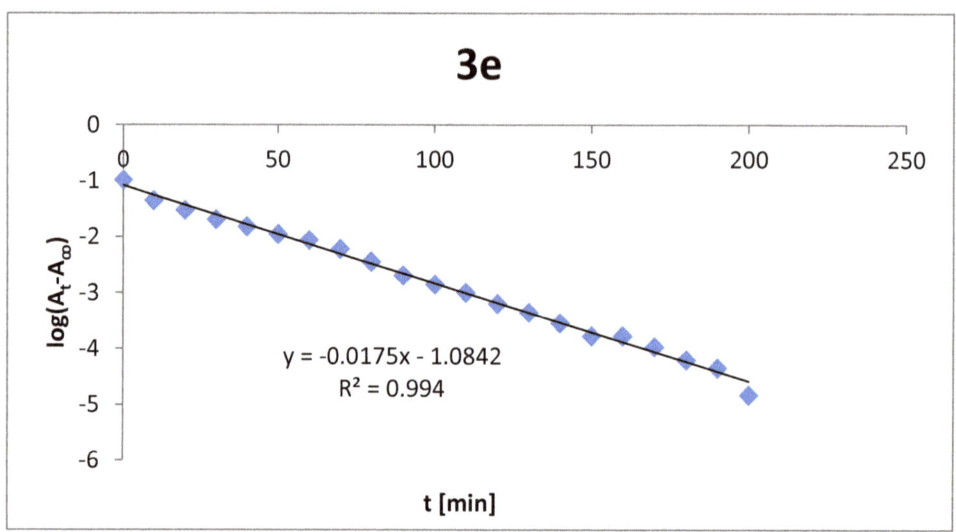

Figure 7. Semilogarithmic plots used to determine rate constants for spontaneous hydrolysis of compound **3e**.

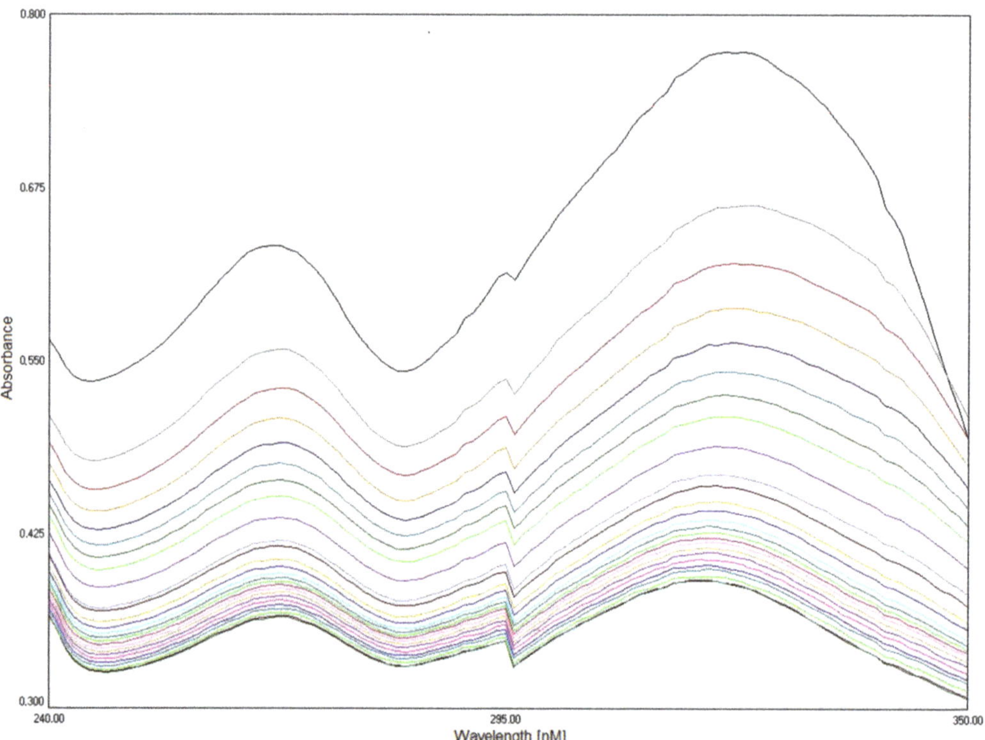

Figure 8. The changes in absorbance spectra of compound **3e** during spontaneous hydrolysis. The spectrum was monitored for 250 min with 10 min intervals in 0.05 M phosphate buffer (pH 7.3, 25 °C).

3. Materials and Methods

3.1. Chemistry

All reactions were performed under a nitrogen atmosphere. All reagents and starting materials were purchased from commercial suppliers and used without further purification. The dichloromethane was dried over calcium hydride. ^{1}H NMR (400 MHz) and ^{13}C NMR (100 MHz) spectra were recorded on a Bruker Avance III multinuclear instrument. High resolution mass spectrometry measurements were performed using a Synapt G2-Si mass spectrometer (Waters) equipped with a quadrupole-Time-of-flight mass analyser. The mass spectrometer was operated in the positive ion detection mode. The results of the measurements were processed using the MassLynx 4.1 software (Waters) incorporated with the instrument. Melting points were determined in open glass capillaries and are uncorrected. Analytical TLC was performed using Macherey-Nagel Polygram Sil G/UV$_{254}$ 0.2 mm plates. 2,2-Diethyl malonyl dichloride and appropriate thioureas were commercial materials (Merck).

3.1.1. 1-(4-(2-Chloroacetyl)phenyl)-3,3-diethylazetidine-2,4-dione (**2**)

2,2-Diethyl malonyl dichloride (1.51 g, 1.32 mL, 7.67 mmol) was added to a stirred solution of 1-(4-aminophenyl)-2-chloroethanone (1) (1.0 g, 5.90 mmol) in dry dichloromethane (25 mL) and then triethylamine (1.55 g, 2.13 mL, 15.34 mmol) was added. The reaction mixture was stirred at room temperature for 20 h. The solvent was removed under reduced pressure, tetrahydrofuran (20 mL) was added and next the triethylamine hydrochloride was filtered off and the solvent was removed under reduced pressure. The product was purified on silica gel column chromatography (230–400 mesh) using (dichloromethane, R_f = 0.58) to reach the desired product: 1.30 g, 76%; mp 76–78 °C. ^{1}H NMR (DMSO-d_6, 400 MHz), δ (ppm): 0.98 (t, 6H, 2CH$_3$, J = 7.7 Hz); 1.86 (q, 4H, 2CH$_2$, J = 7.0 Hz); 5.19 (s, 2H, CH$_2$); 7.88 (d, 2H, 2CH, J = 8.4 Hz); 8.12 (d, 2H, 2CH, J = 7.7 Hz). ^{13}C NMR (DMSO-d_6, 100 MHz), δ (ppm): 9.32 (2CH$_3$); 23.18 (2CH$_2$); 48.02 (C); 71.98 (C); 119.37 (2C); 130.55 (2C); 132.64 (C); 137.54 (C); 171.99 (2C); 190.96 (C).

3.1.2. 1-(4-(2-(4-Chlorophenylamino)thiazol-4-yl)phenyl)-3,3-diethylazetidine-2,4-dione (**3a**) Typical Procedure

N-(4-Chlorophenyl)thiourea (0.187 g, 1.0 mmol) was added to a stirred solution of 1-(4-(2-chloroacetyl)phenyl)-3,3-diethylazetidine-2,4-dione (**2**) (0.294 g, 1.0 mmol) in absolute ethyl alcohol (20 mL). The reaction mixture was stirred under reflux for 20 h. Next, the reaction mixture was added to water (50 mL) and neutralized with NaHCO$_3$ solution. The separate precipitate was collected by filtration and dried in vacuum to afford the desired product: 0.35 g, 82%; mp 164–166 °C, eluent: (dichloromethane/methanol (95:5); R_f = 0.81). ^{1}H NMR (DMSO-d_6, 400 MHz), δ (ppm): 0.99 (t, 6H, 2CH$_3$, J = 7.7 Hz); 1.85 (q, 4H, 2CH$_2$, J = 7.7 Hz); 7.39 (d, 2H, 2CH, J = 9.1 Hz); 7.42 (s, 1H, CH); 7.76–7.79 (m, 4H, 4CH); 8.05 (d, 2H, 2CH, J = 8.4 Hz); 10.47 (s, 1H, NH). ^{13}C NMR (DMSO-d_6, 100 MHz), δ (ppm): 9.39 (2CH$_3$); 23.31 (2CH$_2$); 71.64 (C); 104.62 (C); 118.86 (2C); 119.91 (2C); 125.06 (C); 127.17 (2C); 129.30 (2C); 132.63 (C); 133.44 (C); 140.46 (C); 149.53 (C); 163.36 (C); 172.08 (2C). ESI-HRMS (*m/z*) calculated for $C_{22}H_{21}ClN_3O_2S$: 426.1043 [M + H]$^+$. Found: 426.1039 [M + H]$^+$.

3.1.3. 3,3-Diethyl-1-(4-(2-(4-fluorophenylamino)thiazol-4-yl)phenyl)azetidine-2,4-dione (**3b**)

Yield: 0.14 g, 68%, (dichloromethane/methanol (95:5), R_f = 0.81); mp 155–157 °C. ^{1}H NMR (DMSO-d_6, 400 MHz), δ (ppm): 0.97 (t, 6H, 2CH$_3$, J = 8.0 Hz); 1.83 (q, 4H, 2CH$_2$, J = 7.6 Hz); 7.17 (t, 2H, 2CH, J = 8.8 Hz); 7.37 (s, 1H, CH); 7.18 (q, 2H, 2CH, J = 4.8 Hz); 7.75 (d, 2H, 2CH, J = 9.2 Hz); 8.02 (d, 2H, 2CH, J = 9.6 Hz); 10.31 (s, 1H, NH). ^{13}C NMR (DMSO-d_6, 100 MHz), δ (ppm): 9.40 (2CH$_3$); 23.30 (2CH$_2$); 71.64 (C); 104.20 (C); 115.99 (d, 2C, J = 20.9 Hz); 118.95 (d, 2C, J = 9.4 Hz); 119.93 (2C); 127.14 (2C); 132.49 (C); 133.53 (C); 138.13 (C); 149.45 (C); 157.37 (d, C, J = 227.8 Hz); 163.88 (C); 172.21 (2C). ESI-HRMS (*m/z*) calculated for $C_{22}H_{21}FN_3O_2S$: 410.1339 [M + H]$^+$. Found: 410.1335 [M + H]$^+$.

3.1.4. 3,3-Diethyl-1-(4-(2-(3,4,5-trimethoxyphenylamino)thiazol-4-yl)phenyl)azetidine-2,4-dione (**3c**)

Yield: 0.34 g, 71%, (dichloromethane/methanol (95:5), R_f = 0.61); mp 188–190 °C. ^{1}H NMR (DMSO-d_6, 400 MHz), δ (ppm): 0.97 (t, 6H, 2CH$_3$, J = 7.6 Hz); 1.83 (q, 4H, 2CH$_2$, J = 6.8 Hz); 3.61 (s, 3H, CH$_3$); 3.80 (s, 6H, 2CH$_3$); 7.11 (s, 2H, 2CH); 7.37 (s, 1H, CH); 7.75 (d, 2H, 2CH, J = 8.4 Hz); 8.01 (d, 2H, 2CH, J = 8.8 Hz); 10.24 (s, 1H, NH). ^{13}C NMR (DMSO-d_6, 100 MHz), δ (ppm): 9.40 (2CH$_3$); 23.34 (2CH$_2$); 56.10 (2CH$_3$); 60.61 (CH$_3$); 71.68 (C); 95.19 (2C); 104.07 (C); 120.03 (2C); 126.95 (2C); 132.43 (C); 132.49 (C); 133.59 (C); 137.75 (C); 149.35 (C); 153.48 (2C); 163.37 (C); 172.16 (2C). ESI-HRMS (*m/z*) calculated for C$_{25}$H$_{28}$N$_3$O$_5$S: 482.1750 [M + H]$^+$. Found: 482.1749 [M + H]$^+$.

3.1.5. 3,3-Diethyl-1-(4-(2-(perfluorophenylamino)thiazol-4-yl)phenyl)azetidine-2,4-dione (**3d**)

Yield: 0.43 g, 89%, (dichloromethane/methanol (95:5), R_f = 0.80); mp 162–164 °C. ^{1}H NMR (DMSO-d_6, 400 MHz), δ (ppm): 0.96 (t, 6H, 2CH$_3$, J = 7.6 Hz); 1.82 (q, 4H, 2CH$_2$, J = 7.6 Hz); 7.42 (s, 1H, CH); 7.70 (d, 2H, 2CH, J = 8.4 Hz); 7.87 (d, 2H, 2CH, J = 8.8 Hz); 10.10 (s, 1H, NH). ^{13}C NMR (DMSO-d_6, 100 MHz), δ (ppm): 9.35 (2CH$_3$); 23.29 (2CH$_2$); 71.62 (C); 106.24 (C); 119.87 (2C); 120.80 (C); 126.34 (C); 127.10 (2C); 132.62 (C); 133.14 (C); 141.44 (2C); 143.86 (2C); 149.13 (C); 164.64 (C); 172.10 (2C). ESI-HRMS (*m/z*) calculated for C$_{22}$H$_{17}$F$_5$N$_3$O$_2$S: 482.0962 [M + H]$^+$. Found: 482.0963 [M + H]$^+$.

3.1.6. 4-(4-(4-(3,3-Diethyl-2,4-dioxoazetidin-1-yl)phenyl)thiazol-2-ylamino)benzenesulfonamide (**3e**)

Yield: 0.36 g, 82%, (dichloromethane/methanol (95:5), R_f = 0.39); mp 130–132 °C. ^{1}H NMR (DMSO-d_6, 400 MHz), δ (ppm): 1.00 (t, 6H, 2CH$_3$, J = 7.7 Hz); 1.86 (q, 4H, 2CH$_2$, J = 7.7 Hz); 7.22 (s, 2H, NH$_2$); 7.51 (s, 1H, CH); 7.80 (d, 2H, 2CH, J = 6.3 Hz); 7.82 (d, 2H, 2CH, J = 6.3 Hz); 7.88 (d, 2H, 2CH, J = 9.1 Hz); 8.09 (d, 2H, 2CH, J = 8.4 Hz); 10.74 (s, 1H, NH). ^{13}C NMR (DMSO-d_6, 100 MHz), δ (ppm): 9.40 (2CH$_3$); 23.31 (2CH$_2$); 71.66 (C); 105.37 (C); 116.68 (2C); 119.97 (2C); 127.26 (2C); 127.65 (2C); 132.67 (C); 133.34 (C); 136.54 (C); 144.22 (C); 149.66 (C); 163.00 (C); 172.16 (2C). ESI-HRMS (*m/z*) calculated for C$_{22}$H$_{23}$N$_4$O$_4$S$_2$: 471.1161 [M + H]$^+$. Found: 471.1157 [M + H]$^+$.

3.1.7. 1-(4-(2-(3,5-Dimethylphenylamino)thiazol-4-yl)phenyl)-3,3-diethylazetidine-2,4-dione (**3f**)

Yield: 0.20 g, 99%, (dichloromethane/methanol (95:5), R_f = 0.87); mp 140–143 °C. ^{1}H NMR (DMSO-d_6, 400 MHz), δ (ppm): 0.97 (t, 6H, 2CH$_3$, J = 8.0 Hz); 1.83 (q, 4H, 2CH$_2$, J = 8.0 Hz); 2.62 (s, 6H, 2CH$_3$); 6.61 (s, 1H, CH); 7.29 (s, 2H, 2CH); 7.35 (s, 1H, CH); 7.76 (d, 2H, 2CH, J = 9.2 Hz); 8.01 (d, 2H, 2CH, J = 9.2 Hz); 10.15 (s, 1H, NH). ^{13}C NMR (DMSO-d_6, 100 MHz), δ (ppm): 9.41 (2CH$_3$); 21.75 (2CH$_3$); 23.33 (2CH$_2$); 71.68 (C); 104.02 (C); 115.30 (2C); 120.00 (2C); 123.56 (C); 127.08 (2C); 132.49 (C); 133.58 (C); 138.48 (2C); 141.48 (C); 149.36 (C); 163.93 (C); 172.19 (2C). ESI-HRMS (*m/z*) calculated for C$_{24}$H$_{26}$N$_3$O$_2$S: 420.1746 [M + H]$^+$. Found: 420.1747 [M + H]$^+$.

3.1.8. 1-(4-(2-(1*H*-Indazol-5-ylamino)thiazol-4-yl)phenyl)-3,3-diethylazetidine-2,4-dione (**3g**)

Yield: 0.16 g, 83%, (dichloromethane/methanol (95:5), R_f = 0.42); mp 186–188 °C. ^{1}H NMR (DMSO-d_6, 400 MHz), δ (ppm): 0.98 (t, 6H, 2CH$_3$, J = 7.2 Hz); 1.84 (q, 4H, 2CH$_2$, J = 7.6 Hz); 7.33 (s, 1H, CH); 7.39–7.43 (m, 1H, CH); 7.48–7.53 (m, 1H, CH); 7.77 (d, 2H, 2CH, J = 8.8 Hz); 8.05–8.07 (m, 3H, 3CH); 8.30 (d, 1H, CH, J = 2.0 Hz); 10.23 (s, 1H, NH); 12.94 (s, 1H, NH). ^{13}C NMR (DMSO-d_6, 100 MHz), δ (ppm): 9.43 (2CH$_3$); 23.32 (2CH$_2$); 71.67 (C); 103.56 (C); 107.14 (C); 111.10 (C); 119.87 (C); 119.94 (2C); 123.55 (C); 127.17 (2C); 132.46 (C); 133.67 (C); 133.74 (C); 135.08 (C); 136.75 (C); 149.53 (C); 164.58 (C); 172.16 (2C). ESI-HRMS (*m/z*) calculated for C$_{23}$H$_{22}$N$_5$O$_2$S: 432.1494 [M + H]$^+$. Found: 432.1491 [M + H]$^+$.

3.1.9. 3,3-Diethyl-1-(4-(2-(naphthalen-1-ylamino)thiazol-4-yl)phenyl)azetidine-2,4-dione (**3h**)

Yield: 0.14 g, 63%, (dichloromethane/methanol (95:5), R_f = 0.89); mp 69–71 °C. ^{1}H NMR (DMSO-d_6, 400 MHz), δ (ppm): 0.98 (t, 6H, 2CH$_3$, J = 7.2 Hz); 1.83 (q, 4H, 2CH$_2$,

J = 7.6 Hz); 7.31 (s, 1H, CH); 7.50–7.58 (m, 3H, 3CH); 7.65–7.69 (m, 1H, CH); 7.75 (d, 2H, 2CH, *J* = 7.6 Hz); 7.91–7.96 (m, 1H, CH); 8.01 (d, 2H, 2CH, *J* = 8.8 Hz); 8.25–8.33 (m, 2H, 2CH); 10.20 (s, 1H, NH). ^{13}C NMR (DMSO-d_6, 100 MHz), δ (ppm): 9.40 (2CH$_3$); 23.31 (2CH$_2$); 71.64 (C); 104.65 (C); 117.06 (C); 119.95 (2C); 122.62 (C); 123.85 (C); 126.17 (C); 126.45 (C); 126.59 (C); 126.64 (C); 127.12 (2C); 128.76 (C); 132.52 (C); 133.70 (C); 134.48 (C); 137.14 (C); 149.44 (C); 166.18 (C); 172.18 (2C). ESI-HRMS (*m/z*) calculated for $C_{26}H_{24}N_3O_2S$: 442.1589 [M + H]$^+$. Found: 442.1581 [M + H]$^+$.

3.1.10. 3,3-Diethyl-1-(4-(2-(4-(trifluoromethyl)phenylamino)thiazol-4-yl)phenyl)-azetidine-2,4-dione (3i)

Yield: 0.24 g, 52%, (dichloromethane/methanol (95:5), R$_f$ = 0.92); mp 170–172 °C. ^{1}H NMR (DMSO-d_6, 400 MHz), δ (ppm): 0.99 (t, 6H, 2CH$_3$, *J* = 7.0 Hz); 1.85 (q, 4H, 2CH$_2$, *J* = 7.7 Hz); 7.50 (s, 1H, CH); 7.70 (d, 2H, 2CH, *J* = 8.4 Hz); 7.79 (d, 2H, 2CH, *J* = 8.4 Hz); 7.94 (d, 2H, 2CH, *J* = 8.4 Hz); 8.08 (d, 2H, 2CH, *J* = 8.4 Hz); 10.76 (s, 1H, NH). ^{13}C NMR (DMSO-d_6, 100 MHz), δ (ppm): 9.37 (2CH$_3$); 23.29 (2CH$_2$); 71.66 (C); 105.37 (C); 117.05 (2C); 119.82 (2C); 123.78 (C); 126.46 (C); 126.80 (2C); 127.22 (2C); 132.72 (C); 133.32 (C); 144.79 (C); 149.62 (C); 162.96 (C); 172.15 (2C). ESI-HRMS (*m/z*) calculated for $C_{23}H_{21}F_3N_3O_2S$: 460.1307 [M + H]$^+$. Found: 460.1302 [M + H]$^+$.

3.1.11. 3,3-Diethyl-1-(4-(2-(2,4,6-trichlorophenylamino)thiazol-4-yl)phenyl)azetidine-2,4-dione (3j)

Yield: 0.22 g, 89%, (dichloromethane/methanol (95:5), R$_f$ = 0.91); mp 78–80 °C. ^{1}H NMR (DMSO-d_6, 400 MHz), δ (ppm): 0.96 (t, 6H, 2CH$_3$, *J* = 7.2 Hz); 1.82 (q, 4H, 2CH$_2$, *J* = 7.6 Hz); 7.30 (s, 1H, CH); 7.70 (d, 2H, 2CH, *J* = 8.0 Hz); 7.82 (s, 2H, 2CH); 7.85 (d, 2H, 2CH, *J* = 8.4 Hz); 9.86 (s, 1H, NH). ^{13}C NMR (DMSO-d_6, 100 MHz), δ (ppm): 9.39 (2CH$_3$); 23.28 (2CH$_2$); 71.59 (C); 104.84 (C); 119.88 (2C); 127.07 (2C); 129.36 (4C); 132.28 (C); 132.53 (C); 133.40 (C); 134.71 (C); 149.46 (C); 166.16 (C); 172.12 (2C). ESI-HRMS (*m/z*) calculated for $C_{22}H_{19}Cl_3N_3O_2S$: 494.0264 [M + H]$^+$. Found: 494.0261 [M + H]$^+$.

3.2. Human Neutrophil Elastase Inhibition Assay

Kinetic Analysis of the Inhibition of Human Neutrophil Elastase

Kinetic studies were conducted in the presence of different concentrations of substrate (5, 25, 50 and 100 µM) and test compounds (in the range of 0–500 nM). The reaction conditions were the same as in the HNE inhibition assay and the reaction was monitored for 10 min at 25 °C with the same E$_x$ an E$_m$ wavelengths. By Lineweaver–Burk plots, kinetic values such as the Michaelis–Menten constants and maximum velocity were determined. Next, two inhibition constants for inhibitor binding with free enzyme or enzyme–substrate complex (K$_i$ and K$_{is}$) were obtained by plotting the secondary plot of the slopes of the determined straight lines or vertical intercept (1/V$_{max}$) versus inhibitor concentration [44].

3.3. Molecular Docking Study

The ligand molecules were prepared manually by using the Avogadro 1.1.1 software [45] and optimized within the UFF force field [46] (3000 steps, conjugate gradient algorithm). The flexible, optimized ligands were docked into the binding pockets of the high-resolution elastase structures found in the following five PDB entries: 1bma (X-ray resolution: 0.192 nm), 1hv7 (X-ray resolution: 0.17 nm), 1qnj (X-ray resolution: 0.11 nm), 2de9 (X-ray resolution: 0.13 nm) and 2h1u (X-ray resolution: 0.16 nm). The AutoDock Vina software [47] was applied in all docking simulations. The procedure of docking was carried out within the cuboid region which covered the whole co-crystallized ligand present in the 2h1u PDB structure as well as the closest amino-acid residues that exhibit contact with that ligand (of dimensions: 20 × 20 × 20 Å^3). In order to provide a sufficient number of conformationally diverse ligand arrangements, the number of possible poses generated during a single run was increased to 15, whereas the energy threshold between the highest- and lowest-ranked poses was increased to 5 kcal/mol. Apart from that, all the default procedures and algorithms implemented in AutoDock Vina were applied during the docking procedure. The predicted binding energies were averaged over all protein structures applied for docking, whereas only the ligand poses exhibiting the same desired structural features regarding the distance with respect to Ser195 were subjected to calculations of such average energy. The more favorable binding mode is associated with the lower binding energy value; only the lowest energy values corresponding to the given ligand and poses fulfilling the cutoff criteria were accounted for during subsequent analysis. The visual inspection of the location and orientation of the docked ligands, in order to control the uniformity of the binding pattern, was performed. The docking methodology was initially validated by the docking simulation of the non-covalently-bound ligand molecule originally included in one of the studied protein structures (i.e. PDB:2h1u). Details of the procedure are given in our previous work [25].

3.4. Antiproliferative Activity

Cell preparation, an in vitro antiproliferative assay, and MTT and SRB cytotoxic tests were performed according to the literature [35,48–50].

3.5. Apoptosis and Cell Cycle Assays

The fluorescence-activated cell sorting (FACS) technique was employed to analyze the 3c- and 3e-induced cell death, the activation of executive caspase-3, and the cell cycle in the A549 cells. The quantitative analysis of apoptosis and necrosis was performed using an Annexin V-fluorescein isothiocyanate (FITC)/ propidium iodide (PI) apoptosis kit (BD Biosciences, BD Pharmingen™, San Jose, CA, USA) as previously described [36]. To determine the active form of caspase-3, the A549 cells were plated into 6-well plates at a density of 6×10^5 cells/well. The next day, the growth medium was replaced with a fresh one supplemented with compound **3c** or **3e** (10, 20, and 40 μM) or the DMSO vehicle (0.1%; control cells). After 48-h incubation, the samples were harvested and the level of active caspase-3 was determined using the phycoerythrin (PE) Active Caspase-3 Apoptosis Kit (BD Pharmingen™) according to the manufacturer's instructions. The stained cells were analyzed using FACS Calibur, and data were analyzed using Cell Quest Pro Version 6.0 (BD Biosciences, San Jose, CA, USA) for the Macintosh operating system. The results were calculated as a percent of cells with active caspase-3 among all the analyzed cells. The cell cycle analysis was performed by determination of the DNA contents on the basis of PI staining as previously described [36].

3.6. Analysis of Compounds Stability

Spontaneous hydrolysis of most active compounds was determined as previously described [51]. The kinetics hydrolysis of the compounds was evaluated at 25 °C in 0.05 M phosphate buffer, pH 7.3, by measuring changes in absorbance spectra during incubation using a T60U spectrophotometer (PG Instruments). The absorbance (A_t) at characteristic absorption maxima for each compound was measured at 10 min intervals until there was no further decrease in absorbance (A_∞). Using these measurements, semilogarithmic plots of the dependence of $\log(A_t - A_\infty)$ on time were prepared, and k' values were determined from the slope of these plots according to first order reaction kinetics. Half-conversion times were calculated using equation: $t_{1/2} = 0.693/k'$.

4. Conclusions

In the present work, we developed an efficient method for the synthesis of new thiazole derivatives containing 3,3-diethylazetidine-2,4-dione moiety. The resulting compounds are promising human neutrophil elastase inhibitors. The best results were obtained introducing the trimethoxyphenyl, benzenesulfonamide and naphthyl groups, which resulted in corresponding HNE inhibitors with IC_{50} values of 35.02–44.59 nM. Additionally, the tested compounds demonstrate high stability under physiological conditions. The molecular docking study showed good correlation of the binding energies with the IC_{50} values, which, as a consequence of accepted logical constraint, suggests that the potency of compounds is determined during ligand alignment in the binding cavity prior to its covalent binding via Ser195. Additionally, the obtained derivatives show a high and broad spectrum of antiproliferative activity, which suggests that these new drugs, viaHNE inhibition, will be active against the cancers that produce immunoreactive HNE. Additionally, compounds **3c** and **3e** can induce cell cycle arrest at the G2/M phase and apoptosis via caspase-3 activation, leading to the inhibition of A549 cell proliferation.

Supplementary Materials: The following supporting information can be downloaded at: https://www.mdpi.com/article/10.3390/10.3390/ijms23147566/s1.

Author Contributions: Conceptualization, K.Z.Ł.; organic synthesis, K.Z.Ł.; validation K.Z.Ł., B.D., M.Ś., J.W., M.M.-K., B.Z. and W.P.; formal analysis, B.D., M.Ś., W.P., M.M.-K., B.Z. and K.Z.Ł.; investigation, B.D., M.Ś., W.P., M.M.-K., B.Z. and K.Z.Ł.; writing—original draft preparation, K.Z.Ł., B.Z. and W.P.; writing—review and editing, K.Z.Ł.; visualization, B.D., W.P. and K.Z.Ł.; supervision, K.Z.Ł.; funding acquisition, K.Z.Ł. All authors have read and agreed to the published version of the manuscript.

Funding: This research was funded by the Nicolaus Copernicus University, The Excellence Initiative—Research University (IDUB 2020-1 NZ).

Informed Consent Statement: Not applicable.

Data Availability Statement: Not applicable.

Conflicts of Interest: The authors declare that they have no conflict of interest.

References

1. Lee, W.L.; Downey, G.P. Leukocyte elastase: Physiological functions and role in acute lung injury. *Am. J. Respir. Crit. Care Med.* **2001**, *64*, 896–904. [CrossRef] [PubMed]
2. Stevens, T.; Ekholm, K.; Gränse, M.; Lindahl, M.; Kozma, V.; Jungar, C.; Ottosson, T.; Falk-Håkansson, H.; Churg, A.; Wright, J.L.; et al. AZD9668: Pharmacological characterization of a novel oral inhibitor of neutrophil elastase. *J. Pharmacol. Exp. Ther.* **2011**, *339*, 313–320. [CrossRef] [PubMed]
3. Dengler, V.; Downey, G.P.; Tuder, R.M.; Eltzschig, H.K.; Schmidt, E.P. Neutrophil intercellular communication in acute lung injury: Emerging roles of microparticles and gap junctions. *Am. J. Respir. Cell Mol. Biol.* **2013**, *49*, 1–5. [CrossRef] [PubMed]
4. Travis, J.; Dubin, A.; Potempa, J.; Watorek, W.; Kurdowska, A. Neutrophil proteinases. *Ann. N. Y. Acad. Sci.* **1991**, *624*, 81–86. [CrossRef]
5. Zhou, X.; Dai, Q.; Huang, X. Neutrophils in acute lung injury. *Front. Biosci.* **2012**, *17*, 2278–2283. [CrossRef]
6. Carrell, R.W.; Jeppsson, J.O.; Laurell, C.B.; Brennan, S.O.; Owen, M.C.; Vaughan, L.; Boswell, D.R. Structure and variation of human-1-antitrypsin. *Nature* **1982**, *298*, 329–334. [CrossRef]
7. Thomson, R.C.; Ohlsson, K. Isolation, properties, and complete amino acid sequence of human secretory leukocyte protease inhibitor: A potent inhibitor of leukocyte elastase. *Proc. Natl. Acad. Sci. USA* **1986**, *83*, 6692–6696. [CrossRef]
8. Reilly, C.F.; Travis, J. The degradation of human lung elastin by neutrophil proteinases. *Biochim. Biophys. Acta* **1980**, *621*, 147–157. [CrossRef]
9. Moroy, G.; Alix, A.J.; Sapi, J.; Hornebeck, W.; Bourguet, E. Neutrophil elastase as a target in lung cancer. *Anticancer Agents Med. Chem.* **2012**, *12*, 565–579. [CrossRef]
10. Sato, T.; Takahashi, S.; Mizumoto, T.; Harao, M.; Akuziki, M.; Takasugi, M.; Fukutomi, T.; Yamashita, J.I. Neutrophil elastase and cancer. *Surg. Oncol.* **2006**, *15*, 217–222. [CrossRef]
11. Akizuki, M.; Fukutomi, T.; Takasugi, M.; Takahashi, S.; Sato, T.; Harao, M.; Mizumoto, T.; Yamashita, J. Prognostic significance of immunoreactive neutrophil elastase in human breast cancer: Long-term follow-up results in 313 patients. *Neoplasia* **2007**, *9*, 260–264. [CrossRef] [PubMed]
12. Lerman, I.; Garcia-Hernandez, M.L.; Rangel-Moreno, J.; Chiriboga, L.; Pan, C.; Nastiuk, K.L.; Krolewski, J.J.; Sen, A.; Hammes, S.R. Infiltrating myeloid cells exert protumorigenic actions via neutrophil elastase. *Mol. Cancer Res.* **2017**, *15*, 1138–1152. [CrossRef] [PubMed]
13. Lerman, I.; Hammes, S.R. Neutrophil elastase in the tumor microenvironment. *Steroids* **2017**, *133*, 96–101. [CrossRef] [PubMed]
14. Vaguliene, N.; Zemaitis, M.; Lavinskiene, S.; Miliauskas, S.; Sakalauskas, R. Local and systemic neutrophil inflammation in patients with lung cancer and chronic obstructive pulmonary disease. *BMC Immunol.* **2013**, *14*, 36. [CrossRef] [PubMed]
15. Hunt, K.K.; Wingate, H.; Yokota, T.; Liu, Y.; Mills, G.B.; Zhang, F.; Fang, B.; Su, C.-H.; Zhang, M.; Yi, M.; et al. Elafin, an inhibitor of elastase, is a prognostic indicator in breast cancer. *Breast Cancer Res.* **2013**, *15*, R3. [CrossRef]
16. Goulet, B.; Markovic, Y.; Leduy, L.; Nepveu, A. Proteolytic processing of cut homeobox 1 by neutrophil elastase in the MV4; 11 myeloid leukemia cell line. *Mol. Cancer Res.* **2008**, *6*, 644–653. [CrossRef]

17. Houghton, A.M.; Rzymkiewicz, D.M.; Ji, H.; Gregory, A.D.; Egea, E.E.; Metz, H.E.; Stolz, D.B.; Land, S.R.; Marconcini, L.A.; Kliment, C.R.; et al. Neutrophil elastase-mediated degradation of IRS-1 accelerates lung tumor growth. *Nat. Med.* **2010**, *16*, 219–223. [CrossRef]

18. Inada, M.; Yamashita, J.; Ogawa, M. Neutrophil elastase inhibitor (ONO-5046-Na) inhibits the growth of human lung cancer cell lines transplanted into severe combined immunodeficiency (scid) mice. *Res. Commun. Mol. Pathol. Pharmacol.* **1997**, *97*, 229–232.

19. Crocetti, L.; Schepetkin, I.A.; Cilibrizzi, A.; Graziano, A.; Vergelli, C.; Giomi, D.; Khlebnikov, A.I.; Quinn, M.T.; Giovannoni, M.P. Optimization of *N*-benzoylindazole derivatives as inhibitors of human neutrophil elastase. *J. Med. Chem.* **2013**, *56*, 6259–6272. [CrossRef]

20. Crocetti, L.; Schepetkin, I.A.; Ciciani, G.; Giovannoni, M.P.; Guerrini, G.; Iacovone, A.; Khlebnikov, A.I.; Kirpotina, L.N.; Quinn, M.T.; Vergelli, C. Synthesis and pharmacological evaluation of indole derivatives as deaza analogues of potent human neutrophil elastase inhibitors (HNE). *Drug Dev. Res.* **2016**, *77*, 285–299. [CrossRef]

21. Crocetti, L.; Giovannoni, M.P.; Schepetkin, I.A.; Quinn, M.T.; Khlebnikov, A.I.; Cantini, N.; Guerrini, G.; Iacovone, A.; Teodori, E.; Vergelli, C. 1*H*-pyrrolo[2,3-b]pyridine: A new scaffold for human neutrophil elastase (HNE) inhibitors. *Bioorg. Med. Chem.* **2018**, *26*, 5583–5595. [CrossRef]

22. Vergelli, C.; Schepetkin, I.A.; Crocetti, L.; Iacovone, A.; Giovannoni, M.P.; Guerrini, G.; Khlebnikov, A.I.; Ciattini, S.; Ciciani, G.; Quinn, M.T. Isoxazol-5(2*H*)-one: A new scaffold for potent human neutrophil elastase (HNE) inhibitors. *J. Enzyme Inhib. Med. Chem.* **2017**, *32*, 821–831. [CrossRef] [PubMed]

23. Hasdemir, B.; Sacan, O.; Yasa, H.; Kucuk, H.B.; Yusufoglu, A.S.; Yanardag, R. Synthesis and elastase inhibition activities of novel aryl, substituted aryl, and heteroaryl oxime ester derivatives. *Arch. Pharm. Chem. Life Sci.* **2018**, *351*, 1700269. [CrossRef] [PubMed]

24. Zheng, Q.; Woehl, J.L.; Kitamura, S.; Santos-Martins, D.; Smedley, C.J.; Li, G.; Forli, S.; Moses, J.E.; Wolan, D.W.; Sharpless, K.B. SuFEx-enabled, agnostic discovery of covalent inhibitors of human neutrophil elastase. *Proc. Natl. Acad. Sci. USA* **2019**, *116*, 18808–18814. [CrossRef]

25. Donarska, B.; Świtalska, M.; Płaziński, W.; Wietrzyk, J.; Łączkowski, K.Z. Effect of the dichloro-substitution on antiproliferative activity of phthalimide-thiazole derivatives. Rational design, synthesis, elastase, caspase 3/7, and EGFR tyrosine kinase activity and molecular modeling study. *Bioorg. Chem.* **2021**, *110*, 104819. [CrossRef] [PubMed]

26. Ebnöther, A.; Jucker, E.; Rissi, E.; Rutschmann, J.; Schreier, E.; Steiner, R.; Süess, R.; Vogel, A. Über Azetidin-2,4-dione (Malonimide). *Helv. Chim. Acta* **1959**, *42*, 918–955. [CrossRef]

27. Mulchande, J.; Guedes, R.C.; Tsang, W.; Page, M.I.; Moreira, R.; Iley, J. Azetidine-2,4-diones (4-Oxo-β-lactams) as scaffolds for designing elastase inhibitors. *J. Med. Chem.* **2008**, *51*, 1783–1790. [CrossRef]

28. Mulchande, J.; Oliveira, R.; Carrasco, M.; Gouveia, L.; Guedes, R.C.; Iley, J.; Moreira, R. 4-Oxo-β-lactams (azetidine-2,4-diones) are potent and selective inhibitors of human leukocyte elastase. *J. Med. Chem.* **2010**, *53*, 241–253. [CrossRef]

29. Mulchande, J.; Simões, S.I.; Gaspar, M.M.; Eleutério, C.V.; Oliveira, R.; Cruz, M.E.M.; Moreira, R.; Iley, J. Synthesis, stability, biochemical and pharmacokinetic properties of a new potent and selective 4-oxo-β-lactam inhibitor of human leukocyte elastase. *J. Enzyme Inhib. Med. Chem.* **2011**, *26*, 169–175. [CrossRef]

30. Areias, L.R.P.; Ruivo, E.F.P.; Gonçalves, L.M.; Duarte, M.T.; André, V.; Moreira, R.; Lucas, S.D.; Guedes, R.C. A unified approach toward the rational design of selective low nanomolar human neutrophil elastase inhibitors. *RSC Adv.* **2015**, *5*, 51717–51721. [CrossRef]

31. Ruivo, E.F.P.; Gonçalves, L.M.; Carvalho, L.A.R.; Guedes, R.V.; Hofbauer, S.; Brito, J.A.; Archer, M.; Moreira, R.; Lucas, S.D. Clickable 4-oxo-β-lactam-based selective probing for human neutrophil elastase related proteomes. *ChemMedChem* **2016**, *11*, 2037–2042. [CrossRef] [PubMed]

32. Nunes, A.; Marto, J.; Gonçalves, L.M.; Simões, S.; Félix, R.; Ascenso, A.; Lopes, F.; Ribeiro, H.M. Novel and modified neutrophil elastase inhibitor loaded in topical formulations for psoriasis management. *Pharmaceutics* **2020**, *12*, 358. [CrossRef] [PubMed]

33. de Siqueiraa, L.R.P.; de Moraes Gomes, P.A.T.; de Lima Ferreira, L.P.; de Melo Rêgo, M.J.B.; Leite, A.C.L. Multi-target compounds acting in cancer progression: Focus on thiosemicarbazone, thiazole and thiazolidinone analogues. *Eur. J. Med. Chem.* **2019**, *170*, 237–260. [CrossRef] [PubMed]

34. Chhabria, M.T.; Patel, S.; Modi, P.; Brahmkshatriya, P.S. Thiazole: A Review on Chemistry, Synthesis and Therapeutic Importance of its Derivatives. *Curr. Top. Med. Chem.* **2016**, *16*, 2841–2862. [CrossRef]

35. Piechowska, K.; Świtalska, M.; Cytarska, J.; Jaroch, K.; Łuczykowski, K.; Chałupka, J.; Wietrzyk, J.; Misiura, K.; Bojko, B.; Kruszewski, S.; et al. Discovery of tropinone-thiazole derivatives as potent caspase 3/7 activators, and noncompetitive tyrosinase inhibitors with high antiproliferative activity: Rational design, one-pot tricomponent synthesis, and lipophilicity determination. *Eur. J. Med. Chem.* **2019**, *175*, 162–171. [CrossRef]

36. Piechowska, K.; Mizerska-Kowalska, M.; Zdzisińska, B.; Cytarska, J.; Baranowska-Łączkowska, A.; Jaroch, K.; Łuczykowski, K.; Płaziński, W.; Bojko, B.; Kruszewski, S.; et al. Tropinone-derived alkaloids as potent anticancer agents: Synthesis, tyrosinase inhibition, mechanism of action, DFT calculation, and molecular docking studies. *Int. J. Mol. Sci.* **2020**, *21*, 9050. [CrossRef]

37. Łączkowski, K.Z.; Anusiak, J.; Świtalska, M.; Dzitko, K.; Cytarska, J.; Baranowska-Łączkowska, A.; Plech, T.; Paneth, A.; Wietrzyk, J.; Białczyk, J. Synthesis, molecular docking, ctDNA interaction, DFT calculation and evaluation of antiproliferative and anti-Toxoplasma gondii activities of 2,4-diaminotriazine-thiazole derivatives. *Med. Chem. Res.* **2018**, *27*, 1131–1148. [CrossRef]

38. Łączkowski, K.Z.; Misiura, K.; Biernasiuk, A.; Malm, A.; Siwek, A.; Plech, T.; Ciok-Pater, E.; Skowron, K.; Gospodarek, E. Synthesis, in vitro biological screening and molecular docking studies of novel camphor-based thiazoles. *Med. Chem.* **2014**, *10*, 600–608. [CrossRef]

39. Crocetti, L.; Giovannoni, M.P.; Cantini, N.; Guerrini, G.; Vergelli, C.; Schepetkin, I.A.; Khlebnikov, A.I.; Quinn, M.T. Novel sulfonamide analogs of sivelestat as potent human neutrophil elastase inhibitors. *Front. Chem.* **2020**, *8*, 795. [CrossRef]

40. Nevozhay, D. Cheburator software for automatically calculating drug inhibitory concentrations from in vitro screening assays. *PLoS ONE* **2014**, *9*, e106186. [CrossRef]

41. Stark, G.R.; Taylor, W.R. Analyzing the G2/M checkpoint. *Methods Mol. Biol.* **2004**, *280*, 51–82. [PubMed]

42. Hustedt, N.; Durocher, D. The control of DNA repair by the cell cycle. *Nat. Cell Biol.* **2016**, *19*, 1–9. [CrossRef] [PubMed]

43. Yadav, P.; Yadav, R.; Jain, S.; Vaidya, A. Caspase-3: A primary target for natural and synthetic compounds for cancer therapy. *Chem. Biol. Drug Des.* **2021**, *98*, 144–165. [CrossRef] [PubMed]

44. Kim, J.Y.; Lee, J.H.; Song, Y.H.; Jeong, W.M.; Tan, X.; Uddin, Z.; Park, K.H. Human neutrophil elastase inhibitory alkaloids from *Chelidonium majus* L. *J. Appl. Biol. Chem.* **2015**, *58*, 281–285. [CrossRef]

45. Hanwell, M.D.; Curtis, D.E.; Lonie, D.C.; Vandermeersch, T.; Zurek, E.; Hutchison, G.R.J. Avogadro: An advanced semantic chemical editor, visualization, and analysis platform. *Cheminformatics* **2012**, *4*, 17. [CrossRef]

46. Rappe, A.K.; Casewit, C.J.; Colwell, K.S.; Goddard, W.A.; Skiff, W.M. UFF, a full periodic table force field for molecular mechanics and molecular dynamics simulations. *J. Am. Chem. Soc.* **1992**, *114*, 10024–10035. [CrossRef]

47. Trott, O.; Olson, A.J. AutoDock Vina: Improving the speed and accuracy of docking with a new scoring function, efficient optimization, and multithreading. *J. Comput. Chem.* **2010**, *31*, 455–461. [CrossRef]

48. Rubinstein, L.V.; Shoemaker, R.H.; Paul, K.D.; Simon, R.M.; Tosini, S.; Skehan, P.; Sudiero, D.A.; Monks, A.; Boyd, M.R. Comparison of in vitro anticancer-drug-screening data generated with a tetrazolium assay versus a protein assay against a diverse panel of human tumor cell lines. *J. Nat. Cancer Inst.* **1990**, *82*, 1113–1118. [CrossRef]

49. Bramson, J.; McQuillan, A.; Aubin, R.; Alaoui-Jamali, M.; Batist, G.; Christodoulopoulos, G.; Panasci, L.C. Nitrogen mustard drug resistant B-cell chronic lymphocytic leukemia as an in vivo model for crosslinking agent resistance. *Mut. Res.* **1995**, *336*, 269–278. [CrossRef]

50. Sidoryk, K.; Świtalska, M.; Wietrzyk, J.; Jaromin, A.; Piętka-Ottlik, M.; Cmoch, P.; Zagrodzka, J.; Szczepek, W.; Kaczmarek, L.; Peczyńska-Czoch, W. Synthesis and biological evaluation of new amino acid and dipeptide derivatives of neocryptolepine as anticancer agents. *J. Med. Chem.* **2012**, *55*, 5077–5087. [CrossRef]

51. Schepetkin, I.A.; Khlebnikov, A.I.; Quinn, M.T. *N*-Benzoylpyrazoles are novel small-molecule inhibitors of human neutrophil elastase. *J. Med. Chem.* **2007**, *50*, 4928–4938. [CrossRef] [PubMed]

International Journal of
Molecular Sciences

Article

Micro-RNAs in Response to Active Forms of Vitamin D$_3$ in Human Leukemia and Lymphoma Cells

Justyna Joanna Gleba [1,*], Dagmara Kłopotowska [1], Joanna Banach [1], Karolina Anna Mielko [1,2], Eliza Turlej [1,3], Magdalena Maciejewska [1], Andrzej Kutner [4] and Joanna Wietrzyk [1]

[1] Department of Experimental Oncology, Hirszfeld Institute of Immunology and Experimental Therapy, Polish Academy of Sciences, Weigla 12, 53-114 Wroclaw, Poland; dagmara.klopotowska@hirszfeld.pl (D.K.); joanna.banach@hirszfeld.pl (J.B.); karolina.mielko@pwr.edu.pl (K.A.M.); eliza.turlej@upwr.edu.pl (E.T.); magdalena.maciejewska@hirszfeld.pl (M.M.); joanna.wietrzyk@hirszfeld.pl (J.W.)

[2] Department of Biochemistry, Molecular Biology and Biotechnology, Faculty of Chemistry, Wroclaw University of Science and Technology, Norwida 4/6, 50-373 Wroclaw, Poland

[3] Department of Experimental Biology, The Wroclaw University of Environmental and Life Sciences, Norwida 27 B, 50-375 Wroclaw, Poland

[4] Department of Bioanalysis and Drug Analysis, Faculty of Pharmacy, Medical University of Warsaw, 1 Banacha, 02-097 Warsaw, Poland; andrzej.kutner@wum.edu.pl

* Correspondence: justyna.gleba@hotmail.com; Tel.: +48-1-904-207-2571

Abstract: Non-coding micro-RNA (miRNAs) regulate the protein expression responsible for cell growth and proliferation. miRNAs also play a role in a cancer cells' response to drug treatment. Knowing that leukemia and lymphoma cells show different responses to active forms of vitamin D$_3$, we decided to investigate the role of selected miRNA molecules and regulated proteins, analyzing if there is a correlation between the selected miRNAs and regulated proteins in response to two active forms of vitamin D$_3$, calcitriol and tacalcitol. A total of nine human cell lines were analyzed: five leukemias: MV-4-1, Thp-1, HL-60, K562, and KG-1; and four lymphomas: Raji, Daudi, Jurkat, and U2932. We selected five miRNA molecules—miR-27b, miR-32, miR-125b, miR-181a, and miR-181b— and the proteins regulated by these molecules, namely, CYP24A1, Bak1, Bim, p21, p27, p53, and NF-kB. The results showed that the level of selected miRNAs correlates with the level of proteins, especially p27, Bak1, NFκB, and CYP24A1, and miR-27b and miR-125b could be responsible for the anticancer activity of active forms of vitamin D$_3$ in human leukemia and lymphoma.

Keywords: leukemia; lymphoma; miRNAs; vitamin D$_3$ analogs; calcitriol; tacalcitol; cell cycle; aneuploids; miR-27b; miR-125b

Citation: Gleba, J.J.; Kłopotowska, D.; Banach, J.; Mielko, K.A.; Turlej, E.; Maciejewska, M.; Kutner, A.; Wietrzyk, J. Micro-RNAs in Response to Active Forms of Vitamin D$_3$ in Human Leukemia and Lymphoma Cells. *Int. J. Mol. Sci.* **2022**, 23, 5019. https://doi.org/10.3390/ijms23095019

Academic Editor: Cheng-Chia Yu

Received: 9 March 2022
Accepted: 29 April 2022
Published: 30 April 2022

Publisher's Note: MDPI stays neutral with regard to jurisdictional claims in published maps and institutional affiliations.

1. Introduction

Disorders of crucial stages of hematopoiesis, such as self-renewal, proliferation, and differentiation, are the leading causes of hematopoietic and lymphatic neoplasms. An essential element is unblocking the differentiation process and directing the cells to apoptosis. Therapies lead to complete remission and thus extend the patient's survival time. One of the therapies that offer such a possibility is differential therapy using a retinoic acid derivative (ATRA) [1]. The use of ATRA was initiated to treat patients with acute promyelocytic leukemia, achieving complete disease remission. Thanks to differential therapy, this fatal disease became utterly curable. Further clinical trials confirmed the high effectiveness of ATRA, also in combination with conventional chemotherapy [2–4]. The appearance of the so-called differentiation syndrome (DS), i.e., the retinoic acid syndrome, is a severe and prevalent complication after such treatment [5]. Therefore, it constantly emphasizes developing new, combined differential therapies that target specific abnormalities underlying cancer pathogenesis. Attempts are being made to use the active form of vitamin D$_3$—calcitriol—and its derivatives in patients with myeloid neoplasms [6,7].

Oncogenic microRNAs play a key role in the development of leukemia and lymphoma [8,9]. Moreover, they could be biomarkers and molecular factors in diagnosing and treating leukemias and lymphomas [10–12]. It turns out that miRNA molecules can directly regulate the actions of active forms of vitamin D_3. In the 3'UTR region of the classic vitamin D receptor (VDR) mRNA, there is an element recognized by miR-125b [12]. Studies have shown that its overexpression reduces the level of vitamin D receptor protein in MCF-7 breast cancer cells by 40% compared to the control [12]. miR-125b is not the only molecule that regulates the level of the vitamin D receptor, as miR-27b also binds at the 3'UTR site of the gene for the classic vitamin D receptor, reducing its expression, which was observed in cancer cells of the pancreas and colon [13]. Similarly, the expression of the CYP24A1 enzyme, which converts calcitriol, is regulated by miR-125b, suggesting that the level of this enzyme may be correlated with the level of miR-125b expression [14]. The miR-125b molecule, in addition to regulating the expression of proteins directly responsible for the action of calcitriol, also controls other key proteins of the cell [15,16]. The participation of this molecule in many essential processes, e.g., in apoptosis, suggests that it could become a target in anti-cancer therapy [17,18]. Other reports indicate that VDR may upregulate the expressions of miR-125b and miR-27b. This shows a more complex mechanism of interaction of these miRNAs and vitamin D_3 [19–21]. miR-125b regulates apoptosis in mature hematopoietic cells by blocking the pro-apoptotic Bak1 protein, is also responsible for the resistance to chemotherapeutic agents used to treat leukemia, such as vincristine and daunorubicin, and plays a significant role in the regulation of p53 protein expression and influences the activity of the NF-κB factor [22–24]. It is known that the use of calcitriol may affect the expression profile of miRNA molecules in cells. Experiments on the HL-60 acute promyelocytic leukemia and U937 lymphoma cells proved that calcitriol reduced the expression of miR-181a and miR-181b [25]. Decreasing the level of these non-coding structures increased the level of mRNA and p27 protein [26]. The p27 protein is directly related to monocyte differentiation [27]. Calcitriol also decreases miRNAs such as miR-302c and miR-520c in Kasumi-1 and K562 human leukemia cells, increasing their sensitivity to NK cell activity [28]. The expression of miR-32 is also reduced after the use of calcitriol, which is associated with an increase in the level of the pro-apoptotic protein Bim and sensitizes AML cells to the cytostatic cytosine arabinoside [29]. Similarly, the effect of calcitriol was noticed against miR-26a in HL-60 acute promyelocytic leukemia cells, which was associated with the regulation of factor E2F7, which causes an increase in p21 protein expression [30]. Moreover, calcitriol caused a significant reduction in the levels of miR-17-5p/20a/106a, miR-125b, and miR-155, which are repressors for AML1, VDR, and C/EBPβ in acute promyelocytic leukemia HL-60 cells [31]. In the conducted research, we analyzed the level of the miRNA molecules and the regulated proteins in human leukemia and lymphoma cells. The level of tested molecules was analyzed in response to calcitriol and tacalcitol.

2. Results

2.1. Calcitriol and Tacalcitol Effect on the Cell Cycle in Human Leukemia and Lymphoma

Calcitriol and tacalcitol increased the percentage of cells in the G_1/G_0 phase of the cell cycle. The changes were observed only in the MV-4-11, Thp-1, and HL-60 cell lines. These cell lines have been defined as sensitive to calcitriol and tacalcitol. In the case of the MV-4-11 cell line, an increase in cells in the G_1/G_0 phase was observed after 48 h, and this effect lingered until 120 h after calcitriol and tacalcitol use. At the same time, the population of cells in the S phase of the cell cycle decreased. The most significant reduction in the percentage of cells in the S phase was observed after 72 h, as compared to the control cells. There were no significant changes in the percentage of cells in the G_2/M phase of the cell cycle (Figure 1a). A similar effect of an increase in the percentage of cells in the G_1/G_0 phase of the cell cycle was observed in the Thp-1 cell line. Ninety-six hours after using the tested compounds, an increase in the percentage of cells in the G_1/G_0 phase by about 20% compared to control cells was observed, with a simultaneous decrease in the number of cells in the S phase of the cell cycle. However, there were no significant changes in the percentage of cells in the G_2/M phase of the cell cycle (Figure 1b).

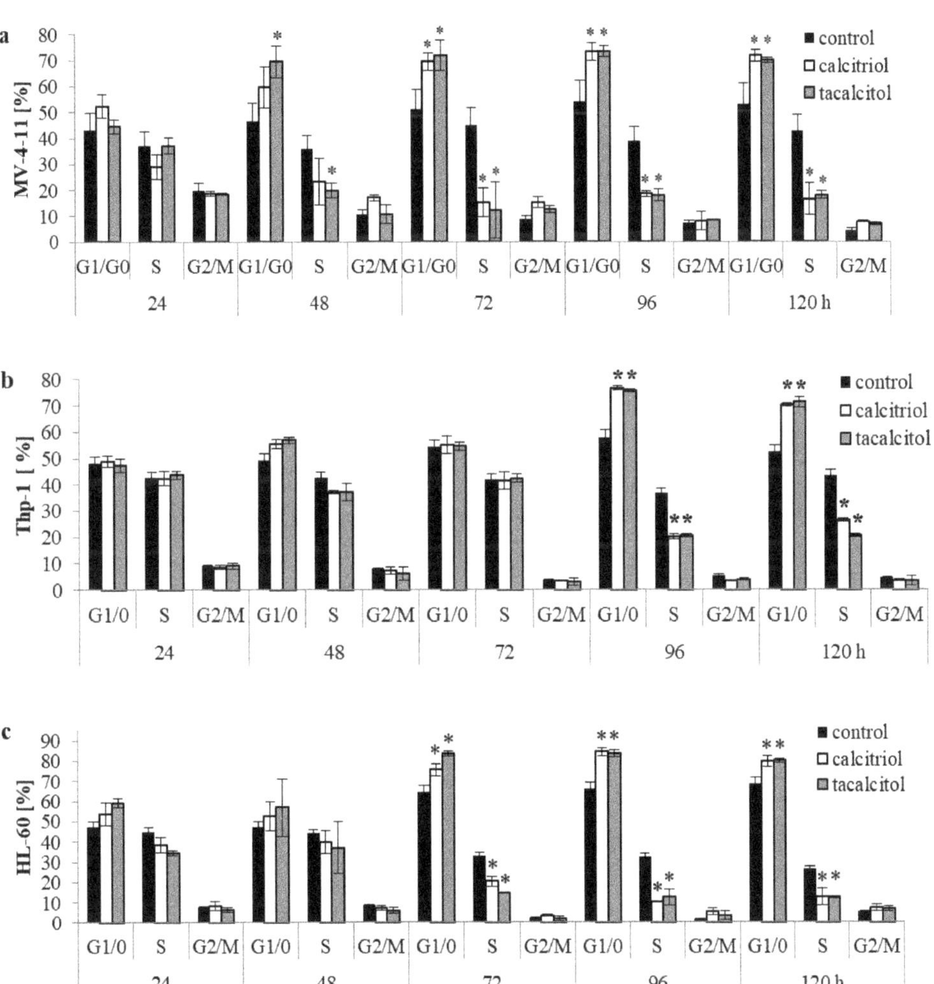

Figure 1. Cell cycle analysis of leukemia cells: acute biphenotypic myelomonocytic leukemia MV-4-11 (**a**), acute monocytic leukemia Thp-1 (**b**), and acute promyelocytic leukemia HL-60 (**c**) after the use of calcitriol and tacalcitol. The cell cycle was analyzed 24, 48, 72, 96, and 120 h after incubation with 10 nM calcitriol and tacalcitol. The graph shows the mean ± standard deviation. *—statistical significance ($p < 0.05$) compared to the control.

The use of calcitriol and tacalcitol showed a similar effect against the third sensitive cell line HL-60. Incubation with the tested compounds increased the percentage of cells in the G_1/G_0 phase. A slight effect of the substances used was observed after 24 h of incubation, but the most remarkable differences were noted after 72 h. Around a 20% increase in the percentage of cells in the G_1/G_0 phase was observed compared to the control cells. In addition, the population of cells in the S phase of the cell cycle was reduced. There were no changes in the percentage of cells in the G_2/M phase (Figure 1c). After calcitriol and tacalcitol use, there were no statistically significant changes in the other tested leukemia and lymphoma cells (Supplementary Materials, Figure S1).

2.2. Population of Aneuploid Cells and Response to Calcitriol and Tacalcitol

The population of aneuploid cells was analyzed in human leukemia and lymphoma cells. Studies have shown that the highest percentage of aneuploid cells was present in the two sensitive to calcitriol and tacalcitol cell lines: MV-4-11 and Thp-1. About 14% of aneuploid cells were observed in the MV-4-11 cell line, while about 17% were observed in the Thp-1 cell line. Interestingly, the remaining leukemia cell lines were 10% aneuploids (Figure 2a). However, analysis of the lymphoma cells showed a low percentage of aneuploid cells. In Raji, about 2% of aneuploid cells were observed. Around 3% of Daudi cells were aneuploid, and only 1% in Jurkat and U2932 (Figure 2b).

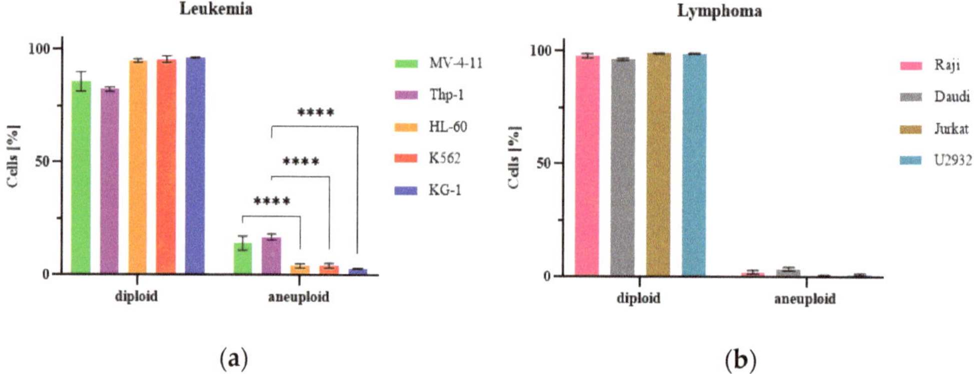

Figure 2. Diploids and aneuploids in human leukemia (**a**) and lymphoma (**b**) cells. The highest percentage of cells with abnormal chromosome numbers (aneuploids) was observed in the two sensitive to calcitriol and tacalcitol leukemia cell lines: MV-4-11 and Thp-1. The graph shows the mean $\pm$ standard deviation (n = 3). **** $p < 0.0001$.

Interestingly, the use of calcitriol and tacalcitol reduced the number of aneuploid cells in three sensitive cell lines: MV-4-11, Thp-1, and HL-60. Incubation of the MV-4-11 cells with calcitriol and tacalcitol reduced the number of aneuploids. The most significant effect of the tested substances was observed after 72 h when there was an approximately 10% decrease in the population of aneuploid cells. Lowering the number of aneuploid cells was observed after 96 and 120 h (Figure 3a). A similar effect was observed in Thp-1 cells. A decrease in the percentage of aneuploids was observed after 24 h, while significant differences were observed after 48 h of incubation with the tested compounds. Tacalcitol reduced the number of aneuploids by about 10% and calcitriol by about 5% (Figure 3b). A downward trend in the aneuploid population was also observed in HL-60 cells. A slight decrease in the percentage of aneuploids was observed 48 h after the treatment. However, a significant reduction occurred after 72 h, and this effect continued after 96 and 120 h. Interestingly, in the case of HL-60 cells, the percentage of aneuploids increased over time in the cell culture, and after 120 h, it was the highest, at about 15% (Figure 3c).

In the case of two other leukemia cell lines of KG-1 and K562 and all lymphoma cells, no changes in the percentage of aneuploid cells were observed after the calcitriol and tacalcitol treatment (Supplementary Materials, Figure S2).

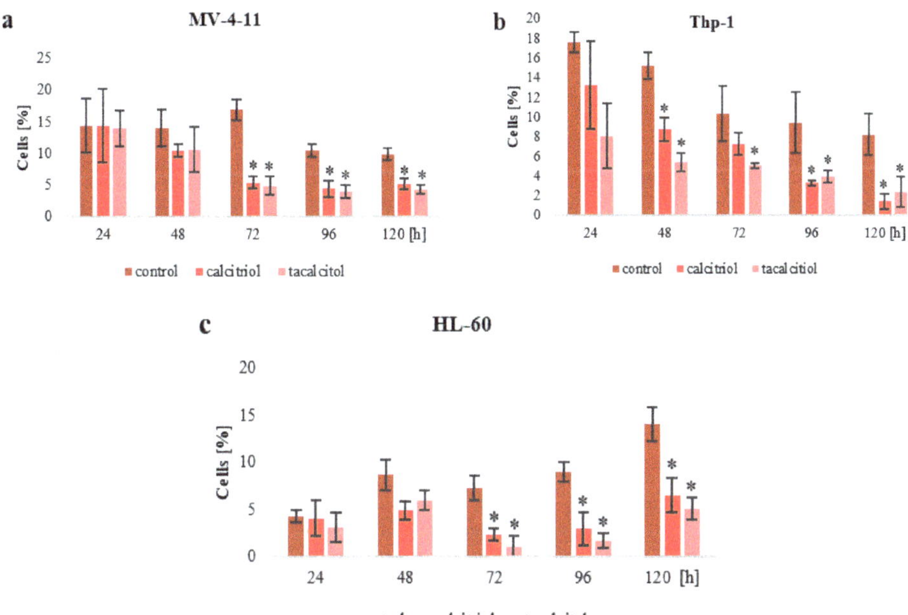

Figure 3. The population of aneuploids in leukemia cells after calcitriol and tacalcitol use. Active forms of vitamin D_3 reduced the percentage of aneuploids in MV-4-11 (**a**), Thp-1 (**b**), and HL-60 (**c**). The graph shows the mean $\pm$ standard deviation. *—statistical significance ($p < 0.05$) compared to the corresponding time point.

2.3. miR-181b and p27 Protein Involved in Cell Cycle

The basic level of p27 protein was significantly higher in the Raji and Daudi cells than PBMCs. However, the level of remaining cells was comparable to the control cells (Figure 4a). The use of calcitriol and tacalcitol resulted in an increased level of p27 in MV-4-11, HL-60, and Thp-1 cells. Conversely, calcitriol significantly reduced p27 in Raji and tacalcitol in KG-1, Daudi, and Jurkat cells (Figure 4b). miR-181a can impact the level of p21 protein by regulating the p53 protein, while miR-181b directly regulates the level of p27 [32,33]. It turned out that most cell lines have similar levels of the miR-181b molecule compared to PBMC. The exception was the HL-60 cell line, which had a much higher miR-181b than the control cells (Figure 4c). Interestingly, miR-181b decreased after using calcitriol and tacalcitol in three sensitive cell lines, MV-4-11, Thp-1, and HL-60. However, it increased in K562, Raji, Daudi, and Jurkat cells (Figure 4d). The miR-181a and p21 protein analysis did not show any correlation between the basal level and after treatment (Figure S3).

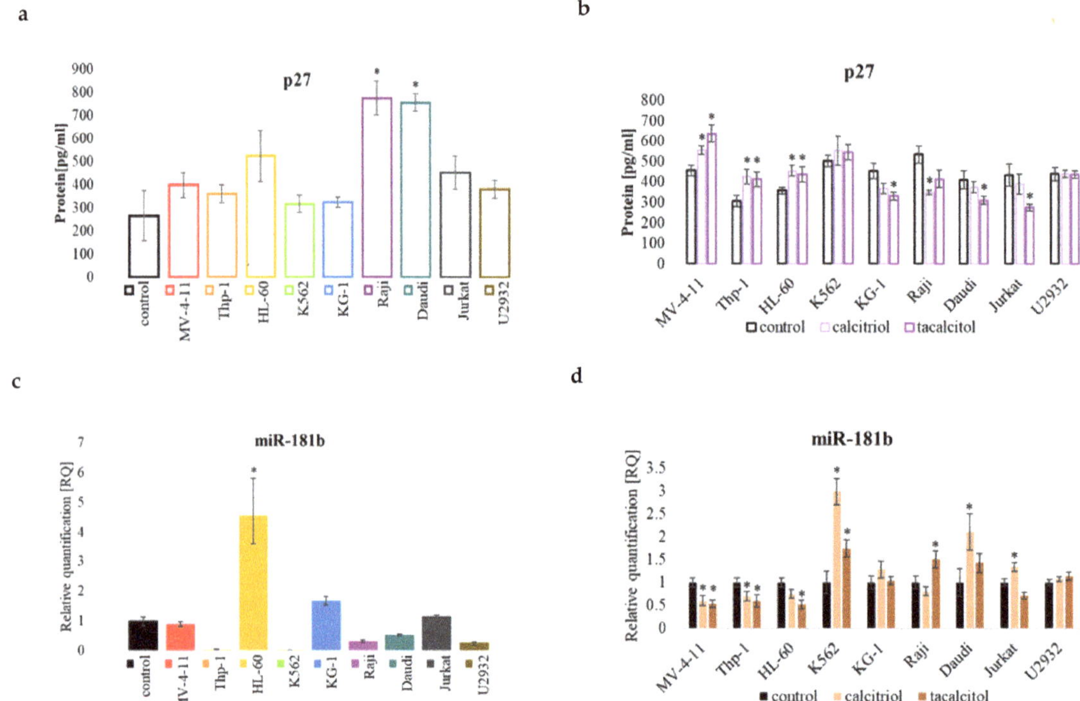

Figure 4. The p27 (**a**) protein (ELISA test) in human leukemia and lymphoma cells compared to the control: (**a**) PBMCs; (**b**) untreated cells of the appropriate cell line. miR-181b (**c**) in human leukemia and lymphoma cells compared to the control: (**c**) PBMCs; (**d**) untreated cells of the appropriate cell line. The relative quantification (RQ) was calculated using as control: (**a**) and (**c**) PBMCs; (**d**) and untreated cells of the appropriate cell line. *—statistical significance ($p < 0.05$) compared to the control.

2.4. miR-32b and Pro-Apoptotic Bim Protein

The miR-32 molecule and the Bim protein levels were determined in human leukemia and lymphoma cells. The miR-32 level in all tested cell lines was significantly lower than in the PBMCs used as the control. In all leukemia cell lines, the level of Bim protein was significantly higher than the level observed in the control cells. Lymphoma cells had a similar level of Bim protein to the normal blood cells (Figure 5a). Cell incubation with calcitriol or tacalcitol caused a decrease in the Bim level in three sensitive cell lines: MV-4-11, Thp-1, HL-60, KG-1, Daudi, and Jurkat (Figure 5b). The analyzed level of the miR-32 molecule showed that all the tested cell lines had significantly lower levels compared to the control (Figure 5c). Interestingly, the active forms of vitamin D$_3$ increased the level of miR-32 in three sensitive cell lines: MV-4-11, Thp-1, HL-60, K562, and Daudi. Calcitriol caused an increase in Jurkat, while tacalcitol decreased the level in this cell line (Figure 5d).

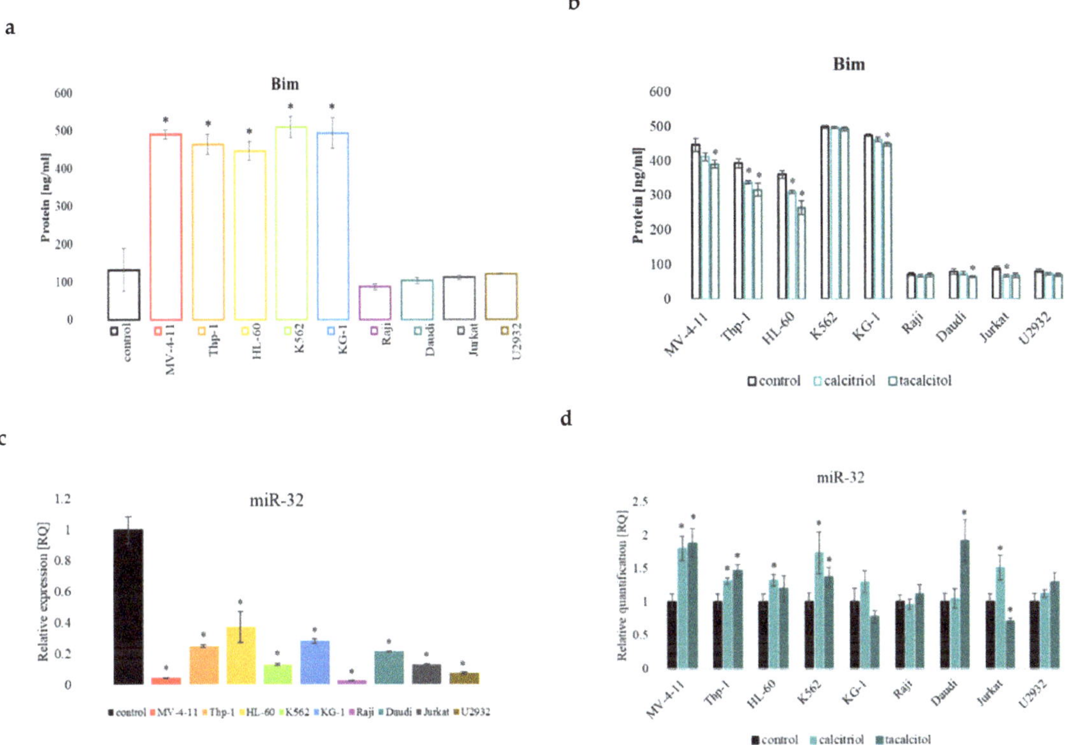

Figure 5. Pro-apoptotic protein Bim (ELISA test) level (**a**), and miR-32 molecule (**c**) in human leukemia and lymphoma compared to the control: (**a**) and (**c**) PBMCs; (**b**,**d**) untreated cells of the appropriate cell line. The relative quantification (RQ) was calculated using as the control: (**a**,**c**) PBMCs; (**d**) untreated cells of the appropriate cell line. The Bim (**b**) and miR-32 (**d**) after use of calcitriol and tacalcitol. *—statistical significance ($p < 0.05$) compared to the control.

2.5. miR-125b and Regulated Proteins: Pro-Apoptotic Protein Bak1, NFκB, and p53, and CYP24A1

2.5.1. miR-125b

The miR-125b in the tested cell lines showed that six cell lines: MV-4-11, Thp-1, K562, Raji, Daudi, and Jurkat, had a lower level of this molecule compared to the PBMCs. HL-60 had a similar level to the control cells, and two cell lines—KG-1 and U2932—had levels higher than the PBMCs (Figure 6a). The use of calcitriol and tacalcitol caused the increase of miR-125b in seven tested cell lines. No changes were observed in KG-1 and Raji. No significant differences also were observed after calcitriol use in the Daudi and U2932 cell lines (Figure 6b).

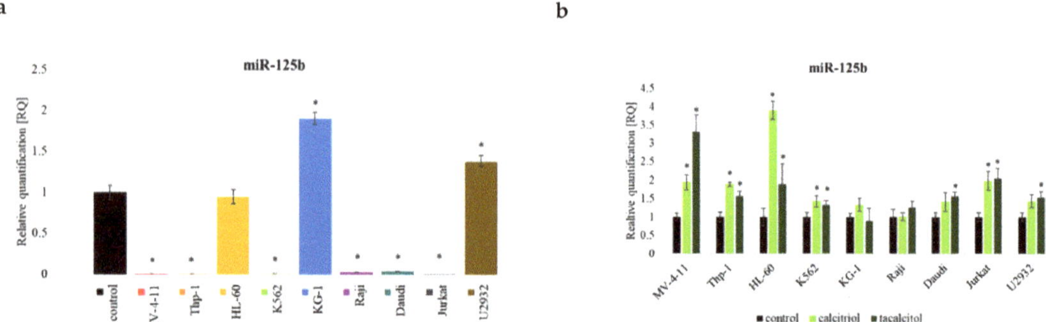

Figure 6. The miR-125b in human leukemia and lymphoma cells compared to the control: (**a**) PBMCs; (**b**) untreated cells of the appropriate cell line. The relative quantification (RQ) was calculated using as control: (**a**) PBMCs; (**b**) untreated cells of the appropriate cell line. The miR-125b in tested cell lines (**a**), and after incubation with calcitriol and tacalcitol (**b**). The graph shows the mean $\pm$ standard deviation. *—statistical significance ($p < 0.05$).

2.5.2. Bak1

It turned out that the level of Bak1 protein in all myeloid neoplasms and Daudi and Jurkat cells was similar to the level present in PBMCs. A significantly higher pro-apoptotic protein Bak1 than in the control cells was observed in Raji and U2932 lymphoma cells (Figure 7a). Calcitriol and tacalcitol caused significant upregulation of the Bak1 protein in the sensitive cells MV-4-11, Thp-1, and HL-60. Interestingly, calcitriol itself caused a reduction in the level of Bak1 protein in K562 and Daudi cells (Figure 7b).

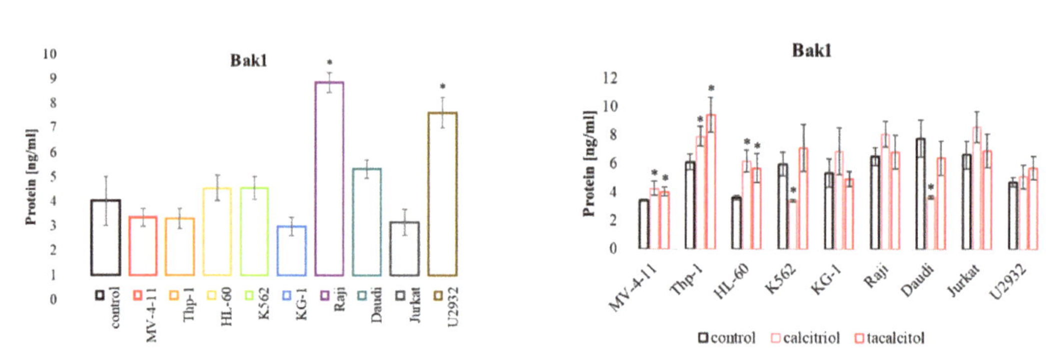

Figure 7. Bak1 in human leukemia and lymphoma cells compared to the control: (**a**) PBMCs; (**b**) untreated cells of the appropriate cell line. The relative quantification (RQ) was calculated using as the control: (**a**) PBMCs. The Bak1 level in the tested cell lines (**a**), and after incubation with calcitriol and tacalcitol (**b**). The graph shows the mean $\pm$ standard deviation. *—statistical significance ($p < 0.05$).

2.5.3. NFκB Protein

A protein that the miR-125b molecule regulates is also the transcription factor NFκB. The analysis of the level of this molecule in cells not treated with calcitriol and tacalcitol showed that MV-4-11 and Daudi cells had levels similar to those observed in normal cells. On the other hand, Thp-1, Jurkat, and U2932 cells have much more significant NFκB protein levels than normal cells. The remaining cells of the HL-60, K562, KG-1, and Raji lines selected for the study had lower levels compared to the control cells (Figure 8a). After calcitriol and tacalcitol treatment, the level of NFκB was downregulated in three sensitive cell lines: MV-4-11, Thp-1, and HL-60. There were no significant changes in the other cell lines (Figure 8b).

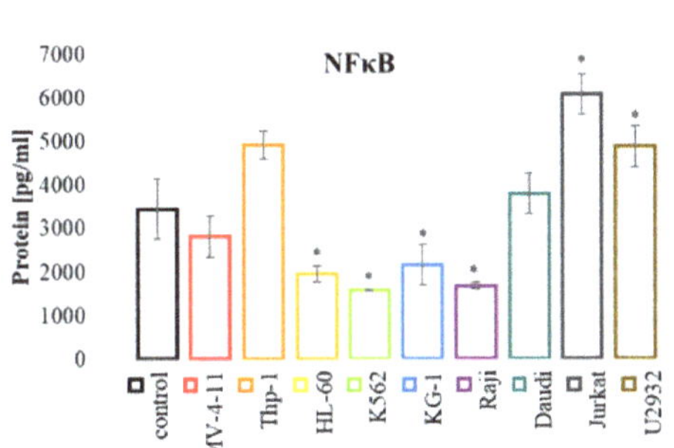

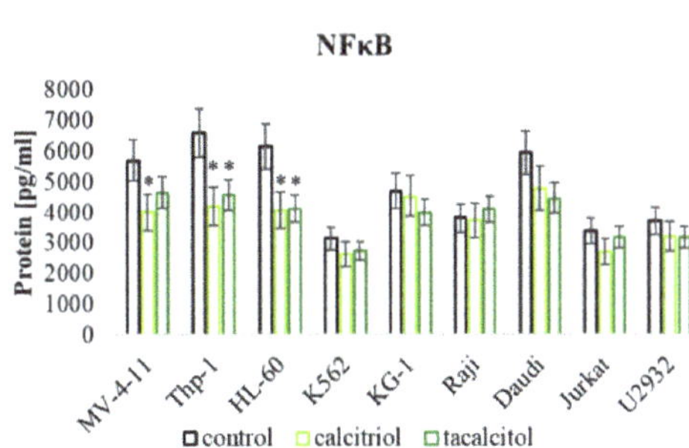

Figure 8. NFκB in human leukemia and lymphoma cells compared to the control: (**a**) PBMCs; (**b**) untreated cells of the appropriate cell line. The relative quantification (RQ) was calculated using as control: (**a**) PBMCs. NFκB in the tested cell lines (**a**), and after incubation with calcitriol and tacalcitol (**b**). The graph shows the mean ± standard deviation. *—statistical significance ($p < 0.05$).

2.5.4. CYP24A1

Analysis of the amount of protein in cells not treated with calcitriol and tacalcitol showed that all the cell lines tested had a significantly higher level of the CYP24A1 compared to the control PBMCs (Figure 9a). However, only in the sensitive cells, namely, MV-4-11, Thp-1, and HL-60, an increase in the level of the CYP24A1 protein was observed after the use of calcitriol and tacalcitol (Figure 9b).

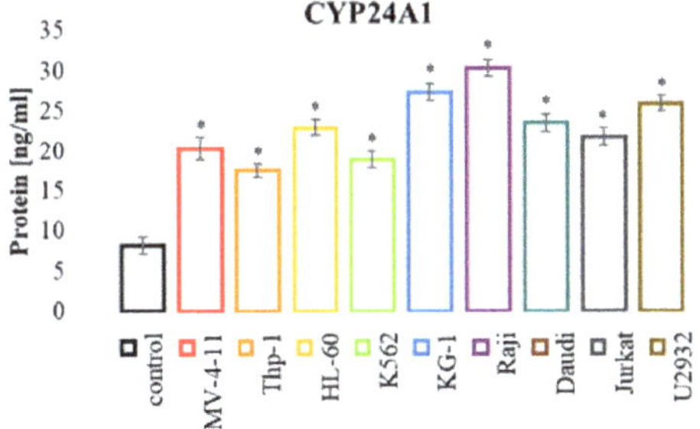

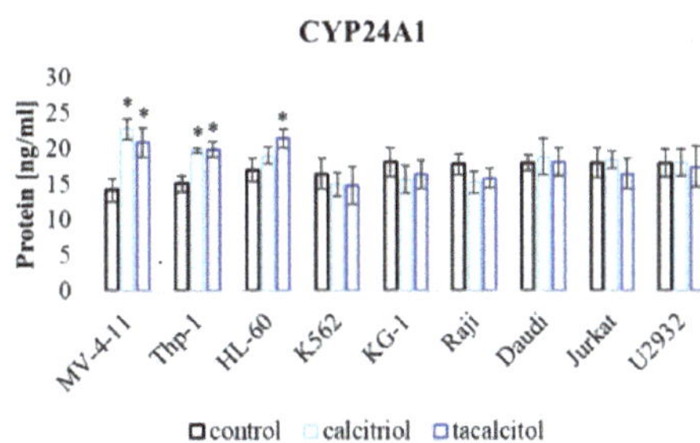

Figure 9. CYP24A1 in human leukemia and lymphoma cells compared to the control: (**a**) PBMCs; (**b**) untreated cells of the appropriate cell line. The relative quantification (RQ) was calculated using as control: (**a**) PBMCs. CYP24A1 in the tested cell lines (**a**), and after incubation with calcitriol and tacalcitol (**b**). The graph shows the mean ± standard deviation. *—statistical significance ($p < 0.05$).

2.6. miR-27b

These studies analyzed two microRNA molecules that regulate the classical vitamin D receptor levels: miR-27b (Figure 10) and miR-125b (Figure 6). The level of the miR-27b molecule was lower in all tested cell lines compared to the PBMCs (Figure 10a). Cell incubation with calcitriol and tacalcitol increased miR-27b in almost all tested cell lines. The K562 cell line was the only cell line where no significant changes were observed. In addition, there were no statistically significant changes in Thp-1, KG-1, Raji, and Daudi after using calcitriol (Figure 10b).

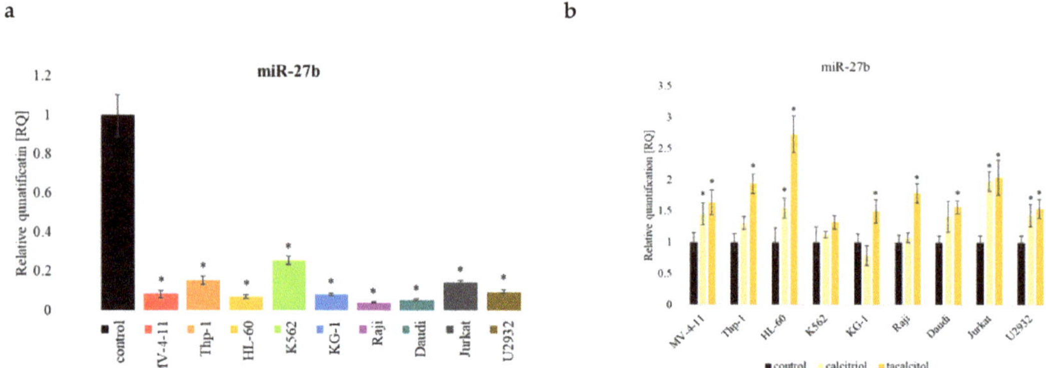

Figure 10. miR-27b in human leukemia and lymphoma cells compared the control: (**a**) PBMCs; (**b**) untreated cells of the appropriate cell line. The relative quantification (RQ) was calculated using as control: (**a**) PBMCs; (**b**) untreated cells of the appropriate cell line. The miR-27b in the tested cell lines (**a**), and after incubation with calcitriol and tacalcitol (**b**). The graph shows the mean ± standard deviation. *—statistical significance ($p < 0.05$).

2.7. Selected mRNA and miRNA Expression Level and Overall Survival

Using The Cancer Genome Atlas (TCGA) data, we correlated the expression level of selected mRNAs and miRNAs with overall survival (OS). The analyses were performed for acute myeloid leukemia. No data on the expression of the studied genes in patients with lymphoma were found. The analysis showed that the high level of the five examined genes are a poor prognosis factor in AML. These genes included VDR, Bak1, Bim, p21, and NFκB. Interestingly, the high expression level of p27 and p53 and all selected miRNAs was a good prognostic marker, extending the life span for patients with AML (Figure 11). There was no data about the CYP24A1 expression level and OS in AML patients.

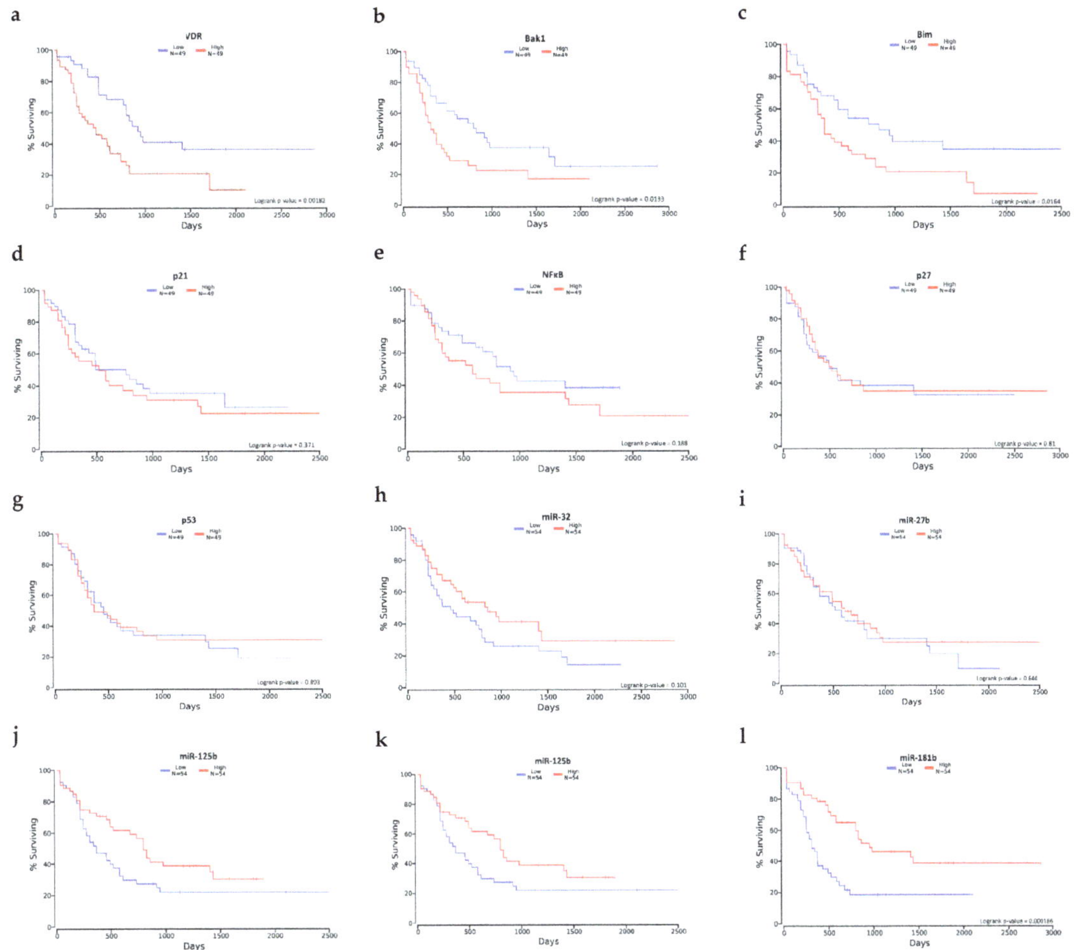

Figure 11. Kaplan–Meier overall survival (OS) for AML patients with high (red) and low (blue) expression of tested mRNA and miRNA. The VDR (**a**), Bak1 (**b**), Bim (**c**), p21 (**d**), and NFκB (**e**) high expression showed shorter overall survival (OS) compared to the low expression. On the other hand, high expression of p27 (**f**), p53 (**g**), miR-32 (**h**), miR-27b (**i**), miR-125b (**j**), miR-181a (**k**), and miR-181b (**l**) showed longer overall survival compared to low expression.

3. Discussion

Accumulating evidence indicates that small, non-coding microRNAs (miRNAs) are involved in the cellular response to calcitriol [34]. MicroRNAs are RNA molecules (usually 18 to 25 nucleotides) that play a regulatory role in plant and animal cells. They attach to the messenger RNA of target genes and control their expression and basic cell processes. Mature miRNA molecules anneal complementarily to the 3 'UTR region of the target gene mRNA, blocking and reducing its expression [35–37]. Using computational predictions, studies published by Calin et al. showed that one cluster containing miR-15a/16-1 holds about 14% of the genes in the human genome [38]. Studies confirm that microRNAs can directly regulate tumor growth. In 2006, Costinean et al., using transgenic mice overexpressing miR-155, proved that this molecule is responsible for developing B-cell neoplasms [39]. This experiment shows that miRNAs are essential in understanding the complex process of cancer formation. Many studies confirm that changes in the miRNA expression profile correlate with cancer development. Moreover, numerous experiments show that these molecules are responsible for cancer cells' sensitivity to chemotherapeutic agents [40], suggesting that they may be an effective therapeutic target [36,37,41]. Analysis of the microRNA expression profile in cancer cells can provide valuable information on the stage of the disease [42,43].

3.1. Active Forms of Vitamin D_3 Affect the Cell Cycle of Human Leukemia and Lymphoma Cells

The antiproliferative effect of calcitriol, through the arrest of cells in the G_0/G_1 phase, was observed in colon cancer cells. In these studies, this effect was caused by an increase in the cyclin p21 and p27-dependent kinase inhibitors and a simultaneous decrease in the amount of cyclin A and cyclin E [44]. In pancreatic cancer cells, calcitriol and its analogs also inhibited cells in the G_1 phase, associated with an increase in p21 and p27 proteins [45]. The p21 protein belongs to the family of cyclin-dependent kinase inhibitors, including the p27 and p57 proteins. The p21 protein is crucial in arresting cell growth following DNA damage. Increasing the amount of this protein leads to an accumulation of cells in the G_1 phase [46]. Our previous studies also showed that active forms of vitamin D_3 increased the number of human leukemia and lymphoma cells in G_0/G_1 [47]. The use of calcitriol and tacalcitol increased the percentage of cells in the G_0/G_1 phase and decreased the number in the S phase in sensitive MV-4-11, Thp-1, and HL-60 leukemia cells. No changes in the cell cycle were observed in the remaining leukemia and lymphoma cells used in the studies. Interestingly, all tested cell lines showed a significantly higher level of p21 protein than normal blood cells. At the same time, calcitriol and tacalcitol did not cause significant changes in the level of p21 protein in all tested cell lines. On the other hand, the amount of p27 protein was comparable in most cell lines to that observed in normal PBMCs. However, two cell lines, Raji and Daudi, expressed significantly higher levels of p27 than control cells. The use of calcitriol and tacalcitol increased the amount of p27 protein in sensitive cells, while in insensitive cells, no changes were observed, or the amount of this molecule was reduced. The results may indicate that the inhibition of sensitive cells in the G_1/G_0 phase is due to an increase in p27 protein by calcitriol and tacalcitol. Increasing the expression of p27 by calcitriol and tacalcitol may also be very beneficial in terms of the overall survival in patients with AML. It turns out that high p27 levels extend the OS of AML patients. The use of calcitriol and tacalcitol increases the level of p27, which may be an effective therapeutic tool.

Analysis of the miR-181 molecules and p21 and p27 levels showed that miR-181a was significantly lower in all tested cell lines. In contrast, the level of miR-181b was similar in most of the cell lines tested to that observed in control PBMCs. The use of calcitriol and tacalcitol caused an increase in miR-181a and a decrease in miR-181b in sensitive cells, while in insensitive cells, mainly a decrease in miR-181a and an increase in miR-181b were observed. The obtained results suggest that the p21 protein is not involved in the blockade of sensitive to calcitriol and tacalcitol cells in the G_1/G_0 phase, but p27 and its level may depend on the amount of miR-181b molecule. The pioneers who discovered the role of the miR-181b molecule in the regulation of the cell cycle believe that the transition of cells between these phases requires the joint operation of several mechanisms and that interference with only one of the factors regulating this complicated process may have only a moderate impact on its course. They also emphasize that it is unrealistic to expect that the p27 protein is fully controlled only by the miR-181b molecule [25]. Cell cycle control is a very complex process involving many molecular factors, including many miRNA molecules [32]. Interestingly, the induction of the p21 or p27 protein leads to an increase in the expression of the Bak1 protein and the activation of caspase 7 [48]. In the conducted studies, we observed a simultaneous increase in the levels of the Bak1 and p27 proteins. For a full explanation of the mechanism of action of the active forms of vitamin D_3, it will be rational to determine the activity of caspases.

3.2. Population of Aneuploid Cells in Human Leukemia and Lymphoma Cells and Response to Active Forms of Vitamin D₃

The number of cells with an abnormal number of chromosomes—aneuploids—was also estimated. In leukemia cell lines, the percentage of aneuploids was significantly higher compared to lymphomas. Calcitriol and tacalcitol caused a significant reduction in the rate of aneuploid cells in the sensitive cell lines MV-4-11, Thp-1, and HL-60. The role of aneuploid cells in tumor development has not yet been elucidated. It turns out that the presence of aneuploid cells is responsible for the more aggressive phenotype of cancer [49]. In vitro studies confirm that aneuploid cells have a different growth rate, metabolism, cell cycle kinetics, and size than diploid cells. Numerous studies show that aneuploidy is a hallmark of developing cancer cells. Many mechanisms are responsible for their formation, opening the door for developing new therapeutic strategies against aneuploid cells [50]. Interestingly, studies show that the presence of aneuploid cells increases the sensitivity of tumors to cytostatics, which may also be an excellent target in anti-cancer therapy [51]. Calcitriol and tacalcitol showed the most significant activity against cell lines with the highest percentage of aneuploid cells, suggesting that the sensitivity of human leukemia cells to these substances may depend on the presence of aneuploid cells. On the other hand, knowing that aneuploid cells arise through disturbances in the mechanisms of cell division, it may be suggested that calcitriol and its analogs show selective activity against cells in which these processes are disturbed. However, additional and detailed studies are necessary to explain these phenomena.

3.3. Pro-Apoptotic Proteins Bim and Bak1 in the Response of Human Leukemia and Lymphoma Cells to Calcitriol and Tacalcitol

Programmed cell death or apoptosis occurs in response to developmental changes and disease states, and it is required to maintain a state of homeostasis in the body. Cell resistance to apoptosis is one of the hallmarks of cancer cells [52,53]. Proteins of the Bcl-2 family play a key role in regulating apoptosis and modulating death signaling through the internal or mitochondrial pathway [54]. While the Bcl-2 proteins and several close relatives promote cell survival, other family members, such as Bax and Bak, induce apoptosis. For example, the BH3 family members (Bim, Puma, Noxa, Bid, Bad, and Bik) induce apoptosis by activating the Bax and Bak proteins [55]. Interestingly, a low concentration of the Bim protein can activate Bak1. However, to obtain the same level of both proteins, a high concentration (μM) of the Bim protein is necessary [56]. In the conducted studies, the use of calcitriol and tacalcitol resulted in a decrease in the pro-apoptotic protein Bim and an increase in the amount of Bak1 protein. Therefore, examining the status of other pro-apoptotic BH3 proteins can explain the role in regulating Bak1 levels under the influence of calcitriol and tacalcitol.

These results indicate that cells sensitive to calcitriol and its analog may undergo apoptosis by increasing the amount of the Bak1 protein. It is known that miRNA molecules can regulate both tested pro-apoptotic proteins. Studies have shown that the miR-125b molecule regulates Bak1 protein expression. In vivo studies confirmed that inhibition of miR-125b reduced the tumor growth of CML cells in mice. Low miR-125b levels affected the expression of the Bak1 protein, promoting apoptosis and inhibition of cell proliferation [57]. The reduction in Bak1 levels inhibited paclitaxel-induced apoptosis and led to an increase in paclitaxel resistance. In turn, the restoration of Bak1 expression with the miR-125b inhibitor restored the sensitivity of these cells. [58]. The conducted studies that indicated that the miR-125b molecule also significantly affects the chemotherapy of ovarian cancer cells. The increase in the expression of this molecule contributes to cisplatin resistance by inhibiting the amount of the Bak1 [59]. The pro-apoptotic protein Bim is regulated by the miR-32 molecule and miR-181a [60]. This molecule is crucial in the apoptosis of cancer cells of epithelial origin and the response to paclitaxel chemotherapy. Lowering the amount of Bim protein by the miR-32 molecule may contribute to the resistance of cancer cells to the induction of apoptosis in the tumor environment [61,62]. Interestingly, the level of Bim protein in all leukemia cell lines was higher than that observed in normal cells. However, this level in lymphoma cells was similar to the level present in control cells. Furthermore, the use of calcitriol and tacalcitol resulted in a decrease in Bim and an increase in the amount of the miR-32 molecule, which suggests that the amount of this pro-apoptotic protein may be regulated by the miR-32 molecule. Moreover, very interesting observations were made based on the analysis of the survival of patients with AML and the level of expression of miR-32, Bak1, and Bim. The high level of Bak1 and Bim expression correlates with a shorter overall survival. On the other hand, a high level of miR-32 is a good prognostic marker and extends the survival rate. Moreover, the use of calcitriol and tacalcitol causes an increase in miR-32 and a simultaneous reduction if Bim in sensitive cells. These observations suggest that calcitriol and tacalcitol may be useful treatment for regulating gene expression and increase the overall survival prognostic factor in AML patients AML. On the other hand, the level of the pro-apoptotic Bak1 protein in cells not treated with calcitriol and tacalcitol was similar to the control cells (except for Raji and Daudi cells). There were also no differences in the basal level of this protein between sensitive and insensitive cells. The use of calcitriol and tacalcitol increased the amount of Bak1 and miR-125b proteins, indicating that this molecule probably does not affect the level of Bak1 protein in human leukemia and lymphoma cells.

3.4. miR-125b and Proteins Involved in Cell Proliferation and Survival: NFκB and p53
NFkB

One of the critical factors that initiate, or block, apoptosis is the transcription factor NF-κB, which can be affected by p53. In the conducted studies, no relationship was observed between the basal amount of the NF-κB protein and the sensitivity of cells to calcitriol and tacalcitol. Studies indicate that blocking the NF-κB activity plays a crucial role in p53-induced cell death. It turns out that inhibition of NF-κB in cancer cells with an unmutated form of p53 leads to a decrease in response to cytostatics [63]. The nuclear factor-kappa B belongs to the family of transcription proteins, responsible for regulating the inflammatory process, immune response and proliferation, and apoptosis. Constitutive activation of the NF-κB pathway is often present in many types of cancer and chronic inflammatory diseases, such as multiple sclerosis, enteritis, rheumatoid arthritis, or asthma [64]. The regulation of cellular metabolism by NF-κB depends on the protein p53. Many mutations, including those involving p53, contribute to the activation of NF-κB in cancer cells. The p53 protein is antagonistic to the factor NF-κB, which may inhibit cancer development. A mutation in the p53 blocks the NF-κB pathway [65]. In addition, this factor can interfere with p53 transcriptional activity by lowering the p53 levels and inhibiting apoptosis by increasing the anti-apoptotic proteins [66].

In the conducted studies, it was observed that the level of NF-κB factor is reduced, in both sensitive and insensitive cells, after using calcitriol. The miR-125b molecule is known to be involved in regulating the expression of NF-κB [67]. It turns out that the level of miR-181b can also indirectly affect the amount of this transcription factor by regulating the CYDL protein. This protein is a tumor suppressor that lowers the NF-κB levels. It turns out that an increase in the amount of miR-181b causes a slowdown in CYDL production and an increase in the amount of NF-κB [68]. Interestingly, the transcription factor NF-κB can induce apoptosis and increase the amount of p21 protein in Ewing sarcoma cells through a pathway independent of the p53 protein. Based on the conducted research, it was found that an increase in the amount of the p21 protein protected cancer cells from apoptosis induced by TNF-α [69]. In other studies, it has also been found that monocyte apoptosis induced by TNF-α can be inhibited by an increase in p21 via NF-κB [70]. In these studies, a decrease in the level of transcription factor NF-κB was related to an increase in the amount of the miR-125b molecule, which may indicate that this molecule may affect the level of the NF-κB protein. We did not observe significant changes in the p21 and p53 protein levels (Figures S3 and S4). In cells sensitive to calcitriol and tacalcitol, a decrease in the amount of miR-181b was observed, and in insensitive cells, an increase in miR-181b, which may indicate that this molecule is not involved in the regulation of the NF-κB levels in human leukemia and lymphoma cells by calcitriol or its analog. Interestingly, calcitriol can block the NF-κB. Studies showed that VDR interacts with IKKβ, which is necessary for NF-κB activation [71,72]. On the other hand, cells without VDR are characterized by an increased level of NF-κB [73]. The simultaneous increase in the VDR and miR-125b levels is responsible for the lowering of the NF-kB level, but further analyses are required. Interestingly, high levels of NFkB are associated with a reduced life expectancy in AML patients. The action of calcitriol and tacalcitol in the human leukemias and lymphomas is presented in the Figure 12.

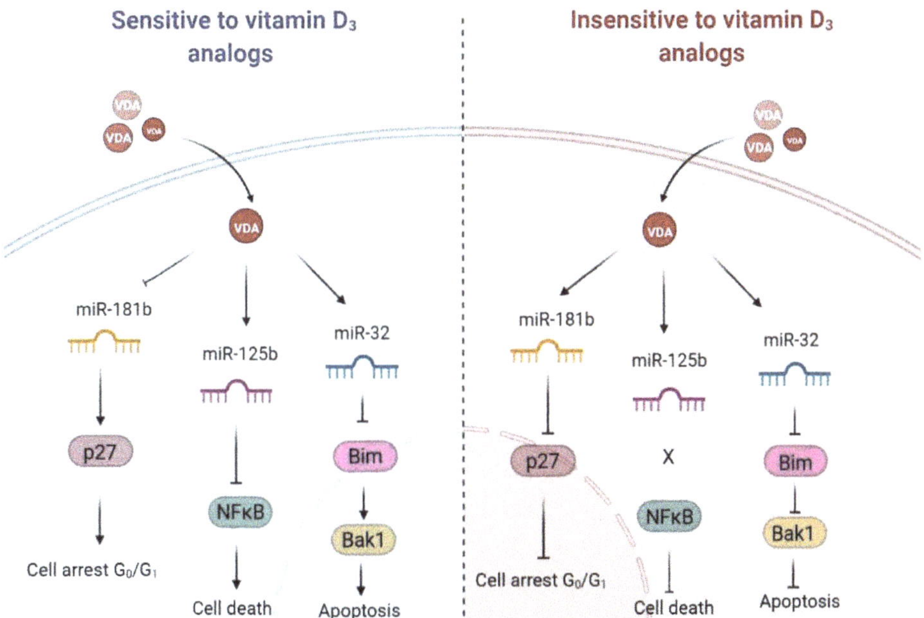

Figure 12. Vitamin D₃ analogs (VDA) biological activity in leukemia and lymphoma cells. The effect on miRNA and protein expression in sensitive and insensitive to VDAs leukemia and lymphoma cells.

3.5. miR-125b and miR-27b in the Regulation of Classic Vitamin D Receptor and CYP24A1

Our recent studies showed that the level of two receptors, the classic vitamin D receptor (VDR) and 1,25D₃-MARRS, increased at the same time under the influence of active forms of vitamin D₃. This effect was observed after calcitriol and tacalcitol use in three cell lines, MV-4-11, Thp-1, and HL-60. In the case of the remaining cells used in the studies, an increase in one of the tested receptors or a decrease in the level of these molecules was observed, and a simultaneous lack of sensitivity to calcitriol and tacalcitol [74]. It turns out that the response to calcitriol and tacalcitol via the vitamin D-VDR receptor may depend on the level of miRNA molecules that regulate the level of this protein. Research shows that two miRNAs, miR-27b and miR-125b, can control the amount of the vitamin D receptor— VDR [75]. The analysis of the level of these molecules showed that almost all leukemia and lymphoma cells selected for the study had significantly lower levels of miR-27b and miR-125b. The higher miR-125b characterized only KG-1 and U2932 compared to normal cells. In turn, the use of calcitriol and tacalcitol resulted in a significant increase in the amount of these molecules in all cell lines selected for research. The results indicate that the increase in miR-27b and miR-125b molecules did not reduce the level in cells sensitive to calcitriol and tacalcitol. Research indicates that the expression of miR-27b and miR-125b can be activated by the action of VDR [20,21]. However, the detailed molecular mechanism has not been known so far. Therefore, a thorough understanding of miR-27b and miR-125b, and vitamin D$_3$, should be of interest to future research. Interestingly, high levels of miR-27b and miR-125b extend survival in AML patients. The use of calcitriol and tacalcitol increased the expression of these molecules. It seems that the use of these substances in patient treatment could significantly extend the survival rate. Therefore, further work is needed on the precise understanding of their molecular interactions. Significantly lower levels of CYP24A1 mRNA characterized human leukemia and lymphoma cells selected for the study compared to normal cells. In contrast, the level of CYP24A1 protein was significantly higher in all neoplastic cells than in control cells. No differences in the level of this enzyme were observed between sensitive and insensitive cells, which may suggest that the basic level of this enzyme does not determine the sensitivity of leukemia and lymphoma cells to calcitriol and tacalcitol. The use of calcitriol and tacalcitol increased the amount of CYP24A1 mRNA in all cancer cell lines used in the study. In turn, the protein level analysis showed that the increase in the level of this molecule was observed only in cells sensitive to calcitriol and tacalcitol. In contrast, in non-sensitive cells, a decrease in the level of the CYP24A1 protein was observed. A vitamin D response element in the gene's promoter region encodes the CYP24A1 enzyme. Accumulating evidence suggests that high levels of CYP24A1 affect the biological activity of calcitriol in cancer cells. The combined use of calcitriol and CYP24A1 inhibitors increases the antiproliferative activity of calcitriol and its analogs [76]. The amount of CYP24A1 protein may depend on the level of the miR-125b molecule [77]. In the conducted studies, an increase in the level of the miR-125b molecule was observed in all tested cell lines. In cells sensitive to calcitriol and tacalcitol, an increase in the level of the CYP24A1 protein was observed. In contrast, in insensitive cells, a decrease in the amount of this enzyme was mainly observed. In cells sensitive to calcitriol and tacalcitol, along with an increase in the CYP24A1 protein, an increase in the miR-125b molecule was noticed, suggesting that this molecule may not affect the level of this protein in these cells. In contrast, in cells insensitive to calcitriol and tacalcitol, the level of CYP24A1 may be decreased by the miR-125b molecule. Interestingly, only in cells sensitive to the action of calcitriol and tacalcitol, did we observe an increase in VDR and CYP24A1 expression [74]. VDR is known for activating the CYP24A1 promoter region and initiating CYP24 transcription. It is also known that calcitriol actively induces CYP24 splicing [78]. Further research is required to identify the mechanisms responsible for this differential response of sensitive and insensitive cells to vitamin D$_3$ derivatives.

4. Materials and Methods

4.1. Human Leukemia and Lymphoma Cell Lines

The following human cancer cell lines were used: leukemia: KG-1, K562, HL-60, MV-4-11, Thp-1; lymphoma–Jurkat, U2932, Daudi, and Raji. All cells were derived from the Hirszfeld Institute of Immunology and Experimental Therapy repository, Polish Academy of Sciences in Wroclaw, Poland. All tested cell lines were cultured in RPMI 1640w/GLUTAMAX-I media (Gibco, Gaithersburg, MD, USA) containing 10% fetal bovine serum (Thermo Fisher Scientific, Waltham, MA, USA), streptomycin (0,1 mg/mL), and penicillin (100 U/mL) (Polfa Tarchomin, Warsaw, Poland). In addition, media for Thp-1 were supplemented with 0.05 mM β-mercaptoethanol, MV-4-11, and HL-60 with sodium pyrogroniate 1 mM and HL-60 additionally with 3.5 g/L glucose (all Sigma-Aldrich, Steinheim, Germany). Cells were incubated at 37 °C with a 5% CO$_2$ in a humid atmosphere incubator (NuAire, Plymouth, MN, USA).

4.2. Human Normal Blood Cells

Samples of peripheral blood from 10 healthy people who consciously and voluntarily submitted them for testing and signed individual statements were collected after obtaining the consent of the Bioethics Committee at the Medical University of Wroclaw, No. 71/2017. First, whole blood was collected into test tubes with EDTA anticoagulant by specialized personnel of the Diagnostics laboratory at the Institute of Immunology and Experimental

Therapy. According to the manufacturer's instructions, blood mononuclear cells were then isolated using the Ficoll-Paque PLUS kit (GE Healthcare, Amersham, UK).

4.3. Calcitriol and Tacalcitol

The active forms of vitamin D, calcitriol (1,25-$(OH)_2D_3$) and tacalcitol (PRI-2191, 1,24-$(OH)_2D_3$, were used. Compounds were synthesized at the Pharmaceutical Research Institute (PRI, Warsaw, Poland). Stock solutions of calcitriol and tacalcitol were obtained by dissolving 50 µg lyophilizate in 99.8% ethanol (Avantor, Gliwice, Poland), obtaining a 10^{-4} nM solution, and was stored at $-20\,^{\circ}$C.

4.4. Real-Time PCR

Cells suspended in the media were centrifuged for 5 min at $300\times g$ and washed with PBS. The cell pellet was resuspended in 700 µL of TRIzol (Invitrogen, Carlsbad, CA, USA). Then 0.2 mL of chloroform (Avantor, Gliwice, Poland) was added and mixed for 15 s. Samples were incubated for 5 min at room temperature and centrifuged for 15 min at $12,000\times g$ at 4 $^{\circ}$C. The aqueous phase was transferred to new tubes, and then 0.5 mL of isopropanol (Avantor, Gliwice, Poland) was added. Samples were incubated for 5 min at room temperature and centrifuged for 15 min at $12,000\times g$ at 4 $^{\circ}$C. The supernatant was removed, and the RNA pellet was resuspended in 1 mL of 75% ethanol with DEPC. Samples were centrifuged for 5 min $7500\times g$ at 4 $^{\circ}$C. The supernatant was removed, and the samples were air-dried for 5 min at room temperature. Each sample was dissolved in 30 µL of molecular biology-grade water (Sigma-Aldrich, Steinheim, Germany) and left on ice until completely dissolved. The RNA concentration was measured using a NanoDrop 2000 spectrophotometer (ThermoFisher Scientific, Waltham, MA, USA).

A total of 1 µg of RNA was used for purification. The following mixture was prepared: 1 µg RNA, 1 µL RNAz inhibitors, 1.5 µL DNAseI buffer, 1 µL DNAseI enzyme (all reagents from ThermoFisher Scientific, Waltham, MA, USA), made up to 15 µL with molecular biology water (Sigma-Aldrich, Steinheim, Germany). DNA digestion was carried out in a Veriti thermal cycler (Applied Biosystems, Foster, CA, USA) at 37 $^{\circ}$C for 15 min. After incubation, 2 µL EDTA (ThermoFisher Scientific, Waltham, MA, USA) was added to the samples and incubated for 10 min at 65 $^{\circ}$C to inactivate the enzyme. Electrophoresis was performed on both samples after and before cleaning.

For mRNA, 15 µL of the mixture after purification, 4 µL of a $5\times$ concentrated buffer for reverse transcriptase, and 1 µL of reverse transcriptase enzyme from the iScript kit (BioRad, Hercules, CA, USA) were added. The reverse transcription process was carried out in a Veriti thermal cycler under the following conditions: 5 min at 25 $^{\circ}$C, 30 min at 42 $^{\circ}$C, and 5 min at 85 $^{\circ}$C. Samples were stored at $-20\,^{\circ}$C. For miRNA, 1 µg of RNA was used for reverse transcription. The reaction mixture was then prepared using the TaqMan MicroRNA Reverse Transcription Kit (ThermoFisher Scientific, Waltham, MA, USA). The following reverse transcription microRNA probes were used: TaqMan MicroRNA Assay RT hsa-miR-27b, TaqMan MicroRNA Assay RT has-miR-32-5p, TaqMan MicroRNA Assay RT hsa-miR-125b-5p, TaqMan MicroRNA Assay RT hsa-miR-181a-5p, TaqMan MicroRNA Assay RT hsa-miR-181b-3p. The reverse transcription reaction was performed in a Veriti thermocycler under the following conditions: 30 min at 16 $^{\circ}$C, 30 min at 42 $^{\circ}$C, and 5 min at 85 $^{\circ}$C. After reverse transcription, the samples were stored at $-20\,^{\circ}$C. To analyze vitamin D receptor (VDR) (Hs01045840_m1), 1.25D3-MARRS (Hs00607126_m1), and CYP24A1 (Hs00167999_m1) expression, 50 ng of cDNA was used. The final reaction volume was 10 µL, and TaqMan Universal PCR Master Mix (Applied Biosystems, Foster, CA, USA) and TaqMan probes were used. 50 ng of cDNA per reaction was used to examine the expression of selected microRNA molecules. The final volume of the reaction mixture was 10 µL. The template also contained TaqMan Universal PCR Master Mix, TaqMan Small RNA probes (Applied Biosystems, Foster, CA, USA), and 4.8 µL of reverse transcription product in addition to the template. The following probes were used to perform the Real-Time PCR reaction: TaqMan MicroRNA Assay TM hsa-miR-27b, TaqMan MicroRNA Assay

TM hsa-miR-32-5p, TaqMan MicroRNA Assay TM hsa-miR-125b-5p, TaqMan MicroRNA Assay TM hsa-miR-181a-5p, and TaqMan MicroRNA Assay TM hsa-miR-181b-3p. The reaction was run under the following conditions: 95 °C for 10 min, 40 cycles consisting of denaturation at 95 °C for 15 sec, and annealing at 60 ° C for 1 min in a Real-Time PCR Via 7 Reaction Machine (Applied Biosystems, Foster, CA, USA). Expression of the tested genes was estimated using the comparative method $\Delta\Delta$Ct. Data were analyzed using Data Assist v. 3.01 (Applied Biosystems, Foster, CA, USA). At least three independent RNA isolations were performed for all cell lines used in the study.

4.5. ELISA Tests

Commercial ELISA assays (Elabscience, Houston, TX, USA) were used to determine the CYP24A1, p53, NFκB, Bak1, Bim, p21, and p27 proteins. The manufacturer's instructions were followed in all steps. Standard curves were prepared by diluting the standard for each protein to be assayed in 1 mL Reference Standard & Sample Dilution. The standard dilutions were prepared as follows: CYP24A1 (10; 5; 2.5; 1.25; 0.63; 0.31; 0.16; 0 ng/mL), p53 (5000; 2500; 1250; 625; 312.5; 156.25; 78.13; 0 pg/mL), NFκB (4000; 2000; 1000; 500; 250; 125; 62.5; 0 pg/mL), Bak1 (10; 5; 2.5; 1.25; 0.63; 0.31; 0.16; 0 ng/mL), Bim (20; 10; 5; 2.5; 1.25; 0.625; 0.313; 0 ng/mL), p21 (1000; 500; 250; 125; 62.5; 31.25; 15.63; 0 pg/mL), p27 (20; 10; 5; 2.5; 1.25; 0.625; 0.313; 0 ng/mL). In total, 100 ng of each sample were diluted 25× in Reference Standard & Sample Dilution. Then 100 μL of the sample and the standard were added to the wells of a 96-well coated plate. Samples were incubated for 90 min at 37 °C. The solution in the well was then carefully removed, and 100 μL of biotinylated Biotinylated Detection Ab was immediately added and incubated for 1 h at 37 °C. After incubation, the wells were washed three times with 350 μL Wash Buffer, and 100 μL of the HRP conjugate solution was added to each well and incubated for 30 min at 37 °C. Then the plates were washed 5 times with Wash Buffer. In total, 90 μL of Substrat Solution were added, and plates were incubated in the dark for 15 min at 37 °C. After a time, 50 μL of Stop Solution was added. The color of the solution turned from blue to yellow. Spectrophotometric measurement was performed at a wavelength of 450 nm using a Synergy H4 Hybrid Multi-Mode Microplate Reader (BioTek Instruments, Inc., Winooski VT, USA). Standard curves were plotted using the Curve Expert software (Hyams, D.G., CurveExpert software, http://www.curveexpert.net, 2010 (accessed on 10 August 2017)).

4.6. Cell Cycle Analysis and Aneuploids Population

Cells (10^5 cells/well) were plated in 24-well plates (Corning Incorporated, Corning, NY, USA) in 2 mL of media. After 24 h, 10 nM of calcitriol and tacalcitol were added. Cells were harvested at 24, 48, 72, 96, and 120 h in 15 mL tubes (Corning Incorporated, Corning, NY, USA), centrifuged at 400× g at 4 °C for 10 min. The supernatant was removed, and cells were washed with PBS and counted. One million cells were resuspended in 0.7 mL of 70% cold ethanol and stored at −20 ° C for a minimum of 24 h. On the day of the cell cycle analysis, the 70% ethanol was removed by centrifuging the samples at 400× g at 4 °C for 10 min. Then the cell pellet was washed by adding 0.5 mL PBS and resuspended in an 8 μg/mL RNAse solution (ThermoFisher Scientific, Waltham, MA, USA), transferred to cytometer tubes (Corning Incorporated, Corning, NY, USA), and incubated for 1 h at 37 °C. After incubation, 50 μL of 0.1 mg/mL propidium iodide (ThermoFisher Scientific, Waltham, MA, USA) was added to the cells. After 30 min incubation at room temperature, samples were analyzed using a BD LSRFortessa flow cytometer using FACS Diva software (Becton Dickinson, San Jose, CA, USA). The percentage of cells in particular phases of the cell cycle was determined using ModFit LT software (Verity Software House, Topsham, ME, USA).

4.7. Statistical Analysis

Statistical analysis of the obtained results was performed using the Statistica 7.1 program (StatSoft Inc., Tulsa, OK, USA). Before selecting the test, the normality of the distribution was analyzed using Levene's test. Next, data showing a normal distribution were analyzed using one-way ANOVA. Finally, non-parametric data were analyzed with the Kruskal–Wallis test. Results were statistically significant when $p < 0.05$. The charts were prepared using Excel (Microsoft Office) and GraphPad Prism version 6.04 for Windows (GraphPad Software, La Jolla, CA, USA).

4.8. Overall Survival Analysis

TCGA survival data was linked to VDR, Bak1, Bim, p21, p27, p53, miR-27b, miR-32, miR-125b, miR-181a, and miR-181b expression in patients with acute myeloid leukemia by downloading data from http://www.oncolnc.org/ (accessed on 18 April 2022). Survival data up to 3000 days was used for the survival analysis. Kaplan–Meier curves and log-rank tests were performed using the survival package [79].

5. Conclusions

The studies indicate that the antitumor activity of calcitriol and tacalcitol results from the interaction of several miRNA molecules and the level of the targeted protein. Active forms of vitamin D_3 caused an increase in the p27 protein that blocked cells in the G_0/G_1 phase of the cell cycle. At the same time, there was an increase in the pro-apoptotic protein Bak1. It correlated with the decreased level of miR-181b regulating p27. Interestingly, the use of calcitriol and tacalcitol resulted in an increase in miR-27b and miR-125b as well as CYP24A1, with a simultaneous reduction in NF-kB. These two miRNAs' (miR-27b and miR-125b) roles should be studied further to understand the molecular mechanism of vitamin D_3 action.

Supplementary Materials: The following supporting information can be downloaded at: https://www.mdpi.com/article/10.3390/ijms23095019/s1.

Author Contributions: Conceptualization, J.J.G., D.K. and J.W.; methodology, J.J.G.; formal analysis, J.J.G., J.B. and K.A.M.; resources, A.K.; data curation J.J.G., J.B., K.A.M., E.T., M.M.; writing, J.J.G. and D.K.; writing—review and editing, J.W., D.K. and A.K.; visualization, J.J.G.; supervision, J.W. and D.K.; project administration, J.J.G. and J.W.; funding acquisition, J.J.G. and J.W. All authors have read and agreed to the published version of the manuscript.

Funding: This research was funded by the National Science Center, Poland, under the PRELUDIUM 9 research project "The role of miRNA in the differentiation of human leukemia and lymphoma cells under the influence of calcitriol and its analog PRI-2191" No. 2015/17/N/NZ4/02833 for the years 2016–2019.

Institutional Review Board Statement: The study was conducted according to the guidelines of the Bioethics Committee at the Medical University of Wroclaw, No. 71/2017.

Informed Consent Statement: Informed consent was obtained from all subjects involved in the study.

Data Availability Statement: The data presented in this study are available in this article (and Supplementary Materials).

Conflicts of Interest: The authors declare no conflict of interest.

References

1. Hamadani, M.; Awan, F.T. Remission induction, consolidation and novel agents in development for adults with acute myeloid leukaemia. *Hematol. Oncol.* **2010**, *28*, 3–12. [CrossRef] [PubMed]
2. Schlenk, R.F.; Fröhling, S.; Hartmann, F.; Fischer, J.T.; Glasmacher, A.; del Valle, F.; Grimminger, W.; Götze, K.; Waterhouse, C.; Schoch, R.; et al. Phase III study of all-trans retinoic acid in previously untreated patients 61 years or older with acute myeloid leukemia. *Leukemia* **2004**, *18*, 1798–1803. [CrossRef] [PubMed]
3. Di Febo, A.; Laurenti, L.; Falcucci, P.; Tosti, M.E.; Fianchi, L.; Pagano, L.; Leone, G. All-trans Retinoic Acid in Association with Low Dose Cytosine Arabinoside in the Treatment of Acute Myeoid Leukemia in Elderly Patients. *Am. J. Ther.* **2007**, *14*, 351–355. [CrossRef] [PubMed]
4. Zhu, J.; Chen, Z.; Lallemand-Breitenbach, V.; de Thé, H. How acute promyelocytic leukaemia revived arsenic. *Nat. Rev. Cancer* **2002**, *2*, 705–714. [CrossRef] [PubMed]

5. Sanz, M.A.; Montesinos, P. How we prevent and treat differentiation syndrome in patients with acute promyelocytic leukemia. *Blood* **2014**, *123*, 2777–2782. [CrossRef]

6. Deeb, K.K.; Trump, D.L.; Johnson, C.S. Vitamin D signalling pathways in cancer: Potential for anticancer therapeutics. *Nat. Rev. Cancer* **2007**, *7*, 684–700. [CrossRef]

7. Woloszynska-Read, A.; Johnson, C.S.; Trump, D.L. Vitamin D and cancer: Clinical aspects. *Best Pract. Res. Clin. Endocrinol. Metab.* **2011**, *25*, 605–615. [CrossRef]

8. Gounaris-Shannon, S.; Newbury, S.; Chevassut, T. MicroRNAs—Key Players in Haematopoiesis. *Curr. Signal Transduct. Ther.* **2013**, *8*, 91–98. [CrossRef]

9. Fazi, F.; Rosa, A.; Fatica, A.; Gelmetti, V.; De Marchis, M.L.; Nervi, C.; Bozzoni, I. A minicircuitry comprised of microRNA-223 and transcription factors NFI-A and C/EBPalpha regulates human granulopoiesis. *Cell* **2005**, *123*, 819–831. [CrossRef]

10. Xiao, C.; Calado, D.P.; Galler, G.; Thai, T.H.; Patterson, H.C.; Wang, J.; Rajewsky, N.; Bender, T.P.; Rajewsky, K. MiR-150 controls B cell differentiation by targeting the transcription factor c-Myb. *Cell* **2007**, *131*, 146–159. [CrossRef]

11. Fontana, L.; Pelosi, E.; Greco, P.; Racanicchi, S.; Testa, U.; Liuzzi, F.; Croce, C.M.; Brunetti, E.; Grignani, F.; Peschle, C. MicroRNAs 17-5p-20a-106a control monocytopoiesis through AML1 targeting and M-CSF receptor upregulation. *Nat. Cell Biol.* **2007**, *9*, 775–787. [CrossRef] [PubMed]

12. Mohri, T.; Nakajima, M.; Takagi, S.; Komagata, S.; Yokoi, T. MicroRNA regulates human vitamin D receptor. *Int. J. Cancer* **2009**, *125*, 1328–1333. [CrossRef] [PubMed]

13. Pan, Y.Z.; Gao, W.; Yu, A.M. MicroRNAs regulate CYP3A4 expression via direct and indirect targeting. *Drug Metab. Dispos.* **2009**, *37*, 2112–2117. [CrossRef] [PubMed]

14. Wang, L.; Gao, Z.; Wang, L.; Gao, Y. Loss of miR-125b contributes to upregulation of CYP24 in uraemic rats. *Nephrology* **2016**, *21*, 1063–1068. [CrossRef]

15. Sun, Y.M.; Lin, K.Y.; Chen, Y.Q. Diverse functions of miR-125 family in different cell contexts. *J. Hematol. Oncol.* **2013**, *6*, 1–8. [CrossRef]

16. Alemdehy, M.F.; Erkeland, S.J. MicroRNAs: Key players of normal and malignant myelopoiesis. *Undefined* **2012**, *19*, 261–267. [CrossRef]

17. Gong, J.; Zhang, J.P.; Li, B.; Zeng, C.; You, K.; Chen, M.X.; Yuan, Y.; Zhuang, S.M. MicroRNA-125b promotes apoptosis by regulating the expression of Mcl-1, Bcl-w and IL-6R. *Oncogene* **2013**, *32*, 3071–3079. [CrossRef]

18. Surdziel, E.; Cabanski, M.; Dallmann, I.; Lyszkiewicz, M.; Krueger, A.; Ganser, A.; Scherr, M.; Eder, M. Enforced expression of miR-125b affects myelopoiesis by targeting multiple signaling pathways. *Blood* **2011**, *117*, 4338–4348. [CrossRef]

19. Yokoi, T.; Nakajima, M. Toxicological Implications of Modulation of Gene Expression by MicroRNAs. *Toxicol. Sci.* **2011**, *123*, 1–14. [CrossRef]

20. Ge, X.; Yuan, L.; Wei, J.; Nguyen, T.; Tang, C.; Liao, W.; Li, R.; Yang, F.; Zhang, F.; Zhao, B.; et al. Vitamin D/VDR signaling induces miR-27a/b expression in oral lichen planus. *Sci. Rep.* **2020**, *10*, 310. [CrossRef]

21. Giangreco, A.A.; Vaishnav, A.; Wagner, D.; Finelli, A.; Fleshner, N.; Van Der Kwast, T.; Vieth, R.; Nonn, L. Tumor Suppressor microRNAs, miR-100 and-125b, Are Regulated by 1,25-dihydroxyvitamin D in Primary Prostate Cells and in Patient Tissue. *Cancer Prev. Res.* **2013**, *6*, 483–494. [CrossRef] [PubMed]

22. Akbari Moqadam, F.; Lange-Turenhout, E.A.M.; Ariës, I.M.; Pieters, R.; den Boer, M.L. MiR-125b, miR-100 and miR-99a co-regulate vincristine resistance in childhood acute lymphoblastic leukemia. *Leuk. Res.* **2013**, *37*, 1315–1321. [CrossRef] [PubMed]

23. Kim, S.W.; Ramasamy, K.; Bouamar, H.; Lin, A.P.; Jiang, D.; Aguiar, R.C.T. MicroRNAs miR-125a and miR-125b constitutively activate the NF-κB pathway by targeting the tumor necrosis factor alpha-induced protein 3 (TNFAIP3, A20). *Proc. Natl. Acad. Sci. USA* **2012**, *109*, 7865–7870. [CrossRef] [PubMed]

24. Le, M.T.N.; The, C.; Shyh-Chang, N.; Xie, H.; Zhou, B.; Korzh, V.; Lodish, H.F.; Lim, B. MicroRNA-125b is a novel negative regulator of p53. *Genes Dev.* **2009**, *23*, 862–876. [CrossRef] [PubMed]

25. Wang, X.; Gocek, E.; Liu, C.-G.; Studzinski, G.P. MicroRNAs181 regulate the expression of p27Kip1 in human myeloid leukemia cells induced to differentiate by 1,25-dihydroxyvitamin D3. *Cell Cycle* **2009**, *8*, 736–741. [CrossRef] [PubMed]

26. Cuesta, R.; Martínez-Sánchez, A.; Gebauer, F. miR-181a regulates cap-dependent translation of p27(kip1) mRNA in myeloid cells. *Mol. Cell. Biol.* **2009**, *29*, 2841–2851. [CrossRef]

27. Rots, N.Y.; Iavarone, A.; Bromleigh, V.; Freedman, L.P. Induced differentiation of U937 cells by 1,25-dihydroxyvitamin D3 involves cell cycle arrest in G1 that is preceded by a transient proliferative burst and an increase in cyclin expression. *Blood* **1999**, *93*, 2721–2729. [CrossRef]

28. Min, D.; Lv, X.B.; Wang, X.; Zhang, B.; Meng, W.; Yu, F.; Hu, H. Downregulation of miR-302c and miR-520c by 1,25(OH)2D3 treatment enhances the susceptibility of tumour cells to natural killer cell-mediated cytotoxicity. *Br. J. Cancer* **2013**, *109*, 723–730. [CrossRef]

29. Gocek, E.; Wang, X.; Liu, X.; Liu, C.G.; Studzinski, G.P. MicroRNA-32 upregulation by 1,25-dihydroxyvitamin D3 in human myeloid leukemia cells leads to Bim targeting and inhibition of AraC-induced apoptosis. *Cancer Res.* **2011**, *71*, 6230. [CrossRef]

30. Salvatori, B.; Iosue, I.; Mangiavacchi, A.; Loddo, G.; Padula, F.; Chiaretti, S.; Peragine, N.; Bozzoni, I.; Fazi, F.; Fatica, A. The microRNA-26a target E2F7 sustains cell proliferation and inhibits monocytic differentiation of acute myeloid leukemia cells. *Cell Death Dis.* **2012**, *3*, e413. [CrossRef]

31. Iosue, I.; Quaranta, R.; Masciarelli, S.; Fontemaggi, G.; Batassa, E.M.; Bertolami, C.; Ottone, T.; Divona, M.; Salvatori, B.; Padula, F.; et al. Argonaute 2 sustains the gene expression program driving human monocytic differentiation of acute myeloid leukemia cells. *Cell Death Dis.* **2013**, *4*, e926. [CrossRef]
32. Bueno, M.J.; Malumbres, M. MicroRNAs and the cell cycle. *Biochim. Biophys. Acta* **2011**, *1812*, 592–601. [CrossRef] [PubMed]
33. Raut, S.K.; Singh, G.B.; Rastogi, B.; Saikia, U.N.; Mittal, A.; Dogra, N.; Singh, S.; Prasad, R.; Khullar, M. miR-30c and miR-181a synergistically modulate p53–p21 pathway in diabetes induced cardiac hypertrophy. *Mol. Cell. Biochem.* **2016**, *417*, 191–203. [CrossRef] [PubMed]
34. Bartel, D.P. MicroRNAs: Genomics, Biogenesis, Mechanism, and Function. *Cell* **2004**, *116*, 281–297. [CrossRef]
35. Costinean, S.; Zanesi, N.; Pekarsky, Y.; Tili, E.; Volinia, S.; Heerema, N.; Croce, C.M. Pre-B cell proliferation and lymphoblastic leukemia/high-grade lymphoma in E(mu)-miR155 transgenic mice. *Proc. Natl. Acad. Sci. USA* **2006**, *103*, 7024–7029. [CrossRef] [PubMed]
36. Esquela-Kerscher, A.; Slack, F.J. Oncomirs—microRNAs with a role in cancer. *Nat. Rev. Cancer* **2006**, *6*, 259–269. [CrossRef]
37. Vasilatou, D.; Papageorgiou, S.; Pappa, V.; Papageorgiou, E.; Dervenoulas, J. The role of microRNAs in normal and malignant hematopoiesis. *Eur. J. Haematol.* **2010**, *84*, 1–16. [CrossRef]
38. Calin, G.A.; Cimmino, A.; Fabbri, M.; Ferracin, M.; Wojcik, S.E.; Shimizu, M.; Taccioli, C.; Zanesi, N.; Garzon, R.; Aqeilan, R.I.; et al. MiR-15a and miR-16-1 cluster functions in human leukemia. *Proc. Natl. Acad. Sci. USA* **2008**, *105*, 5166. [CrossRef]
39. Ma, J.; Dong, C.; Ji, C. MicroRNA and drug resistance. *Cancer Gene Ther.* **2010**, *17*, 523–531. [CrossRef]
40. Calin, G.A.; Croce, C.M. MicroRNA signatures in human cancers. *Nat. Rev. Cancer* **2006**, *6*, 857–866. [CrossRef]
41. O'Connell, R.M.; Chaudhuri, A.A.; Rao, D.S.; Gibson, W.S.J.; Balazs, A.B.; Baltimore, D. MicroRNAs enriched in hematopoietic stem cells differentially regulate long-term hematopoietic output. *Proc. Natl. Acad. Sci. USA* **2010**, *107*, 14235–14240. [CrossRef] [PubMed]
42. Pan, Y.; Meng, M.; Zhang, G.; Han, H.; Zhou, Q. Oncogenic microRNAs in the genesis of leukemia and lymphoma. *Curr. Pharm. Des.* **2014**, *20*, 5260–5267. [CrossRef] [PubMed]
43. Tagawa, H.; Ikeda, S.; Sawada, K. Role of microRNA in the pathogenesis of malignant lymphoma. *Cancer Sci.* **2013**, *104*, 801–809. [CrossRef]
44. Pereira, F.; Larriba, M.J.; Muñoz, A. Vitamin D and colon cancer. *Endocr. Relat. Cancer* **2012**, *19*, 67–81. [CrossRef] [PubMed]
45. Kawa, S.; Nikaido, T.; Aoki, Y.; Zhai, Y.; Kumagai, T.; Furihata, K.; Fujii, S.; Kiyosawa, K. Vitamin D analogues up-regulate p21 and p27 during growth inhibition of pancreatic cancer cell lines. *Br. J. Cancer* **1997**, *76*, 884–889. [CrossRef]
46. Boulaire, J.; Fotedar, A.; Fotedar, R. The functions of the cdk-cyclin kinase inhibitor p21(WAF1). *Pathol. Biol.* **2000**, *48*, 190–202.
47. Trynda, J.; Turlej, E.; Milczarek, M.; Pietraszek, A.; Chodyński, M.; Kutner, A.; Wietrzyk, J. Antiproliferative Activity and in Vivo Toxicity of Double-Point Modified Analogs of 1,25-Dihydroxyergocalciferol. *Int. J. Mol. Sci.* **2015**, *16*, 24873–24894. [CrossRef]
48. Don, M.J.; Chang, Y.H.; Chen, K.K.; Ho, L.K.; Chau, Y.P. Induction of CDK inhibitors (p21(WAF1) and p27(Kip1)) and Bak in the beta-lapachone-induced apoptosis of human prostate cancer cells. *Mol. Pharmacol.* **2001**, *59*, 784–794. [CrossRef]
49. Dey, P. Aneuploidy and malignancy: An unsolved equation. *J. Clin. Pathol.* **2004**, *57*, 1245–1249. [CrossRef]
50. Gordon, D.J.; Resio, B.; Pellman, D. Causes and consequences of aneuploidy in cancer. *Nat. Rev. Genet.* **2012**, *13*, 189–203. [CrossRef]
51. Tang, Y.C.; Williams, B.R.; Siegel, J.J.; Amon, A. Identification of aneuploidy-selective antiproliferation compounds. *Cell* **2011**, *144*, 499–512. [CrossRef] [PubMed]
52. Cory, S.; Huang, D.C.S.; Adams, J.M. The Bcl-2 family: Roles in cell survival and oncogenesis. *Oncogene* **2003**, *22*, 8590–8607. [CrossRef] [PubMed]
53. Hanahan, D.; Weinberg, R.A. Hallmarks of cancer: The next generation. *Cell* **2011**, *144*, 646–674. [CrossRef] [PubMed]
54. Johnstone, R.W.; Ruefli, A.A.; Lowe, S.W. Apoptosis: A link between cancer genetics and chemotherapy. *Cell* **2002**, *108*, 153–164. [CrossRef]
55. Cory, S.; Adams, J.M. The Bcl2 family: Regulators of the cellular life-or-death switch. *Nat. Rev. Cancer* **2002**, *2*, 647–656. [CrossRef]
56. Kale, J.; Osterlund, E.J.; Andrews, D.W. BCL-2 family proteins: Changing partners in the dance towards death. *Cell Death Differ.* **2017**, *25*, 65–80. [CrossRef]
57. Zeng, Q.H.; Xu, L.; Liu, X.D.; Liao, W.; Yan, M. xia [miR-125b promotes proliferation of human acute myeloid leukemia cells by targeting Bak1]. *Zhonghua Xue Ye Xue Za Zhi* **2013**, *34*, 1010–1014. [CrossRef]
58. Zhou, M.; Liu, Z.; Zhao, Y.; Ding, Y.; Liu, H.; Xi, Y.; Xiong, W.; Li, G.; Lu, J.; Fodstad, O.; et al. MicroRNA-125b confers the resistance of breast cancer cells to paclitaxel through suppression of pro-apoptotic Bcl-2 antagonist killer 1 (Bak1) expression. *J. Biol. Chem.* **2010**, *285*, 21496–21507. [CrossRef]
59. Kong, F.; Sun, C.; Wang, Z.; Han, L.; Weng, D.; Lu, Y.; Chen, G. miR-125b confers resistance of ovarian cancer cells to cisplatin by targeting pro-apoptotic Bcl-2 antagonist killer 1. *J. Huazhong Univ. Sci. Technol. Med. Sci.* **2011**, *31*, 543–549. [CrossRef]
60. Taylor, M.A.; Sossey-Alaoui, K.; Thompson, C.L.; Danielpour, D.; Schiemann, W.P. TGF-β upregulates miR-181a expression to promote breast cancer metastasis. *J. Clin. Investig.* **2013**, *123*, 150–163. [CrossRef]
61. Tan, T.T.; Degenhardt, K.; Nelson, D.A.; Beaudoin, B.; Nieves-Neira, W.; Bouillet, P.; Villunger, A.; Adams, J.M.; White, E. Key roles of BIM-driven apoptosis in epithelial tumors and rational chemotherapy. *Cancer Cell* **2005**, *7*, 227–238. [CrossRef] [PubMed]

62. Ambs, S.; Prueitt, R.L.; Yi, M.; Hudson, R.S.; Howe, T.M.; Petrocca, F.; Wallace, T.A.; Liu, C.G.; Volinia, S.; Calin, G.A.; et al. Genomic profiling of microRNA and messenger RNA reveals deregulated microRNA expression in prostate cancer. *Cancer Res.* **2008**, *68*, 6162–6170. [CrossRef] [PubMed]

63. Ryan, K.M.; Ernst, M.K.; Rice, N.R.; Vousden, K.H. Role of NF-kappaB in p53-mediated programmed cell death. *Nature* **2000**, *404*, 892–897. [CrossRef]

64. Hayden, M.S.; West, A.P.; Ghosh, S. NF-κB and the immune response. *Oncogene* **2006**, *25*, 6758–6780. [CrossRef]

65. Xia, Y.; Shen, S.; Verma, I.M. NF-κB, an active player in human cancers. *Cancer Immunol. Res.* **2014**, *2*, 823–830. [CrossRef] [PubMed]

66. Dolcet, X.; Llobet, D.; Pallares, J.; Matias-Guiu, X. NF-kB in development and progression of human cancer. *Virchows Arch.* **2005**, *446*, 475–482. [CrossRef]

67. Zhou, R.; Hu, G.; Gong, A.Y.; Chen, X.M. Binding of NF-kappaB p65 subunit to the promoter elements is involved in LPS-induced transactivation of miRNA genes in human biliary epithelial cells. *Nucleic Acids Res.* **2010**, *38*, 3222. [CrossRef]

68. Ma, X.; Becker Buscaglia, L.E.; Barker, J.R.; Li, Y. MicroRNAs in NF-κB signaling. *J. Mol. Cell Biol.* **2011**, *3*, 159. [CrossRef]

69. Javelaud, D.; Wietzerbin, J.; Delattre, O.; Besançon, F. Induction of p21Waf1/Cip1 by TNFα requires NF-κB activity and antagonizes apoptosis in Ewing tumor cells. *Oncogene* **2000**, *19*, 61–68. [CrossRef]

70. Pennington, K.N.; Taylor, J.A.; Bren, G.D.; Paya, C.V. IkappaB kinase-dependent chronic activation of NF-kappaB is necessary for p21(WAF1/Cip1) inhibition of differentiation-induced apoptosis of monocytes. *Mol. Cell. Biol.* **2001**, *21*, 1930–1941. [CrossRef]

71. Chen, Y.; Zhang, J.; Ge, X.; Du, J.; Deb, D.K.; Li, Y.C. Vitamin D Receptor Inhibits Nuclear Factor κB Activation by Interacting with IκB Kinase β Protein. *J. Biol. Chem.* **2013**, *288*, 19450. [CrossRef] [PubMed]

72. Fekrmandi, F.; Wang, T.T.; White, J.H. The hormone-bound vitamin D receptor enhances the FBW7-dependent turnover of NF-κB subunits. *Sci. Rep.* **2015**, *5*, 13002. [CrossRef] [PubMed]

73. Sun, J.; Kong, J.; Duan, Y.; Szeto, F.L.; Liao, A.; Madara, J.L.; Yan, C.L. Increased NF-κB activity in fibroblasts lacking the vitamin D receptor. *Am. J. Physiol.–Endocrinol. Metab.* **2006**, *291*, 315–322. [CrossRef] [PubMed]

74. Gleba, J.J.; Kłopotowska, D.; Banach, J.; Turlej, E.; Mielko, K.A.; Gębura, K.; Bogunia-Kubik, K.; Kutner, A.; Wietrzyk, J. Polymorphism of VDR Gene and the Sensitivity of Human Leukemia and Lymphoma Cells to Active Forms of Vitamin D. *Cancers* **2022**, *14*, 387. [CrossRef]

75. Signature of VDR miRNAs and Epigenetic Modulation of Vitamin D Signaling in Melanoma Cell Lines–PubMed. Available online: https://pubmed.ncbi.nlm.nih.gov/22213330/ (accessed on 23 January 2022).

76. Luo, W.; Hershberger, P.A.; Trump, D.L.; Johnson, C.S. 24-Hydroxylase in cancer: Impact on vitamin D-based anticancer therapeutics. *J. Steroid Biochem. Mol. Biol.* **2013**, *136*, 252–257. [CrossRef]

77. Craig, T.A.; Zhang, Y.; Magis, A.T.; Funk, C.C.; Price, N.D.; Ekker, S.C.; Kumar, R. Detection of 1α,25-dihydroxyvitamin D-regulated miRNAs in zebrafish by whole transcriptome sequencing. *Zebrafish* **2014**, *11*, 207–218. [CrossRef]

78. Peng, X.; Tiwari, N.; Roy, S.; Yuan, L.; Murillo, G.; Mehta, R.R.; Benya, R.V.; Mehta, R.G. Regulation of CYP24 splicing by 1,25-dihydroxyvitamin D3 in human colon cancer cells. *J. Endocrinol.* **2012**, *212*, 207–215. [CrossRef]

79. Anaya, J. OncoLnc: Linking TCGA survival data to mRNAs, miRNAs, and lncRNAs. *PeerJ Comput. Sci.* **2016**, *2016*, e67. [CrossRef]

International Journal of
Molecular Sciences

Article

Improved Trimethylangelicin Analogs for Cystic Fibrosis: Design, Synthesis and Preliminary Screening

Christian Vaccarin [1,2,†], Daniela Gabbia [1,†], Erica Franceschinis [1], Sara De Martin [1], Marco Roverso [3], Sara Bogialli [3], Gianni Sacchetti [4], Chiara Tupini [4], Ilaria Lampronti [4,5], Roberto Gambari [4,5], Giulio Cabrini [4,5], Maria Cristina Dechecchi [6], Anna Tamanini [6], Giovanni Marzaro [1] and Adriana Chilin [1,5,*]

[1] Department of Pharmaceutical and Pharmacological Sciences, University of Padova, Via Marzolo 5, 35131 Padova, Italy
[2] Center for Radiopharmaceutical Sciences ETH-PSI-USZ, Paul Scherrer Institute, 5232 Villigen, Switzerland
[3] Department of Chemical Sciences, University of Padova, Via Marzolo 1, 35131 Padova, Italy
[4] Department of Life Sciences and Biotechnology, University of Ferrara, Via Fossato di Mortara 74, 44121 Ferrara, Italy
[5] Center of Innovative Therapies for Cystic Fibrosis (InnThera4CF), University of Ferrara, Via Fossato di Mortara 74, 44121 Ferrara, Italy
[6] Department of Neurosciences, Biomedicine and Movement, Section of Clinical Biochemistry, University of Verona, Piazzale Stefani 1, 37126 Verona, Italy
* Correspondence: adriana.chilin@unipd.it
† These authors contributed equally to this work.

Abstract: A small library of new angelicin derivatives was designed and synthesized with the aim of bypassing the side effects of trimethylangelicin (TMA), a promising agent for the treatment of cystic fibrosis. To prevent photoreactions with DNA, hindered substituents were inserted at the 4 and/or 6 positions. Unlike the parent TMA, none of the new derivatives exhibited significant cytotoxicity or mutagenic effects. Among the synthesized compounds, the 4-phenylderivative **12** and the 6-phenylderivative **25** exerted a promising F508del CFTR rescue ability. On these compounds, preliminary in vivo pharmacokinetic (PK) studies were carried out, evidencing a favorable PK profile *per se* or after incorporation into lipid formulations. Therefore, the selected compounds are good candidates for future extensive investigation to evaluate and develop novel CFTR correctors based on the angelicin structure.

Keywords: trimethylangelicin; synthesis; CFTR; pharmacokinetics; SEDDS

Citation: Vaccarin, C.; Gabbia, D.; Franceschinis, E.; De Martin, S.; Roverso, M.; Bogialli, S.; Sacchetti, G.; Tupini, C.; Lampronti, I.; Gambari, R.; et al. Improved Trimethylangelicin Analogs for Cystic Fibrosis: Design, Synthesis and Preliminary Screening. *Int. J. Mol. Sci.* **2022**, *23*, 11528. https://doi.org/10.3390/ijms231911528

Academic Editor: Hidayat Hussain

Received: 11 September 2022
Accepted: 26 September 2022
Published: 29 September 2022

Publisher's Note: MDPI stays neutral with regard to jurisdictional claims in published maps and institutional affiliations.

1. Introduction

Cystic fibrosis (CF) is an autosomal recessive disorder caused by mutations in the gene for the cystic fibrosis transmembrane conductance regulator (CFTR) protein, an anion-conducting transmembrane channel that regulates electrolyte transport and mucociliary clearance in the airways [1,2]. The recently approved drugs ivacaftor, lumacaftor, tezacaftor, and elexacaftor have changed the approach to CF therapy, improving the quality of life of CF patients [3]. These new drugs work on the CFTR defect, correcting or potentiating CFTR function [4–6]. TMA (4,6,4′-trimethylangelicin) (Figure 1) is a small molecule known to modulate the function of CFTR and control the inflammatory status of CF by targeting NF-κB and reducing IL-8 expression, demonstrating to be a promising agent for the treatment of CF [7–9]. Thus, TMA received an Orphan Drug Designation for the treatment of CF (EU/3/13/1137) by the EMA in 2013 [10].

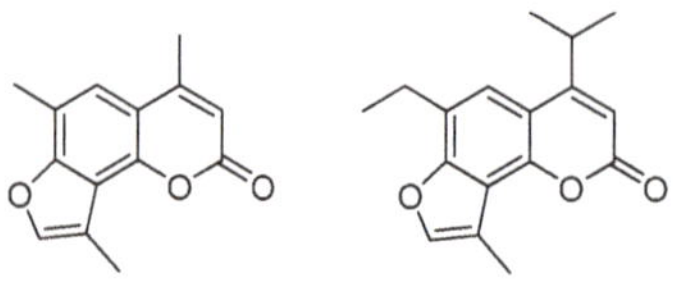

Figure 1. Structure of 4,6,4′-trimethylangelicin (TMA) and 4-isopropyl-6-ethyl-4′-methylangelicin (IPEMA).

Unluckily, TMA suffers from serious disadvantages, such as poor solubility, phototoxicity and mutagenicity. These drawbacks aroused a growing interest in investigating the structural modification of the TMA nucleus to identify new-generation TMA analogs with a comparable or better activity profile for the original hit compound but without the undesirable effects.

In previous work, we defined some structural determinants of the angelicin scaffold to maintain biological activities and eliminate side effects, thus identifying 4-isopropyl-6-ethyl-4′-methylangelicin (IPEMA) (Figure 1) as a promising candidate with TMA-like inhibitory activity on NF-kB/DNA interactions but lacking mutagenicity and DNA damaging properties [11]. Further investigations revealed that IPEMA significantly rescued F508del CFTR-dependent chloride efflux at nanomolar concentrations and maintained the potentiation activity of CFTR in FRT-YFP-G551D cells, acting by the same mechanism of TMA [12].

Going on modifying the TMA scaffold, we report here the design, synthesis and preliminary screening of a small library of new TMA analogs (Figure 2). The pharmacokinetic profile of the most active derivatives, together with the previously reported IPEMA, was also investigated to identify the drug-likeness features to move TMA derivatives forward.

Table 1. R substituents for the A series of TMA analogs.

Compounds	R$_4$	R$_6$	ID Final Product [1]
1	-	Me	-
2	-	Et	-
3, 8, 11	*c*Pr	Me	**CPDMA**
4, 9, 12	Ph	Me	**4-PhDMA**
5, 10, 13	Ph	Et	**4-PhEMA**
3, 14, 17	*c*Pr	Me	**CPMA**
6, 15, 18	*i*Pr	Me	**IPMA**
7, 16, 19	*i*Pr	Et	**IPEA**

[1] The ID is derived from the chemical name for TMA: CP, cyclopropyl; DMA, dimethylangelicin; Ph, phenyl; EMA, ethylmethylangelicin; MA, methylangelicin; IP, isopropyl; EA, ethylangelicin.

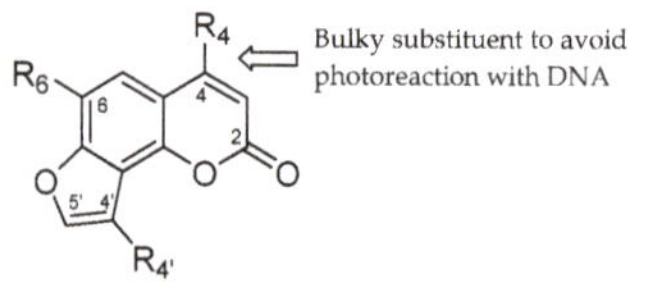

R$_4$ — Bulky substituent to avoid photoreaction with DNA

R$_6$

R$_4$'

Figure 2. General structure of TMA analogs. See Tables 1–3 for the R specification.

Table 2. R substituents for the B series of TMA analogs.

Compounds	R$_4$	R$_6$	ID Final Product [1]
24	Me	Ph	6-PhDMA
25	Me	*p*-NH$_2$Ph	pANDMA
26	Me	*m*-NH$_2$Ph	mANDMA
27	Me	Pyridyl	PyDMA
28	Me	Thienyl	ThiDMA

[1] The ID is derived from the chemical name as for TMA: Ph, phenyl; DMA, dimethylangelicin; pAN, para-aniline; mAN, meta-aniline; Py, pyridyl; Thi, thienyl.

Table 3. R substituents for the three series of TMA analogs.

Compounds	R$_4$	R$_6$	ID Final Product [1]
29	-	Ph	-
30, 32, 34	*i*Pr	Ph	IPPhMA
31, 33, 35	*c*Pr	Ph	CPPhMA

[1] The ID is derived from the chemical name for TMA: IP, isopropyl; Ph, phenyl; MA, methylangelicin; CP, cyclopropyl.

2. Results and Discussion

2.1. Chemistry

As previously described, a hindered substituent at position 4 impairs DNA intercalation, thus preventing the unwanted photoreaction of the lactone or furan double bond of the furocoumarin scaffold with the nucleic acid [11,13]. For this reason, new analogs with cyclopropyl and aryl substituents at position 4 were designed (Figure 3, A series), in addition to the ethyl, propyl, and isopropyl derivatives already reported [11]. To also investigate the role of steric hindrance at position 6, we designed a further series of analogs with aryl substituents at this position (Figure 3, B series). Finally, derivatives with steric hindrances at both positions 4 and 6 were also considered (Figure 3, C series). In all cases, a 4′-methyl group or no substituent was inserted in the furan ring.

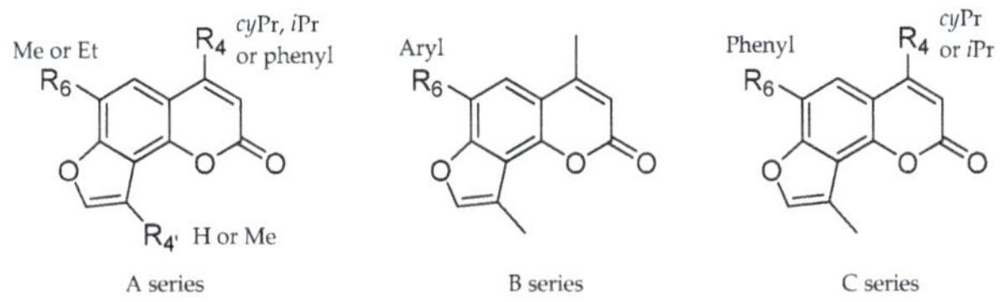

Figure 3. Series of TMA analogs.

Depending on the series of compounds, three different synthetic pathways were required. The A series, which included the analogs with steric hindrance only at position 4 was synthesized according to Scheme 1, adopting the previously reported synthetic strategies [11,14].

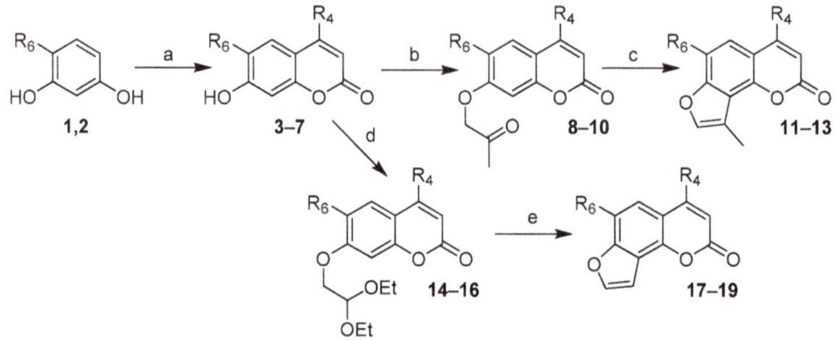

Scheme 1. Synthesis of the A series of TMA analogs. Reaction conditions: a. Ethyl-3-cyclopropyl-3-oxopropionate (for **3**) or ethyl benzoylacetate (for **4,5**) or ethyl isobutyrylacetate (for **6,7**), H_2SO_4, 1 h, 40–95%; b. Chloroacetone, K_2CO_3, KI, acetone, reflux, 12 h, 40–95%; c. KOH, EtOH abs., reflux, 2 h, 10–33%. d. Chloroacetaldehyde diethylacetale, K_2CO_3, KI, DMF, reflux, 18 h, 36–83%; e. TFA, MW, 120 °C, 5 min, 24–29%. See Table 1 for the R specification.

The starting 7-hydroxybenzopyran-2-ones (**3–7**) were synthesized by condensing the opportune resorcinol derivative (**1, 2**) with the appropriate acetoacetic ester derivatives. Compounds **3–5** were functionalized with α-chloroacetone, obtaining the O-α-ketoethers **8–10**. The intermediate ethers were finally cyclized in an anhydrous alkaline medium that provided the desired 4′-methylangelicins **11–13**. Alternatively, starting benzopyranones **3**, **6**, and **7** were condensed with chloroacetaldehyde diethylacetale, affording ethers **14–16**, which were finally cyclized to angelicins **17–19** characterized by an unsubstituted furan ring.

The B series, which included analogs with steric hindrance at position 6, was synthesized according to Scheme 2, using bromoangelicin **23** as a common intermediate.

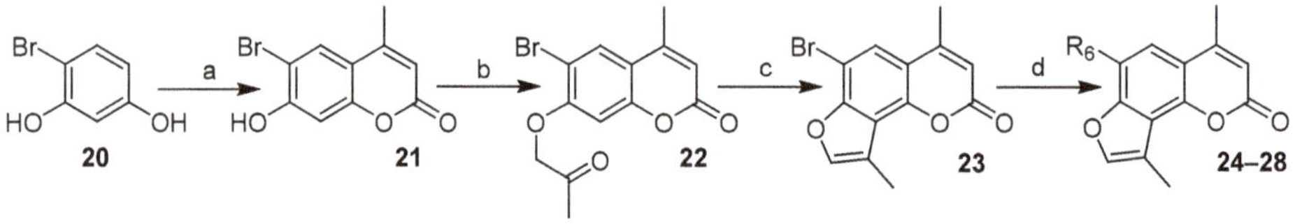

Scheme 2. Synthesis of the B series of TMA analogs. Reaction conditions: a. Ethyl acetoacetate, H_2SO_4, 1 h, 82%; b. Chloroacetone, K_2CO_3, KI, acetone, reflux, 12 h, 69%; c. KOH, EtOH abs., reflux, 1 h, 34%; d. Arylboronic acid, Na_2CO_3, $Pd(PPh_3)_4$, DME, reflux, 2 h, 33–76%. See Table 2 for the R specification.

The starting 6-bromo-7-hydroxybenzopyran-2-one (**21**) was synthesized by condensing bromoresorcinol (**20**) with ethyl acetoacetate, which was then functionalized with α-chloroacetone to O-α-ketoether **22** and finally cyclized to 4-methyl-6-bromo-4′-methylangelicins **23** in the same conditions as previously reported. The resulting 6-bromoangelicin was submitted to Suzuki coupling with opportune arylboronic acids to obtain the desired 6-arylangelicin (**24-28**). Benzene, pyridine and thiophene were chosen as representative aryl substituents; *p*- and *m*-aniline as more hydrophilic and salifiable groups were also introduced at position 6. Compound **25** was also prepared as a mesylate salt (**25a, pANDMA mesylate** or *p*ANDMAa) to improve the poor solubility of the parent TMA and analogs.

The C series, which included analogs with steric hindrance at both positions 4 and 6, was synthesized according to Scheme 3, using phenylresorcinol as a key intermediate.

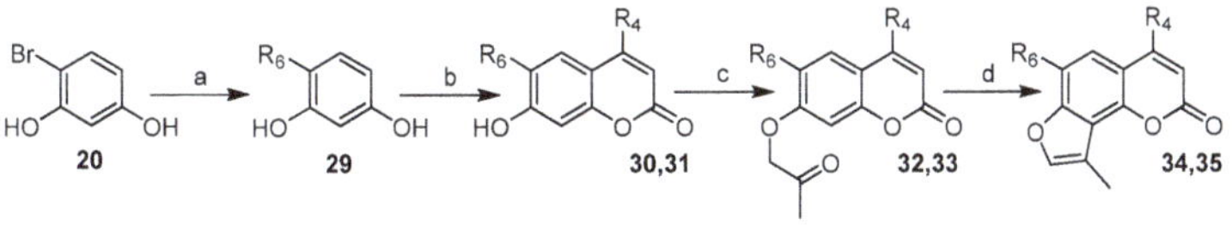

Scheme 3. Synthesis of the C series of TMA analogs. Reaction conditions: a. Phenylboronic acid, Na_2CO_3, $Pd(PPh_3)_4$, DME, MW, 130 °C, 5 min, 26%; b. Ethyl isobutyrylacetate (for **30**) or ethyl-3-cyclopropyl-3-oxopropionate (for **31**), H_2SO_4, 1 h, 25%; c. Chloroacetone, K_2CO_3, KI, acetone, reflux, 6 h, 95%; d. KOH, EtOH abs., reflux, 1 h, 20–22%. See Table 3 for the R specification.

In this case, the 6-phenyl substituent was inserted via Suzuki coupling on the starting resorcinol **20**, since we noted that Pechmann condensation on bromoresorcinol with hindered acetoacetic esters did not occur. The obtained 6-phenylresorcinol (**29**) was condensed with the opportune derivatives of ethyl acetoacetate, then functionalized with α-chloroacetone and finally cyclized to 4-alkyl-6-phenyl-4′-methylangelicins (**34, 35**).

2.2. Biological Activity

2.2.1. Functional Rescue of Mutant F508del CFTR Protein in CF Airway Cells

To investigate the corrective properties of the new TMA derivatives, we performed functional experiments to test whether some analogs maintain the F508del CFTR rescue property of the parent TMA.

To verify the possibility that the title compounds may act as correctors in CFBE41o-, overexpressing F508del and the high sensitivity halide-sensing yellow fluorescent protein (HS-YFP) YFP-H148Q/I152L, we examined the effect of 48 h preincubation with vehicle or the TMA analogs or TMA or VX809 (Lumacaftor), as a positive control.

At the time of the assay, the correction of the F508del-CFTR function was evaluated in single-cell analysis by measuring the decrease in YFP fluorescence after the concomitant addition of extracellular iodide and the activating cocktail (20 μM forskolin and 5 μM VX770), in the presence or absence of the 10 μM CFTR inhibitor, CFTRinh-172. Figure 4 shows the activity of the TMA analogs compared to VX809 and TMA. Analogs not represented in the figure were less active than the vehicle. At least two derivatives show CFTR activity similar to that of the parent TMA.

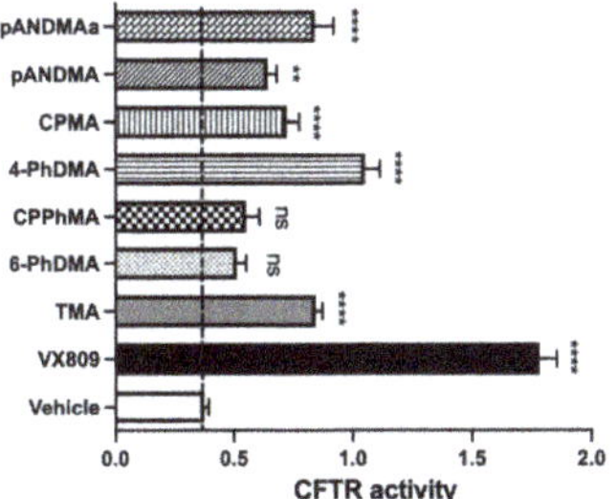

Figure 4. Evaluation of different TMA analogs as F508del CFTR correctors in CFBE41o- YFP-F508del cells. Correction of F508del-CFTR function was assessed, in a single cell, by measuring the decrease in YFP fluorescence upon addition of extracellular iodide and with the activating cocktail, in the presence or absence of CFTR inhibitor, CFTRInh-172. Iodide influx was tested under control condition (vehicle) or after 48 h incubation with the correctors 200 nM or 1 μM VX809. The YFP fluorescence decay rate was calculated by fitting the fluorescence data with an exponential function of time courses. CFTR activity was obtained by the difference between the fluorescence decay rate of YFP in the presence and absence of CFTRInh-172. Fluorescence coming from every single cell of at least five cells per field was recorded. Each bar corresponds to the mean ± SEM of data points coming from at least three different experiments (5–10 different cells/each experiment). Statistical comparisons were made using a nonparametric Kruskall–Wallis test (** $p < 0.01$, and **** $p < 0.0001$, ns: not significant). **pANDMAa = pANDMA mesylate.**

Combining these data and analogous data obtained from the previous set of TMA derivatives [11,12], it can be seen that rescue activity was maintained, also hindering the 4 or 6 positions alternatively, taking TMA as a reference. The steric hindrances at positions 4 and 6 simultaneously have a detrimental effect depending on the dimension of the two substituents. If one of the substituents is a phenyl group, the other must be at most a methyl group, as can be seen by comparing compound **4-PhDMA** (active) with **4-PhEMA** (inactive, Figure 4). The presence of a 6-aryl substituent is poorly tolerated, as in the case of **6-PhDMA, mANDMA, PyDMA, ThiDMA, IPPhMA**, and **CPPhMA** (inactive), unless an amine is present specifically at the para position (compare **pANDMA** with **mANDMA**). A moderate steric hindrance at both positions 4 and 6 is allowed only if the substituents are small alkyl groups: this is the case of the previously identified **IPEMA** [12], bearing an isopropyl and an ethyl group, respectively. Almost all the derivatives without substituents in the furan ring exhibited activities lower than those of the vehicle.

As shown in the representative experiments, correction of F508del-CFTR upon treatment with **4-PhDMA** and **pANDMA mesylate** (Figure 5, panels A and B, respectively) resulted in a greater reduction in YFP fluorescence induced by increased iodide influx compared to cells treated with vehicle alone.

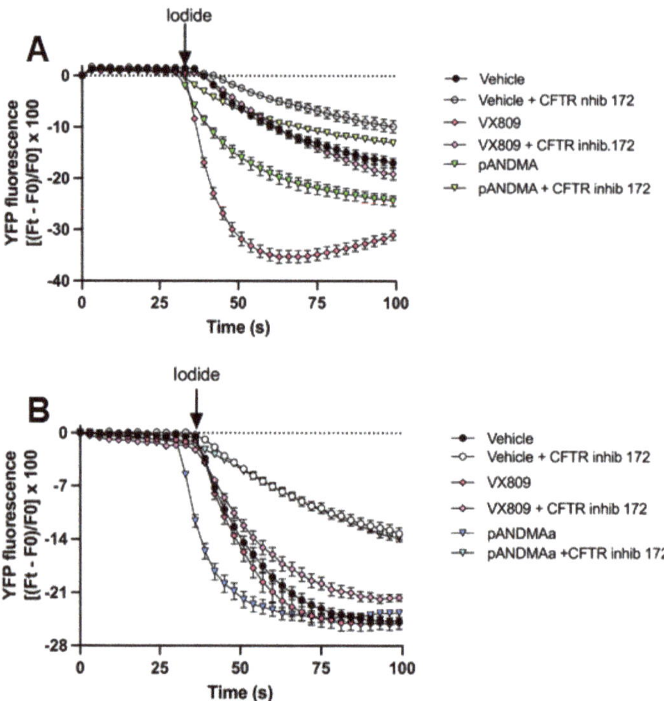

Figure 5. Representative time courses. Cells were treated for 48 h with vehicle or VX809 or **4-PhDMA** (**A**) or pANDMAa (**pANDMA mesylate**) (**B**). CFTR activity was measured as described in Figure 3. The arrow indicates the addition of iodide.

Based on this preliminary screening, compounds **4-PhDMA** and **pANDMA mesylate** were the most interesting of the new series.

Differently from **IPEMA** [12], none of these TMA analogs were able to potentiate the CFTR function (data shown in Supplementary Materials, Figure S1).

2.2.2. In Vitro Toxicity Studies: Photochemical Behavior

To confirm that the phototoxicity of the parent TMA was abolished, the photoadduct formation with DNA under UVA irradiation was investigated in the new TMA analogs by means of TLC analysis upon UVA irradiation of DNA/compound mixture.

Typically, if photoadducts between the lactone ring and thymine took place upon irradiation, the photoadducts appear as a violet fluorescent band at lower Rf values when observed under UV light, whereas the unchanged compounds (not photoreacted) appear at higher Rf values.

As shown in the Supplementary Materials, only TMA, used as a reference, showed photoadducts (Figure S2A) upon UVA irradiation, while the other compounds exhibited only the unchanged bands (Figure S2B–D, on **4-PhDMA**, **6-PhDMA**, and **pANDMA**, as representative compounds), thus demonstrating that they do not react with DNA under UVA irradiation.

Interestingly, the abolition of photoreaction is achieved by hindering positions 4 or 6, despite the distance of the latest one from the photoinducible double bonds on the furocoumarins ring. This is probably due to the non-planar shape of the 6-arylderivatives, preventing the intercalation into DNA and the subsequent addition.

2.2.3. In Vitro Toxicity Studies: Evaluation of Mutagenic Effects (Ames Test)

To investigate whether changes in the structure of TMA abolish the mutagenic effect observed with this class of compounds, the most promising TMA analogs were evaluated by the Ames test. Compounds **4-PhDMA** and **pANDMA** were tested for genotoxicity evaluation using the histidine-requiring *Salmonella typhimurium* mutant TA97A, TA98, TA100, and TA1535 strains that allow checking frameshift mutation and base pair substitution, with and without metabolic activation induced by S9 Mix.

The histidine-requiring *Salmonella typhimurium* mutant strains described are particularly appropriate to check the genotoxic potential of new chemical compounds because they both have a plasmid that induces an increase in error-prone DNA damage repair (pKM101), and mutations that induce increased permeability to larger molecules through a defective lipopolysaccharide layer (*rfa* mutations) and avoid the possibility to repair excision DNA damage (*uvrB* mutations) [15,16].

The number of spontaneous revertants is relatively constant, but it dramatically increases—generally in a dose-dependent way—when a mutagen is added to the plate. In general, the cut-off to discriminate between a mutagenic and nonmutagenic result is considered as a two-fold revertants increase over the response of the negative control (spontaneous revertants).

For the above-reported considerations, the Ames test is generally considered an appropriate preliminary screening tool to determine the mutagenic potential of new chemicals for different industrial applications, from agrochemicals, pharmaceuticals, and medical devices to health products (e.g., for personal care, cosmetics, nutraceuticals, etc.). For the high significant predictive capacity, the regulatory agencies require the Ames test for the registration or acceptance of new compounds [16].

The results obtained for the compounds **4-PhDMA** and **pANDMA** are reported in the Supplementary Materials (Tables S1–S3): data were obtained employing the cited mutant, according to the method reported by Maron & Ames [15]. None of the samples had genotoxic activity at all concentrations tested with all the *Salmonella* strains. In conclusion, all of the samples can be considered non-genotoxic for the Ames test.

2.2.4. In Vitro Toxicity Studies: Evaluation of the Cytotoxic Activity toward HepG2 Cells

Since the compounds are designed for oral administration and are likely to undergo the hepatic first-pass effect, we tested their cytotoxic potential toward HepG2 hepatoma cells. We could exclude the significant cytotoxicity of **4-PhDMA, pANDMA, and pANDMA mesylate,** as well as of the other compounds we screened, as cell viability was greater than 75% at the highest concentration tested (50 μM) (Figure S3 in the Supplementary Materials).

2.3. Pharmacokinetic Profile

The main pharmacokinetic parameters of **4-PhDMA, pANDMA,** and **pANDMA mesylate** were evaluated in c57BL/6 male mice, using water as dispersing vehicle. In addition, previously reported **IPEMA** was also tested, as it appeared to be the most active compound of all the synthesized TMA analogs.

Figure 6 shows the plasma concentration versus time curves obtained after administration of the selected compounds.

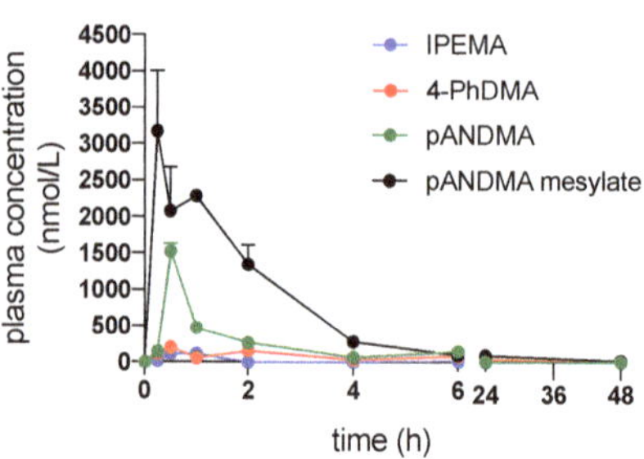

Figure 6. Plasma concentration versus time curves obtained after administration of the selected compounds **IPEMA, 4-PhDMA, pANDMA,** and **pANDMA mesylate** suspended or dissolved in water to c57BL/6 male mice.

The results highlight that the oral bioavailability of lipophilic TMA derivatives (**IPEMA, 4-PhDMA,** and **pANDMA**) was very low, due to their poor water solubility, while the hydrophilic salt **pANDMA mesylate** was nearly four-times more bioavailable than **pANDMA** (Table 4).

Table 4. Main PK parameters of the selected compounds.

Compound	$t_{1/2}$ (h)	C_{max} (nmol/L)	T_{max} (h)	AUC 0-inf (nmol/L*h)	CL/F (mL/h/g)
IPEMA	9.01 ± 1.34	120.4 ± 55.3	1 ± 0	394.50 ± 98.72	0.0633 ± 0.005
4-PhDMA	21.13 ± 0.98	221.77 ± 184.09	1.25 ± 1.06	2747.36 ± 513.61	0.0014 ± 0.0014
pANDMA	1.91 ± 0.71	1523.5 ± 260.9	0.5 ± 0	2405.4 ± 208.2	0.0083 ± 0.0007
pANDMA mesylate	1.87 ± 0.43	3868.3 ± 1065.2	0.375 ± 0.177	9402.6 ± 742.2	0.0091 ± 0.0002

The experiment was performed in c57BL/6 male mice. $t_{1/2}$: elimination half-life; C_{max}: maximum plasma concentration, T_{max}: time to maximum plasma concentration, AUC0-inf: area under the plasma concentration vs. time curve, CL/F: apparent oral clearance.

2.3.1. Formulation of Lipophilic TMA Analogs

Since the oral bioavailability of lipophilic molecules is limited due to their poor water solubility, the most promising lipophilic TMA derivatives (**4-PhDMA** and **IPEMA**) were formulated in self-emulsifying formulations (self-emulsifying drug delivery systems, SEDDS). SEDDS are mixtures containing drugs, an oil phase, surfactant, and co-surfactant/cosolvents that spontaneously self-emulsify upon contact with the aqueous environment in the gastrointestinal tract producing a milky or transparent emulsion able to improve drug solubilization. Furthermore, other phenomena, such as reduced efflux mediated by P-glycoproteins, along with improved lymphatic transport and intestinal wall permeability, can also contribute to an increase in oral bioavailability [17]. Thus, the most promising lipophilic derivatives of TMA (**4-PhDMA** and **IPEMA**) were formulated in two self-emulsifying formulations, previously built for the parent TMA [18], and characterized by the presence of two different oil phases. The first formulation, named A, contains a mixture of monoglycerides, diglycerides, and triglycerides of long-chain fatty acids, mainly linoleic and oleic acid (Maisine® CC) as the oil phase, instead the second formulation, named B, contains a mixture of medium-chain triglycerides (mainly capric and caprylic acid derivatives (LabrafacTM Lipophile WL1349) as the oil phase.

The formulations were characterized by measuring the droplet size and PDI. The results are reported in Table 5.

Table 5. Properties of the SEDDS developed for **IPEMA** and **4-PhDMA**.

Compound	Formulation	Droplet Size (nm)	PDI (-)
IPEMA	A	14.62 ± 2.80	0.24
IPEMA	B	15.10 ± 1.50	0.18
4-PhDMA	A	13.23 ± 2.10	0.31
4-PhDMA	B	14.56 ± 1.57	0.35

The results show that both formulations are produced after dispersion emulsion with a droplet size smaller than 20 nm with a narrow droplet size distribution (PDI < 0.7); therefore, after oral administration, diffusion and lipid digestion processes can be facilitated.

2.3.2. Bioavailability of IPEMA and 4-PhDMA Formulated in SEDDS

Figure 7 shows the comparison between the plasma concentration versus time curves obtained after the administration of the selected lipophilic compounds **IPEMA** and **4-PhDMA** suspended in water and formulated in SEDDS.

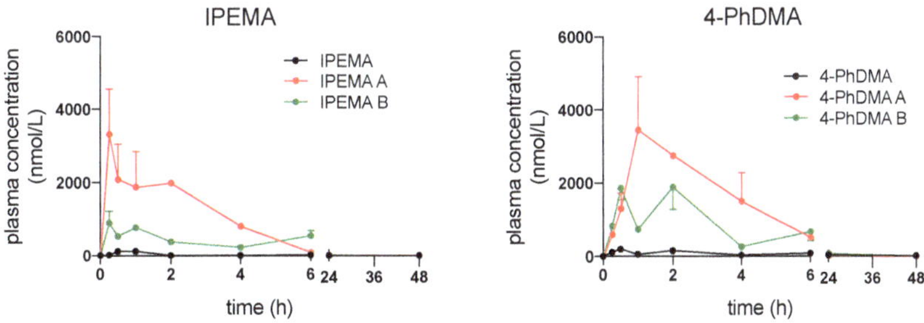

Figure 7. Plasma concentration versus time curves obtained after administration of the selected compounds **IPEMA** and **4-PhDMA**, suspended in water or formulated in SEDDS to c57BL/6 male mice.

The PK profiles clearly indicate that the SEDDS formulations were able to significantly increase the oral bioavailability of lipophilic derivatives.

In particular, the bioavailability of **IPEMA** formulated in SEDDS was increased 20-fold and 16-fold for A and B, respectively, while for **4-PhDMA** it was increased 6.4-fold and 4.8-fold for A and B, respectively (Table 6). Formulations A containing a mixture of long-chain mono, di, and triglycerides as oil phase were more effective than formulations B containing medium-chain triglycerides; this is probably due to the higher solvent capacity of the mixture of long-chain mono, di and triglycerides, in particular, after oral administration [19].

Table 6. Main PK parameters of the selected compounds formulated in SEDDS.

Compound/. Formulation	$t_{1/2}$ (h)	C_{max} (nmol/L)	T_{max} (h)	AUC 0-inf (nmol/L*h)	CL/F (mL/h/g)
IPEMA A	5.99 ± 0.87	$3184,582 \pm 897,45$	0.25 ± 0	$12,456.1 \pm 1456.3$	0.0021 ± 0.007
IPEMA B	6.15 ± 1.09	845.73 ± 76.54	0.25 ± 0	6671.16 ± 432.13	0.0037 ± 0.002
4-PhDMA A	6.08 ± 4.78	4374.8 ± 2296.6	1.5 ± 0.71	$17,699.4 \pm 6170.2$	0.0012 ± 0.0005
4-PhDMA B	7.31 ± 4.29	2683.0 ± 370.4	1.25 ± 1.06	$13,337.8 \pm 6463.4$	0.0017 ± 0.0008

The experiment was performed in c57BL/6 male mice. $t_{1/2}$: elimination half-life; C_{max}: maximum plasma concentration, T_{max}: time to maximum plasma concentration, AUC0-inf: area under the plasma concentration vs. time curve, CL/F: apparent oral clearance.

3. Materials and Methods

3.1. Chemistry

All the general and detailed synthetic procedures, characterization data, and mass spectra can be found in Supplementary Materials.

3.2. Biology

3.2.1. CFTR Functional Rescue

The effect of TMA analogs on CFTR function was tested in CFBE41o- overexpressing F508del and the high sensitivity halide-sensing yellow fluorescent protein (HS-YFP) YFP-H148Q/I152L (a generous gift from N. Pedemonte, Gaslini Institute, Genova, Italy). YFP CF cells were seeded on round glass coverslips and preincubated for 48 h with vehicle or test compounds or VX809 or TMA. At the time of the assay, cells were washed in Dulbecco's PBS (in mM: 137 NaCl, 2.7 KCl, 8.1 Na_2HPO_4, 1.5 KH_2PO_4, 1 $CaCl_2$, and 0.5 $MgCl_2$, pH 7.4) and subsequently incubated with a stimulation cocktail (20 µM forskolin and 5 µM VX770) in the presence and absence of 10 µM CFTR inhibitor, CFTRInh-172 (Sigma), for 30 min. The coverslips were then transferred on a Nikon TMD inverted microscope through a Nikon Fluor 40 objective. The signal was acquired with a Hamamatsu C2400–97 charge-coupled intensified video camera at a rate of 1 frame/3 s. The fluorescence coming from every single cell was analyzed by custom software (Spin, Vicenza, Italy). The results are presented as transformed data to obtain the percentage signal variation (Fx) relative to the time of addition of the stimulus, according to the equation: $Fx = ([Ft - Fo]/Fo) \times 100$, where Ft and Fo are the fluorescence values at time t and at the time of stimulus addition, respectively. The assay of each sample consists of a continuous 99-s fluorescence reading with 30 s before and 69 s after injection of the iodide-rich Dulbecco's PBS (in mM: 137 NaI, 2.7 KCl, 8.1 Na_2HPO_4, 1.5 KH_2PO_4, 1 $CaCl_2$, and 0.5 $MgCl_2$, pH 7.4) to reach a final iodide concentration of 50 mM [20]. The different sensitivity of HS-YFP toward iodide and chloride allows assays to be performed by measuring the transport of anions through the plasma membrane as changes in cell fluorescence. For this assay, cells expressing HS-YFP are equilibrated in a physiological chloride-rich saline solution (e.g., Dulbecco's PBS). During fluorescence reading, cells are exposed to a high concentration of iodide. Iodide influx quenches the cell fluorescence with a rate that depends on the permeability of the cell membrane to halides, and therefore, on the activity of the anion channels or transporters [21].

3.2.2. Phototoxicity

UVA irradiation was carried out by using Philips HPW 125 lamps, mainly emitting at 365 nm. The total energy that struck the sample was monitored by means of a radiometer (Variocontrol, Waldmann, Villingen-Schwenningen, Germany), equipped with a Vario-control UV Sensor (Waldmann). The radiant power emitted by the UVA lamp was about 8 mW/cm^2. The samples were maintained at room temperature during irradiation.

The test compounds were dissolved in ethanol (2 mg/mL) and salmon testes DNA (Sigma-Aldrich) was dissolved in 10 mM NaCl and 1 mM EDTA solution (1.5×10^{-3} M). Each test solution (250 µL) was added dropwise to the DNA solution (5 mL). The mixture was irradiated with 15 J/cm2 UVA light in a glass dish for 1 h. After irradiation, the DNA was precipitated with 1 M NaCl and two volumes of cold ethanol; the precipitated DNA was collected, washed with 80% ethanol, dried, and then dissolved in water. To obtain hydrolysis, the final solution was acidified with 1 M HCl, heated at 100 °C for 1 h, neutralized with NaOH and extracted three times with chloroform. Then, all of the organic layers were collected, dried under a high vacuum, and dissolved in ethanol. The hydrolyzed mixtures were separated by thin layer chromatography (TLC; F_{254} plates, 0.25 mm, Merck, Darmstadt, Germany) eluting with EtOAc/EtOH (9:1). The plates were analyzed with a Mineral Light Lamp VL-4LC VILBERT-LOURMAT (Vilber, Marne-la-Vallée, France).

3.2.3. Mutagenicity Assay

The mutagenicity assay was performed for the test compounds following the plate incorporation method with the histidine-requiring Salmonella typhimurium mutant TA97A, TA98, TA100, and TA1535 strains purchased by Molecular Toxicology Inc. (Boone, NC, USA; moltox.com). All strains (100 μL per plate of fresh overnight cultures) were checked with and without the addition of 0.5 mL of a 5% S9 exogenous metabolic activator (S9 mix). The lyophilized post-mitochondrial supernatant S9 mix (Aroclor 1254-induced, Sprague–Dawley male rat liver in 0.154 M KCl solution), commonly used for the activation of pro-mutagens in mutagenic metabolites (Molecular Toxicology, Inc., Boone, NC, USA), was stored at −80 °C before use. The concentration tested for all the samples were 5, 10, 20, 50, and 100 μL/plate from a stock solution of 500 nM. A fully grown culture of the appropriate tester strain (0.1 mL) was added to 2 mL molten top agar (0.6% agar, 0.6% NaCl, 0.5 mM L-histidine/biotin solution) at 46 °C, together with 0.1 mL of each sample solution at different concentrations, and 0.5 mL S9 mix for assays with metabolic activation. The ingredients were thoroughly mixed and poured onto minimal glucose agar plates (1.5% agar in 2% Vogel–Bonner medium E with 5% glucose solution). DMSO was used as negative a control (100 μL/plate). Positive controls were prepared as follows: 2-aminoanthracene (2 μg/plate) and 2-nitrofluorene (2 μg/plate) for TA 97A, TA98 and TA1535 with and without metabolic activator (S9 mix), respectively; 2-aminoanthracene (2 μg/plate) and sodium azide (2 μg/plate) for TA100, with and without metabolic activator (S9 mix), respectively. The plates were incubated at 37 °C for 72 h and then his+ revertants were checked and counted using a Colony Counter 560 Suntex (Antibioticos, Italy). A sample was considered mutagenic when the observed number of colonies was at least twofold over the spontaneous level of revertants [15,16]. All determinations were made in triplicate. Statistical analysis: relative standard deviations were calculated using the statistical software STATISTICA 6.0 (StatSoft, Italia srl, Vigonza, Italy).

3.2.4. Cytotoxicity Assay

The potential toxicity of TMA analogs on hepatic tissue was assessed by evaluating the viability of HepG2 cells (ATCC, HB-8065) using the MTT assay, as previously reported [22]. Cells were maintained in DMEM complete medium (10% FBS, 1% l-glutamine and 1% PEN-STREP) and seeded in a 96-well plate (5×10^4 cells/mL). They were treated for 24 h with TMA analogs (concentration range from 3.125 μM to 50 μM). At the end of the treatment, cells were washed and an MTT solution (0.5 mg/mL in complete medium) was added to each well and incubated for 4 h at 37 °C. The solution was discarded, and the produced blue formazan crystals were dissolved with acidified isopropanol (2% HCl). Absorbance was measured at 570 nm using a multiplate reader (Victor Nivo, Perkin Elmer). The percentage of viable cells was calculated considering untreated cells as control.

3.3. Formulation of Lipophilic TMA Derivatives

Self-emulsifying formulations (SEDDS) were prepared as described in previous work [18], in particular, medium-chain triglycerides (LabrafacTM Lipophile WL1349) or glycerol/glyceryl monolinoleate (Maisine® CC) were selected as the oil phase, diethylene glycol-monoethyl ether (Transcutol® HP) as the cosolvent, and polyoxyl-35-castor oil (Cremophor®

EL) as the surfactant. Ten percent (*w*/*w*) of the oil phase, 40% (*w*/*w*) of cosolvent and 50% (*w*/*w*) of surfactant were well mixed, then the TMA derivative (4-PhDMA or IPEMA) was introduced to obtain formulations with a drug concentration of 4 μg/μL. Loaded formulations were characterized by measuring the droplet size, and the polydispersity index (PDI), using a laser light scattering analyzer (Zetasizer ZS 90, Malvern Instruments, Worcestershire, UK). For the measurement, SEDDS were diluted in a ratio of 1:250 (*v*/*v*) with distilled water and mixed for 1 min prior to testing. From the light scattering signal (monitored at 25 °C and 173° in automatic mode), the intensity-weighed diameter of dispersion droplets (reported as the z-average) and the PDI were calculated using the manufacturer's software

3.4. Pharmacokinetic Profile

3.4.1. In Vivo Pharmacokinetics

All experimental procedures involving animals were conducted in accordance with the 3R principle to minimize their pain and discomfort. The experimental design and the procedures have been authorized by the Italian Ministry of Health (Authorization no. 853-2018-PR obtained on 11 November 2018). All mice were maintained in IVC cages with free access to food and drink with a 12/12 h of dark/light. For the PK analysis, C57Bl6 male mice (n = 3 per time point) were gavaged with selected compounds (20 mg/kg), suspended in water, or dissolved in SEDDS formulations, after an overnight fast. Plasma samples were collected from the submandibular plexus before and 0.5, 1, 2, 3, 4, 6 24 and 48 h after the administration of the compounds or their formulations. The main PK parameters have been calculated with standard formulas using the PK Solver plug-in for Microsoft Excel and analyzed by means of Student's t-test or one-way ANOVA using GraphPad Prism ver. 8.0 software (GraphPad Software Inc., San Diego, CA, USA), when appropriate. If not otherwise stated, the PK data are expressed as mean ± SD. $p < 0.05$ was considered statistically significant.

3.4.2. Quantification of TMA Analogs in Plasma Samples by UHPLC-HRMS

The TMA analogs were extracted and analyzed from mouse plasma as already reported [18]. Briefly, 100 μL of plasma were treated with 300 μL of cold acetonitrile spiked with 80 nM of internal standard (IS, 6,4′-dimethylangelicin) [23], and then the samples were centrifuged for 10 min at 12,000× g (Hettich Mikro 120 Benchtop Centrifuge) and 25 °C. Ten microliters of the supernatant were injected into the UHPLC system.

The UHPLC-HRMS system was equipped with an Agilent 1260 Infinity II LC liquid chromatographer coupled to an Agilent 6545 LC/Q-TOF mass analyzer (Agilent Technologies, Palo Alto, CA, USA). The analytical column was a Kinetex 2.6 μm C18 Polar, 100 A, 100 × 2.1 mm (Phenomenex, Bologna, Italy), thermostated at 30 °C. The mobile phase components A and B were water and methanol, respectively, both containing 10 mM ammonium formate. The eluent flow rate was 0.25 mL/min. The mobile phase gradient profile was as follows (t in min): t0-3 2% B; t3-18 2-100% B, t18-20 100% B; t20-30 0% B. The MS conditions were: electrospray (ESI) ionization in positive mode, Gas Temp 325 °C, Drying Gas 10 L/min, Nebulizer 20 psi, Sheath Gas Temp 400 °C, Sheath Gas Flow 12 L/min, VCap 4000 V, Nozzle Voltage 0 V, Fragmentor 180 V. Centroid full scan mass spectra were recorded in the range 100–1000 m/z with a scan rate of 1 spectrum/s. MS data were analyzed using the Mass Hunter Qualitative Analysis software (Agilent Technologies, Palo Alto, CA, USA).

Chromatographic peaks of the TMA analog were identified and integrated by considering the Extracted Ion Chromatogram (EIC) of the $[M+H]^+$ species selected with a window of 5 ppm and normalized by the IS. Quantification of TMA analogs was carried out by external calibration of normalized signals using a seven-point calibration curve obtained by spiking blank plasma with each TMA analog in the range of 20–3000 nM. Linearity showed an R2 > 0.98 and the limit of detection of the method was assessed to be approximately 10 nM.

4. Conclusions

A small library of TMA analogs was designed and synthesized to identify new agents with improved safety profiles and bioavailability with respect to the parent TMA. From preliminary screening data on CFTR correction activity, together with previously obtained data on **IPEMA** [11], we try to outline the structural determinants that allow maintaining the correction potential. A certain steric hindrance at either positions 4 or 6 is essential to abolish photoreactivity towards DNA. However, the simultaneous presence of bulky substituents in both positions compromised the maximal CFTR functional rescue achieved. A balanced combination was found to be the presence of a phenyl group at positions 4 or 6, and a methyl substituent at positions 6 or 4, respectively, as exhibited by **4-PhDMA** and **p-ANDMA**, the most active derivatives of the series. Only small alkyl groups are tolerated together at the two positions, as is the case for **IPEMA**. In the furan ring, the presence of a methyl group at position 4′ leads to a better in vitro pharmacological profile compared to the unsubstituted derivatives. All of the analogs were demonstrated to be nongenotoxic for the Ames test.

In terms of pharmacokinetic profile, the more lipophilic compounds showed poor bioavailability and should be opportunely formulated as SEDDS. On the other hand, the introduction of a hydrophilic group in the *para* position of the 6-phenyl substituent gives rise not only to an improvement in bioavailability but also to the opportunity to make a soluble salt that demonstrated appropriate C_{max} and AUC, together with low interindividual variability in absorption.

In conclusion, by modifying the parent TMA at position 4 or 6, we obtained a large series of analogs with an improved safety profile and bioavailability, per se or after incorporation into suitable delivery systems. Furthermore, a CFTR functional screening assay selected some of these analogs to undergo future extensive electrophysiological experiments in CF primary bronchial epithelial cells, to evaluate their ability to functionally rescue the F508del CFTR protein.

Supplementary Materials: The following supporting information can be downloaded at: https: //www.mdpi.com/article/10.3390/ijms231911528/s1.

Author Contributions: Conceptualization, A.C. and G.M.; writing—original draft preparation, review, and editing, A.C.; methodology, investigation, data curation, writing—review and editing, C.V., D.G., S.D.M., E.F., M.R., S.B., C.T., G.S., I.L., R.G., G.C., M.C.D. and A.T.; resources, supervision, project administration, and funding acquisition, A.C. All authors have read and agreed to the published version of the manuscript.

Funding: This research was funded by the FFC (Fondazione Ricerca Fibrosi Cistica) Projects FFC#1/ 2016 (to A.C.) and FFC#22/2019 (to I.L.) and by University of Padova PRID 2018 (to A.C.).

Institutional Review Board Statement: The procedures involving animals were approved by the Ethics Committee for studies involving laboratory animals of the University of Padua and the Italian Ministry of Health (Authorization no. 853-2018-PR, 11 November 2018).

Data Availability Statement: The data presented in this study are available in this article and in the Supplementary Materials.

Acknowledgments: A.C thanks Giorgia Miolo (University of Padova) for phototoxicity experiments and Susanna Vogliardi and Alessia Forgiarini (University of Padova) for mass spectra. A.C. thanks "Delegazione FFC di Rovigo", "Delegazione FFC di Boschi Sant'Anna Minerbe", "Guadagnin srl", and "Delegazione FFC del Lago di Garda con i Gruppi di Sostegno di Arezzo, dell'Isola Bergamasca, di Chivasso" for the adoption of the project "FFC#1/2016".

Conflicts of Interest: The authors declare no conflict of interest. The funders had no role in the design of the study; in the collection, analyses, or interpretation of data; in the writing of the manuscript; or in the decision to publish the results.

References

1. Elborn, J.S. Cystic fibrosis. *Lancet* **2016**, *388*, 2519–2531. [CrossRef]
2. Cutting, G.R. Cystic fibrosis genetics: From molecular understanding to clinical application. *Nat. Rev. Genet.* **2015**, *16*, 45–56. [CrossRef] [PubMed]
3. King, J.A.; Nichols, A.L.; Bentley, S.; Carr, S.B.; Davies, J.C. An Update on CFTR Modulators as New Therapies for Cystic Fibrosis. *Paediatr. Drugs* **2022**, *24*, 321–333. [CrossRef]
4. Deeks, E.D. Lumacaftor/Ivacaftor: A Review in Cystic Fibrosis. *Drugs* **2016**, *76*, 1191–1201. [CrossRef]
5. Guerra, L.; Favia, M.; Di Gioia, S.; Laselva, O.; Bisogno, A.; Casavola, V.; Colombo, C.; Conese, M. The preclinical discovery and development of the combination of ivacaftor + tezacaftor used to treat cystic fibrosis. *Expert Opin. Drug Discov.* **2020**, *15*, 873–891. [CrossRef]
6. Hoy, S.M. Elexacaftor/Ivacaftor/Tezacaftor: First Approval. *Drugs* **2019**, *79*, 2001–2007. [CrossRef]
7. Laselva, O.; Molinski, S.; Casavola, V.; Bear, C.E. The investigational Cystic Fibrosis drug Trimethylangelicin directly modulates CFTR by stabilizing the first membrane-spanning domain. *Biochem. Pharmacol.* **2016**, *119*, 85–92. [CrossRef]
8. Laselva, O.; Molinski, S.; Casavola, V.; Bear, C.E. Correctors of the Major Cystic Fibrosis Mutant Interact through Membrane-Spanning Domains. *Mol. Pharmacol.* **2018**, *93*, 612–618. [CrossRef]
9. Tamanini, A.; Borgatti, M.; Finotti, A.; Piccagli, L.; Bezzerri, V.; Favia, M.; Guerra, L.; Lampronti, I.; Bianchi, N.; Dall'Acqua, F.; et al. Trimethylangelicin reduces IL-8 transcription and potentiates CFTR function. *Am. J. Physiol. Lung Cell Mol. Physiol.* **2011**, *300*, L380–L390. [CrossRef]
10. Europe Medicinal Agency. EU/3/13/1137: Orphan Designation for the Treatment of Cystic Fibrosis. 2020. Available online: https://www.ema.europa.eu/en/medicines/human/orphan-designations/eu3131137 (accessed on 22 August 2022).
11. Marzaro, G.; Lampronti, I.; D'Aversa, E.; Sacchetti, G.; Miolo, G.; Vaccarin, C.; Cabrini, G.; Dechecchi, M.C.; Gambari, R.; Chilin, A. Design, synthesis and biological evaluation of novel trimethylangelicin analogues targeting nuclear factor kB (NF-kB). *Eur. J. Med. Chem.* **2018**, *151*, 285–293. [CrossRef]
12. Laselva, O.; Marzaro, G.; Vaccarin, C.; Lampronti, I.; Tamanini, A.; Lippi, G.; Gambari, R.; Cabrini, G.; Bear, C.E.; Chilin, A.; et al. Molecular Mechanism of Action of Trimethylangelicin Derivatives as CFTR Modulators. *Front. Pharmacol.* **2018**, *9*, 719. [CrossRef]
13. Miolo, G.; Dall'Acqua, F.; Moustacchi, E.; Sage, E. Monofunctional angular furocoumarins: Sequence specificity in DNA photobinding of 6,4,4′-trimethylangelicin and other angelicins. *Photochem. Photobiol.* **1989**, *50*, 75–84. [CrossRef]
14. Chimichi, S.; Boccalini, M.; Cosimelli, B.; Viola, G.; Vedaldi, D.; Dall'Acqua, F. A convenient synthesis of psoralens. *Tetrahedron* **2002**, *58*, 4859–4863. [CrossRef]
15. Maron, D.M.; Ames, B.N. Revised methods for the Salmonella mutagenicity test. *Mutat. Res.* **1983**, *113*, 173–215. [CrossRef]
16. Mortelmans, K.; Zeiger, E. The Ames Salmonella/microsome mutagenicity assay. *Mutat. Res.* **2000**, *455*, 29–60. [CrossRef]
17. Holm, R. Bridging the gaps between academic research and industrial product developments of lipid-based formulations. *Adv. Drug Deliv. Rev.* **2019**, *142*, 118–127. [CrossRef]
18. Franceschinis, E.; Roverso, M.; Gabbia, D.; De Martin, S.; Brusegan, M.; Vaccarin, C.; Bogialli, S.; Chilin, A. Self-Emulsifying Formulations to Increase the Oral Bioavailability of 4,6,4′Trimethylangelicin as a Possible Treatment for Cystic Fibrosis. *Pharmaceutics* **2022**, *14*, 1806. [CrossRef]
19. Yin, H.F.; Yin, C.-M.; Ouyang, T.; Sun, S.-D.; Chen, W.-G.; Yang, X.-L.; He, X.; Zhang, C.-F. Self-Nanoemulsifying Drug Delivery System of Genkwanin: A Novel Approach for Anti-Colitis-Associated Colorectal Cancer. *Drug Des. Devel. Ther.* **2021**, *15*, 557–576. [CrossRef]
20. Favia, M.; Mancini, M.T.; Bezzerri, V.; Guerra, L.; Laselva, O.; Abbattiscianni, A.C.; Debellis, L.; Reshkin, S.J.; Gambari, R.; Cabrini, G.; et al. Trimethylangelicin promotes the functional rescue of mutant F508del CFTR protein in cystic fibrosis airway cells. *Am. J. Physiol. Lung Cell Mol. Physiol.* **2014**, *307*, L48–L61. [CrossRef]
21. Pedemonte, N.; Zegarra-Moran, O.; Galietta, L.J. High-throughput screening of libraries of compounds to identify CFTR modulators. *Methods Mol. Biol.* **2011**, *741*, 13–21. [CrossRef]
22. Frión-Herrera, Y.; Gabbia, D.; Cuesta-Rubio, O.; De Martin, S.; Carrara, M. Nemorosone inhibits the proliferation and migration of hepatocellular carcinoma cells. *Life Sci.* **2019**, *235*, 116817. [CrossRef] [PubMed]
23. Guiotto, A.; Rodighiero, P.; Manzini, P.; Pastorini, G.; Bordin, F.; Baccichetti, F.; Carlassare, F.; Vedaldi, D.; Dall'Acqua, F.; Tamaro, M.; et al. 6-Methylangelicins: A new series of potential photochemotherapeutic agents for the treatment of psoriasis. *J. Med. Chem.* **1984**, *27*, 959–967. [CrossRef] [PubMed]

International Journal of *Molecular Sciences*

Article

Antibacterial and Cytotoxicity Evaluation of New Hydroxyapatite-Based Granules Containing Silver or Gallium Ions with Potential Use as Bone Substitutes

Kamil Pajor [1], Anna Michalicha [2], Anna Belcarz [2], Lukasz Pajchel [1], Anna Zgadzaj [3], Filip Wojas [3] and Joanna Kolmas [1,*]

[1] Department of Analytical Chemistry, Chair of Analytical Chemistry and Biomaterials, Medical University of Warsaw, Faculty of Pharmacy, 02-097 Warsaw, Poland; kamil.pajor@wum.edu.pl (K.P.); lukasz.pajchel@wum.edu.pl (L.P.)

[2] Chair and Department of Biochemistry and Biotechnology, Medical University of Lublin, 20-093 Lublin, Poland; anna.michalicha@gmail.com (A.M.); anna.belcarz@umlub.pl (A.B.)

[3] Department of Environmental Health Sciences, Medical University of Warsaw, Faculty of Pharmacy, 02-097 Warsaw, Poland; anna.zgadzaj@wum.edu.pl (A.Z.); filip.wojas@gmail.com (F.W.)

* Correspondence: joanna.kolmas@wum.edu.pl

Abstract: The aim of the current work was to study the physicochemical properties and biological activity of different types of porous granules containing silver or gallium ions. Firstly, hydroxyapatites powders doped with Ga^{3+} or Ag^+ were synthesized by the standard wet method. Then, the obtained powders were used to fabricate ceramic microgranules (AgM and GaM) and alginate/hydroxyapatite composite granules (AgT and GaT). The ceramic microgranules were also used to prepare a third type of granules (AgMT and GaMT) containing silver or gallium, respectively. All the granules turned out to be porous, except that the AgT and GaT granules were characterized by higher porosity and a better developed specific surface, whereas the microgranules had very fine, numerous micropores. The granules revealed a slow release of the substituted ions. All the granules except AgT were classified as non-cytotoxic according to the neutral red uptake (NRU) test and the MTT assay. The obtained powders and granules were subjected to various antibacterial test towards the following four different bacterial strains: *Staphylococcus aureus*, *Staphylococcus epidermidis*, *Pseudomonas aeruginosa* and *Escherichia coli*. The Ag-containing materials revealed high antibacterial activity.

Keywords: silver; gallium; calcium phosphates; biomaterials; antibacterial activity

Citation: Pajor, K.; Michalicha, A.; Belcarz, A.; Pajchel, L.; Zgadzaj, A.; Wojas, F.; Kolmas, J. Antibacterial and Cytotoxicity Evaluation of New Hydroxyapatite-Based Granules Containing Silver or Gallium Ions with Potential Use as Bone Substitutes. *Int. J. Mol. Sci.* **2022**, *23*, 7102. https://doi.org/10.3390/ijms23137102

Academic Editors: Geoffrey Brown, Andrzej Kutner and Enikö Kallay

Received: 6 May 2022
Accepted: 22 June 2022
Published: 26 June 2022

Publisher's Note: MDPI stays neutral with regard to jurisdictional claims in published maps and institutional affiliations.

1. Introduction

A significant increase in the development of biomaterials for use in bone disease treatment has been recorded in recent years. One of the main reasons is the increasing number of orthopaedical surgeries and the need to replace bone tissue with an appropriate multifunctional biomaterial [1–5].

Currently, materials are being sought that can act as carriers for delivering drugs to the bone as a poorly vascularized tissue. Particular importance is attached to the administration of antibacterial agents (antibiotics) in this way, due to the high risk of potential bacterial infection during surgery, known as surgical site infections (SSIs) [6–10].

Calcium phosphates (CaPs) are used to a great extent in orthopaedic surgery and dentistry, in the form of cements, scaffolds, granules or coatings [11–13]. Among the CaPs, synthetic hydroxyapatite (HA), with the formula $Ca_{10}(PO_4)_6(OH)_2$, can be distinguished, due to its beneficial properties, such as its similarity to bone mineral, bioactivity, osteoconductivity and non-toxicity [14–17].

It is worth mentioning that HA can be easily modified by various ionic substitution, in order to obtain additional biological or physicochemical properties [13,15,17]. For example, the incorporation of carbonates (CO_3^{2-}) into the HA structure leads to an increased solubility of material and a great tendency to nanocrystallinity, whereas substitution with fluorides (F^-) causes better thermal stability of HA [18–22]. Moreover, a feature of HA that deserves attention is the ability to adsorb many biologically active substances; therefore, HA can potentially be used as a carrier for drugs [15,16,23].

Silver has been well known for its antibacterial properties for many years (its beneficial role in the treatment of infection dates back to at least 4000 B.C.) and its mechanism of action is one of the best understood [24,25]. Silver ions mainly affect thiol groups (-SH), present in bacterial proteins' structure, by substituting hydrogen atoms and the arising of S-Ag binding. Such modifications in the protein structure of bacteria cells cause a denaturation, deactivation and malfunction of membrane pumps. As a consequence, membrane cells shrink and detach from the cell wall, then the content of the cell leaks outside the membrane and finally, the cell wall is torn apart [26].

As an antibacterial agent, silver is currently used in hospitals to reduce nosocomial infections in the treatment of burns and open wounds, but also in water cleaning systems [24,27]. Recently, silver has been identified as a promising agent for potential use in treating multidrug-resistant bacteria infections [28,29]. Many studies focusing on the synthesis of silver-substituted HA have been reported, which highlight its abilities in inhibiting bacterial growth and simultaneously intensifying osseointegration, consequently resulting in silver-doped HA being regarded as a very promising biomaterial [30–33].

Gallium ions also exhibit antibacterial activity; however, their use is not as common as silver. Their main advantage over silver ions is significantly lower cytotoxicity in higher concentrations with regard to human cells. The Ga^{3+} antibacterial mechanism of action is mainly based on the substitution of iron ions with gallium ions in the bacteria protein metabolism ("the Trojan horse strategy"), which causes the impairment of bacteria cell functions [26,34]. Moreover, there are many studies that outline the beneficial effect of gallium on bone tissue. It should be noted that Ga ions exhibit antiresorptive and antiosteoporotic properties, as well as antitumor, anti-inflammatory and immune suppressive properties [26,35,36]. As a result, HA doped with gallium can be used as biomaterial, which on the one hand enhances bone growth and on the other hand protects from bacterial infection [37–40]. In the available literature, the use of the aforementioned biomaterial as a drug delivery system is reported [41–43].

In the present work, the new bone substitutes, based on HA modified with Ag^+ or Ga^{3+} ions as antibacterial agents, were prepared in the form of three types of granules. The obtained materials were subjected to physicochemical analysis, followed by cytotoxicity and antimicrobial evaluation.

2. Results

2.1. Chemical Structure and Elemental Analysis of the Synthesized Powders

The powder X-ray diffractometry (PXRD) patterns of the samples are presented in Figure 1a. The revealed reflections in all the diffractograms indicated that the materials are composed of hydroxyapatite. The obtained powders are homogenous, with no additional crystalline phase.

It is worth mentioning that the reflections were broad and poorly resolved, illustrating the poorly crystalline feature of the synthesized powders. In order to determine crystallite sizes along *c* and *a* axes, the Scherrer formula was used for the reflections at approx. 25.9° and 39.8°, respectively [44], which is as follows

$$d = \frac{0.94\lambda}{\beta \cos \theta},$$ (1)

where

d—crystallite size (nm)
λ—wavelength of used radiation (nm)
β—full width at half maximum (FWHM) of the peak (radians)
θ—the diffraction angle of the corresponding reflection (°).

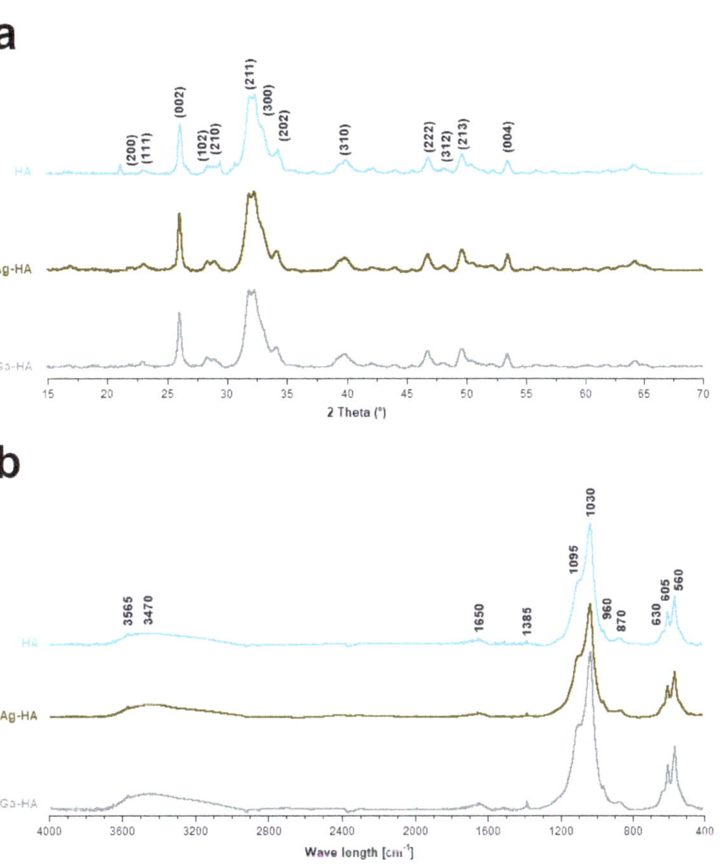

Figure 1. Powder X-ray diffractograms (**a**) and FT-IR spectra (**b**) of the synthesized powder samples.

It was also possible to evaluate the crystallinity of the materials, using the following formula [45]:

$$\chi = \left(\frac{K}{\beta_{(002)}}\right)^3 \tag{2}$$

where

χ—the degree of crystallinity

K—constant (for hydroxyapatite it is equal to 0.24)

$\beta_{(002)}$—full width in a half minimum (FWHM)for (002) reflection (°).

The calculated parameters are shown in Table 1. The values of the crystal dimensions confirm that apatitic crystals in all the samples were elongated along the *c* axis. The samples are nanocrystalline with a low degree of crystallinity.

Table 1. Parameters of the obtained HA powders.

	Ag-HA	Ga-HA
Crystallinity index (CI)	0.45	0.39
Crystallite size—*c*-axis (nm)	27 ± 2	26 ± 3
Crystallite size—*a*-axis (nm)	6 ± 2	7 ± 2
Ionic dopant content (wt%)	0.46	0.41

The Fourier transform infrared spectroscopy (FT-IR) spectra presented in Figure 1b show a band corresponding to the apatitic chemical structure. One can observe three intensive bands within the range of 1095–960 cm^{-1} and two bands within the range of 605–560 cm^{-1}, originating from the stretching and bending vibrations of phosphate P–O bonds, respectively. At approximately 3565 cm^{-1} and 630 cm^{-1}, respectively, the stretching and libration vibrations of the characteristic structural hydroxyl groups can be observed. In addition, a wide band at approx. 3470 cm^{-3} (stretching) and a band at 1650 cm^{-3} (bending) confirm the presence of adsorbed water in the samples, which is common for the wet method. The slightly detectable band at 1385 cm^{-3} originates from the nitrate residues, while the band at 872 cm^{-1} comes from the carbonates [46–49].

The Ag^+ and Ga^{3+} ion content in the Ag-HA and Ga-HA samples, respectively, were measured by the atomic absorption spectrometry (AAS) method. The elemental analysis shows that the dopant's concentration is slightly lower than the theoretical value (0.54 and 0.45% for Ag and Ga, respectively), indicating only a partial substitution of these elements into the apatitic structure (see Table 1).

2.2. Ultrastructure, Porosity and Mechanical Strength of the Granules

The representative scanning electron microscopy (SEM) images of various types of granules are shown in Figure 2a–i. The ceramic AgM and GaM granules comprised micro-sized, regular spheres with an approximate diameter of 0.2–1 mm (see Figure 2a). The composite granules (AgT, GaT, AgMT and GaMT) were significantly larger and their average diameter was around 3.5 mm (see Figure 2d,g). As observed, all the samples revealed a porous feature; however, the outer surface of the ceramic microgranules seemed to be more porous than the outer surface of the composite, which in turn was rough and undulating. In Figure 2b,c,e,f,h,i, the cross-sections through all the types of granules are presented. It should be noted that the large number of macropores and mesopores in the internal structure of the granules can be observed, especially in the composite granules, which may indicate a high porosity of the materials. In addition, the cross-sections through the AgMT and GaMT granules show the microgranules, which were used during their preparation (indicated with white arrows in Figure 2h,i).

The porosity measurements using the mercury intrusion were only available for the composite granules (see Table 2). As observed, the total volume of pores and the degree of porosity is significantly higher for the AgT and GaT granules than for the AgMT and GaMT granules. It is worth mentioning that the specific surface area (SSA) of the pores is also better developed for the AgT and GaT granules, which is in accordance with the SEM results (see Figure 2d–f). The high SSA of these granules may be caused by their well-developed mesoporous structure. This is also an important factor in the case of the average diameter of pores; highly developed mesopores, as well as the presence of numerous macropores, may be the reason for a larger diameter of pores in the case of the AgT and GaT granules (see Table 2).

Table 2. Results of porosity and mechanical strength studies.

Sample	Total Volume of Pores (cm³/g)	Porosity of Granules (%)	SSA of Pores (m²/g)	Volume of Mesopores (cm³/g)	Percentage of Mesopores (%)	Average Diameter of Pores (nm)	Apparent Density of Granules (g/cm³)	Average Mechanical Strength (N/granule)
GaT	0.87	65	79	0.32	37	40	0.8	34
GaMT	0.28	36	21	0.06	22	35	1.3	110
AgT	0.94	65	55	0.23	25	65	0.7	37
AgMT	0.56	55	42	0.12	22	45	1.0	15

It should be noted that the average mechanical strength of the synthesized granules was vastly different in two analyzed types of granules (see Table 2). The AgT and GaT granules exhibited mechanical strength of around 35 N, which was approximately three times lower than the mechanical strength of the GaMT granules. This may be explained by the higher apparent density of the GaMT granules. Surprisingly, the AgMT granules showed significantly lower mechanical strength, together with great apparent density.

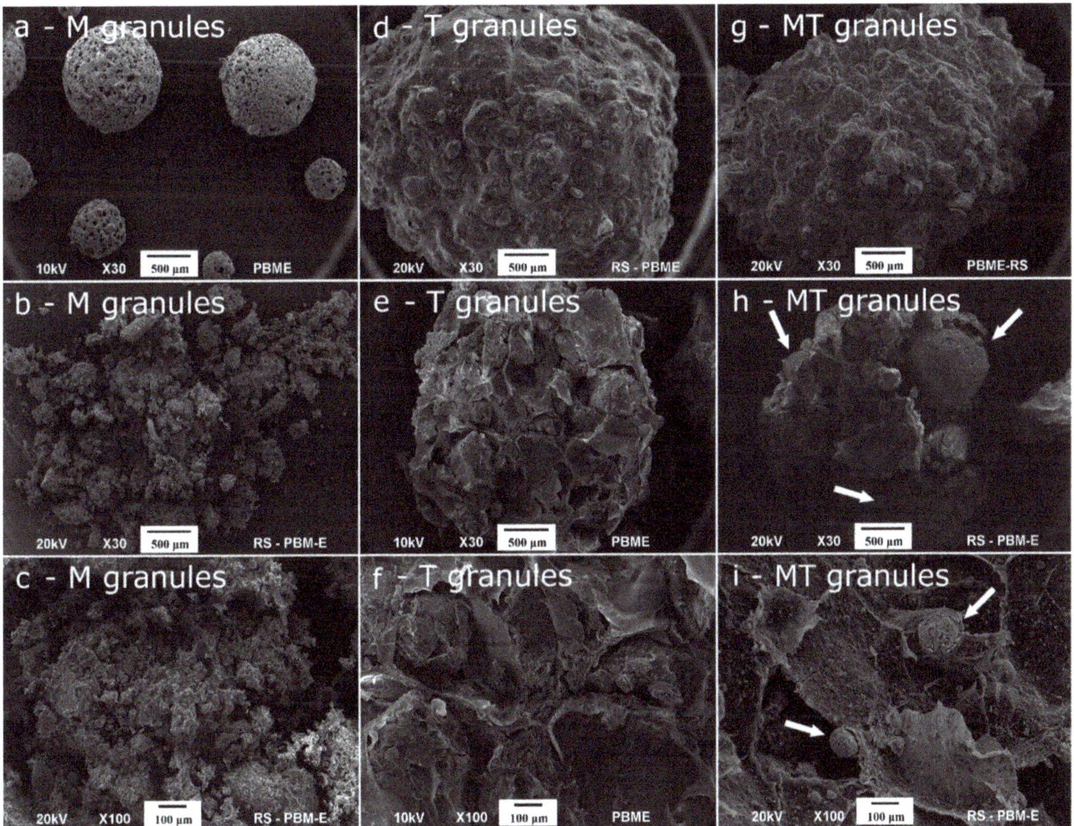

Figure 2. Representative SEM images of the samples: ceramic microgranules (AgM and GaM) (**a–c**); composite granules (AgT and GaT) (**d–f**); composite granules AgMT and GaMT (**g–i**).

In the case of the ceramic microgranules (AgM and GaM), it was not possible to conduct an evaluation of the mechanical compressive strength and mercury intrusion porosimetry method, due to the very small size of the microgranules and their pores. Therefore, we decided to estimate the sizes of the pores using the SEM pictures (data not shown) and Olympus software measure IT.

According to this evaluation, ceramic microgranules AgM and AgT have two types of pores, including larger, irregular ones with a diameter of about 25 ± 3 μm and 15 ± 3 μm for AgM and GaM, respectively, and smaller, oval ones with a diameter of about 5 ± 2 μm.

2.3. Study of Silver and Gallium Ions Release from Granules

In Figure 3a,b, the results of the release of silver and gallium ions are presented. It can be observed that in all of the cases (for the "T" granules, as well as the "TM" granules), the amount of released silver ions was fairly low. The lowest release was observed for GaMT granules. It is also worth noting that in the case of the AgMT and GaMT granules, silver or gallium ions started to be released after a longer time than from the AgT and GaT granules, respectively. Regarding the AgT and GaT granules, the presence of the doped ions could be observed in the sample after only 1 h, whereas in the case of the AgMT and GaT granules, the ions were detected in the samples after 12 h. However, in the case of the silver-containing samples, this did not influence the final concentration of silver ions, which was similar in both cases.

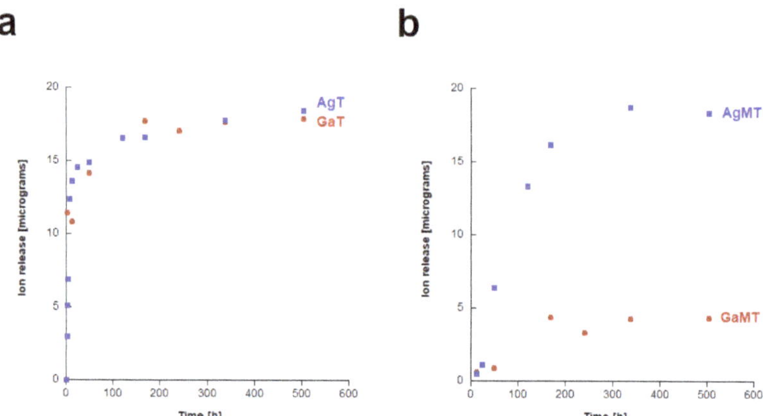

Figure 3. The results of the release study of silver and gallium ions from: AgT and GaT granules (**a**); AgMT and GaMT granules (**b**).

2.4. Cytotoxicity Studies of Powders and Granules

Figure 4 and Figure S1 (in Supplementary Materials) show the results obtained in the cytotoxicity tests. As can clearly be observed, the BALB/c 3T3 cells viability or mitochondrial metabolic activity did not fall below 70% in comparison to the untreated control, using the AgM, GaM, GaT, AgMT and GaMT granules in the NRU; neither was this the case in the MTT test across the whole range of tested dilutions (Figure 4). Therefore, all these materials were classified as non-cytotoxic in both assays (Tables 3 and 4). On the other hand, Ag-HA powder (see Figure S1) and AgT reduced 3T3 cells viability and enzymatic activity to below 70% after exposition to the undiluted extracts (100 mg/mL) and were classified as cytotoxic in both assays (Tables 3 and 4). The Ga-HA powder was not classified as cytotoxic, according to the methodology of the NRU assay, but it significantly decreased the mitochondrial metabolic activity of cells treated with the undiluted extract (see Figure S1 and Table 3). However, in the first of the dilutions in the twofold dilution series, none of these samples (Ag-HA, Ga-HA powders and AgT) negatively affected the cell culture condition.

Table 3. Results of the neutral red uptake test for the highest concentrations of tested extracts (100 mg/mL) in comparison to the untreated control.

Sample	Cells Viability ± SD (%)	IC50 (mg/mL)	Classification
HA powder	101 ± 2	N	Non-cytotoxic
Ag-HA powder	**0 ± 0**	75	Cytotoxic
Ga-HA powder	80 ± 10	N	Non-cytotoxic
AgM granules	100 ± 3	N	Non-cytotoxic
GaM granules	98 ± 4	N	Non-cytotoxic
AgT granules	**0 ± 0**	61	Cytotoxic
GaT granules	80 ± 4	N	Non-cytotoxic
AgMT granules	93 ± 7	N	Non-cytotoxic
GaMT granules	96 ± 5	N	Non-cytotoxic
LT	**0 ± 0**	<10	Cytotoxic
PE	102 ± 7	N	Non-cytotoxic

LT—latex, reference cytotoxic material. PE—polyethylene foil, reference non-cytotoxic material. N—calculation was not possible due to the lack of cytotoxicity in the whole range of tested concentrations. SD—standard deviation. Results with cells viability decreased under 70% are bolded.

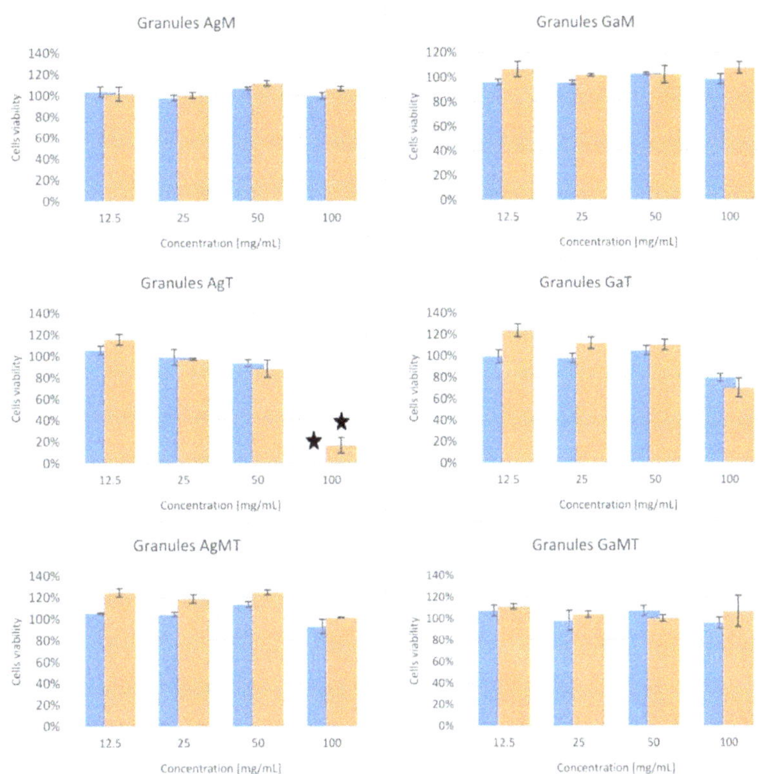

Figure 4. The NRU (blue bars) and MTT (orange bars) tests results obtained for granules in the whole range of tested extracts concentrations. Black stars indicate the decrease in the cells' viability under 70%, which classifies the sample as cytotoxic.

Table 4. Results of the MTT assay for the highest concentrations of tested extracts (100 mg/mL) in comparison to the untreated control.

Sample	Cells Viability $\pm$ SD (%)	IC50 (mg/mL)	Classification
Ag-HA powder	0 ± 0	59	Cytotoxic
Ga-HA powder	37 ± 9	86	Cytotoxic
AgM granules	107 ± 2	N	Non-cytotoxic
GaM granules	108 ± 5	N	Non-cytotoxic
AgT granules	17 ± 7	74	Cytotoxic
GaT granules	70 ± 9	N	Non-cytotoxic
AgMT granules	101 ± 0	N	Non-cytotoxic
GaMT granules	107 ± 15	N	Non-cytotoxic
LT	0 ± 0	<10	Cytotoxic
PE	102 ± 7	N	Non-cytotoxic

2.5. Results of Antibacterial Activity Studies

2.5.1. Preliminary Studies

The pilot test of antibacterial activity was based on the migration of Ag and Ga ions from the prepared samples to the surrounding agar medium and inhibition of bacterial growth within the migration zone. The results obtained in this experiment showed that only the Ag-HA powder and porous AgT granules exhibited antibacterial activity (Table 5), confirming the release of Ag^+ from the samples. Larger growth inhibition zones were observed for Ag-HA powder (11–13 mm) than for AgT granules (8–10 mm). This difference can be explained by the fact that the granules are composed not only of Ag-HA powder but also of sodium alginate and chondroitin sulphate; therefore, the content of Ag-enriched powder was lower in the granules. The lowest susceptibility to silver ions was noted with regard to the *E. coli* strain (Table 5). In the case of the AgM ceramic microgranules, no antibacterial activity was observed for all tested strains. This was expected, due to the low porosity of the microgranules, which reduced the Ag^+ release. No effect of Ga^{3+} ions on the antibacterial activity of the samples was observed under the test conditions. This could have been caused by the lower antibacterial activity of the Ga ions or their slower release from the samples, in comparison with the Ag-enriched samples.

Table 5. Results of agar plate test. Diameter of the well drilled in agar medium: 7.5 mm.

Sample	Zones of Bacterial Growth Inhibition (mm)			
	Staphylococcus aureus	*Staphylococcus epidermidis*	*Escherichia coli*	*Pseudomonas aeruginosa*
HA	0	0	0	0
Ag-HA	12	13	11	12
Ga-HA	0	0	0	0
AgM	0	0	0	0
GaM	0	0	0	0
AgT	10	10	8	10
GaT	0	0	0	0
AgMT	0	0	0	0
GaMT	0	0	0	0

The agar plate test has limitations, which are related to the restricted release and migration of antibacterial ions in viscous and solid agar medium. Therefore, to gain a deeper insight into the antibacterial properties of the tested samples, an antibacterial activity assessment of porous materials and a bacterial adhesion test were performed. Due to the different form of the samples (powders and granules), the tests were performed separately for the powders pressed into the tablets and for the porous granules.

2.5.2. Antibacterial Activity of Powders

The assessment of the antibacterial activity of the tested materials is based on a comparison of the number of viable bacteria eluted from the pure hydroxyapatite and Ag- and Ga-doped hydroxyapatite. The results of this assessment for pressed Ag-HA and Ga-HA powders were compared with those of the pressed HA powder and the positive controls (amount of bacteria in 50 μL of working bacterial suspensions). The tablets pressed from HA powder served as a reference, which allowed us to estimate the amount of viable bacteria eluted from pure non-doped material. As shown in Figure S2a, approximately 3% and 48% of bacteria introduced into the tablets were eluted from the HA reference samples inoculated with *S. aureus* and *S. epidermidis*, respectively. No viable Gram-positive bacteria were eluted from the Ag-HA and Ga-HA tablets (in the case of *S. epidermidis*, the results were statistically different), suggesting the antibacterial activity of both silver- and gallium-doped hydroxyapatite. However, for both Gram-negative strains, no bacteria were eluted from the HA reference tablets or from the Ag-HA and Ga-HA tablets (Figure S2a). Thus, the evaluation of antibacterial activity of the Ag and Ga ions was impossible in the case of the *E. coli* and *P. aeruginosa* strains. This observation was possibly caused by the relatively high density and low porosity of the pressed tablets, which enabled the bacteria to enter into the tablets but did not allow them to be eluted after the incubation. In turn, the bacterial adhesion test allows us to evaluate the ability of tested materials to prevent the adhesion of bacteria, which is a crucial starting point in the biofilm formation process. In this test, the tablets of pressed HA powder served as a reference for the Ag-HA tablets and the Ga-HA tablets, similar to the AATCC test method 100-2004. The results are presented in Figure S2b.

For all the tested bacterial strains, Ag-HA significantly reduced the number of adhered bacteria in comparison with pure HA by 10–82%, depending on the strain (Figure S2b). A higher rate of bacterial viability reduction was observed for both Gram-negative strains (by 49–82% compared with pure HA) in comparison with both Gram-positive strains (by 10–33% compared with pure HA) (Figure S2b). The effect of gallium ions doping on the antiadhesive properties of HA was less distinct, which is probably related to the lower rate of Ga^{3+} release from the Ga-HA tablets. A statistically significant reduction in the number of adhered bacterial cells in the case of the Ga-HA tablets was found for all strains (by 12.5–64% compared with pure HA), with the exception of *S. epidermidis* (Figure S2b).

Difficulties in the interpretation of the results of antibacterial activity assessment for the reference and ions-doped HA powders (as shown in Figure S2a) necessitated the testing of the antibacterial activity of the powders in another mode. Thus, powdered samples were subjected to additional tests, based on the direct contact between the powders and bacteria, followed by an evaluation of their survival rate. In this test, the impact of bacterial adhesion on the tested powder was eliminated, as both free and powder-adhered cells were plated and counted. Two concentrations of powder suspensions, 0.1 mg/mL and 1 mg/mL, were tested, while the final titre of all bacterial strains was 3.0×10^6 CFU/mL. A statistically different antibacterial effect was found even for pure HA powder in both tested concentrations for *S. epidermidis* and *P. aeruginosa*, causing the reduction in bacterial viability to 40–60% of the control. However, the effect of HA powder for *S. aureus* and *E. coli* was not detected (Figure 5).

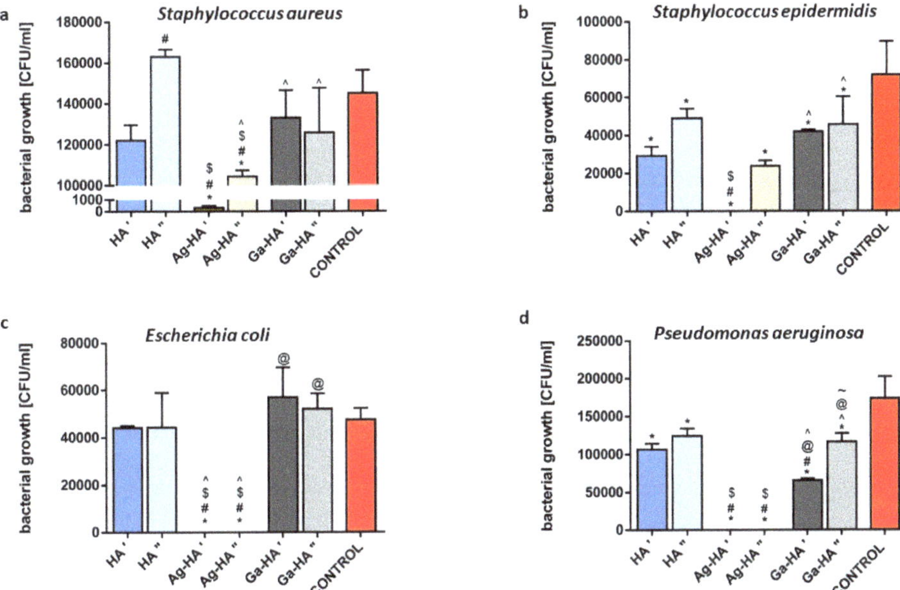

Figure 5. Antibacterial activity of hydroxyapatite powders against 4 bacterial strains: Staphylococcus aureus (**a**), Staphylococcus epidermidis (**b**), Escherichia coli (**c**) and Pseudomonas aeruginosa (**d**). (′) and (″) symbols accompanying powders designation on X axes indicate their concentration in mixtures (0.1 mg/mL and 1 mg/mL, respectively). (*) symbol indicates statistically significant differences between the samples and control, (#) symbol indicates statistically significant results between HA′ and the samples, ($) symbol indicates statistically significant results between HA″ and the samples, (^) symbol indicates statistically significant results between Ag-HA′ and the samples, (@) symbol indicates statistically significant results between Ag-HA″ and the samples, (~) symbol indicates statistically significant results between Ga-HA′ and the samples; according to one-way ANOVA with post-hoc Dunnett's test or post-hoc Tukey's test ($p < 0.05$).

Although it was recognized that hydroxyapatite was biologically inert and did not act toxically on bacteria, there were some observations indicating the antibacterial activity of nanohydroxyapatite, both artificial (obtained by the wet precipitation method or by the microwave-assisted method) [50,51] and natural source-derived [52]. Our results are, to some extent, confirmation of previous reports. The effect of the Ag-HA powder was the most striking; in the case of all strains, 1 mg/mL of Ag-HA powder reduced the bacterial viability completely or almost completely. The same was observed for the 0.1 mg/mL Ag-HA powder concentration by comparison with both Gram-negative strains, while in the case of the Gram-positive strains, the reduction in bacterial viability reached 25–35% of the control (Figure 5). The statistical analysis also confirmed that the presence of Ag ions caused a statistically different antibacterial effect in comparison with pure HA powder (with the exception of a 0.1 mg/mL concentration, tested against *S. epidermidis* strain). In turn, the effect of Ga ions in the HA powder was much less obvious. It did reduce the viability of *S. epidermidis* and *P. aeruginosa* by comparison with the control (to approx. 40% of the control) (Figure 5b,d). However, in the case of 1 mg/mL of Ga-HA and *P. aeruginosa*, the antibacterial effect was statistically different compared with the same concentration of HA powder (Figure 5d). Moreover, this effect was dose-dependent (Figure 5d). This observation is the more important in light of the fact that *P. aeruginosa* is a critically dangerous bacterial strain, causing morbidity and mortality in many patients and is remarkably resistant to antibacterial agents, thus is difficult to eradicate [53]. These observations are in accordance with the results of the bacterial adhesion test, which showed that *P. aeruginosa* is the most susceptible to Ga-HA antibacterial activity among all tested strains. To summarize, the direct contact test revealed more details than the antibacterial activity test, based on the AATCC test method 100-2004, and confirmed the strong antibacterial activity of Ag-HA and the moderate antibacterial activity of Ga-HA.

2.5.3. Antibacterial Activity of Granules

For porous granules, the assessment of the antibacterial activity required a reference sample in the form of granules prepared from pure HA powder (HAT), which was prepared in an analogous way as AgT, GaT, AgMT and GaMT. The physicochemical properties of the HAT granules (size, morphology, porosity) were similar to the AgT and GaT granules (data not shown). The results of the antibacterial activity for these granules, based on the AATCC test method 100-2004, showed that amount of both Gram-positive bacteria significantly increased after incubation with the reference HAT granules, in comparison with the control (Figure 6a).

This phenomenon might have been caused by the presence of organic polymers (sodium alginate and chondroitin sulphate), which served as a nutrient for bacterial cell propagation. Alginate oligosaccharides are known for their antibacterial activity exhibited in relation to *S. aureus* and other staphylococci [54]. However, hydrogel wound dressings, based on alginate polymers, require additional antibacterial agents to reveal antibacterial activity [55]. In turn, the number of *E. coli* and *P. aeruginosa* was reduced after incubation with HAT, which can especially be explained in the case of the latter strain, as *P. aeruginosa* cannot use alginate as a carbon source [56]. Silver-containing AgT and AgMT granules caused the most significant mortality of all the tested strains (Figure 6a), confirming their strong antibacterial activity. Gallium-doped materials (GaT and GaMT) also caused a decrease in bacterial viability by 70–100% by comparison with pure HAT (Figure 6a); only in the case of *S. epidermidis* and GaT granules was this effect less notable (only a 20% decrease). The bacterial adhesion test related to the second aspect of the bacteria-biomaterial interactions showed the strong impact of AgT on bacterial adhesion (although for *S. epidermidis*, it was less and was not significant) (Figure 6b). AgMT granules exhibited a much weaker effect on bacterial adhesion than AgT, which can be explained by the lower content of silver in the granules. The effect of gallium presence in GaT and GaMT on bacterial adhesion limitation was much less distinct than that observed for the presence of silver in the tested granules (Figure 6b). Surprisingly *E. coli* showed a stronger reaction to GaMT than to GaT, although the latter contained a higher Ga concentration (Figure 6b).

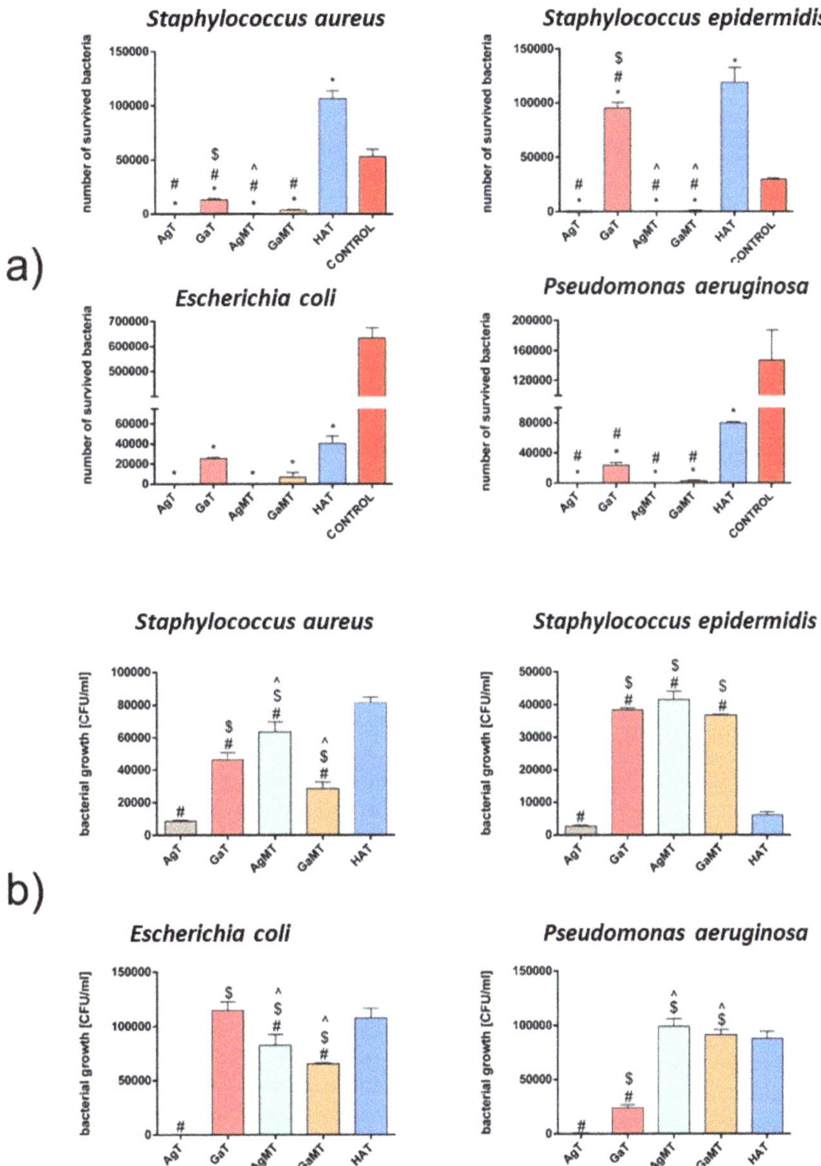

Figure 6. Antibacterial activity according to the AATCC 100-2004 test method (**a**) and bacterial adhesion (**b**) of hydroxyapatite-based granules against 4 bacterial strains. (*) symbol indicates statistically significant differences between the samples and control, (#) symbol indicates statistically significant results between the samples and HAT granules, ($) symbol indicates statistically significant results between AgT and all doped samples, (^) symbol indicates statistically significant results between GaT and all doped samples; according to one-way ANOVA with post-hoc Dunnett's test or post-hoc Tukey's test ($p < 0.05$).

An unexpected observation was made in the case of *S. epidermidis*, namely, the bacterial adhesion to AgMT, GaT and GaMT was significantly higher than that of the reference HAT granules (Figure 6b). This phenomenon is difficult to explain; however, this may be on account of the different surface topography of these granules.

AgM and GaM microgranules, due to the low porosity resulting from the high sintering temperature (15–25 μm and 5 μm; as mentioned above) and micro-dimensional form, could not be evaluated for antibacterial activity, using the same tests as other materials. Therefore, extracts obtained after microgranule incubation (24 h) with bacterial culture broth were inoculated with four reference bacterial strains. Then, the bacterial growth in the extracts was monitored. However, no antibacterial effect was observed. The rate of bacterial growth in both collected extracts was on a comparable level with the bacterial growth of the control (data not shown). The above results may be explained by the low porosity of microgranules, which negatively affects the release of Ag^+ and Ga^+ ions from the samples in this particular test. However, we have indirect proof that AgM and GaM microgranules reveal antibacterial activity, namely, that AgMT and GaMT granules (composed of 50% of AgM and GaM microgranules, respectively) exhibit notable antibacterial activity, according to the AATCC 100-2004 test method (Figure 6a). Therefore, their antibacterial activity must have been related to the presence of AgM and GaM microgranules in the granules' structure.

3. Materials and Methods

3.1. Synthesis of Silver- or Gallium-Containing Hydroxyapatite Powders

Hydroxyapatite powders enriched with gallium (Ga-HA) or silver ions (Ag-HA) with 0.45 wt% and 0.54 wt% nominal value of Ga or Ag, respectively, were synthesized using the conventional wet method (coprecipitation in an aqueous solution), which was described in detail in our previous work [57]. The following reagents were used in the aforementioned synthesis: calcium nitrate tetrahydrate $Ca(NO_3)_2 \cdot 4H_2O$ (Sigma-Aldrich, Bangalore, India), ammonium dibasic phosphate $(NH_4)_2HPO_4$ (Chempur, Piekary Śląskie, Poland), silver nitrate $AgNO_3$ (Avantor Performance Materials, Gliwice, Poland) and gallium nitrate trihydrate $Ga(NO_3)_3 \cdot 3H_2O$ (Sigma-Aldrich, Burlington, MA, USA) as sources of calcium, phosphorus, silver and gallium, respectively. Briefly, an aqueous solution of $(NH_4)_2HPO_4$ was added dropwise into an aqueous solution of $Ca(NO_3)_2 \cdot 4H_2O$ and one of the aforementioned reagents ($AgNO_3$ or $Ga(NO_3)_3$ $3H_2O$) and stirred gently at pH 10 and at a temperature of 60 °C for 2 h. Then, the obtained precipitate was left for 24 h for ageing. Next, the precipitates were filtered, soaked several times in distilled water and dried at 120 °C in air. For comparison, pure, unsubstituted hydroxyapatite (HA) was synthesized by the same method.

3.2. Preparation of Microgranules

Microgranules were obtained using Ag-HA and Ga-HA powders according to the method adapted from [58]. Additional reagents, used in the preparation of microgranules by using camphene emulsion, were as follows: gelatine, 20 mesh pure p.a. (Chempur, Piekary Śląskie Poland); poly(vinyl alcohol) PVA, average molecular weight 130,000 (Sigma-Aldrich, USA); poly(acrylic acid sodium salt) (Sigma-Aldrich, St. Louis, MO, USA); Triton™ X-100 (Sigma-Aldrich, St. Louis, MO, USA); camphene (Sigma-Aldrich, Madrid, Spain).

Firstly, a 10% aqueous gelatine solution was prepared, with the addition of 2% PVA. Afterwards, 0.2% Triton X-100 and 0.3% poly(acrylic acid sodium salt) were added to the solution as dispersants. Meanwhile, camphene and Ag-HA or Ga-HA powder were mixed together at 60 °C in a 0.5:1 ratio. A 10% gelatine solution was added to the camphene/Ag-HA (Ga-HA) mixture (2 mL of solution per 1 g of hydroxyapatite powder) and the obtained slurry was dispersed in oil in a beaker using a magnetic stirrer (150–250 rpm). Subsequently, the beaker was kept in an ice-cooled bath for 5 min, then the obtained microgranules were separated from the oil, rinsed with ethanol and finally dried at room temperature.

At the next stage, microgranules were sintered at a high temperature (initially 500 °C for 1 h at a heating rate of 3 °C/min, then at 1250 °C for 3 h at a heating rate of 5 °C/min). The granules were then sieved using test sieves (Ø 0.2 mm and Ø 1 mm), in order to separate those with a 0.2–1 mm diameter.

The obtained granules were named AgM and GaM for silver- and gallium-containing samples, respectively. For comparison, pure HA was synthesized according to the aforementioned procedure.

3.3. Preparation of Composite Granules

In order to fabricate composite granules, the following reagents were used: sodium alginate (Sigma Aldrich, US), chondroitin sulphate sodium salt (TCI, Belgium), calcium chloride anhydrous $CaCl_2$ (Sigma-Aldrich, China) and the following previously synthesized powders and microgranules: HA, Ag-HA, Ga-HA, AgM and GaM.

At first, a 4% aqueous sodium alginate solution was prepared at 40 °C and chondroitin sulphate sodium salt was added to obtain a 0.5% suspension. Then, two types of composite granules were prepared.

In the first type of granules, 1 g of Ag-HA or Ga-HA (or HA) powder was added to the suspension (10 mL) and mixed vigorously, resulting in a milky, dense slurry. Meanwhile, the cross-linking solution (1.5% $CaCl_2$) was prepared. Finally, the slurry was added dropwise to a $CaCl_2$ solution, stirred using a magnetic stirrer and granules were formed. The granules obtained were left in the cross-linking agent for 10 min, rinsed with distilled water, dried in air and then lyophilized.

During the preparation of the second type of granules, pure, unsubstituted HA and AgM or GaM microgranules (ratio 1:1) were used instead of Ag-HA or Ga-HA powders. The other stages of production remained unchanged. All the obtained granules are listed in Table 6.

Table 6. List of all the obtained granules from Ag-HA or Ga-HA materials.

Granules	Fabrication method	Comment
AgM	Camphene emulsion	Microgranules containing Ag-HA
GaM	Camphene emulsion	Microgranules containing Ga-HA
HAT	Alginate cross-linking	Composite granules containing HA
AgT	Alginate cross-linking	Composite granules containing Ag-HA
GaT	Alginate cross-linking	Composite granules containing Ga-HA
AgMT	Alginate cross-linking	Composite granules containing AgM and HA
GaMT	Alginate cross-linking	Composite granules containing GaM and HA

3.4. Physicochemical Analysis of Ag-HA and Ga-HA

The phase composition of the powder samples was determined by powder X-ray diffractometry (PXRD, Bruker D8 Advance, Billerica, MA, USA). The diffractometer was equipped with a LYNEXEYE position sensitive detector and with Cu-Kα radiation (λ = 0.15418 nm). The measurements were carried out in the Bragg–Brentano (θ/θ) horizontal geometry (flat reflection mode) between 15° and 60° (2θ) in a continuous scan, using 0.03° steps and 2 s/step (total time 384 s/step). Phase identification was achieved by comparing the obtained diffractograms of HA, Ag-HA and Ga-HA samples with the JCPDS 09-0432 standard pattern.

Fourier-transform infrared studies (FT-IR) were conducted using the Spectrum 1000 spectrometer (Perkin Elmer, Llantrisant, UK). The data were collected with a 2 cm^{-1} resolution over a range of 4000–400 cm^{-1} at 30 scans, using the standard KBr pellet technique.

PXRD patterns and FT-IR spectra were processed using GRAM/AI 8.0 software (Thermo Scientific, Burlington, ON, USA) and subsequently, graphs were prepared with KaleidaGraph 3.5 software (Synergy Software, Reading, PA, USA).

The gallium and silver content in the Ga-HA and Ag-HA samples, respectively, was measured by atomic absorption spectrometry (AAS). Briefly, $AgNO_3$ (Avantor Performance

Materials, Poland) and Ga(NO$_3$)$_3$·3H$_2$O (Sigma-Aldrich, USA) were weighed out, dissolved in distilled water (separately) and then diluted several times, in order to prepare the solutions necessary for the calibration curves. Then, the known quantities of synthesized Ag-HA and Ga-HA powders were weighted out, dissolved in suprapure 63% HNO$_3$ and adequately diluted with distilled water. Finally, the obtained solutions were measured by AAS spectrometry (ANALYST 400, Perkin Elmer, Llantrisant, UK), with detection at a wavelength λ = 328.07 nm for silver and λ = 287.42 nm for gallium.

3.5. Physicochemical Analysis of the Granules

In order to determine the morphology of the prepared granules, a microscopical study was performed using scanning electron microscopy (SEM) JSM 6390 LV (JEOL, Tokyo, Japan) at 20 or 30 kV accelerating voltage. The survey was based on taking images of granules (previously covered with a gold layer in a vacuum chamber) from both the outer and the inner surface (after the cross-section). The cross-sections were prepared by carefully cutting the granules with a surgical lancet.

The porosity and specific surface area of the granules were evaluated using the mercury intrusion porosimetry method with the Autopore IV 9510 instrument (Micromeritics, Norcross, GA, USA). The measurements were conducted with Hg intrusion pressure in the range of 0.0035–400 MPa. The dried fragments of the tested sample were degassed in a penetrometer to a pressure of 50 mmHg. Finally, the volumes and size distributions of the pores were calculated using the Washburn equation [59], which is as follows:

$$P_C = \frac{2\sigma \cos\theta}{r} \tag{3}$$

where

P_C—capillary pressure
σ—mercury interfacial tension
θ—contact angle
r—pore radius.

The samples were also tested in order to evaluate the mechanical compressive strength, by measuring the strength needed for destruction of the granules. The study was performed using the Tinius Olsen H 10K-S instrument (Tinius Olsen, Horsham, PA, USA). Briefly, the granules were placed between the stationary plate and the measuring head, then were put under a pressure test, while moving the head at a speed of 5 mm/s. The mechanical compressive strength is a ratio of pressure used for the destruction of the granule (N) and the diameter of the granule (mm).

In order to evaluate in vitro Ga^{3+} and Ag$^+$ release from the granules, 0.5 g of each sample was placed in a conical tube with a volume of 50 mL, then 50 mL of phosphate buffered saline (PBS) of pH = 7.4 was added. Afterwards, the tubes were placed in a water bath at 37 °C and stirred gently. The release study of silver and gallium ions was carried out for three weeks. Sample aliquots of 10 mL were taken at specific time intervals, namely 1 h, 2 h, 3 h, 6 h, 12 h, 24 h, 2 days, 5 days, 1 week, 2 weeks and 3 weeks, and replaced with the same amount of fresh PBS.

The amount of released ions was determined by inductively coupled plasma mass spectrometry (ICP-MS), using an ICP mass spectrometer, Thermo Electron X Series II (Thermo Electron Corporation, Allentown, PA, USA).

3.6. In Vitro Cytotoxicity Studies

In order to evaluate the cytotoxicity to mammalian cells, the materials (Ag-HA, Ga-HA powders and six types of granules), were tested with the following two assays: the neutral red uptake (NRU) test performed on the basis of the ISO 10993 guideline Annex A [60] and the MTT assay, based on the reduction in 3-(4,5-dimethylthiazol-2-yl)-2,5-diphenyltetrazolium bromide by the mitochondrial succinate dehydrogenase. In the NRU assay, the quantitative estimation of viable cells in the tested cultures was based on their neutral red uptake in comparison to the results obtained for untreated cells. Dead cells have no ability to accumulate the dye in their lysosomes. The MTT assay allowed us to evaluate the mitochondrial metabolic activity of the tested cultures. Only cellular oxidoreductase enzymes in living cells have the ability to reduce the tetrazolium MTT dye into an insoluble, purple formazan. Both tests were performed with the BALB/c 3T3 clone A31 mammalian cell line (mouse embryonic fibroblasts from American Type Culture Collection).

For the NRU and MTT assays, the BALB/c 3T3 cells were seeded in 96-well microplates (15,000 cells/100 µL) in DMEM (Lonza) culture medium (supplemented with 10% of calf bovine serum, 100 IU/mL penicillin and 0.1 mg/mL streptomycin) and incubated for 24 h (5% CO_2, 37 °C, >90% humidity). At the end of the incubation, each well was examined under a microscope to ensure that the cells formed a confluent monolayer. Subsequently, the culture medium was replaced by the tested extracts of materials. The extracts were prepared by incubation of the tested materials in the cell culture medium (100 mg/mL) with reduced serum concentration (5%) at 37 °C for 24 h, then shaken and sterilized by filtration. The cells were treated with four dilutions of each extract in a twofold dilution series for 24 h (three data points for each). Subsequently, the treatment medium was removed. The cells were washed with PBS and treated with the neutral red medium or MTT medium for 2 h. Then, the medium was discarded, the cells were washed with PBS and treated with desorbing fixative (ethanol and acetic acid water solution or isopropanol). The amount of neutral red medium accumulated by the cells was evaluated colorimetrically at 540 nm. The amount of insoluble purple formazan was evaluated colorimetrically at 570 nm. Polyethylene film and latex were used as the reference materials (with no cytotoxicity and high cytotoxicity, respectively). The percentage of viable cells in each well was calculated by comparing its OD_{540} or OD_{570} result with the mean result obtained for untreated cells (incubated in the same conditions with fresh culture medium). Samples were considered cytotoxic if they reduced cell survival or mitochondrial metabolic activity below 70%, compared to the untreated cells (a baseline cell viability and enzymatic activity). When the BALB/c 3T3 cell viability was not decreased below 70% across the whole range of tested dilutions of the samples, it was considered non-cytotoxic in this range of concentrations.

3.7. Antibacterial Activity Studies

3.7.1. Strains and Maintenance

The following bacterial strains from ATCC were used in this study: *Staphylococcus aureus* ATCC 25923, *Staphylococcus epidermidis* ATCC 12228, *Pseudomonas aeruginosa* ATCC 27853 and *Escherichia coli* ATCC 25922. Bacteria, maintained in microbanks at −80 °C, were cultured in Mueller–Hinton (M–H) agar medium (Biomaxima, Lublin, Poland), at 37 °C for 20–24 h and then transferred into a Mueller–Hinton (M–H) broth (Biomaxima, Lublin, Poland) for a further 24 h incubation at 37 °C. Depending on the particular experiment, bacterial inoculates were obtained either by bacterial cell scrapping from agar medium to sterile saline or by the dilution of the suspension in the culture broth prior to the experiments. Bacterial titres were selected individually for each test. The tested biomaterials were sterilized by EthO (ethylene oxide) before the tests. All tests of antibacterial activity were performed in triplicate.

3.7.2. Agar Plate Test

Wells (Ø = 7.5 mm) were drilled in Mueller–Hinton agar (Biomaxima, Lublin, Poland) in 90 mm Petri dishes (thickness: approx. 5 mm). A number of 25 ± 1 mg samples, sterilized in EtOh, were placed inside the wells to evenly cover the surface of the wells. Then, liquid Mueller–Hinton agar medium (approx. 40 °C) was poured into the wells and allowed to set. Next, 100 µL of bacterial inoculate in sterile 0.9% NaCl (0.1 McFarland standard, an equivalent of approx. 3.0×10^7 CFU (colony forming units)/mL) was placed onto the agar and evenly spread. The test was performed individually for each bacterial strain. Subsequently, plates were incubated at 37 °C for 20–24 h. Afterwards, bacterial growth inhibition zones (in mm) were measured.

3.7.3. Antibacterial Activity Test (AATCC Test Method 100-2004 "Antibacterial Finishes on Textile Materials: Assessment of Developed from American Association of Textile Chemists and Colorists")

Pellets for antibacterial activity studies were obtained by weighting 0.2 g of powder (HA, Ag-HA or Ga-HA) and forming it into a tablet, using a manual hydraulic pellet press. An evaluation of the antibacterial activity of the tested samples was performed on the basis of the AATCC test method 100-2004 for textile materials and adapted for porous ceramic materials. Briefly, samples (200 $\pm$ 15 mg of pellets formed from the obtained powders and 50 $\pm$ 2 mg of each synthesized granule) were sterilized by ethylene oxide. AgM and GaM microgranules were not subjected to this test, due to their small size and low porosity caused by high sintering temperature, which significantly reduced the accuracy of this particular measurement. A working bacterial suspension (3.0×10^7 CFU/mL) of each strain was prepared in M–H broth and diluted 250-fold in sterile 0.9% NaCl. Samples were placed on sterile microscope slides (Chemland, Stargard Szczeciński, Poland) and inoculated with a working bacterial suspension, which was completely absorbed by the samples, leaving no remaining liquid (50 µL both for 200 mg tablets and 50 mg granule portions). Then, all samples were transferred to sterile 50 mL Falcon tubes (Corning, Union City, CA, USA), which were screwed to prevent evaporation and then incubated at 37 °C for 24 h. An amount of 50 µL of working bacterial suspension of each strain was placed in another sterile 50 mL Falcon tube and treated as above, to control the viability of bacteria, without contact with the tested materials (control+). Afterwards, 5 mL of sterile 0.9% NaCl was added to all samples and vigorously shaken (1 min) to elute the bacterial cells. Samples of the collected eluate were plated onto M–H agar Petri dishes using an EasySpiral Dilute (Interscience, Saint Nom La Bretèche, France) automatic plater (each sample in triplicate). M–H agar plates with plated bacteria eluted from the samples were incubated at 37 °C for 20–28 h. CFUs were then counted for each plate using a Scan 300 colony counter (Interscience, Saint Nom La Bretèche, France). The reduction in the number of bacteria was calculated as a percentage of CFU in control+ for each bacterial strain individually.

3.7.4. Bacterial Adhesion Test

The samples (200 $\pm$ 15 mg of pellets formed from the obtained powders and 50 $\pm$ 2 mg of each synthesized granule) were sterilized by ethylene oxide in the wells of a 12-well plate (Corning, Union City, CA, USA). AgM and GaM microgranules were also not subjected to this test for the same reason as mentioned in Section 3.7.3, namely due to the difficulty in the test performance. Then, 2 mL of bacterial suspensions (3.0×10^8 CFU/mL) in M–H broth were added to each well individually for each bacterial strain. Next, the plates were protected with stripes of parafilm (Bemis, Neenah, WI, USA) to prevent evaporation at the liquid phase, then incubated in an Innova 42 incubator (New Brunswick Scientific, Enfield, CT, USA) at 37 °C, 2 h, 100 rpm. Afterwards, tablets and granules were aseptically transferred onto a sterilized Whatman filter membrane to remove the excess of the bacterial suspension, placed in sterile 50 mL Falcon tubes (Corning, Union City, CA, USA) and washed carefully 4 times with 20 mL of 0.9% NaCl, to remove all unbound bacterial cells. Then, the pieces were transferred again to the sterilized Whatman filter membrane to remove excess saline, moved to another set of 50 mL Falcon tubes and treated with 1 mL of 0.25% trypsin-EDTA solution (Sigma-Aldrich, St. Louis, MO, USA) to digest the proteins, which enable the cells to adhere to the prostheses (15 min at 37 °C). After intense vortexing (1 min), trypsin was inactivated with 4 mL of M–H broth and diluted with 5 mL of 0.9% NaCl. Finally, 50 µL of each resulting liquid was plated in triplicate on M–H agar plates using the EasySpiral Dilute plater (Interscience, Saint Nom La Bretèche, France), then the plates were incubated at 37 °C for 20–24 h and the CFU were counted on each plate using a Scan 300 counter (Interscience, Saint Nom La Bretèche, France).

3.7.5. Direct Contact with Powders Test

Sterilized HA, Ag-HA and Ga-HA powders were suspended in sterile PBS pH 7.4 to obtain the suspensions 2 mg/mL and 0.2 mg/mL, which were sonicated (Sonic-6, Polsonic, Warsaw, Poland) for 15 min for better dispersion. Bacterial working solutions were prepared for each bacterial strain, including 6.0×10^6 CFU/mL of M–H broth diluted 125-fold in sterile 0.9% NaCl. Then, the powder suspensions and bacterial working solutions were mixed in sterile 5 mL screwed tubes in the proportion 1:1 (resulting in a mixture of 1 mg/mL or 0.1 mg/mL of powder with 3.0×10^6 CFU/mL in M–H broth, diluted 250-fold in sterile 0.9% NaCl). Positive controls (control+) for each bacterial strain were prepared by mixing PBS pH 7.4 and bacterial working solutions under the same conditions as the powder suspensions. The tubes were then incubated at 37 °C, for 24 h at 5 rpm using a RM 5–30 V CAT roller mixer (Ingenieurbüro CAT M.Zipperer, Ballrechten-Dottingen, Germany). Finally, samples of the mixtures (50 µL) were plated onto M–H agar Petri dishes using an EasySpiral Dilute (Interscience, Saint Nom La Bretèche, France) automatic plater (each sample in triplicate). M–H agar plates with plated bacteria eluted from the samples were incubated at 37 °C for 20–28 h. CFUs were then counted for each plate, using a Scan 300 colony counter (Interscience, Saint Nom La Bretèche, France).

3.7.6. Antibacterial Activity in Sample Extracts

Samples of AgM and GaM microgranules were sterilized by ethylene oxide in 50 mL Falcon tubes and were then immersed in sterile PBS pH 7.4 (proportion: 0.1 g of granules/1 mL PBS) at 37 °C for 24 h in an Innova 42 incubator (New Brunswick Scientific, Enfield, CT, USA) at 100 rpm. An amount of 100 µL of the extracts was collected under sterile conditions and transferred into the wells of a 96-well plate, mixed with 100 µL of sterile M–H broth and inoculated with 10 µL of bacterial working suspension to produce a final titre of 0.75×10^5 CFU/mL. An amount of 100 µL PBS with pH 7.4, mixed with 100 µL of sterile M–H broth and inoculated with bacteria as above, served as a positive control, while the non-inoculated variant served as an assay control. Then, the plates were incubated at 37 °C and 200 rpm for 24 h in an Innova 42 incubator (New Brunswick Scientific, Enfield, CT, USA) and absorbance of the extracts was measured at 660 nm using a Synergy H4 Hybrid Microplate Reader (Thermo Electron Corporation, Allentown, PA, USA). The absorbance of the assay control was subtracted from the absorbance of the samples.

3.7.7. Statistical Analysis

Statistically significant differences between the various samples were calculated according to a one-way ANOVA with post-hoc Dunnett's test or post-hoc Tukey's test, or according to a Student's *t*-test, using GraphPad Prism 8.0.0 Software (San Diego, CA, USA). Samples were used in different numbers for various tests but at least in triplicate (details in appropriate sections).

4. Conclusions

In our work, we investigated the antibacterial activity of various materials with potential application as bone substitutes. We focused on two ionic dopants with antibacterial properties, Ag^+ and Ga^{3+}, due to their different mechanisms of action and efficiency.

Three types of granules containing silver or gallium ions (ceramic microgranules, hydroxyapatite/alginate composite granules, and granules made of ceramic microgranules and alginate) were examined. The studied samples differed in morphology, porosity and mechanical properties. The granules were tested for the release of silver and gallium ions and for their antibacterial properties. It should be noted that the granules exhibited different silver and gallium release. Two cytotoxicity tests showed that the majority of the materials are not toxic, except the Ag-HA powder and AgT granules that were found toxic in both assays. However, in the first of the dilutions in the twofold dilution series, none of these samples negatively affected the cell culture condition. The results of the antibacterial tests turned out to be promising, as all the silver-containing materials caused significant mortality of the tested bacterial strains. The antibacterial efficacy of gallium-containing materials were significantly lower.

The aim of our future work is to investigate the porous granules doped with silver or gallium as antibiotic delivery systems targeting bone tissue.

We assume that the presence of silver or gallium ions, together with an antibiotic, may improve the effectiveness of the prevention and treatment of intraoperative infections of osseous tissue.

Supplementary Materials: The following supporting information can be downloaded at: https://www.mdpi.com/article/10.3390/ijms23137102/s1.

Author Contributions: Conceptualization, J.K. and A.B.; methodology, K.P., A.Z. and A.B.; software, K.P. and L.P.; investigation, K.P., A.M., F.W., L.P. and A.B.; resources, K.P. and F.W.; writing—original draft preparation, K.P., A.B. and A.Z.; writing—review and editing, J.K. and A.B.; visualization, K.P.; supervision, J.K.; funding acquisition, J.K. and A.B. All authors have read and agreed to the published version of the manuscript.

Funding: Studies were supported by the Ministry of Education and Science in Poland within statutory activity of the Medical University of Warsaw (FW232/N/2022). Synthesis and physicochemical studies of the powders were supported by NCN Sonata Bis grant (Project NCN UMO-2016/22/E/ST5/00564). Antibacterial activity studies were supported by the Ministry of Education and Science in Poland within statutory activity of the Medical University of Lublin (DS6/2022).

Institutional Review Board Statement: Not applicable.

Informed Consent Statement: Not applicable.

Data Availability Statement: The data that support the findings of this study are available from the corresponding author upon reasonable request.

Conflicts of Interest: The authors declare no conflict of interest.

References

1. Barrere, F.; Mahmood, T.; De Groot, K.; Van Blitterswijk, C. Advanced biomaterials for skeletal tissue regeneration: Instructive and smart functions. *Mater. Sci. Eng. R Rep.* **2008**, *59*, 38–71. [CrossRef]
2. Chen, Z.-Y.; Gao, S.; Zhang, Y.-W.; Zhou, R.-B.; Zhou, F. Antibacterial biomaterials in bone tissue engineering. *J. Mater. Chem. B* **2021**, *9*, 2594–2612. [CrossRef] [PubMed]
3. Alvarez Echazú, M.I.; Perna, O.; Olivetti, C.E.; Antezana, P.E.; Municoy, S.; Tuttolomondo, M.V.; Galdopórpora, J.M.; Alvarez, G.S.; Olmedo, D.G.; Desimone, M.F. Recent Advances in Synthetic and Natural Biomaterials-Based Therapy for Bone Defects. *Macromol. Biosci.* **2022**, *22*, 2100383. [CrossRef]
4. Ambrosio, L.; Raucci, M.G.; Vadalà, G.; Ambrosio, L.; Papalia, R.; Denaro, V. Innovative Biomaterials for the Treatment of Bone Cancer. *Int. J. Mol. Sci.* **2021**, *22*, 8214. [CrossRef] [PubMed]
5. Wang, W.; Yeung, K.W. Bone grafts and biomaterials substitutes for bone defect repair: A review. *Bioact. Mater.* **2017**, *2*, 224–247. [CrossRef]
6. Borchardt, R.A.; Tzizik, D. Update on surgical site infections: The new CDC guidelines. *J. Am. Acad. Pas* **2018**, *31*, 52–54. [CrossRef] [PubMed]
7. Andersson, R.; Søreide, K.; Ansari, D. Surgical Infections and Antibiotic Stewardship: In Need for New Directions. *Scand. J. Surg.* **2021**, *110*, 110–112. [CrossRef]
8. Spina, N.T.; Aleem, I.S.; Nassr, A.; Lawrence, B.D. Surgical site infections in spine surgery: Preoperative prevention strategies to minimize risk. *Glob. Spine J.* **2018**, *8*, 31S–36S. [CrossRef]
9. European Centre for Disease Prevention Control. *Healthcare-Associated Infections: Surgical Site Infections*; ECDC: Stockholm, Sweden, 2019.
10. Li, H.-K.; Rombach, I.; Zambellas, R.; Walker, A.S.; McNally, M.A.; Atkins, B.L.; Lipsky, B.A.; Hughes, H.C.; Bose, D.; Kümin, M. Oral versus intravenous antibiotics for bone and joint infection. *N. Engl. J. Med.* **2019**, *380*, 425–436. [CrossRef]
11. Eliaz, N.; Metoki, N. Calcium phosphate bioceramics: A review of their history, structure, properties, coating technologies and biomedical applications. *Materials* **2017**, *10*, 334. [CrossRef]
12. Dorozhkin, S.V. Bioceramics of calcium orthophosphates. *Biomaterials* **2010**, *31*, 1465–1485. [CrossRef] [PubMed]
13. Samavedi, S.; Whittington, A.R.; Goldstein, A.S. Calcium phosphate ceramics in bone tissue engineering: A review of properties and their influence on cell behavior. *Acta Biomater.* **2013**, *9*, 8037–8045. [CrossRef] [PubMed]
14. Aminzare, M.; Eskandari, A.; Baroonian, M.; Berenov, A.; Hesabi, Z.R.; Taheri, M.; Sadrnezhaad, S. Hydroxyapatite nanocomposites: Synthesis, sintering and mechanical properties. *Ceram. Int.* **2013**, *39*, 2197–2206. [CrossRef]

15. Haider, A.; Haider, S.; Han, S.S.; Kang, I.-K. Recent advances in the synthesis, functionalization and biomedical applications of hydroxyapatite: A review. *Rsc Adv.* **2017**, *7*, 7442–7458. [CrossRef]

16. Prakasam, M.; Locs, J.; Salma-Ancane, K.; Loca, D.; Largeteau, A.; Berzina-Cimdina, L. Fabrication, properties and applications of dense hydroxyapatite: A review. *J. Funct. Biomater.* **2015**, *6*, 1099–1140. [CrossRef]

17. Koutsopoulos, S. Synthesis and characterization of hydroxyapatite crystals: A review study on the analytical methods. *J. Biomed. Mater. Res. Off. J. Soc. Biomater. Jpn. Soc. Biomater. Aust. Soc. Biomater. Korean Soc. Biomater.* **2002**, *62*, 600–612. [CrossRef]

18. Boanini, E.; Gazzano, M.; Bigi, A. Ionic substitutions in calcium phosphates synthesized at low temperature. *Acta Biomater.* **2010**, *6*, 1882–1894. [CrossRef]

19. Basu, S.; Basu, B. Doped biphasic calcium phosphate: Synthesis and structure. *J. Asian Ceram. Soc.* **2019**, *7*, 265–283. [CrossRef]

20. Singh, G.; Singh, R.P.; Jolly, S.S. Customized hydroxyapatites for bone-tissue engineering and drug delivery applications: A review. *J. Sol-Gel Sci. Technol.* **2020**, *94*, 505–530. [CrossRef]

21. Ratnayake, J.T.; Mucalo, M.; Dias, G.J. Substituted hydroxyapatites for bone regeneration: A review of current trends. *J. Biomed. Mater. Res. Part B Appl. Biomater.* **2017**, *105*, 1285–1299. [CrossRef]

22. Šupová, M. Substituted hydroxyapatites for biomedical applications: A review. *Ceram. Int.* **2015**, *41*, 9203–9231. [CrossRef]

23. Shirtliff, M.E.; Calhoun, J.H.; Mader, J.T. Experimental osteomyelitis treatment with antibiotic-impregnated hydroxyapatite. *Clin. Orthop. Relat. Res. (1976–2007)* **2002**, *401*, 239–247. [CrossRef] [PubMed]

24. Barras, F.; Aussel, L.; Ezraty, B. Silver and antibiotic, new facts to an old story. *Antibiotics* **2018**, *7*, 79. [CrossRef] [PubMed]

25. Abram, S.L.; Fromm, K.M. Handling (nano) silver as antimicrobial agent: Therapeutic window, dissolution dynamics, detection methods and molecular interactions. *Chem.–A Eur. J.* **2020**, *26*, 10948–10971. [CrossRef]

26. Kolmas, J.; Groszyk, E.; Kwiatkowska-Różycka, D. Substituted hydroxyapatites with antibacterial properties. *BioMed Res. Int.* **2014**, *2014*, 178123. [CrossRef]

27. McRee, A.E. Therapeutic review: Silver. *J. Exot. Pet Med.* **2015**, *2*, 240–244. [CrossRef]

28. Kędziora, A.; Speruda, M.; Krzyżewska, E.; Rybka, J.; Łukowiak, A.; Bugla-Płoskońska, G. Similarities and differences between silver ions and silver in nanoforms as antibacterial agents. *Int. J. Mol. Sci.* **2018**, *19*, 444. [CrossRef]

29. Slavin, Y.N.; Asnis, J.; Häfeli, U.O.; Bach, H. Metal nanoparticles: Understanding the mechanisms behind antibacterial activity. *J. Nanobiotechnol.* **2017**, *15*, 65. [CrossRef]

30. Trujillo, N.A.; Oldinski, R.A.; Ma, H.; Bryers, J.D.; Williams, J.D.; Popat, K.C. Antibacterial effects of silver-doped hydroxyapatite thin films sputter deposited on titanium. *Mater. Sci. Eng. C* **2012**, *32*, 2135–2144. [CrossRef]

31. Dubnika, A.; Zalite, V. Preparation and characterization of porous Ag doped hydroxyapatite bioceramic scaffolds. *Ceram. Int.* **2014**, *40*, 9923–9930. [CrossRef]

32. Dubnika, A.; Loca, D.; Salma, I.; Reinis, A.; Poca, L.; Berzina-Cimdina, L. Evaluation of the physical and antimicrobial properties of silver doped hydroxyapatite depending on the preparation method. *J. Mater. Sci. Mater. Med.* **2014**, *25*, 435–444. [CrossRef] [PubMed]

33. Gokcekaya, O.; Ueda, K.; Narushima, T.; Ergun, C. Synthesis and characterization of Ag-containing calcium phosphates with various Ca/P ratios. *Mater. Sci. Eng. C* **2015**, *53*, 111–119. [CrossRef] [PubMed]

34. Łapa, A.; Cresswell, M.; Campbell, I.; Jackson, P.; Goldmann, W.H.; Detsch, R.; Boccaccini, A.R. Gallium-and cerium-doped phosphate glasses with antibacterial properties for medical applications. *Adv. Eng. Mater.* **2020**, *22*, 1901577. [CrossRef]

35. Verron, E.; Masson, M.; Khoshniat, S.; Duplomb, L.; Wittrant, Y.; Baud'Huin, M.; Badran, Z.; Bujoli, B.; Janvier, P.; Scimeca, J.C. Gallium modulates osteoclastic bone resorption in vitro without affecting osteoblasts. *Br. J. Pharmacol.* **2010**, *159*, 1681–1692. [CrossRef]

36. Strazic Geljic, I.; Melis, N.; Boukhechba, F.; Schaub, S.; Mellier, C.; Janvier, P.; Laugier, J.P.; Bouler, J.M.; Verron, E.; Scimeca, J.C. Gallium enhances reconstructive properties of a calcium phosphate bone biomaterial. *J. Tissue Eng. Regen. Med.* **2018**, *12*, e854–e866. [CrossRef]

37. Melnikov, P.; Matos, M.d.F.C.; Malzac, A.; Teixeira, A.R.; de Albuquerque, D.M. Evaluation of in vitro toxicity of hydroxyapatite doped with gallium. *Mater. Lett.* **2019**, *253*, 343–345. [CrossRef]

38. Cassino, P.C.; Rosseti, L.S.; Ayala, O.I.; Martines, M.A.U.; Portugal, L.C.; Oliveira, C.G.d.; Silva, I.S.; Caldas, R.d.A. Potencial of different hydroxyapatites as biomaterials in the bone remodeling. *Acta Cir. Bras.* **2018**, *33*, 816–823. [CrossRef]

39. Kurtjak, M.; Vukomanović, M.; Krajnc, A.; Kramer, L.; Turk, B.; Suvorov, D. Designing Ga (iii)-containing hydroxyapatite with antibacterial activity. *RSC Adv.* **2016**, *6*, 112839–112852. [CrossRef]

40. Melnikov, P.; Teixeira, A.; Malzac, A.; Coelho, M.d.B. Gallium-containing hydroxyapatite for potential use in orthopedics. *Mater. Chem. Phys.* **2009**, *117*, 86–90. [CrossRef]

41. Dubnika, A.; Loca, D.; Rudovica, V.; Parekh, M.B.; Berzina-Cimdina, L. Functionalized silver doped hydroxyapatite scaffolds for controlled simultaneous silver ion and drug delivery. *Ceram. Int.* **2017**, *43*, 3698–3705. [CrossRef]

42. Sampath Kumar, T.; Madhumathi, K.; Rubaiya, Y.; Doble, M. Dual mode antibacterial activity of ion substituted calcium phosphate nanocarriers for bone infections. *Front. Bioeng. Biotechnol.* **2015**, *3*, 59. [CrossRef] [PubMed]

43. Nie, L.; Deng, Y.; Zhang, Y.; Zhou, Q.; Shi, Q.; Zhong, S.; Sun, Y.; Yang, Z.; Sun, M.; Politis, C. Silver-doped biphasic calcium phosphate/alginate microclusters with antibacterial property and controlled doxorubicin delivery. *J. Appl. Polym. Sci.* **2021**, *138*, 50433. [CrossRef]

44. Klug, H.P.; Alexander, L.E. *X-ray Diffraction Procedures: For Polycrystalline and Amorphous Materials*; Wiley: New York, NY, USA, 1974.
45. Landi, E.; Tampieri, A.; Celotti, G.; Sprio, S. Densification behaviour and mechanisms of synthetic hydroxyapatites. *J. Eur. Ceram. Soc.* **2000**, *20*, 2377–2387. [CrossRef]
46. Okada, M.; Furuzono, T. Hydroxylapatite nanoparticles: Fabrication methods and medical applications. *Sci. Technol. Adv. Mater.* **2012**, *13*, 064103. [CrossRef]
47. Kolmas, J.; Kalinowski, E.; Wojtowicz, A.; Kolodziejski, W. Mid-infrared reflectance microspectroscopy of human molars: Chemical comparison of the dentin–enamel junction with the adjacent tissues. *J. Mol. Struct.* **2010**, *966*, 113–121. [CrossRef]
48. Geng, Z.; Wang, R.; Zhuo, X.; Li, Z.; Huang, Y.; Ma, L.; Cui, Z.; Zhu, S.; Liang, Y.; Liu, Y. Incorporation of silver and strontium in hydroxyapatite coating on titanium surface for enhanced antibacterial and biological properties. *Mater. Sci. Eng. C* **2017**, *71*, 852–861. [CrossRef]
49. Gopi, D.; Shinyjoy, E.; Kavitha, L. Synthesis and spectral characterization of silver/magnesium co-substituted hydroxyapatite for biomedical applications. *Spectrochim. Acta Part A Mol. Biomol. Spectrosc.* **2014**, *127*, 286–291. [CrossRef]
50. Ragab, H.; Ibrahim, F.; Abdallah, F.; Al-Ghamdi, A.A.; El-Tantawy, F.; Radwan, N.; Yakuphanoglu, F. Synthesis and in vitro antibacterial properties of hydroxyapatite nanoparticles. *IOSR J. Pharm. Biol. Sci.* **2014**, *9*, 77–85. [CrossRef]
51. Lamkhao, S.; Phaya, M.; Jansakun, C.; Chandet, N.; Thongkorn, K.; Rujijanagul, G.; Bangrak, P.; Randorn, C. Synthesis of hydroxyapatite with antibacterial properties using a microwave-assisted combustion method. *Sci. Rep.* **2019**, *9*, 4015. [CrossRef]
52. Jahangir, M.U.; Islam, F.; Wong, S.Y.; Jahan, R.A.; Matin, M.A.; Li, X.; Arafat, M.T. Comparative analysis and antibacterial properties of thermally sintered apatites with varied processing conditions. *J. Am. Ceram. Soc.* **2021**, *104*, 1023–1039. [CrossRef]
53. Pang, Z.; Raudonis, R.; Glick, B.R.; Lin, T.-J.; Cheng, Z. Antibiotic resistance in Pseudomonas aeruginosa: Mechanisms and alternative therapeutic strategies. *Biotechnol. Adv.* **2019**, *37*, 177–192. [CrossRef] [PubMed]
54. Asadpoor, M.; Ithakisiou, G.-N.; Van Putten, J.P.; Pieters, R.J.; Folkerts, G.; Braber, S. Antimicrobial Activities of Alginate and Chitosan Oligosaccharides Against Staphylococcus aureus and Group B Streptococcus. *Front. Microbiol.* **2021**, *12*, 700605. [CrossRef] [PubMed]
55. Aderibigbe, B.A.; Buyana, B. Alginate in wound dressings. *Pharmaceutics* **2018**, *10*, 42. [CrossRef] [PubMed]
56. Ertesvåg, H. Alginate-modifying enzymes: Biological roles and biotechnological uses. *Front. Microbiol.* **2015**, *6*, 523. [PubMed]
57. Pajor, K.; Pajchel, Ł.; Zgadzaj, A.; Piotrowska, U.; Kolmas, J. Modifications of hydroxyapatite by gallium and silver ions—physicochemical characterization, cytotoxicity and antibacterial evaluation. *Int. J. Mol. Sci.* **2020**, *21*, 5006. [CrossRef] [PubMed]
58. Yang, J.H.; Kim, K.H.; You, C.K.; Rautray, T.R.; Kwon, T.Y. Synthesis of spherical hydroxyapatite granules with interconnected pore channels using camphene emulsion. *J. Biomed. Mater. Res. Part B Appl. Biomater.* **2011**, *99*, 150–157. [CrossRef]
59. Liu, M.; Wu, J.; Gan, Y.; Hanaor, D.A.; Chen, C. Evaporation limited radial capillary penetration in porous media. *Langmuir* **2016**, *32*, 9899–9904. [CrossRef]
60. *EN ISO 10993-5:2009*; Biological Evaluation of Medical Devices—Part 5: Tests for In Vitro Cytotoxicity (ISO 10993-5:2009), Annex A Neutral Red Uptake (NRU) Cytotoxicity Test. International Organization for Standardization: Geneva, Switzerland, 2009.

International Journal of
Molecular Sciences

Review

Current Knowledge on Interactions of Plant Materials Traditionally Used in Skin Diseases in Poland and Ukraine with Human Skin Microbiota

Natalia Melnyk [1], Inna Vlasova [1,2], Weronika Skowrońska [3], Agnieszka Bazylko [3], Jakub P. Piwowarski [1] and Sebastian Granica [1,*]

[1] Microbiota Lab, Department of Pharmacognosy and Molecular Basis of Phytotherapy, Medical University of Warsaw, Banacha 1, 02-097 Warsaw, Poland
[2] Department of Pharmacognosy, National University of Pharmacy, 53 Pushkinska Str., 61002 Kharkiv, Ukraine
[3] Department of Pharmacognosy and Molecular Basis of Phytotherapy, Medical University of Warsaw, Banacha 1, 02-097 Warsaw, Poland
*   Correspondence: sgranica@wum.edu.pl; Tel.: +48-225-720-9053

Abstract: Skin disorders of different etiology, such as dermatitis, atopic dermatitis, eczema, psoriasis, wounds, burns, and others, are widely spread in the population. In severe cases, they require the topical application of drugs, such as antibiotics, steroids, and calcineurin inhibitors. With milder symptoms, which do not require acute pharmacological interventions, medications, dietary supplements, and cosmetic products of plant material origin are gaining greater popularity among professionals and patients. They are applied in various pharmaceutical forms, such as raw infusions, tinctures, creams, and ointments. Although plant-based formulations have been used by humankind since ancient times, it is often unclear what the mechanisms of the observed beneficial effects are. Recent advances in the contribution of the skin microbiota in maintaining skin homeostasis can shed new light on understanding the activity of topically applied plant-based products. Although the influence of various plants on skin-related ailments are well documented in vivo and in vitro, little is known about the interaction with the network of the skin microbial ecosystem. The review aims to summarize the hitherto scientific data on plant-based topical preparations used in Poland and Ukraine and indicate future directions of the studies respecting recent developments in understanding the etiology of skin diseases. The current knowledge on investigations of interactions of plant materials/extracts with skin microbiome was reviewed for the first time.

Keywords: dermatology; skin; microbiota; interaction; topical; skin diseases; plant materials; phytotherapy; keratinocytes; fibroblasts

Citation: Melnyk, N.; Vlasova, I.; Skowrońska, W.; Bazylko, A.; Piwowarski, J.P.; Granica, S. Current Knowledge on Interactions of Plant Materials Traditionally Used in Skin Diseases in Poland and Ukraine with Human Skin Microbiota. *Int. J. Mol. Sci.* **2022**, *23*, 9644. https://doi.org/10.3390/ijms23179644

Academic Editors: Geoffrey Brown, Enikö Kallay and Andrzej Kutner

Received: 25 July 2022
Accepted: 21 August 2022
Published: 25 August 2022

Publisher's Note: MDPI stays neutral with regard to jurisdictional claims in published maps and institutional affiliations.

1. Introduction

More and more drugs, dietary supplements, and cosmetic products appear on the pharmaceutical and cosmetic market, which contain medicinal plant materials or substances of plant origin [1,2]. In recent years, phytotherapeutic preparations are gaining greater importance in solving many problems in dermatology and cosmetology [3]. Severe skin ailments require the application of antibiotics, steroids, or calcineurin inhibitors. However, in mild skin disorders, the topical application of plant-based remedies in various pharmaceutical forms, such as raw infusions, creams, ointments, balms, and tinctures can be effective and successfully prevent further disease development [4,5].

Human skin is a complex organ that accounts for about 15% of the total body weight of an adult and has a surface area of 1.5–2 m^2. This organ is responsible for many vital functions [6]. It protects against external factors; participates in thermoregulation, metabolism, the regulation of fluid balance, and body shape maintenance; and eliminates toxins from the body by sweat excretion [7]. It consists of several layers, such as the living tissue of the dermis, epidermis, and the outer-facing layer, which is called the stratum corneum [8].

On the microanatomic level, skin is composed of keratinocytes, Langerhans cells, fibroblasts, mast cells, macrophages, endothelial cells, and lymphocytes, which form a complicated and fine-tuned organization such as the skin immune system [9,10]. Likewise, the superficial layer of the skin is home to millions of bacteria, fungi, and viruses that compose the skin microbiota [8]. This microbial ecosystem supports many skin functions, including metabolism, vitamin synthesis, protection against pathogen invasion, immunity development, and regulation [11].

Currently, an increase in the incidence of skin diseases worldwide is observed, which is mostly linked to the adverse factors of modern civilization [12]. For instance, 42 of the 145 surveyed people in Poland reported past or present skin disorders [13]. In Ukraine, the morbidity rate of skin diseases increases every year. The currently reported problems mostly relate to dermatitis, atopic dermatitis, eczema, psoriasis, wounds, and others. Many factors influence skin well-being, with the highest contribution attributed to domestic detergents, cosmetics, and antiseptics.

The present review aims to summarize the current information on the plant materials that are contained in pharmaceutical and cosmetic preparations available in Poland and Ukraine and marketed as non-prescription medicines or medical products. The paper presents the current knowledge regarding the traditionally used plant raw materials related to their influence on the skin and skin microbiota. Medicinal plants used in treating wounds, burns, dermatitis, atopic dermatitis, eczema, and other skin inflammatory diseases have been included.

Based on the scientific database Scopus, a graphical representation of the publication trend in the field of the present review was generated. The trend in the number of papers focused on plant extracts or materials plants and skin shows a significant increase over the last 20 years. Around 190 reports were published in 2001 according to Scopus, and in 2021, it was over 1300 original papers and reviews. Similarly, a dynamic increase in the interest in research devoted to plant preparations and microbiota was observed (Figure 1) in 2001 when 4 papers matching chosen keywords were recorded compared to 377 in 2021. This analysis confirms that the general interest in the different aspects of the interaction of plant materials with skin increases, and the subject is worth further investigations. This trend can be explained by the fact that natural therapies are gaining much interest in the context of changes in the lifestyle of the population around the world.

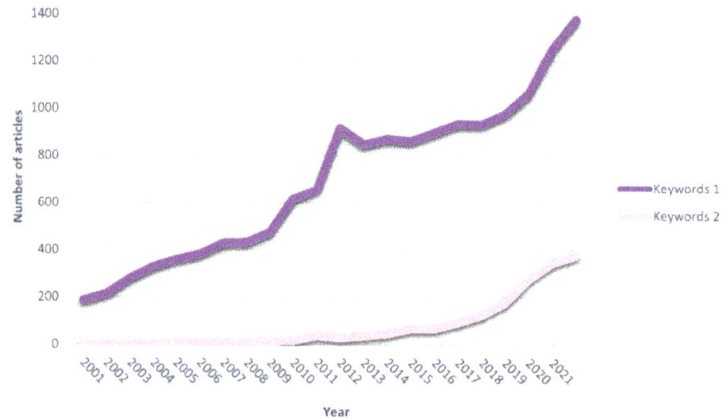

Figure 1. Graphical representation of the publication trend in the field of using plants related to their influence on the skin and microbiota. A graph was created based on the Scopus database, keywords 1 (plant material or plant extract and skin) and keywords 2 (plant material or plant extract and microbiota) were used.

2. Methods Used in the Review

A literature search was conducted using scientific databases such as PubMed, Scopus, and Wiley Online Library for relevant studies with the keywords "microbiome", "microbiota", "commensal", "topical", "plant materials", and "phytotherapy", "dermatology", "skin", "skin disorders", "keratinocytes", "fibroblasts". All search terms were used in various combinations, and studies were screened for relevance based on their abstracts. Studies written in English, Polish, and Ukrainian languages were considered.

The selection of non-prescription medicines and medical products was based on the search of topically used remedies containing material of plant origin in the Ukrainian directory of drugs "Compendium" and on the Polish pharmacy websites. Moreover, the Ukrainian online service "Tabletki.ua", which provides information on the availability of medical preparations and other pharmaceutical products in pharmacies, was considered.

3. Human Microbiota in the Skin Inflammation Process

The skin is the habitat and a source of nutrients for various symbiotic, commensal, and pathogenic microorganisms described as skin microbiota [7]. Early studies showed that abundant bacterial genera on the surface layers of the human skin include *Staphylococcus*, *Propionibacterium*, *Micrococcus*, and *Corynebacterium* [14]. The composition of the human microbiome similarly varies quantitatively and qualitatively according to the body site on which it is located, depending on the distinctive characteristics (pH, moisture, salinity, and sebum content), and may also vary due to other factors (e.g., genotype, age, and sex) [5,12]. The recent surveys have a greatly advanced understanding of the host-symbiont and host-pathogen relationships and established that the skin microbiota plays a beneficial role, much like the gut microbiota, indicating that the bacteria present on our skin have similar functions in immune regulation and disease pathogenesis [11,13,14].

Commensal bacteria can passively occupy a similar ecological niche to a pathogenic microbe, thus impeding its skin colonization. Additionally, commensal bacteria on human skin can selectively induce antimicrobial peptides production and provide a protective effect in vivo when administered before the infectious challenge. The human microbiome may also modulate the immune system, directing it to eliminate the disease-causing factor [14,15].

4. Interaction of Plant Origin Material with Microbiota

The interaction of microorganisms with plants can be considered from two sides. On the one hand, there is the influence of plants on microbes' growth and metabolic functions, as medicinal plants contain extractable biochemical and bioactive compounds, which can target certain viruses, bacteria, or fungi [16]. On the other hand, a metabolically active microbial community can alter the chemical structure and composition of natural products applied to the skin.

To date, there are a lot of known natural products which act effectively against bacteria, fungi, viruses, or protozoa. For instance, phenolics have antifungal and antiviral properties, and it is established that their overall harmfulness to microorganisms is associated with the measure of hydroxyl bunches and their locations on the phenol bunch. Low dosages of phenols (0.032%, 320 g/mL) destroyed fast-developing cultures of *Staphylococci* and *Streptococci* [17]. Quinones have been found to frame irreversible edifices with nucleophilic amino acids in proteins, leading to protein inactivation and function loss resulting in antibacterial effects. Quinones may also make it difficult for bacteria to obtain substrates [18]. The antimicrobial mode of action of tannins is linked to their ability to inactivate microbial adhesins, enzymes, and cell envelope transport proteins [19].

Several in vivo and in vitro examinations demonstrated that different plant preparations repress bacterial species found in cutaneous diseases. Chamomile essential oil and bisabolol were found to have activity mostly against Gram-positive bacteria, *Staphylococcus aureus*, *Bacillus subtilis*, and the fungus *Candida albicans*. Aqueous extracts from *Allium cepa* showed antifungal activity against *Malassezia furfur*, *Candida albicans*, other *Candida* sp., and other dermatophyte species [17].

In terms of the influence of microbiota on plant materials, it should be mentioned that the literature review showed a lack of information about the skin microbiota metabolism. However, nowadays, many investigations on microbiota residing in the gut have been described. For instance, human and swine microbiota transformed natural products in the goldenrod infusion into smaller molecules, mainly phenylpropanoid acid derivatives [20].Twenty metabolites were detected and characterized after incubating the linden flower extract with human gut microbiota [21]. All changes in the chemical composition of the raw plant material caused by the interaction with microbiota can lead to potentially active compounds responsible for their bioactivity in vivo. This is also evidenced by experiments on the gut microbiota-derived metabolites of ellagitannins-urolithins. It was established that urolithins inhibit proinflammatory cytokines expression in RAW 264.7 macrophages, which has an important role in inflammatory bowel diseases [22].

5. Phytotherapy in Skin Diseases

Patients and physicians widely use phytotherapy throughout the whole world. This type of therapy is as old as humankind. Plant-derived drugs are in demand because of several advantages, such as often having fewer side effects and better patient tolerance. Apart from this, treatment with natural products is more affordable to patients than chemical medicines.

Clinical trials and in vitro and in vivo experiments were conducted for many plant materials, showing their effectiveness in inhibiting the formation of cytokines and eicosanoids, preventing the inflammatory reaction cascade. However, still, the use of most herbal medicines is based solely on their longstanding traditional use in folk medicine [23].

Over the years, plants that deserve special attention in treating skin diseases have been identified. Below is a list of plant materials that are the most popular in the traditional treatment of skin diseases in Poland and Ukraine (Table 1).

Table 1. Traditionally used plant material in skin disorders in Poland and Ukraine.

Botanical Name	Common Name	Part Used	Application Properties [24–26]	Traditional Use/Therapeutic Area [24–26]
Achillea millefolium L.	yarrow	herb	antimicrobial, anti-inflammatory, antioxidant, antiproliferative, and cytotoxic	small superficial wounds
Agrimonia eupatoria L.	agrimony	herb	anti-inflammatory and antioxidant	minor inflammation and small, superficial wounds
Allium cepa L.	onion	bulbs	anti-inflammatory, antioxidant, and antimicrobial	insect bites treatment, wounds, minor burns, boils, warts, and treatment of bruises
Aloe vera (L.) Burm.f.	aloes	leaves	antiproliferative	bedsores
Artemisia absinthium L.	wormwood	herb	antiseptic and anti-inflammatory	boils, wounds, and bruises
Arctium lappa L.	burdock	root	antibacterial, antiviral, antioxidant, anti-inflammatory, antiallergic, antimutagenic, and antiproliferative	seborrheic skin conditions
Arnica montana L.	arnica	flowers	anti-inflammatory, antimicrobial activity, and antioxidant, immunotoxic, cytotoxic, and anti-platelet	inflammation
Bidens tripartita L.	three-lobe beggartick	herb	anti-inflammatory, bactericidal, and hemostatic	diathesis
Calendula officinalis L.	calendula	flowers	wound healing, antiviral, antimicrobial, anti-inflammatory, antioxidant, photoprotective, repellent, and anti-irritative	sunburn, minor wounds, and minor inflammations
Chelidonium majus L.	greater celandine	herb	antiseptic and fungicidal	warts, callus, corns, pimples, shingles, eczema, and skin tumors
Equisetum arvense L.	field horsetail	herb	antibacterial, antioxidant, anti-inflammatory, and wound healing	superficial wounds

Table 1. *Cont.*

Botanical Name	Common Name	Part Used	Application Properties [24–26]	Traditional Use/Therapeutic Area [24–26]
Hamamelis virginiana L.	hamamelis	leaves, bark	antibacterial, anti-inflammatory, antiviral, and radical-scavenging	minor inflammation and dryness
Hippophae rhamnoides L.	sea-buckthorn	fruits	anti-inflammatory, bactericidal, analgesic, and epithelializing properties	rashes, eczema, burns, bedsores, frostbite, ulcers that do not heal well, and radiation skin diseases
Hypericum perforatum L.	st. John's wort	herb	anti-inflammatory wound healing	skin disorders and minor wounds
Linum usitatissimum L.	linseed	seeds	antiproliferative	trophic ulcers, burns, and radiation damage
Melissa officinalis L.	melissa	leaves	anti-inflammatory, antiviral, antimicrobial, antioxidant, and anti-inflammatory	external remedies for herpes
Matricaria chamomilla L.	chamomile	flowers	anti-inflammatory, antimicrobial, and wound healing	irritation, minor inflammation, sunburn, superficial wounds, and furuncles
Plantago lanceolata L.	ribwort plantain	leaves, herb	anti-inflammatory, antibacterial, antiviral, antioxidant, and analgesic	Boils, edema, and insect bites
Potentilla erecta L.	tormentil	rhizomes	antibacterial, antioxidant, and antitumor	minor inflammations
Quercus robur L.	common oak	bark	astringent and anti-inflammatory	minor inflammation
Salvia officinalis L.	sage	leaves	antimicrobial, antioxidant, anti-inflammatory, cytoprotective, and wound healing	minor inflammations
Sophora japonica L.	Japanese pagoda tree	fruits	hemostatic, antiseptic, and wound healing	wounds, trophic ulcers, and seborrheic dermatitis
Symphytum officinale L.	comfrey	roots	anti-inflammatory, wound healing, and antibacterial	pain and inflammation
Trigonella foenum-graecum L.	fenugreek	seeds	anti-inflammatory and antiulcer	minor inflammations
Urtica dioica L.	nettle	leaves, herb	anti-inflammatory, analgesic, and local anesthetic	seborrheic skin conditions
Viola tricolor L.	wild pansy	herb	antibacterial, antioxidant, cytotoxic activity against cancer cells, anti-nociceptive, and anti-inflammatory	skin disorders, minor wounds, and mild seborrheic skin conditions

Different herbal parts are used in treating skin ailments, but the most usable are the above-ground parts, such as herbs, leaves, or flowers. Among all plant effects are frequently noted anti-inflammatory, bactericidal, and wound healing effects. The mostly described plants are traditionally used for minor superficial wounds, burns, minor inflammations of different etiology, irritations, bedsores, and ulcers. However, for example, *Chelidonii herba* is applied to warts, corns, and pimples due to their antiseptic and fungicidal effects, and *Melissae folium*, like antiviral, antimicrobial external remedies for herpes.

Polish and Ukrainian pharmaceutical markets have been analyzed and preparations marketed as non-prescription medicines or medical products containing plant material from Table 1 as an active ingredient are presented in Table 2.

In Poland, as in Ukraine, the most popular plant material for treating skin diseases is marigold flowers (*Calendulae flos*). It is in pharmaceutical forms, such as ointments or tinctures, and is produced by different manufacturers. It is used externally in treating mild inflammatory conditions of the skin and ancillary in treating minor skin injuries. At Polish pharmacies, a large number of remedies from linseed, chamomile, and tormentil are found. Tormentil is presented mostly in ointments ("Tormentiol", "Tormentile Forte", and "Tormentillae unguentum compositum"), linseed in ointments and creams ("Linomag"), and chamomile in ointments, gels, and tinctures ("Kamagel" and "Azulan"). Ukrainian pharmacies have some distinguishing topical remedies of *Hippophae rhamnoides*, such as "Sea buckthorn ointment", "Olasol spray", and some balms.

The plant materials were selected for the literature surveys based on the above approach. The plants which are considered effective in the treatment of inflammatory skin diseases are discussed below. Additionally, to information on the chemical composition, activity studies, and traditional use, data were also searched for in studies of the effect of skin microbiota on natural products contained in the listed plant materials and/or the analyses of changes in the composition of the microbiome as a result of the use of preparations from the raw materials mentioned above.

5.1. Allium cepa L. (Onion Bulbs)

The plant substance of *Allii cepae bulbus* consists of thick and fleshy leaf sheaths and leaf approaches from *Allium cepa* L. (Amaryllidaceae). Biologically active compounds of onion bulbs are flavonoids and sulfur-containing compounds. In traditional medicine, onions have been used externally to treat insect bites, wounds, minor burns, boils, warts, and bruises [27]. Many studies have demonstrated the antioxidant and antimicrobial activity of the extracts [28]. The study using human fibroblasts showed that the onion extract inhibited their proliferation, induced apoptosis, and decreased expression of β1 integrin, which may be beneficial in treating keloid and hypertrophic scars [29]. However, data on the activity of onion extracts in treating hypertrophic scars are inconclusive.

The effect of Mederma (Merz Pharma, Frankfurt, Germany) gel, containing *Allium cepa* as the active ingredient, was investigated in a rabbit hypertrophic scar model. No significant effect of the gel on the reduction in scar hypertrophy, vascularization, or inflammation was determined, but an improvement in the organization of dermal collagen was observed [30]. The effectiveness of the same gel was investigated in patients with new surgical scars compared to petrolatum emollient. No significant differences were found in the activity of these formulations in the treatment of scars as assessed by redness, itching, burning, pain, thickness, or overall cosmetic appearance [31]. Another study comparing the activity of Mederna with a petroleum-based emollient assessed the appearance and symptomatology of postoperative scars. The onion extract gel was ineffective, no difference was found in the evaluation of the redness and itching of the scar after one month of use. In contrast, a reduction in redness has been observed in patients using the emollient [32].

Table 2. Selected preparations marketed as non-prescription medicines or medical products containing plant material as an active ingredient in Poland (PL) and Ukraine (UA).

Plant Material	Country Code	Marketed Product	Active Ingredient(s)	Manufacturer	Pharmaceutical Form	Registered Indications
Allium cepa **L.** **(onion bulbs)**	PL	Cepan	in 100 g of cream: 20.0 g ethanolic extract of *Allium cepa*, 5.0 g extract of *Matricaria chamomilla*, 5000.0 IU sodium heparin, and 1.0 g allantoin	Unia	cream	scars and keloids after burns and surgery; treatment of contractures; treatment of scarring of the eyelids; treatment of scars from boils, ulcers, and acne
	PL	Contractubex	in 100 g of gel: 10.0 g extract of *Allium cepa*, 5000.0 IU sodium heparin, and 1.0 g allantoin	Merz Pharma	gel	scars restricting movement, enlarged (hypertrophic, swollen, and keloid-shaped), unaesthetic postoperative scars, amputation scars, burn and accident scars, contractures of, e.g., fingers (Dupuytren's contracture), tendon contractures caused by injuries, and scar shrinkage
Arnica montana **L.** **(arnica flowers)**	PL	Arnithei	in 100 g of gel: 24.0 g arnica tincture	Dr. Theiss Naturwaren	gel	relieve bruises, sprains, and local muscle pain
	PL	Uzarin	in 100 g of gel: 1.0 g extract of *Arnica montana*, 1.0 g extract of *Calendula officinalis*, and 1.0 g aluminum acetate	Nes Pharma	gel	bruises, swellings, first-degree burns, and insect bites
Calendula officinalis **L. (marigold flowers)**	UA	Marigold ointment	tincture of calendula flowers	Viola DKP Pharmaceutical Factory Lubnyfarm Ternopharm FITOPHARM	ointment	burns, cuts, cracks in the skin, and purulent wounds
	PL	Marigold ointment	ethanolic extract of *Calendula officinalis*	Elissa	ointment	mild inflammation of the skin, as an aid to the healing of minor wounds

Table 2. *Cont.*

Plant Material	Country Code	Marketed Product	Active Ingredient(s)	Manufacturer	Pharmaceutical Form	Registered Indications
	PL	Marigold ointment	extract of *Calendula officinalis* (extraction solvent: liquid paraffin)	Ziaja	ointment	symptomatic treatment of mild skin inflammations and as an auxiliary in the treatment of minor wounds (abrasions of the epidermis)
	UA	Marigold tincture	tincture of calendula flowers	Lubnypharm / FITOPHARM / Viola / Vishpa	tincture	drugs that promote wound healing
	PL	Marigold tincture	tincture of marigold flowers	FITOPHARM	tincture	symptomatic treatment of mild skin inflammations (such as sunburn) and as an adjunct in the treatment of minor skin wounds and mild inflammation of the mouth and throat
Chelidonium majus **L.** **(greater celandine herb)**	UA	CHYSTOTIL	celandine extract	Khimpharmzavod Chervona Zirka	cream	bactericidal and wound healing
	UA	CHYSTOTIL	oil extracts of flowers, leaves, and roots of celandine	NATURE LIFE	ointment	analgetic, anti-inflammatory, and bactericidal
Hippophae rhamnoides **L.** **(sea-buckthorn fruits)**	UA	Sea buckthorn ointment	sea buckthorn oil	Fitolic	ointment	healing (scarring) of wounds
	UA	Olasol spray	sea buckthorn oil—5.40 g; chloramphenicol—1.62 g; benzocaine—1.62 g; and boric acid—0.27 g	STOMA	spray	infected wounds, including long-term non-healing burns, trophic ulcers, and skin grafts

Table 2. *Cont.*

Plant Material	Country Code	Marketed Product	Active Ingredient(s)	Manufacturer	Pharmaceutical Form	Registered Indications
	UA	Mintalon	propolis, mummy, sea buckthorn oil, wheat germ oil, natural honey, geranium oil, terpene oil, vaseline oil, lecithin, pine resin, birch tar, camphor, and vitamin E	MINTA	balm	wounds, mechanical damage to the body surface, bruises; thermal damage (burns and frostbite); inflammation and purulent processes; bedsores; bumps; animal and insect bites; dryness and cracks of the skin; skin irritation; prevention of negative effects in frost, sun, wind; and moisturizing and normalizing skin nutrition
	UA	Reskinol	terpene oil, propolis, mummy, lecithin, sea buckthorn oil, pine resin, wheat germ oil, birch tar, natural honey, camphor, geranium oil, and vitamin E	Botany	balm	from pain and inflammation
Linum usitatissimum L. (linseed)	PL	Linomag	virgin linseed oil	Ziołolek	Ointment, cream, or liquid	eczema, blemishes, and nappy rash: in states of excessive dryness of the skin; and to relieve the symptoms of psoriasis
	PL	Poldermin Hydro	*Linum usitatissimum* seed extract, *Avena sativa* extract, xylitol, and β-glucan	Polfa Tarchomin	Cream	intensively moisturizes the skin, softens it, and eliminates itching. In addition, it creates a protective film on the skin surface, accelerates the healing process of irritations, and reduces skin peeling
	PL	Aquastop Radioterapia	linseed oil and allantoin	Ziołolek	Cream	for skin care during and after radiotherapy

Table 2. *Cont.*

Plant Material	Country Code	Marketed Product	Active Ingredient(s)	Manufacturer	Pharmaceutical Form	Registered Indications
Matricaria chamomilla **L.** **(chamomile flowers)**	PL	Chamomile ointment	ethanolic extract of *Matricaria chamomilla*	Elissa	ointment	skin inflammation
	PL	Kamagel	glycolic extract of *Matricaria chamomilla* and aluminum acetate	KRKA	Gel	inflammatory symptoms in various inflammations of the skin
	PL	Azulan	ethanolic extract of *Matricaria chamomilla*	Herbapol	Tinctura	in inflammation of the skin and mucous membranes, e.g., for rinsing in inflammations of the mouth and throat and inflammation of the gums
Potentilla erecta **L.** **(tormentil roots)**	PL	Tormentil complex ointment (Tormentillae unguentum compositum)	in 100 g of ointment: 3.0 g liquid extract of the tormentil and 2.0 g ammonium bituminosulphonate 20.0 g zinc oxide	Unia / Amara / Ziaja / Prolab	Ointment	treatment of minor skin lesions, such as skin abrasions and scrapes
	PL	Tormentile forte	in 100 g of ointment: 3.0 g liquid extract of the tormentil, 2.0 g ammonium bituminosulphonate, 20.0 g zinc oxide, and 1.0 g borax (sodium tetraborate decahydrate)	Farmina	Ointment	skin lesions such as eczema lesions, first-degree burns, and mild acne vulgaris
	PL	Tormentiol	in 100 g of ointment: 2.0 g liquid extract of the tormentil, 2.0 g ammonium bituminosulphonate, 20.0 g zinc oxide, and 1.0 g borax (sodium tetraborate decahydrate)	Omega pharma	Ointment	minor skin damage such as abrasions and scratches. Incidentally, in purulent lesions and skin inflammations
Quercus robur **L.** **(common oak bark)**	UA	BIOFLORIN	oak bark extract, coriander essential oil, and jojoba oil	Khimpharmzavod Chervona Zirka	cream	anti-inflammatory and wound healing

Table 2. *Cont.*

Plant Material	Country Code	Marketed Product	Active Ingredient(s)	Manufacturer	Pharmaceutical Form	Registered Indications
Salvia officinalis **L. (sage leaves)**	UA	Tinctura Salviae	tincture of leaves salviae	DKP Pharmaceutical Factory	tincture	inflammation of the mucous membranes of the mouth, gums (stomatitis, gingivitis, periodontitis), pharynx, tonsils (pharyngitis, sore throat), upper respiratory tract, and infected wounds, cuts, and skin burns
	PL	Sage ointment	ethanolic extract of *Salvia officinalis*	Elissa	ointment	skin inflammation
Sophora japonica **(Japanese pagoda tree fruits)**	UA	Tinctura sophorae japonicae	tinctura of Japanese sophora fruit	FITOPHARM	tinctura	antiseptics and disinfectants; purulent inflammatory processes (wounds and trophic ulcers)
Symphytum officinale **L. (comfrey roots)**	UA	Ointment Dr. Taissa with Comrey	tincture of comfrey and tocopherol acetate	Dr. Theiss Naturwaren GmbH	ointment	the anti-inflammatory, analgesic effect, and promote the formation of calluses
	UA	Bainvel Comfrey Dr. Theiss	tincture of comfrey	Dr. Theiss Naturwaren GmbH	cream	degenerative-dystrophic and inflammatory diseases of the joints, as well as for recovery after sports and excessive or prolonged physical exertion
	UA	Ointment with Comfrey	comfrey root tincture and tocopherol acetate (vitamin E)	DKP Pharmaceutical Factory	ointment	pain in the joints, back, lower back with radiculitis, osteochondrosis, and arthritis. Sports, domestic injuries and bruises, sprains, and closed bone fractures; and dryness and cracked skin
Another/Complex	UA	Wundehil ointment	propolis, calendula, gooseberry foxglove, Japanese sophora fruit, and yarrow herb	AEM	ointment	remedies for wounds and ulcers

Table 2. *Cont.*

Plant Material	Country Code	Marketed Product	Active Ingredient(s)	Manufacturer	Pharmaceutical Form	Registered Indications
	UA	TRAUMEEL® S	*Achillea millefolium* 0.09 g, *Aconitum napellus* D1 0.05 g, *Arnica montana* D3 1.5 g, *Atropa belladonna* D1 0.05 g, *Bellis perennis* 0.1 g, *Calendula officinalis* 0.45 g, *Echinacea* 0.15 g, *Echinacea purpurea* 0.15 g, *Hamamelis virginiana* 0.45 g, Hepar sulfuris D6 0.025 g, *Hypericum perforatum* D6 0.09 g, *Matricaria recutita* 0.15 g, Mercurius solubilis Hahnemanni D6 0.04 g, and *Symphytum officinale* D4 0.1 g	Biologische Heilmittel Heel GmbH	ointment	bedsores, burns, pityriasis, and trophic ulcers
	UA	Express BITE	*Chamomila recutita extract, Melaleuca alternifolia.* pantpenol, *Aloe arborescens extract, and Eugenia caryophyllus oil*	Georg BioSystems	cream	to eliminate skin itching
	UA	Express Burn	Chamomila extract Aloe extract Shea butter Tea tree cil Colloidal silver D-panthenol	Georg BioSystems	cream	treatment wounds and scars
	UA	UGRIN	*Millefolii herb, Menthae folia, Calendulae officinalis flores, Tanaceti flores, Lavandulae herba, Chelidoni herba,* and *Chamomillae recutitae flores,*	Khimpharmzavod Chervona Zirka	tincture	wound healing, anti-inflammatory, and antimicrobial action

The effectiveness of this plant material in treating scars after minor dermatological procedures has been proven for the occlusive overnight intensive patch medical device containing onion bulbs extract and allantoin. After 24 weeks, a clear improvement in the appearance of the scar was observed, assessed with the Patient and Observer Scar Assessment Scale and a Global Aesthetic Improvement Scale [33]. The beneficial effect in the treatment of scars has also been confirmed for other combined preparations containing *Allium cepa* extract. The gel enriched with allantoin and heparin (Contractubex, Merz Pharma, Frankfurt, Germany) improved vascularization, pigmentation, and the overall appearance of the scar according to the Vancouver Scar Scale [34]. The use of a gel containing *Allium cepa* extract, allantoin, and pentaglycan (Kaloidon gel, Laboratorio Farmacologico Milanese SRL, Caronno Pertusella, Italy) for 24 weeks successfully reduced neoangiogenesis in patients with hypertrophic scars and keloids, resulting in the clinical improvement of skin lesions [35]. Moreover, using a patch containing 10% *Allium cepa* extract, 1% allantoin, and 4% pentaglycan (Kaloidon patch, Laboratorio Farmacologico Milanese SRL, Caronno Pertusella, Italy), after 24 weeks, showed beneficial effects according to the Patient and Observer Scar Assessment Scale. In addition, significantly improved skin scar thickness and vascularization were observed after 12 weeks [36].

There are two medicinal products available on the Polish market, containing *Allium cepa* extract, allantoin, and heparin. It is a Contractubex (Merz Pharma, Frankfurt, Germany) gel and Cepan (Unia, Warsaw, Poland) cream, which additionally contains chamomile extract. In the literature, no reports on the interactions of extracts from onion bulbs with skin microbiota were found.

5.2. Aloe vera L. (Aloes Leaves)

Aloe vera is a plant that belongs to the Asphodelaceae family. *Aloe vera* leaf gel is mainly used in dermal ailments. This gel is rich in polysaccharides. Glucomannan, acetylated glucomannan, galactogalacturan, glucogalactomannan, and acemannan had been extracted and described from this *Aloe* species [37]. Moreover, such sterols as lupeol, campesterol, and β-sitosterol were also found [37].

In the past, the influence on skin cells was investigated. In vitro pharmacological studies determined that the effect of *Aloe vera* gel and its compounds on HaCaT cells lies in decreasing photodamage; maintaining membrane integrity; reducing the levels of TNF-α, IL-8, IL-12 and p65; and increasing IκB-α protein expression. *Aloe vera* gel, in turn, increases the wound healing ability, number of cells, keratinocyte proliferation and differentiation, and cell surface expression of adhesion molecules (β1-integrin, α6-integrin, β4-integrin, and E-cadherin) in HEKa [38]. In vivo pharmacological studies on *Aloe vera* were also conducted and showed an increasing amount of fibroblasts, TGF-β gene expression, wound closure and skin tensile strength, collagen deposition, and wound healing activity as re-epithelialization and angiogenesis [38]. Moreover, it was established that *Aloe* sterols reduce skin dryness, epidermal thickness, wrinkle formation, and pro-inflammatory cytokines levels, and lupeol, campesterol, and β-sitosterol are significantly anti-inflammatory in wounded mice [37]. Polysaccharides isolated from *A. vera* help to regulate the wound healing activity, inducing matrix metallopeptidase (MMP)-3 and metallopeptidase inhibitor-2 gene expression during the skin wound repair in rats [39].

In most cases, aloes occur in complex medical products as additional ingredients. On the Ukrainian market, it is presented in creams "Express BITE" and "Express Burn" (Georg BioSystems, Kirovograd, Ukraine).

5.3. Arnica montana L. (Arnica Flowers)

Arnica montana is a widely used therapeutic plant belonging to the Asteraceae family. This plant possesses numerous medicinal activities due to such constituents as flavonoids, sesquiterpene lactones (metacryl, isobutyryl, tygloyl, methacryloyl, and isovaleryl helenalin derivatives), acetylenes, hydroxycoumarines (umbelliferone and scopoletin), phenyl acrylic acids, essential oil components, and phenolic acids (chlorogenic and caffeic acid). It is also known to contain pyrrolizidine alkaloids (tussilagin and isotussilagin) [40]. It has been used for centuries in dermatology as an antiphlogistic, antibiotic, and anti-inflammatory remedy [41]. The pharmaceutical form in traditional topical use is presented as herbal preparations in semi-solid and liquid dosage forms for cutaneous use [42].

Lyss et al. investigation shows that the main anti-inflammatory sesquiterpene lactone from arnica, helenalin, modifies the NF-κB/IκB complex, preventing the release of IκB [38,40]. Some results present that arnica reduced the UVB-induced inflammatory response as demonstrated by the inhibition of myeloperoxidase activation, a decrease in NF-κB levels, and a reduction in proinflammatory cytokines levels (IL-1β, IL-6, TNF-α, and IFN-γ) in in vivo studies [43]. In some in vitro experimental models, the production of IL-6, IL-8, and TNF-α pro-inflammatory cytokines was also measured. The secretion of IL-6, IL-8, and TNF-α in an H_2O_2-stressed fibroblast cell culture decreased, which indicates the cytoprotective effect against cell membrane oxidative damage and higher anti-inflammatory activity [42,44].

Arnica is a popular and characteristic plant material for the Polish market and two OTC medicines with extracts from this plant material are available at Polish pharmacies. "Arnithei" gel (Dr. Theiss Naturwaren, Homburg, Germany) contains an arnica tincture and is used for relieving bruises, sprains, and local muscle pain. "Uzarin" gel (Nes Pharma, Tarnów, Poland) contains the extract of *Arnica montana* and *Calendula officinalis* and is applied for bruises, swellings, first-degree burns, and after insect bites. In some sources, the beneficial effect of arnica extracts on the composition of skin microbiome is mentioned; however, there are no fine basic studies supporting these statements [45].

5.4. Calendula officinalis L. (Marigold Flowers)

Calendula officinalis (Asteraceae) is one of the most popular plants used clinically throughout the world [46]. This plant contains triterpene saponins (2–10%), mainly oleanolic acid glycosides; free and esterified triterpene alcohols, especially faradiol 3-mono- and diesters; carotenoids (up to 3%): α- and β-carotene, lutein, and rubixanthin; flavonoids (0.3–0.8%) based on quercetin, quercitrin and isorhamnetin; polysaccharides; sterols; sesquiterpenoids (aloaromadendrol and epicubebol); phenolcarbonic acids; fatty and amino acids; tocopherols; and essential oil (0.2–0.3%) with α-cadinol and β-cadinen as the major components [24,25,47,48]. Frequently, the infusion, tincture, and ointment of marigold are used as a wound healing remedy for inflammation of the skin and mucous membranes and externally in the treatment of long-healing wounds, cuts, boils, burns, and ulcers [25,48,49].

Pharmacological studies had confirmed that extracts from this plant exhibit a broad range of biological effects, such as antibacterial, antifungal [47,48,50,51], antioxidant [47,52], anti-inflammatory [47,48,51,52], spasmolytic, anticancer [24,53], anti-HIV, and hepatoprotective activities [51], and stimulate the proliferation and migration of fibroblasts in vitro [54]. In fact, most of the studies on marigold focus on its anti-inflammatory property. Some results present that *C. officinalis*, with other plants, reduced cutaneous inflammation at the price of the downregulation of inflammatory IL-1β, IL-6, and IL-8 and suppressed an increase in stratum corneum dehydration through the upregulation of AQP3 [55]. Marigold flowers extracts protect HaCaT skin cells against an oxidative stress challenge in the form of H_2O_2 [56]. Moreover, other scientists found that the *n*-hexane and the ethanolic extracts modulated the inflammatory phase of wound healing by activating the transcription factor NF-κB and increasing the amount of the chemokine IL-8 [57].

Nowadays, marigold flowers are one of the most famous plant materials among manufacturers and patients as they are presented in large amounts in Polish and Ukrainian pharmacies. For patients, it is available mainly in two forms (ointments and tincture) and is used in mild inflammations of the skin, minor wounds, burns, and cuts. One study on the potential influence of the marigold extract on the composition of skin microbiota was performed. The 90% hydroethanolic extract from marigold was shown to inhibit the growth of *P. acens* and *S. epidermidis*. The results suggest that this plant material can be considered a skin prebiotic important in treating acne [58].

5.5. Chelidonium majus L. (Greater Celandine Herb)

Chelidonium majus L. is also known as greater celandine (family Papaveraceae). The herb of *Chelidonium majus* L. contains over 20 different alkaloids, including chelerythrine, chelidonine, sanguinarine, isochelidonine, and protoberberines (berberine, coptisine, dihydrocoptisine, and stylopine) protopine [59]. Several flavonoids were found in the aerial parts in low amounts. Among them are derivatives of kaempferol and quercetin. Moreover, other phenolic compounds such as hydroxycinnamic acids, hydroxybenzoic acids, and their derivatives were identified. Additionally, organic acids (chelidonic, malic, citric, and succinic acids), biogenic amines (histamine, methylamine, and tyramine), essential oil constituents, triterpenoids, saponins, vitamins A and C, and nicotinic acid were found in *C. majus* extracts [60].

N. Cordes et al.'s investigation shows that ukrain, an alkaloid thiophosphoric acid derivative of *C. majus.*, demonstrates a protective effect in normal human fibroblasts in modulating radiation toxicity [61]. Vavrecková et al. determined the antiproliferative activity of the extract on human keratinocytes, showing IC_{50} was lowest for sanguinarine (2.26 µM), extract (as chelidonine) ca. 5.68 µM, chelidonine, and chelerythrine ca. 28 µM, and poor activity of berberine and hydrastinine. The lactate dehydrogenase assay showed the cytostatic activity of the *C. majus* extract rather than cytotoxic activity which can be considered beneficial in treating wards [62]. Some medical products with celandine extracts are available on the Ukrainian market and presented in such preparations as "CHYSTOTIL" cream (Khimpharmzavod Chervona Zirka, Ukraine) and "CHYSTOTIL" ointment (NATURE LIFE, Ukraine). Bactericidal, wound healing, anti-inflammatory, and analgesic properties are indicated.

5.6. Hamamelis virginiana L. (Witch Hazel Leaves, Bark)

Hamamelis virginiana L., also known as witch hazel, is a shrub that belongs to the Hamamelidaceae family [63,64]. Preparations from *Hamamelis* leaves, bark, and twigs, present in extracts, tinctures, creams, and salves, are utilized to treat dermatological (sunburn, irritated skin, and atopic eczema) and vascular disorders (hemorrhoids, varicose veins, and phlebitis), highlighting the fact that this plant has a wide range of biologically active substances [64–66].

Witch hazel bark contains up to 10% tannins (hamamelitannin and catechins), free gallic acid, and a small amount of flavonols, fats, and waxes. Leaves contain 3–10% of tannins (a mixture of gallotannins and condensed catechins–procyanidins); notably a small amount of hamamelitannin; phenolic acids (caffeic and gallic acids); flavonoids such as kaempferol, quercetin, quercitrin, and isoquercitrin; and essential oil [26,64,67–69]. Numerous in vitro and in vivo studies have shown that this plant has antitumoral, antioxidant, anti-inflammatory, antibacterial, antiviral, and antimutagenic activity [63,64,68–70]. For instance, the extract of witch hazel leaves and small twigs can decrease the amount of IL-8 produced by fibroblast cells [71]. The hazel extract and its component—hexagallloylglucose—regulated the inflammatory response via inhibiting NF-κB and PAR-2 pathways in human keratinocytes [72].

Witch hazel, one of the active ingredients, was found in the combined homeopathic ointment TRAUMEEL® S (Biologische Heilmittel Heel GmbH, Baden-Baden, Germany), which is available on the Ukrainian market and applied for the treatment of bedsores, burns, pityriasis, and trophic ulcers. No papers reporting the possible interactions of hazel extracts with human sin microbiome were found.

5.7. Hippophae rhamnoides L. (Sea-Buckthorn Fruits)

Hippophae rhamnoides L., (Elaeagnaceae family), is commonly known as sea buckthorn. Fruits and seeds contain fatty oil (about 8% in fruits and about 12% in seeds), which contains a significant amount of carotene (up to 250 mg%), vitamins E, F, and K, phospholipids (up to 1%), and fatty acids (linoleic, oleic, palmitic, palmitoleic, and stearic). Fruits contain mono- and disaccharides, mucus, vitamins (C, B_1, B_2, B_6, B_9, P, and PP), organic acids (malic, tartaric, oxalic, and succinic), sulfur-containing substances, including betaine and choline, tannins, flavonoids (rutin, quercetin, kaempferol, and isorhamnetin), phenolic acids (chlorogenic and caffeic), and coumarin [25]. Patients have used sea buckthorn for a long time due to its rich composition, which provides a wound healing effect, modification of sebum characteristics, and improvement of atopic skin [73].

To date, many investigations have been conducted on seeds and their non-polar compounds. It was shown that the palmitic acid-enriched fraction supported cell proliferation properties on normal human keratinocytes (NHEK) and normal human dermal fibroblasts (HDFa). However, some fractions did not alter the cellular morphology of normal keratinocytes and did not influence the inflammatory response [73]. There are also studies confirming that sea buckthorn seed oil stimulated the proliferation of dysplastic cells, while it also impaired the ability of both normal and dysplastic cells to migrate over a denuded area [74]. 1,5-dimethyl citrate isolated from *Hippophae rhamnoides* was demonstrated to prevent LPS-induced NO production and inhibited the expression of IKK-α/β, IκB-α, NF-κB p65, iNOS, and COX-2 and the activities of IL-6 and TNF-α [75].

Sea buckthorn ointment (Fitolic, Ivano-Frankivsk, Ukraine) and Olasol spray (STOMA, Kharkiv, Ukraine) are drugs popular in Ukraine and contain sea buckthorn oil, used for healing infected wounds, including long-term non-healing burns, trophic ulcers, and skin grafts. Moreover, some other medical products consisting of *Hippophae rhamnoides* extract are available, such as the balms Mintalon (Minta, Kharkiv, Ukraine) and Reskinol (Botany, Kramatorsk, Ukraine). To the best of our knowledge, the interaction of sea buckthorn extracts with skin microbiota has never been investigated.

5.8. Linum usitatissimum L. (Linseed)

Linseed is usually defined as the ripe dried seeds of *Linum usitatissimum* L. (Linaceae), which contain 30–45% of fixed oil, 25% protein, 3–9% polysaccharides, and 0.1–1.5% cyanogenic glycosides. In addition, it also contains lignans, mainly secoisolariciresinol, and its glycosides [76]. It is a herbal medicinal product with well-established use in treating habitual constipation [77]. In traditional medicine, flax seeds have been used to relieve inflammation of the upper respiratory tract and gastrointestinal tract and externally in skin inflammation, eczema, ulcers, burns, chilblains, hard-to-heal wounds, skin drying, and cracking [78].

On the Polish market, some drugs contain virgin linseed oil, indicated in the treatment of skin diseases such as eczema and rash, as well as in conditions of excessive dryness and symptoms of psoriasis. In addition, there are preparations containing flax in medical products that intensely moisturize, soften, and nourish the skin during or after radiotherapy.

In vivo studies have shown that the external application of linseed oil has an anti-inflammatory effect. In rats with carrageenan-induced paw edema, a reduction in clinical signs of inflammation, infiltration of inflammatory cells, vascular congestion, and an improvement in biochemical parameters were observed [79].

The effectiveness of treating burns with linseed oil has been proven in animal models. After applying linseed oil to second-degree burns in rats, a higher wound closure rate was observed, and the performed biopsy showed better tissue regenerative properties and higher angiogenesis than the control [80]. In another study, also on the model of second-degree burns in rats, it was confirmed that after 21 days of linseed oil application, the severity of the inflammatory process decreased, and collagen synthesis, re-epithelialization, and angiogenesis increased [81]. In a burn model in rabbits, after 12 days of treatment with linseed oil, the degree of wound closure was significantly higher than in the control group; moreover, complete wound closure was 9 days earlier than in the control group. A histopathological study showed that in the group treated with linseed oil, the wound contained fewer inflammatory cells and had complete re-epithelialization with reduced thickness compared to the non-treated control. In addition, an increased number of new capillaries, collagen fibers, and fibroblasts was observed [82]. The effectiveness in treating burns was also confirmed for a gel containing flax seed polysaccharides (composed of glucose, mannose, xylose, and arabinose in glycerol) in the rats' burns model. The group treated with the gel showed the best results. The skin was naturally colored, and a histological evaluation showed epidermal regeneration without inflammation, growth in connective tissue, and increased collagen production [83].

The beneficial effects of flax have also been confirmed in wound healing. In wound models made in rats, it was shown that the application of linseed oil accelerates wound closure, increases re-epithelialization, and reduces inflammation [84,85]. The histopathological examination also showed the increased synthesis of collagen fibers, vascularization, and hair follicles [86]. Linseed oil used on wounds made with a scalpel in rabbits increased the skin's elasticity and firmness and stimulated microcirculation and the influx of fibroblasts, as well as the growth of collagen fibers [87]. No information on the research focusing on interactions of flax extracts with skin microbiota was found in the literature.

5.9. Matricaria chamomilla L. (Chamomile Flowers)

Flowers from *Matricaria chamomilla* L. (Asteraceae) contain not less than 4 mL/kg of essential oil and 0.25% of apigenin-7-glucoside. In addition to essential oil and flavonoids, they contain coumarins, phenolic acids, and polysaccharides [88]. The indications for external use include the treatment of minor ulcers and inflammation of the mouth and throat, inflammation of the skin (sunburn), superficial wounds and small boils (furuncles), and as an adjunct to the treatment of skin and mucosa irritation around the anus and genital region [89].

The anti-inflammatory properties of the essential oil and water extract of chamomile flowers have been confirmed in animal models. In a model of rats with carrageenan-induced paw edema, they reduced swelling and decreased prostaglandin E_2 secretion and NO levels. In a mouse model with xylene-induced ear swelling, they reduced the swelling and lowered the allergic reaction. Additionally, the essential oil reduced the duration and frequency of scratching in mice with dextran-induced itching [90].

The acceleration of the second-degree burn regeneration in rats was observed for the chamomile flower oil extract, which significantly reduced the lesion area after 20 days [91]. In addition, another study showed that the same extract reduced the size of the incision made on the back of rats after just 5 days. Complete healing was observed after 11 days, while for olive oil alone, the healing process took 20 days [92].

The ability to regenerate wounds was also confirmed for ethanolic and methanolic extracts from *M. chamomilla* flowers. The daily use of the gel containing 5 or 10% ethanolic extract improved wound healing in diabetic rats by increasing fibroblast proliferation and revascularization. A higher wound closure ratio was noted after three days of treatment compared to the control group. No significant differences were observed between the treatment with the gel containing 5 and 10% of the extract [93]. The wound healing capacity of the methanolic extracts was tested in an excision wound model on the rats' dorsum, using concentrations of 2.5, 5, and 10%. Seven days after the injury, a difference in wound regeneration was observed between the test and control groups, the ointment containing 10% of the extract had the strongest healing effect. After 11 days, the differences in wound healing after the application of the 2.5, 5, and 10% ointments were insignificant, and still, all of them aided the healing more strongly than the control. Histopathological examinations carried out after 14 days showed that using the ointments with the chamomile flower extract increased the number of fibroblasts, basal epidermal cells, and the amount of collagen. In contrast, the number of neutrophils in the wound decreased [94]. The activity of the ethanolic extract from chamomile flowers was also tested in a model of wounds infected with *Staphylococcus aureus* in mice. After 14 days of daily use, the wound was completely healed, and the hair was present, without scarring, while scar tissue appeared in the gentamycin-treated group and severe inflammation in the control group. Moreover, in the group treated with chamomile extract, an increase in granulation tissue production, fibroblast density, keratinization on the wound surface, and thickness of collagen fibers was observed [95].

The clinical efficacy of the *M. chamomilla* flower extract applied to the forearm and face of healthy volunteers was tested. Skin physiology was assessed after 2 h and 2 and 4 weeks of daily use. The use of the extract significantly increased the hydration of the corneum, and after prolonged use, it reduced the transepidermal water loss by 27% [96]. Another study looked at the activity of chamomile gel in preventing acute radiation dermatitis in head and neck cancer patients. The effect of a gel containing 8.35% of chamomile flower extract was compared with urea cream. The use of the gel delayed the onset of the inflammatory reaction to the radiation. In addition, its use was associated with a lesser incidence of itching, burning, and discoloration among patients, which were seen in the group using urea cream [97].

The activity of 3% of *M. chamomilla* essential oil was tested in a BALB/c mouse model in which atopic dermatitis was induced with dinitrochlorobenzene. The daily use of the oil for four weeks contributed to a decrease in the levels of Ig E and Ig G1 and histamine in the blood of the animals. In addition, it reduced the frequency of scratching [98]. Although several basic and clinical studies were performed with chamomile extract as a skin medicine, no research considering its influence on skin microbiota has been reported so far.

5.10. Potentilla erecta L. (Tormentil Root)

Tormentillae rhizoma is a whole or cut dried rhizome of *Potentilla erecta* (syn. *Potentilla tormentilla*, Rosaceae), containing not less than 7% of tannins expressed as pyrogallol. The composition is dominated by condensed tannins (up to 22%), but ellagitannins (including agrimoniin and pedunculagin) are also present. In addition, there are phenolic acids (coumaric, sinapic, caffeic, and gallic acids and their derivatives), flavonoids (kaempferol and quercetin and their derivatives), as well as triterpene saponins [99].

The indications for use exclusively based on longstanding use only include the symptomatic treatment of mild diarrhea and mild inflammation of the oral mucosa [100].

On the Polish market, ointments containing a liquid extract from the rhizome of common tormentil are available. It is used together with zinc oxide and ichthammol. The indication for the use of these preparations is the treatment of minor skin lesions, such as scratches or abrasions of the epidermis. In the case of ointments that additionally contain borax, the indications for use are extended to the treatment of purulent and acne lesions.

Studies on in vitro models have shown the astringent, antimicrobial, antioxidant, and anti-inflammatory effects of tormentil rhizomes [101].

The potent anti-inflammatory properties of the agrimoniin-rich fraction were confirmed in in vitro and in vivo models of UVB-induced inflammation. The investigation using HaCaT keratinocyte cells showed that the abovementioned fraction reduced the production of prostaglandin PGE_2 by inhibiting COX-2 [102]. In similar studies on the HaCaT cell model, it was demonstrated that the tormentil ethanolic extract inhibits the activation of NF-κB and, in addition to inhibiting PGE_2 production, also inhibits the production of IL-6 [103]. In the in vivo model of erythema induction on the skin of healthy volunteers, a significant reduction in inflammation and redness was observed after the use of the agrimoniin-rich fraction in the concentration of 100 mg/mL [102]. The effect of the methanolic extract from *P. erecta* rhizomes on the healing of diabetic wounds was investigated in Wistar rats with streptozocin-induced diabetes. Studies have shown that the extract significantly accelerates wound contraction compared to control and increases nitric oxide, glutathione, and collagen levels, while the thiobarbituric-acid reactive substances levels decreased [104]. No studies on the influence of the tormentil rhizome extract on skin microbiota or the metabolism of natural products contained in this plant material by skin microorganisms have been reported.

5.11. Quercus robur L. (Common Oak Bark)

Common oak belongs to the family Fagaceae and contains a highly variable amount of tannins (8–20%). *Quercus cortex* contains hydrolyzable tannins (gallotannins, ellagitannins, and flavonols-ellagitannins) and condensed tannins (proanthocyanidins). More than 20 compounds (catechins and low-molecular-mass, oligomeric, and polymeric proanthocyanidins) have been isolated from the bark [105,106]. Triterpenes, insoluble lipid polyesters, and volatile acids are also presented in oak bark. It is considered a traditional herbal medicinal product for the symptomatic treatment of minor inflammation of the oral mucosa or skin in pharmaceutical forms, such as infusion and decoction [105,106].

Ji-Ae Hong et al. found that *Quercus fruits* rescued UVB-induced cytotoxicity and substantially inhibited cellular ROS production in human keratinocytes [107]. Likewise, this study showed that *Quercus fruits* effectively prevent skin photoaging by enhancing collagen deposition and inhibiting MMP-1 via the ERK/AP-1 signaling pathway. *Quercus mongolica* and its isolated compounds have shown inhibitory activities toward inflammatory cytokines and chemokines. Potent activities against MCP-1, TARC, IL-6, IL-8, IL-10, and IL-13 in keratinocytes irradiated with UVB were determined [108]. Chang Seok Lee et al.'s study revealed that oak wood vinegar has anti-inflammatory and antiproliferative effects in a 2,4-dinitrochlorobenzene-induced contact dermatitis mice model. Furthermore, they showed that the mechanism by which oak wood vinegar most likely inhibits epithelial proliferation is through STAT3 inactivation [109].

In most cases, oak bark is used in decoctions by humankind, but in the Ukrainian market, this plant material was also found in a combined medical product as cream, "Bioflorin" (Khimpharmzavod Chervona Zirka, Kharkiv, Uktaine), which has anti-inflammatory and wound healing properties. No information on the interactions of oak bark extracts with sin microbiota was found in the available literature.

5.12. Salvia officinalis L. (Sage Leaf)

Salviae folium, obtained from *Salvia officinalis* L. (Lamiaceae), has a monograph in the European Pharmacopoeia (Ph. Eur. 10th Edition), European Medicines Agency (EMA), and European Scientific Cooperative on Phytotherapy (ESCOP). The medicinal plant material consists of whole or cut dried sage leaves containing not less than 12 or 10 mL/kg of essential oil, respectively. In addition to the essential oil containing monoterpenes and sesquiterpenes, the chemical composition includes diterpenoids, triterpenoids, flavonoids, hydroxycinnamic acid derivatives, and phenolic glycosides [110]. Traditionally, sage leaf can be used to relieve dyspeptic disorders such as heartburn and flatulence, reduce hyperhidrosis, relieve inflammation of the mouth and throat, and treat inflammation of the skin [111].

Medicinal products available on the Polish market are intended primarily for treating inflammations in the mouth and throat. These are concentrates for preparing rinse solutions (Dentosept, Salviasept, Tinctura Salviae, and Tymsal) and gels applied directly to the lesions within the oral cavity (Aperisan, Dentosept A, and Mucosit). In addition, sage leaf is available as a single herb for making infusions (Salviae folium), as well as in the form of herbal mixtures for making infusions for gargling (Septosan) and for use in mild inflammatory conditions of the female genitalia (Vagosan). The only medicinal product to be applied directly to the skin is sage ointment containing the ethanolic extract of *Salvia officinalis* leaves.

The traditional use of sage leaf has been partially supported by scientific research. Studies conducted on in vitro and in vivo models have confirmed its anti-inflammatory, antioxidant, and antimicrobial properties as well as the beneficial effects on wound healing [112].

Strong anti-inflammatory properties after topical application on a model of mouse ear edema induced by croton oil have been demonstrated for *n*-hexane and chloroform extracts from sage leaves. The component responsible for the anti-inflammatory activity was ursolic acid [113].

The essential oil of *S. officinalis* showed antifungal activity against clinical strains of dermatophytes isolated from skin and nails and also inhibited NO production by LPS-stimulated macrophages [114]. In in vitro studies, sage leaf oil showed stronger antibacterial properties against *S. aureus* and *P. aeruginosa* than penicillin and mupirocin. Studies on an in vivo model of infected wounds in BALB/c mice showed that the ointment containing 4% of the *S. officinalis* oil statistically reduced the number of bacteria compared to the control group and the mupirocin-treated group. Moreover, compared to the control group, there was a shortening of the inflammatory phase, acceleration of cell proliferation, increased collagen accumulation, revascularization, and re-epithelialization. The content of pro-inflammatory cytokines (IL-1β, IL-6, and TNF-α) decreased, and the content of growth factors (FGF-2 and VEGF) increased [115].

In another in vivo study in a Wistar rat model, the wound healing capacity of an ointment containing 3 and 5% hydroethanolic extract of sage leaves was tested. The time of wound closure and re-epithelialization was significantly shortened compared to the control group. Moreover, the formation of new blood vessels and the number of fibroblasts increased in the wound, improving the proliferation phase of healing [116]. The methanolic extract of *S. officinalis* leaves inhibited hyaluronidase, elastase, and collagenase activity in vitro. In vivo studies on Swiss albino mice inhibited the formation of wrinkles induced by UV exposure [117]. In the prospective randomized, double-blind placebo-controlled study, an area of the skin of healthy volunteers was irradiated to induce local erythema. An ointment containing 2% sage extract has been shown to reduce local inflammation and erythema to a similar extent as 1% hydrocortisone ointment [118]. Although sage leaves extracts and essential oil are widely used for skin problems, no reports on the interactions of this plant material with skin microbiome were published.

5.13. Sophora japonicum L. (Japanese Pagoda Tree Fruits)

The Japanese pagoda tree is also named sophora and belongs to the *Fabaceae* family. At least 153 constituents, including flavonoids, isoflavonoids, triterpenoids, alkaloids, mineral elements, and amino acids, were identified and isolated from *S. japonica*. The most important and abundant components of the dried flower buds and ripe fruits are rutin and sophoricoside [119]. These substances have been used to control the quality of medicinal products and determine the high medicinal value of this plant material. Based on the chemical composition, this plant material has anti-inflammatory, antibacterial, antiviral, and antioxidant effects and is traditionally used in treating wounds and trophic ulcers.

Sophora's polysaccharides protect HaCaT keratinocytes from UVB irradiation-induced skin injuries and may involve the MAPK signaling pathway, which contributes to apoptotic cell death [120]. Sophoricoside ameliorates contact dermatitis due to the inhibition of the phosphorylation and degradation of IκB-α/β and the nuclear translocation of NF-κB p65 in B cells [121]. In turn, sophoricoside exhibited a potent inhibitory effect in the IL-5 bioassay in a dose-dependent manner [122]. This isoflavone glycoside inhibited the IL-6 bioactivity with an IC$_{50}$ value of 6.1 μM. In contrast, it had no effects on IL-1β and TNF-α production and was established as a selective inhibitor of cyclooxygenase COX-2 activity [123].

In Ukraine, a tincture of Japanese sophora fruit (FITOPHARM, Kyiv, Ukraine) is very popular as an antiseptic and wound healing preparation and is used in purulent inflammatory processes (wounds and trophic ulcers).

5.14. Symphytum officinale L. (Comfrey Root)

Symphytum officinale is a plant belonging to the *Boraginaceae* family, rich in allantoin, phenolic acids (e.g., rosmarinic, *p*-hydroxybenzoic, caffeic, chlorogenic, and p-coumaric acids), pyrrolizidine alkaloids, triterpene saponins, tannins, amino acids, flavonoids, triterpenes, terpenoids, saponins, sterols, and mucopolysaccharides [124]. The comfrey plant material therapeutic properties include anti-inflammatory, analgesic, granulation-promoting, and anti-exudative effects [125,126].

The previous investigation determined the in vivo wound healing effects of *Symphytum officinale* L. leaves extract. The results showed that comfrey extract modulates the inflammatory process and stimulates collagen production [126]. Proliferative and antioxidant studies demonstrate a beneficial effect on human skin fibroblasts. It is non-toxic and simultaneously expresses the high ability to reduce ROS [125]. Several investigations with isolated compounds from comfrey were conducted, and it was established that crude comfrey polysaccharides possess the antioxidative activity and revealed that, by efficiency, it is superior to allantoin ointment in burn wound healing. Moreover, poly[3-(3,4-dihydroxyphenyl) glyceric acid from *S. asperum* and *S.caucasicum* roots inhibits the TNF-α production by human macrophages [127]. Furthermore, rosmarinic acid isolated from *Symphytum officinale* L. was shown to inhibit the formation of inflammation mediators of the arachidonic acid cascade in vitro [128]. Moreover, researchers conclude that comfrey root extract inhibits NF-κB by interfering with the activation pathway, at least in part at the level of IκB-α phosphorylation and possibly of IKK activation.

Mostly, medicines from comfrey root are presented in ointments ("Ointment Dr. Taissa with Comfrey", Dr. Theiss Naturwaren GmbH, Homburg, Germany, and "Ointment with Comfrey", DKP Pharmaceutical Factory, Zhytomyr, Ukraine) and used in the treatment of pain in the joints, back, lower back with radiculitis, osteochondrosis, arthritis, domestic injuries and bruises, sprains, dryness, and cracked skin. No research on the interaction of comfrey extracts with skin microbiome is available.

6. Conclusions

Plants have been used to prevent and treat skin diseases of various etiologies since ancient times. Due to their longstanding use, people have gained information regarding their effectiveness, active ingredients, as well as associated side effects. However, very often, it is not clear what mechanisms are responsible for the observed therapeutic effects. Moreover, recent advances in understanding the contribution of the skin microbiota in the maintenance of skin homeostasis can put new light on understanding the activity of topically applied plant-based products. Although the influence of various plants on skin-related ailments is well documented in vivo and in vitro, little is known about the interaction with the network of the skin microbial ecosystem, especially considering the prolonged treatment. It is also unclear whether skin microbiota can alter the chemical composition of herbal drugs applied directly on the skin surface. It was shown that some of the reported plant materials (e.g., sage leaves preparations) could have antimicrobial potential. However, available reports are strictly limited to investigating plant preparations influencing the growth of single strains of the chosen microorganisms. The analysis of the number of studies reported in Scopus between 2001 and 2021 using plant extract/material and skin microbiota as keywords showed that only 48 reports were found. Skin microbiota is a complex ecosystem that can certainly be modulated by plant extracts in many ways. Without solid basic studies involving human skin microbiota, the interaction of plant materials with the microbiome will remain unknown. Recently, there has been an arising interest in an investigation of the interaction between drugs and gut microbiota [129,130]. Based on the present review, it can be suspected that one of the major problems related to the lack of proper studies is the lack of well-described and reliable models that can be used for the investigation of the interactions between skin microbiome and plant extracts in vitro. More focus on this aspect of the problem is needed. For sure, future studies devoted to the investigation of skin microbiota with topically used plant materials or extracts are essential for the complex understating of the mechanism of action of those natural drugs in the prevention and treatment of various skin diseases.

Author Contributions: Conceptualization, S.G.; data writing—original draft preparation, N.M., W.S. and I.V.; writing—review and editing, N.M., W.S., S.G., A.B. and J.P.P.; supervision, S.G. All authors have read and agreed to the published version of the manuscript.

Funding: The presented research was financially supported by the NCN research grant Preludium Bis 2 No. 2020/39/O/NZ7/01109.

Institutional Review Board Statement: Not applicable.

Informed Consent Statement: Not applicable.

Data Availability Statement: Not applicable.

Conflicts of Interest: The authors declare no conflict of interest.

References

1. Bilia, A.R. Herbal Medicinal Products *versus* Botanical-Food Supplements in the European market: State of Art and Perspectives. *Nat. Prod. Commun.* **2015**, *10*, 125–131. [CrossRef]
2. Aberer, W. Contact Allergy and Medicinal Herbs. *JDDG J. Ger. Soc. Dermatol.* **2007**, *6*, 15–24. [CrossRef]
3. Gurib-Fakim, A. Medicinal Plants: Traditions of Yesterday and Drugs of Tomorrow. *Mol. Asp. Med.* **2006**, *27*, 1–93. [CrossRef] [PubMed]
4. Shedoeva, A.; Leavesley, D.; Upton, Z.; Fan, C. Wound Healing and the Use of Medicinal Plants. *Evid. Based Complement. Altern. Med.* **2019**, *2019*, 2684108. [CrossRef]
5. Wyszkowska-Kolatko, M.; Koczurkiewicz, P.; Wójcik, K.; Pękala, E. Rośliny Lecznicze w Terapii Chorób Skóry. *Postępy Fitoter.* **2011**, *12*, 191–196.
6. Bos, J.D.; Kapsenberg, M.L. The Skin Immune System Its Cellular Constituents and their Interactions. *Immunol. Today* **1986**, *7*, 235–240. [CrossRef]
7. Bos, J.D.; Kapsenberg, M.L. The Skin Immune System: Progress in Cutaneous Biology. *Immunol. Today* **1993**, *14*, 75–78. [CrossRef]
8. Byrd, A.L.; Belkaid, Y.; Segre, J.A. The Human Skin Microbiome. *Nat. Rev. Microbiol.* **2018**, *16*, 143–155. [CrossRef]
9. Lai-Cheong, J.E.; McGrath, J.A. Structure and Function of Skin, Hair and Nails. *Medicine* **2021**, *49*, 337–342. [CrossRef]
10. Атопічний дерматит. Available online: https://www.dec.gov.ua/mtd/atopichnyj-dermatyt/ (accessed on 18 June 2022). (In Ukrainian)
11. Spiewak, R. Częstość Występowania Chorób Skóry w Losowej Grupie Dorosłych Polaków. *Estetologia Med. I Kosmetol.* **2012**, 50–53. [CrossRef]
12. Christensen, G.J.M.; Brüggemann, H. Bacterial Skin Commensals and their Role as Host Guardians. *Benef. Microbes* **2014**, *5*, 201–215. [CrossRef] [PubMed]
13. Zeeuwen, P.; Kleerebezem, M.; Timmerman, H.M.; Schalkwijk, J. Microbiome and Skin Diseases. *Curr. Opin. Allergy Clin. Immunol.* **2013**, *13*, 514–520. [CrossRef] [PubMed]
14. Belkaid, Y.; Segre, J.A. Dialogue Between Skin Microbiota and Immunity. *Science* **2014**, *346*, 954–959. [CrossRef]
15. Yu, Y.; Dunaway, S.; Champer, J.; Kim, J.; Alikhan, A. Changing Our Microbiome: Probiotics in Dermatology. *Br. J. Dermatol.* **2019**, *182*, 39–46. [CrossRef]
16. Dhama, K.; Karthik, K.; Khandia, R.; Munjal, A.; Tiwari, R.; Rana, R.; Khurana, S.K.; Ullah, S.; Khan, R.U.; Alagawany, M.; et al. Medicinal and Therapeutic Potential of Herbs and Plant Metabolites/Extracts Countering Viral Pathogens-Current Knowledge and Future Prospects. *Curr. Drug Metab.* **2018**, *19*, 236–263. [CrossRef]
17. Rahman, M.; Rahaman, S.; Islam, R.; Hossain, E.; Mithi, F.M.; Ahmed, M.; Saldías, M.; Akkol, E.K.; Sobarzo-Sánchez, E. Multifunctional Therapeutic Potential of Phytocomplexes and Natural Extracts for Antimicrobial Properties. *Antibiotics* **2021**, *10*, 1076. [CrossRef]
18. El-Najjar, N.; Gali-Muhtasib, H.; Ketola, R.A.; Vuorela, P.; Urtti, A.; Vuorela, H. The Chemical and Biological Activities of Quinones: Overview and Implications in Analytical Detection. *Phytochem. Rev.* **2011**, *10*, 353–370. [CrossRef]
19. Farha, A.K.; Yang, Q.-Q.; Kim, G.; Li, H.-B.; Zhu, F.; Liu, H.-Y.; Gan, R.-Y.; Corke, H. Tannins as an Alternative to Antibiotics. *Food Biosci.* **2020**, *38*, 100751. [CrossRef]
20. Popowski, D.; Czerwińska, M.E.; Kruk, A.; Pawłowska, K.A.; Zentek, J.; Melzig, M.F.; Piwowarski, J.P.; Granica, S. Gut Microbiota Metabolism and the Permeability of Natural Products Contained in Infusions from Herb of European Goldenrod *Solidago virgaurea* L. *J. Ethnopharmacol.* **2021**, *273*, 113924. [CrossRef]
21. Kruk, A.; Granica, S.; Popowski, D.; Malinowska, N.; Piwowarski, J.P. Tiliae Flos Metabolites and their Beneficial Influence on Human Gut Microbiota Biodiversity ex vivo. *J. Ethnopharmacol.* **2022**, *294*, 115355. [CrossRef]
22. Piwowarski, J.P.; Kiss, A.K.; Granica, S.; Moeslinger, T. Urolithins, Gut Microbiota-Derived Metabolites of Ellagitannins, Inhibit LPS-Induced Inflammation in RAW 264.7 Murine Macrophages. *Mol. Nutr. Food Res.* **2015**, *59*, 2168–2177. [CrossRef]
23. Dawid-Pac, R. Medicinal Plants Used in Treatment of Inflammatory Skin Diseases. *Adv. Dermatol. Allergol.* **2013**, *3*, 170–177. [CrossRef]
24. ESCOP. *ESCOP Monographs: The Scientific Foundation for Herbal Medicinal Products*, 2nd ed.; European Scientific Cooperative on Phytotherapy: Exeter, UK, 2019. Available online: https://escop.com/ (accessed on 17 August 2022).
25. Кисличенко, В.С.; Журавель, І.О.; Марчишин, С.М.; Мінарченко, В.М.; Хворост, О.П. *фармакогнозія, Частина 1*; Національний фармацевтичний Університет: Харків, Україна, 2015; pp. 48–548. ISBN 978-966-400-365-7. (In Ukrainian)

26. Кисличенко, В.С.; Журавель, І.О.; Марчишин, С.М.; Мінарченко, В.М.; Хворост, О.П. *фармакогнозія Частина 2*; Національний фармацевтичний Університет: Харків, Україна, 2015; pp. 549–669. ISBN 978-966-400-365-7. (In Ukrainian)

27. HMPC. *Assessment Report on Allium cepa L., Bulbus*; EMA/HMPC/347195/2011; European Medicines Agency: London, UK, 2011. Available online: https://www.ema.europa.eu/en/documents/herbal-report/draft-assessment-report-allium-cepa-l-bulbus_en.pdf (accessed on 17 August 2022).

28. Teshika, J.D.; Zakariyyah, A.M.; Zaynab, T.; Zengin, G.; Rengasamy, K.R.; Pandian, S.K.; Fawzi, M.M. Traditional and Modern Uses of Onion Bulb (*Allium cepa* L.): A Systematic Review. *Crit. Rev. Food Sci. Nutr.* **2018**, *59*, S39–S70. [CrossRef]

29. Pikuła, M.; Żebrowska, M.E.; Poblocka-Olech, L.; Krauze-Baranowska, M.; Sznitowska, M.; Trzonkowski, P. Effect of Enoxaparin and Onion Extract on Human Skin Fibroblast Cell Line—Therapeutic Implications for the Treatment of Keloids. *Pharm. Biol.* **2013**, *52*, 262–267. [CrossRef]

30. Saulis, A.S.; Mogford, J.H.; Mustoe, T.A. Effect of Mederma on Hypertrophic Scarring in the Rabbit Ear Model. *Plast. Reconstr. Surg.* **2002**, *110*, 177–183. [CrossRef]

31. Chung, V.Q.; Kelley, L.; Marra, D.; Jiang, S.B. Onion Extract Gel Versus Petrolatum Emollient on New Surgical Scars: A Prospective Double-Blinded Study. *Dermatol. Surg.* **2006**, *32*, 193–197. [CrossRef]

32. Jackson, B.A.; Shelton, A.J. Pilot Study Evaluating Topical Onion Extract as Treatment for Postsurgical Scars. *Dermatol. Surg.* **1999**, *25*, 267–269. [CrossRef]

33. Prager, W.; Gauglitz, G.G. Effectiveness and Safety of an Overnight Patch Containing Allium cepa Extract and Allantoin for Post-Dermatologic Surgery Scars. *Aesthetic Plast. Surg.* **2018**, *42*, 1144–1150. [CrossRef]

34. Güngör, E.S.; Güzel, D.; Zebitay, A.G.; Ilhan, G.; Verit, F.F. The Efficacy of Onion Extract in the Management of Subsequent Abdominal Hypertrophic Scar Formation. *J. Wound Care* **2020**, *29*, 612–616. [CrossRef]

35. Campanati, A.; Savelli, A.; Sandroni, L.; Marconi, B.; Giuliano, A.; Giuliodori, K.; Ganzetti, G.; Offidani, A. Effect of Allium Cepa-Allantoin-Pentaglycan Gel on Skin Hypertrophic Scars: Clinical and Video-Capillaroscopic Results of an Open-Label, Controlled, Nonrandomized Clinical Trial. *Dermatol. Surg.* **2010**, *36*, 1439–1444. [CrossRef]

36. Campanati, A.; Ceccarelli, G.; Brisigotti, V.; Molinelli, E.; Martina, E.; Talevi, D.; Marconi, B.; Giannoni, M.; Markantoni, V.; Gregoriou, S.; et al. Effects of in Vivo Application of an Overnight Patch Containing *Allium Cepa*, Allantoin, and Pentaglycan on Hypertrophic Scars and Keloids: Clinical, Videocapillaroscopic, and Ultrasonographic Study. *Dermatol. Ther.* **2020**, *34*, e14665. [CrossRef] [PubMed]

37. Reynolds, T.; Dweck, A. Aloe vera leaf gel: A review update. *J. Ethnopharmacol.* **1999**, *68*, 3–37. [CrossRef]

38. Sánchez, M.; González-Burgos, E.; Iglesias, I.; Gómez-Serranillos, M.P. Pharmacological Update Properties of Aloe Vera and its Major Active Constituents. *Molecules* **2020**, *25*, 1324. [CrossRef] [PubMed]

39. Radha, M.H.; Laxmipriya, N.P. Evaluation of Biological Properties and Clinical Effectiveness of Aloe Vera: A Systematic Review. *J. Tradit. Complement. Med.* **2014**, *5*, 21–26. [CrossRef] [PubMed]

40. Kriplani, P.; Guarve, K.; Baghael, U.S. *Arnica montana* L.—A Plant of Healing: Review. *J. Pharm. Pharmacol.* **2017**, *69*, 925–945. [CrossRef]

41. Oberbaum, M.; Galoyan, N.; Lerner-Geva, L.; Singer, S.R.; Grisaru, S.; Shashar, D.; Samueloff, A. The Effect of the Homeopathic Remedies Arnica Montana and Bellis Perennis on Mild Postpartum Bleeding—A Randomized, Double-Blind, Placebo-Controlled Study—Preliminary Results. *Complement. Ther. Med.* **2005**, *13*, 87–90. [CrossRef]

42. HMPC. *Assessment Report on Arnica montana L., Flos*; EMA/HMPC/198794/2014; European Medicines Agency: London, UK, 2011. Available online: https://www.ema.europa.eu/en/documents/herbal-report/final-assessment-report-arnica-montana-l-flos_en.pdf (accessed on 17 August 2022).

43. Prade, J.D.S.; Bálsamo, E.C.; Machado, F.R.; Poetini, M.R.; Bortolotto, V.C.; Araújo, S.M.; Londero, L.; Boeira, S.P.; Sehn, C.P.; de Gomes, M.G.; et al. Anti-inflammatory effect of Arnica montana in a UVB radiation-induced skin-burn model in mice. *Cutan. Ocul. Toxicol.* **2020**, *39*, 126–133. [CrossRef]

44. Huber, R.; Bross, F.; Schempp, C.; Gründemann, C. Arnica and stinging nettle for treating burns—A self-experiment. *Complement. Ther. Med.* **2011**, *19*, 276–280. [CrossRef]

45. Hillebrand, G.; Ike, R.B.; Dimitriu, P.; Monh, W. Method and Topical Composition for Modification of Skin Microbiome. Patent PCT/US2018/0629387, 29 November 2018. Available online: https://patents.google.com/patent/US20190160117A1 (accessed on 17 August 2022).

46. Arora, D.; Rani, A.; Sharma, A. A Review on Phytochemistry and Ethnopharmacological Aspects of Genus Calendula. *Pharmacogn. Rev.* **2013**, *7*, 179–187. [CrossRef]

47. Muley, B.P.; Khadabadi, S.S.; Banarase, N. Phytochemical Constituents and Pharmacological Activities of Calendula officinalis Linn (Asteraceae): A Review. *Trop. J. Pharm. Res.* **2009**, *8*, 48090. [CrossRef]

48. Fialová, S.B.; Rendeková, K.; Mučaji, P.; Nagy, M.; Slobodníková, L. Antibacterial Activity of Medicinal Plants and Their Constituents in the Context of Skin and Wound Infections, Considering European Legislation and Folk Medicine—A Review. *Int. J. Mol. Sci.* **2021**, *22*, 10746. [CrossRef] [PubMed]

49. HMPC. *Assessment Report on Calendula officinalis L., Flos*; EMA/HMPC/603409/2017; European Medicines Agency: London, UK, 2018. Available online: https://www.ema.europa.eu/en/documents/herbal-report/final-assessment-report-calendula-officinalis-l-flos-revision-1_en.pdf (accessed on 17 August 2022).

50. Parente, L.M.L.; Júnior, R.D.S.L.; Tresvenzol, L.M.F.; Vinaud, M.C.; de Paula, J.R.; Paulo, N.M. Wound Healing and Anti-Inflammatory Effect in Animal Models of *Calendula officinalis* L. Growing in Brazil. *Evid. Based Complement. Altern. Med.* **2012**, *2012*, 375671. [CrossRef] [PubMed]

51. Tresch, M.; Mevissen, M.; Ayrle, H.; Melzig, M.; Roosje, P.; Walkenhorst, M. Medicinal Plants as Therapeutic Options for Topical Treatment in Canine Dermatology? A Systematic Review. *BMC Veter. Res.* **2019**, *15*, 174. [CrossRef] [PubMed]

52. Bernatoniene, J.; Masteikova, R.; Davalgiene, J.; Peciura, R.; Gauryliene, R.; Bernatoniene, R.; Majiene, D.; Lazauskas, R.; Civinskiene, G.; Velziene, S.; et al. Topical Application of *Calendula officinalis* (L.): Formulation and Evaluation of Hydrophilic Cream with Antioxidant Activity. *J. Med. Plants Res.* **2011**, *5*, 868–877. [CrossRef]

53. Hu, J.J.; Cui, T.; Rodriguez-Gil, J.L.; Allen, G.O.; Li, J.; Takita, C.; Lally, B.E. Complementary and Alternative Medicine in Reducing Radiation-Induced Skin Toxicity. *Radiat. Environ. Biophys.* **2014**, *53*, 621–626. [CrossRef]

54. Serra, M.B.; Barroso, W.A.; Da Silva, N.N.; Silva, S.D.N.; Borges, A.C.R.; Abreu, I.C.; Borges, M.O.D.R. From Inflammation to Current and Alternative Therapies Involved in Wound Healing. *Int. J. Inflamm.* **2017**, *2017*, 3406215. [CrossRef]

55. Wu, X.X.; Siu, W.S.; Wat, C.L.; Chan, C.L.; Koon, C.M.; Li, X.; Cheng, W.; Ma, H.; Tsang, M.S.M.; Lam, C.W.-K.; et al. Effects of Topical Application of a Tri-Herb Formula on Inflammatory Dry-Skin Condition in Mice with Oxazolone-Induced Atopic Dermatitis. *Phytomedicine* **2021**, *91*, 153691. [CrossRef]

56. Alnuqaydan, A.M.; Lenehan, C.E.; Hughes, R.R.; Sanderson, B.J. Extracts from *Calendula officinalis* Offer in Vitro Protection Against H_2O_2 Induced Oxidative Stress Cell Killing of Human Skin Cells. *Phytotherapy Res.* **2014**, *29*, 120–124. [CrossRef]

57. Nicolaus, C.; Junghanns, S.; Hartmann, A.; Murillo, R.; Ganzera, M.; Merfort, I. In Vitro Studies to Evaluate the Wound Healing Properties of Calendula Officinalis Extracts. *J. Ethnopharmacol.* **2017**, *196*, 94–103. [CrossRef]

58. Pelletier-ahmed, L. Selective Inhibition of Propionibacterium Acnes by Calendula Officinalis: A Potential Role in Acne Treatment through Modulation of the Skin Microbiome. Master's. Thesis, Middlesex University, London, UK, 2016. [CrossRef]

59. HMPC. *Public Statement on Chelidonium majus L., Herba*; EMA/HMPC/743927/2010; European Medicines Agency: London, UK, 2011. Available online: https://www.ema.europa.eu/en/documents/public-statement/final-public-statement-chelidonium-majus-l-herba_en.pdf (accessed on 17 August 2022).

60. Zielińska, S.; Jezierska-Domaradzka, A.; Wójciak-Kosior, M.; Sowa, I.; Junka, A.; Matkowski, A.M. Greater Celandine's Ups and Downs—21 Centuries of Medicinal Uses of Chelidonium majus From the Viewpoint of Today's Pharmacology. *Front. Pharmacol.* **2018**, *9*, 299. [CrossRef]

61. Cordes, N.; Plasswilm, L.; Bamberg, M.; Rodemann, H.P. Ukrain ®, an Alkaloid Thiophosphoric Acid Derivative of Chelidonium Majus L. Protects Human Fibroblasts but not Human Tumour Cells in Vitro Against Ionizing Radiation. *Int. J. Radiat. Biol.* **2002**, *78*, 17–27. [CrossRef] [PubMed]

62. Vavrečková, C.; Gawlik, I.; Müller, K. Benzophenanthridine Alkaloids of Chelidonium majus; II. Potent Inhibitory Action Against the Growth of Human Keratinocytes. *Planta Medica* **1996**, *62*, 491–494. [CrossRef] [PubMed]

63. Raposo, A.; Saraiva, A.; Ramos, F.; Carrascosa, C.; Raheem, D.; Bárbara, R.; Silva, H. The Role of Food Supplementation in Microcirculation—A Comprehensive Review. *Biology* **2021**, *10*, 616. [CrossRef] [PubMed]

64. HMPC. *Assessment Report on Hamamelis virginiana L., Cortex, Hamamelis virginiana L. Folium, Hamamelis virginiana L., Folium et Cortex Aut Ramunculus Destillatum*; EMA/HMPC/114585/2008; European Medicines Agency: London, UK, 2009. Available online: https://www.ema.europa.eu/en/documents/herbal-report/assessment-report-hamamelis-virginiana-l-cortex-hamamelis-virginiana-l-folium-hamamelis-virginiana-l_en.pdf (accessed on 17 August 2022).

65. Natella, F.; Guantario, B.; Ambra, R.; Ranaldi, G.; Intorre, F.; Burki, C.; Canali, R. Human Metabolites of Hamaforton™ (*Hamamelis virginiana* L. Extract) Modulates Fibroblast Extracellular Matrix Components in Response to UV-A Irradiation. *Front. Pharmacol.* **2021**, *12*, 747638. [CrossRef] [PubMed]

66. Smeriglio, A.; Barreca, D.; Bellocco, E.; Trombetta, D. Proanthocyanidins and Hydrolysable Tannins: Occurrence, Dietary Intake and Pharmacological Effects. *Br. J. Pharmacol.* **2017**, *174*, 1244–1262. [CrossRef]

67. Wang, H.; Provan, G.J.; Helliwell, K. Determination of Hamamelitannin, Catechins and Gallic Acid in Witch Hazel Bark, Twig and Leaf by HPLC. *J. Pharm. Biomed. Anal.* **2003**, *33*, 539–544. [CrossRef]

68. Duckstein, S.M.; Stintzing, F.C. Investigation on the Phenolic Constituents in Hamamelis Virginiana Leaves by HPLC-DAD and LC-MS/MS. *Anal. Bioanal. Chem.* **2011**, *401*, 677–688. [CrossRef]

69. Masaki, H.; Atsumi, T.; Sakurai, H. Hamamelitannin as a New Potent Active Oxygen Scavenger. *Phytochemistry* **1994**, *37*, 337–343. [CrossRef]

70. Theisen, L.L.; Erdelmeier, C.A.J.; Spoden, G.A.; Boukhallouk, F.; Sausy, A.; Florin, L.; Muller, C.P. Tannins from Hamamelis virginiana Bark Extract: Characterization and Improvement of the Antiviral Efficacy against Influenza A Virus and Human Papillomavirus. *PLoS ONE* **2014**, *9*, e88062. [CrossRef]

71. Thring, T.S.; Hili, P.; Naughton, D.P. Antioxidant and Potential Anti-inflammatory Activity of Extracts and Formulations of White Tea, Rose, and Witch Hazel on Primary Human Dermal Fibroblast Cells. *J. Inflamm.* **2011**, *8*, 27. [CrossRef]

72. Choi, J.; Yang, D.; Moon, M.Y.; Han, G.Y.; Chang, M.S.; Cha, J. The Protective Effect of *Hamamelis virginiana* Stem and Leaf Extract on Fine Dust-Induced Damage on Human Keratinocytes. *Cosmetics* **2021**, *8*, 119. [CrossRef]

73. Dudau, M.; Codrici, E.; Tarcomnicu, I.; Mihai, S.; Popescu, I.D.; Albulescu, L.; Constantin, N.; Cucolea, I.; Costache, T.; Rambu, D.; et al. A Fatty Acid Fraction Purified From Sea Buckthorn Seed Oil Has Regenerative Properties on Normal Skin Cells. *Front. Pharmacol.* **2021**, *12*, 737571. [CrossRef] [PubMed]

74. Dudau, M.; Vilceanu, A.; Codrici, E.; Mihai, S.; Popescu, I.; Albulescu, L.; Tarcomnicu, I.; Moise, G.; Ceafalan, L.; Hinescu, M.; et al. Sea-Buckthorn Seed Oil Induces Proliferation of both Normal and Dysplastic Keratinocytes in Basal Conditions and under UVA Irradiation. *J. Pers. Med.* **2021**, *11*, 278. [CrossRef] [PubMed]

75. Balkrishna, A.; Sakat, S.S.; Joshi, K.; Joshi, K.; Sharma, V.; Ranjan, R.; Bhattacharya, K.; Varshney, A. Cytokines Driven Anti-Inflammatory and Anti-Psoriasis Like Efficacies of Nutraceutical Sea Buckthorn (*Hippophae rhamnoides*) Oil. *Front. Pharmacol.* **2019**, *10*, 1186. [CrossRef]

76. ESCOP. Lini Semen (Linseed). Available online: https://escop.com/lini-semen-linseed-online/ (accessed on 17 August 2022).

77. HMPC. *Assessment report on Linum usitatissimum L., Semen*; EMA/HMPC/377674/2014; European Medicines Agency: London, UK, 2015. Available online: https://www.ema.europa.eu/en/documents/herbal-report/assessment-report-linum-usitatissimum-l-semen_en.pdf (accessed on 17 August 2022).

78. Sederski, M.E. *Prawie Wszystko o Ziołach i Ziołolecznictwie*; Mateusz, E., Ed.; Senderski: Wroclaw, Poland, 2017; p. 383. ISBN 978-83-945-5401-9.

79. Bardaa, S.; Turki, M.; Ben Khedir, S.; Mzid, M.; Rebai, T.; Ayadi, F.; Sahnoun, Z. The Effect of Prickly Pear, Pumpkin, and Linseed Oils on Biological Mediators of Acute Inflammation and Oxidative Stress Markers. *BioMed Res. Int.* **2020**, *2020*, 5643465. [CrossRef]

80. Bardaa, S.; Moalla, D.; Ben Khedir, S.; Rebai, T.; Sahnoun, Z. The Evaluation of the Healing Proprieties of Pumpkin and Linseed Oils on Deep Second-Degree Burns in Rats. *Pharm. Biol.* **2015**, *54*, 581–587. [CrossRef]

81. Naseri, S.; Golpich, M.; Roshancheshm, T.; Joobeni, M.G.; Khodayari, M.; Noori, S.; Zahed, S.A.; Razzaghi, S.; Shirzad, M.; Salavat, F.S.; et al. The Effect of Henna and Linseed Herbal Ointment Blend on Wound Healing in Rats with Second-Degree Burns. *Burns* **2020**, *47*, 1442–1450. [CrossRef]

82. Beroual, K.; Agabou, A.; Abdeldjelil, M.-C.; Boutaghane, N.; Haouam, S.; Hamdi-Pacha, Y. Evaluation of Crude Flaxseed (*Linum usitatissimum* L.) Oil in Burn Wound Healenig New Zeland Rabbits. *Afr. J. Tradit. Complement. Altern. Med.* **2017**, *14*, 280–286. [CrossRef]

83. Trabelsi, I.; Ben Slima, S.; Ktari, N.; Bardaa, S.; Elkaroui, K.; Abdeslam, A.; Ben Salah, R. Purification, Composition and Biological Activities of a Novel Heteropolysaccharide Extracted from *Linum usitatissimum* L. Seeds on Laser Burn Wound. *Int. J. Biol. Macromol.* **2019**, *144*, 781–790. [CrossRef]

84. Franco, E.D.S.; de Aquino, C.M.F.; de Medeiros, P.L.; Evêncio, L.B.; Góes, A.J.D.S.; Maia, M.B.D.S. Effect of a Semisolid Formulation of *Linum usitatissimum* L. (Linseed) Oil on the Repair of Skin Wounds. *Evid.-Based Complement. Altern. Med.* **2011**, *2012*, 270752. [CrossRef]

85. Farahpour, M.R. Wound Healing Activity of Flaxseed *Linum usitatissimum* L. in Rats. *Afr. J. Pharm. Pharmacol.* **2011**, *5*, 258. [CrossRef]

86. Rafiee, S.; Nekouyian, N.; Sarabandi, F.; Chavoshi-Nejad, M.; Mohsenikia, M.; Yadollah-Damavandi, S.; Seifaee, A.; Jangholi, E.; Eghtedari, D.; Najafi, H.; et al. Effect of Topical Linum usitatissimum on Full Thickness Excisional Skin Wounds. *Trauma Mon.* **2016**, *22*, 39045. [CrossRef]

87. Abdul Jabbar, O.; Kashmoola, M.A.; Mustafa Al-Ahmad, B.E.; Mokhtar, K.I.; Muhammad, N.; Abdul Rahim, R.; Qouta, L.A. The Effect of Flaxseed Extract on Skin Elasticity of The Healing Wound In Rabbits. *IIUM Med. J. Malays.* **2019**, *18*, 230. [CrossRef]

88. ESCOP. Matricariae flos (Matricaria flower). Available online: https://escop.com/downloads/matricariae-flos-matricaria-flower/ (accessed on 17 August 2022).

89. HMPC. *Assessment Report on Matricaria recutita L., Flos*; EMA/HMPC/55837/2011; European Medicines Agency: London, UK, 2015. Available online: https://www.ema.europa.eu/en/documents/herbal-report/final-assessment-report-matricaria-recutita-l-flos-matricaria-recutita-l-aetheroleum-first-version_en.pdf (accessed on 17 August 2022).

90. Wu, Y.-N.; Xu, Y.; Yao, L. Anti-Inflammatory and Anti-allergic Effects of German Chamomile (*Matricaria chamomilla* L.). *J. Essent. Oil Bear. Plants* **2011**, *14*, 549–558. [CrossRef]

91. Jarrahi, M. An Experimental Study of the Effects of *Matricaria Chamomilla* Extract on Cutaneous Burn Wound Healing in Albino Rats. *Nat. Prod. Res.* **2008**, *22*, 422–427. [CrossRef]

92. Jarrahi, M.; Vafaei, A.A.; Taherian, A.A.; Miladi, H.; Pour, A.R. Evaluation of Topical *Matricaria Chamomilla* Extract Activity on Linear Incisional Wound Healing in Albino Rats. *Nat. Prod. Res.* **2010**, *24*, 697–702. [CrossRef]

93. Nematollahi, P.; Aref, N.M.; Meimandi, F.Z.; Rozei, S.L.; Zareé, H.; Mirlohi, S.M.J.; Rafiee, S.; Mohsenikia, M.; Soleymani, A.; Ashkani-Esfahani, S.; et al. Matricaria Chamomilla Extract Improves Diabetic Wound Healing in Rat Models. *Trauma Mon.* 2019; *in press*. [CrossRef]

94. Niknam, S.; Tofighi, Z.; Faramarzi, M.A.; Abdollahifar, M.A.; Sajadi, E.; Dinarvand, R.; Toliyat, T. Polyherbal Combination for Wound Healing: *Matricaria chamomilla* L. and *Punica granatum* L. *DARU J. Pharm. Sci.* **2021**, *29*, 133–145. [CrossRef]

95. Khashan, A.A.; Hamad, M.A.; Jadaan, M.S. In Vivo Antimicrobial Activity of Matricaria Chamomilla Extract against Pathogenic Bacteria Induced Skin Infections in Mice. *Syst. Rev. Pharm.* **2020**, *11*, 672–676. [CrossRef]

96. Nobrega, A.T.; Wagemaker, T.A.L.; Campos, P.M.B.G.M. Antioxidant Activity of *Matricaria chamomilla* L. Extract and Clinical Efficacy of Cosmetic Formulations Containing This Extract and Its Isolated Compounds. *J. Biomed. Biopharm. Res.* **2013**, *10*, 249–261. [CrossRef]

97. Ferreira, E.B.; Ciol, M.A.; De Meneses, A.G.; Bontempo, P.D.S.M.; Hoffman, J.M.; Dos Reis, P.E.D. Chamomile Gel versus Urea Cream to Prevent Acute Radiation Dermatitis in Head and Neck Cancer Patients: Results from a Preliminary Clinical Trial. *Integr. Cancer Ther.* **2020**, *19*, 1534735420962174. [CrossRef] [PubMed]

98. Lee, S.-H.; Heo, Y.; Kim, Y.-C. Effect of German Chamomile Oil Application on Alleviating Atopic Dermatitis-Like Immune Alterations in Mice. *J. Veter Sci.* **2010**, *11*, 35–41. [CrossRef] [PubMed]

99. ESCOP. Tormentillae Rhizoma. Available online: https://escop.com/wp-content/uploads/edd/2015/09/Tormentil.pdf (accessed on 17 August 2022).

100. HMPC. *Assessment Report on Potentilla erecta (L.) Raeusch., Rhizoma*; EMA/HMPC/5511/2010; European Medicines Agency: London, UK, 2010. Available online: https://www.ema.europa.eu/en/documents/herbal-report/final-assessment-report-potentilla-erecta-l-raeusch-rhizoma_en.pdf (accessed on 17 August 2022).

101. Melzig, M.F.; Böttger, S. Tormentillae rhizoma—Review for an Underestimated European Herbal Drug. *Planta Medica* **2020**, *86*, 1050–1057. [CrossRef] [PubMed]

102. Hoffmann, J.; Casetti, F.; Bullerkotte, U.; Haarhaus, B.; Vagedes, J.; Schempp, C.M.; Wölfle, U. Anti-Inflammatory Effects of Agrimoniin-Enriched Fractions of Potentilla erecta. *Molecules* **2016**, *21*, 792. [CrossRef]

103. Wölfle, U.; Hoffmann, J.; Haarhaus, B.; Mittapalli, V.R.; Schempp, C.M. Anti-Inflammatory and Vasoconstrictive Properties of Potentilla Erecta—A Traditional Medicinal Plant From the Northern Hemisphere. *J. Ethnopharmacol.* **2017**, *204*, 86–94. [CrossRef]

104. Kaltalioglu, K.; Balabanli, B.; Coskun-Cevher, S. Phenolic, Antioxidant, Antimicrobial, and In-vivo Wound Healing Properties of *Potentilla erecta* L. *Root Extract in Diabetic Rats.* **2020**, *19*, 264–274. [CrossRef]

105. HMPC. *Assessment Report on Quercus robur L., Quercus petraea (Matt.) Liebl., Quercus pubescens Willd, Cortex*; EMA/HMPC/3206/2009; European Medicines Agency: London, UK, 2010. Available online: https://www.ema.europa.eu/en/documents/herbal-report/final-assessment-report-quercus-robur-l-quercus-petraea-matt-liebl-quercus-pubescens-willd-cortex_en.pdf (accessed on 17 August 2022).

106. HMPC. *Community Herbal Monograph on Quercus robur L., Quercus petraea.(Matt.) Liebl., Quercus pubescens Willd, Cortex*; EMA/HMPC/3203/2009; European Medicines Agency: London, UK, 2010. Available online: https://www.ema.europa.eu/en/documents/herbal-monograph/final-community-herbal-monograph-quercus-robur-l-quercus-petraea-matt-liebl-quercus-pubescens-willd_en.pdf (accessed on 17 August 2022).

107. Hong, J.-A.; Bae, D.; Oh, K.-N.; Oh, D.-R.; Kim, Y.; Kim, Y.; Im, S.J.; Choi, E.-J.; Lee, S.-G.; Kim, M.; et al. Protective Effects of Quercus Acuta Thunb. Fruit Extract Against UVB-induced Photoaging through ERK/AP-1 Signaling Modulation in Human Keratinocytes. *BMC Complement. Med. Ther.* **2022**, *22*, 6. [CrossRef]

108. Yin, J.; Kim, H.H.; Hwang, I.H.; Kim, D.H.; Lee, M.W. Anti-Inflammatory Effects of Phenolic Compounds Isolated from Quercus Mongolica Fisch. ex Ledeb. on UVB-Irradiated Human Skin Cells. *Molecules* **2019**, *24*, 3094. [CrossRef]

109. Lee, C.S.; Yi, E.H.; Kim, H.-R.; Huh, S.-R.; Sung, S.-H.; Chung, M.H.; Ye, S.-K. Anti-dermatitis Effects of Oak Wood Vinegar on the DNCB-induced Contact Hypersensitivity via STAT3 Suppression. *J. Ethnopharmacol.* **2011**, *135*, 747–753. [CrossRef]

110. ESCOP. Salviae Folium (Sage Leaf). Available online: https://escop.com/downloads/salviae-folium-sage-leaf/ (accessed on 17 August 2022).

111. HMPC. *European Union Herbal Monograph on Salvia officinalis L., Folium*; EMA/HMPC/277152/2015; European Medicines Agency: London, UK, 2016. Available online: https://www.ema.europa.eu/en/documents/herbal-monograph/final-european-union-herbal-monograph-salvia-officinalis-l-folium-revision-1_en.pdf (accessed on 17 August 2022).

112. Grdiša, M.; Jug-Dujaković, M.; Lončarić, M.; Carović-Stanko, K.; Ninčević, T.; Liber, Z.; Radosavljević, I.; Šatović, Z. Dalmatian Sage (*Salvia officinalis* L.): A Review of Biochemical Contents, Medical Properties and Genetic Diversity. *Agric. Conspec. Sci.* **2015**, *80*, 69–78.

113. Baricevic, D.; Sosa, S.; Della Loggia, R.; Tubaro, A.; Simonovska, B.; Krasna, A.; Zupancic, A. Topical Anti-inflammatory Activity of *Salvia officinalis* L. Leaves: The Relevance of Ursolic Acid. *J. Ethnopharmacol.* **2001**, *75*, 125–132. [CrossRef]

114. Abu-Darwish, M.S.; Cabral, C.; Ferreira, I.V.; Gonçalves, M.J.; Cavaleiro, C.; Cruz, M.T.; Al-Bdour, T.H.; Salgueiro, L. Essential Oil of Common Sage (*Salvia officinalis* L.) from Jordan: Assessment of Safety in Mammalian Cells and Its Antifungal and Anti-Inflammatory Potential. *BioMed Res. Int.* **2013**, *2013*, 538940. [CrossRef] [PubMed]

115. Farahpour, M.R.; Pirkhezr, E.; Ashrafian, A.; Sonboli, A. Accelerated Healing by Topical Administration of Salvia Officinalis Essential Oil on Pseudomonas Aeruginosa and Staphylococcus Aureus Infected Wound Model. *Biomed. Pharmacother.* **2020**, *128*, 110120. [CrossRef] [PubMed]

116. Karimzadeh, S.; Farahpour, M.R. Topical Application of Salvia Officinalis Hydroethanolic Leaf Extract Improves Wound Healing Process. *Indian J. Exp. Biol.* **2017**, *55*, 98–106.

117. Khare, R.; Upmanyu, N.; Jha, M. Exploring the Potential Effect of Methanolic Extract of Salvia officinalis Against UV Exposed Skin Aging: In vivo and In vitro Model. *Curr. Aging Sci.* **2021**, *14*, 46–55. [CrossRef]

118. Reuter, J.; Jocher, A.; Hornstein, S.; Mönting, J.S.; Schempp, C.M. Sage Extract Rich in Phenolic Diterpenes Inhibits Ultraviolet-Induced Erythema in Vivo. *Planta Medica* **2007**, *73*, 1190–1191. [CrossRef]

119. He, X.; Bai, Y.; Zhao, Z.; Wang, X.; Fang, J.; Huang, L.; Zeng, M.; Zhang, Q.; Zhang, Y.; Zheng, X. Local and Traditional Uses, Phytochemistry, and Pharmacology of *Sophora japonica* L.: A Review. *J. Ethnopharmacol.* **2016**, *187*, 160–182. [CrossRef]

120. Li, L.; Huang, T.; Lan, C.; Ding, H.; Yan, C.; Dou, Y. Protective Effect of Polysaccharide from Sophora Japonica L. Flower Buds Against UVB Radiation in a Human Keratinocyte Cell Line (HaCaT Cells). *J. Photochem. Photobiol. B Biol.* **2018**, *191*, 135–142. [CrossRef]

121. Lee, H.K.; Kim, H.S.; Kim, Y.J.; Kim, J.S.; Park, Y.S.; Kang, J.S.; Yuk, D.Y.; Hong, J.T.; Kim, Y.; Han, S.-B. Sophoricoside isolated from Sophora japonica ameliorates contact dermatitis by inhibiting NF-κB signaling in B cells. *Int. Immunopharmacol.* **2013**, *15*, 467–473. [CrossRef]

122. Min, B.; Oh, S.R.; Lee, H.-K.; Takatsu, K.; Chang, I.-M.; Min, K.R.; Kim, Y. Sophoricoside Analogs as the IL-5 Inhibitors from Sophora japonica. *Planta Medica* **1999**, *65*, 408–412. [CrossRef]

123. Kim, B.H.; Chung, E.Y.; Ryu, J.-C.; Jung, S.-H.; Min, K.R.; Kim, Y. Anti-inflammatory Mode of Isoflavone Glycoside Sophoricoside by Inhibition of lnterleukin-6 and Cyclooxygenase-2 in Inflammatory Response. *Arch. Pharmacal Res.* **2003**, *26*, 306–311. [CrossRef] [PubMed]

124. Salehi, B.; Sharopov, F.; Tumer, T.B.; Ozleyen, A.; Rodríguez-Pérez, C.; Ezzat, S.M.; Azzini, E.; Hosseinabadi, T.; Butnariu, M.; Sarac, I.; et al. Symphytum Species: A Comprehensive Review on Chemical Composition, Food Applications and Phytopharmacology. *Molecules* **2019**, *24*, 2272. [CrossRef] [PubMed]

125. Sowa, I.; Paduch, R.; Strzemski, M.; Zielińska, S.; Rydzik-Strzemska, E.; Sawicki, J.; Kocjan, R.; Polkowski, J.; Matkowski, A.; Latalski, M.; et al. Proliferative and Antioxidant Activity of *Symphytum Officinale* Root Extract. *Nat. Prod. Res.* **2017**, *32*, 605–609. [CrossRef] [PubMed]

126. Silva-Barcellos, N.M.; Araujo, L.U.; Reis, P.G. In Vivo Wound Healing Effects of *Symphytum officinale* L. Leaves extract in different topical formulations. *Die Pharm. Int. J. Pharm. Sci.* **2012**, *67*, 355–360. [CrossRef]

127. Mulkijanyan, K.; Barbakadze, V.; Novikova, Z.; Sulakvelidze, M.; Gogilashvili, L.; Amiranashvili, L.; Merlani, M. Burn Healing Compositions from Caucasian Species of Comfrey (*Symphytum* L.). *Chemistry* **2009**, *3*, 2–5.

128. Matta, K.L.; Ochrymowycz, A.; Neplyuev, V.M.; Sinenko, A.; Pel, S.; Miiller, C.; Gracza, L.; Koch, H.; Löffler, E. Isolierung von Rosmarinsaure Aus Symphytum Officinale Und Ihre Anti-Inflammatorische Wirksamkeit in Einem In-Vitro-Model. *Arch. Pharm.* **1985**, *318*, 1090–1095.

129. Piwowarski, J.P.; Granica, S.; Kiss, A.K. Influence of Gut Microbiota-Derived Ellagitannins' Metabolites Urolithins on Pro-Inflammatory Activities of Human Neutrophils. *Planta Medica* **2014**, *80*, 887–895. [CrossRef]

130. Piwowarski, J.P.; Granica, S.; Zwierzyńska, M.; Stefańska, J.; Schopohl, P.; Melzig, M.F.; Kiss, A.K. Role of Human Gut Microbiota Metabolism in the Anti-inflammatory Effect of Traditionally Used Ellagitannin-Rich Plant Materials. *J. Ethnopharmacol.* **2014**, *155*, 801–809. [CrossRef]

International Journal of
Molecular Sciences

Review

The Relevance of Crystal Forms in the Pharmaceutical Field: Sword of Damocles or Innovation Tools?

Dario Braga *, Lucia Casali and Fabrizia Grepioni

Department of Chemistry G. Ciamician, University of Bologna, Via Selmi 2, 40126 Bologna, Italy
* Correspondence: dario.braga@unibo.it

Abstract: This review is aimed to provide to an "educated but non-expert" readership and an overview of the scientific, commercial, and ethical importance of investigating the crystalline forms (polymorphs, hydrates, and co-crystals) of active pharmaceutical ingredients (API). The existence of multiple crystal forms of an API is relevant not only for the selection of the best solid material to carry through the various stages of drug development, including the choice of dosage and of excipients suitable for drug development and marketing, but also in terms of intellectual property protection and/or extension. This is because the physico-chemical properties, such as solubility, dissolution rate, thermal stability, processability, etc., of the solid API may depend, sometimes dramatically, on the crystal form, with important implications on the drug's ultimate efficacy. This review will recount how the scientific community and the pharmaceutical industry learned from the catastrophic consequences of the appearance of new, more stable, and unsuspected crystal forms. The relevant aspects of hydrates, the most common pharmaceutical solid solvates, and of co-crystals, the association of two or more solid components in the same crystalline materials, will also be discussed. Examples will be provided of how to tackle multiple crystal forms with screening protocols and theoretical approaches, and ultimately how to turn into discovery and innovation the purposed preparation of new crystalline forms of an API.

Keywords: crystal polymorphism; hydrates; co-crystals of active pharmaceuticals

Citation: Braga, D.; Casali, L.; Grepioni, F. The Relevance of Crystal Forms in the Pharmaceutical Field: Sword of Damocles or Innovation Tools? *Int. J. Mol. Sci.* **2022**, *23*, 9013. https://doi.org/10.3390/ijms23169013

Academic Editors: Geoffrey Brown, Enikö Kallay and Andrzej Kutner

Received: 14 July 2022
Accepted: 5 August 2022
Published: 12 August 2022

Publisher's Note: MDPI stays neutral with regard to jurisdictional claims in published maps and institutional affiliations.

1. Introduction

Crystal polymorphism has been known and studied since the early days of solid-state chemistry and crystallography [1,2], but it is only in the recent past that it has emerged as a strategic research area involving the use of molecular crystalline materials (pharmaceuticals, nutraceuticals, fertilizers, pigments, high-energy materials, etc.) [3–6].

Although the unexpected appearance of a new crystal form of a known active principle is often a threat for an API on the market (see below), it is also true that the urge for a careful pre-screening and form selection is a potent stimulus for research in various areas, and provides opportunities for innovation and new discoveries, especially in the burgeoning subfield of molecular co-crystals. This latter class of crystalline compounds is proving particularly apt to innovation and the development of new drugs and/or of new formulations of old ones [7,8].

In order to help the reader to put the current academic and industrial interest on crystal forms into a wide perspective, this review will move from the early days awareness of the importance of searching for crystal forms to the current impact and consequences for the scientific community and for the industrial sector of the discovery of polymorphs, solvates, and co-crystals of APIs.

2. Polymorphism: The Awareness

The existence of a substance in more than one crystal form has been known since 1822 [1], but we have to wait until 1962 for Walter McCrone, unanimously recognized as the father of this field of research, to provide the definition of a polymorph as "a solid crystalline phase of a given compound resulting from the possibility of at least two crystalline arrangements of the molecules of that compound in the solid state … every compound has different polymorphic forms … " He also opinioned that "the number of forms known for a given compound is proportional to the time and money spent in research on that compound" [2]. This "prediction" was published in 1962, but it took more than three decades before the phenomenon of polymorphism would hit the pharmaceutical field in a rather shocking way (see below the Section 5).

In addition to polymorphs, i.e., crystals having the same chemical composition but different structures, the term "crystal forms" nowadays encompasses not only the association of the molecule of interest with solvents (solvates), but also the association with molecules that form solids at room temperature (co-crystals) or with salts (ionic co-crystals). It is important to stress that all these crystal forms can be polymorphic. Therefore, understanding polymorphism is of primary importance when embarking on any drug development/authorization/manufacture/formulation process.

The effort is by no means only theoretical or academic and has important implications both in terms of the drug ultimate efficacy and of the protection of the intellectual property rights associated with the final pharmaceutical product [9].

A number of statistical analyses of the literature have been carried out in an attempt to estimate the extent of polymorphism. A search of the Cambridge Structural Database on the keywords "polymorph", "form", "modification", or "phase" indicates that approximately 4.2% of the ~1,200,000 entries fall into this category. Approximately 25% of the entries are either solvates or hydrates. Other studies based on different selection criteria reveal results falling somewhere between these two extremes [10–13].

It is worthy of note that the "International Conference on Harmonization" [14] includes under the heading of "polymorphs": "single entity polymorphs; molecular adducts (solvates, hydrates), amorphous forms". The FDA currently requires that pharmaceutical manufacturers investigate the polymorphism of the active ingredients before clinical tests and that polymorphism is continuously monitored during scale-up and production processes [15]. The European Patent Office also demands the characterization of solid drugs by means of X-ray diffraction to ensure the integrity of the crystal form [16].

3. Polymorphism: The Implications. Different Crystals. Different Properties

A truly significant contribution to the understanding of crystal polymorphism in all its numerous facets and industrial implications (from polymorph detection, screening, and assessment to the impact on intellectual property rights in the case of active principles) is due to the work of the late Joel Bernstein and to his intense dissemination efforts [3].

Polymorphs, although possessing exactly the same chemical composition, may differ in a number of properties (see Table 1). The analogy with molecular isomers is strong: if a crystal is seen as a supermolecule then its polymorphic modifications are solid state super-isomers [17]. These isomers may show physico-chemical differences that, in some cases, are as large as to make them behave as practically different species altogether.

Table 1. Physico-chemical properties that may depend on the crystal form.

Physical and Thermodynamic Properties	Density and refractive index, thermal and electrical conductivity, hygroscopicity, melting points, free energy and chemical potential, heat capacity, vapor pressure, solubility, thermal stability, and color and shape of crystals.
Spectroscopic Properties	Electronic, vibrational, and rotational properties, and nuclear magnetic resonance spectral features.
Kinetic Properties	Rate of dissolution, kinetics of solid-state reactions, and kinetic inertness.
Surface Properties	Surface free energy, crystal habit, surface area, and particle size distribution.
Mechanical Properties	Hardness, compression, and thermal expansion.
Chemical Properties	Chemical and photochemical reactivity.

Even shape and color may differ in a significant manner from form to form with important implications at the manufacturing and processing levels. A striking example is provided by ROY (ROY = red, orange, yellow polymorphs of 5-methyl-2-[(2-nitrophenyl)amino]-3-thiophene carbonitrile), the most polymorphic compound in the Cambridge Structural Database, with its crystal forms differing in color and morphology as shown in Figure 1 [18,19]. The palette of polymorphs of ROY has been recently enriched by the discovery of new ways to search for polymorphs and increase polymorphic diversity, based on crystallization induced by suitably designed mixed-crystal seeds (see also below) [20].

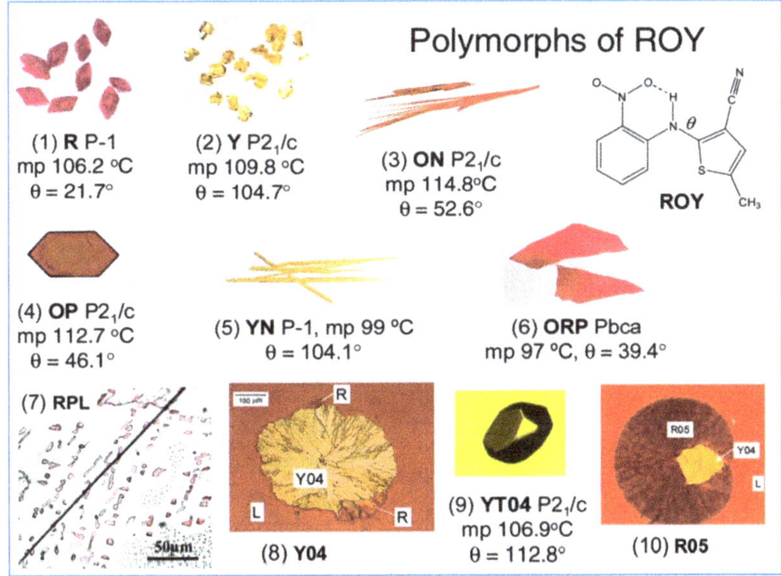

Figure 1. Differences in shape and color between the first ten (out of 13) discovered polymorphs of ROY (ROY = red, orange, yellow polymorphs of 5-methyl-2-[(2-nitrophenyl) amino]-3-thiophene carbonitrile). Reprinted with permission from Ref. [21]. 2010, American Chemical Society.

4. Polymorphism: The Rationale

Crystal polymorphism is a manifestation of the perpetual thermodynamic–kinetic dualism ruling the physical world. The thermodynamic stability of a polymorph is strictly dependent on pressure and temperature; however, due to kinetic considerations, metastable forms can exist or coexist in the presence of more stable forms [22,23].

For crystals of organic molecules, such as most APIs, the energy difference between different polymorphic forms is usually of the order of few kJ/mol, mainly because of the entropic contribution to the free energy. Polymorphs can be grouped in two major categories depending on whether there is a transition point between two solid phases at a given temperature, i.e., the two phases interconvert via a phase transition, or the two phases do not share a point of identical free energy before melting, i.e., the two phases do not interconvert via a phase transition. In the first case, the two phases are said to be enantiotropically related, while in the second case the two phases are said to be monotropically related (see Figure 2) and will be discussed briefly in the following.

When polymorphs are enantiotropically related, there is a transition temperature at a temperature below the melting point of the lower melting form. The two crystalline phases are in equilibrium at the transition temperature. The transition temperature is real (Burger–Ramberger Rule 1) and corresponds to $\Delta G_{trans} = 0$, i.e., $\Delta H = T_{trans} \Delta S$. Melting is observed only for the polymorph that is stable at a higher temperature (mpI in Figure 2, left).

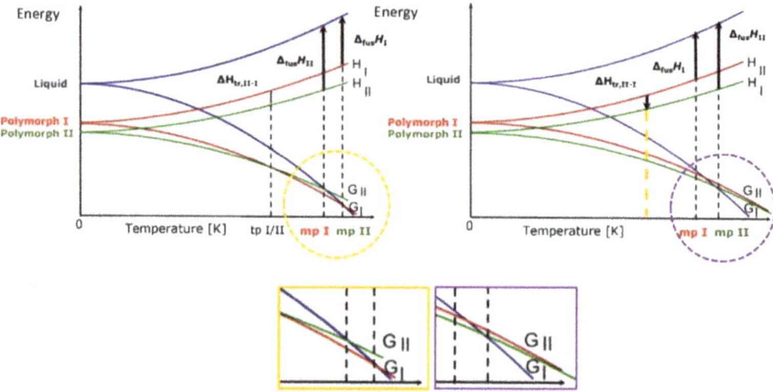

Figure 2. E/T diagrams, with G and H vs. temperature in the enantiotropic (**left**) and monotropic cases (**right**).

In the case of a monotropic system, the transition temperature between two crystals is only virtual, since the two G curves only cross in the field of stability of the liquid phase. The two polymorphs have independent melting points (mpII and mpI in the Figure 2, right), and they cannot interconvert in the solid state, as there is no point in the phase diagram where $\Delta G_{trans} = 0$. The transformation can only occur in one direction, from the metastable to the stable form, and cannot be predicted on a thermodynamic ground but may be activated for kinetic reasons. As it will be shown in the following, it is often the case that a thermodynamically metastable crystal form is kinetically favored at the nucleation stage and is crystallized in preference to the thermodynamic form. Therefore, it is often possible to make intentional use of thermodynamically metastable crystal forms to take advantage of very special properties (see below).

The transformation of a stable to a metastable polymorph in a monotropic system can occur only if it is mediated by a liquid or gas phase, as in fast recrystallization from melt, crystallization from solution, or in vapour digestion processes. Conversion from the metastable to the stable form can be obtained via slurry, or may occur because of changes in pressure, as during a mechanical treatment; it can also be triggered by the presence of impurities.

In this respect, monotropic systems are the true Sword of Damocles for the pharmaceutical industry, because the interplay of kinetic and thermodynamic factors in a crystallization process is often unpredictable, with consequences that are well known to the practitioners in the area, as it will be discussed below.

As efficaciously pointed out by Bernstein "it is sometimes difficult to comprehend why and how new polymorphs still emerge (while others disappear) long after crystal-form screens presumably have been completed. [...] The point is that it can never be stated with certainty that the most stable form has been found; at best it can be determined which of the known forms is the most stable. [...] a new (and most often more stable) form can appear at any stage in the history of a compound (or life-cycle of a drug)" [24].

These are the reasons why the "quest for polymorphs" has become a central point in the development of a substance that is administered in the solid form, whether this is a drug, a nutraceutical, a fertilizer, etc. This will be clarified in the following.

4.1. Examples of Polymorphism in Single Component (Unary) Systems

An early textbook example of polymorphism affecting a very commonly used drug is provided by paracetamol.

The crystallization of paracetamol from methanol affords a so-called Form I, which although poorly compressible, is the marketed phase. Form I [25] is the thermodynamically stable form at all temperatures, and melts at 440–445 K with a ΔH_{fus} of 26–34 kJ mol^{-1}.

Crystallization from benzyl alcohol yields the highly compressible Form II [26], which is metastable, and melts in the range 427–433 K with a ΔH_{fus} of 26.4–33.5 kJ mol^{-1}. The difference in the hydrogen bonded chains, constituting the basic packing motif in both forms, is shown in Figure 3. A Form III has also been discovered, which can only be stabilized under certain conditions [27].

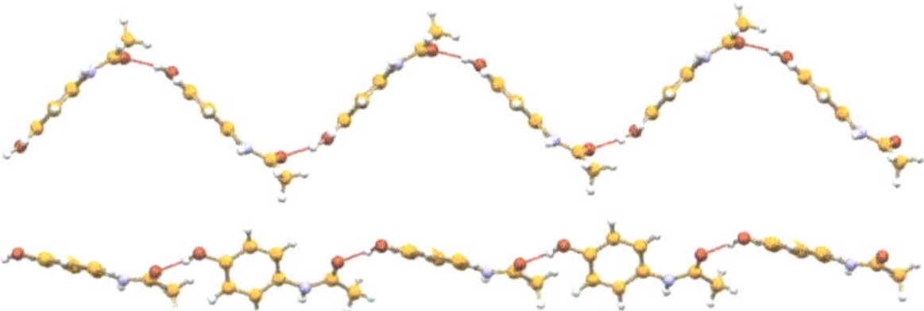

Figure 3. Hydrogen bonded chains in paracetamol Form I (**top**, refcode HXACAN04) and Form II (**bottom**, refcode HXACAN21).

The two forms have different crystal shapes. Moreover, Form II is elusive and hard to crystallize, unless selective impurities are present in solution. Recently, it has been shown that metacetamol, used as an additive, inhibits the growth of Form I and favors the growth of Form II (see Figure 4) [28].

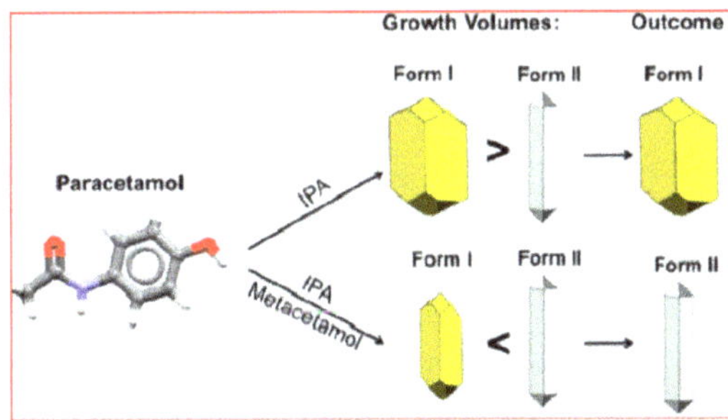

Figure 4. Crystal habits and shapes of the two polymorphs of paracetamol. The outcome of the crystallization can be changed by doping the crystallization with metacetamol. Reprinted with permission from Ref. [28]. 2020, American Chemical Society.

Importantly, the two types of paracetamol crystals have also been shown to possess different wettability properties, with consequences on the way the drug is processed and formulated [29].

An example of the dependence of bioavailability on polymorphic form is provided by the antibiotic chloramphenicol palmitate. Chloramphenicol palmitate exists in three polymorphic forms [30], which recently have been fully characterized thanks to advances in analytical methods. Of these, the so-called forms A and B are monotropically related. Form B is pharmacologically active and is used in suspensions (see Figure 5), while Form A is inactive as an antibiotic. However, Form B is metastable under ambient conditions and, due to its better solubility, in suspension it slowly recrystallizes into form A. Using a thermodynamically metastable modification in the production of tablets, creams, suspensions, and solutions, is sometimes the reason why unwanted changes take place upon storage, caused by transition into the thermodynamically stable modification at ambient conditions. Chloramphenicol palmitate shows that it is possible to make use of thermodynamically unstable crystal forms taking advantage of the considerable kinetic inertness [31].

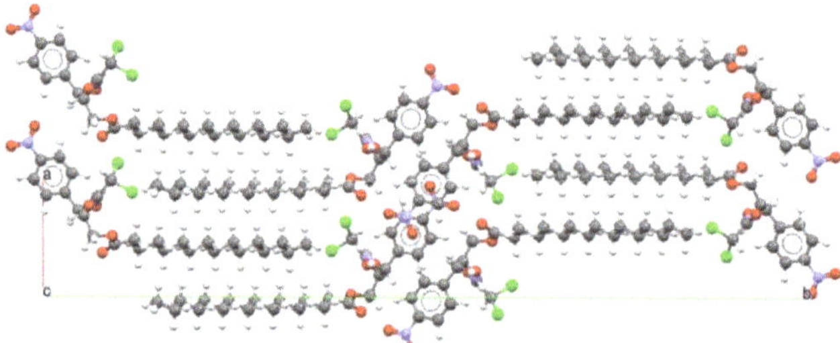

Figure 5. Crystal packing of chloramphenicol palmitate Form B (view down the crystallographic *c*-axis).

Another example is provided by the drug Bitopertin, which has been shown to possess three unsolvated, non-hygroscopic crystalline forms, designated as form A, form B, and form C [32]. Form A is enantiotropically related to form B, with form A being the thermodynamically stable polymorph below the transition temperature (~83 °C). Form A and form C are also enantiotropically related, with form A being the thermodynamically stable form below the calculated transition temperature (~70 °C). Form B and form C are otherwise monotropically related. The complex relationship between the three phases is shown in Figure 6.

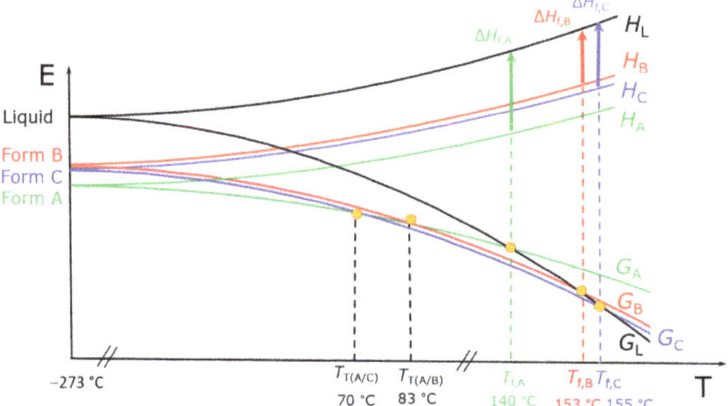

Figure 6. E/T (energy vs. temperature) diagram showing the complex phase relationship between the three unsolvated, non-hygroscopic crystalline forms, designated as form A, form B, and form C of Bitopertin. Figure adapted from [32].

The reader, by now, will have appreciated that the nomenclature of crystal forms is a relevant problem. There is no convention on the naming of polymorphs, with consequences in the understanding of differences and properties of one or another form. This aspect may become particularly relevant when tackling intellectual property issues related to polymorphism.

4.2. Conformational Polymorphism

Conformational polymorphism, viz. polymorphism originated by different molecular conformations in different crystals, is a widespread phenomenon. It has been estimated that ca. 39% of the flexible organic molecules in the CSD exhibit conformational polymorphism [33]. As pointed out by Cruz-Cabeza and Bernstein, however, conformational polymorphism results from "conformational changes", which should not be confused with "conformational adjustments". Conformational adjustments occur for any structurally non-rigid molecule in the solid state as a compromise between the optimization of packing energy and the optimization of molecular structure (with respect to the gas-phase unconstrained environment); no energy barrier is involved. Conformational changes, on the contrary, are observed only if an energy barrier separates distinct minima in the intramolecular energy conformational curve, as shown in Figure 7.

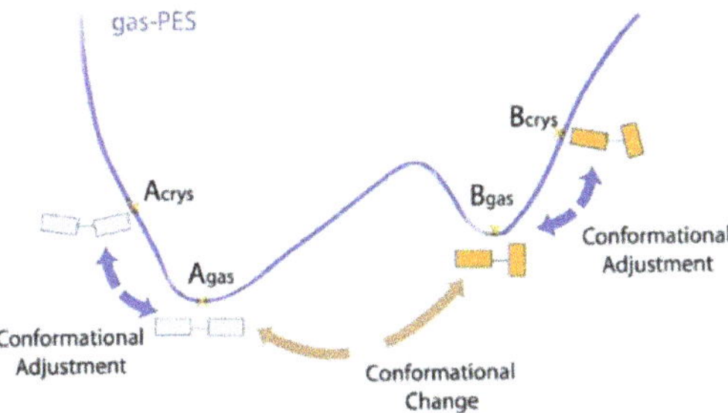

Figure 7. Schematic representation of the concepts of "conformation change" and "conformational adjustment" in a PES (potential energy surface) profile. Reprinted with permission from Ref. [33]. 2014, American Chemical Society.

L-glutamic acid is an example of conformational polymorphism. L-glutamic acid, in its zwitterionic form, crystallizes in the two forms α and β, both orthorhombic $P2_12_12_1$ (see Figure 8). A detailed thermodynamic investigation of the temperature dependence of the two forms has shown that α-glutamic acid is the preferred form at low temperatures and the β form is most stable at ambient temperatures [34].

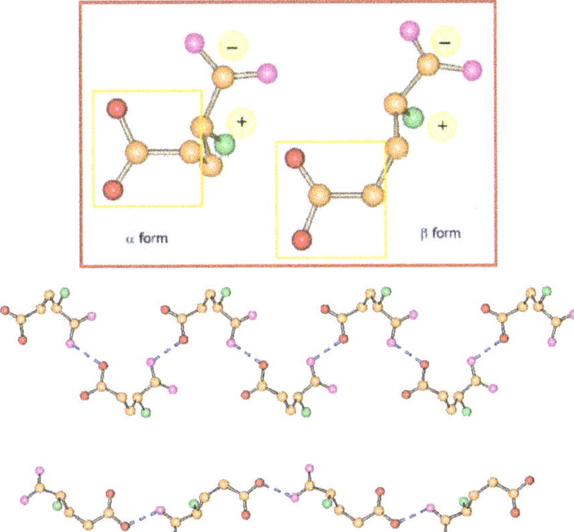

Figure 8. (**Top**) The α and β forms of L-glutamic acid and a comparison of the hydrogen bonded chains in the α (**middle**) and β (**bottom**) polymorphs.

4.3. Tautomeric Polymorphism

Tautomeric polymorphism may occur due to the crystallization of different tautomers. A textbook example is provided by barbituric acid. The keto form, used in all common representations of this important chemical, is the preferred tautomer in solution; it is also found in two polymorphs and one hydrated form. The stable thermodynamic form, however, as shown by both solid-state NMR and X-ray diffraction [35,36], turned out to be the enol form IV, obtained by the grinding of commercial barbituric acid, or via extremely slow solid-state conversion upon storage at ambient conditions. See Figure 9 for a view of the packing of form IV.

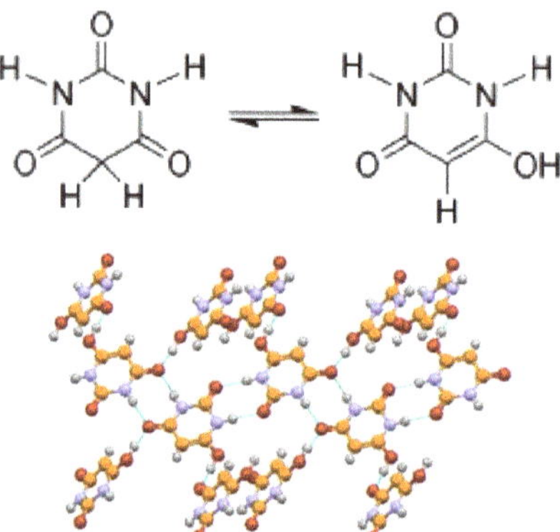

Figure 9. (**Top**): keto/enol tautomerism of barbituric acid; (**bottom**): crystal structure at room temperature of the enol form.

The isostructural compound 2-thiobarbituric acid possesses five polymorphs and one hydrated form. In both the crystalline form II and the hydrate form, the 2-thiobarbituric acid molecules are present in the enol form, whereas only the keto isomer is present in crystalline forms I, III, V, and VI. The stable form IV can also be obtained mechanochemically, as in the case of barbituric acid, and has been shown, by single-crystal X-ray diffraction and 1D and 2D (1H, 13C, and 15N) solid-state NMR spectroscopy, to contain both tautomers in a 50:50 ordered distribution (see Figure 10).

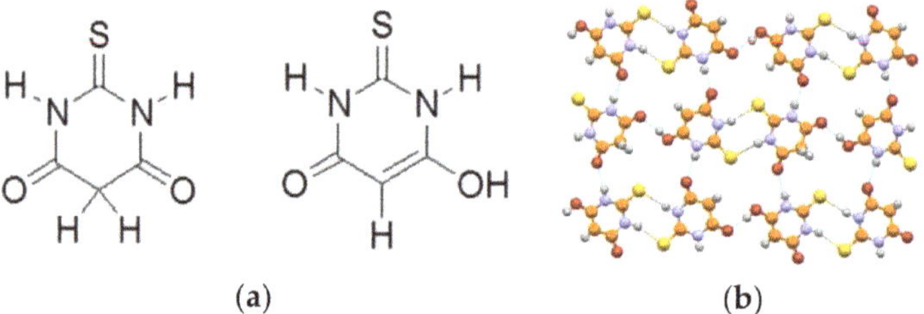

(a)   (b)

Figure 10. (**a**) Keto/enol tautomerism of thiobarbituric acid and (**b**) the crystal structure of the most stable form at room temperature, showing the presence of both the keto and enol forms [37].

5. Polymorphism: The Impact

The examples provided in the previous sections were intended only to give an idea of the spread and complexity of the phenomenon. It was only when some major "polymorphism incidents" severely hit the pharmaceutical industry, however, that the community at large became aware of the "sword of Damocles".

Undoubtedly, the case of the drug Ritonavir (Norvir®) is one of the most striking, also because it had a huge impact on a particularly fragile typology of patients. Norvir® was produced by Abbott and administered for the treatment of HIV. After many years of research, production, and distribution, in 1998, drug production lines begun to show problems related to "undesirable" crystal formation in a series of production batches that failed the dissolution test. An investigation of the reason for the failure showed the unexpected appearance of a new crystalline form of ritonavir that affected the way the drug dissolved, hence its absorption [38]. In spite of all the efforts, Abbott was not able to avoid formation of what turned out to be the thermodynamically more stable, much less soluble, form of ritonavir, form II. Form I and form II are monotropically related with no thermodynamic solid-to-solid transition point at any temperature, hence crystallization of form II could not be predicted.

In terms of crystal structure, the two forms differ in the relative arrangement of the molecules, which affects the hydrogen bonding pattern, as is shown in Figure 11.

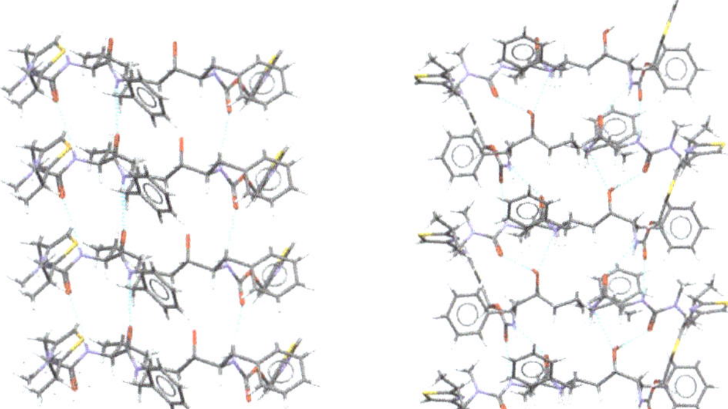

Figure 11. The hydrogen bonding networks in crystals of Ritonavir form I (**left**) and form II (**right**) [38].

This dramatic incident (the drug was not available for patients till Abbott found an alternative formulation based on soft-gel capsules) was a shock for the pharmaceutical industry and prompted a more thorough investigation of the relative stability of crystal forms. The episode has been thoroughly described in a review by Bučar, Lancaster, and Bernstein [24]. Subsequent investigations led to the discovery of several additional crystalline forms of ritonavir [39,40], all less thermodynamically stable than the "unwanted" Form II.

Another important case of unexpected (and unwanted) appearance of a more stable monotropic crystal form of a drug is that of Rotigotine (Neupro®), a Parkinson drug produced by UCB and administered to patients as skin patches. In 2008, a new form suddenly appeared, which crystallised in the patches (see Figure 12), reducing the drug efficacy [41]. The product had to be withdrawn from the market with considerable impact on the patients and on the company. The new crystal form was described in a patent filed in November 2008 and granted in July 2012 [42,43].

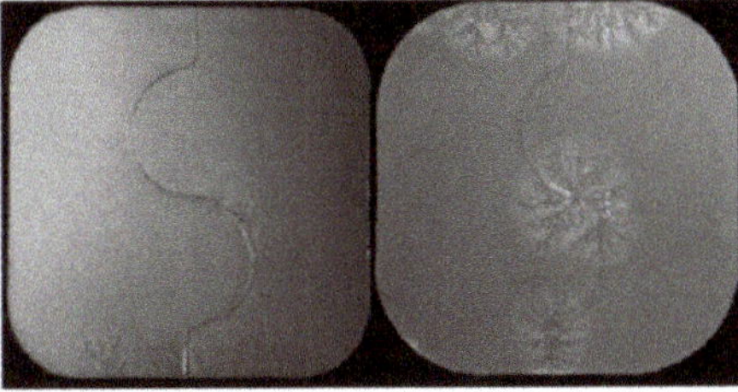

Figure 12. The appearance of the stable crystalline form of Rotigotine on the patches used for administering the drug. Reprinted with permission from Ref. [43]. 2015, Elsevier.

There are, of course, several other examples of "disappearing (and reappearing) polymorphs". The interested reader is addressed to the review published by J. Bernstein and others in 2015 [24,44].

5.1. Polymorphism: The Reaction of the Scientific Community

Unsurprisingly, the events described above motivated renewed efforts from both the academic and industrial communities. Polymorphism began to be systematically investigated and a number of spinoff companies or new research branches within large pharmaceutical companies were launched. Nonetheless, despite the knowledge of the factors that can cause polymorphs to "appear" (or to disappear), our ability to predict the real occurrence of polymorphism is still embryonic. In most cases, the crystallization of a new crystal form of a substance is still an unexpected event (for example, accidental seeding with impurities may trigger the crystallization of a new, more stable polymorph) rather than the result of a controlled process.

It became clear that the investigation of the crystalline phase(s) of a new API was not simply the characterization of the new pharmaceutical in its solid form, but, rather, the beginning of a long journey in the quest for crystalline materials with controllable properties. This journey required specific skill and training and also access to a variety of solid-state methods and techniques to be used in combination. The objective is that of minimizing, if not eliminating, the chances of an unexpected appearance of unknown new crystal forms of a drug at later stages of its development, or even when the drug is already on the market.

It is now clear that it is not only necessary to explore as thoroughly as possible the "crystal space domain" of the molecule of interest, but also to be able to follow the production, storage, and distribution of the product to guarantee persistence of the solid form, hence of the selected properties.

Polymorph assessment has indeed become part of the system of quality control in the pharmaceutical industry [4–6]. It is necessary to make sure that the scaling-up from laboratory preparation to industrial production does not introduce variations in crystal forms. Polymorph assessment also guarantees that the product conforms to the guidelines of the appropriate regulatory agencies and does not infringe on the intellectual property protection that may cover other crystal forms. The schematic diagram in Figure 13 shows how polymorph screening can lead to relevant patenting in the process of drug development, while continuous polymorph assessment will be required once the drug is on the market.

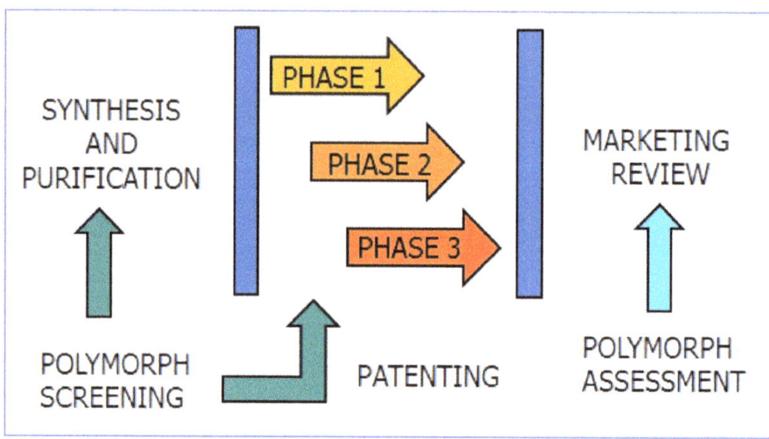

Figure 13. A flow-chart showing how polymorph screening ought to be associated with the API selection, before entering trial stage, and the quality control process, once the API is on the market.

Both initial polymorph screening and continuous crystal form assessment require the combined use of several solid-state techniques, among them (not exclusively or in any preferential order): microscopy and hot stage microscopy (HSM), differential scanning calorimetry (DSC), thermogravimetric analysis (TGA), infrared and Raman spectroscopy (IR and Raman), single crystal/powder X-ray diffraction (SCXRD, PXRD), and solid state nuclear magnetic resonance spectroscopy (SSNMR) [45]. The discussion of these techniques and of the pros and cons and pitfalls is well beyond the scope of this review.

Clearly, in the pharmaceutical field, the screening of crystal forms is motivated by safety and commercial necessities and is relevant in terms of patenting and in general of intellectual property protection. Interestingly, this has had repercussion also for fundamental science. Litigations over polymorphs and hydrates, in addition to the polymorphism "incidents" mentioned above, have fueled research in solid-state chemistry, and oriented the experience and competence of many academic research groups worldwide. The birth of many spinoff companies, responding to the demand of industrial outsourcing of accurate solid-state investigation, ought also to be mentioned.

5.2. Crystal Structure Prediction

On closing this section, it is also important to mention the increasing importance that is being acquired by computational crystal structure prediction (CSP). The term "crystal structure prediction" (CSP) comprises computational methods to explore the thermodynamic domain of molecular crystals. CSP aims to find the most thermodynamically stable crystal structure of a given molecule by evaluating the crystal energy landscape. In this respect, crystal structure prediction is complementary to experimental screening and provides information on the existence and relative energies of polymorphs [46,47].

As we have seen, if polymorphs are enantiotropically related, the stable structure depends on the temperature, i.e., polymorphs may interconvert. If the polymorphs are monotropically related, thermodynamically metastable polymorphs may be kinetically inert and persist indefinitely because of the difficulty in transforming into the stable form. The cases discussed above of ritonavir and rotigotine are emblematic examples of the consequences of the late appearance of more stable crystal forms. It would thus be of paramount importance in the development of a new drug, or of any new material where the solid form properties are essential for its utilization, to reach a good level of confidence on the relative thermodynamic stability of the crystal phase under examination. Even more so it would be of great value to be able to design ab initio the most stable crystal structure of a given molecule.

Because of polymorphism, computed crystal energy landscapes invariably contain several crystal structures separated by small differences in energy. Hence, the main use of CSP is to explore the range of packings and relative energies of the thermodynamically feasible crystal structures. A successful CSP would always generate the most stable crystal structure; the same structure, however, would invariably be obtained experimentally only if the crystallisation process were completely under thermodynamic control.

A detailed discussion of CSP is beyond the scope of this review. The interested reader is addressed to a more specialized literature [46,48]. It is worth mentioning, however, that crystal structure prediction methods are periodically assessed via the Blind Tests organized by the Cambridge Crystallographic Data Centre [49]. The Blind Test is based on a collection of unpublished crystal structures, which are sent out in the form of chemical diagrams to those developing CSP methods, with the challenge to submit predictions of the crystal structure by a given deadline.

The sixth blind test of organic crystal structure prediction was held in 2016 over five target systems constituted of a small nearly rigid molecule, a polymorphic drug candidate, a chloride salt hydrate, a co-crystal, and a bulky flexible molecule (see Table 2). The challenge saw the participation of 25 teams [50].

All targets, apart from a single potentially disordered polymorph of the drug candidate, were predicted by at least one participating group, each group being allowed to propose up to 100 predicted structures ranked in order of likelihood [50].

Overall, the results of the 2016 Blind Test demonstrates the increased maturity of CSP methods and shows how CSP calculations can guide and complement our understanding and experimental studies of organic solid forms.

The seventh crystal structure prediction blind test, organized by the CCDC, was launched in October 2020 and registration closed on 14 June 2022: results will be presented at the end of September 2022 [51].

Table 2. The five systems used in the sixth blind test.

Target	Chemical Diagram	Crystallization Conditions, Remarks, and Clarifications
(XXII)		Crystallized from an acetone/water mixture; chiral-like character due to potential flexibility of the six-membered ring, but no chiral precursors used in synthesis.
(XXIII)		Five known polymorphs (*A–E*). The most stable polymorphs at 257 and 293 K. Crystallization conditions include slow evaporation of acetone solution and of an ethyl acetate:water mixture.
(XXIV)		Crystallized from 1 M HCl solution. The substituents of the C=C double bond are in the *cis* configuration.
(XXV)		Slow evaporation of a methanol solution, which contained a racemic mixture of the enantiomers of Tröger's base.
(XXVI)		Slow evaporation from 1:1 mixture of hexane and dichloromethane. No chiral precursors used in synthesis.

The fact that the most stable structure in CSP is not always observed reflects the limitations of the thermodynamic modelling of crystallization. Kinetic effects on the nucleation and growth of less stable crystal forms are not taken in account. Moreover, a crystallization in the "real world" (in vitro, not in silico) implies the use of solvents and compounds with a purity profile determined by the detection methods and also of hardware (glassware and instruments) with the possible release of microparticles, all implying the possibility of unintentional seeding and other physical effects that may favor the nucleation of less thermodynamically stable forms.

However, CSP is progressing rapidly. The increasing success of the Blind Tests indicate that the future will show a wider utilization of CSP to guide the experimental work in the quest for "missing" polymorphs [48].

6. Solvates and, Especially, Hydrates

An API can form polymorphs, i.e., different crystals of the same chemical entity; however, an API can also form different crystal structures with solvent molecules, i.e., solvates [52]. Crystalline solvates may present different stoichiometries, i.e., mono-, di-, tri-solvates, etc. or they can be non-stoichiometric. In this latter case, they adsorb/release a variable number of solvent molecules depending on the temperature, relative humidity (in the case of hydrates), or other physical conditions. Solvate formation is also largely unpredictable. Solvates are sometimes called "pseudo-polymorphs", but this practice ought to be discouraged, since solvates have a different chemical composition from the pure API. Moreover, crystalline solvates can show polymorphism, i.e., the same compositions, the same API/solvent stoichiometric ratio, but different crystal structures.

When crystalline materials are being used for living beings, the permitted solvents are often restricted to water and very few bio-compatible (GRAS: generally regarded as safe) solvents [53]. Clearly, in the case of an API, hydrates are not only common but are also amply manageable and fully acceptable in formulation.

For this reason, we will hereafter focus on hydrates, with the understanding that solvates and hydrates share much in terms of the methods of characterization and analysis.

In fact, precipitation from a solution either by solvent evaporation or by a temperature gradient is the most common way to obtain crystals. In these conditions, the formation of a solvate is an unsurprising event. In the case of hydrates, it is also very common that water is taken up from glassware, reactants, and solvents. The situation is further complicated by the ubiquity of water [54]. The formation of hydrates, though not certain, is indeed very common [52,55].

A statistical analysis based on the crystal structures deposited in the CSD [56] (until 2016) showed that approximately 7–8% of the organic crystal structures are in the form of hydrates, whereas only 1.4% form single entity polymorphs, as listed in Table 3.

Table 3. Occurrence in the CSD of various crystal forms [56].

CRYSTAL FORMS	% of All Organic Structures
Organic crystal structures	100
Single component molecular organic structures	72.1
Single component polymorphic structures	1.4
Hydrates	7.4
Molecular organic hydrates	2.7
Polymorphic molecular organic hydrates	1.0
Cocrystals	1.1
Polymorphic cocrystals	1.9

The association of water with a crystalline material can take different forms [52]. Water may form stoichiometric hydrates, whereby water molecules are linked, generally via a hydrogen bond and/or via coordination of the oxygen atoms to other atoms in the crystal, or may be absorbed in disordered regions or cracks and cavities within the crystal mosaic or adsorbed on the crystal surface (Figure 14). This is an important notion to keep in mind when evaluating the amount of water present in a crystalline material, especially when the extent of hydration is a relevant aspect for the utilization of the crystalline material, as in the formulation of pharmaceuticals.

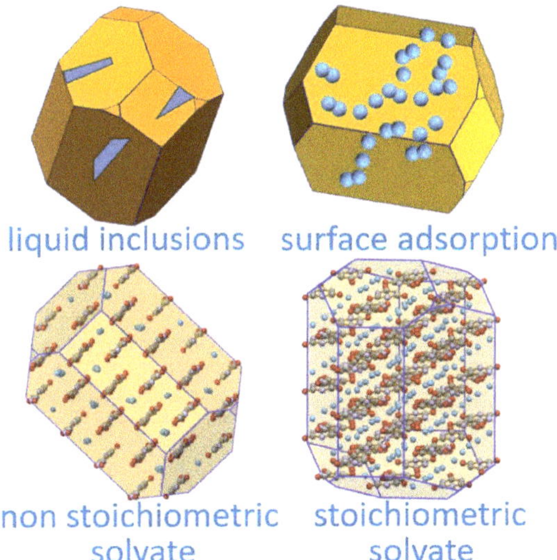

Figure 14. A representation of the types of association of a solvent with a crystalline solid.

Stoichiometric versus Non-Stoichiometric Hydrates

As mentioned above, crystalline hydrates can be non-stoichiometric, i.e., may absorb a variable number of water molecules in the crystal structure, depending on the environmental conditions such as humidity and temperature. The most relevant structural difference between a stoichiometric and a non-stoichiometric hydrate is that, in the former case, hydrate and anhydrate usually possess different crystal structures, whereas in the non-stoichiometric case the features of the crystal structure are retained (almost) unaltered upon absorption and release various amounts of water. This is made possible by the presence, in the structure of non-stoichiometric hydrates, of channels or cavities that allow the entrance, transit, and exit of water molecules without significantly affecting the overall packing.

The stoichiometric versus non-stoichiometric nature of the hydrate can be assessed by dynamic vapor sorption (DVS), a technique that allows us to establish whether water uptake is continuous as the humidity increases or passes through an abrupt change, which accompanies the crystal structure transformation from anhydrate to hydrate. In the stoichiometric case, anhydrous and hydrated crystals show distinctly different physicochemical properties (e.g., aqueous solubility) and hysteresis between hydration and dehydration. If the structures are investigated by diffraction, they will very likely show different unit cells and a different organization of the molecules. In the non-stoichiometric case, on the contrary, the difference in the physico-chemical properties will be small and the solid phases will show a highly variable composition in terms of water content, but a very small variation in cell parameters and overall structure organization.

Ampicillin and theophylline are good examples of the dependence of hydrate formation on water activity in organic solvents for the compounds.

Anhydrous crystalline ampicillin is kinetically stable for several days in MeOH/H_2O mixtures over the whole range of water activity. Even though at ambient conditions the crystalline trihydrate is more thermodynamically stable and less soluble than the anhydrate, conversion to the hydrated form occurs only with the addition of seeds at a water activity >0.381 [57] (see Figure 15).

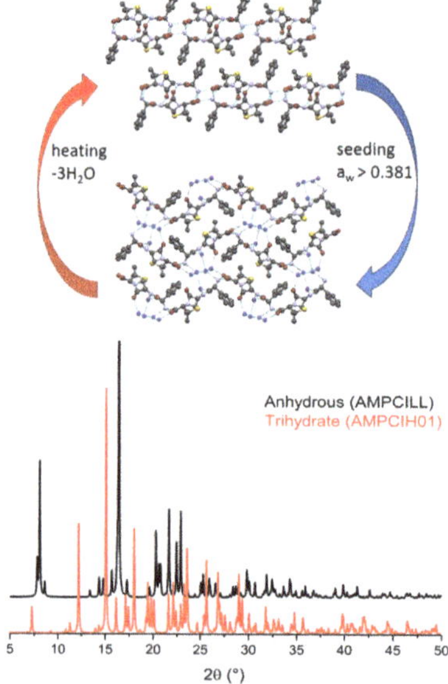

Figure 15. (**Top**): packing comparison and interconversion process between anhydrous and trihydrate ampicillin (refcodes AMCILL and AMPCIH01, respectively). (**Bottom**): comparison between the powder diffractograms calculated on the basis of the single-crystal data, showing the difference between the two patterns.

Theophylline forms an anhydrate and a monohydrate that interconvert in organic solvents, depending on the water activity. At a low water activity (<0.25) the anhydrate is the only species present, whereas at a higher water activity the monohydrate is the most stable form [57,58] (see Figure 16).

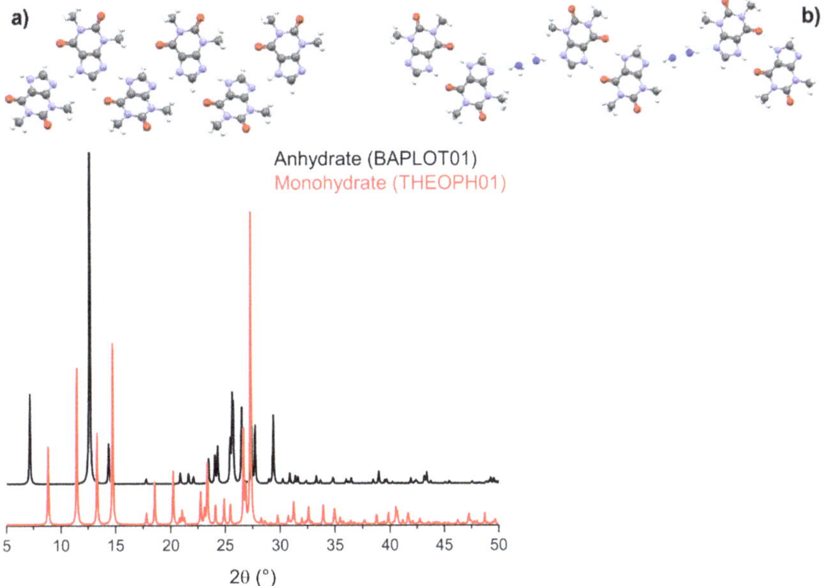

Figure 16. (Top): The structures of the anhydrate ((**a**), refcode BAPLOT01) and of the monohydrate ((**b**), refcode THEOPH01) crystals of theophylline. The anhydrate is obtained from a mixture of an organic solvent and water at low water activity. (**Bottom**): comparison of the the powder diffractograms calculated on the basis of the single-crystal data, showing the difference between the two patterns.

Although hydrate formation can sometimes be reversed by the drying process, the dehydration of hydrates can lead to the formation of amorphous material or crystal defects that can have potential deleterious effects on physical and chemical stability [59,60]. Hydrates are generally expected to be thermodynamically more stable, hence less soluble and slower to dissolve than anhydrates above the critical water activity for hydrate formation. Hence, dehydrated hydrates tend to be metastable with an easy uptake of water (or other solvents).

An interesting notion, albeit slightly counterintuitive, is that most organic hydrates are less soluble in water than the corresponding anhydrous compounds at the same temperature, as shown by the solubility data compared in Table 4. As an example, the solubilities of anhydrous and hydrated crystals of caffeine are 49.7 and 21.8 mg/mL, those of carbamazepine 0.424 and 0.139 mg/mL, and those of sulfaguanidine 1.38 and 1.07 mg/mL, respectively [59]. The reason is that in hydrates, some of the water–molecule interactions, mainly taking place via hydrogen bonds, are already satisfied in the crystals, thus decreasing the solvation energy contributions to dissolution.

Hydrates are also patentable because they often meet the required non-obviousness and innovativeness criteria. In terms of novelty, anhydrates and hydrates, often because of their difference in solubility (see above), have properties that are distinct from those of polymorphs and other solvates of the same API. One should keep in mind that, even when crystallization takes place from water, it is difficult if not impossible to predict whether a stable hydrate might precipitate out, let alone its physico-chemical properties.

Table 4. Comparison of solubility data for three examples of hydrated and anhydrous compounds. Data reprinted with permission from Ref. [59]. 2008, Elsevier.

Material	Molecular Structure	Solid State Form	Solubility (mg/mL)
Caffeine (CAF)		Anh = triclinic β-phase form Hyd = monoclinic 4/5 hydrate	Anh = 49.7 Hyd = 21.8
Carbamazepine (CBZ)		Anh = monoclinic low temp. form Hyd = orthorhombic dihydrate	Anh = 0.424 Hyd = 0.139
Sulfaguanidine (SFG)		Anh = monoclinic form II Hyd = monoclinic monohydrate	Anh = 1.38 Hyd = 1.07

Rifaximin: An Example of a Multiple Non-Stoichiometric Hydrate.

A cautionary word is in order: the water content, as determined from diffraction data, is averaged over all unit cells forming the crystals used for that specific experiment. In the case of polycrystalline material, however, adsorbed water might be present, and this may impact on the water content established by analytical methods (e.g., Karl Fischer titration) [55]. Care should thus be taken when comparing information coming from diffraction and analytical methods, especially in the cases of non-stochiometric hydrates (see below).

A good example of a commercial drug with important properties depending on the degree of hydration is provided by Rifaximin. Rifaximin, (4-deoxy-4′-methylpyrido[1′,2′-1,2]imidazo-[5,4-c] rifamycin SV) is a synthetic antibiotic; its mechanism of action relies on the inhibition of bacterial RNA synthesis by binding the β-subunit of the deoxyribonucleic acid (DNA)-dependent ribonucleic acid (RNA) polymerase [61,62]. Several distinct crystalline forms of Rifaximin have been reported [63,64]. They are basically non-stoichiometric hydrates that interconvert depending on the amount of water and on the way dehydration/hydration is carried out. What is important is that they exhibit remarkable differences in water solubility ranging from 2 mg/L up to more than 40 mg/L, which deeply affects the Rifaximin bioavailability. It is thus of paramount importance to be able to clearly identify the different hydrates, and this can be done via powder X-ray diffraction (see Figure 17).

Form α, a form with a low water content, can only be obtained by the dehydration of form β. Figure 18 shows that transition β → α-form on heating on a hot stage microscope a single crystal of the β-form. It is noteworthy how the β-form at 25 °C (monoclinic β angle, shown in red, ca. 91°) transforms at 50 °C into the α-form with the monoclinic β angle changing to ca. 110°.

The relationship between the different crystal phases can be better appreciated by comparing the structures determined by single-crystal X-ray diffraction (see Figure 19). The packing in all forms of Rifaximin can be seen as obtained by a juxtaposition of "molecular rods" formed by pairs of Rifaximin molecules. In addition to a change in the relative orientation of Rifaximin molecules within a dimeric pair, forms α, β, δ, and ε present a different relative arrangement of the molecular rods, which seem to "slide" on passing from one form to the other.

Crystal forms of Rifaximin solvates with organic solvents are also known in the patent literature [65,66].

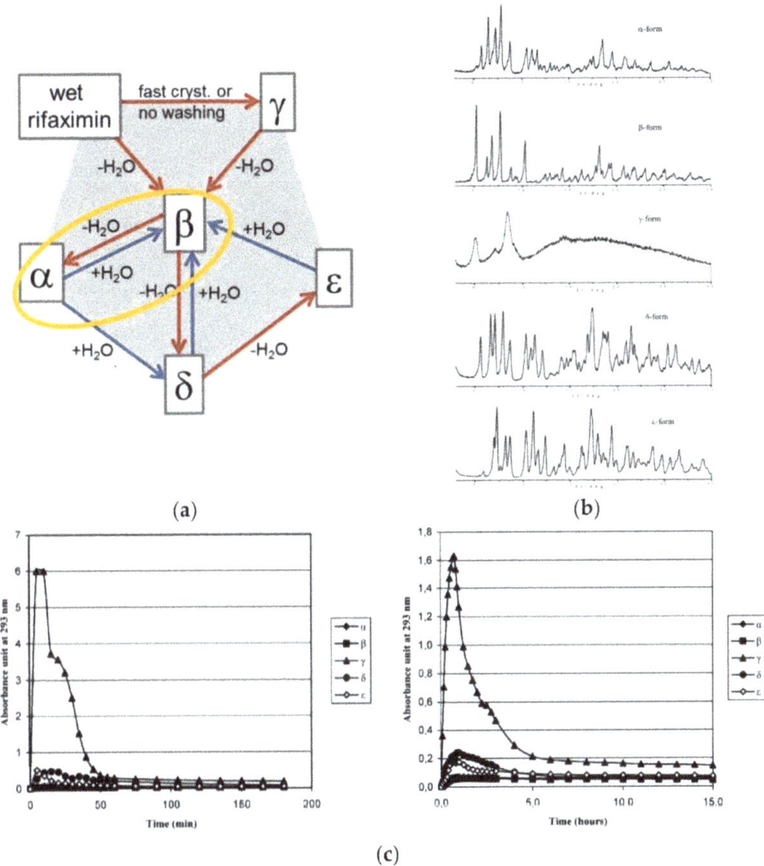

Figure 17. (**a**) The intricate hydration/dehydration phase relationship between the various forms of Rifaximin, (**b**) the powder X-ray diffraction patterns five forms of Rifaximin, (**c**) dissolution rates of Rifaximin hydrates at 250 (**left**) and 100 (**right**) rpm of mixing rate. Time vs. absorbance at 293 nm. Reproduced with permission from [63,64].

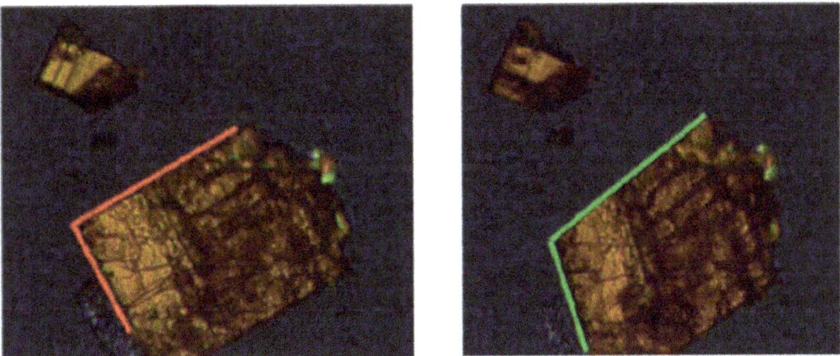

Figure 18. The single crystal to single crystal form β → form α transformation. The red and green lines evidence the change in the monoclinic β-angle from 91° in Rifaximin form α to 110° in Rifaximin form β. Reprinted with permission from Ref. [63]. 2019, Royal Society of Chemistry.

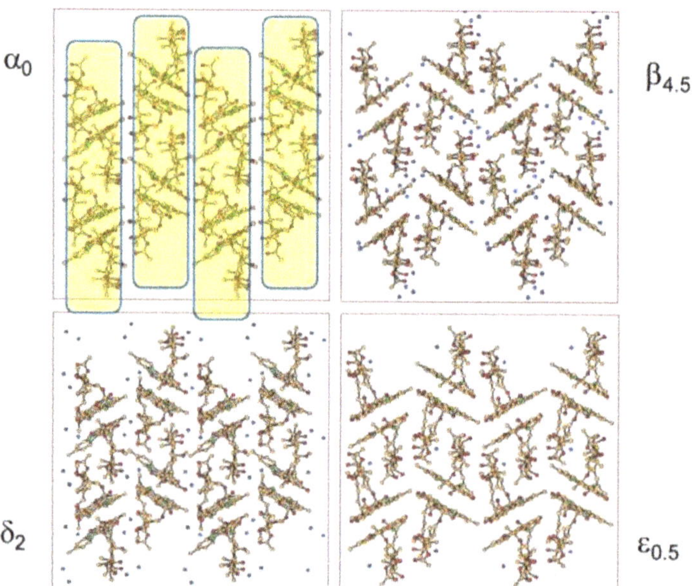

Figure 19. A comparison of molecular packings in four hydrate forms of the antibiotic Rifaximin. Reproduced with permission from [63].

An interesting example of how dissolution rate profiles may be altered, and hygroscopic behavior improved by the formation of solvates is that of the recently reported and patented Rifaximin τ, a transcutol® (transcutol® IUPAC name 2-(2-ethoxyethoxy)-ethanol) solvate crystal shown in Figure 20.

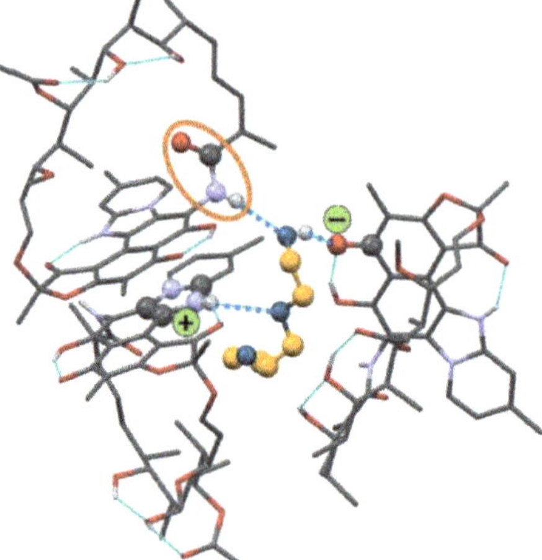

Figure 20. The solid-state structure of Rifaximin τ, showing the interaction of Rifaximin with transcutol® [67,68]. Reprinted with permission from Ref. [67]. 2019, Royal Society of Chemistry.

The comparison of the dissolution rate profiles in Figure 21 clearly shows how the dissolution rate of Rifaximin τ is higher than the one observed for amorphous Rifaximin.

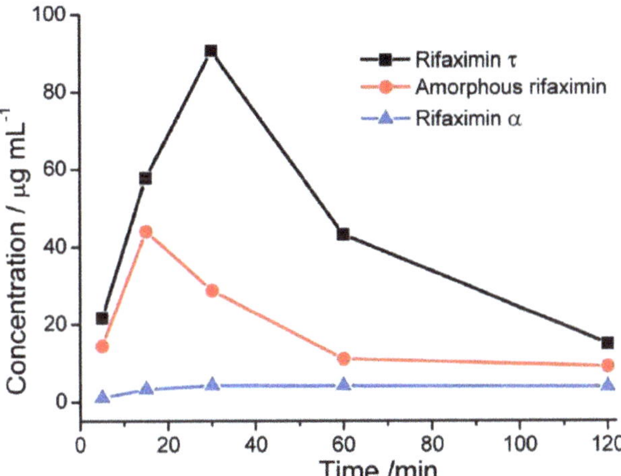

Figure 21. Comparison of dissolution rates for Rifaximin τ, amorphous Rifaximin, and Rifaximin α at neutral pH. Reprinted with permission from Ref. [67]. 2019, Royal Society of Chemistry.

7. Molecular Co-Crystals

After discussing polymorphs and hydrates, we will now review the topic of co-crystals. Undoubtedly co-crystals have become one of the major attractions for all those interested in altering the physico-chemical properties of APIs or in finding new ones. This is because the association in the solid state of two or more chemically distinct entities, each forming stable solid phases at ambient conditions, is proving to be one of the most fruitful ways to modify solid state as well as biological properties of the crystals of active molecules. This can be applied to pharmaceuticals already in use and/or to access different pharmacological properties by combining different drugs in one crystalline material (co-drugs).

The term co-crystal was used for the first time in 1963 by Hoogsteen when he reported the structure of an adduct between 1-methyl adenine and 1-methyl thymine (CSD refcode MTHMAD, see Figure 22) [69].

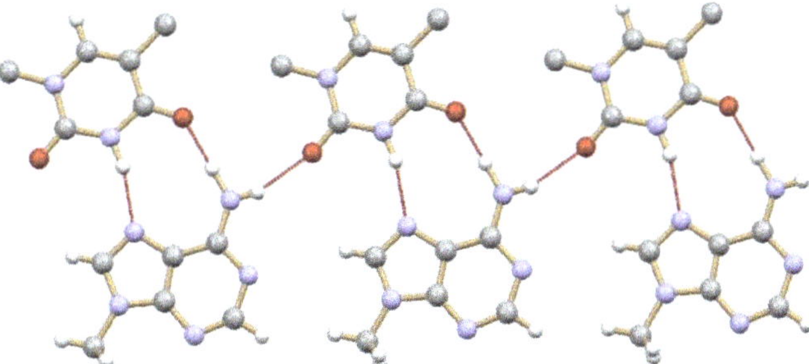

Figure 22. Hydrogen bonding pattern in the 1:1 co-crystal of 1-methyl adenine and 1-methyl thymine, first reported in 1963 by Hoogsteen (CSD refcode MTHMAD).

The first hint to the possibility of using co-crystallization as an instrument to modify solid state properties can probably be found in a 1991 paper by Margaret Etter [70], where she discussed "ways to prepare organic crystals and to use co-crystallization to probe the forces involved in aggregation phenomena" and "how molecular aggregation can impart unexpected new properties to organic compounds".

As a matter of fact, the very definition of a co-crystal is not straightforward, and has been addressed in slightly different ways by various authors. In 2003, J. Dunitz defined a co-crystal as "a crystal containing two or more components together" [71]. In 2004, M.

Zaworotko and O. Almarsson provided a definition of a pharmaceutical co-crystal as "a stoichiometric multiple component crystal in which at least one component is molecular and a solid at room temperature (the co-crystal former) and forms a supramolecular synthon with a molecular or ionic API" [72]. In 2005, C. Aakeröy and D. J. Salmon preferred "compounds constructed from neutral molecular species [...] that are solids at ambient conditions [...] and [...] present in definite stoichiometric amounts" [73].

In the context of this review, the definition of a co-crystal as "a multicomponent crystal formed by two or more compounds that are solid at RT and that interact via non-covalent bonding" has been adopted. A corollary of this definition is that a co-crystal is not a solvate (solvent molecules are not solid at RT) and is not a salt (ions do not have separate identities) but could be the association of a neutral molecule with a coordination compound or with an organic or inorganic salt. In this latter case, the definition of ionic co-crystals is adopted.

Indeed, co-crystals may offer new ways to design or to alter the properties of solid active ingredients including the thermal stability, the shelf life, the solubility, the dissolution rate, the compressibility, etc., by linking the co-crystal former with a suitable ancillary molecule, a co-former. Obviously, these ought to be GRAS molecules for pharmaceuticals [53]. If the co-former happens to be another API, the co-crystal is a co-drug, with all the implications for the regulatory process and authorizations. It is important to appreciate that the differences in physico-chemical properties between a co-crystal and the parent single-molecule crystal are usually larger than those between polymorphs and often also than those between the active ingredient and its solvates/hydrates.

Co-crystals may also be polymorphic. Figure 23 shows a schematic representation of two polymorphs of a co-crystal of an API and a conformer.

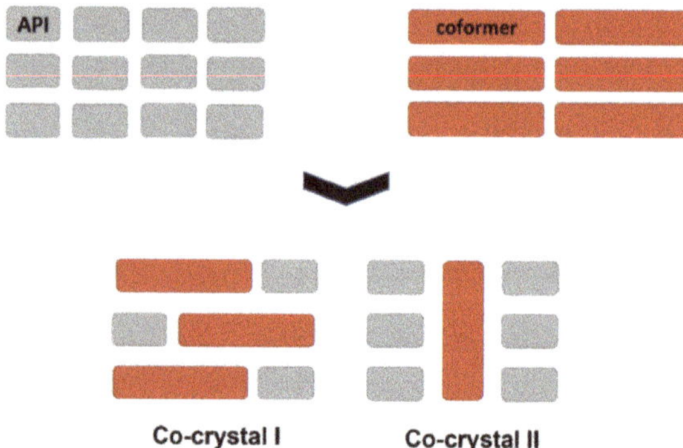

Figure 23. A schematic representation of two polymorphs of a co-crystal of the same API and a conformer.

7.1. Co-Crystals in Pharmaceutics

For the reasons outlined above, it should be clear why the topic of pharmaceutical co-crystals is being extensively investigated. The subject has been addressed in a number of books, publications, and reviews. Good entry points are the books cited above [6,7], while reviews on the subject matter, with a focus on pharmaceutical co-crystals, were written by Zaworotko et al. in 2016 [74] and, very recently, by Nangia et al. [75]. The occurrence of polymorphism in multicomponent systems, including co-crystals, has also been reviewed recently [76].

As an example, it is worth citing the thorough exploration of the co-crystal domain of carbamazepine with a series of pharmaceutically acceptable carboxylic acids, as reported by Childs et al. [77]. The authors not only explored a large number of co-formers, but also tested and compared four different screening techniques to form co-crystals. Out of this screening, 27 co-crystals with 18 carboxylic acids were generated and characterized both by XRD (see Figure 24).

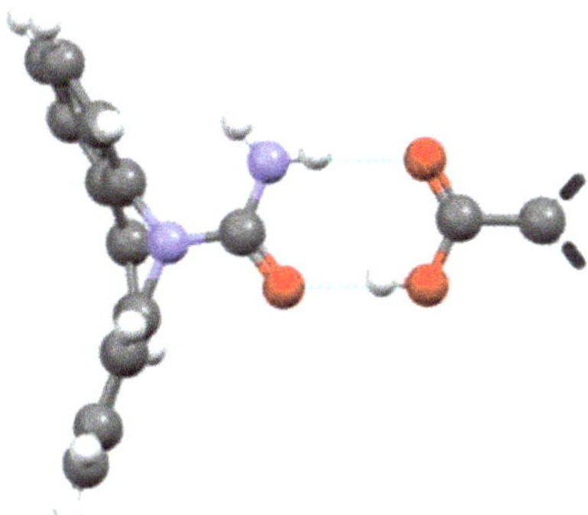

Figure 24. The acid–amide synthon frequently occurring in the co-crystals of carbamazepine with carboxylic acids.

Within the broad family of co-crystals, ionic co-crystals deserve a special mention, as they contain ionized components together with an API or an API precursor [78]. The ionic co-former may be either a salt of an ionizable molecule (carboxylic acid, amine, etc.), or a metal salt, e.g., LiCl. The stability of the ionic co-crystals depends on the interactions established by the organic moiety with cations and anions; usually oxygen or nitrogen atoms donate electrons towards the cation, while hydrogen bonds are formed between hydrogen donor groups on the organic moiety and the anions. Therefore, they resemble the interactions that a solvent molecule might establish with ions in solution.

An example of an ionic co-drug co-crystal co-drug is provided by the co-crystals [79] obtained by reacting LiCl (a drug used as a mood stabilizer in patients affected by bipolar disorder) together with piracetam, a cognition enhancing medicine (Nootropil®). The preparation of these materials can be easily attained by mechanical mixing of the API (in this case piracetam) with LiCl, but also with other salts such as LiBr as well as with other bio-compatible inorganic salts [80,81].

It is worth mentioning, on passing, that ionic co-crystals have also been used in the food sector long before the subject of co-crystals gained popularity. The preparation of compounds based on the association of sugars with inorganic salts dates back more than a century. Since combining NaCl with carbohydrates allows for the introduction of a combined source of sodium and calories, the idea of ionic co-crystals with sugars is rather interesting from a nutraceutical point of view. A number of carbohydrates have been found to form stable co-crystals with NaCl, in particular pentoses (e.g., ribose, arabinose, and xylose), hexoses (e.g., glucose, fructose, galactose, and mannose), as well as disaccharides (e.g., sucrose, lactose, and trehalose). The subject of ionic co-crystals with carbohydrates has been reviewed by Oertling [82].

7.2. Co-Crystals Properties

Compared to polymorphs, solvates and co-crystals are more likely to induce significant changes in the solid-state properties of the active ingredient. Clearly, whether additional pharmaceutical and clinical tests might be required will depend on the nature of the coformer, especially if it does not belong to the molecules admitted as GRAS.

In many pharmaceutical applications, the key issues are often, but not exclusively, those related to the solubility and/or dissolution rate of the active molecules of interest. This aspect is also of interest from a patenting point of view, because it may allow an extension of the intellectual property protection of an active ingredient.

A number of authors have explored the preparation of pharmaceutical co-crystals to attain better solubility or better dissolution rates [83,84]. The reader may refer to Good and Rodriguez-Hornedo (2010) [85] for an example of evaluation of the factors controlling and affecting co-crystal solubility. Several studies have been reported on the intrinsic dissolution rates of co-crystals. For instance, in the case of the co-crystals of poorly soluble 2-[4-(4-chloro-2-fluorophenoxy)phenyl]pyrimidine-4-carboxamide with glutaric acid, a dissolution rate 18 times higher than that of the pure API was observed [86].

The co-crystal of melatonin with pimelic acid is another example (Figure 25) [87]. The oral bioavailability of melatonin in humans is limited and efforts to improve the dissolution rate and solubility are ongoing. Co-crystal formation occurs when a 1:1 molten mixture of melatonin and pimelic acid is allowed to crystallize in the temperature range 50–70 °C. The co-crystal displays a significantly higher apparent solubility and acceptable stability as compared to the original melatonin. Figure 26 also shows that the melatonin concentration in the end drops back to the equilibrium solubility level of the pure drug, but the 3 h of increased solubility can be crucial for obtaining a higher bioavailability.

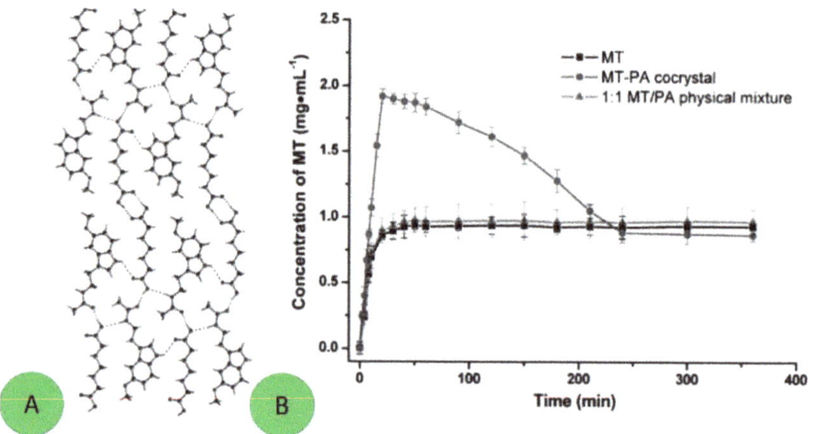

Figure 25. Packing diagram of the melatonine/pimelic acid co-crystal showing the hydrogen bonding (**A**). Powder dissolution profiles of melatonine, melatonine-pimelic acid co-crystal, and of a 1:1 physical mixture in PBS (pH 6.8) at 37 °C (**B**). Adapted with permission from Ref. [87]. 2015, Royal Society of Chemistry.

Dissolution enhancement can also be found in a number of different crystal forms of niclosamide, an active ingredient belonging to the salicylamide class [88], obtained by solvent-free synthesis and characterized by diffraction, thermal methods, and solid-state NMR spectroscopy (1H, 15N, 13C CPMAS). Since niclosamide has a very poor aqueous solubility in water, it has proven extremely important to increase its dissolution rate via the formation of ionic co-crystals.

The effect of co-crystallization on the dissolution rate has been studied in the case of the co-crystals of hesperetin with picolinic acid, nicotinamide, and caffeine, shown in Figure 26a. The dissolution rate experiments in Figure 26b show the parachute effect accompanying dissolution of the three co-crystals, with solubility reaching a maximum concentration in a short time then dropping down, first abruptly, then slowly but steadily, approaching the levels shown by pure hesperetin. This change in solubility is explained by the breakdown of co-crystals to the starting molecules on extended exposure to an aqueous medium, as is confirmed by FT-IR analyses of residues.

7.3. Chiral Resolution via Co-Crystallization

Co-crystallization is a viable route to the resolution of racemic mixtures. There are, basically, two main strategies for chiral resolution via co-crystal formation. The first approach is based on the preparation of a chiral host compound with cavities capable of forming inclusion complexes with only one of the enantiomers of the guest molecule. In the second approach, an enantiopure co-former is co-crystallized with a target molecule resulting in the formation of "diastereoisomeric" co-crystals. Broadly speaking, a supramolecular heterosynthon should favor chirality with respect to homosynthons. This is because a homosynthon is inherently centrosymmetric, while a heterosynthon is inherently non-centrosymmetric and should favor polarity/chirality.

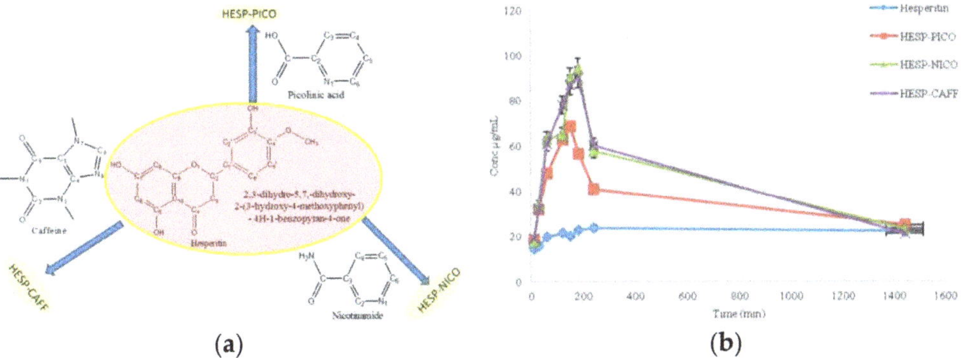

Figure 26. (**a**) The effect of co-crystallization on the dissolution rate for co-crystals of hesperetin with picolinic acid, nicotinamide, and caffeine. (**b**) Equilibrium solubility (24 h) of hesperetin, HESP-PICO, HESP-NICO, and HESP-CAFF. Reprinted with permission from Ref. [89]. 2017, American Chemical Society.

If the chosen coformer is enantiopure, chiral resolution could be achieved via either enantiospecific co-crystal formation, when the co-crystal is formed with only one enantiomer of the target molecule or a diastereomeric co-crystal pair—when the co-crystals are formed with both enantiomers. Unlike for salts, for which diastereomeric salt pair formation appears to be a general rule, co-crystals more frequently behave enantiospecifically.

The basic idea is that the reaction (whether in solution or in the solid state) of a racemic R,S molecule (R,S-M) with an enantiopure coformer, say an R-coformer (R-C), might lead to the formation of a "diastereoisomeric co-crystal", namely R-M/R-C and S-M/R-C aggregates. These co-crystals will possess definitely different crystal structures and different physicochemical properties that could be used for resolution. Analogously, if R,S-M is capable of salt formation, say via an acid-base reaction with an enantiopure acid/base capable of hydrogen bonding donor/acceptor interaction via proton transfer, "diastereoisomeric salts" will be formed, e.g., R-MH$^+$—R-C$^-$ and S-MH$^+$—R-C$^-$.

There are several examples in the literature. What is probably the first utilization of the "diastereoisomeric co-crystal" approach dates back to 1939 when Eisenlohr and Meier reported on the chiral resolution of racemic 5-(1-hydroxyethyl)benzene-1,3-diol (resorcylmethylcarbinol) with an alkaloid brucine (see Figure 27).

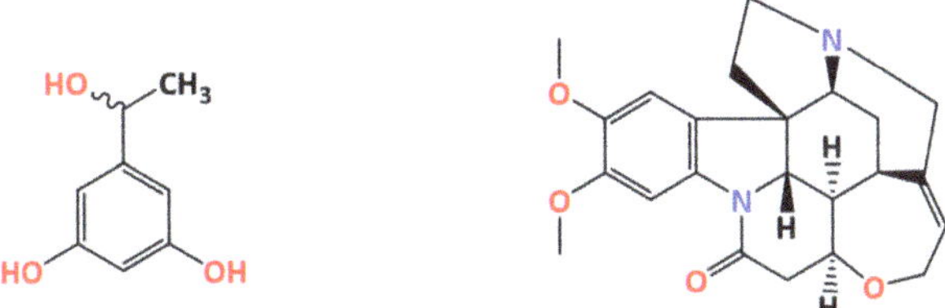

Figure 27. If RS-resorcylmethylcarbinol (**left**) and brucine (**right**) are co-crystallized from methanol, only dextro-resorcylmethyl carbinol forms a co-crystal with brucine [90].

Resolution of optical isomers of 4-amino-p-chlorobutyric acid lactam (Baclofen®) by co-crystallization with (2R,3R)-(+)-tartaric acid has been obtained by forming a co-crystal in which only the (R) enantiomer is present. This can be attributed to a supramolecular heterosynthon (see Figure 28) [91].

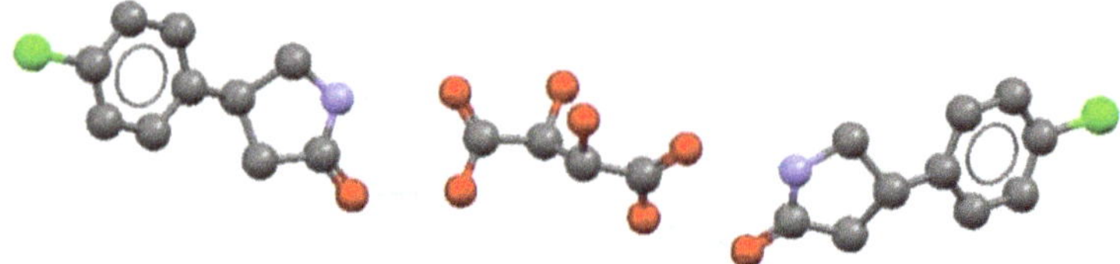

Figure 28. Resolution of optical isomers of 4-amino-p-chlorobutyric acid lactam (Baclofen®) by co-crystallization 4-amino-p-chlorobutyric acid lactam lactam [91].

The unsubstituted (2R,3R)-tartaric acid has been successfully applied for chiral resolution of 4-amino-p-chlorobutyric acid lactam. It was established that only (R-)-4-amino-p-chlorobutyric acid lactam co-crystallizes with (2R,3R)-tartaric acid, resulting in the formation of a 2:1 co-crystal (Figure 29) [91].

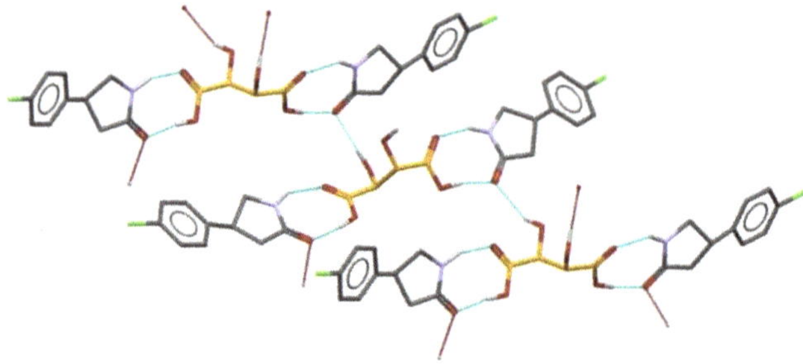

Figure 29. The 2:1 co-crystal of (R-)-4-amino-p-chlorobutyric acid lactam with (2R,3R)-tartaric acid [91].

Leyssens et al. have shown enantioselective co-crystal formation in the case of 2-(2-oxopyrrodin-1-yl)butanamide (etiracetam) with S-mandelic and S-tartaric acid (see Figure 30), while co-crystals are not formed with the R-etiracetam enantiomer [92]. Conglomerate formation was also reported as a result of crystallization from a racemate via formation of molecular co-crystals [93].

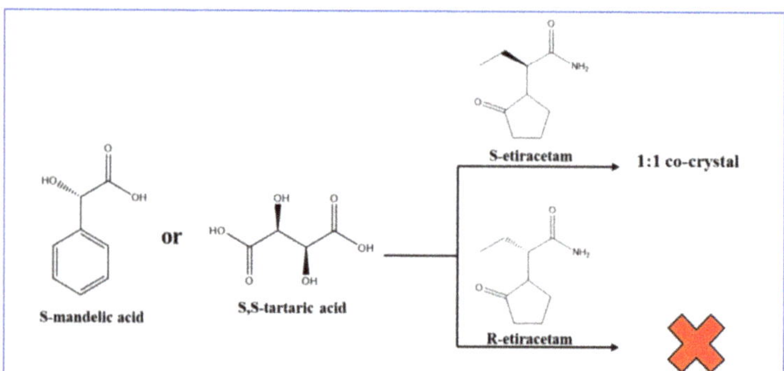

Figure 30. The enantioselective co-crystal formation in the case of S-enantiomer of 2-(2-oxopyrrodin-1-yl)butanamide (S-etiracetam) with S-mandelic and S-tartaric acid. The same reaction with R-etiracetam does not lead to co-crystal formation [92].

An alternative route to chiral resolution via crystal engineering is that based on inclusion compounds, a strategy pioneered by F. Toda [94]. Co-crystallization with derivatives of lactic acid led to the enantiomeric excess (ee), including an "inclusion complex" with 3-methylcyclohexanone obtained with an ee > 99% (KUCJAT, Figure 31) [95].

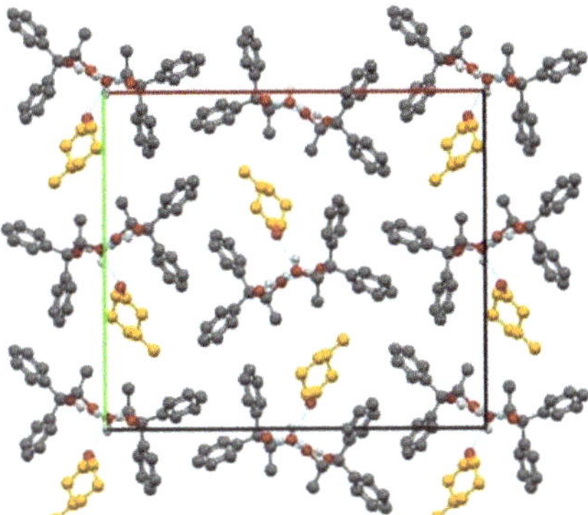

Figure 31. KUCJAT. The carbons of 3-methylcyclohexane are marked in orange and H_{CH} have been removed for the sake of clarity [95].

A similar approach has been applied for the optical resolution of racemic N-R-3-hydroxypyrrolidines using ((2R,3R)-1,4-dioxaspiro[4.4]nonane-2,3-diyl)bis (diphenylmethanol) as an optically active host compound [96]. Three racemic N-R-3-hydroxypyrrolidine were tested (R=H, benzyl, or ethyl) and it was established that the capability of the optical resolution of ((2R,3R)-1,4-dioxaspiro[4.4]nonane-2,3-diyl)bis(diphenylmethanol) was very high for only N-benzyl-3-hydroxypyrrolidine (ee 100%), whereas for N-ethyl- and 3-hydroxylpyrrolidine, enantioselectivity was low.

8. Conclusions

The purpose of this article was to provide an overview of the main issues concerning drugs in the crystalline state.

There is little doubt that the design, synthesis, characterization, and biological and pharmacological evaluation of new active ingredients are the first and paramount objectives of research in the pharmaceutical field. However, obtaining a new drug is not enough: the road from bench to market is often long, steep, and tortuous. One aspect that has emerged in all its relevance is that of the impact of the solid-state form of the active ingredient on its utilization as a drug. The vast majority of drugs are constituted of organic molecules. Organic molecules are often structurally non-rigid and conformationally adaptable, and often carry functional groups that form strong and directional intermolecular interactions (hydrogen bonds, halogen bonds, etc.) or that confer anisotropy and polarity, hence impose orientational preference to surrounding molecules, whether of the same or of different type. This variability has consequences when the active ingredient is crystallized from a solution, melt, or vapor phase, because the formation of a solid from a less condensed phase is the outcome of a competition between kinetics and thermodynamic aspects. The isolation of stable or metastable crystal forms, whether enantiotropically or monotropically related, is a manifestation of such dualism, as it is the precipitation of a solvate or of an unsolvated crystal of the API from solution. As we have seen in this review, an active ingredient can be brought in the solid state in an impressive number of alternative forms: solvates, hydrates, salts, co-crystals, and all these forms can be polymorphic.

These are the reasons why the discovery of a drug, whether destined to treat a new disease or to replace old and less efficacious pharmaceuticals, is undoubtedly a fantastic achievement, but it is also the beginning of a journey in another domain, that of the aggregate form of the new drug, the crystal domain. As a matter of fact, the drug will have to be isolated, its preparation scaled up, purified, stored, packed, distributed, and ultimately administered to patients and all these steps will be carried out, in the vast majority of cases, with the drug in the solid form, sometimes amorphous, most often a polycrystalline material.

To summarize:

(i)   An API can take many solid forms (polymorphs, solvates, salts, co-crystals) all containing exactly the same active principle but in different structural arrangements and/or associated to different molecules (solvents, coformers, counterions, salts, etc.);

(ii)   Different solid forms will possess different properties, which might affect a drug's processing, formulation, distribution, storage, administration to the patient and, ultimately, its therapeutic efficacy;

(iii)   Since there is no way to guarantee that a transformation to a more stable form will occur, or that a more stable form will appear, a preliminary, thorough exploration of the crystal domain is the only way to minimize the chances of the undesired appearance of a new and more stable form at a later stage of development;

(iv)   Besides maximising control on the crystal form of the API, the investigation of hydrates and co-crystals may afford alternative, often improved or innovative properties of the drug;

(v)   Polymorphs, hydrates, and co-crystals may meet the requisites of novelty, utility, and non-obviousness useful for extending the patenting life of a drug.

The examples provided in the article have been chosen with the intention of stressing why the quest for new crystal forms of any given API can be both "joy and sorrow" for the academic and industrial researcher. This article has a provocative title, the same title as the lecture delivered by one of us (D.B.) at the ACCORD22 meeting.

Author Contributions: Conceptualization, D.B.; writing—original draft preparation, D.B.; writing—review and editing, F.G. and L.C.; supervision, D.B.; funding acquisition, D.B. All authors have read and agreed to the published version of the manuscript.

Funding: D.B., F.G. and L.C acknowledge financial support from the University of Bologna and from MUR, project PRIN2020 "Nature inspired crystal engineering".

Acknowledgments: This review article is dedicated to the memory of Joel Bernstein (Beer Sheva University, Israel) and of Giuseppe Claudio Viscomi (Alfasigma S.p.A., Bologna, Italy) for having shared with us many "joys and sorrows" of the research on crystal forms of pharmaceuticals, and for their friendship and continuous scientific encouragement.

Conflicts of Interest: The authors declare no conflict of interest.

References

1.   Mitscherlich, E. Sur la relation qui existe entre la forme cristalline e le proportions chimiques, I. Memoires sur les arseniates et les phosphates. *Ann. Chim. Phys.* **1822**, *19*, 350–419.

2.   McCrone, W.C. Physics and Chemistry of the Organic Solid State. In *Physics Today*, 2nd ed.; Fox, D., Labes, M.M., Weissenberg, A., Eds.; Interscience: New York, NY, USA, 1965; Volume 2, 725p.

3.   Bernstein, J. *Polymorphism in Molecular Crystals*, 2nd ed.; Oxford University Press: Oxford, UK, 2020.

4.   Brittain, H.G. (Ed.) *Polymorphism in Pharmaceutical Solids*, 2nd ed.; Informa Healthcare: London, UK, USA, 1999.

5.   Hilfiker, R.; Von Raumer, M. (Eds.) *Polymorphism in the Pharmaceutical Industry: Solid Form and Drug Development*, 2nd ed.; Hilfiker, R.; Von Raumer, M. (Eds.) Wiley-VCH: Weinheim, Germany, 2019.

6.   Gruss, M. (Ed.) *Solid State Development and Processing of Pharmaceutical Molecules: Salts, Cocrystals, and Polymorphism*; John Wiley & Sons: Hoboken, NJ, USA, 2019.

7.   Wouters, J.; Quéré, L. (Eds.) *Pharmaceutical Salts and Co-Crystals*, 1st ed.; Royal Society of Chemistry: Cambridge, UK, 2011.

8.   Aakeröy, C.B.; Sinha, A.S. (Eds.) *Co-Crystals: Preparation, Characterization and Applications*, 1st ed.; Royal Society of Chemistry: Piccadilly, UK, 2018.

9.   Braga, D.; Grepioni, F.; Gavezzotti, A.; Bernstein, J. Re: "Crystal Engineering in the Regulatory and Patent Literature of Pharmaceutical Solid Forms". *Cryst. Growth Des.* **2017**, *17*, 933–939. [CrossRef]

10. Seddon, K.R.; Zaworotko, M.J. *Crystal Engineering: The Design and Application of Functional Solids*; Kluwer: Dordrecht, The Netherlands, 1999.

11. Bruno, I.J.; Cole, J.C.; Edgington, P.R.; Kessler, M.; Macrae, C.F.; McCabe, P.; Pearson, J.; Taylor, R. New software for searching the Cambridge Structural Database and visualizing crystal structures. *Acta Cryst. B* **2002**, *58*, 389–397. [CrossRef] [PubMed]

12. Görbitz, C.H.; Hersleth, H.P. On the Inclusion of Solvent Molecules in the Crystal Structures of Organic Compounds. *Acta Cryst. B* **2000**, *56*, 526–534. [CrossRef]

13. Werner, J.E.; Swift, J.A. Data mining the Cambridge Structural Database for hydrate-anhydrate pairs with SMILES strings. *CrystEngComm* **2020**, *22*, 7290–7297. [CrossRef]

14. ICH Official Website. Available online: https://www.ich.org/ (accessed on 14 June 2022).

15. U.S. Food & Drug Administration. Guidance Document, Pharmaceutical Solid Polymorphism: Chemistry, Manufacturing, and Controls Information. Available online: https://www.fda.gov/regulatory-information/search-fda-guidance-documents/andaspharmaceutical-solid-polymorphism-chemistry-manufacturing-and-controls-information (accessed on 14 June 2022).

16. European Patent Office. Available online: http://www.european-patent-office.org/ (accessed on 14 June 2022).

17. Dunitz, J.D. Phase Transitions in Molecular Crystals from a Chemical Viewpoint. *Pure Appl. Chem.* **1991**, *63*, 177–185. [CrossRef]

18. Chen, S.; Guzei, I.A.; Yu, L. New Polymorphs of ROY and New Record for Coexisting Polymorphs of Solved Structures. *J. Am. Chem. Soc.* **2005**, *127*, 9881–9885. [CrossRef]

19. Yu, L.; Stephenson, G.A.; Mitchell, C.A.; Bunnell, C.A.; Snorek, S.V.; Bowyer, J.J.; Borchardt, T.B.; Stowell, J.G.; Byrn, S.R. Thermochemistry and Conformational Polymorphism of a Hexamorphic Crystal System. *J. Am. Chem. Soc.* **2000**, *122*, 585–591. [CrossRef]

20. Leévesque, A.L.; Maris, T.; Wuest, J.D. ROY Reclaims Its Crown: New Ways to Increase Polymorphic Diversity. *J. Am. Chem. Soc.* **2020**, *142*, 11873–11883. [CrossRef]

21. Yu, L. Polymorphism in Molecular Solids: An Extraordinary System of Red, Orange, and Yellow Crystals. *Acc. Chem. Res.* **2010**, *43*, 1257–1266. [CrossRef]

22. Burger, A. *Topics in Pharmaceutical Sciences*; Breimer, D.D., Speiser, P., Eds.; Elsevier: Amsterdam, The Netherlands, 1983.

23. Threlfall, T.L. Analysis of organic polymorphs. A review. *Analyst* **1995**, *120*, 2435–2460. [CrossRef]

24. Bučar, D.K.; Lancaster, R.W.; Bernstein, J. Disappearing Polymorphs Revisited. *Angew. Chem.* **2015**, *54*, 6972–6993. [CrossRef]

25. Haisa, M.; Kashino, S.; Kawai, R.; Maeda, H. The Monoclinic Form of P-Hydroxyacetanilide. *Acta Cryst. B* **1976**, *32*, 1283–1285. [CrossRef]

26. Haisa, M.; Kashino, S.; Maeda, H. The Orthorhombic Form of P-Hydroxyacetanilide. *Acta Crystallogr. Sect. B* **1974**, *30*, 2510–2512. [CrossRef]

27. Martino, P.D.; Conflant, P.; Drache, M.; Huvenne, J.P.; Guyot-Hermann, A.M. Preparation and Physical Characterization of Forms H and HI of Paracetamol. *J. Therm. Anal. Cal.* **1997**, *48*, 447–458. [CrossRef]

28. Liu, Y.; Gabriele, B.; Davey, R.J.; Cruz-Cabeza, A.J. Concerning Elusive Crystal Forms: The Case of Paracetamol. *J. Am. Chem. Soc.* **2020**, *142*, 6682–6689. [CrossRef] [PubMed]

29. Heng, J.Y.Y.; Williams, D.R. Wettability of Paracetamol Polymorphic Forms I and II. *Langmuir* **2006**, *22*, 6905–6909. [CrossRef]

30. Mishra, R.; Srivastava, A.; Sharma, A.; Tandon, P.; Baraldi, C.; Gamberini, M.C. Structural, Electronic, Thermodynamical and Charge Transfer Properties of Chloramphenicol Palmitate Using Vibrational Spectroscopy and DFT Calculations. *Spectrochim. Acta Part A Mol. Biomol. Spectrosc.* **2013**, *101*, 335–342. [CrossRef] [PubMed]

31. Szulzewsky, K.; Kulpe, S.; Schulz, B.; Kunath, D. The structure of the modification of chloramphenicol palmitate—A redetermination. *Acta Crystallogr. Sect. B* **1981**, *37*, 1673–1676. [CrossRef]

32. Diodone, R.; Hidber, P.C.; Kammerer, M.; Meier, R.; Schwitter, U.; Thun, J. Industry Case Studies. In *Polymorphism in the Pharmaceutical Industry: Solid Form and Drug Development*, 2nd ed.; Hilfiker, R., Von Raumer, M., Eds.; Wiley-VCH: Weinheim, Germany, 2019; Chapter 15, pp. 453–456.

33. Cruz-Cabeza, A.J.; Bernstein, J. Conformational Polymorphism. *Chem. Rev.* **2014**, *114*, 2170–2191. [CrossRef] [PubMed]

34. Ruggiero, M.T.; Sibik, J.; Zeitler, J.A.; Korter, T.M. Examination of L-Glutamic Acid Polymorphs by Solid-State Density Functional Theory and Terahertz Spectroscopy. *Phys. Chem. A* **2016**, *120*, 7490–7495. [CrossRef] [PubMed]

35. Chierotti, M.R.; Gobetto, R.; Pellegrino, L.; Milone, L.; Venturello, P. Mechanically Induced Phase Change in Barbituric Acid. *Cryst. Growth Des.* **2008**, *8*, 1454–1457. [CrossRef]

36. Schmidt, M.U.; Brüning, J.; Glinnemann, J.; Hützler, M.W.; Mörschel, P.; Ivashevskaya, S.N.; Van De Streek, J.; Braga, D.; Maini, L.; Chierotti, M.R.; et al. The Thermodynamically Stable Form of Solid Barbituric Acid: The Enol Tautomer. *Angew. Chem. Int. Ed.* **2011**, *50*, 7924–7926. [CrossRef] [PubMed]

37. Chierotti, M.R.; Ferrero, L.; Garino, N.; Gobetto, R.; Pellegrino, L.; Braga, D.; Grepioni, F.; Maini, L.; Ferrero, L.; Garini, N.; et al. The Richest Collection of Tautomeric Polymorphs: The Case of 2-Thiobarbituric. *Acid. Chem. Eur. J.* **2010**, *16*, 4347–4358.

38. Chemburkar, S.R.; Bauer, J.; Deming, K.; Spiwek, H.; Patel, K.; Morris, J.; Henry, R.; Spanton, S.; Dziki, W.; Porter, W.; et al. Dealing with the Impact of Ritonavir Polymorphs on the Late Stages of Bulk Drug Process Development. *Org. Process Res. Dev.* **2000**, *4*, 413–417. [CrossRef]

39. Bauer, J.; Spanton, S.; Henry, R.; Quick, J.; Dziki, W.; Porter, W.; Morris, J. Ritonavir: An Extraordinary Example of Conformational Polymorphism. *J. Pharm. Res.* **2001**, *6*, 59–66.

40. Morissette, S.L.; Soukasene, S.; Levinson, D.; Cima, M.J.; Almarsson, Ö. Elucidation of crystal form diversity of the HIV protease inhibitor ritonavir by high-throughput crystallization Polymorphism and Crystalforms. *Proc. Natl. Acad. Sci. USA* **2003**, *100*, 2180–2184. [CrossRef]

41. Georgi, C. European Patent Specification. *Supply Manag.* **2006**, *1*, 1–11.

42. Wolff, H.-M.; Queré, L.; Riedner, J. Polymorphic Form of Rotigotine. EP2215072B1, 25 November 2008.

43. Rietveld, I.B.; Céolin, R. Rotigotine: Unexpected Polymorphism with Predictable Overall Monotropic Behavior. *J. Pharm. Sci.* **2015**, *104*, 4117–4122. [CrossRef]

44. Dunitz, J.D.; Bernstein, J. Disappearing Polymorphs. *Acc. Chem. Res.* **1995**, *28*, 193–200. [CrossRef]
45. Bernstein, J. Polymorphism and Patents from a Chemist's Point of View. In *Polymorphism in the Pharmaceutical Industry*; Elsevier: Amsterdam, The Netherlands, 2006; pp. 365–384.
46. Price, S.L. Predicting Crystal Structures of Organic Compounds. *Chem. Soc. Rev.* **2014**, *43*, 2098–2111. [CrossRef] [PubMed]
47. Price, L.S.; Price, S.L. Packing Preferences of Chalcones: A Model Conjugated Pharmaceutical Scaffold. *Cryst. Growth Des.* **2022**, *22*, 1801–1816. [CrossRef] [PubMed]
48. Price, S.L. Is Zeroth Order Crystal Structure Prediction (CSP_0) Coming to Maturity? What Should We Aim for in an Ideal Crystal Structure Prediction Code? *Faraday Discuss.* **2018**, *211*, 9–30. [CrossRef]
49. CSP Blind Tests—The Cambridge Crystallographic Data Centre (CCDC). Available online: https://www.ccdc.cam.ac.uk/Community/initiatives/cspblindtests/past-csp-blind-tests (accessed on 14 June 2022).
50. Reilly, A.M.; Cooper, R.I.; Adjiman, C.S.; Bhattacharya, S.; Boese, A.D.; Brandenburg, J.G.; Bygrave, P.J.; Bylsma, R.; Campbell, J.E.; Car, R.; et al. Report on the Sixth Blind Test of Organic Crystal Structure Prediction Methods. *Acta Crystallogr. Sect. B Struct. Sci. Cryst. Eng. Mater.* **2016**, *72*, 439–459. [CrossRef]
51. Available online: https://www.ccdc.cam.ac.uk/Community/initiatives/cspblindtests/csp-blind-test-7/ (accessed on 14 June 2022).
52. Griesser, U.J. Importance of Solvates. In *Polymorphism in the Pharmaceutical Industry: Solid Form and Drug Development*; Hilfiker, R., Ed.; Wiley-VCH: Weinheim, Germany, 2006; pp. 211–234.
53. Generally Recognized as Safe (GRAS). Available online: https://www.fda.gov/food/food-ingredients-packaging/generally-recognized-safe-gras (accessed on 15 June 2022).
54. Braun, D.E.; Koztecki, L.H.; Mcmahon, J.A.; Price, S.L.; Reutzel-Edens, S.M. Navigating the Waters of Unconventional Crystalline Hydrates. *Mol. Pharm.* **2015**, *12*, 3069–3088. [CrossRef]
55. Jurczak, E.; Mazurek, A.H.; Szeleszczuk, L.; Pisklak, D.M.; Zielinska-Pisklak, M. Pharmaceutical Hydrates Analysis-Overview of Methods and Recent Advances. *Pharmaceutics* **2020**, *12*, 959. [CrossRef] [PubMed]
56. Groom, C.R.; Bruno, I.J.; Lightfoot, M.P.; Ward, S.C. The Cambridge Structural Database. *Acta Crystallogr. Sect. B Struct. Sci. Cryst. Eng. Mater.* **2016**, *72*, 171–179. [CrossRef]
57. Zhu, H.; Yuen, C.; Grant, D.J.W. Influence of Water Activity in Organic Solvent + Water Mixtures on the Nature of the Crystallizing Drug Phase. 1. Theophylline. *Int. J. Pharm.* **1996**, *135*, 151–160. [CrossRef]
58. Braga, D.; Grepioni, F. Crystal Engineering: State of the Art and Open Challenges in Intermolecular Interactions in Crystals. In *Intermolecular Interactions in Crystals: Fundamentals of Crystal Engineering*; Novoa, J.J., Ed.; The Royal Society of Chemistry: London, UK, 2018.
59. Gift, A.D.; Luner, P.E.; Luedeman, L.; Taylor, L.S. Influence of Polymeric Excipients on Crystal Hydrate Formation Kinetics in Aqueous Slurries. *J. Pharm. Sci.* **2008**, *97*, 5198–5211. [CrossRef]
60. Fang, T.; Haiyan, Q.; Zimmermann, A.; Mtunk, T.; Joergensen, A.C.; Rantanen, J. Factors affecting crystallization of hydrates. *J. Pharm. Pharmac.* **2010**, *62*, 1534–1546.
61. Marchi, E.; Montecchi, L. Imidazo-Rifamycin Derivatives with Antimicrobial Utility. U.S. Patent 4341785A, 27 July 1982.
62. Gillis, J.C.; Brogden, R.N. Rifaximin: A Review of Its Antibacterial Activity, Pharmacokinetic Properties and Therapeutic Potential in Conditions Mediated by Gastrointestinal Bacteria. *Drugs* **1995**, *49*, 467–484. [CrossRef] [PubMed]
63. Braga, D.; Grepioni, F.; Chelazzi, L.; Campana, M.; Confortini, D.; Viscomi, G.C. The Structure-Property Relationship of Four Crystal Forms of Rifaximin. *CrystEngComm* **2012**, *14*, 6404–6411. [CrossRef]
64. Viscomi, G.C.; Campana, M.; Barbanti, M.; Grepioni, F.; Polito, M.; Confortini, D.; Rosini, G.; Righi, P.; Cannata, V.; Braga, D. Crystal Forms of Rifaximin and Their Effect on Pharmaceutical Properties. *CrystEngComm.* **2008**, *10*, 1074–1081. [CrossRef]
65. Viscomi, G.C.; Campana, M.; Braga, D.; Confortini, D.; Cannata, V.; Severini, D.; Righi, P.; Rosini, G. Polymorphous Forms of Rifaximin, Processes for Their Production and Use thereof in Medicinal Preparations. U.S. Patent US7045620, 5 June 2006.
66. Viscomi, G.C.; Campana, M.; Confortini, D.; Barbanti, M.M.; Braga, D. New polymorphous forms of rifaximin, processes for their production and use thereof in the medicinal preparations. PCT Patent WO2006/094662, 14 September 2006.
67. Grepioni, F.; Braga, D.; Chelazzi, L.; Shemchuk, O.; Maffei, P.; Sforzini, A.; Viscomi, G.C. Improving solubility and storage stability of Rifaximin via solid-state solvation with Transcutol®. *CrystEngComm* **2019**, *21*, 5278–5283. [CrossRef]
68. Viscomi, G.C.; Maffei, P.; Sforzini, A.; Grepioni, F.; Chelazzi, L. Solvated Crystal Form of Rifaximin, Production, Compositions and Uses. U.S. Patent 9,938,298 B2 (45), 10 April 2018.
69. Hoogsteen, K. The crystal and molecular structure of a hydrogen-bonded complex between 1-methylthymine and 9-methyladenine. *Acta Crystallogr.* **1963**, *16*, 907–916. [CrossRef]
70. Etter, M.C. Hydrogen bonds as design elements in organic chemistry. *J. Phys. Chem.* **1991**, *95*, 4601–4610. [CrossRef]
71. Dunitz, J.D. Crystal and cocrystal: A second opinion. *CrystEngComm* **2003**, *5*, 506. [CrossRef]
72. Almarsson, Ö.; Zaworotko, M.J. Crystal engineering of the composition of pharmaceutical phases. Do pharmaceutical cocrystals represent a new path to improved medicines? *Chem. Comm.* **2004**, *17*, 1889–1896. [CrossRef]
73. Aakeröy, C.B.; Salmon, D.J. Building cocrystals with molecular sense and supramolecular sensibility. *CrystEngComm* **2005**, *7*, 439–448. [CrossRef]
74. Duggirala, N.K.; Perry, M.L.; Almarsson, O.; Zaworotko, M.J. Pharmaceutical cocrystals: Along the path to improved medicines. *Chem. Commun.* **2016**, *52*, 640. [CrossRef]

75. Bolla, G.; Sarma, B.; Nangia, A.K. Crystal Engineering of Pharmaceutical Cocrystals in the Discovery and Development of Improved Drugs. *Chem. Rev.* **2022**, *122*, 11514–11603. [CrossRef] [PubMed]

76. Cruz-Cabeza, A.J.; Reutzel-Edens, S.M.; Bernstein, J. Facts and fictions about polymorphism. *Chem. Soc. Rev* **2015**, *44*, 8619. [CrossRef]

77. Childs, S.L.; Rodriguez-Hornedo, N.; Reddy, L.S.; Jayasankar, A.; Maheshwari, C.; McCausland, L.; Shipplett, R.; Stahly, B.C. Screening strategies based on solubility and solution composition generate pharmaceutically acceptable cocrystals of carbamazepine. *CrystEngComm* **2008**, *10*, 856–864. [CrossRef]

78. Braga, D.; Grepioni, F.; Maini, L.; Prosperi, S.; Gobetto, R.; Chierotti, M.R. From unexpected reactions to a new family of ionic co-crystals: The case of barbituric acid with alkali bromides and caesium iodide. *Chem. Commun* **2010**, *46*, 7715–7717. [CrossRef] [PubMed]

79. Braga, D.; Grepioni, F.; Maini, L.; Capucci, D.; Nanna, S.; Wouters, J.; Aerts, L.; Quéré, L. Combining piracetam and lithium salts: Ionic cocrystals and co-drugs? *Chem. Commun.* **2012**, *48*, 8219–8221. [CrossRef] [PubMed]

80. Grepioni, F.; Wouters, J.; Braga, D.; Nanna, S.; Fours, B.; Coquerel, G.; Geraldine Longfils, G.; Rome, S.; Aerts, L.; Quéré, L. Ionic cocrystals of racetams: Solid-state properties enhancement of neutral active pharmaceutical ingredients via addition of Mg^{2+} and Ca^{2+} chlorides. *CrystEngComm* **2014**, *16*, 5887–5896. [CrossRef]

81. Ong, T.T.; Kavuru, P.; Nguyen, T.; Cantwell, R.; Wojtas, Ł.; Zaworotko, M.J. 2:1 cocrystals of homochiral and achiral amino acid zwitterions with Li^+ salts: Water-stable zeolitic and diamondoid metal-organic materials. *J. Am. Chem. Soc.* **2011**, *133*, 9224–9227. [CrossRef]

82. Oertling, H. Interactions of alkali- and alkaline earth-halides with carbohydrates in the crystalline state—The overlooked salt and sugar cocrystals. *CrystEngComm* **2016**, *18*, 1676–1692. [CrossRef]

83. Aakeröy, C.B.; Forbes, S.; Desper, J. Using cocrystals to systematically modulate aqueous solubility and melting behavior of an anticancer drug. *J. Am. Chem. Soc.* **2009**, *131*, 17048–17049. [CrossRef]

84. Shiraki, K.; Takata, N.; Takano, R.; Hayashi, Y.; Terada, K. Dissolution Improvement and the Mechanism of the Improvement from Cocrystallization of Poorly Water-soluble Compounds. *Pharm. Res.* **2008**, *25*, 2581–2592. [CrossRef] [PubMed]

85. Good, D.J.; Naır, N.; Rodrıguez-Hornedo, R. Cocrystal Eutectic Constants and Prediction of Solubility Behavior. *Cryst. Growth Des.* **2010**, *10*, 1028–1032. [CrossRef]

86. McNamara, D.P.; Childs, S.L.; Giordano, J.; Iarriccio, A.; Cassidy, J.; Shet, M.S.; Mannion, R.; O'Donnell, E.; Park, A. Use of a Glutaric Acid Cocrystal to Improve Oral Bioavailability of a Low Solubility API. *Pharm. Res.* **2006**, *23*, 1888–1899. [CrossRef] [PubMed]

87. Yan, Y.; Chen, J.M.; Geng, N.; Lu, T.B. Thermodynamics and preliminary pharmaceutical characterization of a melatonin–pimelic acid cocrystal prepared by a melt crystallization method. *CrystEngComm* **2015**, *17*, 612–620. [CrossRef]

88. Grifasi, F.; Chierotti, M.R.; Gaglioti, K.; Gobetto, R.; Maini, L.; Braga, D.; Dichiarante, E.; Curzi, M. Using Salt Cocrystals to Improve the Solubility of Niclosamide. *Cryst. Growth Des.* **2015**, *15*, 1939–1948. [CrossRef]

89. Chadha, K.; Karan, M.; Bhalla, Y.; Chadha, R.; Khullar, S.; Mandal, S.; Vasisht, K. Cocrystals of Hesperetin: Structural, Pharmacokinetic, and Pharmacodynamic Evaluation. *Cryst. Growth Des.* **2017**, *17*, 2386–2405. [CrossRef]

90. Eisenlohr, F.; Meier, G. Die Aufspaltung von Racematen mit Hilfe von Molekülverbindungen. *Ber. Der Dtsch. Chem. Ges. A B Ser.* **1938**, *71*, 1005–1013. [CrossRef]

91. Caira, M.R.; Nassimbeni, L.R.; Scott, J.L.; Wildervanck, A.F. Resolution of optical isomers of 4-amino-p-chlorobutyric acid lactam by cocrystallization. *J. Chem. Cryst.* **1996**, *26*, 117–122. [CrossRef]

92. Springuel, G.; Leyssens, T. Innovative chiral resolution using enantiospecific cocrystallization in solution. *Cryst. Growth Des.* **2012**, *12*, 3374–3378. [CrossRef]

93. George, F.; Norberg, B.; Robeyns, K.; Wouters, J.; Leyssens, T. Peculiar Case of Levetiracetam and Etiracetam α-Ketoglutaric Acid Cocrystals: Obtaining a Stable Conglomerate of Etiracetam. *Cryst. Growth Des.* **2016**, *16*, 5273–5282. [CrossRef]

94. Toda, F. Isolation and optical resolution of materials utilizing inclusion crystallization. In *Molecular Inclusion and Molecular Recognition—Clathrates I*; Springer: Berlin/Heidelberg, Germany, 2005; pp. 43–69.

95. Weber, E.; Wimmer, C.; Llamas-Saiz, A.L.; Foces-Foces, C. New chiral selectors derived from lactic acid. Cocrystalline and sorptive optical resolutions, and the crystal structure of an inclusion complex with 3-methylcyclohexanone. *J. Chem. Soc. Chem. Commun.* **1992**, 733–735. [CrossRef]

96. Nishikawa, K.; Tsukada, H.; Abe, S.; Kishimoto, M.; Yasuoka, N. Mode of molecular recognition during optical resolution: A structural study of the molecular complex involving both 3-hydroxypyrrolidines and (R,R)-(-)-trans-bis(hydroxydiphenylrnethyl)-1,4-dioxaspiro[4,5]decane. *Chirality* **1999**, *11*, 166–171. [CrossRef]

International Journal of
Molecular Sciences

Article

Development of a Sensitive Screening Method for Simultaneous Determination of Nine Genotoxic Nitrosamines in Active Pharmaceutical Ingredients by GC-MS

Anna B. Witkowska [1,2,*], Joanna Giebułtowicz [3], Magdalena Dąbrowska [1] and Elżbieta U. Stolarczyk [1,4]

[1] Research Analytics Team, Analytical Department, Łukasiewicz Research Network—Industrial Chemistry Institute, 8 Rydygiera Street, 01-793 Warsaw, Poland
[2] Department of Bioanalysis and Drugs Analysis, Doctoral School, Medical University of Warsaw, 61 Żwirki i Wigury, 02-091 Warsaw, Poland
[3] Department of Bioanalysis and Drugs Analysis, Faculty of Pharmacy, Medical University of Warsaw, 1 Banacha, 02-097 Warsaw, Poland
[4] Spectrometric Methods Department, National Medicines Institute, Chełmska 30/34, 00-725 Warsaw, Poland
* Correspondence: anna.witkowska@wum.edu.pl

Abstract: A worldwide crisis with nitrosamine contamination in medical products began in 2018. Therefore, trace-level analysis of nitrosamines is becoming an emerging topic of interest in the field of quality control. A novel GC-MS method with electron ionization and microextraction was developed and validated for simultaneous determination of nine carcinogenic nitrosamines (NDMA, NMEA, NDEA, NDBA, NMOR, NPYR, NPIP, NDPA, and *N*-methyl-npz) in active pharmaceutical ingredients (APIs): cilostazol, sunitinib malate, and olmesartan medoxomil. The method was validated according to the International Council for Harmonisation of Technical Requirements for Pharmaceuticals for Human Use (ICH) guidelines, demonstrating good linearity in the range of LOQ up to 21.6 ng/mL (120% of specification limit). The limits of detection for the nine nitrosamines were determined to be in the range 0.15–1.00 ng/mL. The developed trace level GC-MS method turned out to be specific, accurate, and precise. The accuracy of all the tested APIs ranged from 94.09% to 111.22% and the precision evaluated by repeatability, intermediate precision, and system precision was RSD ≤ 7.65%. Nitrosamines were not detected in cilostazol and sunitinib, whereas in olmesartan medoxomil NDEA was detected at the level of LOQ. The novel protocol was successfully applied for nitrosamines determination in selected APIs and can be used for the routine quality control of APIs under Good Manufacturing Practices rules, ensuring the safety and effectiveness of pharmaceutical products.

Keywords: nitrosamines; gas chromatography–mass spectrometry; active pharmaceutical ingredient; ionization; microextraction; validation

Citation: Witkowska, A.B.; Giebułtowicz, J.; Dąbrowska, M.; Stolarczyk, E.U. Development of a Sensitive Screening Method for Simultaneous Determination of Nine Genotoxic Nitrosamines in Active Pharmaceutical Ingredients by GC-MS. *Int. J. Mol. Sci.* **2022**, 23, 12125. https://doi.org/10.3390/ijms232012125

Academic Editors: Geoffrey Brown, Enikö Kallay and Andrzej Kutner

Received: 31 August 2022
Accepted: 4 October 2022
Published: 12 October 2022

Publisher's Note: MDPI stays neutral with regard to jurisdictional claims in published maps and institutional affiliations.

1. Introduction

A worldwide crisis with nitrosamine contamination in medical products began in 2018 [1]. Since then, widespread investigations by regulatory agencies, including the European Medicines Agency (EMA) and United States Food and Drug Administration (US FDA), were undertaken. In drugs such as angiotensin II receptor blockers (ARBs), ranitidine, metformin, rifampin, rifapentine, and, recently, verenicline, *N*-nitrosodimethylamine (NDMA) and other nitrosamines have been detected [2,3]. Nitrosamines are known as probable human carcinogens (IARC 2A group for NDMA), potent genotoxic agents, and "cohort of concern" compounds according to ICH M7 guidance [4,5]. They can form in reaction of amine source (e.g., 2°, 3° amines, amidines, hydrazines) and nitrosating agents (e.g., nitrite in acidic conditions, nitrogen oxide, nitrous acid). According to newly established guidelines, every marketing authorization holder has to identify the risk of nitrosamine formation or cross-contamination in the manufacturing/storage of chemical and biological medicines. Root causes of nitrosamine impurities in APIs (active pharmaceutical ingredients) and drug products can be manufacturing process related as well as stability of the drug substance/product or excipient compatibility related (Figure 1) [6–8].

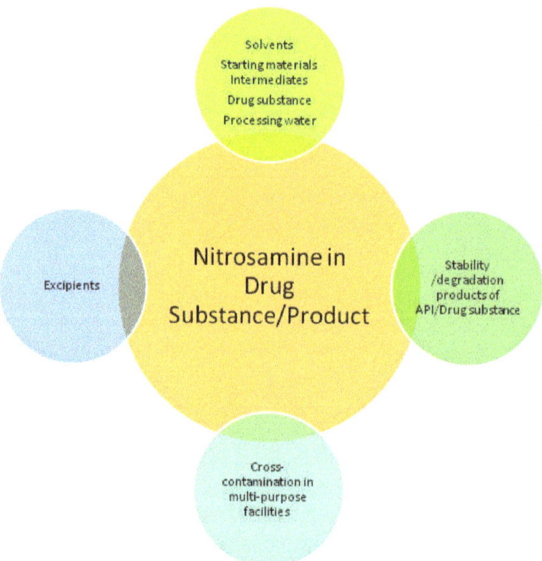

Figure 1. Root causes of nitrosamine impurities in APIs and drug products.

The scale of the problem is revealed by the fact that, till now, more than one thousand batches of medical products have been recalled from the market due to the detection of harmful nitrosamines [9].

A large number of studies also resulted in the introduction of new, more restrictive legislation that regulated the concentration of nitrosamines in drug products. Therefore, trace-level analysis of nitrosamines is becoming an emerging topic of interest in the field of quality control. Till now, a wide range of quantitative analytical methods to measure nitrosamine traces in pharmaceuticals have been established, including gas chromatography with single quadrupole or tandem mass spectrometers (GC-MS/MS) with a headspace system or direct injection [10–16], liquid chromatography–tandem mass spectrometers (LC-MS/MS) or high-resolution mass spectrometers [17–20], and liquid chromatography with UV detectors and supercritical fluid chromatography (SFC) [21–23]. Solid-phase extraction (SPE), solid-phase microextraction (SPME), and dispersive liquid–liquid microextraction (DLLME) are also commonly used with GC-MS to produce comparable results of the LOQ (range 0.5–1.5 ppb) to LC-MS/MS [24–26]. Many already established methods are complex and can be applied to only one drug product/API, such as ranitidine and metformin. Our newly established, sensitive analytical screening procedure involving microextraction is fast, simple, environmentally friendly, and with limits of quantitation (LOQs) close to multi-step methods involving SPME or SPE extraction. Moreover GC-MS offers low operational costs, and is commonly used for analysis of volatile impurities in APIs. We demonstrate, for the first time, that our analytical procedure can be successfully applied for various APIs selected for quantitative testing after risk assessment.

2. Results and Discussion

2.1. Risk Assessment

Figure 2 presents the structures of tested APIs and N-nitrosamines examined in this study.

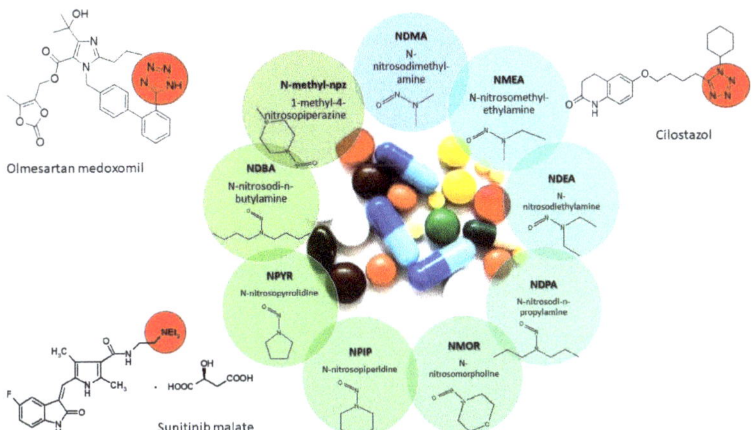

Figure 2. Structures of the APIs and nitrosamines presented in this study. Red circles—types of nitrosamine-related substructures present in APIs.

Since 2019, national regulatory agencies (US FDA, EMA) have obligated every marketing authorization holder to review their manufacturing process in order to identify the risk of nitrosamine formation or cross-contamination and, if necessary, to perform confirmatory testing. Kao et al. [27] in their recent study reported a substructure-based approach as a method for screening and investigating nitrosamines in medical products. According to their work, there are eight substructures present in APIs: DBA (dibutylamine), DEA (dietylamine), DIPA (diisopropylamine), DMA (dimethylamine), IPEA (isopropylethylamine), MPA (methylphenulamine, NMP (N-methyl-2-pyrrolidone), and a tetrazole ring can be suspected to be precursors of nitrosamines. A tetrazole ring is present in the structure of olmesartan medoxomil and cilostazol, while DEA in sunitinib malate (Figure 2). Another risk factor for nitrosamine contamination of sunitinib malate is the use of pyrrolidine in the synthesis. Many suppliers of raw materials for the synthesis of APIs do not provide information about the reagents used in the synthesis, which is why it is so important to selectively check batches with risk elements. Regulatory agencies have implemented analytical methods for sartans, antihistamines, and antidiabetic drugs, but they also continue to test other drug products; so, our new, sensitive screening method for determination of the nine nitrosamines in cilostazol and sunitinib malate is a very useful and suitable tool.

2.2. Method Development

A key parameter in the development of the microextraction method for the samples was optimization of the sample solvent. The solubility studies of olmesartan medoxomil, sunitinib malate, and cilostazol are presented in Supplementary Tables S1–S3. The choice of solvent involved the selection of the medium in which centrifugal extraction of the analytes under study was most efficient. Finally, methanol was chosen as the solvent in which cilostazol, olmesartan medoxomil, and sunitinib malate were slightly soluble. During method development, the centrifuge speed, centrifugation time, and amount of solvent taken for extraction were optimized. The best sensitivity was obtained for a centrifugation speed of 15,000 rpm, time of 10 min, and 250 µL of solvent. At lower centrifugation speeds and shorter times (5000 rpm, 10,000 rpm, 2 and 5 min), there was no complete extraction of the analytes (Table S4). At longer centrifugation times, no increase in method sensitivity was observed. During the extraction, no evaporation of the solvent in the centrifuge chamber was noticed. From the capillary columns commonly used in the analysis of nitrosamines, such as DB-5 ms, DB-624 ms, and VF-WAXms [28–30], the best chromatographic separation was obtained on the column VF-WAXms (30 m × 0.32 mm, film thickness of 1.0 µm). For the quantification analysis on a single quadruple mass spectrometer, we used electron ionization in selected ion monitoring mode (SIM) of particular ions, ensuring the highest sensitivity was chosen (see Section 3.2).

2.3. Validation Results

During the development of the method, satisfactory separation of nine nitrosamines was obtained in selected APIs. The standard added to the sample did not cause peaks splitting and the matrix did not affect the result (Figures S1–S27). Figure 3 shows a chromatogram of the standard solution.

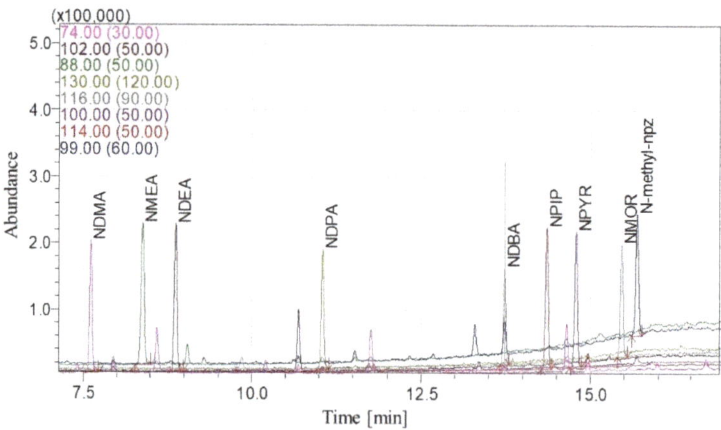

Figure 3. SIM mode chromatogram of the N-nitrosamines standard—18 ng/mL.

Results for linearity, limit of detection (LOD), and LOQ are summarized in Table 1. The determination coefficients (R2) of the linear regression for the nine nitrosamines were over 0.995, which clearly demonstrates that the method was suitable for quantitative analysis.

Table 1. Validation results for linearity, LOD, and LOQ.

Analyte	Range (ng/mL)	*Y = ax + b ($R^2 \geq$ 0.990) ** t_a ** t_b	LOD (ng/mL)		LOQ (ng/mL)	
			S/N $\geq$ 3	(3.3× σ/S)	S/N $\geq$ 10	(10× σ/S)
NDMA	0.45–21.6	Y = 0.0127x + 0.0005 (R^2 = 0.9996) 158.53 0.5695	0.15	0.47	0.45	1.43
NMEA	1.5–21.6	Y = 0.0104x + 0.0011 (R^2 = 0.9997) 159.85 1.4050	0.50	0.40	1.50	1.20
NDEA	1.0–21.6	Y = 0.0088x + 0.0009 (R^2 = 0.9997) 153.81 1.3608	0.30	0.45	1.00	1.37
NDPA	3.0–21.6	Y = 0.0029x + 0.0007 (R^2 = 0.9988) 70.521 1.4061	1.00	0.78	3.00	2.38

Table 1. *Cont.*

Analyte	Range (ng/mL)	*Y = ax + b ($R^2 \geq 0.990$) ** t_a ** t_b	LOD (ng/mL)		LOQ (ng/mL)	
			S/N $\geq$ 3	(3.3$\times$ σ/S)	S/N $\geq$ 10	(10$\times$ σ/S)
NMOR	3.0–21.6	Y = 0.0039x−0.0002 (R^2 = 0.9991) 83.655 0.7650	1.00	0.67	3.00	2.03
NPYR	1.5–21.6	Y = 0.0085x + 0.0007 (R^2 = 0.9994) 105.24 0.7193	0.50	0.60	1.50	1.82
NPIP	1.5–21.6	Y = 0.0087x + 0.0005 (R^2 = 0.9992) 91.615 0.4482	0.50	0.69	1.50	2.10
NDBA	1.5–21.6	Y = 0.0032x + 0.0002 (R^2 = 0.9986) 71.346 0.4260	0.50	0.88	1.5	2.66
N-methyl-npz	1.5–21.6	Y = 0.0065x + 0.0013 (R^2 = 0.9996) 137.02 2.2741	0.50	0.46	1.50	1.40

* a—slope; b—y-intercept. ** t_a, t_b—parameters of the Student's *t*-test; t_b = b/S_b, t_a = a/S_a, where S_a—the standard deviation of a, and S_b—the standard deviation of b.

The LODs and LOQs calculated both ways from the linearity and ratio peak/noise height were very similar. LODs were determined to be within the range 0.15–1.00 ppb (Table 2); such low LODs are comparable to the results obtained with GC-MS/MS by Liu at al. [11] (Table S5).

The accuracy of the method was verified at three levels—80%, 100%, and 120%—of the specification limit. The results for accuracy for all tested APIs are summarized in Table 2. The results exhibited that recoveries for all nitrosamines in sunitinib malate ranged from 94.09% to 111.22%, for cilostazol from 95.91% to 104.87%, and for olmesartan medoxomil from 92.20% to 108.72%, respectively.

Method precision was evaluated by repeatability, intermediate precision, and system precision. As summarized in Table 3, this GC-MS method exhibited satisfactory RSD% values for repeatability in all APIs in the range of 0.76–5.75%, and for intermediate precision and system precision in the range of 1.13–7.65% and 1.03–3.65% (area), respectively. Since the acceptance criteria for recovery and precision for all APIs were established for 60–115% and RSD < 21%, all the results meet the criteria according to the Association of Official Analytical Chemists [31].

The robustness of the GC-MS method was evaluated on standard solution to determine the impact on the result in case of small variations in the chromatographic parameters: oven initial column temperature ± 5 °C; first temperature ramp ± 1 °C/min; and gas pressure ± 10%. The robustness results (Table S6) showed that the retention time for all nitrosamines change within a range of ± 1.1 min and the separation of the nitrosamines was not affected.

Table 2. Validation results of the accuracy for sunitinib, cilostazol, and olmesartan medoxomil.

Analyte	Accuracy Mean Recovery 60–115% RSD $\leq$ 21% ($n = 9$)		
	Sunitinib Malate	Cilostazol	Olmesartan Medoxomil
NDMA	94.45 1.98	98.26 2.11	104.66 5.68
NMEA	94.09 2.10	99.14 1.78	95.85 4.24
NDEA	97.97 3.57	97.15 2.02	97.34 4.44
NDPA	97.07 5.05	95.91 2.15	92.20 6.50
NMOR	111.22 2.89	97.10 3.66	103.60 5.16
NPYR	95.87 3.82	100.62 3.00	108.72 5.29
NPIP	99.37 5.13	96.66 6.02	ND
NDBA	99.30 6.10	104.87 2.38	102.17 8.98
N-methyl-npz	102.94 7.53	99.90 5.17	ND

ND—not detected.

Table 3. Validation results of precision for sunitinib malate, cilostazol, and olmesartan medoxomil.

Matrix / Analyte	Sunitinib		Cilostazol		Olmesartan Medoxomil		System Precision	
	Repeatability RSD (%)	Intermediate Precision RSD (%)	Repeatability RSD (%)	Intermediate Precision RSD (%)	Repeatability RSD (%)	Intermediate Precision RSD (%)	Area Nitrosamine/Area NDMA-d6 RSD (%)	Retention Time RSD (%)
NDMA	2.12	2.05	0.76	1.13	3.07	2.51	1.31	0.03
NMEA	2.25	1.99	1.27	1.42	2.47	2.21	1.08	0.02
NDEA	2.48	3.37	1.10	1.37	3.00	4.04	1.31	0.02
NDPA	3.95	6.26	2.97	2.25	5.75	4.24	1.03	0.01
NMOR	3.18	5.55	2.73	4.49	3.07	3.97	3.52	0.01
NPYR	2.51	5.08	1.77	3.91	2.55	7.50	2.42	0.01
NPIP	3.67	4.06	3.91	4.60	ND	ND	2.01	0.01
NDBA	4.32	6.81	1.68	1.46	4.51	4.71	3.07	0.01
N-methyl-npz	2.22	6.45	5.60	7.65	ND	ND	3.65	0.01

ND—not detected.

Moreover, during validation, the stability of the standard solution and sample solution was investigated. Standard solution stability studies for all nitrosamines in methanol demonstrated that after 24 and 48 h in 25 °C, the RSD% (area) ranged from 0.63 to 4.79%, demonstrating that the standard solution was stable up to 48 h after preparation (Tables S7–S15). Stability of the sample solution of cilostazol, olmesartan medoxomil, and sunitinib malate spiked with nitrosamines at 100% of a specification level was evaluated after 24 h in 25 °C by measuring the peak area quotient of nitrosamines and NDMA-d6. The RSD% for all the nitrosamines in olmesartan medoxomil was in range of 1.36–18.33%, in cilostazol, 0.69–9.14%, and in sunitinib malate, 0.38–10.63%, which indicated the increase in RSD with time and the necessity of sample solution preparation directly before analysis (Tables S16–S24).

Supplementary Material Table S5 shows the comparison of the results of this publication with various reported GC-MS and GC-MS/MS methods for the detection of N-nitrosamines. The novel GC-MS method with microextraction showed superior results in the validation test compared with other GC-MS and GC-MS/MS methods. The developed method offers several benefits over previously established methodologies, including enhanced sensitivity (LOD below 1 ppb), with no requirements for derivatization and multistep sample preparation (SPE, SPME or DLLEM); a reduction in solvent usage (only 250 µL); and application to a wide variety of APIs. We have also demonstrated that the developed method using a single quadrupole is as good as the method with tandem mass spectrometry.

2.4. Application of the Method for Nitrosamine Determination in Selected APIs

The validated GC-MS method was applied to determine nine nitrosamines in three active pharmaceutical ingredients: cilostazol, sunitinib malate, and olmesartan medoxomil. In cilostazol and sunitinib, no nitrosamines were found. NDEA was detected in olmesartan medoxomil at a concentration of 1.66 ng/mL. Therefore, based on the proposed European Pharmacopoeia Commission limit (30 ppb), the acceptance criteria for nitrosamines in APIs have been meet. Olmesartan medoxomil have never been recalled from the market due to detection of nitrosamines beyond acceptable intake limits [2].

3. Materials and Methods

3.1. Materials

Cilostazol (99.8%), sunitinib malate (99.7%), and olmesartan medoxomil (99.7%) were purchased from Pharmaceutical Research Institute (Poland). The nitrosamines mix (containing NDMA, NMEA, NDEA, NDPA, NMOR, NPYR, NPIP, and NDBA) EPA 8270 (Appendix IX), certified reference material; 2000 µg/mL in methanol) was obtained from Sigma-Aldrich (USA). 1-Methyl-4-nitrosopiperazine was obtained from Angene International Limited (London, England). N-Nitrosodimethylamine-d6 (1000 µg/mL in methanol) was purchased from LGC (Kiełpin, Poland). Methanol LC-MS reagent was obtained from J.T Baker (Oslo, Norway).

3.2. Equipment and Method

The GC-MS analysis was performed using a Shimadzu gas chromatograph GCMS-QP2010 ULTRA with MS detector. The separation of nitrosamines was performed on a VF-WAXms polyethylene glycol column (30 m × 0.32 mm, film thickness of 1.0 µm) from Agilent Technologies. In the study, centrifuge tubes with a 0.2 µm PVDF filter from Anchem and an M-Science centrifuge from MPW Med. Instruments were used. All the GC-MS method's experimental conditions are presented in Table 4. The ions selected for each nitrosamine for quantification and confirmation are shown in Table 5.

3.3. Standard Solution Preparation

The internal standard (IS) solution (70 ng/mL in methanol) was prepared from the internal stock solution NDMA-d6 (1000 µg/mL in methanol) and diluted with methanol. The nitrosamine intermediate dilution was prepared by dissolving the appropriate amount of the nitrosamines mix (certified reference material) to reach 10 µg/mL of the nitrosamine mix in MeOH. A standard solution of nine nitrosamines at the concentration level of 18.0 ng/mL was prepared from the nitrosamine intermediate dilution in a 10 mL volumetric flask with methanol.

Table 4. GC-MS method experimental conditions.

GC Parameters	
Carrier gas	Helium
Column flow rate	2.1 mL/min
Oven initial temperature	75 °C (hold time: 2 min)
First temperature ramp	9.0 °C/min
Second temperature	200 °C/min (hold time: 2 min)
Second temperature ramp	15.0 °C/min
Final temperature	240 °C (hold time: 3 min)
Pressure	45.5 kPa
Injection mode	Splitless, high-pressure injection 130 kPa
Injection port temperature	150 °C
MS Parameters	
Ion Source Temperature	240
Interface Temperature	240
Acquisition type	SIM
Solvent delay	6.5 min

Table 5. GC-MS SIM for the detected nitrosamines.

Compound	Quantification Ion (*m/z*)	Confirmation Ion (*m/z*)
NDMA	74	44
NMEA	88	56
NDEA	102	56
NDPA	130	70
NMOR	116	56
NPYR	100	68
NPIP	114	55
NDBA	116	158, 84
N-methyl-npz	99	100, 56
NDMA-d6	80	50

3.4. Sample Solution Preparation

A total of 150 mg of API was weighed into 2 mL centrifuge tubes with filtration equipment and 250 μL of the internal standard solution (70 ng/mL NDMA-d6) was added. Then, the solution was centrifuged at 15,000 rpm for 10 min. A clear organic phase was transferred to a sample vial because the selected APIs were practically insoluble in methanol.

3.5. Blank Preparation

The blank solution was prepared as described in Section 3.4., but without addition of the sample.

3.6. Validation of the GC-MS Method

The GC-MS method for the nine nitrosamines in API was validated according to ICH guidelines. The specificity of the method was examined using a blank solution, standard solution, sample solution, and reference solution (sample solution spiked with standard solution at a 100% concentration level of the specified limit).

The calibration curve was prepared in methanol at concentrations from LOQ to 120% (0.45–21.6 ng/mL) of the specified limit for nitrosamines with respect to sample preparation, the internal standard solution was added in a concentration of 70 ng/mL NDMA-d6. The linearity for every nitrosamine was evaluated by linear regression and statistical evaluation. Linearity experiments included the determination coefficient, y-intercept, slope of regression line, and residual sum of squares (Sr). The determination coefficients (R2) of the linear regression for each calibration curve should be ≥0.995.

The accuracy of the method was established on three different concentrations: a sample spiked with nitrosamines at 80%, 100%, and 120% of the specification limit; the acceptance criteria were 60–115% and the precision at each concentration should not exceed RSD < 21%. In 2020, the European Pharmacopoeia Commission adopted a new general chapter for trace-level analysis of nitrosamine impurities where the recommended maximum allowed limit for nitrosamines in active substances was set at 0.03 ppm (30 ppb) [32]; therefor, in the present study, the same limit was implemented.

Limit of detection (LOD) and limit of quantitation (LOQ) were calculated in two different ways: using peak height (S) and noise height (N), where the LOQ was stated as $S/N \geq 10$ and LOD as $S/N \geq 3$, and using linearity parameters such as the standard deviation of the response (σ) and the slope of the curve (S); then, LOD and LOQ were determined from the following equations: $LOD = (3.3 \times \sigma/S)$ and $LOQ = (10 \times \sigma/S)$.

The precision of the method was examined as repeatability, intermediate precision, and system precision. Repeatability for all nitrosamines was obtained from analysis of the six sample solutions. Intermediate precision was assessed as a result of the intra-laboratory variations, i.e., a different day or analyst (the RSD should not exceed 21%), while the difference between the results were tested by statistical Horwitz's test. System precision was evaluated by six replicate injections of the standard solution (calculation—Supplementary Note S1).

During validation, the stability of the solutions and robustness of the method also were evaluated. The stability of the solutions was established by measuring the peak area quotient of the nitrosamines and NDMA-d6 for the reference solution, with nitrosamines at a 100% concentration level of the specified limit and for a sample solution of API (cilostazol, sunitinib, and olmesartan medoxomil) spiked with nitrosamines at 100% of the specification level (18.0 ng/mL) after 24 h under autosampler conditions (20 °C ± 5 °C). Robustness of the developed method was investigated by injecting the standard solution in different chromatographic conditions: column temperature, ±5 °C; rate, ±1 °C/min; and carrier gas pressure, ±10%.

4. Conclusions

A new, sensitive, simple, environmentally friendly method for extracting nine nitrosamines from active pharmaceutical ingredients, using the high-pressure direct injection GC-MS method, was developed in this study. Based on the performed risk evaluation in the manufacturing process, three APIs were examined. This is the first described method for the determination of nitrosamines in cilostazol and sunitinib malate. The implemented GC-MS method for cilostazol and sunitinib malate showed no influence of the manufacturing process on the development of nitrosamine impurities. The quantitative method for each API was validated according to the requirements of the ICH guideline. The validation of the GC-MS method proved that the method was sensitive, accurate, precise, and suitable for nitrosamine determination in multiple APIs. The results obtained during the validation showed that the LOD and LOQ were in the range from 0.15 to 1.00 ng/mL. In conclusion, the presence of nine nitrosamines was determined as below the acceptable intake limits in selected APIs; however, it is necessary to test other APIs containing specific amine substructures in their structures to ensure the safety of the drugs for patients.

Supplementary Materials: The following supporting information can be downloaded at: https://www.mdpi.com/article/10.3390/ijms232012125/s1.

Author Contributions: Conceptualization, supervision, E.U.S.; methodology, validation, formal analysis, investigation, visualization, A.B.W.; writing—original draft, A.B.W.; writing—review and editing, A.B.W., E.U.S., J.G., M.D.; funding acquisition A.B.W. All authors have read and agreed to the published version of the manuscript.

Funding: The project was financially supported by the Ministry of Science and Higher Education: contract number DWD/5/0401/2021.

Institutional Review Board Statement: Not applicable.

Informed Consent Statement: Not applicable.

Data Availability Statement: Not applicable.

Acknowledgments: The study was supported by the Polish Ministry of Science and Higher Education, statutory grant no. 841343A. The authors would like to thank Marta Zezula for her support in this study.

Conflicts of Interest: The authors declare no conflict of interest.

References

1. US Food and Drug Administration. FDA Announces Voluntary Recall of Several Medicines Containing Valsartan Following Detection of an Impurity. Available online: https://www.fda.gov/news-events/press-announcements/fda-announces-voluntary-recall-several-medicines-containing-valsartan-following-detection-impurity (accessed on 30 August 2022).
2. US Food and Drug Administration. Recalls, Market Withdrawals, & Safety Alerts. Available online: https://www.fda.gov/safety/recalls-market-withdrawals-safety-alerts (accessed on 30 August 2022).
3. Tao, X.; Tian, Y.; Liu, W.-H.; Yao, S.; Yin, L. Trace Level Quantification of 4-Methyl-1-Nitrosopiperazin in Rifampicin Capsules by LC-MS/MS. *Front Chem* **2022**, *10*, 834124. [CrossRef] [PubMed]
4. European Medicines Agency; Committee for Medicinal Products for Human Use. *ICH Guideline M7 on Assessment and Control of DNA Reactive (Mutagenic) Impurities in Pharmaceuticals to Limit Potential Carcinogenic Risk-Addendum*; European Medicines Agency: Amsterdam, The Netherlands, 2021; Volume 31.
5. International Agency for Research on Cancer. *IARC Publications Website-Overall Evaluations of Carcinogenicity: An Updating of IARC Monographs*; International Agency for Research on Cancer: Lyon, France, 1987; Volume 1–42.
6. Wu, Y.; Levons, J.; Narang, A.S.; Raghavan, K.; Rao, V.M. Reactive Impurities in Excipients: Profiling, Identification and Mitigation of Drug–Excipient Incompatibility. *AAPS PharmSciTech* **2011**, *12*, 1248–1263. [CrossRef]
7. Yokoo, H.; Yamamoto, E.; Masada, S.; Uchiyama, N.; Tsuji, G.; Hakamatsuka, T.; Demizu, Y.; Izutsu, K.I.; Goda, Y. N-Nitrosodimethylamine (Ndma) Formation from Ranitidine Impurities: Possible Root Causes of the Presence of Ndma in Ranitidine Hydrochloride. *Chem. Pharm. Bull.* **2021**, *69*, 872–876. [CrossRef]
8. Sedlo, I.; Kolonić, T.; Tomić, S. Presence of Nitrosamine Impurities in Medicinal Products. *Arh. Hig. Rada Toksikol.* **2021**, *72*, 1–5. [CrossRef] [PubMed]
9. Bharate, S.S. Critical Analysis of Drug Product Recalls Due to Nitrosamine Impurities. *J. Med. Chem.* **2021**, *64*, 2923–2936. [CrossRef] [PubMed]
10. Parr, M.K.; Joseph, J.F. NDMA Impurity in Valsartan and Other Pharmaceutical Products: Analytical Methods for the Determination of N-Nitrosamines. *J. Pharm. Biomed. Anal.* **2019**, *164*, 536–549. [CrossRef]
11. Liu, J.; Xie, B.; Mai, B.; Cai, Q.; He, R.; Guo, D.; Zhang, Z.; Fan, J.; Zhang, W. Development of a Sensitive and Stable GC-MS/MS Method for Simultaneous Determination of Four N-Nitrosamine Genotoxic Impurities in Sartan Substances. *J. Anal. Sci. Technol.* **2021**, *12*, 3. [CrossRef]
12. US Food and Drug Administration. *Combined Headspace N-Nitrosodimethylamine (NDMA), N-Nitrosodiethylamine (NDEA), N-Nitrosoethylisopropylamine (NEIPA), and N-Nitrosodiisopropylamine (NDIPA) Impurity Assay by GC-MS/MS*; US Food and Drug Administration: Silver Spring, MD, USA, 2019; pp. 1–7.
13. Tsutsumi, T.; Akiyama, H.; Demizu, Y.; Uchiyama, N.; Masada, S.; Tsuji, G.; Arai, R.; Abe, Y.; Hakamatsuka, T.; Izutsu, K.; et al. Analysis of an Impurity, N-Nitrosodimethylamine, in Valsartan Drug Substances and Associated Products Using GC-MS. *Biol. Pharm. Bull.* **2019**, *42*, 547–551. [CrossRef]
14. New OMCL Method for Simultaneous Determination of NDMA and NDEA in Sartan. Available online: https://www.edqm.eu/documents/52006/71923/Ad-hoc-projects-OMCL-Network-ranitidine.pdf/f775fbf3-705e-ce82-d1e1-8eb3c06002d4?t=1628667875462 (accessed on 30 August 2022).
15. Wichitnithad, W.; Sudtanon, O.; Srisunak, P.; Cheewatanakornkool, K.; Nantaphol, S.; Rojsitthisak, P. Development of a Sensitive Headspace Gas Chromatography–Mass Spectrometry Method for the Simultaneous Determination of Nitrosamines in Losartan Active Pharmaceutical Ingredients. *ACS Omega* **2021**, *6*, 11048–11058. [CrossRef]
16. Lee, D.H.; Hwang, S.H.; Park, S.; Lee, J.; Oh, H.B.; Han, S.B.; Liu, K.-H.; Lee, Y.-M.; Pyo, H.S.; Hong, J. A Solvent-Free Headspace GC/MS Method for Sensitive Screening of N-Nitrosodimethylamine in Drug Products. *Anal. Methods* **2021**, *13*, 3402–3409. [CrossRef] [PubMed]
17. Chang, S.-H.; Chang, C.-C.; Wang, L.-J.; Chen, W.-C.; Fan, S.-Y.; Zang, C.-Z.; Hsu, Y.-H.; Lin, M.-C.; Tseng, S.-H.; Wang, D.-Y. A Multi-Analyte LC-MS/MS Method for Screening and Quantification of Nitrosamines in Sartans. *J. Food Drug Anal.* **2020**, *28*, 292–301. [CrossRef]
18. Yang, J.; Marzan, T.A.; Ye, W.; Sommers, C.D.; Rodriguez, J.D.; Keire, D.A. A Cautionary Tale: Quantitative LC-HRMS Analytical Procedures for the Analysis of N-Nitrosodimethylamine in Metformin. *AAPS J.* **2020**, *22*, 89. [CrossRef]
19. Schmaler-ripcke, J.; Hannes, C. *Test Method for the Determination of NDMA by LC-MS/MS in Ranitidine Drug Substance and Film Coated Tablets*; EDQM: Strasbourg, France, 2019; pp. 1–7.

20. European Directorate for the Quality of Medicines and HealthCare. LC-MS/MS Method for the Determination of NDEA and NMDA in Valsartan, Irbesartan and Losartan APIs and Finished Dosage Forms. Available online: https://www.edqm.eu/sites/default/files/medias/fichiers/Sartans/de-by-lc-ms_ls.pdf (accessed on 30 August 2022).
21. Schmidtsdorff, S.; Schmidt, A.H. Simultaneous Detection of Nitrosamines and Other Sartan-Related Impurities in Active Pharmaceutical Ingredients by Supercritical Fluid Chromatography. *J. Pharm. Biomed. Anal.* **2019**, *174*, 151–160. [CrossRef] [PubMed]
22. French National Agency for Medicines and Health Products Safety. Determination of NDMA in Valsartan Active Substances and Finished Products by HPLC/UV. Available online: https://www.edqm.eu/sites/default/files/omcl-method-determination-ndma-valsartan-ansm-september2018.pdf (accessed on 30 August 2022).
23. Maziarz, M.; Rainville, P. Reliable HPLC/UV Quantification of Nitrosamine Impurities in Valsartan and Ranitidine Drug Substances [TECHNOLOGY BRIEF]. Available online: https://www.waters.com/webassets/cms/library/docs/720006775en.pdf (accessed on 30 August 2022).
24. Giménez-Campillo, C.; Pastor-Belda, M.; Campillo, N.; Hernández-Córdoba, M.; Viñas, P. Development of a New Methodology for the Determination of N-Nitrosamines Impurities in Ranitidine Pharmaceuticals Using Microextraction and Gas Chromatography-Mass Spectrometry. *Talanta* **2021**, *223*, 121659. [CrossRef] [PubMed]
25. Alshehri, Y.M.; Alghamdi, T.S.; Aldawsari, F.S. HS-SPME-GC-MS as an Alternative Method for NDMA Analysis in Ranitidine Products. *J. Pharm. Biomed. Anal.* **2020**, *191*, 113582. [CrossRef]
26. Lim, H.H.; Oh, Y.S.; Shin, H.S. Determination of N-Nitrosodimethylamine and N-Nitrosomethylethylamine in Drug Substances and Products of Sartans, Metformin and Ranitidine by Precipitation and Solid Phase Extraction and Gas Chromatography–Tandem Mass Spectrometry. *J. Pharm. Biomed. Anal.* **2020**, *189*, 113460. [CrossRef]
27. Kao, Y.-T.; Wang, S.-F.; Wu, M.-H.; Her, S.-H.; Yang, Y.-H.; Lee, C.-H.; Lee, H.-F.; Lee, A.-R.; Chang, L.-C.; Pao, L.-H. A Substructure-Based Screening Approach to Uncover N-Nitrosamines in Drug Substances. *J. Food Drug Anal.* **2022**, *30*, 150–162. [CrossRef] [PubMed]
28. Sieira, B.J.; Carpinteiro, I.; Rodil, R.; Quintana, J.B.; Cela, R. Determination of N-Nitrosamines by Gas Chromatography Coupled to Quadrupole–Time-of-Flight Mass Spectrometry in Water Samples. *Separations* **2020**, *7*, 3. [CrossRef]
29. Jurado-Sánchez, B.; Ballesteros, E.; Gallego, M. Comparison of the Sensitivities of Seven N-Nitrosamines in Pre-Screened Waters Using an Automated Preconcentration System and Gas Chromatography with Different Detectors. *J. Chromatogr. A* **2007**, *1154*, 66–73. [CrossRef] [PubMed]
30. ChaoYe, S.; HongQuan, C.; SaiFeng, P.; Yun, Z. Determination of Nine Nitrosamines in Water by Solid Phase Extraction and Gas Chromatography-Mass Spectrometry. *J. Environ. Occup. Med.* **2019**, *36*, 1060–1065. [CrossRef]
31. Association of Official Analytical Chemists. *Appendix F: Guidelines for Standard Method Performance Requirements*; Association of Official Analytical Chemists: Rockville, MD, USA, 2016.
32. European Pharmacopoeia. *2.5.42 N-Nitrosamines in Active Substances*; European Pharmacopoeia: Strasbourg, France, 2020.

International Journal of
Molecular Sciences

Article

Comparative Study of Chemical Stability of a PK20 Opioid–Neurotensin Hybrid Peptide and Its Analogue [Ile^9]PK20—The Effect of Isomerism of a Single Amino Acid

Barbara Żyżyńska-Granica [1], Adriano Mollica [2], Azzurra Stefanucci [2], Sebastian Granica [3] and Patrycja Kleczkowska [4,5,*]

[1] Chair and Department of Biochemistry, Faculty of Medicine, Medical University of Warsaw, 02-097 Warsaw, Poland
[2] Department of Pharmacy, G. d'Annunzio University of Chieti-Pescara, 66100 Chieti, Italy
[3] Microbiota Lab, Department of Pharmacognosy and Molecular Basis of Phytotherapy, Medical University of Warsaw, 02-097 Warsaw, Poland
[4] Military Institute of Hygiene and Epidemiology, 01-163 Warsaw, Poland
[5] Maria Sklodowska-Curie Medical Academy in Warsaw, 03-411 Warsaw, Poland
* Correspondence: hazufiel@wp.pl; Tel.: +48-690-888-774

Abstract: Chemical stability is one of the main problems during the discovery and development of potent drugs. When ignored, it may lead to unreliable biological and pharmacokinetics data, especially regarding the degradation of products' possible toxicity. Recently, two biologically active drug candidates were presented that combine both opioid and neurotensin pharmacophores in one entity, thus generating a hybrid compound. Importantly, these chimeras are structurally similar except for an amino acid change at position 9 of the peptide chain. In fact, isoleucine ($C_6H_{13}NO_2$) was replaced with its isomer *tert*-leucine. These may further lead to various differences in hybrids' behavior under specific conditions (temperature, UV, oxidative, acid/base environment). Therefore, the purpose of the study is to assess and compare the chemical stability of two hybrid peptides that differ in nature by way of one amino acid (*tert*-leucine vs. isoleucine). The obtained results indicate that, opposite to biological activity, the substitution of *tert*-leucine into isoleucine did not substantially influence the compound's chemical stability. In fact, neither hydrolysis under alkaline and acidic conditions nor oxidative degradation resulted in spectacular differences between the two compounds—although the number of potential degradation products increased, particularly under acidic pH. However, such a modification significantly reduced the compound's half-life from 204.4 h (for PK20 exposed to 1M HCl) to 117.7 h for [Ile^9]PK20.

Keywords: hybrid peptides; stability; degradation; chemical structure

Citation: Żyżyńska-Granica, B.; Mollica, A.; Stefanucci, A.; Granica, S.; Kleczkowska, P. Comparative Study of Chemical Stability of a PK20 Opioid–Neurotensin Hybrid Peptide and Its Analogue [Ile^9]PK20—The Effect of Isomerism of a Single Amino Acid. *Int. J. Mol. Sci.* **2022**, *23*, 10839. https://doi.org/10.3390/ijms231810839

Academic Editor: Ross Bathgate

Received: 25 August 2022
Accepted: 14 September 2022
Published: 16 September 2022

Publisher's Note: MDPI stays neutral with regard to jurisdictional claims in published maps and institutional affiliations.

1. Introduction

Drug degradation, either enzymatic or chemical, is crucial for its clinical response and efficacy. However, most drugs available tend to have a short half-life or are unstable as a result of their exposure to various environmental factors. In addition to enzymatic stability, the chemical stability of the molecule is another equally important feature of all compounds tested for potential medical use. It determines the sensitivity of the compound to degradation by various non-enzymatic processes. In fact, several chemical reactions may affect structures in aqueous solutions, including hydrolysis, deamidation, isomerization or oxidation [1]. Each of these processes may lead to the transformation of the physicochemical properties of the compound, which may occasionally be manifested by a change in the action profile and therapeutic potential.

The most important factors that can often cause inactivation of drugs are temperature (both too low and too high), environmental pH and UV radiation. This property is important due to the exposure of peptide drugs to unfavorable conditions during the production process and subsequent storage and use of the finished preparation. Furthermore, the resistance of the drug substance to the degrading effect of acidic pH is important in the case of oral formulations that enter the stomach. High chemical stability is necessary for the drug substance to exhibit adequate biological activity throughout the shelf-life of the preparation and to be able to produce desired effects. Indeed, as the drug undergoes degradation, it becomes less effective. In addition, drug decomposition may yield toxic byproducts that are harmful to the patient.

All of the abovementioned factors are especially true for peptide compounds, reducing their therapeutic application. In fact, peptides, compared to other structures formed by antibodies or proteins, are more susceptible to enzymatic and chemical degradation. They owe this property to their chemical structure, as each amino acid in the sequence is linked to the other by way of peptide bonds (amide bonds), which undergo spontaneous degradation through hydrolysis. Other reactions may occur, such as diketopiperazine formation as a consequence of the degradation of the *N*-terminus [2,3]. In addition, oxidation appears to be the most important type of chemical degradation of protein/peptides, which can be affected by pH [4] or even the flexibility of the peptide backbone [5].

Since hybrid compounds have gained attention due to their potent multifunctional behavior with a more favorable profile in terms of possible side effects, much work devoted to such compounds can be found. PK20 and [Ile^9]PK20, which are hybrid structures encompassing both opioid and neurotensin pharmacophores, although modified, were designed and synthesized to reduce pain. Indeed, a modified pharmacophore of an opioid endorphin-2 (Tyr-Pro-Phe-Phe-NH$_2$) has been designed by an incorporation of known elements that possess the ability to increase its resistance to metabolic degradation: Tyr was replaced by Dmt in position 1 and a *D*-amino acid residue (*D*-Lys) was inserted in position 2 of the peptide sequence. In addition, a neurotensin pharmacophore (*pyr*Glu-Leu-Tyr-Glu-Asn-Lys-Pro-Arg^8-Arg^9-Pro-Tyr-Ile-Leu-OH) was strongly reduced and modified, especially by the replacement of Arg8-Arg9 with Lys-Lys. Nonetheless, the structures of these two hybrid peptides are similar (Figure 1), if not the same, except for one amino acid: *tert*-leucine → isoleucine (Tle → Ile). An additional substitution such as Ile12 by Tle was suggested to improve the metabolic stability [6]; PK20 revealed its resistance to degradation as the exact half-life of the peptide was calculated to be 31 h 45 min [7]. However, most of the drug biological activity was affected, which was further confirmed in our studies focused on antinociceptive effect [8,9] and neuroprotective effects. For example, [Ile^9]PK20-induced analgesia was significantly lower: 0.005 at 120 min and 0.02 nmol/rat after specific time-points of 30, 60 and 120 min, respectively, when compared to PK20. PK20 was also significantly stronger than its structural analogue at a dose of 0.02 nmol/rat and at 30 min drug post-administration when compared to morphine (3 nmol/rat) [8,9]. These were also confirmed by differences in the intrinsic activity of the receptor which was targeted ([Ile^9]PK20 had lower efficacy (Emax = 151.2% $\pm$ 74.5) and potency at mu opioid receptor (EC50 = 1244 nM) relative to PK20 (Emax = 149.17% $\pm$ 2.9 and EC50 = 79 nM), respectively) [9]. Additionally, the protective properties of PK20 and [Ile^9]PK20, assessed in an in vitro model of excitotoxic injury in organotypic hippocampal slice cultures subjected to NMDA, demonstrated PK20 hybrid as a more potent agent when compared to its [Ile^9]-analogue. The extent of damage to the CA1 region of the hippocampus, in the case of a combined administration of PK20 and NMDA, was equal: 1.47% $\pm$ 1.20 for the dose of 25 ng/mL, 5.18% $\pm$ 2.84 for the dose of 50 ng/mL and 6.82% $\pm$ 6.08 for the dose of 100 ng/mL, respectively [10]. In contrast, for [Ile^9]PK20 administered simultaneously with NMDA, the values were as follows: 8.53% $\pm$ 2.65 for the dose of 25 ng/mL, 19.04% $\pm$ 8.52 for the dose of 50 ng/mL and 8.62% $\pm$ 0.62 for the dose of 100 ng/mL, respectively (data not published).

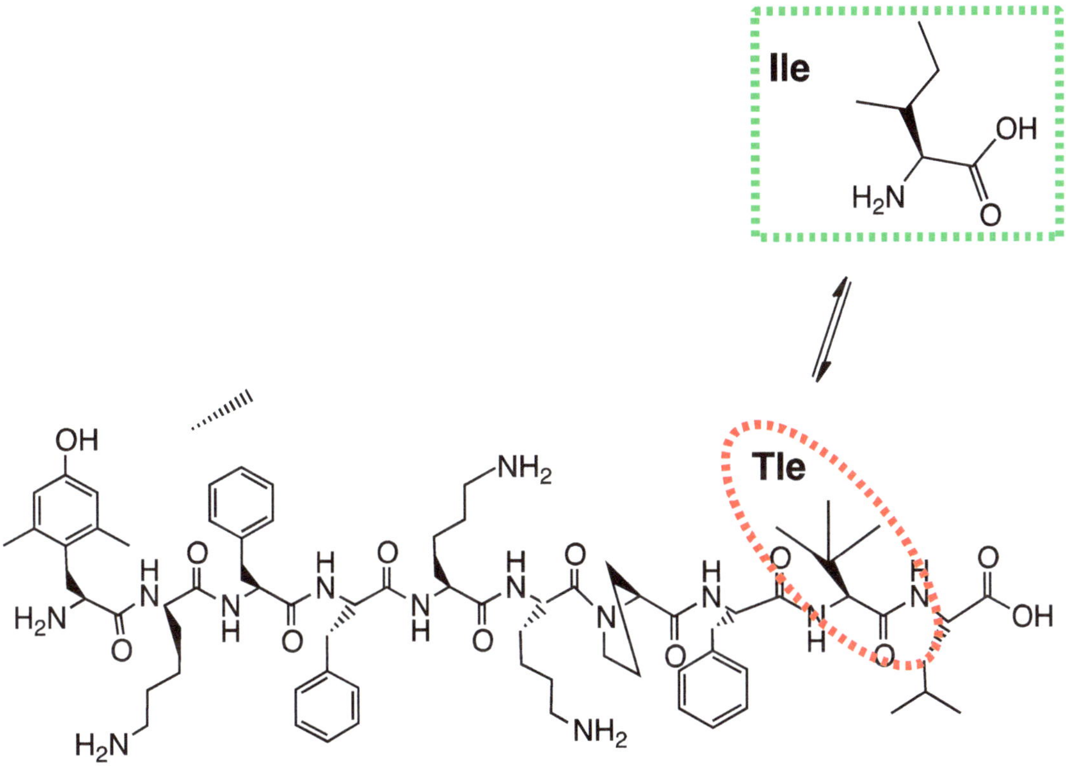

Dmt-D-Lys-Phe-Phe-Lys-Lys-Pro-Phe-**X**-Leu

PK20: X = Tle

[Ile⁹]PK20: X = Ile

Figure 1. Chemical structure of the PK20 opioid–neurotensin hybrid peptide with the indication of a performed modification resulting in the production of [Ile⁹]PK20.

Tle and Ile are structural isomers belonging to the same class of leucines with the same atomic composition ($C_6H_{13}NO_2$) but they have different chemical structures, including different bond coordination and stereochemistry (Figure 1). Thus, their presence in the peptide structure led to diverse results. In fact, incorporation of Ile into the 9th position of the peptide chain caused a decrease in the analgesic activity in vivo when compared with PK20, having a tert-leucine (Tle9) in the corresponding position [8]. Nonetheless, [Ile⁹]PK20 produced the maximal pain-relieving effect, although with a distinct pharmacokinetic profile in comparison with PK20 [9]. Furthermore, in the case of [Ile⁹]PK20, a replacement of Tle with Ile led to reduced binding at the target receptors (i.e., mu opioid receptor and NTS1 neurotensin receptor) [9].

Considering the aforementioned, this paper aims to present whether such a slight structural difference may influence the chemical stability of both drugs when exposed to thermal, acidic/basic, oxidative and UV factors.

2. Results

PK20 and [Ile^9]PK20 Degradation Depending on Conditions

PK20 opioid–neurotensin hybrid peptide was found to be quite stable, particularly in thermal and acidic degradation (Table 1, Figures 2A and 3A—left panel). At varying temperatures ranging from 22 °C (room temperature) through 37 °C to +80 °C, the remaining concentration of the peptide was approximately 50 µg/mL, which corresponds to 100%. Furthermore, PK20 behaved similarly when incubated at room temperature or −80 °C (Table 1), resulting in a 100% recovery. In contrast, alkaline stress led to significant peptide degradation (Table 1, Figure 2—left panel). In fact, after 24 h of hydrolysis in 1 M NaOH and at 37 °C, the percentage of PK20 remaining was 30.32%, while after 24 h of hydrolysis in 1M hydrochloric acid, the recovery reached almost 80% (78.19%). In addition, the calculated half-life of PK20 in 1M HCl was 204.4 h, while in 1M NaOH, $t_{1/2}$ was 11.36 h.

Table 1. Summarized results obtained for the chemical degradation of PK20 and [Ile^9]PK20 opioid–neurotensin hybrid peptides.

Condition	Drug Concentration Injected (µg/mL)	Concentration Found (Mean ± SD, µg/mL)		Recovery (%)		RSD (%) *	
		PK20	**[Ile^9]PK20**	**PK20**	**[Ile^9]PK20**	**PK20**	**[Ile^9]PK20**
Acidic degradation (1 M HCl)							
37 °C (24 h)	50	39.09 ± 6.33	44.39 ± 4.48	78.19	88.79	16.18	10.10
80 °C (12 h)	50	27.22 ± 2.75	26.12 ± 1.52	54.43	52.25	10.09	5.83
Basic degradation (1 M NaOH)							
37 °C (24 h)	50	15.16 ± 0.34	18.56 ± 0.59	30.32	37.13	2.27	3.19
80 °C (12 h)	50	5.71 ± 1.67	5.61 ± 1.83	11.44	11.23	29.17	32.69
Oxidative degradation (30% H$_2$O$_2$)							
37 °C (24 h)	50	30.52 ± 0.49	31.39 ± 0.27	61.04	62.79	1.60	0.87
80 °C (12 h)	50	9.80 ± 0.2	8.24 ± 6.69	19.60	16.48	2.07	81.14
Thermal degradation							
Room temperature (24 h)	50	51.36 ± 1.64	49.71 ± 2.24	102.72	99.41	3.19	4.50
37 °C (24 h)	50	49.98 ± 0.35	44.16 ± 0.99	99.97	88.32	0.70	2.24
80 °C (12 h)	50	48.67 ± 2.74	43.27 ± 1.18	97.34	86.54	5.63	2.73
−80 °C (24 h)	50	51.90 ± 0.20	51.87 ± 0.40	103.80	103.74	0.38	0.77
Two freeze–thaw cycles (24 h (−80 °C), −2 h (22 °C), 24 h (−20 °C))	50	43.25 ± 9.42	49.04 ± 1.03	86.50	98.08	21.78	2.09
Photolytic degradation (365 nm UV light 7 h)	50	18.72 ± 0.94	9.63 ± 0.11	37.44	19.25	5.04	1.14

* RSD—relative standard deviation.

Importantly, neither acidic nor basic degradation provided at the higher temperature of +80 °C for 12 h showed any of PK20's stability improvement; the percentage of PK20 remaining was 54.43 and 11.44%, respectively. It is noteworthy that, in all of the studied stress conditions, PK20 degraded to similar degradants, as it was revealed in Figure 2A,B (left panel). However, one of the newly produced compounds was characteristic for alkaline conditions only: Compound J (Figure 4) with a corresponding m/z of 487.42 (Figure 2B; left panel).

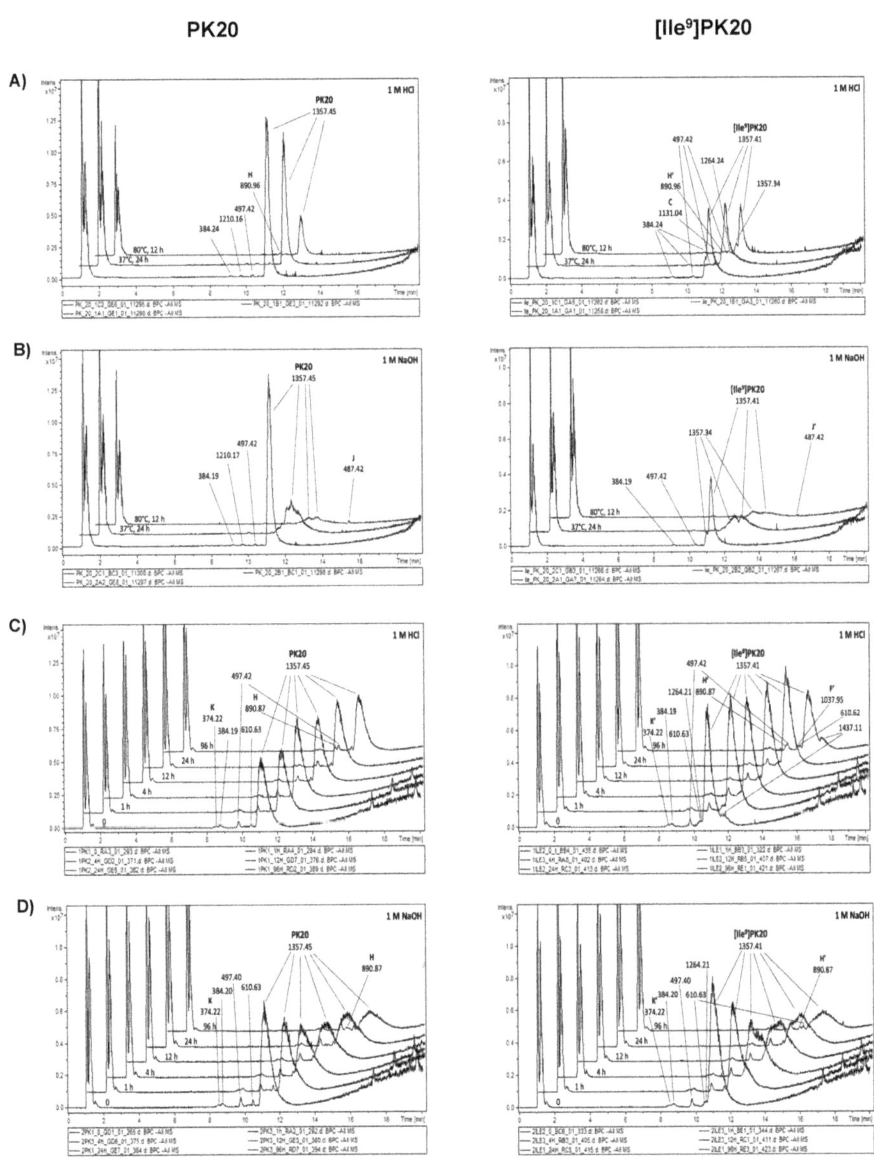

Figure 2. PK20 (panel left) and [Ile⁹]PK20 (panel right) degradation in acidic (1M HCl) and basic (1M NaOH) conditions depends on: the temperature and time of exposure (panel (**A**) and (**B**), respectively) and time of exposure (panel (**C**) and (**D**), respectively). Chromatograms show degradation products with corresponding molecular weights.

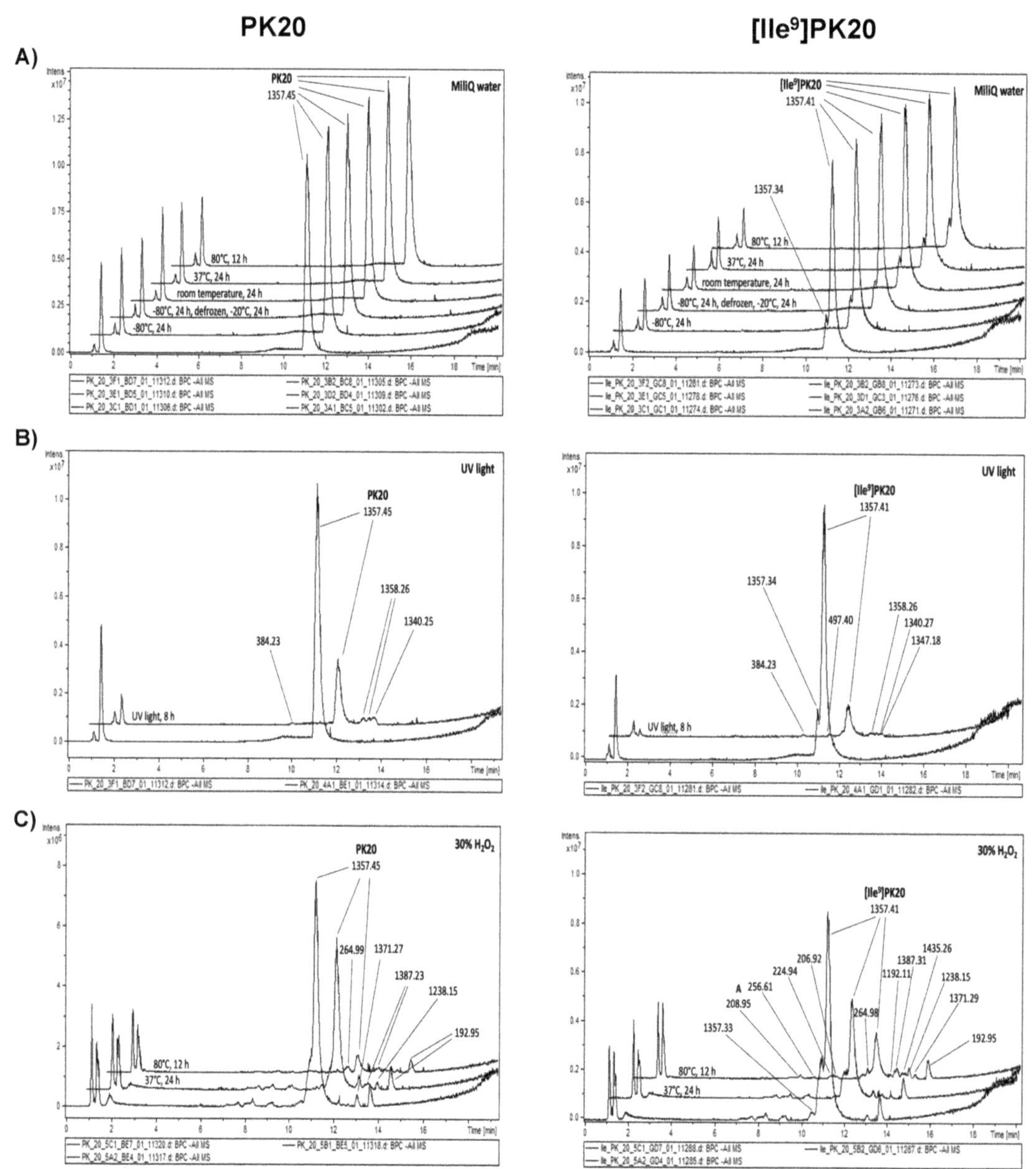

Figure 3. Thermal (**A**) and photolytic (**B**), and oxidative (**C**) PK20 (panel left) and [Ile⁹]PK20 (panel right) chimera degradation.

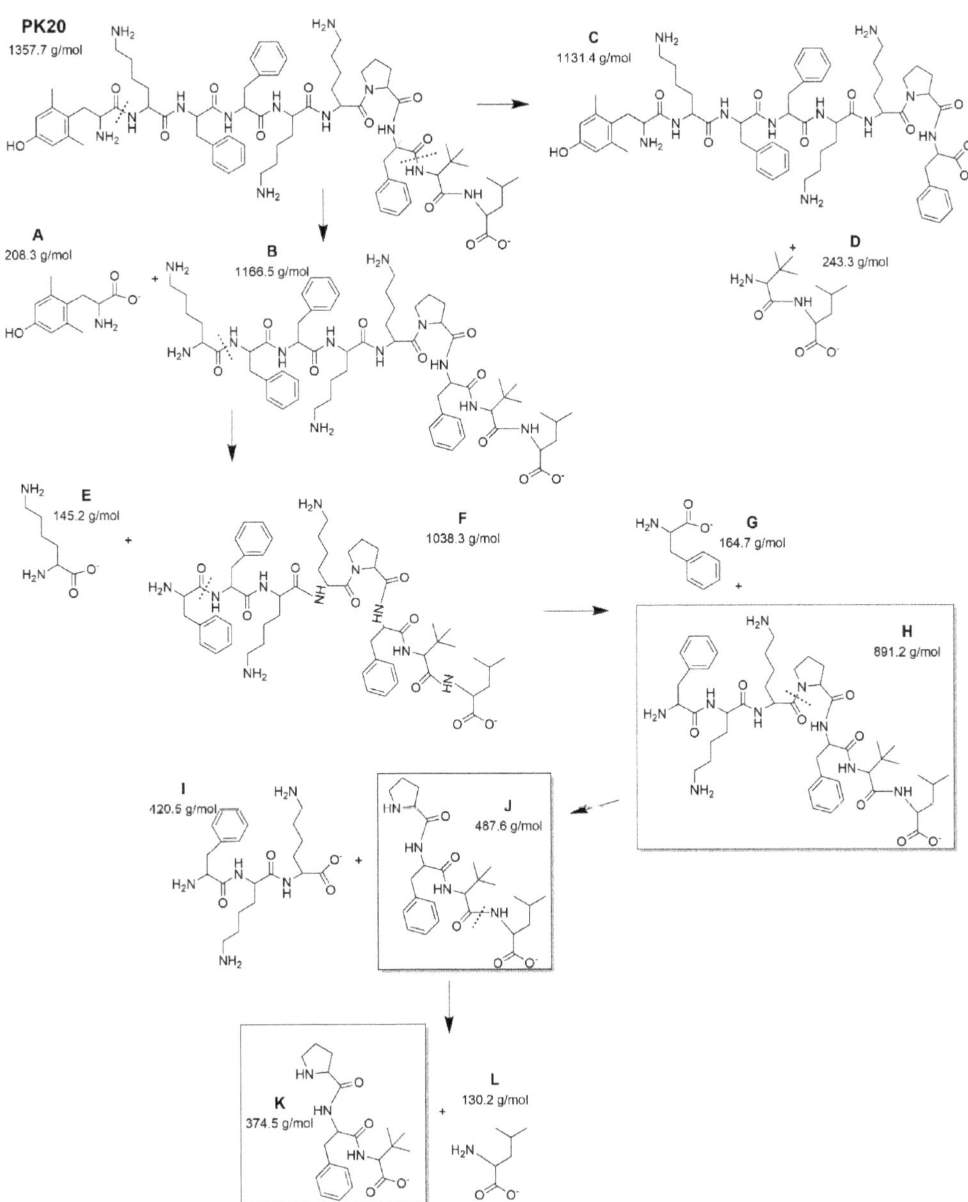

Figure 4. Representative structures of potential PK20′s degradation products based on the molecular weight obtained. Products commonly observed in Figure 2 are given in brackets, and these are as follows: compound K—Pro-Phe-Tle [m/z observed = 374.22 g/mol vs. MW calculated = 374.5 g/mol]; compound J—Pro-Phe-Tle-Leu m/z observed = 487.42 g/mol vs. MW calculated = 487.6 g/mol]; compound H—Phe-Lys-Lys-Pro-Phe-Tle-Leu [m/z observed = 890.96 g/mol vs. MW calculated = 891.2 g/mol] with PK20 [m/z observed = 1357.45 g/mol vs. MW calculated = 1357.7 g/mol].

The replacement of Tle in PK20 with Ile in [Ile^9]PK20 did not significantly change the compound's chemical stability. In fact, neither hydrolysis under alkaline and acidic conditions nor oxidative degradation resulted in spectacular differences between the two compounds–although the number of potential degradation products increased, particularly under acidic pH (Figure 2A,C; right panel). Interestingly, only the [Ile^9]PK20 exposure to UV light led to an observable decrease in the stability, as the percentage of the compound remaining was only 19.25% (in comparison with 37.44% for PK20) (Table 1). Some additional differences were noted when the compound was treated with temperature, especially under the two freeze–thaw cycles. Surprisingly, [Ile^9]PK20 was much more resistant than its mature hybrid compound PK20, as the recovery reached approximately 100% (Table 1).

Even though there were no changes in the recovery values for both compounds examined under acidic and basic conditions, the [Ile^9]PK20 half-life deteriorated significantly, estimated at 117.7 h for acidic and 4.69 h for alkaline conditions.

The exposure of an aqueous solution of PK20 to UV light for 7 h resulted in a slight degradation (Table 1, Figure 3B—left panel). A similar result was found for PK20 under oxidative stress (Table 1, Figure 3C—left panel), although an increase in temperature worsened the obtained results. For PK20 treatment with 30% H_2O_2, either at a temperature of 37 °C for 24 h or 80 °C for 12 h, the remaining compound was 61.04 and 19.60%, respectively (Table 1). Importantly, some novel degradants were detected as a result of H_2O_2 treatment, with molecular weights exceeding the value of PK20 (m/z = 1357.45 g/mol) (Figure 3C; left panel). This was true for the peaks with the corresponding m/z of 1371.27 and 1387.23 g/mol.

As was presented in Figure 2, the PK20 chimera is stable under specific conditions, although some degradation products can be distinguished (Figure 4). According to the time of appearance on the chromatogram (retention time), one of the first degradation products (T_R = 8.6 min) was the compound **K** with m/z of 374.22 g/mol corresponding with the amino acid structure Pro-Phe-Tle (Figure 2C,D (left panel) and Figure 4). In both acidic and basic stress, PK20 degradation resulted in the formation of a compound **H** with a molecular mass of 890.96 g/mol (T_R = 10.0 min) (Figure 2; left panel). Furthermore, in both conditions, PK20 was degraded to unknown products and/or impurities at the retention time of PK20 peak (11.2 min): 8.6 (MW = 374.22 g/mol), 9.2 (MW = 384.24 g/mol), 9.7 (MW = 1210.16 g/mol) and 10.6 min (MW = 497.40 g/mol). However, only under alkaline stress, at 80 °C, and at the time-point of 13.6 min, did the chimera produce the compound **J** (Pro-Phe-Tle-Leu, MW = 487.42 g/mol) (Figure 2B (left panel) and Figure 4).

Although both PK20 and its analogue [Ile^9]PK20 are quite similar in terms of their stability under thermal, photolytic and oxidative stress (Figure 3, Table 1), where the differences occurred. These relate to the quantity and quality of the degradation products. Indeed, as it was presented, the hybrid peptide [Ile^9]PK20 breaks down into a much larger amount of degradation products (Figure 3—right panel)—though some are the same for PK20 (e.g., compounds with m/z of 1340 and 384 for UV light exposure; Figure 3B, right and left panels).

Possible structures of several new degradants are presented in Figure 5.

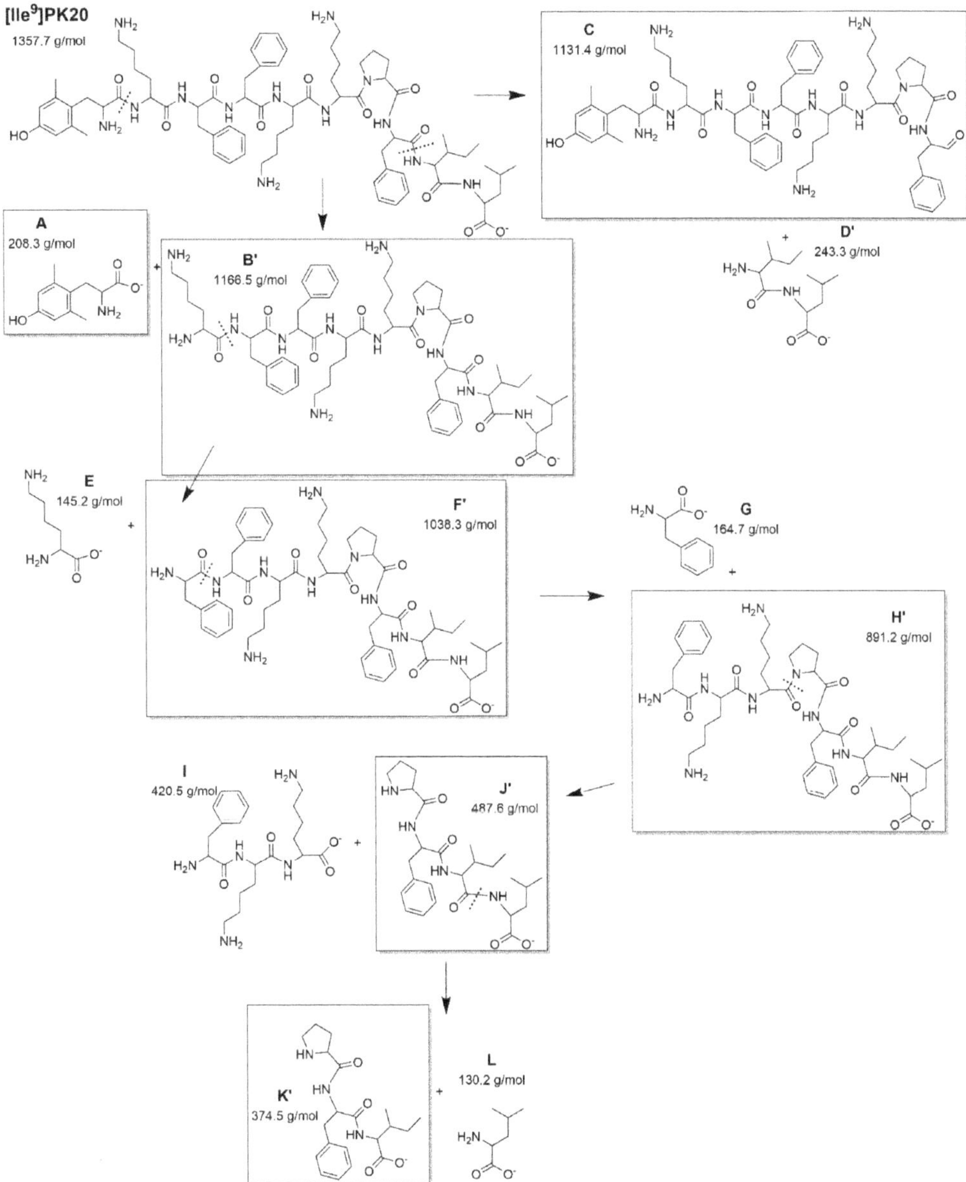

Figure 5. Representative structures of potential [Ile⁹]PK20's degradation products based on the molecular weight obtained, and identified as: compound K'—Pro-Phe-Tle [m/z observed = 374.22 g/mol vs. MW calculated = 374.5 g/mol]; compound J'—Pro-Phe-Tle-Leu [m/z observed = 487.42 g/mol vs. MW calculated = 487.6 g/mol]; compound H'—Phe-Lys-Lys-Pro-Phe-Tle-Leu [m/z observed = 890.96 g/mol vs. MW calculated = 891.2 g/mol] with [Ile⁹]PK20 [m/z observed = 1357.45 g/mol vs. MW calculated = 1357.7 g/mol].

3. Discussion

Peptides are known for their extreme instability under enzymatic and chemical conditions, which are particularly important in terms of drug storage and the route of its administration into the body. This, in turn, is critically important to both the safety and efficacy of drugs. Such stability, or lack thereof, is due to the structure of the molecule. The prediction of the possible pathway of degradation enables an understanding of labile functionalities crucial in designing less reactive and more stable analogues [11]. Therefore, herein we decided to determine whether substituting one amino acid with its structural isomer may influence the compound stability when exposed to different stress conditions.

Based on previous studies demonstrating [Ile^9]PK20 as a much weaker analgesic peptide in comparison to PK20, with a completely different pharmacokinetic profile [8,9] it was probable that the insertion of Ile instead of Tle into the compound's peptide chain could also result in dramatic changes and dissimilarities in chemical stability. However, none of the factors used, such as acidic or basic pH, oxidative conditions and low or high temperature, resulted in substantial changes between either peptide. Moreover, these two compounds were found to be stable, particularly under acidic conditions. Additionally, thermal treatment did not affect the recovery of the peptides. However, when exposed to UV light for 7 h, [Ile^9]PK20 turned out to be more rapidly degraded than its mature compound, as the percentage of [Ile^9]PK20 remaining was only 19.25% (in comparison with 37.44% for PK20).

Although there were no important changes in the behavior of both compounds, the replacement of Tle with Ile in the peptide chain influenced their half-life.

The half-life is a key element in the therapeutic efficacy and potency of drug molecules at the site where it is administered. In addition, it is well known that incorporating a non-natural amino acid into the peptide results in the extension of its half-life [12]. This is consistent with our results demonstrating Tle as a crucial element that possibly induces conformational changes of the entire molecule. Thus, the final effect is the improvement in the resistance to the action of numerous stress factors. Indeed, the presence of Tle in PK20 led to the achievement of long-term stability, both in acidic and basic conditions. The PK20 half-life values were 204.4 h for 1 M HCl and 11.36 h for 1 M NaOH. At the same time, Tle substitution for Ile resulted in a significant reduction in the half-life: 117.7 h for acidic and 4.69 h for alkaline conditions [13].

When analyzing the mass peaks from the LC-MS/MS and, further, the type of the assigned structure that were formed under the stress conditions, significantly more peaks were shown for [Ile^9]PK20 compared with PK20. The degradation reactions, and ultimately the efficiency of degradation processes, vary as a function of the environmental conditions and reaction types, and are dependent on the structure of the compound exposed. In line with this, apart from the identified degradation products, the used factors (i.e., UV, temperature, acidic or basic pH, etc.) possibly induced modifications on the amino acids, as some products with the molecular weight exceeding the output value were detected. The situations mentioned above are well known in the literature [14,15]. For example, as provided by Alsant et al. [11], changes in the MW of +16 and +32 amu occur frequently and correspond to the addition of one and two oxygen atoms, respectively. Likewise, a change in the MW of +18 or −18 amu can readily be explained by the addition or loss of water. Additionally, other products have been described in the literature, such as side chain ring opening or isobaric conversion between amino acids by way of the loss of side chain groups [16]. Furthermore, UV irritation is well known for its ability to produce reactive species, including hydroxyl radicals (OH$^\bullet$), superoxides (O$_2^-$), solvated free electrons (e$_{aq}^-$), hydroperoxyls (HO$_2^\bullet$) and hydrogen peroxide (H$_2$O$_2$) [16,17]. Similarly, several free or bond amino acids can be found to undergo photolytic oxidation when exposed to UV light or H$_2$O$_2$ solely [18,19]. Our studies revealed numerous unidentified degradation products, although we can suspect the modification type from the change of m/z. For instance, the decrease in mass of −17 amu under UV light suggests detachment of the OH$^\bullet$ radical (1340.25). However, most of the degradation products cannot be identified only by LC-MS/MS analysis. Nonetheless, additional studies are needed to determine the exact structures of every potential degradation product or impurities, as well as to predict the pathway of degradation.

4. Materials and Methods

4.1. Drugs and Chemicals

PK20 (Dmt-D-Lys-Phe-Phe-Lys-Lys-Pro-Phe-Tle-Leu-OH) and its analogue [Ile[9]]PK20 (Dmt-D-Lys-Phe-Phe-Lys-Lys-Pro-Phe-Ile-Leu-OH) were synthesized as previously described using Fmoc-based solid-phase peptide synthesis (SPPS) [20]. Methanol (LC-MS grade) and acetonitrile (LC-MS grade) were purchased from Merck KGaA (Darmstadt, Germany). Deionized water was produced with a Simplicity UV system (Merck-Millipore (Burlington, MA, USA). Hydrogen peroxide (H_2O_2), sodium hydroxide (NaOH) and hydrochloric acid (HCl) were purchased from POCH (Gliwice, Poland).

4.2. LC Apparatus and LC-MS Conditions

The LC-MS/MS was performed using a Dionex Ultimate 3000RS device (Dionex, Sunnyvale, CA, USA) equipped with an autosampler, column oven and degasser. The chromatograph was coupled with a Bruker Amazon SL ion trap mass spectrometer (Bruker Daltonik, Bremen, Germany) without splitting.

The separation was carried out using Kinetex XB-C_{18} column (150 mm × 2.1 mm × 1.7 μm; Phenomenex, Torrance, CA, USA). The mobile phase consisted of A: 0.1% formic acid in deionized water and B: 0.1% formic acid in acetonitrile (MeCN). The two-step gradient was used from 5% B to 50% B in 13 min and from 50% B to 65% B up to 20 min. The column temperature was maintained at 25 °C with a flow rate equal to 0.3 mL/min. The eluate was introduced directly to the mass spectrometer. The ion trap setting was as follows: capillary voltage 4500 V, endplate offset 500 V, nebulizer pressure 40 psi, drying gas temperature 145 °C and gas flow rate 9 L/min. The instrument used a smart parameter setting (SPS) fixed at 1000 amu. The scan range was from m/z = 70 to m/z = 2200. Compounds were analyzed in negative ion mode.

4.3. The Stock Solution of Investigated Compounds

Accurately weighed, around 1 mg of each compound was dissolved in deionized water to reach the stock solution's concentration of 1 mg/mL. The stock solution was prepared before each experiment and used immediately.

4.4. Quantification of Peptides Using LC-MS

Before LC analysis, the ionization of each quantified compound in the positive ion mode was optimized by the direct injection of standard (50 μg/mL in 0.1% HCOOH in H_2O; 0.3 mL/min) into the ESI source of the mass spectrometer. The intensity of the most abundant ion for each compound was monitored in order to choose optimal conditions. The quantification of investigated compounds was performed using the dominating ion (for both PK20 and [Ile[9]] PK20 the ion at m/z =1357.40 ± 0.3, retention time ca. 11.2 min).

The calibration curves were plotted as the amount of injected compound (ng) vs. detector response (peak area) using extracted ion chromatograms (EIC) for the characteristic ion for each compound. The linear range for each quantified compound was between 5 and 200 ng per injection. Five amount levels were used for the plotting of curves. Samples at each level were analyzed in triplicate.

4.5. Method Validation

The method for quantifying peptides detected during the LC-MS analysis was validated according to ICH guidelines [21]. Method selectivity, sensitivity, linearity and precision were evaluated. The developed method met ICH criteria in each of the evaluated aspects. The method was selective, sensitive, linear and precise and could be used to assess the degradation processes of investigated peptides. Calibration curves, recorded MS spectra for standards, samples chromatogram of standards and method precision and recovery data are given in the supplementary materials (Figures S1–S3 and Table S1).

4.6. Degradation and Analysis Procedures of the Stressed Compounds

The stability of PK20 and [Ile[9]]PK20 was tested under acidic, basic, oxidative, heat, freeze and ultraviolet light conditions. All experiments were conducted in triplicate (n = 3). The comparison of peak areas in degradative and non-degradative conditions for both compounds can be observed in Figures 3 and 4.

4.6.1. Acid and Base Hydrolysis

Drug solution (50 μg/mL) was prepared in HCl (1 M) or NaOH (1 M). Aliquots were kept at a temperature of 80 °C (LBK type water bath, SWL Bytom, Poland) for 12 h and 37 °C (medical water bath, LW102, Auritronic, Krakow, Poland) for 24 h. At a specific time-point, acidic and basic solutions of both PK20 and [Ile[9]]PK20 were neutralized with an equal volume of 1 M NaOH or 1 M HCl, respectively, and diluted with an equal volume of 0.2% formic acid in acetonitrile and analyzed using LC-MS.

4.6.2. Thermal Degradation

Solutions (50 μg/mL) of both hybrid peptides were prepared in Mili-Q water, and aliquots were kept at room temperature 22 °C, 37 °C and −80 °C for 24 h and at 80 °C for 12 h. One aliquot was also subjected to two freeze–thaw cycles with the sample first frozen at −80 °C for 24 h, defrosted under running tap water at room temperature for 2 h and second frozen at −20 °C for 24 h and defrosted under running tap water at room temperature for 2 h. Then, the solutions of PK20 and [Ile[9]]PK20 were diluted with an equal volume of 0.2% formic acid in acetonitrile and analyzed with LC-MS.

4.6.3. UV Degradation

PK20 and [Ile[9]]PK20 solution (50 μg/mL) was prepared in Mili-Q water, and aliquots were kept in clear plastic vials exposed to UV light (366 nm, Camag, UV Lamp 4, Camag, Switzerland) for 7 h. After the incubation, the drug solution was diluted with an equal volume of 0.2% formic acid in acetonitrile and analyzed with LC-MS.

4.6.4. Oxidative Degradation

PK20 and [Ile[9]]PK20 solution (50 μg/mL) was prepared in 30% (*v/v*) hydrogen peroxide (H_2O_2). Aliquots were incubated at 37 °C for 24 h and 80 °C for 12 h. After the incubation, each drug solution was diluted with an equal volume of 0.2% formic acid in acetonitrile and analyzed with LC-MS.

5. Conclusions

Drug degradation is crucial for the clinical response of a compound. Here, we compared the chemical stability of two hybrid peptides that differ in nature by way of one amino acid (*tert*-leucine vs. isoleucine in PK20 and [Ile[9]]PK20, respectively). Our studies indicated that, although the difference in chemical stability is not substantial, the half-time of PK20 is significantly higher than for [Ile[9]]PK20 for both basic as well as acidic conditions. Moreover, the recovery of PK20 is higher for both UV and oxidative conditions. These differences may be important for future in vivo use of both compounds. Additionally, our study shows, for the first time, that the simple change of one amino acid can influence the chemical stability of a whole molecule. Thus, it is important to perform stability analysis for every analogue of a known compound, even if the modification is small and seems to be negligible.

Supplementary Materials: The supporting information can be downloaded at: https://www.mdpi.com/article/10.3390/ijms231810839/s1.

Author Contributions: B.Ż.-G. conducted the experiments, performed data analysis, reviewed and edited the manuscript; A.M. and A.S. conducted the experiment (peptide synthesis) and contributed reagents or analytical tools, reviewed and edited the manuscript; S.G. conducted the experiments, performed data analysis, reviewed and edited the manuscript; P.K. provided supervision and designed the research, wrote the original draft, reviewed and edited the manuscript, and conducted the experiment (peptide synthesis). All authors have read and agreed to the published version of the manuscript.

Funding: This research received no external funding.

Institutional Review Board Statement: Not applicable.

Informed Consent Statement: Not applicable.

Data Availability Statement: All data in this article is available from the corresponding author upon reasonable request.

Conflicts of Interest: The authors declare no competing interests.

References

1. Powell, M.F. Chapter 30. Peptide Stability in Drug Development: In vitro Peptide Degradation in Plasma and Serum. *Ann. Rep. Med. Chem.* **1993**, *28*, 285–294. [CrossRef]
2. Battersby, J.E.; Hancock, W.S.; Canova-Davis, E.; Oeswein, J.; O'Onnor, B. Diketopiperazine formation and N-terminal degradation in recombinant human growth hormone. *Int. J. Pept. Protein Res.* **2009**, *44*, 215–222. [CrossRef] [PubMed]
3. Kertscher, U.; Bienert, M.; Krause, E.; Sepetov, N.F.; Mehlis, B. Spontaneous chemical degradation of substance P in the solid phase and in solution. *Int. J. Pept. Protein Res.* **2009**, *41*, 207–211. [CrossRef] [PubMed]
4. Chu, J.; Yin, J.; Brooks, B.R.; Wang, D.I.; Ricci, M.S.; Brems, D.N.; Trout, B.L. A comprehensive picture of non-site specific oxidation of methionine residues by peroxides in protein pharmaceuticals. *J. Pharm. Sci.* **2004**, *93*, 3096–3102. [CrossRef] [PubMed]
5. Pogocki, D.; Ghezzo-Schöneich, E.; Schöneich, C. Conformational Flexibility Controls Proton Transfer between the Methionine Hydroxy Sulfuranyl Radical and the N-Terminal Amino Group in Thr−(X)n−Met Peptides. *J. Phys. Chem. B* **2001**, *105*, 1250–1259. [CrossRef]
6. García-Garayoa, E.; Bläuenstein, P.; Bruehlmeier, M.; Blanc, A.; Iterbeke, K.; Conrath, P.; Tourwé, D.; Schubiger, P.A. Preclinical evaluation of a new, stabilized neurotensin(8–13) pseudopeptide radiolabeled with (99m)Tc. *J. Nucl. Med.* **2002**, *43*, 374–383. [PubMed]
7. Kleczkowska, P.; Bojnik, E.; Lesniak, A.; Kosson, P.; Eynde, I.V.D.; Ballet, S.; Benyhe, S.; Tourwé, D.; Lipkowski, A.W. Identification of Dmt-D-Lys-Phe-Phe-OH as a highly antinociceptive tetrapeptide metabolite of the opioid-neurotensin hybrid peptide PK20. *Pharmacol. Rep.* **2013**, *65*, 836–846. [CrossRef]
8. Kleczkowska, P.; Kosson, P.; Ballet, S.; Van den Eynde, I.; Tsuda, Y.; Tourwé, D.; Lipkowski, A.W. PK20, a New Opioid-Neurotensin Hybrid Peptide That Exhibits Central and Peripheral Antinociceptive Effects. *Mol. Pain* **2010**, *6*, 86. [CrossRef] [PubMed]
9. Kleczkowska, P.; Hermans, E.; Kosson, P.; Kowalczyk, A.; Lesniak, A.; Pawlik, K.; Bojnik, E.; Benyhe, S.; Nowicka, B.; Bujalska-Zadrozny, M.; et al. Antinociceptive effect induced by a combination of opioid and neurotensin moieties vs. their hybrid peptide [Ile(9)]PK20 in an acute pain treatment in rodents. *Brain Res.* **2016**, *1648*, 172–180. [CrossRef] [PubMed]
10. Kleczkowska, P.; Kawalec, M.; Bujalska-Zadrozny, M.; Filip, M.; Zablocka, B.; Lipkowski, A.W. Effects of the Hybridization of Opioid and Neurotensin Pharmacophores on Cell Survival in Rat Organotypic Hippocampal Slice Cultures. *Neurotox. Res.* **2015**, *28*, 352–360. [CrossRef] [PubMed]
11. Alsante, K.M.; Huynh-Ba, K.; Baertschi, S.; Reed, R.A.; Landis, M.S.; Kleinman, M.H.; Foti, C.; Rao, V.M.; Meers, P.; Abend, A.; et al. Recent Trends in Product Development and Regulatory Issues on Impurities in Active Pharmaceutical Ingredient (API) and Drug Products. Part 1: Predicting Degradation Related Impurities and Impurity Considerations for Pharmaceutical Dosage Forms. *AAPS PharmSciTech* **2013**, *15*, 198–212. [CrossRef] [PubMed]
12. Stefanucci, A.; Pinnen, F.; Feliciani, F.; Cacciatore, I.; Lucente, G.; Mollica, A. Conformationally Constrained Histidines in the Design of Peptidomimetics: Strategies for the χ-Space Control. *Int. J. Mol. Sci.* **2011**, *12*, 2853–2890. [CrossRef] [PubMed]
13. Wu, L. Regulatory considerations for peptide therapeutics. In *Peptide Therapeutics: Strategy and Tactics for Chemistry, Manufacturing, and Controls*; Srivastava, V., Ed.; Royal Society of Chemistry: London, UK, 2019; pp. 1–30.
14. Baertschi, S.W.; Jansen, P.J.; Alsante, K.M. Stress testing: A predictive tool (Chapter 2). In *Pharmaceutical Stress Testing: Predicting Drug Degradation*; Baertschi, S.W., Alsante, K.M., Reed, R.A., Eds.; Informa Life Sciences: London, UK, 2011; pp. 10–48.
15. Alsante, K.M.; Baertschi, S.W. Reviewing advances in knowledge of drug degradation chemistry. In Proceedings of the Forced Degradation for Pharmaceuticals: Conference Proceedings, London, UK, 18–19 January 2012.
16. Xu, G.; Chance, M.R. Hydroxyl Radical-Mediated Modification of Proteins as Probes for Structural Proteomics. *Chem. Rev.* **2007**, *107*, 3514–3543. [CrossRef] [PubMed]
17. Stadtman, E.R.; Levine, R.L. Free radical-mediated oxidation of free amino acids and amino acid residues in proteins. *Amino Acids* **2003**, *25*, 207–218. [CrossRef] [PubMed]

18. Meddah, B.; Kamel, S.; Giroud, C.; Brazier, M. Effects of ultraviolet light on free and peptide-bound pyridinoline and deoxypyridinoline cross-links. Protective effect of acid pH against photolytic degradation. *J. Photochem. Photobiol. B Biol.* **2000**, *54*, 168–174. [CrossRef]

19. Sakura, S.; Fujimoto, D.; Sakamoto, K.; Mizuno, A.; Motegi, K. Photolysis of pyridinoline, a cross-linking amino acid of collagen, by ultraviolet light. *Can. J. Biochem.* **1982**, *60*, 525–529. [CrossRef] [PubMed]

20. Mollica, A.; Pinnen, F.; Stefanucci, A.; Constante, R. The evolution of peptide synthesis: From early days to small molecular machines. *Curr. Bioact. Comp.* **2013**, *9*, 184–202. [CrossRef]

21. Available online: https://somatek.com/wp-content/uploads/2014/06/sk140605h.pdf (accessed on 20 August 2022).

International Journal of
Molecular Sciences

Article

Implementation of QbD Approach to the Development of Chromatographic Methods for the Determination of Complete Impurity Profile of Substance on the Preclinical and Clinical Step of Drug Discovery Studies

Lidia Gurba-Bryśkiewicz *, Urszula Dawid, Damian A. Smuga, Wioleta Maruszak *, Monika Delis, Krzysztof Szymczak, Bartosz Stypik, Aleksandra Moroz, Aleksandra Błocka, Michał Mroczkiewicz, Krzysztof Dubiel and Maciej Wieczorek

Celon Pharma S.A., ul. Marymoncka 15, 05-152 Kazuń Nowy, Poland
* Correspondence: authors: lidia.gurba@celonpharma.com (L.G.-B.); wioleta.maruszak@celonpharma.com (W.M.)

Citation: Gurba-Bryśkiewicz, L.;
Dawid, U.; Smuga, D.A.; Maruszak, W.;
Delis, M.; Szymczak, K.; Stypik, B.;
Moroz, A.; Błocka, A.; Mroczkiewicz, M.;
et al. Implementation of QbD
Approach to the Development of
Chromatographic Methods for the
Determination of Complete Impurity
Profile of Substance on the Preclinical
and Clinical Step of Drug Discovery
Studies. *Int. J. Mol. Sci.* **2022**, *23*,
10720. https://doi.org/10.3390/
10.3390/ijms231810720

Academic Editors: Geoffrey Brown,
Enikö Kallay and Andrzej Kutner

Received: 8 August 2022
Accepted: 8 September 2022
Published: 14 September 2022

Publisher's Note: MDPI stays neutral
with regard to jurisdictional claims in
published maps and institutional affiliations.

Abstract: The purpose of this work was to demonstrate the use of the AQbD with the DOE approach to the methodical step-by-step development of a UHPLC method for the quantitative determination of the impurity profile of new CPL409116 substance (JAK/ROCK inhibitor) on the preclinical and clinical step of drug discovery studies. The critical method parameters (CMPs) have been tested extensively: the kind of stationary phase (8 different columns), pH of the aqueous mobile phase (2.6, 3.2, 4.0, 6.8), and start (20–25%) and stop (85–90%) percentage of organic mobile phase (ACN). The critical method attributes (CMAs) are the resolution between the peaks ($\geq$2.0) and peak symmetry of analytes ($\geq$0.8 and $\leq$1.8). In the screening step, the effects of different levels of CMPs on the CMAs were evaluated based on a full fractional design 2^2. The robustness tests were established from the knowledge space of the screening step and performed by application fractional factorial design $2^{(4-1)}$. Method operable design region (MODR) was generated. The probability of meeting the specifications for the CMAs was calculated by Monte-Carlo simulations. In relation to literature such a complete AQbD approach including screening, optimization, and validation steps for the development of a new method for the quantitative determination of the full profile of nine impurities of an innovative pharmaceutical substance with the structure-based pre-development pointed out the novelty of our work. The final working conditions were as follows: column Zorbax Eclipse Plus C18, aqueous mobile phase 10 mM $\pm$ 1 mM aqueous solution of HCOOH, pH 2.6, 20% $\pm$ 1% of ACN at the start and 85% $\pm$ 1% of ACN at the end of the gradient, and column temperature 30 °C $\pm$ 2 °C. The method was validated in compliance with ICH guideline Q2(R1). The optimized method is specified, linear, precise, and robust. LOQ is on the reporting threshold level of 0.05% and LOD at 0.02% for all impurities.

Keywords: Analytical Quality by Design (AQbD); design of experiment (DOE); method operable design region (MODR); pharmaceutical impurity profiling; analytical method development; CHI logD; CPL409116; JAK/ROCK inhibitor

1. Introduction

The pharmaceutical industry is closely regulated by the quality control system. The quality of pharmaceutical products and substances (API) should be assured and documented at every stage of the manufacturing and product lifecycle. According to the International Council for Harmonization ICH Q9 guideline [1], it is not enough to ensure quality, it is also necessary to identify and control the potential quality problems arising during the process of development and manufacture. In order to fulfil this requirement, a systematic approach to the pharmaceutical process and product development is necessary.

This approach is known as quality by design (QbD) and is outlined in the ICH Q8 guideline [2]. Originally, the QbD methodology focused on the development of pharmaceutical manufacturing but recently it has been implemented into the development and optimization of analytical methods and is called Analytical Quality by Design (AQbD) [3–8].

Appropriate analytical methods are necessary to ensure the effectiveness and safety of a pharmaceutical product and to meet regulatory requirements, as well as QC needs. The inefficient analytical methods can lead to inaccurate results and misleading information, significantly affecting the drug development process. The application of the AQbD approach ensures that analytical procedures are well understood, robust, and consistently deliver the intended performances throughout their lifecycle.

The application of QbD into analytical methods development has been well adopted by the pharmaceutical and biopharmaceutical analysts [3–10] and has progressively increased over the last years to achieve a higher quality of analytical methods and thus a higher quality of the pharmaceutical product. Most of the applications are focused on the use of experimental design (DOE) and statistical screening of spaces of the method's operating parameters to account for method robustness, especially for separation techniques [11–20]. Also, there have been published papers describing a general, systematic life-cycle approach to the development of analytical procedures. [6,8,21–25]. However, much more effort is still required to improve the AQbD procedures and target the concept towards all kinds of methods from clinical to commercial and manage their lifecycle management as well as updated regulations. Therefore, it is expected that the update of ICHQ2(R1) [25] and the development of a new ICHQ14 [26], guidelines for the development of current analytical procedures will be released soon and is an excellent opportunity to define how analytical methods should be developed, described, and validated in regulatory submissions.

The AQbD methodology is quite similar to the QbD process as described in ICH Q8 [2]. The starting point is to define the intended purpose of the method and the analytical target profile (ATP), while in QbD it is the step to define the quality target product profile (QTPP). Each analytical method should have its intended purpose, taking into account the area of its application, for example, to support research, manufacturing process, formulation development, release, and stability testing for clinical or commercial drugs, quantitative, qualitative, or limit tests. The method ATP is determined by combining the intended purpose and the ICH Q2(R1) quality requirements [25], such as specificity, precision, accuracy, linearity, range, quantitation limit, and detection limit. Furthermore, the potential critical method attributes (CMAs) could be determined as an analytical equivalent of the critical quality attributes (CQAs) in the QbD approach. CMAs should fulfil the specification requirements as well as the quality limits of the measured parameters. The next step of AQbD workflow is risk-assessed to identify and prioritize factors (critical method parameters, CMP), that can affect the method attributes [11,19,24]. The Ishikawa diagram is a useful tool to identify critical method parameters. CMPs are further investigated for robustness using the statistical DOE and multivariate analysis. DOE is a structured, cost-effective, and cost-efficient method to organize, limit the number of experiments, and determine the simultaneous effects and interactions of multiple CMPs on the CMAs. The objective of DOE is the definition of the analytical DS—MODR (method operable design region), which is the operating range of CMPs that guarantee quality results. [9,20,21]. Figure 1 shows the scheme of the AQbD methodology.

The aim of the study was to apply the AQbD approach to the methodical step-by-step development of a chromatographic method for determining the full profile of impurities, process, and degradations, an innovative pharmaceutical substance of the CPL409116 (Figure 2), dual JAK (Janus kinase) and ROCK (Rho-associated kinase) inhibitor [28–30].

Control of the purity of the CPL409116 substance was crucial at the stage of development of the process of large-scale synthesis and the production of the substance for preclinical and clinical trials.

To meet the quality requirements of the formal ICH regulatory guidelines [2,25], but bearing in mind the cost and time effectiveness of the preclinical stage of drug development main goal was to design a quick, simple, and robust analytical method that allows easy application for control of manufacturing of finished dosage form of a pharmaceutical product.

Analitical Quality by Design (AQbD)

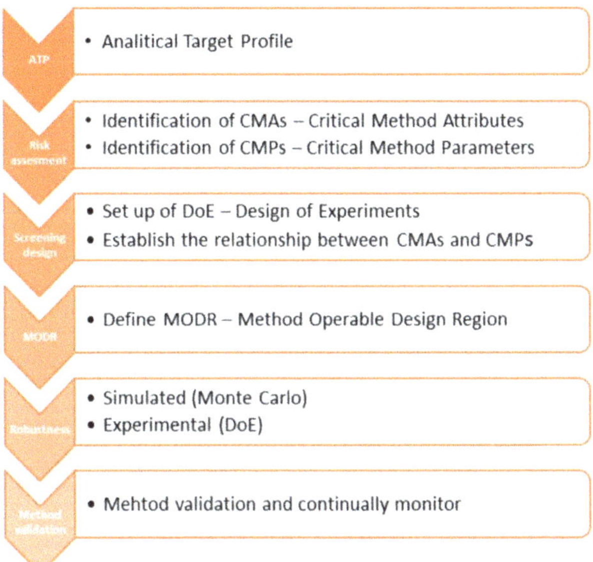

Figure 1. AQbD methodology workflow on the base of Tome at al. [27].

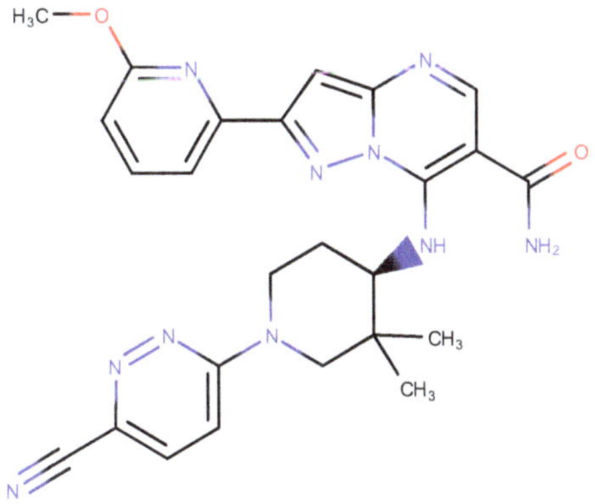

Figure 2. Chemical structure of CPL409116.

Based on our best knowledge, the literature presented works related to the partial introduction of QbD to the correction, verification, and improvement of pharmacopeial methods or methodologies of known pharmaceutical substances [11,12,19,31,32]. Our studies were focused on the implementation of the complete AQbD approach including screening, optimization, and validation steps, for the development of a new method for the quantitative determination of the full profile of nine impurities from the process and degradation of an innovative pharmaceutical substance. In our case, structure-based analysis was a very important step due to the greater differences in the structures of the tested compounds than in the case of the presented optimizations of the methods for the separation of related compounds. For these reasons, the structure-based pre-development studies based on the calculated physicochemical properties such as logP, logD, pKa including CHI logD—experimentally determined chromatographic hydrophobicity index were also performed.

Our work concerns a wider range of simultaneously optimized parameters of the chromatographic method compared to those described in the literature. The development presented in [31–33] was carried out on one column, while our study tested eight different stationary phases. We also used a much wider screening of pH values of the aqueous mobile phase (from 2.6 to 6.8).

Moreover, apart from the commonly considered resolution between neighbors peaks as a critical method attribute (CMA), we also took into account the peaks symmetry factor, which is important in terms of quality criteria.

2. Results and Discussion

2.1. Analytical Target Profile (ATP)

Controlling the impurity profile of a pharmaceutical active substance and pharmaceutical formulation is a key element for drug development and quality assurance since the provided data can directly influence the safety of drug therapy, reducing the impurity-related adverse effects of drug products.

There are two main steps of the control of impurities during the manufacture of drugs, the control of raw materials before drug manufacturing and the control of finished products before batch release. This two-stage control process requires selective, specific, and sensitive analytical techniques. In this context, the most current techniques coupled to AQbD for the quantitation of impurities in active pharmaceuticals ingredients (APIs) are high-performance liquid chromatography (HPLC) and ultrahigh-pressure liquid chromatography (UHPLC) in reversed-phase mode with UV or mass spectrometer detectors [9,11,12,14,15]. In this work, the UHPLC method with UV detection was optimized.

The first step in the AQbD approach is to define ATP [11–13]. The ATP conceptualizes the objective of the analytical method, but also the quality requirements for the reportable result, which include both performance characteristics associated with one or more CMAs and criteria for validation parameters for demonstrating that the method is fit for purpose [11,20]. When dealing with quantitative determination of impurities, the ATP is mainly focused on method selectivity to ensure a complete separation between API, related and unknown impurities, and at a pinch also—excipients. The second objective is also to achieve the required sensitivity of the method (limit of quantitation (LOD) equal or lower than the 0.05%) [11,13]. The method under development was intended to be used in routine quality control, hence the criteria for ATP related to the validation parameters were generally taken from the ICH Q2 (R1) guidelines [9,25].

Therefore, in the present study, ATP is set as: (a) a robust, selective, and specific method, (b) good linearity, precision, and accuracy of impurities determination, and (c) statistically challenged validation [11,12].

2.2. Risk Assessment, Critical Method Attributes CMAs, Critical Method Parameters CMPs

CMAs are an element of method performance that must be measured to access whether a method can produce fit for purpose data. The CMAs are directly connected with the ATP and are response variables that can give information on the quality of the chromatogram. In the present study, the CMAs are: (a) resolution between the peaks should be ≥ 2.0, and (b) peak symmetry of analytes should be ≥ 0.8 and ≤ 1.8. Many parameters could influence the chromatographic performances (of CMAs), which involve aspects related to the chromatographic system, the column, and the mobile phase. In order to indicate the risk factors of the chromatographic method, a fishbone diagram was performed as shown in Figure 3 [11,14,34].

From this diagram, it was possible to identify the CMPs that could potentially affect the selected CMAs and, therefore, required an in-depth investigation using DOE methodologies. The following parameters were considered for the multivariate optimization: stationary phase chemistry, pH of aqueous eluent of the mobile phase, start and stop percentage of organic eluent of the mobile phase, and oven temperature.

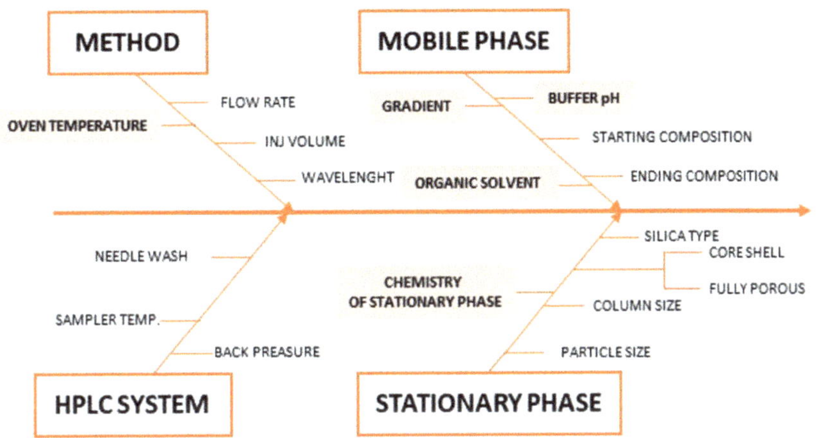

Figure 3. Ishikawa diagram for critical method parameters (CMP).

The settings of the other factors were selected based on preliminary experiments. In particular, following common practice, the injection volume was set to a low value (1 µL), gradient time was set to 7 min (to obtain optimum analysis time, elution time for all of the compounds is lower than 4 min), and the flow rate of mobile phase was set to 0.5 mL min^{-1}.

In this study linear gradient elution has been chosen. This is simple and typically the default elution method ("first choice" method) during chromatography methods development. The linear gradient minimizes peak broadening against isocratic elution and produces more accurate and repeatable results. This second factor is particularly relevant when a transfer method to another instrument is an option, thus linear elution allows for the development of a more universal methodology. Different gradient speeds have occurred by changing the initial and final amount of organic phase, at a range from 20% to 25% and from 85% to 90%, respectively.

ACN was chosen as a type of organic solvent in the mobile phase. MeOH as an organic solvent has been disregarded on a base of preliminary studies, due to high retention and no elution of some of the tested compounds.

Moreover, the wavelength of UV detection was determined and set to 297 nm as a maximum of absorbance JAK01 (CPL409116) and eight of its impurities and 230 nm to control the JAK SM-05 impurity.

2.3. Structure-Based Pre-Development Study

One of the first steps during the development of a chromatographic method is to a knowledge of the chemical structure of analyzed compounds to assess opportunities and identify initial risks during optimization of the analytical method, and also to determine correctly CMPs, like pH of the mobile phase. The basic molecular descriptors, which strongly affect the retention factors, are hydrophobicity (lipophilicity) of molecules or ions, expressed by the partition coefficient logP, and distribution coefficient (logD) for ionizable compounds [35–37]. In addition, most of the new drug molecules that are currently undergoing preclinical research or clinical trials contain ionized groups. Acid-base chemistry strongly influences not only retention and chromatography resolution but also affects band broadening [38,39], which further complicates the development of an effective separation method.

For the initial characterization and determination of the differences between the CPL409116 (JAK01) compound and its impurities, calculations of the physicochemical properties (logP, logD, and pKa) have been performed using the ACD Labs Percepta program (Table 1) [40]. On this basis, it was found that most of the molecules have basic dissociation constants pKa in the range 1.98–2.70 except for JAK-07 (0.90 and 3.98) and JAK-08 (0.79 and 2.70), and JAK ImpB characterized by basic pKa 2.70 and 3.56, but in particular, JAK-09, having a different basic nature (pKa = 2.68 and 9.92). In the pH range of 5.5–8.0, all compounds exist in a single, unprotonated form, which definitely facilitates the development of the chromatographic method.

Table 1. Calculated physico-chemical properties and determined the chromatographic hydrophobicity index CHI logD of CPL409116 and its impurities.

Compound Name	logP	clogD 2.6	clogD 4.5	clogD 7.4	clogD 10.5	pKa (Acid)	pKa (Base)	CHILogD pH 2.6	CHILogD pH 7.4	CHILogD pH 10.5
SM-05	0.46	0.46	0.46	0.46	0.46	–	−2.50	−0.12	−0.03	0.99
JAK-09	1.27	−2.03	−1.83	−1.16	1.16	12.52	−1.12 2.68 9.92	0.71	1.22	2.38
JAK ImpC	−4.22	−4.29	−4.22	−4.22	−4.22	13.41	−0.95 2.00	1.25	3.41	3.23
JAK ImpA	0.54	0.1	0.53	0.43	−1.91	7.95 12.58	−0.53 2.00 2.68	1.49	1.49	1.99
JAK ImpB	1.07	−0.26	1.02	1.07	1.07	12.52 15.09	−1.11 2.70 3.56	2.40	2.21	2.45
CPL409116 (JAK01)	1.44	0.98	1.43	1.44	1.44	12.52	−1.11 2.00 2.70	3.51	3.09	3.05
JAK-07	3.45	0.96	0.62	0.3	0.3	–	−0.78 0.90 3.98	3.80	2.55	2.57
JAK-08	2.55	2.3	2.55	2.55	2.54	12.52	−1.12 0.79 2.70	4.63	4.04	3.61
JAK ImpE	3	2.71	3	3	3	–	−1.20 1.98 2.40	4.86	4.82	4.07
JAK ImpD	2.65	2.36	2.65	2.65	2.65	–	−1.20 1.99 2.40	5.12	4.46	3.83

A much more complicated situation was concerned with the descriptors determining the lipophilicity of compounds (logP and clogD). Generally, logP factors show a promising distinction of retentions in chromatographic conditions. But the data obtained for the compound JAK ImpC (logP/logD = −4.22) suggested very poor retention that was inconsistent with the observed retention times of the compound in chromatographic conditions.

Due to this observation and to better characterise elution conditions for method development, it was decided to determine the chromatographic hydrophobicity index CHI logD of CPL409116 and its impurities by the methodology described by Valko et al. [41]. Chromatographic hydrophobicity index CHI logD has been originally used for high-throughput physicochemical property profiling for rational drug design [42], where determined values strongly correlate to lipophilicity (logP and/or logD) of molecules [37,41].

On a basis of determining lipophilicity described by CHI logD at pH 2.6, 7.4 and 10.5 on ACE Ultracore SuperC18 column more valuable data were obtained in a contest of chromatography interactions molecules with stationary phase extended with their acid-base properties (Table 1). Particularly, strong pH dependence on retention factors has been observed for JAK ImpC and also for JAK07, JAK08, and JAK09. Greater differences and a wider range of CHI logD values, calculated on a base of peaks retention times of compounds, in the lower pH range than in the pH 7.4–10.5 range have been observed. The use of mobile phases with a lower pH would allow obtaining higher chromatographic resolutions and higher selectivity of the analytical method.

Moreover, CPL409116 is practically insoluble in aqueous solutions (e.g., in phosphate buffer with a pH of 7.4 below 0.1 µg ml^{-1}), hence the choice of more acidic separation conditions where the compound may protonate is expected to reduce carryover, problems with clogging of the column and its longer lifetime.

In summary, various dissociation constants of the tested compounds made it necessary to check the elution conditions in more detail in the pH range 2.6–6.8, taking into account peak shape. Using more acidic conditions in the mobile phase should lead to the preparation of more robust and effective analytical methods. As a final CMP, six different buffers at pH 2.6, 3.2, 3.8, 4.0, 4.2, and 6.8 as an aqueous mobile phase has been chosen.

Impurities generally have molecular structures that are close to or related to the API, they also tend to show similar chromatographic behavior. In the case of impurities of CPL409116 JAK ImpC and JAK ImpA have very close hydrophobic index CHI logD at pH 2.6 that same like main compound have similar retention to JAK07 or JAK ImpD and JAK ImpE at pH 7.4 (Table 1).

It is possible to modulate the effectivity of separation and chromatographic resolution by changing the pH of the mobile phase or temperature. But a much more effective way is modulation of selectivity of stationary phase that is strongly important if analysed compounds have a very close molecular structure with comparable hydrophobicity. In this manner, the most effective idea is to explore the selectivity space and chemistry of stationary phases. Based on the work of Snyder and Dolan [43], [44] five terms describing selectivity are: (a) the hydrophobic interaction between the solute and the column; (b) the resistance of the bonded phase to penetration by bulky molecules; (c) hydrogen bond (H-B) interactions between basic solutes and acidic sites (silanols) in the column; (d) H-B interactions between acidic solutes and a vicinal-silanol pair on the column surface; and (e) cation exchange between ionized bases and ionized silanol groups in the column. On this base the Hydrophobic-Subtraction Model (HSM) has been prepared for quantitative comparison of differences in chromatographic columns and the resulting data are publicly available for free through different websites, i.e., www.hplconline.org [45] (accessed on 29 January 2021 or USP chromatographic column database [46].

For the present work, eight columns that differ in chemistry and parameters of stationary phase have been chosen (Table 2).

Table 2. List of the tested column and their properties: Fs—factor describes column similarity; H—parameter is a measure of the phase hydrophobicity; S—a measure of the resistance of the stationary phase to penetration by a solute molecule; A—measure of the hydrogen-bond acidity of the phase; B—a measure of the hydrogen-bond basicity of the phase; C—a measure of the interaction of the phase with ionized solute molecules.

L.p.	Column Name	Fs	H	S	A	B	C (pH 2.8)	C (pH 7.0)	EB	USP Type	Phase Chemistry
1.	ZORBAX Eclipse Plus C18	0	1.03	0	−0.0700	−0.0200	0	0.0200	7.8	L1	C18
2.	Kinetex EVO C18	9.65	1.01	−0.00600	−0.170	−0.0240	−0.110	−0.0100	4.38	L1	C18
3.	ACQUITY UPLC BEH C18	15.51	1.00	0.0280	−0.366	0.00700	0.142	0.0880	6.4	L1	C18
4.	InfinityLab Poroshell 120 Phenyl Hexyl	18.23	0.752	−0.0830	−0.394	0.0180	0.136	0.140	3.59	L11	Phenyl−Hexyl
5.	Kinetex Biphenyl	26.75	0.697	−0.173	−0.583	0.0340	0.122	0.817	2.63	L11	Biphenyl
6.	Kinetex PFP	78.79	0.680	0.0800	−0.270	−0.0300	0.940	1.53	2.4	L43	Pentafluorophenyl
7.	ACQUITY UPLC CSH C18	21.15	0.954	−0.00200	−0.179	0.118	0.0820	0.171	5.62	L1	C18
8.	ACQUITY UPLC CSH Fluoro−Phenyl	25.63	0.708	−0.0590	−0.435	0.129	−0.0680	0.223	2.865	L11	Fluoro−Phenyl

Columns have been chosen on a base similarity factor Fs and also taking into account differences in acid-base interaction of stationary phase (C(pH 2.8) and C(pH 7.0). CPL409116 contains a few different heterocyclic rings, aliphatic amine functionality, amide, ether and nitrile substituents chemistry of stationary phases has been diversified between different chemical modifications, octadecyl, phenyl-hexyl, biphenyl, and pentafluorophenyl.

Therefore, during screening, the authors advise the fixing of qualitative factors such as the type of stationary and mobile phases based on scientific knowledge of the molecules and any relevant impurities. These parameters are very unlikely to change during routine use. Subsequently, quantitative factors such as pH, and the composition of the mobile phase or temperature, can be examined when optimizing a method for robustness. If necessary, pre-tests on qualitative factors can be conducted. Usually, the chosen columns will be those optimizing peak shapes, time of analysis, and selectivity. Knowledge of the interactions between the stationary phase, the mobile phase, and the molecule of interest should of course be included in such studies.

At the stage of preclinical and clinical development, a very important factor is the assessment of the potential genotoxicity and mutagenicity of new compounds in accordance with the requirements of the ICH M7 (R1) guidelines [47]. In the case of risk or analysis of mutagenic compounds, it is necessary to use very sensitive analytical methods, which in most cases means the use more sophisticated methodology or mass spectrometry detection [48,49].

Compound CPL409116 and their related substances, processing by-products and impurities, have been evaluated by two complementary (Q)SAR prediction methodologies to assess the potential mutagenicity of impurities. One methodology should be expert rule-based (DEREK Nexus, module in Star Drop software) [50–52] and the second methodology should be statistically-based (Toxicity Estimation Software Tool (T.E.S.T)) [53,54].

No structural alters and no mutagenicity risks were found for the CPL409116 and its related substances (Table 3) and it did not require changes to the previous assumptions regarding the level of impurity quantification.

Table 3. Summary of evaluation of mutagenicity. (Q)SAR results for CPL409116 and related substances.

Compound Name	DEREK Nexus Results (Q)SAR Rule-Based	T.E.S.T Results (Q)SAR Statistic-Based	ICH M7 Class
CPL409116 (JAK-01)	Mutagenicity is INACTIVE	Mutagenicity NEGATIVE Predicted value $p = 0.25$	Class 5
JAK-07	Mutagenicity is INACTIVE	Mutagenicity NEGATIVE Predicted value $p = 0.26$	Class 5
JAK-08	Mutagenicity is INACTIVE	Mutagenicity NEGATIVE Predicted value $p = 0.23$	Class 5
JAK-09	Mutagenicity is INACTIVE	Mutagenicity NEGATIVE Predicted value $p = 0.39$	Class 5
JAK SM-5	Mutagenicity is INACTIVE	Mutagenicity NEGATIVE Predicted value $p = 0.22$	Class 5
JAK ImpA	Mutagenicity is INACTIVE	Mutagenicity NEGATIVE Predicted value $p = 0.16$	Class 5
JAK ImpB	Mutagenicity is INACTIVE	Mutagenicity NEGATIVE Predicted value $p = 0.28$	Class 5
JAK ImpC	Mutagenicity is INACTIVE	Mutagenicity NEGATIVE Predicted value $p = 0.29$	Class 5
JAK ImpD	Mutagenicity is INACTIVE	Mutagenicity NEGATIVE Predicted value $p = 0.15$	Class 5
JAK ImpE	Mutagenicity is INACTIVE	Mutagenicity NEGATIVE Predicted value $p = 0.16$	Class 5

2.4. Screening Study

The Screening Phase Method development was initiated with the screening activity using an Agilent scouting system equipped with an 8-position column manager and solvent selection valve. For screening experiments, eight column chemistries with a wide range of selectivity difference, wide pH range, different bonding, and with different similarity factors (Fs) [12,43] were used. A large difference in Fs factor indicates that the two columns are very different. During this study, columns' Fs values were compared with Agilent

ZORBAX Eclipse Plus C18. Columns selected, operating pH range of the columns, bonding details, and Fs values were mentioned in Table 2.

The buffer pH selection was carried out in the range of pH 2.6–6.8. Formic acid solution (10 mM) was used to prepare a buffer solution of pH 2.6, and ammonium formate (10 mM) buffer with different concentrations of additives was used to prepare buffer solutions of pH 3.2, 4.0, and 6.8. The initial screening experiments were carried out using different columns, different pH buffer solutions, and different components of a start (from 20% to 25%) and end gradient (from 85% to 90%) organic modifier compositions as variables.

The statistical experimental design was performed based on full fractional design 2^2 (two factors: start and end gradient composition, two-level) with full repetition for each tested column and difference pH of buffer solutions [13,14]. Flow rate, gradient time, and column temperature were kept constant. A total of 128 experiments were performed by statistical design at the screening step.

After running the screening experiments, processing was done to integrate all the peaks properly and then the generated chromatographic responses were transferred to the STATISTICA software [55–57] to generate knowledge space for linear additive effects, curvilinear effects, and complex effects of different variables.

Multiple linear regression was applied for the calculation of the coefficients of the nine models for resolution between all pairs of peaks (R_s) and 10 models of the symmetry factors for all peaks (A_s), then the models were refined to improve their quality by removing some of the non-significant and entangled effects. The evaluation of statistical analysis tools like ANOVA for each response was used to determine the significance of each method parameter selected for the study using the p-value (significance level $p < 0.05$). For all models, good fits were obtained; coefficients of determination R^2 were above 0.99, and lack of fits is not statistically significant ($p > 0.05$).

The results are evaluated based on the desirability factor used in the STATISTICA software [55–57]. The desirability factor approach is a very useful method for the optimization of multiple response processes. The relationship between the approximated (predicted) baseline responses and the utility of the response is called the desirability function. The idea behind this approach is the "quality" of a product or process that has multiple quality characteristics, with one of them outside of some "desirability" limits, is completely unacceptable. The method finds operating conditions that provide the "most desirability" response values. 0.0 represents a completely undesirable value and 1.0 represents a completely desirable or ideal response value. In this study, for resolutions between all pairs of peaks, desirability equals 0.0 when all resolution factors are less than 2.0, for symmetry factors desirability equals 1.0 when all symmetry factors are more than 0.8, and desirability equals 0.0 when all symmetry factors are more than 1.8.

Cumulative desirability plots of results and multiple response curves are used to identify the optimum pH of the aqueous mobile phase, range of start and end % of acetonitrile, and a suitable column for further optimization.

The trellis graph (Figures 4 and 5) shows the results (desirability plot) of screening experiments using different columns, different pH, different start, and final % of acetonitrile, separately for the resolutions and symmetry factors. The red tint in the graphs shows the "knowledge space", the area where all the critical method attributes (CMAs), i.e., the resolution between peaks, and symmetry factors are within the expected range ($R_s \geq 2.0$, and $0.8 \leq A_s \leq 1.8$), and green tint shows the area where no requirements of CMAs are not met. Figure 6 presented an example of chromatograms corresponding to the selected conditions of separation from screening tests.

Based on the evaluation of the screening experiments results calculated by the STATISTICA Software (version 13.3) TIBCO Inc. (Palo Alto, CA, USA), the values of the critical parameters of the methods (CMPs) allow for obtaining the optimal conditions, meeting the acceptance criteria (CMAs) for the separation of the test substance CPL409116, and its nine impurities are shown in Table 4.

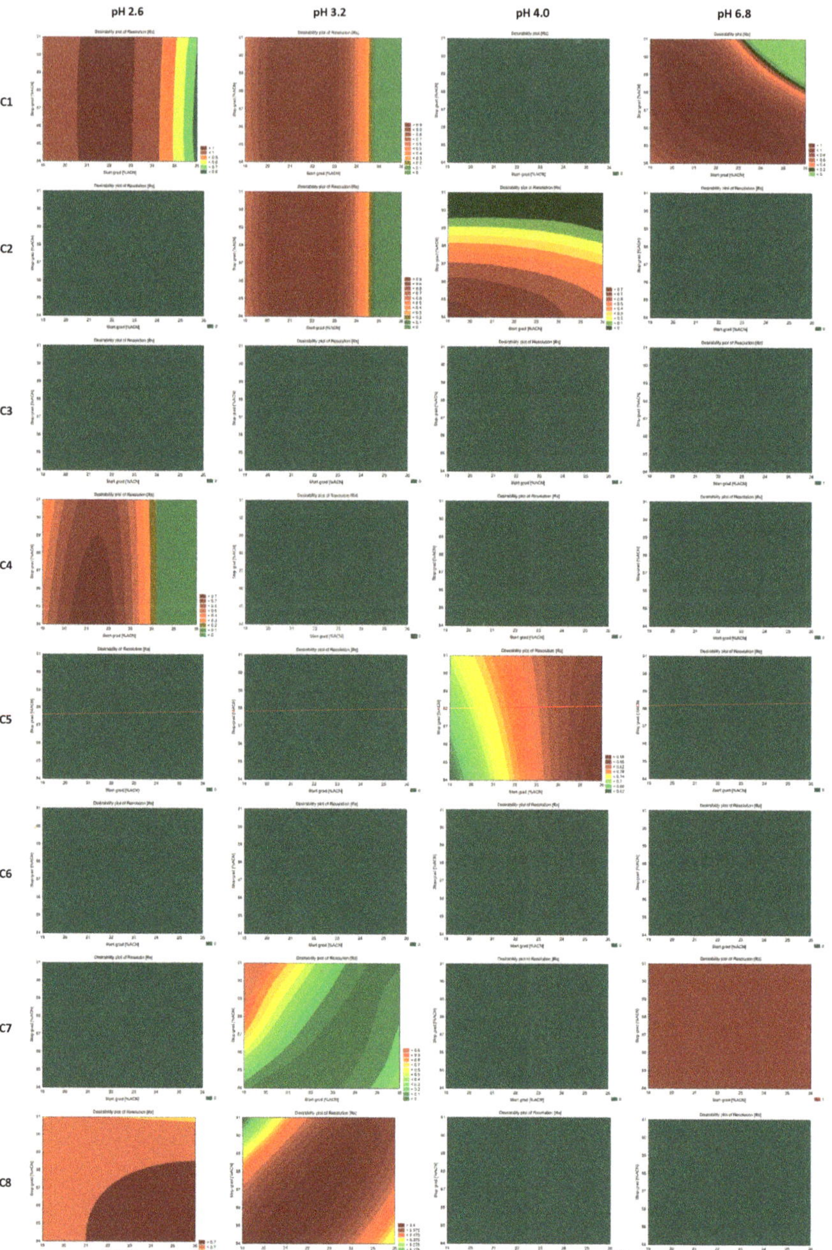

Figure 4. Screening experiment results showing the interaction effect of columns, % of organic solvent at the start and end gradient, and pH with CMAs as a desirability plot of resolutions (R_S). C1—ZORBAX Eclipse Plus C18 (2.1 × 50 mm, 1.8 µm, Agilent Technologies), C2—Kinetex EVO C18 (2.1 × 50 mm, 1.7 µm, Phenomenex), C3—ACQUITY UPLC BEH C18 (2.1 × 50 mm, 1.7 µm, Waters), C4—InfinityLab Poroshell 120 Phenyl Hexyl (2.1 × 50 mm, 1.9 µm, Agilent Technologies), C5—Kinetex Biphenyl (2.1 × 50 mm, 1.7 µm, Phenomenex), C6—Kinetex PFP (2.1 × 50 mm, 1.7 µm, Phenomenex), C7—ACQUITY UPLC CSH C18 (2.1 × 50 mm, 1.7 µm, Waters), and C8—ACQUITY UPLC CSH Fluoro-Phenyl (2.1 × 100 mm, 1.7 µm, Waters).

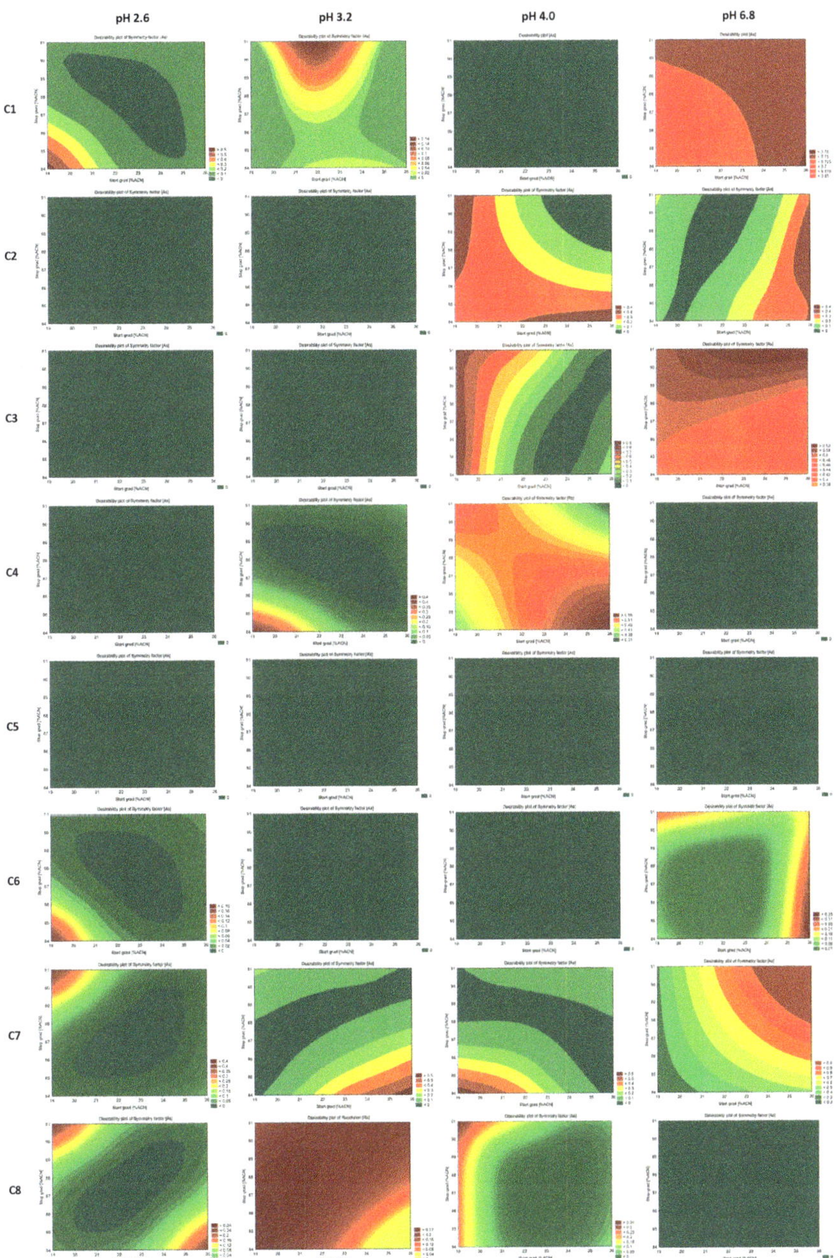

Figure 5. Screening experiment results showing the interaction effect of columns, % of organic solvent at the start and end gradient, and pH with CMAs as a desirability plot of symmetry factors (A_S). C1—ZORBAX Eclipse Plus C18 (2.1 × 50 mm, 1.8 μm, Agilent Technologies), C2—Kinetex EVO C18 (2.1 × 50 mm, 1.7 μm, Phenomenex), C3—ACQUITY UPLC BEH C18 (2.1 × 50 mm, 1.7 μm Waters), C4—InfinityLab Poroshell 120 Phenyl Hexyl (2.1 × 50 mm, 1.9 μm, Agilent Technologies), C5—Kinetex Biphenyl (2.1 × 50 mm, 1.7 μm, Phenomenex), C6—Kinetex PFP (2.1 × 50 mm, 1.7 μm, Phenomenex), C7—ACQUITY UPLC CSH C18 (2.1 × 50 mm, 1.7 μm, Waters), and C8—ACQUITY UPLC CSH Fluoro-Phenyl (2.1 × 100 mm, 1.7 μm, Waters).

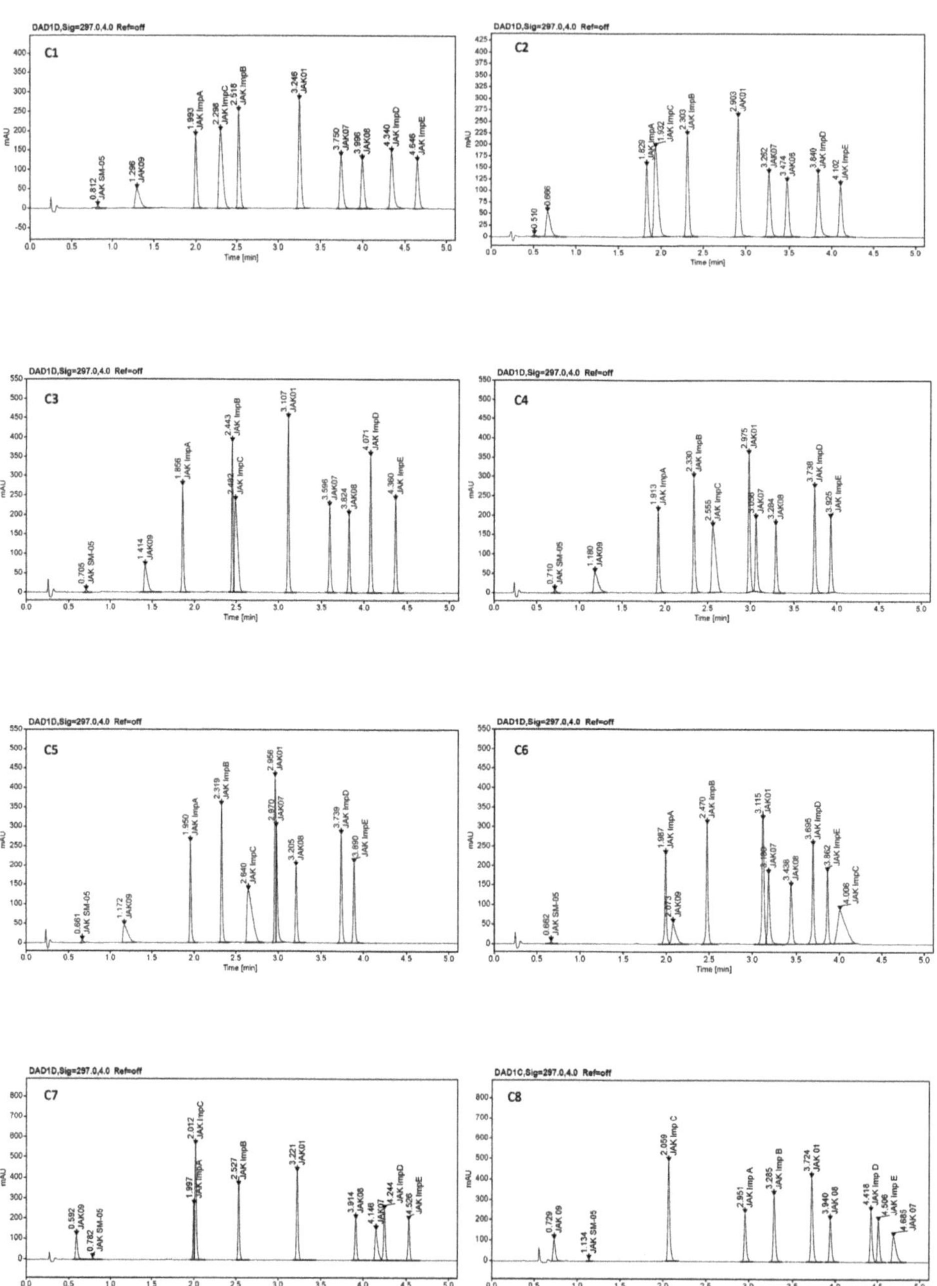

Figure 6. An example of chromatograms from screening experiments on the eight tested columns: C1—ZORBAX Eclipse Plus C18 (2.1 × 50 mm, 1.8 µm, Agilent Technologies), C2—Kinetex EVO C18 (2.1 × 50 mm, 1.7 µm, Phenomenex), C3—ACQUITY UPLC BEH C18 (2.1 × 50 mm, 1.7 µm, Waters), C4—InfinityLab Poroshell 120 Phenyl Hexyl (2.1 × 50 mm, 1.9 µm, Agilent Technologies), C5—Kinetex Biphenyl (2.1 × 50 mm, 1.7 µm, Phenomenex), C6—Kinetex PFP (2.1 × 50 mm, 1.7 µm, Phenomenex), C7—ACQUITY UPLC CSH C18 (2.1 × 50 mm, 1.7 µm, Waters), and C8—ACQUITY UPLC CSH Fluoro-Phenyl (2.1 × 100 mm, 1.7 µm, Waters).

Table 4. Summary of the screening experiments results.

Column	pH of Buffer	Critical Method Parameters (CMAs)	Optimal Conditions of CMPs	
			Start Gradient [% ACN]	End Gradient [% ACN]
ZORBAX Eclipse Plus C18 (2.1 × 50 mm, 1.8 μm, Agilent Technologies)	2.6	Symmetry factor of JAK09 and JAK ImpD Resolution of JAK SM-05/JAK09	19–21	84–86
	3.2	Symmetry factor of JAK 09 Symmetry factor of JAK ImpC (>1.77) Symmetry factor of JAK ImpD Resolution of JAK ImpC/JAK ImpB (<1.5)	21–23	88–90
	4.0	Symmetry factor of JAK ImpD (>2.0) Symmetry factor of JAK ImpE (>1.9)	not found	not found
	6.8	Resolution of JAK ImpC—JAK Imp E	20–23	85–90
Kinetex EVO C18 (2.1 × 50 mm, 1.7 μm, Phenomenex)	2.6	Resolution of JAK ImpA/JAK ImpB (<1.5) Symmetry factor of JAK09 (>1.8) Symmetry factor of JAK ImpC (>1.8)	not found	not found
	3.2	Symmetry factor of JAK09 (>1.8) Symmetry factor of JAK ImpB (>1.8) Symmetry factor of JAK ImpD (>1.8)	not found	not found
	4.0	Symmetry factor of JAK09 (<1.7) Symmetry factor of JAK ImpC (<1.7)	20–24	85–87
	6.8	Resolution of JAK ImpB/JAK07	not found	not found
ACQUITY UPLC BEH C18 (2.1 × 50 mm, 1.7 μm Waters)	2.6	Resolution of JAK ImpB/JAK ImpC (<0.8)	not found	not found
	3.2	Resolution of JAK ImpC/JAK ImpB (<0.5) Symmetry factor of JAK09 (>2.1) Symmetry factor of JAK ImpC (>1.9)	not found	not found
	4.0	Resolution of JAK ImpB/JAK ImpC (<0.8)	not found	not found
	6.8	Symmetry factor of JAK09 (>1.6)	20–25	85–90
InfinityLab Poroshell 120 Phenyl Hexyl (2.1 × 50 mm, 1.9 μm, Agilent Technologies)	2.6	Symmetry factor of JAK ImpC (>2.0)	not found	not found
	3.2	Resolution of JAK07/JAK08 (<0.5) Symmetry factor JAK07 (>1.8)	not found	not found
	4.0	Resolution of JAK08/JAK01 (<0.5) Symmetry factor JAK07 (<0.6)	not found	not found
	6.8	Symmetry factor JAK09 (>2.0) Symmetry factor JAK07 (>1.8)	not found	not found
Kinetex Biphenyl (2.1 × 50 mm, 1.7 μm, Phenomenex)	2.6	Resolution of JAK07/JAK01 (<0.6) Symmetry factor of JAK09 (>2.3) Symmetry factor of JAK ImpB (>2.4) Symmetry factor of JAK07 (>2.3)	not found	not found
	3.2	Resolution of JAK07/JAK01 (<1.1) Symmetry factor of JAK09 (>1.9) Symmetry factor of JAK ImpB (>1.9)	not found	not found
	4.0	Symmetry factor of JAK01 (>2.5)	not found	not found
	6.8	Resolution of JAK07/JAK ImpB (<0.8) Symmetry factor of JAK07 (>2.4)	not found	not found
Kinetex PFP (2.1 × 50 mm, 1.7 μm, Phenomenex)	2.6	Resolution of JAK09/JAK ImpA (<1.6) Symmetry factor of JAK09 (>1.8) Symmetry factor of JAK ImpC (>1.8)	not found	not found
	3.2	Resolution of JAK08/JAK07 (<0.5) Symmetry factor of JAK09 (>1.8) Symmetry factor of JAK ImpC (>2.1)	not found	not found
	4.0	Resolution of JAK ImpE/JAK ImpD (<0.8) Symmetry factor of JAK ImpC (>1.9)	not found	not found
	6.8	Resolution of JAK08/JAK07 (<0.85)	not found	not found
ACQUITY UPLC CSH C18 (2.1 × 50 mm, 1.7 μm, Waters)	2.6	Resolution of JAK ImpC/JAK ImpA (<1.8) Symmetry factor of JAK07 (>1.8)	not found	not found
	3.2	Resolution of JAK07/JAK08 (<2.0) Symmetry factor of JAK07 (>1.8)	not found	not found
	4.0	Resolution of JAK ImpB/JAK ImpC (<0.6)	not found	not found
	6.8	Symmetry factor of JAK09 Symmetry factor of JAK ImpB Symmetry factor of JAK07	20–23	85–87
ACQUITY UPLC CSH Fluoro-Phenyl (2.1 × 100 mm, 1.7 μm, Waters)	2.6	Resolution JAK ImpE/JAK ImpD (<2.28, >2.00) Symmetry factor of JAK07 (<1.8)	20–21	89–90
	3.2	Resolution JAK ImpE/JAK ImpD (<2.3, >2.00) Symmetry factor of JAK07 (<1.8)	20–23	85–90
	4.0	Resolution JAK07/JAK ImpE (<1.3)	not found	not found
	6.8	Resolution JAK08/JAK01 Symmetry factor of JAK01 (<0.8)	not found	not found

For further study follow chromatographic condition was selected ZORBAX Eclipse Plus C18, pH 2.6, 20% of ACN at start gradient, 85% of ACN at end gradient; Kinetex EVO C18, pH 4.0, 21% of ACN at start gradient, 86% of ACN at end gradient; ACQUITY UPLC CSH Fluoro-Phenyl, pH 2.6, 20% of ACN at start gradient, 85% of ACN at end gradient.

2.5. Optimization and Robustness Testing

In the following step of method development, the conditions chosen during the screening experiments were tested. The real sample of JAK01 substance and a sample of the test substance spiked with impurities at the 0.15% level, apart from the SST solution, were running, and robustness tests were performed. Although the acceptance criteria of CMAs were met for all three conditions selected for the screening study, a chromatogram of the real substance with overload concentrations of JAK01 with impurities spiked at 0.15% level is insufficient. On column Kinetex Evo C18, impurity JAK07 elutes on the slope of the main peak of JAK01 (Figure 7). On the ACQUITY UPLC CSH Fluoro-Phenyl column we observe a similar situation, impurity JAK08 elutes on the slope of the peak of JAK01. Because of this, there may be problems with coelution and peak integration when applying the method to routine testing.

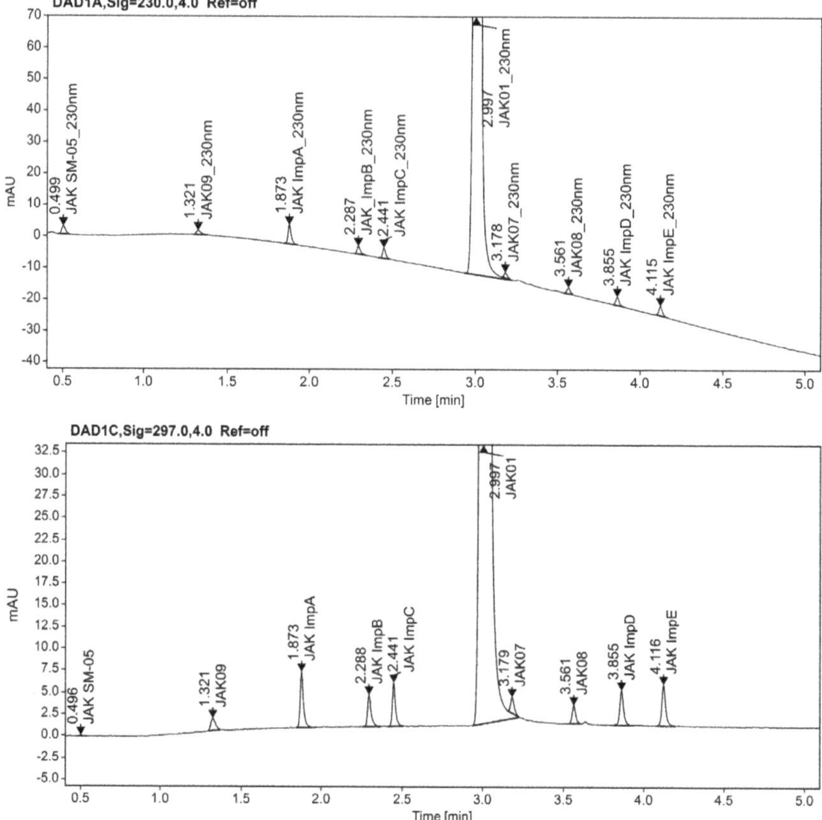

Figure 7. Chromatogram of JAK01 substance solution spiked with impurities at 0.15% level on the Kinetex EVO C18 (2.1 × 50 mm, 1.7 μm, Phenomenex) column.

Therefore, the best choice of chromatographic conditions for the determination of JAK01 impurities seems to be on the Zorbax Eclipse Plus C18 column, pH 2.6, with 20% of ACN at the start and 85% of ACN at the ending gradient. The higher resolution ($R_s > 8.5$) between the main peak of JAK01 and the nearest impurity is observed under this condition on the SST solution, compared to Kinetex Evo C18 and ACQUITY UPLC CSH Fluoro-Phenyl columns. On the Kinetex Evo C18 column, the resolution between JAK01 and the next peak JAK07 equals 2.2, and on the ACQUITY UPLC CSH Fluoro-Phenyl column resolution between JAK01 and JAK08 equals 5.1.

In the next step, separations conditions on a Zorbax Eclipse Plus C18 column with 10 mM formic acid pH 2.6 as an aqueous mobile phase were applied to the robustness test.

Typically, the robustness of a developed method is tested by changing one method parameter at a time, keeping the other variables constant. During our studies, many variables were changed simultaneously in various combinations using the DOE approach, based on the fractional factorial design with central point values, and with full repetition for statistical analysis of robustness.

Apart from variables tested at the screening step, the column temperature, and concentration of formic acid were additionally tested. The range of variable values was established from the knowledge space of the screening step, % of ACN at the start of the gradient (20% ± 1%), % of ACN at the end of the gradient (85% ± 1%). The concentration of formic acid was tested in the range of 10mM ± 1 mM, and the column temperature was 30 °C ± 2 °C. The other parameters of separation are kept at a constant value that was detailed described in Section 3.2.

The application of fractional factorial design ($2^{(4-1)}$) with a center point and repetition instead of full factorial design with center points (3^4) and repetition allows obtaining the same knowledge about the robustness of the method when performing 18 experiments instead of 81. The runs of the experimental plans were carried out in a randomized order with SST solution and a test solution containing 0.5 mg mL^{-1} of JAK01 and 0.75 μg mL^{-1} of JAK01 impurities corresponding to 0.15% level, to assure sufficient selectivity.

The design of the experiment plan and corresponding raw data were presented in Table 5.

Multiple linear regression was applied for the calculation of the coefficients of the nine models for resolution between all pairs of peaks and ten models of the symmetry factors for all peaks, then the models were refined to improve their quality by removing some of the non-significant and entangled effects. The evaluation of statistical analysis tools like ANOVA for each response was used to determine the significance of each method parameter selected for the study using the p-value (significance level $p < 0.05$). The graphical analysis of the effects presents in Figure 8 as an example of the Pareto chart (for R_s between JAK01 and JAK07 and A_s for JAK01), which allows the retained coefficients and the significant terms of the models to be identified the impact of the tested variables on the CMAs.

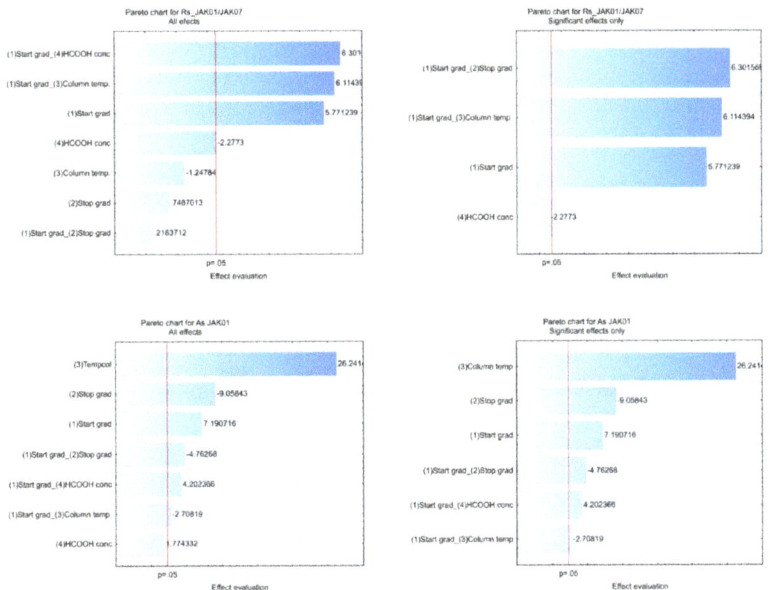

Figure 8. The Pareto chart for Rs between JAK01 and JAK07 and As for JAK01 for robustness tests.

Table 5. Plan of the design and raw data of robustness experiments.

Start Gradient [%ACN]	End Gradient [%ACN]	Column Temperature [°C]	Concentration of HCOOH Solution [mM]	Results of R_s										Results of A_s								
				JAK 09	JAK Imp A	JAK Imp C	JAK Imp B	JAK 01	JAK 07	JAK 08	JAK Imp D	JAK Imp E	JAK SM-05	JAK 09	JAK Imp A	JAK Imp C	JAK Imp B	JAK 01	JAK 07	JAK 08	JAK Imp D	JAK Imp E
19.0	84.0	28.0	9.0	11.20	7.83	5.01	2.79	11.70	7.42	4.20	4.2	4.36	1.57	1.35	1.28	1.40	1.25	1.20	1.26	1.17	1.22	1.23
21.0	86.0	28.0	9.0	3.02	8.27	3.95	3.14	12.15	7.61	3.60	4.57	4.19	1.56	1.79	1.31	1.46	1.26	1.18	1.22	1.18	1.16	1.20
19.0	86.0	28.0	11.0	6.88	7.04	3.63	2.87	10.20	6.75	2.83	3.74	3.54	1.56	1.43	1.15	1.44	1.14	1.16	1.21	1.20	1.20	1.20
21.0	84.0	28.0	11.0	2.70	7.00	4.23	2.96	10.75	6.96	2.96	3.80	3.75	1.60	1.65	1.37	1.45	1.49	1.43	1.39	1.39	1.50	1.51
19.0	86.0	32.0	9.0	6.43	6.71	3.28	3.08	9.50	6.44	2.62	3.40	3.51	1.54	1.47	1.55	1.38	1.55	1.59	1.44	1.48	1.51	1.52
21.0	84.0	32.0	9.0	2.57	8.32	3.77	3.37	11.92	7.52	3.11	4.16	4.31	1.46	1.65	1.71	1.57	1.73	1.69	1.60	1.59	1.60	1.65
19.0	84.0	32.0	11.0	6.09	7.22	3.38	3.02	12.00	6.90	3.18	4.25	4.28	1.42	1.47	1.78	1.49	1.68	1.59	1.51	1.46	1.52	1.57
21.0	86.0	32.0	11.0	2.35	8.28	3.75	3.32	11.69	7.78	3.06	4.14	4.18	1.60	1.64	1.77	1.53	1.63	1.56	1.49	1.45	1.54	1.54
20.0	85.0	30.0	10.0	5.73	9.32	4.54	3.24	14.04	8.53	3.98	5.43	4.76	1.39	1.72	1.31	1.38	1.41	1.32	1.28	1.23	1.40	1.24
19.0	84.0	28.0	9.0	11.6	7.07	5.38	2.84	11.51	7.61	4.28	4.09	4.30	1.55	1.34	1.33	1.38	1.27	1.25	1.25	1.22	1.26	1.29
21.0	86.0	28.0	9.0	3.07	8.15	4.01	3.18	12.05	7.45	3.45	4.85	4.49	1.56	1.79	1.27	1.46	1.23	1.18	1.18	1.15	1.18	1.23
19.0	86.0	28.0	11.0	7.34	7.28	3.81	2.91	10.91	6.99	3.02	3.88	3.61	1.53	1.43	1.12	1.48	1.10	1.11	1.15	1.12	1.18	1.22
21.0	84.0	28.0	11.0	2.65	7.73	4.02	2.88	10.32	6.64	2.78	4.06	3.93	1.63	1.62	1.34	1.42	1.41	1.46	1.40	1.41	1.46	1.50
19.0	86.0	32.0	9.0	6.36	6.79	3.29	3.17	9.51	6.52	2.66	3.39	3.50	1.54	1.44	1.52	1.36	1.51	1.52	1.41	1.42	1.51	1.51
21.0	84.0	32.0	9.0	2.54	8.37	3.75	3.36	11.27	7.19	2.98	4.02	4.20	1.45	1.66	1.73	1.61	1.77	1.67	1.54	1.54	1.58	1.61
19.0	84.0	32.0	11.0	6.37	7.38	3.70	3.18	12.33	6.55	3.20	4.00	4.32	1.40	1.50	1.78	1.52	1.69	1.57	1.48	1.45	1.53	1.55
21.0	86.0	32.0	11.0	2.34	8.20	3.77	3.34	11.77	7.73	3.02	4.10	4.14	1.62	1.66	1.72	1.53	1.65	1.59	1.49	1.46	1.59	1.56
20.0	85.0	30.0	10.0	5.75	9.26	4.53	3.27	14.10	8.51	3.99	5.41	4.70	1.39	1.72	1.35	1.41	1.38	1.30	1.32	1.23	1.55	1.28
		min		2.34	6.71	3.28	2.79	9.50	6.44	2.62	3.39	3.50	1.39	1.34	1.12	1.36	1.10	1.11	1.15	1.12	1.16	1.20
		max		11.60	9.32	5.38	3.37	14.10	8.53	4.28	5.43	4.76	1.63	1.79	1.78	1.61	1.77	1.69	1.60	1.59	1.60	1.65

After verifying the reliability of the regression models, the results were converted to the desirability plots for a specified value of CMAs ($R_s \geq 2.0$ for all pairs of peaks, and the symmetry factor of all peaks $0.8 \leq A_s \leq 1.8$) for values of the CMPs from the robustness testing range. The analysis of these plots was used to estimate the final design space of the proposed method.

Based on the results of the statistical analysis, we can conclude that the assumed requirements for CQAs are met in the whole robustness tested range of the CMPs—desirability factors are well above 0.0, near the maximum value equals 1.0.

2.6. Method Operable Design Region (MODR) and Control Strategy

Based on the results of the design of the experiment from robustness tests, the STATIS-TICA software was also employed for verifying the MODR, i.e., it was checked whether the CMPs selected in the screening phase, in a small range centered on the optimized value, had a significant effect on CMAs.

For this purpose, probability maps were calculated by Monte-Carlo simulations, propagating the predictive error by using the model equation to the CMAs and computing the probability of reaching the desired objectives [11,12]. The threshold for the risk of failure was set to 10%. This means that in the calculated zone the values of the CMAs are satisfied with a probability of 90%. For all variables (CMAs) risk of failure was well below 10% at the robustness tested range of CMPs.

Based on the statistical result evaluations of the screening, optimization, robustness, and risk tests, the MODR corresponds to the following intervals, $20\% \pm 1\%$ of ACN at the start of the gradient, $85\% \pm 1\%$ of ACN at the end of the gradient, the concentration of formic acid 10 mM $\pm$ 1 mM, and the column temperature 30 °C $\pm$ 2 °C.

The control strategy of the method was designed by taking into account the results of robustness testing and identifying system suitability criteria [21]. The intervals for the accepted CMAs values were included between the lowest and the highest values for the CMAs measured when performing a robustness study. The obtained interval resolutions for the SST solution were shown in Table 5. Additionally, based on the screening study, we could define the requirements for the resolution between JAK01 and the nearest impurity peak for SST solution as more than 6.4. Less value of it causes that impurity peak may coelute with the main peak of JAK01 at an overloaded concentration in the test solution.

2.7. Validation

Validation of the method was carried out in compliance with ICH guideline Q2(R1) [25]. The validation data are reported in Table 6, showing adequate performances for the intended purpose. The determined content of all CPL409116 (JAK01) impurities showed evidence of good precision (RSDs were less than 10.0%). LOQ concentration value is on the reporting threshold level of 0.05% for all impurities.

Table 6. Summary of the validation results.

Parameter	Acceptance Criteria	JAK SM-05	JAK 09	JAK Imp A	JAK Imp C	JAK Imp B	JAK 07	JAK 08	JAK Imp D	JAK Imp E	Compliance with Acceptance Criteria (Yes/No)
Linearity, R	LOQ—120% of the specification limit (for each impurity) Concentration levels $n = 5$, $R \geq 0.99$	0.99	0.99	0.99	0.99	0.99	0.99	0.99	0.99	0.99	Yes
Repeatability, RSD (%)	LOQ—%RSD (area) $\leq 15\%$	6.8	8.5	2.7	4.7	4.2	9.1	3.4	2.5	2.0	Yes
	100% of the specification limit—%RSD (area) $\leq 15\%$	4.5	5.2	0.8	1.8	2.0	5.3	1.8	3.3	1.5	Yes
Limit of Detection (LOD),	$S/N \geq 3\ (n = 3)$	6	3	8	9	11	5	5	4	11	Yes
Limit of Quantitation (LOQ)	$S/N \geq 10\ (n = 6)$	21	12	22	29	31	14	16	8	32	Yes
Specificity	Resolution between two neighboring peaks, Rs ≥ 2.0	-	9.6	12.0	8.4	4.6	9.9	4.1	5.8	5.6	Yes

The developed method was finally applied for the analysis of a real sample of CPL409116 substance from a large-scale synthetic process. Overlay chromatograms JAK01 test solution and JAK01 test solution spiked with impurities at 0.15% level shown in Figure 9.

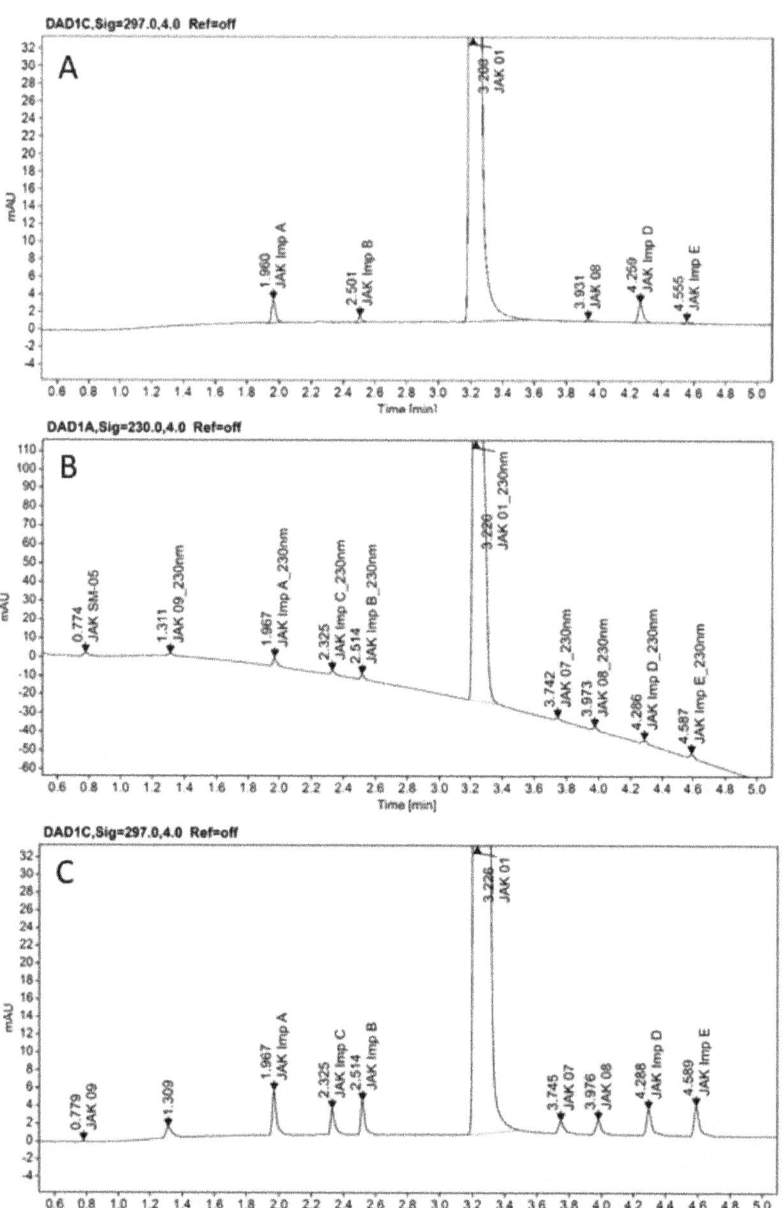

Figure 9. Overlay chromatograms JAK01 test solution (**A**), and JAK01 test solution spiked with impurities at 0.15% level at 230 nm (**B**), and at 297 nm (**C**), on the ZORBAX Eclipse Plus C18 (2.1 × 50 mm, 1.8 µm, Agilent Technologies, Waldbronn, Germany) column.

3. Material and Methods

3.1. Chemicals and Reagents

Reference standard JAK01 (CPL409116) and its impurities (JAK07, JAK08, JAK09, JAK ImpA, JAK ImpB, JAK ImpC, JAK ImpD, JAK ImpE, and JAK SM-05) were manufactured in-house by Celon Pharma S.A. (Lomianki, Poland).

Mix reference substances to the distribution coefficient chromatographically determinations (CHI logD) including paracetamol, acetanilide, acetophenone, propiophenone, butyrophenone, and valerophenone (purity >99.0%) were purchased from Sigma-Aldrich Chemie GmbH (Steinheim, Germany).

Acetonitrile (ACN) (hypergrade for LC-MS), and methanol (MeOH, hypergrade for LC-MS) were purchased from Merck KGaA (Darmstadt, Germany), and dimethyl sulfoxide (DMSO, for HPLC) from POCH (Gliwice, Poland). Formic acid (98–100%, eluent additive for LC-MS) and ammonia (25% solution, eluent additive for LC-MS) were obtained from CHEM-LAB NV (Zedelgem, Belgium). Ultra-pure water for HPLC was obtained from the water purification system Milli-Q IQ 7000 from Merck KGaA (Darmstadt, Germany).

3.2. Solutions and Sample Preparations

A number of buffers (10mM HCOONH$_4$) with different pH levels were prepared for these experiments. Buffers (pH 3.2, 3.8, 4.0, 4.2, and 6.8) were prepared by mixing the formic acid solution with different levels of ammonia (25% solution) to adjust the pH. The individual phases of formic acid were prepared by adding a suitable amount of concentrated acid to water. Stock solutions were prepared for all compounds and stored at 4 °C. A standard stock solution of JAK01 (3 mg mL^{-1}) was prepared using a mixture of DMSO:MeOH (20:80 v/v). Stock solutions (4 mg mL^{-1}) for each impurity (JAK07, JAK08, JAK09, JAK ImpA, JAK ImpB, JAK ImpC, and JAK SM-05) were prepared in a mixture of DMSO: MeOH (20:80 v/v). Stock solutions (4 mg mL^{-1}) for JAK ImpD and JAK ImpE were prepared in DMSO. Working standard solutions were daily prepared using MeOH. A mix of all compounds at the concentration level of 0.1–1 µM (as SST solution) was used to screen the chromatographic conditions. In the next stage of the research, a mixture of JAK01 (0.5 mg mL^{-1}; 100%) with nine of its impurities (each at the level of 0.75 ug mL^{-1}; 0.15%) was used. For the tested impurities, linearity in the range of 0.05% to 0.18% was also demonstrated by preparing mixtures of compounds at the appropriate concentration level and a solution of impurities at 0.02% was prepared for detection limit (LOD) determination.

For the CHI logD determination, separate solutions of all the tested sample were prepared at a concentration of 1 mM in DMSO. Reference mix solutions have a concentration of 1 mM in DMSO.

3.3. Instrumentations and Chromatographic Conditions

Eight different chromatographic columns were evaluated: ZORBAX Eclipse Plus C18 (2.1 × 50 mm, 1.8 µm, Agilent Technologies), Kinetex EVO C18 (2.1 × 50 mm, 1.7 µm, Phenomenex), ACQUITY UPLC BEH C18 (2.1 × 50 mm, 1.7 µm Waters), InfinityLab Poroshell 120 Phenyl Hexyl (2.1 × 50 mm, 1.9 µm, Agilent Technologies), Kinetex Biphenyl (2.1 × 50 mm, 1.7 µm, Phenomenex), Kinetex PFP (2.1 × 50 mm, 1.7 µm, Phenomenex), ACQUITY UPLC CSH C18 (2.1 × 50 mm, 1.7 µm, Waters), and ACQUITY UPLC CSH Fluoro-Phenyl (2.1 × 100 mm, 1.7 µm, Waters). When performing the chromatographic optimization, eluent A consisted of formic acid (10 mM, pH 2.6) and ammonium formate (10 mM) buffer with different concentrations of additives was used to prepare buffer solutions of 3.2, 4.0, and 6.8, and eluent B consisted of the organic solvent (ACN). The elution was performed with linear gradient mode. The starting and ending percentage of eluent B was optimized at a screening phase at a range from 20% to 25% and from 85% to 90%. The ending percentage of eluent B was isocratically maintained for 2 min. The whole gradient time was 7 min and included 4 min of elution step, 1 min of washing and 2 min of stabilization step. The other working conditions were as follows: sample injection volume,

1 µL; flow rate, 0.50 mL min^{-1}; oven temperature, 30 °C (28–32 °C); detection UV 297 nm (230 nm for JAK SM-05), and the autosampler temperature, 15 °C.

The chromatographic analyses were run by an Inifinity II 1290 UHPLC Method Development System (Agilent Technologies, Waldbronn, Germany) equipped with quaternary high-pressure pump, 12-position solvent selection valve, multisampler with thermostat, multicolumn thermostat including 8-position quick-change column selection valve and diode array detector. UHPLC system was controlled by OpenLab Client/Server System (version 2.5) (Agilent Technologies, Waldbronn, Germany).

The CHI logD for JAK01 and its impurities was determined on an UltraCore 2.5 SuperC18 (2.1 × 50 mm, 2.5 µm, ACE) column with eluent A consisting of formic acid (10 mM, pH 2.6), and ammonium acetate (50 mM) buffers with different concentrations of additives were used to prepare buffer solutions of 7.4, and 10.5, and eluent B consisted ACN [58].

3.4. Calculations and Software

Calculations of physico-chemical properties were made with the ACD Labs Percepta software (2020.1.1 release) (Advanced Chemistry Development, Inc., Toronto, ON, Canada) including PhysChem and Drug Profiler modules [40].

For evaluation of potential mutagenicity two complementary (Q)SAR prediction methodologies were used DEREK Nexus module (Lhasa Ltd., Leeds, UK) in the Star Drop software (version 6.6) (Optibrium Ltd., Cambridge, UK) [50–52] was used for rule-based calculations. For statistically-based (Q)SAR methodology the Toxicity Estimation Software Tool T.E.S.T (version 4.2.1) (US Environmental Protection Agency, Cincinnati, OH, USA) [53,54] software was used.

The STATISTICA (version 13.3) TIBCO Software Inc. (Palo Alto, CA, USA) was used for DOE and investigations—generate plans of the experiment, screening, optimization, robustness, and risk study of analytical method conditions, and statistical analysis of obtained results [55–57].

4. Conclusions

The present work has demonstrated the application of AQbD for the selection and development of a chromatographic analytical method for the determination of purity CPL409116 (JAK01) for controlling the large-scale synthetic process of the substance, performing stability tests, and manufacturing of the final drug form for clinical trials.

The ultimate goal of the chromatographic method development is to obtain an acceptable resolution of all components within a reasonable analysis time. To meet the quality requirements of the formal ICH regulatory guidelines [2,25], but bearing in mind the cost and time effectiveness of the preclinical stage of drug development, the main goal was to design a quick, simple, and robust analytical method.

At the beginning of the method development process, physico-chemical properties of CPL409116 and its related substances have been performed. In practice, not all values obtained from calculations give adequate information to use for chromatography method development. However, for this purpose, a very useful strategy is the determination of the hydrophobicity index CHI logD that gives also the possibility to narrow the scope of tested pH of the stationary phases at the screening stage.

In the present work, final method selection has been performed by screening analysis with eight chromatographic columns differing in chemical modifications of the stationary phase (C18, phenyl-hexyl, biphenyl), three linear gradients with various increasing of the organic phase, and aqueous phases differing in pH in acidic range (pH 2.6–6.8).

As a result, we selected three different conditions that meet acceptance criteria (ATP), but as a most universal condition, Agilent Zorbax Eclipse Plus C18 column with 10 mM formic acid pH 2.6 as an aqueous mobile phase, was applied for the robustness test.

Using a DOE approach and application of fractional factorial design with a center point and repetition, instead of full factorial design with center points and repetition, allows for obtaining the same knowledge about the robustness but is limited to 18 experiments based on simultaneous changes of many variables.

In this way, the need to perform a large number of analytical runs was effectively limited, confirming the scope of applicability of the selected analytical method with the AQbD method. Based on the results of the statistical analysis, we can conclude that the assumed requirements for CQAs are met in the whole robustness tested range of the CMPs—desirability factors are well above 0.0, near the maximum value equals 1.0.

Author Contributions: Conceptualization, L.G.-B., D.A.S., U.D. and W.M.; methodology, L.G.-B. and D.A.S.; software, D.A.S. and L.G.-B.; validation, L.G.-B., U.D. and W.M.; formal analysis, L.G.-B. and D.A.S.; investigation, L.G.-B., U.D. and M.D.; resources and chemical synthesis of compounds, M.M., D.A.S., K.D., B.S., K.S., A.M. and A.B.; data curation, D.A.S. and L.G.-B.; writing—original draft preparation, L.G.-B.; writing—review and editing, W.M. and D.A.S.; visualization, L.G.-B., U.D. and D.A.S.; supervision, D.A.S., K.D. and M.W.; project administration, L.G.-B. and U.D.; funding acquisition, M.W. All authors have read and agreed to the published version of the manuscript.

Funding: This research was co-funded by Celon Pharma S.A. and the National Centre for Research and Development (Poland) Grant POIR.01.01.01-00-0382/16.

Institutional Review Board Statement: Not applicable.

Informed Consent Statement: Not applicable.

Data Availability Statement: Not applicable.

Conflicts of Interest: The authors declare the following financial interests/personal relationships, which may be considered as potential competing interests: all contributors to this work at the time of their direct involvement in the project were full-time employees of Celon Pharma S.A. M. Wieczorek is the CEO of Celon Pharma S.A. Some of the authors are the shareholders of Celon Pharma S.A.

References

1. Elder, D.; Teasdale, A. ICH, Q9, Quality Risk Management. In *ICH Quality Guidelines: An Implementation Guide*; Wiley: Hoboken, NJ, USA, 2017.
2. Holm, P.; Allesø, M.; Bryder, M.C.; Holm, R. ICH, Q8(R2), Pharmaceutical Development. In *ICH Quality Guidelines: An Implementation Guide*; Wiley: Hoboken, NJ, USA, 2017.
3. Borman, P.; Chatfield, M.; Nethercote, P.; Thompson, D.; Truman, K. The application of quality by design to analytical methods. *Pharm. Technol.* **2007**, *31*, 142–152.
4. Schweitzer, M.; Pohl, M.; Brown, M.H.; Nethercote, P.; Borman, P.; Smith, K.; Larew, J. Implications and Opportunities of Applying QbD Principles to Analytical Measurements. *Pharm. Technol.* **2010**, *34*, 52–59.
5. Vogt, F.G.; Kord, A.S. Development of quality-by-design analytical methods. *J. Pharm. Sci.* **2011**, *100*, 797–812. [CrossRef] [PubMed]
6. Castle, B.C.; Forbes, R.A. Impact of Quality by Design in Process Development on the Analytical Control Strategy for a Small-Molecule Drug Substance. *J. Pharm. Innov.* **2013**, *8*, 247–264. [CrossRef]
7. Reid, G.; Morgado, J.; Barnett, K.; Harrington, B.; Harwood, J.; Fortin, D. Analytical quality by design (AQbD) in pharmaceutical development. *Am. Pharm. Rev.* **2013**, *16*, 49–59.
8. Parra, M.K.; Schmidt, A.H. Life Cycle Management of Analytical Methods. *J. Pharm. Biomed.* **2018**, *147*, 506–517. [CrossRef]
9. Dispas, A.; Avohou, H.T.; Lebrun, P.; Hubert, P.; Hubert, C. Quality by design approach for the analysis of impurities in pharmaceutical drug products and drug substances. *TrAC Trends Anal. Chem.* **2018**, *101*, 24–33. [CrossRef]
10. Basso, J.; Mendes, M.; Cova, T.F.; Sousa, J.J.; Pais, A.A.; Vitorino, C. Analytical Quality by Design (AQbD) as a multiaddressable platform for co-encapsulating drug assays. The Royal Society of Chemistry 2018. *Anal. Methods* **2018**, *10*, 5659. [CrossRef]
11. Pasquini, B.; Orlandini, S.; Furlanetto, S.; Gotti, R.; Del Bubba, M.; Boscaro, F.; Bertaccini, B.; Douša, M.; Pieraccini, G. Qualit by Design as risk-based strategy in pharmaceutical analysis: Development of a liquid chromatography-tandem mass spectrometry method for determination of nintedanib and its impurities. *J. Chromatogr. A* **2020**, *1611*, 460615. [CrossRef]
12. Pawar, A.; Pandita, N. Statistically Designed, Targeted Profile UPLC Method Development for Assay and Purity of Haloperidol in Haloperidol Drug Substance and Haloperidol 1 mg Tablets. *Chromatographia* **2020**, *83*, 725–737. [CrossRef]
13. Hibbert, D.B. Experimental design in chromatography: A tutorial review. *J. Chromatogr. B* **2012**, *910*, 2–13. [CrossRef] [PubMed]
14. Kochling, J.; Wu, W.; Hua, Y.; Guan, Q.; Castaneda-Merced, J. A platform analytical quality by design (AQbD) approach for multiple UHPLC-UV and UHPLC–MS methods development for protein analysis. *J. Pharm. Biomed. Anal.* **2016**, *125*, 130–139. [CrossRef] [PubMed]
15. Gaudin, K.; Ferey, L. Quality by Design: A Tool for Separation Method Development in Pharmaceutical Laboratories. *LCGC* **2016**, *29*, 16–25.

16. Rácz, N.; Molnár, I.; Zöldhegyi, A.; Rieger, H.-J.; Kormány, R. Simultaneous optimization of mobile phase composition and pH using retention modeling and experimental design. *J. Pharm. Biomed. Anal.* **2018**, *160*, 336–343. [CrossRef] [PubMed]

17. Orlandini, S.; Pinzauti, S.; Furlanetto, S. Application of quality by design to the development of analytical separation methods. *Anal. Bioanal. Chem.* **2013**, *405*, 443–450. [CrossRef] [PubMed]

18. Li, Y.; Terfloth, G.J.; Kord, A.S. A systematic approach to RP-HPLC method development in a pharmaceutical QbD environment. *Am. Pharm. Rev.* **2009**, *12*, 87–95.

19. Bhaskaran, N.A.; Kumar, L.; Reddy, M.S.; Pai, G.K. An analytical "quality by design" approach in RP-HPLC method development and validation for reliable and rapid estimation of irinotecan in an injectable formulation. *Acta Pharm.* **2021**, *71*, 57–79. [CrossRef]

20. Jackson, P.; Borman, P.J.; Campa, C.; Chatfield, M.J.; Godfrey, M.; Hamilton, P.R.; Hoyer, W.; Norelli, F.; Orr, R.; Schofield, T. Using the analytical target profile to drive the analytical method lifecycle. *Anal. Chem.* **2019**, *91*, 2577–2585. [CrossRef]

21. Rathore, A.S.; Winkle, H. Quality by design for biopharmaceuticals. *Nat. Biotechnol.* **2009**, *27*, 26–34. [CrossRef]

22. Yu, L.X. Pharmaceutical quality by design: Product and process development, understanding, and control. *Pharm. Res.* **2007**, *25*, 781–791. [CrossRef]

23. Csoka, I.; Pallagi, E.; Paal, T.L. Extension of quality-by-design concept to the early development phase of pharmaceutical R&D processes. *Drug Discov. Today* **2018**, *23*, 1340–1343. [PubMed]

24. Deidda, R.; Orlandini, S.; Hubert, P.; Hubert, C. Risk-based approach for method development in pharmaceutical quality control context: A critical review. *J. Pharm. Biomed. Anal.* **2018**, *161*, 110–121. [CrossRef] [PubMed]

25. ICH Harmonised Tripartite Guideline. Validation of Analytical Procedures: Text and Methodology Q2(R1). In Proceedings of the International Conference on Harmonisation of Technical Requirements for Registration of Pharmaceuticals for Human Use, Geneva, Switzerland, 9 November 2005.

26. Final Concept Paper: ICH Q14: Analytical Procedure Development and Revision of Q2(R1) Analytical Validation. Available online: https://www.ich.org/fileadmin/Public_Web_Site/ICH_Products/Guidelines/Quality/Q2_Q14/Q2R2Q14EWG_ConceptPaper_2018_1115.pdf (accessed on 25 May 2021).

27. Tome, T.; Žigart, N.; Časar, Z.; Obreza, A. Development and Optimization of Liquid Chromatography Analytical Methods by Using AQbD Principles: Overview and Recent Advances. *Org. Process Res. Dev.* **2019**, *23*, 1784–1802. [CrossRef]

28. Mroczkiewicz, M.; Stypik, B.; Bujak, A.; Szymczak, K.; Gunerka, P.; Dubiel, K.; Wieczorek, M.; Pieczykolan, J. Pyrazole[1,5-a]Pyrimidine Derivatives as Kinase Jak Inhibitors. W.O. Patent WO2018206739A1, 15 November 2018.

29. Mroczkiewicz, M.; Stypik, B.; Bujak, A.; Szymczak, K.; Gunerka, P.; Dubiel, K.; Wieczorek, M.; Pieczykolan, J. Pyrazole[1,5-a]Pyrimidine Derivatives as Kinase Jak Inhibitors. E.P. Patent EP3621966 B1, 18 March 2020.

30. Dulak-Lis, M.; Bujak, A.; Gala, K.; Banach, M.; Kędzierska, U.; Miszkiel, J.; Hucz-Kalitowska, J.; Mroczkiewicz, M.; Stypik, B.; Szymczak, K.; et al. A novel JAK/ROCK inhibitor, CPL409116, demonstrates potent efficacy in the mouse model of systemic lupus erythematosus. *J. Pharmacol. Sci.* **2021**, *145*, 340–348. [CrossRef] [PubMed]

31. Iuliani, P.; Carlucci, G.; Marrone, A. Investigation of the HPLC response of NSAIDs by fractional experimental design and multivariate regression analysis. Response optimization and new retention parameters. *J. Pharm. Biomed. Anal.* **2010**, *51*, 46–55. [CrossRef]

32. Kishore, C.R.P.; Mohan, G.V.K. Development and Validation of Amlodipine Impurities in Amlodipine Tablets Using Design Space Computer Modeling. *Am. J. Anal. Chem.* **2016**, *7*, 918–926. [CrossRef]

33. Prasad, S.S.; Mohan, G.V.K.; Babu, A.N. Orient. *J. Chem.* **2019**, *35*, 140–149.

34. Sahu, P.K.; Ramisetti, N.R.; Cecchi, T.; Swain, S.; Patro, C.S.; Panda, J. An overview of experimental designs in HPLC method development and validation. *J. Pharm. Biomed. Anal.* **2018**, *147*, 590–611. [CrossRef]

35. OECD. *Test No. 107: Partition Coefficient (n-Octanol/Water): Shake Flask Method*; OECD Publishing: Paris, France, 1995.

36. Albert, A. *Selective Toxicity: The Physicochemical Basis of Therapy*; Chapman and Hall: London, UK, 1979.

37. Klose, M.H.M.; Theiner, S.; Varbanov, H.P.; Hoefer, D.; Pichler, V.; Galanski, M.; Meier-Menches, S.M.; Keppler, B.K. Development and Validation of Liquid Chromatography-Based Methods to Assess the Lipophilicity of Cytotoxic Platinum(IV) Complexes. *Inorganics* **2018**, *6*, 130. [CrossRef]

38. Dolan, J.W. Back to Basics: The Role of pH in Retention and Selectivity. *LCGC N. Am.* **2017**, *35*, 22–28.

39. Lewis, J.A.; Lommen, D.C.; Raddatz, W.D.; Dolan, J.W.; Snyder, L.R.; Molnar, I. Computer simulation for the prediction of separation as a function of pH for reversed-phase high-performance liquid chromatography: I. Accuracy of a theory-based model. *J. Chromatogr. A* **1992**, *592*, 183–195. [CrossRef]

40. *ACD/Percepta*; Version 2020.1.1; Advanced Chemistry Development, Inc.: Toronto, ON, Canada, 2020; Available online: http://www.acdlabs.com (accessed on 26 November 2020).

41. Valkó, K.; Bevan, C.; Reynolds, D. Chromatographic Hydrophobicity Index by Fast-Gradient RP-HPLC: A High-Throughput Alternative to log P/log D. *Anal. Chem.* **1997**, *69*, 2022–2029. [CrossRef]

42. Valko, K. *Physicochemical and Biomimetic Properties in Drug Discovery—Chromatographic Techniques for Lead Optimization*, 1st ed.; Wiley: Hoboken, NJ, USA, 2014.

43. Snyder, L.R.; Dolan, J.W.; Carr, P.W. The hydrophobic-subtraction model of reversed-phase column selectivity. *J. Chromatogr. A* **2004**, *1060*, 77–116. [CrossRef]

44. Snyder, L.R.; Dolan, J.W.; Marchand, D.H.; Carr, P.W. The hydrophobic-subtraction model of reversed-phase column selectivity. *Adv. Chromatogr.* **2012**, *50*, 297–376. [PubMed]

45. Available online: http://www.hplccolumns.org (accessed on 29 January 2021).
46. Available online: https://apps.usp.org/app/USPNF/columnsDB.html (accessed on 29 January 2021).
47. Committee for Human Medicinal Products. ICH guideline M7(R1) on assessment and control of DNA reactive (mutagenic) impurities in pharmaceuticals to limit potential carcinogenic risk. *Int. Conf. Harmon.* **2015**, *7*, 1–110.
48. Baertschi, S.; Olsen, B. Chapter 12—Mutagenic impurities. In *Specification of Drug Substances and Products*, 2nd ed.; Riley, C.M., Rosanske, T.W., Reid, G., Eds.; Elsevier: Amsterdam, The Netherlands, 2020; pp. 321–344.
49. Shaikh, T.; Gosar, A.; Sayyed, H. Nitrosamine Impurities in Drug Substances and Drug Products. *J. Adv. Pharm. Pract.* **2020**, *2*, 48–57.
50. Orbitrium. DEREK Nexus—Trial licence. Available online: https://www.optibrium.com (accessed on 2 November 2020).
51. Lhasa Limited. DEREK Nexus. Available online: https://www.lhasalimited.org (accessed on 2 November 2020).
52. Marchant, C.A.; Briggs, K.A.; Long, A. In Silico Tools for Sharing Data and Knowledge on Toxicity and Metabolism: Derek for Windows, Meteor, and Vitic. *Toxicol. Mech. Methods* **2008**, *18*, 177–187. [CrossRef]
53. US EPA Research. Toxicity Estimation Software Tool (TEST). Available online: https://www.epa.gov/chemical-research/toxicity-estimation-software-tool-test (accessed on 30 October 2020).
54. Martin, T. *User's Guide for T.E.S.T. (Version 4.2) (Toxicity Estimation Software Tool) A Program to Estimate Toxicity from Molecular Structure*; EPA: Washington, DC, USA, 2016; Volume 63.
55. StatSoft's Electronic Statistics Textbook. 2006. Available online: http://www.statsoft.pl/textbook/stathome.html (accessed on 27 January 2021).
56. TIBCO Software Inc. Data Science Textbook. 2020. Available online: https://docs.tibco.com/data-science/textbook (accessed on 27 January 2021).
57. Stanisz, A. *Przystępny kurs Statystyki z Zastosowaniem STATISTICA PL na Przykładach z Medycyny—Tom I–III*; StatSoft: Kraków, Poland, 2006.
58. Valko, K.L. Application of biomimetic HPLC to estimate in vivo behavior of early drug discovery compounds. *Future Drug Discov.* **2019**, *1*, FDD11. [CrossRef]

International Journal of
Molecular Sciences

Article

Dialdehyde Starch Nanocrystals as a Novel Cross-Linker for Biomaterials Able to Interact with Human Serum Proteins

Katarzyna Wegrzynowska-Drzymalska [1], Kinga Mylkie [1], Pawel Nowak [1], Dariusz T. Mlynarczyk [2], Dorota Chelminiak-Dudkiewicz [1], Halina Kaczmarek [1], Tomasz Goslinski [2] and Marta Ziegler-Borowska [1,*]

[1] Department of Biomedical Chemistry and Polymer Science, Faculty of Chemistry, Nicolaus Copernicus University in Torun, Gagarina 7, 87-100 Torun, Poland; kasiawd@doktorant.umk.pl (K.W.-D.); kinga.mylkie@doktorant.umk.pl (K.M.); nowak19981411@wp.pl (P.N.); dorotachd@umk.pl (D.C.-D.); halina@umk.pl (H.K.)

[2] Chair and Department of Chemical Technology of Drugs, Poznan University of Medical Sciences, Grunwaldzka 6, 60-780 Poznan, Poland; mlynarczykd@ump.edu.pl (D.T.M.); tomasz.goslinski@ump.edu.pl (T.G.)

* Correspondence: martaz@umk.pl

Abstract: In recent years, new cross-linkers from renewable resources have been sought to replace toxic synthetic compounds of this type. One of the most popular synthetic cross-linking agents used for biomedical applications is glutaraldehyde. However, the unreacted cross-linker can be released from the materials and cause cytotoxic effects. In the present work, dialdehyde starch nanocrystals (NDASs) were obtained from this polysaccharide nanocrystal form as an alternative to commonly used cross-linking agents. Then, 5–15% NDASs were used for chemical cross-linking of native chitosan (CS), gelatin (Gel), and a mixture of these two biopolymers (CS-Gel) via Schiff base reaction. The obtained materials, forming thin films, were characterized by ATR-FTIR, SEM, and XRD analysis. Thermal and mechanical properties were determined by TGA analysis and tensile testing. Moreover, all cross-linked biopolymers were also characterized by hydrophilic character, swelling ability, and protein absorption. The toxicity of obtained materials was tested using the Microtox test. Dialdehyde starch nanocrystals appear as a beneficial plant-derived cross-linking agent that allows obtaining cross-linked biopolymer materials with properties desirable for biomedical applications.

Keywords: dialdehyde starch nanocrystals; cross-linking; Microtox test; HSA and AGP adsorption

Citation:
Wegrzynowska-Drzymalska, K.; Mylkie, K.; Nowak, P.; Mlynarczyk, D.T.; Chelminiak-Dudkiewicz, D.; Kaczmarek, H.; Goslinski, T.; Ziegler-Borowska, M. Dialdehyde Starch Nanocrystals as a Novel Cross-Linker for Biomaterials Able to Interact with Human Serum Proteins. *Int. J. Mol. Sci.* **2022**, 23, 7652. https://doi.org/10.3390/ijms23147652

Academic Editors: Andrzej Kutner, Geoffrey Brown and Enikö Kallay

Received: 21 June 2022
Accepted: 7 July 2022
Published: 11 July 2022

Publisher's Note: MDPI stays neutral with regard to jurisdictional claims in published maps and institutional affiliations.

1. Introduction

Natural polymers are increasingly being used in medical sciences, pharmaceutical, and food industries due to their high biocompatibility and biodegradability. They are also susceptible to numerous functionalizations due to multiple reactive groups, such as hydroxyl and amino groups. Chitosan and gelatin are examples of such materials. Chitosan is non-toxic, bioactive, and antibacterial [1]. This significantly increases the interest in research where chitosan is used in tissue engineering [2], drug delivery systems [3], wound dressing materials [4], packaging materials [5], and as a dietary supplement in weight reduction and preparations lowering cholesterol [6]. Another biomaterial commonly used in biomedical applications is gelatin. Gelatin is one of the main biopolymers widely used in the food, cosmetic, biomedical, and pharmaceutical industries [7]. However, pure gelatin generally has uncontrollable dissolvability and degradability, poor mechanical properties, and thermal instability, which significantly limit its wide practical applications [8]. Polysaccharides require cross-linking to improve their mechanical strength and stability in an aqueous environment. It can be done by physical and chemical methods, e.g., UV irradiation [9], the use of natural and synthetic cross-linkers [10,11], or enzymes [12]. Compared to the other methods, the main advantage of chemical cross-linking is the formation of a strong covalent bond [13]. The standard cross-linker used for biomedical applications is glutaraldehyde.

However, this compound cannot be widely used due to inherent disadvantages, especially in biomedical applications. It was found that glutaraldehyde can lead to undesirable effects of cytotoxicity.

Moreover, glutaraldehyde is corrosive and irritating to the skin, eyes, and respiratory tract and is considered to cause health problems for those who handle it, e.g., occupational asthma [14,15]. Lee et al. [16] showed a decrease in mitochondrial activity in nanofiber scaffolds cross-linked with glutaraldehyde after two days of culture. Rho et al. [17] also observed similar cytotoxicity behavior when using glutaraldehyde to cross-link collagen nanofibers. Current trends aim to look for non-toxic and safe cross-linkers, e.g., based on biopolymers, which can meet high requirements in medicine [18].

In the past few decades, intensive research on the improvement of the properties of polysaccharides resulted in the preparation of nanocrystalline polysaccharides. They are characterized by many valuable properties, such as high tensile strength, differentiated morphology, and large specific surface area, distinguishing them from inorganic nanoparticles [19]. In starch granules, the crystalline and amorphous regions coexist. Starch nanocrystals (NCSs) are crystalline structures resulting from the disruption of the amorphous form of starch granules by acid hydrolysis [20]. Recently, nanostarch has gained much attention as a potential reinforcing material in composites due to its excellent mechanical, biodegradable, renewable, and biocompatible properties. Starch nanocrystals, which combine natural abundance and excellent mechanical properties, are promising candidates as polymer reinforcement agents, e.g., natural rubber [21] and soy protein plastics [22].

Nowadays, numerous studies are focused on the topic of modification of nanocrystalline polysaccharides. The periodate oxidation process is one of the methods of modification of nanocrystalline polysaccharides. In this process, the bond between C2 and C3 in the glycosidic ring is cleaved, and then two aldehyde groups are introduced to these carbon atoms [23]. This simple one-pot method requires only sodium periodate and water as the oxidant and solvent, respectively (Figure 1). The aldehyde groups introduced into the polysaccharide structure can serve as cross-linking agents for biopolymers. The dialdehyde polysaccharides based on pectin [24], chitosan [25], xylan [26], alginate [27], starch [28], hyaluronic acid [29], pullulan [30], galactomannan [31], cellulose [32], carboxymethyl cellulose [33], and xanthan gum [34] are used for cross-linking of polymers containing amino groups.

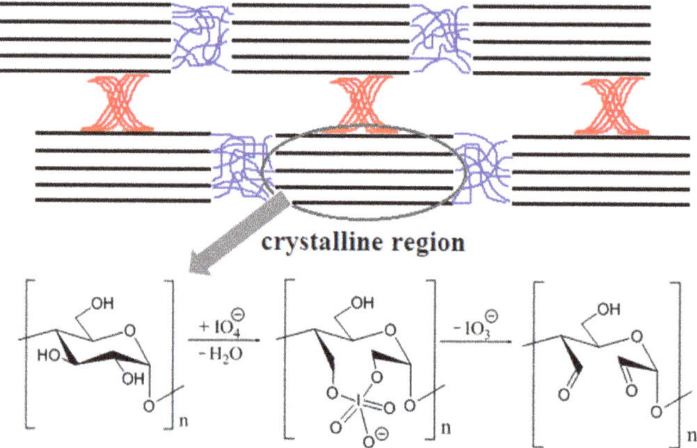

Figure 1. Scheme of the oxidation of nanocrystalline starch with sodium periodate.

Proteins play a crucial role in the functioning of living cells. Research into the adsorption of proteins on the surface of materials has gained widespread attention in various applications [35]. Therefore, it is necessary to study the interaction of multiple types of proteins on the surface of biofilms. Examples of such proteins are human serum albumin

(HSA) and acid glycoprotein (AGP). HSA is the most abundant globular protein in blood plasma, with a physiological concentration of 40–50 g/L. This blood serum protein interacts with many drugs, making it widely used in the clinical treatment of diseases [36]. Acid glycoprotein is a member of the acute phase protein family with a total molecular weight of 41–44 kDa [37]. It has a single polypeptide chain of 183 amino acid residues and is synthesized by the liver and organs such as the heart and lungs [38]. The primary physiological function of AGP is to transport compounds such as basic drugs, heparin, and certain steroid hormones [39].

To the best of our knowledge, there is only published work concerning the application of dialdehyde cellulose nanocrystals as cross-linking agents [40,41], but not about dialdehyde starch nanocrystals. Chen et al. [42] described the synthesis method of oxidized starch nanocrystals. They reported that the oxidation reaction time was two hours in the dark, and the suspensions were washed by centrifugation with distilled water to pH = 7. A subsequent paper by Chen et al. [43] proposed combining nanocrystalline dialdehyde starch with graphene oxide nanosheets to form highly porous aerogels, which can be used as supercapacitor electrodes and efficient adsorbents. However, no research group used nanocrystalline dialdehyde starch as a cross-linking agent. Oxidized starch nanocrystals, like starch nanocrystals, have a large specific surface area [44] of the material, which could affect the efficiency of the cross-linking process. In addition, starch nanocrystals are characterized by good mechanical properties. Hence, they are used as reinforcement agents [45]. Therefore, we consider that dialdehyde starch nanocrystals can also improve the mechanical properties of materials. In addition, the nanocrystalline form of oxidized starch might be a more favorable cross-linking agent for the properties of the obtained material.

The objective of the present study was to prepare dialdehyde starch nanocrystals for cross-linking of biopolymer films. Commonly available methods were applied to perform the detailed characterization of all received materials. In addition, the mechanical properties of the obtained materials and their hydrophilic nature using contact angle measurement were determined. The acute toxicity of the prepared biofilms was preliminarily studied using the Microtox test. The biological properties of designed, cross-linked biopolymer films, including adsorption of protein (human serum albumin (HSA) and acid glycoprotein (AGP)), were tested.

2. Results and Discussion

2.1. Synthesis of Starch Nanocrystals

The acid hydrolysis method of native corn starch was used to synthesize starch nanocrystals. This simple method removes the amorphous region of polysaccharides using sulfuric acid. The differences in the size of the granules, their porosity, and the amylose content affect the hydrolysis efficiency. In addition, it was reported that the pores on the starch surface increase the availability of acid to the inside of the granule, which promotes the hydrolysis process [46]. The reaction yield of the obtained nanocrystalline starch was 26%. The NCS yield from maize starch was the same as that reported for waxy maize [47] and similar to that reported for mango kernel starch [48].

2.2. Characterization of Starch Nanocrystals

The size distributions and particle concentrations of NCSs are shown in Figure 2a. The mean size of the starch nanocrystals was 58 nm and was consistent with previous results described by other research groups [49,50].

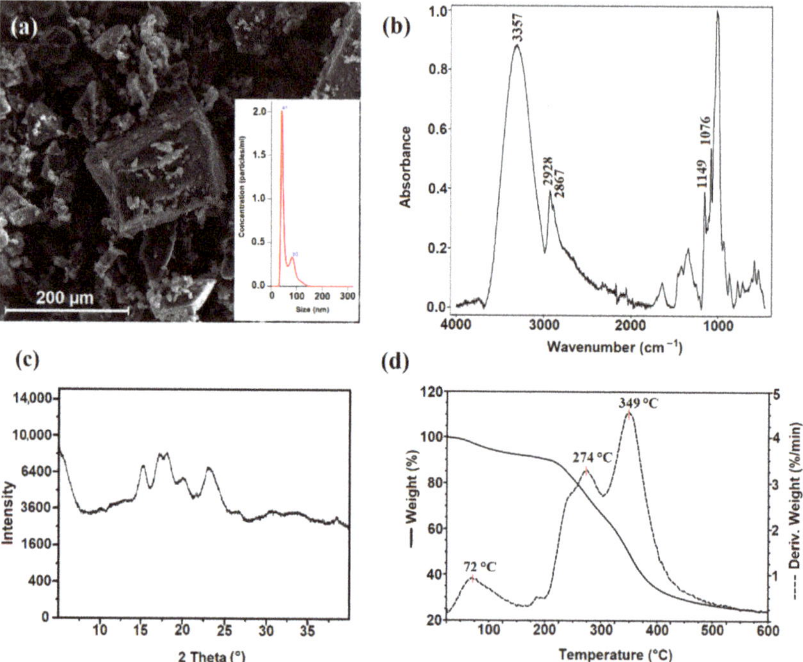

Figure 2. (**a**) SEM image, with inset showing the size distributions and particle concentrations of starch nanocrystals; (**b**) ATR-FTIR spectra; (**c**) X-ray diffraction (XRD) patterns; and (**d**) TGA-DTG curves of starch nanocrystals.

The structure of NCSs was confirmed by ATR-FTIR analysis (Figure 2b). Starch nanocrystals displayed a characteristic FTIR spectrum like that of the native starch. In the spectrum of NCSs, the absorption bands at 3357 cm^{-1} attributed to O–H stretching vibrations, 2867 cm^{-1} to asymmetric CH stretching, 1149 cm^{-1} to the stretching vibration absorption peak of C–O, and 1076 cm^{-1} to C–O stretching were observed. The received spectrum is in agreement with the literature data [51]. The band's intensity at 2928 cm^{-1}, corresponding to the C-H stretching vibration, was found to change with the ratio of amylose to amylopectin [52]. A reduction in the amylose ratio results increased the intensity of this band (2928 cm^{-1}) in the ATR-FTIR spectrum. In the spectrum obtained for starch nanocrystals, an increase in the intensity of this characteristic band was observed, which confirms selective hydrolysis of the amorphous regions, mainly containing amylose.

The X-ray diffraction signal of NCSs (Figure 2c) was observed at 2θ around 15°, a doublet one at 17–18°, and another near 23°, corresponding to a typical A-type starch pattern [53]. Similar observations were previously reported for starch nanocrystals derived from cassava starch [54].

The SEM images of NCSs are shown in Figure 2a. After acid treatment, the corn starch granules were fragmented to nanocrystals, with small, irregular, and square-like structures with larger aggregates. The morphology of starch nanocrystals is in good agreement with the crystalline type, as previously reported [55]. Nanocrystals produced from A-type starches rendered square-like particles.

A TGA-DTG analysis was performed to determine the thermal properties of starch nanocrystals. The TGA curve and its derivative for starch nanocrystals are shown in Figure 2d. The thermogravimetric curve of NCSs shows three degradation stages. In the range between 30 °C and 140 °C, a 7% mass loss was observed due to evaporation of the adsorbed water. The intensive decomposition takes place between 310 °C and 450 °C, with a 41% mass loss, corresponding to the thermal degradation of the starch polymer chain.

The additional decomposition step of nanocrystalline starch in the temperature range of 120–360 °C, with 27% weight loss, due to the partial degradation of the NCS structure was observed.

2.3. Synthesis of Dialdehyde Starch Nanocrystals

The oxidized starch nanocrystals were synthesized by the oxidation reaction of the previously obtained starch nanocrystals. The common oxidation reaction of starch to dialdehyde starch is described. Still, only two reports are available in the literature on the synthesis and potential utility of oxidized starch nanocrystals. Sodium periodate ($NaIO_4$) was used as an oxidation agent in a 1:1 weight ratio with starch nanocrystals to give dialdehyde starch nanocrystals (NDASs) as high as 65% degree of oxidation. The reaction was carried out at 40 °C for 3 h. Oxidized starch nanocrystals were precipitated from the reaction mixture with acetone, which greatly facilitated its separation. The reactions with different sodium periodate:nanocrystal starch ratios (0.5:1, 0.7:1) were also tested, resulting in 30% and 45% oxidation degrees.

2.4. Characterization of Dialdehyde Starch Nanocrystals

The size distributions and particle concentrations of NDASs are shown in Figure 3a. The mean size of the oxidized starch nanocrystals was 212 nm. The oxidation process increased the size of the dialdehyde starch nanocrystals. Opposite results of particle size changes were reported by Chen et al. [42].

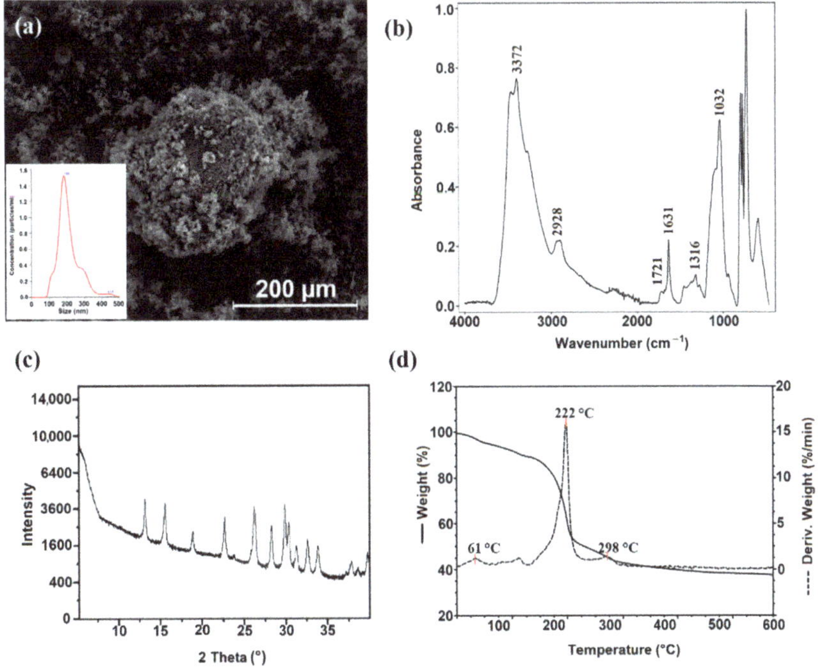

Figure 3. (**a**) SEM image, with inset showing the size distributions and particle concentrations of dialdehyde starch nanocrystals; (**b**) ATR-FTIR spectra; (**c**) the X-ray diffraction (XRD) patterns; and (**d**) the TGA-DTG curves of dialdehyde starch nanocrystals.

The structure of NDASs was confirmed by ATR analysis. After oxidation, the peak appearing at 1721 cm^{-1} was attributed to the characteristic absorption of carbonyl groups (Figure 3b). Its intensity was very weak, possibly because internal cross-links created some hemiacetal linkage during the oxidation process. The new sharp band at 1631 cm^{-1} was observed. It may be derived from the carbonyl group after an effective oxidation process of starch nanocrystals. Additionally, the band at 3372 cm^{-1} representing hydroxyl stretching vibrations was of lower intensity than the NCS spectra. This can be explained by the opening of cyclic structure and oxidation of starch saccharide units. Significant changes were also noticed in the vibration range of 1300–1400 cm^{-1}, which may be caused by the opening and oxidation in the glucoside ring between the second and third carbon atoms.

The X-ray diffraction patterns of NDASs showed many intensive signals (2θ = 13.2°, 15.6°, 22.7°, 26.3°, 30.0°, 31.4°, 32.8°, and 34.1°) proving the presence of the crystalline nature of the obtained product. After careful analysis of the literature data, it was shown that the above diffraction signals could be attributed to the signals derived from NaIO$_3$. It should be pointed out that obtained product was repeatedly washed with deionized water, and despite this, NaIO$_3$ signals in the diffractogram of NDASs were observed. This may indicate the formation of a strong complex between the dialdehyde product and the reduced form of the oxidant. Previous work noted similar mechanisms for dialdehyde chitosan and dialdehyde starch [56,57].

After oxidation, the regular square-like particles of starch nanocrystals disappeared and were transformed into rough shapes with many needle forms (Figure 3a). The needle form length was less than 10 μm, and coarse shapes had an average size of 5–20 μm.

To determine the thermal properties of oxidized starch nanocrystals, a TGA-DTG analysis was performed under the same conditions as in the case of NCSs. The TG curve of NDASs shows three degradation stages (Figure 3d). The first degradation step is observed between 20 °C and 80 °C and shows a 5% loss of weight. This weight loss may be related to the loss of bound water. The intensive destruction takes place in the second stage in the range between 156 and 410 °C and is associated with the 49% mass loss corresponding to the thermal degradation of the dialdehyde starch nanocrystal chain. At the maximum decomposition rate, T_{max} = 222 °C in oxidized starch nanocrystals is lower than that for starch nanocrystals (T_{max} = 365 °C) due to the fact that nanocrystalline oxidized starch has open pyranose rings after oxidation reaction on two and three of the carbon atoms in this structure. The same effect was observed with dialdehyde starch.

2.5. Formation of Chitosan, Gelatin, and Chitosan–Gelatin Films by Cross-Linking with Dialdehyde Starch Nanocrystals

The primary purpose of this research work was to obtain innovative polysaccharide-based material biofilms for biomedical applications. To investigate the effect of oxidized starch nanocrystals as a cross-linking agent, biofilms of chitosan, gelatin, and chitosan–gelatin (1:1) were prepared. The amount of cross-linker was 5 wt%, 10 wt%, and 15 wt% in relation to chitosan, gelatin, and chitosan–gelatin. Biopolymer films were smooth and transparent (Figure S1).

In order to determine the influence of oxidized starch nanocrystals on the properties of biofilms, the structure, morphology, thermal stability, and swelling ability were characterized for obtained materials. The properties of the obtained materials determine their subsequent applications. Materials for biomedical applications should be non-toxic and biocompatible because of their interaction with biological membranes. They should also be characterized by hydrophilicity, adequate strength, and porosity, which will allow the free growth of cells on the material surface.

2.5.1. Determination of Cross-Linking Degree of the Films

The cross-linking of macromolecules causes significant changes in the polymeric material properties. Thus, the degree of cross-linking is an essential and important feature for polymer networks. Cross-linking improves the thermal stability and resistance to cracking effects by liquids. In determining the degree of cross-linking, the non-cross-linked film was dissolved completely in an acetic acid solution (5%), and the degree of cross-linking of studied films is presented in Figure 4a. As expected, the degree of cross-linking of the films increases with the increase in the amount of cross-linking agent. The chemical cross-linking between components of films and dialdehyde starch nanocrystals improved the cross-linking density of polymer films. The highest degree of cross-linking was achieved for chitosan–gelatin films cross-linked with 15% NDAS addition, which may be caused by intermolecular cross-linking between both biopolymers.

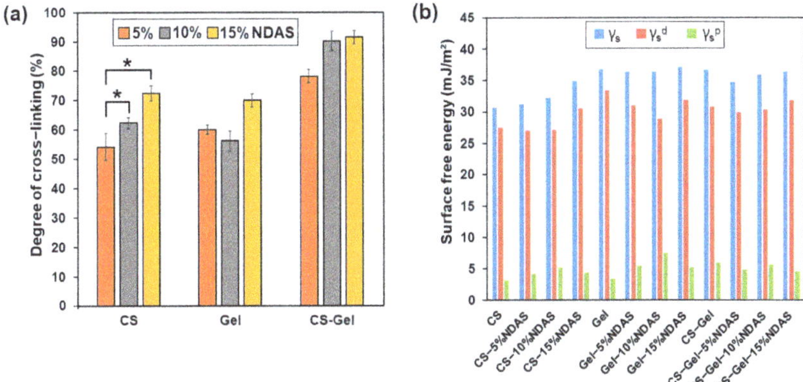

Figure 4. (**a**) Degree of cross-linking of chitosan (CS), gelatin (Gel), and chitosan–gelatin (CS-Gel, 1:1) cross-linked by 5%, 10%, and 15% cross-linker; statistical significance is indicated with asterisks: * $p < 0.05$. (**b**) The surface free energy (γ_s) and its polar ($\gamma_s{}^P$), and dispersive ($\gamma_s{}^d$) components for biofilms based on chitosan (CS), gelatin (Gel), and chitosan–gelatin (CS-Gel) cross-linked with 5%, 10%, and 15% NDASs.

2.5.2. Contact Angle Measurement

Interactions between materials and biological fluids are defined by surface free energy. Measurements of the contact angle of polar (glycerin) and non-polar (diiodomethane) liquids can assess the resulting materials' surface nature and potential applications. Surface free energy is one of the parameters determining the applications of the obtained materials. The dispersive and polar components provide detailed information on the tested surface. The contact angle of glycerin and diiodomethane and the surface free energy and its polar and dispersive components are presented in Table S1 and Figure 4b.

Results prove that all obtained biofilms have average glycerin contact angles lower than 90°, indicating a hydrophilic surface. The value of the surface free energy of pristine chitosan is 30.70 mJ/m^2, which is well confirmed by literature data [58]. Moreover, the chitosan films cross-linked with NDASs were characterized by a lower wettability by glycerol than pure chitosan. Chitosan and gelatin films cross-linked by NDASs were characterized by slightly higher surface free energy values than pure chitosan and pure gelatin. The same situation may be observed for the values of the polar component for neat chitosan and gelatin films after the cross-linking process. Gelatin and chitosan films cross-linked by NDASs were characterized by higher values of $\gamma_s{}^P$ than the neat gelatin and chitosan film. The highest polarity among the selected materials was Gel-10%NDAS due to it having the highest polar component of surface free energy. For chitosan–gelatin materials cross-linked with NDASs, a decrease in surface free energy and its polar component was observed.

Gierszewska et al. [59] investigated the effect of adding glutaraldehyde (0.5 wt% content) to chitosan and chitosan/montmorillonite to determine changes in hydrophilicity. They noticed that the application of the cross-linking agent caused a decrease in the surface free energy, from 33.5 mJ/m^2 (pure chitosan) to 29.5 mJ/m^2 in chitosan with glutaraldehyde. Similar results were obtained for chitosan/montmorillonite cross-linked with dialdehyde starch. The addition of a cross-linker to these samples caused a decrease in surface free energy. In work on chitosan and mixtures of collagen materials, hyaluronic acid, and chitosan cross-linked with oxidized starch, the surface free energy values after the modifications were higher [60].

2.5.3. ATR-FTIR Spectroscopy and X-ray Diffraction (XRD)

The strong and broad bands at 3356 cm^{-1} and 3287 cm^{-1} were attributed to stretching vibrations of O–H and N–H, respectively. A weak band at 2867 cm^{-1} is assigned to stretching vibrations of C–H. The characteristic absorption bands at 1649 cm^{-1} (C=O stretching of amide I) and 1551 cm^{-1} (N–H bending of amide II) were observed in the unmodified chitosan spectrum (Figure 5a). The bands in the region of 1151 cm^{-1} to 1027 cm^{-1} were the characteristic bands of C–O–C asymmetrical stretching and confirmed by literature data [61].

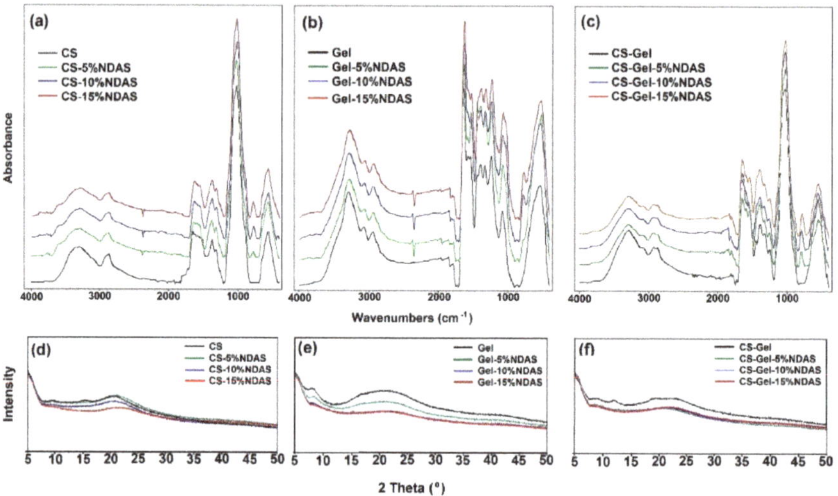

Figure 5. ATR-FTIR spectra of (**a**) chitosan, (**b**) gelatin, and (**c**) chitosan–gelatin films cross-linked by dialdehyde starch nanocrystals and the X-ray diffraction (XRD) patterns of (**d**) chitosan, (**e**) gelatin, and (**f**) chitosan–gelatin films cross-linked with dialdehyde starch nanocrystals.

The changes in the intensity and shape of the hydroxyl band (3356 cm^{-1}) and amine band (3287 cm^{-1}) can be explained by the reaction of an amino group of chitosan with the carbonyl group from a cross-linker. Hoffmann et al. obtained similar results for chitosan films cross-linked with dialdehyde dextran and glutaraldehyde [62]. Furthermore, in chitosan films cross-linked with NDASs, a new band near 778 cm^{-1} is observed, attributed to the =C–H bond (deformation stretching vibrations). It's due to the fact that some aldehyde groups of cross-linker are not linked to chitosan.

In the spectrum of gelatin, the characteristic absorption bands at 3292 cm^{-1} (assigned to the O–H stretching, N–H stretching vibrations, and intramolecular hydrogen bonds), 3076 cm^{-1} (amide B), 2943 cm^{-1} (CH stretching), 1636 cm^{-1} (C=O stretching vibration of amide I), 1537 cm^{-1} (amide II, attributed to a combination of CN stretch and in-plane NH deformation), 1242 cm^{-1} (amide III, corresponds to CN stretching and NH in-plane bending), and 1077 cm^{-1} (C–O stretching) were observed (Figure 5b). The received spectrum is consistent with the literature data [63].

The amino group from gelatin and the aldehyde group of NDASs form a Schiff base, and a new peak can appear in the region 1650–1600 cm^{-1} [64]. In addition, in this bio-film, this cross-linking fingerprint region (1650–1600 cm^{-1}) is covered by the strong amide I peak of gelatin [65]. In addition, the band at 1077 cm^{-1} (C–O stretching vibration) becomes very broad with the addition of NDASs.

The ATR-FTIR spectrum of the CS-Gel composite showed similar characteristic peaks of the pure chitosan and gelatin with some shifts. The peaks related to bending vibrations of the amine groups in pure chitosan were shifted from 1649 cm^{-1} and 1551 cm^{-1} to 1642 cm^{-1} and 1573 cm^{-1}, respectively, in the CS-Gel composite. The shift of amino bands in the CS-Gel composite spectrum showed a complex involving intermolecular hydrogen bonding between chitosan and gelatin [66].

A new band formation at 779 cm^{-1} for CS-Gel cross-linked with NDASs was also observed, as for cross-linked chitosan samples. In addition, no significant changes in the spectrum of the cross-linked CS-Gel film were noticed.

The crystalline structure of CS, Gel, and the CS-Gel films cross-linked by NDASs was characterized by XRD patterns and is shown in Figure 5d–f. The X-ray diffractograms of CS show crystalline diffraction peaks at 2θ = 10.5°, 15.1°, and 20.9°. According to the literature [67], these crystalline peaks of pure CS can be ascribed to the reflection planes of (020), (110), and (200). All the chitosan films cross-linked with NDASs showed diffraction peaks at 2θ = 20.9°, while the intensity of this peak decreased with the increase in the cross-linker amount. The peaks at 2θ = 10.5° and 15.1° for CS-15%NDAS disappeared. For gelatin, the signal located at about 2θ = 8.6° corresponds to the inter-helix distance of the triple helix of gelatin, and a broad diffraction peak in the 2θ range of 15.0–25.0 is typical of the amorphous fraction of gelatin [68]. The obtained pattern is consistent with the literature data [69]. For the gelatin samples, an increase in the amount of cross-linking agents caused a decrease in the intensity of signals. However, the chitosan/gelatin mixture did not show any crystalline peaks, indicating the amorphous structures of chitosan and gelatin. This result also exhibited that the chitosan and gelatin were mixed well, forming the composite [70]. The intensity of the diffraction patterns for CS-Gel cross-linked by NDASs decreases with an increasing amount of cross-linking agent. The same effect was observed for cross-linked gelatin and chitosan films.

2.5.4. Thermal Analysis

The thermogravimetric decomposition process was used to define the thermal properties of the obtained biofilms. The thermal parameters of the obtained biofilms are listed in Table 1, and the TG and DTG curves are presented in Figure S2.

The thermal stability of the films was analyzed using thermogravimetric analysis (TGA) and the derived thermogravimetric analysis (DTG) to observe various stages of degradation. Chitosan film exhibits two degradation stages well explained in the literature [71]. At the first stage of chitosan degradation, approximately 8% of the initial weight loss occurred between 29 and 140 °C due to the loss of moisture content. The main decomposition stage appeared in the temperature range of 180–420 °C. The 51% weight loss in this stage corresponds to main chain scission, side group abstraction, and ring-opening reactions. Gelatin film shows two degradation stages [72]. Neat gelatin shows a main decomposition in the temperature range of 200–450 °C, with 62% weight loss. The 9% initial weight loss between 20 and 120 °C is attributed to the removal of moisture and other volatile impurities. Non-cross-linked chitosan–gelatin film degraded in two steps [73]. The 15% weight loss in the first stage (in the temperature range from 20 to 140 °C) was attributed to water loss. The second stage, with 57% weight loss from 150 to 525 °C, was due to the actual decomposition of the macromolecules.

Three degradation steps were observed in thermogravimetric curves of chitosan, gelatin, and chitosan–gelatin films cross-linked with NDASs. For all cross-linked films, the first decomposition stage ranges between 30 °C and 120 °C. It shows about 8–13% loss in weight due to the evaporation of bound and adsorbed water, similar to unmodified polysaccharides and their mixtures. The main step of thermal degradation of all cross-linked materials is the intensive destruction, which occurs in the temperature range between 160 °C and 350 °C and shows about 55% loss in weight. Furthermore, there is an additional stage of decomposition in the temperature range between 120 °C and 190 °C. The weight loss at this stage is slight (2–4%). It may indicate that the cross-linking bond of biofilms was broken in this step. The content of residue at 600 °C correlates with the range of 23–40%.

Table 1. Thermal parameters of neat chitosan (CS); gelatin (Gel); chitosan–gelatin (CS-Gel, 1:1); and these films cross-linked by 5%, 10%, and 15% addition of NDASs (TGA-DTG analysis in a nitrogen atmosphere).

Sample	First Stage		Second Stage			Third Stage			Residue
	T_{max} (°C)	Δm (%)	T_o (°C)	T_{max} (°C)	Δm (%)	T_o (°C)	T_{max} (°C)	Δm (%)	600 °C (%)
CS	30, 60	8	-	-	-	145	286	51	40
CS-5%NDAS	22, 53	13	134	145	3	162	274	49	34
CS-10%NDAS	31, 52	13	121	142	4	165	277	50	33
CS-15%NDAS	32, 55	12	125	141	4	161	258	47	37
Gel	34, 56	9	-	-	-	168	307	62	27
Gel-5%NDAS	38, 61	11	141	163	3	175	312	64	23
Gel-10%NDAS	35, 59	8	138	164	3	197	312	58	30
Gel-15%NDAS	37, 60	9	124	147	4	185	306	55	32
CS-Gel	26, 70	15	-	-	-	152	290	57	28
CS-Gel-5%NDAS	38, 57	13	135	150	2	168	284	54	31
CS-Gel-10%NDAS	32, 57	12	134	148	3	168	285	51	34
CS-Gel-15%NDAS	33, 57	11	131	144	4	165	270	51	34

Other research groups also investigated the effect of cross-linking agents on the thermal properties of chitosan, gelatin, and their mixtures. Sutirman et al. studied poly (methacrylamide) grafted chitosan beads cross-linked with glutaraldehyde [74], and they also observed three decomposition stages in chitosan cross-linked beads. They observed that the thermal stability of chitosan had been improved through cross-linking and grafting copolymerization. In the work of Kaczmarek-Szczepańska et al. [75], glyoxal (10 wt%) was a cross-linking agent for chitosan hyrogels loaded with tannic acid. After modification, the three steps of degradation of the hydrogel were also noticed. Zhao et al. studied gelatin hydrogels cross-linked with two cross-linking agents (genipin and polyphenol) [76]. They noticed two decomposition stages of samples and observed that the interactions between gelatin and cross-linkers improved the thermal stability. Similar observations have been made by Nieto-Suárez et al. [77] for thermograms of chitosan–gelatin scaffolds cross-linked with glutaraldehyde, and these authors also indicated that chitosan–gelatin materials decompose in three stages. Moreover, they observed that the T_{max} shifts to higher temperatures upon the increase in gelatin concentration, thus indicating higher stability of scaffolds with gelatin.

2.5.5. SEM and AFM

The surface morphology of films was determined by scanning electron microscopy (SEM) for all samples (Figure 6). All the neat films were transparent, smooth-surfaced, and homogeneous. When NDASs were added, the film surfaces became a little rougher in a few samples. However, in most biofilms, the cross-linking process creates a smooth surface, which proves the homogeneity of the samples. The surface structures of the different films depend mainly on the degree of cross-linking and interactions between polymers [78].

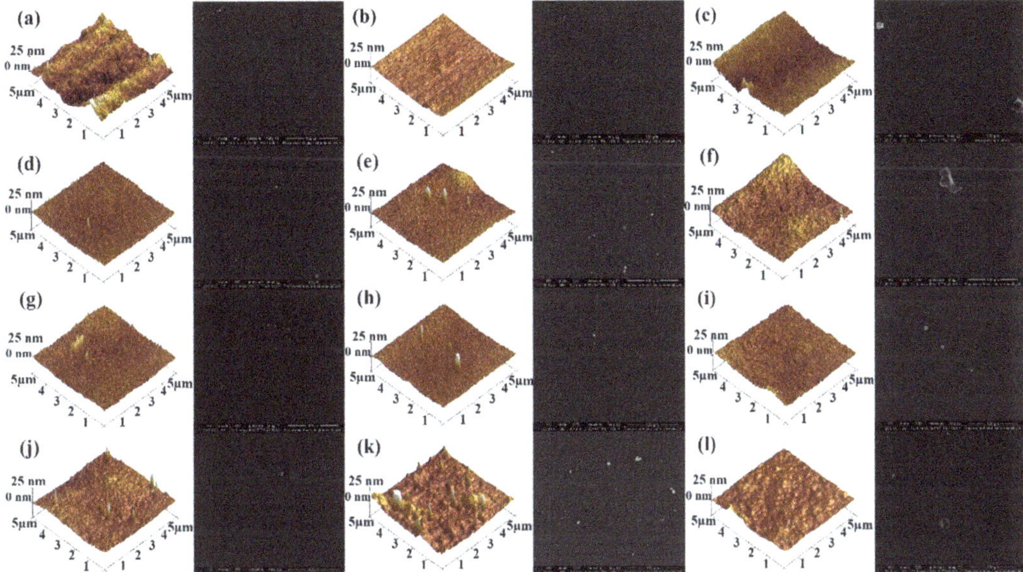

Figure 6. AFM and SEM (1000×) images of (**a**) chitosan films cross-linked by (**d**) 5% NDASs, (**g**) 10% NDASs, and (**j**) 15% NDASs; (**b**) gelatin film cross-linked by (**e**) 5% NDASs, (**h**) 10% NDASs, and (**k**) 15% NDASs; and (**c**) chitosan–gelatin films cross-linked by (**f**) 5% NDASs, (**i**) 10% NDASs, and (**l**) 15% NDASs.

The surface roughness is an essential parameter in subsequent material applications. The surface roughness of obtained biofilms was examined using a topographical atomic force microscope. The 3D and 2D atomic force microscopy (AFM) images of the obtained biofilms are shown in Figure 6 and Figure S3, respectively. The pure gelatin film was smooth and homogenous. After increasing NDAS content, as shown in Table 2, rougher surfaces were observed in the originally smooth gelatin matrix.

Table 2. The roughness parameters of chitosan, gelatin, and chitosan–gelatin films cross-linked with NDASs.

Roughness Parameters (nm)	Sample			
	CS	**CS-5%NDAS**	**CS-10%NDAS**	**CS-15%NDAS**
R_q	9.24	1.73	2.29	6.20
R_a	7.49	1.34	1.78	4.90
R_{max}	89.9	28.1	31.3	33.9
	Gel	**Gel-5%NDAS**	**Gel-10%NDAS**	**Gel-15%NDAS**
R_q	1.83	2.48	3.26	4.70
R_a	1.48	1.61	2.36	2.64
R_{max}	10.2	33.6	86.7	106.0
	CS-Gel	**CS-Gel-5%NDAS**	**CS-Gel-10%NDAS**	**CS-Gel-15%NDAS**
R_q	4.65	3.16	2.06	1.45
R_a	3.14	2.54	1.58	1.07
R_{max}	65.6	27.1	18.3	14.8

Furthermore, the surface of the gelatin film displayed a surface roughness, of R_q value of 1.83 nm, which increased to 4.70 nm as the NDAS addition reached 15%. A relatively smooth morphology was observed for pure chitosan, which is mainly related to the chitosan's physical properties, such as high intrinsic chain stiffness [79]. Biofilms based on chitosan modified with oxidized starch nanocrystals are characterized by a relatively rough surface. The opposite trend was observed for chitosan–gelatin films cross-linked with NDASs, and surface roughness decreases with the rising content of cross-linking agents for these materials.

Other research groups also investigated the effect of cross-linking agents on chitosan and its mixtures. Olewnik-Kruszkowska et al. studied chitosan films cross-linked with 1 wt%, 2 wt%, and 3 wt% squaric acid [80] and noted that roughness parameters increased with an increased amount of the cross-linker. They reported the highest surface roughness parameters' values after cross-linking. In the case of chitosan cross-linked by 3 wt% of squaric acid, Rq and Ra are 3.67 nm and 2.95 nm, respectively. Sionkowska et al. [60] investigated the effect of the addition of dialdehyde starch (5 wt% content) for chitosan and mixtures of collagen, hyaluronic acid, and chitosan on the roughness parameters. They observed that the application of the cross-linking agent promoted an increase in the R_q parameter, from 3.2 nm (pure chitosan) to 4.8 nm in the case of chitosan cross-linked with dialdehyde starch. Opposite results were obtained for collagen/hyaluronic acid/collagen cross-linked with oxidized starch. The addition of a cross-linker caused a decrease in the R_q parameter from 23.0 nm to 7.2 nm. In the work of Abbasi et al. [81] genipin (0.125 wt%, 0.25 wt%, 0.5 wt%, 0.75 wt%, 1 wt%, and 2 wt%) as a cross-linking agent was used for gelatin films. After modification, the observed roughness was lower for materials with genipin than in the case of pure gelatin film.

2.5.6. Mechanical Properties

The mechanical properties are crucial for the applications of biofilms subjected to a certain level of mechanical forces. To investigate the effect of dialdehyde starch nanocrystals on the mechanical properties of biofilms, mechanical tests were performed, and the obtained tensile strength, Young's modulus, and elongation at break are summarized in Figure 7a–c, respectively.

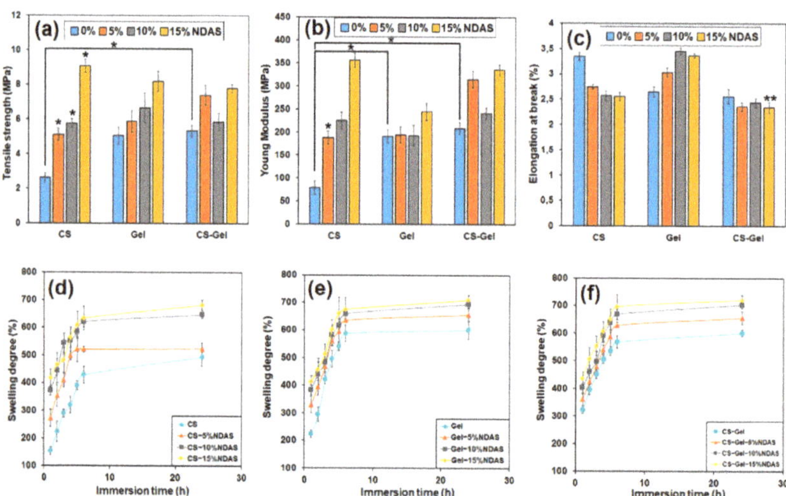

Figure 7. The value of (**a**) tensile strength, (**b**) Young's modulus, and (**c**) elongation at break of chitosan (CS), gelatin (Gel), and chitosan–gelatin (CS-Gel) cross-linked by 5%, 10%, and 15% cross-linking agent; statistical significance is indicated with asterisks: * $p < 0.05$, ** $p < 0.01$. The swelling degree of (**d**) chitosan, (**e**) gelatin, and (**f**) chitosan–gelatin cross-linked by 5%, 10%, and 15% cross-linker (NDASs).

The values of tensile strength for unmodified samples of chitosan, gelatin, and chitosan–gelatin (1:1) were about 3 MPa, 5 MPa, and 5.5 MPa, respectively (Figure 7a). For chitosan and gelatin films cross-linked by NDASs, the tensile strength increases with the increasing addition of cross-linking agents. The sample CS-15%NDAS is characterized by the most significant difference in tensile strength value. This parameter increased to 9.08 MPa, 3 times higher than the value of this parameter for the pure chitosan sample. Moreover, tensile strength increases with rising amounts of cross-linking agents for chitosan and gelatin films cross-linked with NDASs. An increase in strength for all samples with a 5% addition of cross-linking agent in comparison to neat polysaccharides and mixture was observed. Amino groups from chitosan and gelatin form a covalent bond with carbonyl groups from NDASs. The creation of a new cross-linking network in chitosan films is responsible for the increase in mechanical properties.

Young's modulus describes the stiffness of obtained materials; for pure gelatin, it was about 192 MPa. The value of this parameter for a neat gelatin sample is almost the same as that for gelatin films cross-linked with 5% and 10% NDASs (Figure 7b). In this series of measurements, the gelatin sample with a 15% addition of NDASs reached the highest value of Young's modulus—245 MPa. Moreover, for chitosan films cross-linked with NDASs, an increase in the Young's modulus value is observed with an increase in the amount of cross-linking agent. The chitosan samples cross-linked with a 15% addition of NDASs are characterized by the highest value of Young's modulus, which is 4.5 times higher than that of neat chitosan film. In the case of a mixture of chitosan and gelatin, cross-linked with NDASs, an increase the value of the Young modulus is observed with increasing amounts of cross-linking agent, except for the sample with 10% NDAS addition.

The elongation at break is the ratio of the length of the specimen after and before the break. The percentage of elongation at break indicates the stretchability of the samples [82]. The highest value of percentage of elongation at break for the neat samples (CS, Gel, and CS-Gel) was achieved for the chitosan sample (3.35%). This parameter for gelatin and chitosan–gelatin is practically the same and was about 2.65% and 2.56%, respectively (Figure 7c). In the case of chitosan materials cross-linked with NDASs, the percentage of elongation decreases with an increasing amount of cross-linking agent. The opposite trend was observed for cross-linked gelatin samples. Furthermore, for the mixture of chitosan and gelatin, the elongation at break is almost the same for all films—neat and cross-linked.

In our previous research, we obtained chitosan film cross-linked with dialdehyde starch (DAS) [56]. The tensile strength value for the chitosan sample cross-linked with 15% DAS was about 6 MPa, whereas the sample with the same amount of NDASs was above 9 MPa. The most significant difference in Young's modulus value was observed for the CS-15%NDAS sample. This parameter increased to about 350 MPa, above 1.5 times more than that for chitosan film cross-linked with the same amount of DAS. Based on the results, it can be concluded that oxidized starch nanocrystals resulted in a greater improvement in mechanical properties of chitosan compared to the chitosan films cross-linked with DAS.

2.5.7. Swelling Ability

The swelling of biofilms greatly depends not only on the properties of the used polymer but also on the medium of swelling. The polymer structure gives the biofilms the ability to absorb solvent, while the cross-links between the network chains provide the biofilms with the ability to resist uncontrolled biodegradation. The retracting and expanding forces must balance each other to reach swelling equilibrium [83]. The swelling ability of the obtained biofilms in PBS solution was determined and is shown in Figure 7d–f. According to the literature, materials based on biopolymers are easily wettable by polar solvents due to the presence of functional groups capable of interacting with water molecules [84]. This usually results in a high degree of swelling.

The swelling ratio of the pure chitosan sample after the first hour of immersion in the PBS solution was 153%. Moreover, the swelling ratio for chitosan–gelatin films cross-linked with oxidized starch nanocrystals is 3 times higher than that for chitosan film after the same period. For all films cross-linked by NDASs, the percentage swelling ability increases with increasing immersion time in PBS solution. The highest degree of swelling after 24 h is observed for chitosan–gelatin materials cross-linked with NDASs. This is due to the numerous functional groups in chitosan and gelatin macromolecules capable of absorbing polar solvents.

Skopinska-Wisniewska et al. [85] obtained similar results with gelatin materials cross-linked by dialdehyde starch. The swelling ratio increased in the initial stage and stabilized after 6 h of immersion in PBS for these samples. Kaczmarek et al. [86] investigated the swelling ability of 3D chitosan–gelatin hydrogels cross-linked with oxidized starch. Their results showed that the higher cross-linker content resulted in higher water absorption than unmodified scaffolds. For the sample 50Gel/50CTS + 5% ST (dialdehyde starch), the swelling ability is about 842%. This results from the sample structure reorganization after the cross-linker addition and the change of pore sizes. After immersion for 1 h for the sample CS/Gel + 5% NDASs, we obtained a swelling ability of about 360%, which may be related to the different structure of the material in the form of a thin film and the nanocrystalline structure of the cross-linking agent.

2.5.8. Toxicity Assessment

The Microtox test measures the toxicity exerted on Gram-negative *Aliivibrio fischeri* bacteria. The bioluminescence of the species is directly correlated to its metabolism and cell viability. Upon contact with a toxic substance, the bioluminescence decreases in a dose-dependent manner. Additionally, the lyophilized bacteria are supplied with their cell walls broken, which results in higher susceptibility to toxicants. The prepared films were subjected to this test to evaluate their toxicity, and the results are summarized in Figure 8.

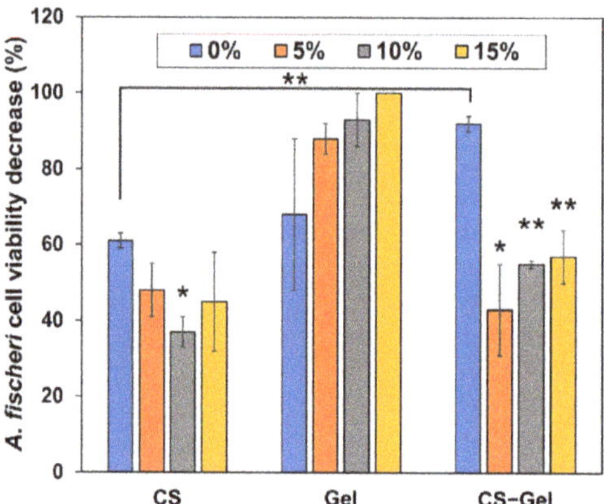

Figure 8. The decrease in *A. fischeri* cell viability upon 5-minute exposure to chitosan (CS), gelatin (Gel), and chitosan–gelatin (CS-Gel) films prepared by cross-linking with 0%, 5%, 10%, or 15% addition of dialdehyde starch nanocrystals (NDASs). Statistical significance is indicated with asterisks: * $p < 0.05$, ** $p < 0.01$.

The toxicity values obtained in this study suggest that non-cross-linked CS-Gel film is the most toxic (non-biocompatible) material among the non-cross-linked films. It can be seen that the toxicity exerted by non-cross-linked chitosan is the lowest as compared to non-cross-linked gelatin and chitosan–gelatin films. However, such a high cell viability decrease for CS (61%) has to be connected to its antimicrobial properties exerted on *A. fischeri* used in the test, as it has been reported as non-toxic [87,88]. In the case of CS and CS-Gel films cross-linked with NDASs, the toxicity of the films decreases as compared to neat biopolymers. However, in the case of gelatin films, the addition of oxidized starch nanocrystals as the cross-linking agent induces dose-dependent toxicity, which increases with an increase in NDASs, reaching almost no viable cells at 15% NDAS addition. This stays in contrast with the toxicity study by Karami Juyani et al., who observed no apparent toxicity of gelatin films towards bone marrow stromal cells while CS films exerted such effects [89]. In the case of the CS-Gel combination, the increasing addition of the cross-linking NDASs also increases the toxicity, but even at 15% NDASs, the cell viability decrease was significantly lower than that for non-cross-linked films.

Other research groups also investigated the toxicity of other types of obtained materials using the Microtox test. Szurkowska et al. conducted preliminary biological tests on the bacteria *Allivibrio fischeri* for hydroxyapatite material co-substituted with manganese(II) and silicate ions [90]. Tammaro et al. [91] examined the potential environmental hazards of photovoltaic panels. In the work of Isidori et al. [92], in situ monitoring of gaseous pollutants at 17 sampling points in two seasons (winter and summer) was performed. However, no research group has used Microtox to analyze the toxicity of biopolymers cross-linked with dialdehyde starch nanocrystals, which is a research novelty.

2.5.9. Protein Adsorption Study

The study of the interaction of the obtained biopolymer films with human serum proteins allows for assessing the suitability of these materials as potential dressings. It is believed that material may be suitable for dressing purposes if it exhibits a high ability to interact with blood serum proteins [93]. When the tested films directly contact the blood, the protein is adsorbed onto the film's surface, resulting in platelet adhesion and activation [93]. Cell adhesion requires the cell adhesion receptors to form cell-anchoring points, and protein adsorption is a crucial step during this process [94]. Human serum albumin (HSA) is the most abundant protein in the blood. It is a negative acute-phase protein. Another important serum protein is an acid glycoprotein (α-AGP), a positive acute-phase protein. The assessment of the material's ability to interact with these proteins allows predicting whether the obtained material is a potential biomaterial for the synthesis of dressings. Therefore, in the present study, as a model, HSA and α-AGP were used to evaluate the protein adsorption behavior of the biopolymer films after different contact times (1–24 h). The obtained results are presented in Table 3 (incubation after 1, 4, and 24 h). Data in the full range of incubation timescales are presented in Supplementary Materials in Table S2 and in the diagram in Figure 9. As can be seen, all the materials show the ability to interact with blood serum proteins. The amount of deposited protein ranges from 0.031 after 1 h to 0.189 mg/cm^2 after 24 h of exposure for HSA and from 0.065 after 1 h to 0.266 mg/cm^2 after 24 h of exposure to α-AGP. The material that shows the ability to bind the largest amount of both proteins is non-cross-linked gelatin (Gel). The amounts of adsorbed HSA and AGP are comparable and amount to 0.111 mg/cm^2 and 0.121 mg/cm^2, respectively, after 24 h of exposure. The properties of gelatin significantly change after it has been cross-linked. The addition of a cross-linking agent in the form of oxidized starch nanocrystals (NDASs) causes the material to preferentially interact with acid glycoprotein from the first hour of incubation. Moreover, increasing the addition of cross-linking agent and thus the degree of cross-linking of the gelatin increases the ability of the material to bind the glycoprotein. Gelatin cross-linked with 15% NDAS addition binds almost 4 times more glycoprotein than albumin.

Table 3. Amount of protein bound on the surface of biopolymers in mg of protein per 1 cm^2 of biopolymer film, after 1, 4, and 24 h of incubation.

Amount of Adsorbed Serum Protein (mg/cm^2)	Incubation Time (h)	Sample			
		CS	CS-5%NDAS	CS-10%NDAS	CS-15%NDAS
HSA	1	0.067	0.116	0.093	0.159
	4	0.067	0.131	0.106	0.167
	24	0.072	0.144	0.119	0.174
AGP	1	0.068	0.073	0.081	0.069
	4	0.070	0.084	0.086	0.081
	24	0.064	0.082	0.085	0.081
		Gel	Gel-5%NDAS	Gel-10%NDAS	Gel-15%NDAS
HSA	1	0.094	0.039	0.036	0.046
	4	0.112	0.042	0.037	0.065
	24	0.126	0.044	0.039	0.069
AGP	1	0.103	0.117	0.136	0.255
	4	0.111	0.129	0.144	0.256
	24	0.111	0.130	0.141	0.256
		CS-Gel	CS-Gel-5%NDAS	CS-Gel-10%NDAS	CS-Gel-15%NDAS
HSA	1	0.031	0.042	0.039	0.052
	4	0.033	0.047	0.040	0.053
	24	0.044	0.049	0.041	0.058
AGP	1	0.066	0.073	0.065	0.079
	4	0.073	0.074	0.067	0.080
	24	0.064	0.072	0.070	0.084

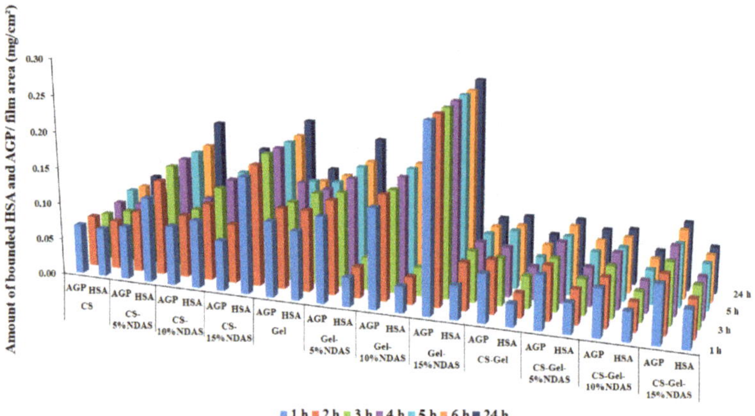

Figure 9. Amount of bound serum proteins AGP and HSA at the surface of obtained biopolymer films.

In the case of the non-cross-linked chitosan film, the preference for binding one of the proteins is no longer apparent. Both HSA and AGP proteins are bound in similar amounts, and the addition of NDASs as a cross-linking agent increases the amount of adsorbed HSA.

A similar situation occurs in the case of the obtained material from mixtures of both chitosan and gelatin biopolymers. For non-cross-linked films, the amounts of bound protein are very similar, with a slight predominance of the amount of glycoprotein. Taking into account that the pure chitosan film does not show such differences, it can be assumed that the greater amount of bound AGP compared to HSA is related to the addition of gelatin, which interacts preferentially with glycoprotein. The addition of NDASs as a cross-linking agent slightly increases the protein binding capacity of this material while reducing the difference between the amounts of adsorbed AGP and HSA.

The obtained results of the ability to bind HSA by the synthesized materials as potential materials for dressings do not differ from the results published in the literature. Singh and Dhiman used bovine serum albumin (BSA) to evaluate the protein adsorption on their polymer films: gum acacia-*cl*-(poly(HEMA-co-carbopol)) (GAHCP) and [poly(HEMA)-co-carbopol] (HCP), where HEMA is 2-hydroxyethyl methacrylate. Researchers observed that after 24 h, GAHCP hydrogel films and HCP films showed albumin adsorption of 0.19 ± 0.02 mg/cm^2 and 0.24 ± 0.02 mg/cm^2, respectively [95]. It should be emphasized that no data on the study of AGP binding capacity of biomaterials or data for similar systems tested with HSA were found in the literature.

3. Materials and Methods

3.1. Materials

Corn starch, chitosan (low molecular weight: MW = 50 kDa, deacetylation degree = 75–85%), sodium periodate, diiodomethane (pure for analysis), glycerol, human serum albumin, and acid glycoprotein were purchased from Sigma-Aldrich (St. Louis, MO, USA) and used without further purification. Acetic acid, sodium hydroxide, concentrated hydrochloric acid (35%), concentrated sulfuric acid (96%), acetone, and phosphate-buffered saline (PBS, pH = 7.4) were purchased from Avantor Performance Materials (Gliwice, Poland). Gelatin was purchased from CHEMPUR (Piekary Slaskie, Poland)—20 mesh pure. Diluent solution 2% NaCl and bacteria *A. fischeri* for toxicity assessment were supplied by the producer of Microtox (Modern Water, Cambridge, UK).

3.2. Starch Nanocrystal Synthesis

The starch nanocrystals were prepared by the hydrolysis process reported previously [96]. Corn starch was hydrolyzed under constant magnetic stirring for five days at 40 °C in a 3.16 M aqueous H$_2$SO$_4$ solution. After the hydrolysis, the nanocrystals were separated from the acid by centrifugation for 10 minutes at 8000 rpm at 4 °C. Subsequent washing and centrifugation with distilled water were performed until neutrality of the eluent and pH were controlled by litmus paper. A homogeneous dispersion of starch nanocrystals was obtained using an Ultra Turrax T25 homogenizer for 5 min at 13,500 rpm. The starch nanocrystals were dried at room temperature for 48 h.

3.3. Dialdehyde Starch Nanocrystal Synthesis

Previously obtained starch nanocrystals (1.5 g) were dissolved in deionized water (30 mL). Subsequently, sodium periodate (0.7 M) with three different weight ratios of oxidant to starch nanocrystals (0.5:1, 0.7:1, and 1:1) was added to the starch nanocrystal suspension. The flask was covered with aluminum foil to prevent light-induced decomposition of sodium periodate. The reaction mixture was stirred with a magnetic stirrer (IKA, Staufen, Germany) in an oil bath at 40 °C for 3 h. After the reaction was complete, the appropriate quantity of acetone was added, and white amorphous powder was formed immediately. Oxidized starch nanocrystals were isolated by filtration, washed with deionized water, and dried at room temperature for 24 h.

3.4. Biopolymer Preparation

Chitosan and gelatin were separately dissolved in acetic acid (C = 1% m/m) to obtain 1% solutions. Both biopolymers were mixed in a volume ratio of 1:1. Chitosan, gelatin, and their blend were mixed with the desired amount of oxidized starch nanocrystals (5%, 10%, and 15%). This amount is relative to dry protein/polysaccharide weight. The mixing of blends was carried out for 2 h with magnetic stirring. Next, all solutions were poured onto the leveled glass plates to allow the solvent to evaporate. Evaporation time was five days at room temperature for the volume 10 mL solutions of polysaccharide and mixture.

3.5. Characterization of Starch Nanocrystals, Dialdehyde Starch Nanocrystals, and Cross-Linked Biopolymers

3.5.1. Determination of the Content of Aldehyde Groups

The aldehyde group content of the samples was determined by acid–base titration [56]. Oxidized starch nanocrystals (0.10 g) and 5 mL 0.25 M sodium hydroxide were added to the Erlenmeyer flask. The mixture was heated in a water bath at 70 °C until the sample was dissolved, and then it was cooled with cold water. Standardized 0.25 M hydrochloric acid (7.5 mL) and distilled water (15 mL) were added. Then, phenolphthalein was added, and the solution was titrated by 0.25 M sodium hydroxide. The procedure was repeated three times. The percentage of dialdehyde units is given by Equation (1):

$$Aldehyde\ content,\ \% = \frac{C_1 V_1 - C_2 V_2}{W} \times 161 \times 100\% \tag{1}$$

where C_1 and C_2 (mol/L) represent the concentrations of NaOH and HCl solutions, respectively; V_1 and V_2 (dm^3) represent the volume of NaOH and HCl solutions, respectively; W is the weight of the sample (g); and 161 (g/mol) is the molecular weight of the repeating unit in dialdehyde starch nanocrystals.

3.5.2. Particle Size Distribution Analysis

The particle size distribution was analyzed using the Malvern Panalytical NanoSight LM10 instrument (sCMOS camera, 405 nm laser) using NTA 3.2 Dev Build 3.2.16 software. Before the measurements, the material dispersions were diluted with deionized water to achieve an operating range of nanoparticle concentration. The temperature of the sample chamber was set and maintained at 25.0 ± 0.1 °C; the syringe pump infusion rate was set to 200. For each sample, three 60-second movies were recorded.

3.5.3. Determination of the Cross-Linking Degree of the Films

The cross-linking degree of films was determined by the extraction method, where the insoluble matrix ("gel") reflected the cross-linked fraction of films, and the non-cross-linked fraction was fully dissolved in the solvent [97]. The acetic acid solution (5%) was chosen as a suitable solvent for chitosan–gelatin-based films. The precise initial weight of each sample (m_1) was determined before measurement. Then, all samples were put in an acetic acid solution and were extracted at 70 °C for 24 h in an oil bath. After this treatment, the insoluble residue was dried at 60 °C for 24 h and then weighed again (m_2). The degree of cross-linking was calculated by using Formula (2):

$$Degree\ of\ the\ cross-linking,\ \% = \frac{m_2}{m_1} \times 100\% \tag{2}$$

3.5.4. Contact Angle Measurement

The contact angle of polar glycerol and non-polar diiodomethane at constant temperature (27 °C) was measured on the biopolymers' surface and cross-linked biofilms. The measure was performed with a DSA10 goniometer (Kruss GmbH, Hamburg, Germany) equipped with a camera. The surface free energy (γ_s) and its polar (γ_s^P) and dispersive (γ_s^d) components were calculated by the Owens–Wendt method using the mean of at least five measurements of the contact angle of two measuring liquids [98].

3.5.5. Attenuated Total Reflectance Fourier Transform Infrared (ATR-FTIR) Spectroscopy and X-ray Diffraction (XRD)

Structures of obtained materials were evaluated by attenuated total reflectance Fourier transform infrared (ATR-FTIR) spectroscopy using a Spectrum Two spectrophotometer (Perkin Elmer, Waltham, MA, USA) equipped with an ATR device (diamond crystal). The spectra were collected in the region from 4000 cm^{-1} to 400 cm^{-1} at a resolution of 4 cm^{-1} and with 64 scans at room temperature.

X-ray diffraction (XRD) measurements were carried out in an X'PertPRO diffractometer (Malvern Panalytical, Almelo, Holland) with CuKα radiation (λ = 1.540 Å). All samples were recorded through the 2θ range of 5–120° with a step size of 0.008° at room temperature.

3.5.6. Thermal Analysis

Thermal gravimetric analysis of dialdehyde starch nanocrystals, starch nanocrystals, and cross-linked biofilms was accomplished on a TA Instruments (SDT 2960 Simultaneous DSC-TGA, New Castle, DE, USA) thermogravimetric analyzer. TGA measurements were performed at a 10 °C/min heating rate in the atmosphere of nitrogen in the range from ambient to 600 °C.

3.5.7. Scanning Electron Microscopy (SEM) and Atomic Force Microscopy (AFM)

Scanning electron microscopy (SEM) analyses of oxidized starch nanocrystals, starch nanocrystals, and cross-linked biofilms were carried out using a 1430 VP LEO (Electron Microscopy Ltd., Cambridge, UK) at an acceleration voltage of 20 kV. The samples were covered with a thin layer of gold before observation, and images were captured at different magnifications.

Topographic images of the obtained biofilms were recorded by the atomic force microscopy (MultiMode Nanoscope IIIa Veeco Metrology Inc., Santa Barbara, CA, USA) technique. Surface images were acquired using a scan width of 1 μm with a scan rate of 1.97 Hz. The samples were scanned with 5 μm × 5 μm areas and analyzed using NanoScope Analysis software. Roughness parameters, namely arithmetic mean, R_a; root mean square, R_q; and the highest peak value, R_{max}, were determined.

3.5.8. Mechanical Properties

In order to perform mechanical testing, the samples were cut with initial dimensions of 50 mm in length and 4.5 mm in width. The samples were inserted between the machine limbs and stretched to break using an EZ-Test E2-LX Shimadzu texture analyzer (Shimadzu, Kyoto, Japan). The measurements were carried out at the speed of 20 mm/min using a 50 N load head. The tensile strength, Young's modulus, and elongation at break for average values of 5 measurements were obtained.

3.5.9. Swelling Ability

The conventional gravimetric methods were used for the determination of the swelling ability [99]. The weight samples were immersed in phosphate-buffered saline (PBS) solution (pH = 7.4) and removed at regular time intervals. The excess buffer on the surface was wiped off, and the samples were weighed until equilibrium was reached. The swelling ability was determined as follows (3):

$$Swelling\ ability,\ \% = \frac{m_s - m_d}{m_d} \times 100\% \tag{3}$$

where m_s and m_d denote the weights of swollen and dry samples, respectively.

3.5.10. Toxicity Assessment

The materials prepared in the study were then assessed in terms of the acute toxicity in the Microtox test. Slight modifications were applied to the standard 81.9% screening test procedure for the testing of films [56]. Briefly, after registering the bioluminescence of the *A. fischeri* bacteria directly before the addition of the sample (time t = 0), 900 μL of 2% sodium chloride solution (Microtox Diluent; Modern Water, Cambridge, UK) precooled to 15 °C was added, and into such a resulting bacterial suspension, a film fragment was submerged. Next, the bioluminescence emitted by the bacteria at 490 nm was registered using Microtox M500 5 minutes and 15 minutes after the start of the exposure to the tested films [100]. The supplied software, Modern Water MicrotoxOmni 4.2 (Modern Water,

Cambridge, UK), was used to calculate the percentage effect (toxicity) exerted by the materials. All the materials were tested in triplicate.

3.5.11. Protein Adsorption Study

Fluorescence emission spectra were obtained on a Jasco FP-8300 spectrofluorometer (Jasco, Tokyo, Japan). Fluorescence spectra were recorded at 298 K, ranging from 290 to 400 nm at excitation wavelength 280 nm for human serum albumin (HSA) (Figure S4) and from 300 nm to 400 nm at excitation wavelength 289 nm for α1-acid glycoprotein (AGP) (Figure S5). Spectrum registration range was 285–400 nm for HSA and 300–400 nm for AGP, scanning speed was 100 nm/min, and Em/Ex bandwidth was 2.5 nm/5 nm. A stock solution of human serum albumin and α1-acid glycoprotein in phosphate buffer (PBS, pH = 7.4, 50 mM) at a concentration of 10 μM was used to prepare a protein standard curve. From the stock protein solution, solutions in PBS, pH = 7.4, 50 mM were prepared with the following concentrations: 1 μM, 2 μM, 3 μM, 4 μM, 5 μM, 6 μM, 7 μM, 8 μM, and 9 μM. Spectra were recorded for each protein solution by excitation at 280 nm for HSA and 289 nm for AGP, and a standard curve was plotted (R^2 = 0.99).

An albumin solution and α1-acid glycoprotein were prepared in phosphate buffer pH = 7.4 (50 mM) at concentrations of 6.62 μM and 9.98 μM, respectively. Further, cut biofilms with a size of 2 $\times$ 2 cm were placed in Eppendorf tubes (Eppendorf, Hamburg, Germany). Two milliliters of the protein suspension were then added to each sample of the biofilms and incubated at 36 °C at 600 rpm. For the supernatant from each tube, fluorescence spectra were recorded by excitation at 280 nm at intervals of 1 h, 2 h, 3 h, 4 h, 5 h, 6 h, and 24 h from the start of incubation. The measurement was repeated three times.

4. Conclusions

In summary, novel biopolymer films based on fully natural and biocompatible materials, gelatin, chitosan, gelatin–chitosan, and NDASs, were prepared by solvent evaporation. The cross-linking agent was obtained by periodate oxidization of starch nanocrystals and characterized by ATR-FTIR, SEM images, NTA, and XRD analysis. Schiff's base bond between NDASs and biopolymers was created. The highest degree of cross-linking was observed for chitosan–gelatin with a 15% addition of NDASs. All the prepared cross-linked films are homogeneous. The values of surface free energy polar component and contact angle for prepared materials allow stating that materials cross-linked with dialdehyde starch nanocrystals are promising for biomedical applications due to the nature of the surface. Creating a new cross-linking network between polysaccharides and oxidized starch nanocrystals leads to better mechanical properties. The swelling ability of all cross-linked biopolymer materials increased with an increase in the cross-linking agent content in the polysaccharide composition. The addition of chitosan into gelatin films caused the lowest toxicity effect on the surface compared to pristine gelatin films. All obtained materials can interact with blood serum proteins: HSA and AGP. Cross-linking of materials with NDASs increases the ability of these interactions, which is beneficial for biomedical applications of materials. Moreover, it has been noticed that NDAS cross-linked gelatin films preferentially bind glycoprotein, which can be used, for example, in pharmaceutical analysis, apart from their use in dressing materials. It can be assumed that the films' properties were improved compared to the properties exhibited by samples without NDASs. The proposed materials can be applied as thin biofilms in biomedical applications.

Supplementary Materials: The following supporting information can be downloaded at: https://www.mdpi.com/article/10.3390/ijms23147652/s1.

Author Contributions: Conceptualization, M.Z.-B.; methodology, M.Z.-B., K.W.-D., K.M., P.N. and D.T.M.; formal analysis, K.W.-D., D.T.M., and D.C.-D.; investigation, M.Z.-B., K.W.-D., D.T.M., D.C.-D., K.M., P.N., T.G. and H.K.; data curation, K.W.-D., K.M., P.N. and D.T.M.; writing—original draft preparation, M.Z.-B. and K.W.-D.; writing—review and editing, M.Z.-B., H.K. and T.G.; funding acquisition, K.W.-D.; supervision, M.Z.-B. All authors have read and agreed to the published version of the manuscript.

Funding: This research was funded by the National Science Centre Poland grant 2016/23/N/ST8/00211.

Institutional Review Board Statement: Not applicable.

Informed Consent Statement: Not applicable.

Data Availability Statement: All data generated or analyzed during this study are included in this manuscript and its Supplementary Materials.

Acknowledgments: Grateful acknowledgment to the SORIMEX company from Torun for the possibility of mechanical testing of the obtained films. K.M., P.N., D.C.-D., and M.Z.-B. are members of Center of Excellence, "Towards Personalized Medicine" operating under Excellence Initiative—Research University. K.W.-D. and H.K. are members of the Emerging Field in the field of Physical Science and Engineering.

Conflicts of Interest: The authors declare no conflict of interest.

References

1. Cheung, R.C.F.; Ng, T.B.; Wong, J.H.; Chan, W.Y. Chitosan: An Update on Potential Biomedical and Pharmaceutical Applications. *Mar. Drugs* **2015**, *13*, 5156–5186. [CrossRef]
2. Rodríguez-Vázquez, M.; Vega-Ruiz, B.; Ramos-Zúñiga, R.; Saldaña-Koppel, D.A.; Quiñones-Olvera, L.F. Chitosan and Its Potential Use as a Scaffold for Tissue Engineering in Regenerative Medicine. *Biomed Res. Int.* **2015**, *2015*, 821279. [CrossRef] [PubMed]
3. Bernkop-Schnürch, A.; Dünnhaupt, S. Chitosan-based drug delivery systems. *Eur. J. Pharm. Biopharm.* **2012**, *81*, 463–469. [CrossRef] [PubMed]
4. Liang, D.; Lu, Z.; Yang, H.; Gao, J.; Chen, R. Novel Asymmetric Wettable AgNPs/Chitosan Wound Dressing: In Vitro and In Vivo Evaluation. *ACS Appl. Mater. Interfaces* **2016**, *8*, 3958–3968. [CrossRef]
5. Cazón, P.; Vázquez, M. Applications of Chitosan as Food Packaging Materials. In *Sustainable Agriculture Reviews 36: Chitin and Chitosan: Applications in Food, Agriculture, Pharmacy, Medicine and Wastewater Treatment*; Crini, G., Lichtfouse, E., Eds.; Springer International Publishing: Cham, Switzerland, 2019; pp. 81–123.
6. Shahidi, F.; Arachchi, J.K.V.; Jeon, Y.-J. Food applications of chitin and chitosans. *Trends Food Sci. Technol.* **1999**, *10*, 37–51. [CrossRef]
7. Song, R.; Murphy, M.; Li, C.; Ting, K.; Soo, C.; Zheng, Z. Current development of biodegradable polymeric materials for biomedical applications. *Drug Des. Devel. Ther.* **2018**, *12*, 3117–3145. [CrossRef]
8. Dash, R.; Foston, M.; Ragauskas, A.J. Improving the mechanical and thermal properties of gelatin hydrogels cross-linked by cellulose nanowhiskers. *Carbohydr. Polym.* **2013**, *91*, 638–645. [CrossRef] [PubMed]
9. Chan, B.P.; Chan, O.C.M.; So, K.F. Effects of photochemical crosslinking on the microstructure of collagen and a feasibility study on controlled protein release. *Acta Biomater.* **2008**, *4*, 1627–1636. [CrossRef]
10. He, L.; Mu, C.; Shi, J.; Zhang, Q.; Shi, B.; Lin, W. Modification of collagen with a natural cross-linker, procyanidin. *Int. J. Biol. Macromol.* **2011**, *48*, 354–359. [CrossRef] [PubMed]
11. Boekema, B.K.H.L.; Vlig, M.; Olde Damink, L.; Middelkoop, E.; Eummelen, L.; Bühren, A.V.; Ulrich, M.M.W. Effect of pore size and cross-linking of a novel collagen-elastin dermal substitute on wound healing. *J. Mater. Sci. Mater. Med.* **2014**, *25*, 423–433. [CrossRef]
12. Chen, R.-N.; Ho, H.-O.; Sheu, M.-T. Characterization of collagen matrices crosslinked using microbial transglutaminase. *Biomaterials* **2005**, *26*, 4229–4235. [CrossRef] [PubMed]
13. Ermis, M.; Calamak, S.; Calibasi Kocal, G.; Guven, S.; Durmus, N.G.; Rizvi, I.; Hasan, T.; Hasirci, N.; Hasirci, V.; Demirci, U. Chapter 15—Hydrogels as a New Platform to Recapitulate the Tumor Microenvironment. In *Handbook of Nanomaterials for Cancer Theranostics*; Conde, J., Ed.; Elsevier: Amsterdam, The Netherlands, 2018; pp. 463–494.
14. Takigawa, T.; Endo, Y. Effects of Glutaraldehyde Exposure on Human Health. *J. Occup. Health* **2006**, *48*, 75–87. [CrossRef] [PubMed]
15. Emmanuel, E.; Hanna, K.; Bazin, C.; Keck, G.; Clément, B.; Perrodin, Y. Fate of glutaraldehyde in hospital wastewater and combined effects of glutaraldehyde and surfactants on aquatic organisms. *Environ. Int.* **2005**, *31*, 399–406. [CrossRef] [PubMed]
16. Lee, S.J.; Yoo, J.J.; Lim, G.J.; Atala, A.; Stitzel, J. In vitro evaluation of electrospun nanofiber scaffolds for vascular graft application. *J. Biomed. Mater. Res. A* **2007**, *83A*, 999–1008. [CrossRef]
17. Rho, K.S.; Jeong, L.; Lee, G.; Seo, B.-M.; Park, Y.J.; Hong, S.-D.; Roh, S.; Cho, J.J.; Park, W.H.; Min, B.-M. Electrospinning of collagen nanofibers: Effects on the behavior of normal human keratinocytes and early-stage wound healing. *Biomaterials* **2006**, *27*, 1452–1461. [CrossRef] [PubMed]
18. Azeredo, H.M.C.; Waldron, K.W. Crosslinking in polysaccharide and protein films and coatings for food contact—A review. *Trends Food Sci. Technol.* **2016**, *52*, 109–122. [CrossRef]

19. Islam, M.S.; Chen, L.; Sisler, J.; Tam, K.C. Cellulose nanocrystal (CNC)–inorganic hybrid systems: Synthesis, properties and applications. *J. Mater. Chem. B* **2018**, *6*, 864–883. [CrossRef] [PubMed]
20. Perez Herrera, M.; Vasanthan, T.; Chen, L. Rheology of starch nanoparticles as influenced by particle size, concentration and temperature. *Food Hydrocoll.* **2017**, *66*, 237–245. [CrossRef]
21. Angellier, H.; Molina-Boisseau, S.; Lebrun, L.; Dufresne, A. Processing and Structural Properties of Waxy Maize Starch Nanocrystals Reinforced Natural Rubber. *Macromolecules* **2005**, *38*, 3783–3792. [CrossRef]
22. Zheng, H.; Ai, F.; Chang, P.R.; Huang, J.; Dufresne, A. Structure and properties of starch nanocrystal-reinforced soy protein plastics. *Polym. Compos.* **2009**, *30*, 474–480. [CrossRef]
23. Potthast, A.; Schiehser, S.; Rosenau, T.; Kostic, M. Oxidative modifications of cellulose in the periodate system—Reduction and beta-elimination reactions 2nd ICC 2007, Tokyo, Japan, October 25–29, 2007. *Holzforschung* **2009**, *63*, 12–17. [CrossRef]
24. Fan, L.; Sun, Y.; Xie, W.; Zheng, H.; Liu, S. Oxidized pectin cross-linked carboxymethyl chitosan: A new class of hydrogels. *J. Biomater. Sci. Polym. Ed.* **2012**, *23*, 2119–2132. [CrossRef] [PubMed]
25. Amirian, J.; Zeng, Y.; Shekh, M.I.; Sharma, G.; Stadler, F.J.; Song, J.; Du, B.; Zhu, Y. In-situ crosslinked hydrogel based on amidated pectin/oxidized chitosan as potential wound dressing for skin repairing. *Carbohydr. Polym.* **2021**, *251*, 117005. [CrossRef] [PubMed]
26. Fu, G.-Q.; Zhang, S.-C.; Chen, G.-G.; Hao, X.; Bian, J.; Peng, F. Xylan-based hydrogels for potential skin care application. *Int. J. Biol. Macromol.* **2020**, *158*, 244–250. [CrossRef]
27. Bajpai, S.K.; Bajpai, M.; Shah, F.F. Alginate dialdehyde (AD)-crosslinked casein films: Synthesis, characterization and water absorption behavior. *Des. Monomers Polym.* **2016**, *19*, 406–419. [CrossRef]
28. Kuchaiyaphum, P.; Chotichayapong, C.; Butwong, N.; Bua-ngern, W. Silk Fibroin/Poly (vinyl alcohol) Hydrogel Cross-Linked with Dialdehyde Starch for Wound Dressing Applications. *Macromol. Res.* **2020**, *28*, 844–850. [CrossRef]
29. Thomas, L.V.; Vg, R.; Nair, P.D. Effect of stiffness of chitosan-hyaluronic acid dialdehyde hydrogels on the viability and growth of encapsulated chondrocytes. *Int. J. Biol. Macromol.* **2017**, *104*, 1925–1935. [CrossRef]
30. Liu, J.; Zhang, L.; Liu, C.; Zheng, X.; Tang, K. Tuning structure and properties of gelatin edible films through pullulan dialdehyde crosslinking. *LWT* **2021**, *138*, 110607. [CrossRef]
31. Lucas de Lima, E.; Fittipaldi Vasconcelos, N.; da Silva Maciel, J.; Karine Andrade, F.; Silveira Vieira, R.; Andrade Feitosa, J.P. Injectable hydrogel based on dialdehyde galactomannan and N-succinyl chitosan: A suitable platform for cell culture. *J. Mater. Sci. Mater. Med.* **2019**, *31*, 5. [CrossRef]
32. Pietrucha, K.; Safandowska, M. Dialdehyde cellulose-crosslinked collagen and its physicochemical properties. *Process Biochem.* **2015**, *50*, 2105–2111. [CrossRef]
33. Dou, Y.; Zhang, L.; Zhang, B.; He, M.; Shi, W.; Yang, S.; Cui, Y.; Yin, G. Preparation and Characterization of Edible Dialdehyde Carboxymethyl Cellulose Crosslinked Feather Keratin Films for Food Packaging. *Polymers* **2020**, *12*, 158. [CrossRef]
34. Ngwabebhoh, F.A.; Zandraa, O.; Patwa, R.; Saha, N.; Capáková, Z.; Saha, P. Self-crosslinked chitosan/dialdehyde xanthan gum blended hypromellose hydrogel for the controlled delivery of ampicillin, minocycline and rifampicin. *Int. J. Biol. Macromol.* **2021**, *167*, 1468–1478. [CrossRef]
35. Terracciano, R.; Zhang, A.; Butler, E.B.; Demarchi, D.; Hafner, J.H.; Grattoni, A.; Filgueira, C.S. Effects of Surface Protein Adsorption on the Distribution and Retention of Intratumorally Administered Gold Nanoparticles. *Pharmaceutics* **2021**, *13*, 216. [CrossRef] [PubMed]
36. Afrin, S.; Riyazuddeen; Rabbani, G.; Khan, R.H. Spectroscopic and calorimetric studies of interaction of methimazole with human serum albumin. *J. Lumin.* **2014**, *151*, 219–223. [CrossRef]
37. Matsumoto, K.; Sukimoto, K.; Nishi, K.; Maruyama, T.; Suenaga, A.; Otagiri, M. Characterization of ligand binding sites on the alpha1-acid glycoprotein in humans, bovines and dogs. *Drug Metab. Pharmacokinet.* **2002**, *17*, 300–306. [CrossRef] [PubMed]
38. Bteich, M. An overview of albumin and alpha-1-acid glycoprotein main characteristics: Highlighting the roles of amino acids in binding kinetics and molecular interactions. *Heliyon* **2019**, *5*, e02879. [CrossRef] [PubMed]
39. Jiao, Q.; Wang, R.; Jiang, Y.; Liu, B. Study on the interaction between active components from traditional Chinese medicine and plasma proteins. *Chem. Cent. J.* **2018**, *12*, 48. [CrossRef]
40. Xu, Q.; Ji, Y.; Sun, Q.; Fu, Y.; Xu, Y.; Jin, L. Fabrication of Cellulose Nanocrystal/Chitosan Hydrogel for Controlled Drug Release. *Nanomaterials* **2019**, *9*, 253. [CrossRef]
41. Kwak, H.W.; Lee, H.; Park, S.; Lee, M.E.; Jin, H.-J. Chemical and physical reinforcement of hydrophilic gelatin film with di-aldehyde nanocellulose. *Int. J. Biol. Macromol.* **2020**, *146*, 332–342. [CrossRef]
42. Chen, Y.; Hao, Y.; Ting, K.; Li, Q.; Gao, Q. Preparation and emulsification properties of dialdehyde starch nanoparticles. *Food Chem.* **2019**, *286*, 467–474. [CrossRef]
43. Chen, Y.; Dai, G.; Gao, Q. Starch Nanoparticles-Graphene Aerogels with High Supercapacitor Performance and Efficient Adsorption. *ACS Sustain. Chem. Eng.* **2019**, *7*, 14064–14073. [CrossRef]
44. Lin, N.; Huang, J.; Dufresne, A. Preparation, properties and applications of polysaccharide nanocrystals in advanced functional nanomaterials: A review. *Nanoscale* **2012**, *4*, 3274–3294. [CrossRef] [PubMed]
45. Condés, M.C.; Añón, M.C.; Mauri, A.N.; Dufresne, A. Amaranth protein films reinforced with maize starch nanocrystals. *Food Hydrocoll.* **2015**, *47*, 146–157. [CrossRef]
46. Jayakody, L.; Hoover, R. The effect of lintnerization on cereal starch granules. *Food Res. Int.* **2002**, *35*, 665–680. [CrossRef]

47. Sanchez de la Concha, B.B.; Agama-Acevedo, E.; Nuñez-Santiago, M.C.; Bello-Perez, L.A.; Garcia, H.S.; Alvarez-Ramirez, J. Acid hydrolysis of waxy starches with different granule size for nanocrystal production. *J. Cereal Sci.* **2018**, *79*, 193–200. [CrossRef]

48. Silva, A.P.M.; Oliveira, A.V.; Pontes, S.M.A.; Pereira, A.L.S.; Souza Filho, M.d.s.M.; Rosa, M.F.; Azeredo, H.M.C. Mango kernel starch films as affected by starch nanocrystals and cellulose nanocrystals. *Carbohydr. Polym.* **2019**, *211*, 209–216. [CrossRef]

49. Xu, Y.; Ding, W.; Liu, J.; Li, Y.; Kennedy, J.F.; Gu, Q.; Shao, S. Preparation and characterization of organic-soluble acetylated starch nanocrystals. *Carbohydr. Polym.* **2010**, *80*, 1078–1084. [CrossRef]

50. Duan, B.; Sun, P.; Wang, X.; Yang, C. Preparation and properties of starch nanocrystals/carboxymethyl chitosan nanocomposite films. *Starch Stärke* **2011**, *63*, 528–535. [CrossRef]

51. Sessini, V.; Raquez, J.-M.; Kenny, J.M.; Dubois, P.; Peponi, L. Melt-processing of bionanocomposites based on ethylene-co-vinyl acetate and starch nanocrystals. *Carbohydr. Polym.* **2019**, *208*, 382–390. [CrossRef]

52. Oyeyinka, S.A.; Singh, S.; Adebola, P.O.; Gerrano, A.S.; Amonsou, E.O. Physicochemical properties of starches with variable amylose contents extracted from bambara groundnut genotypes. *Carbohydr. Polym.* **2015**, *133*, 171–178. [CrossRef]

53. Mukurumbira, A.; Mariano, M.; Dufresne, A.; Mellem, J.J.; Amonsou, E.O. Microstructure, thermal properties and crystallinity of amadumbe starch nanocrystals. *Int. J. Biol. Macromol.* **2017**, *102*, 241–247. [CrossRef] [PubMed]

54. Md Shahrodin, N.S.; Rahmat, A.R.; Arsad, A. Synthesis and Characterization of Cassava Starch Nanocrystals by Hydrolysis Method. *Adv. Mat. Res.* **2015**, *1113*, 446–452. [CrossRef]

55. LeCorre, D.; Bras, J.; Dufresne, A. Influence of botanic origin and amylose content on the morphology of starch nanocrystals. *J. Nanopart. Res.* **2011**, *13*, 7193–7208. [CrossRef]

56. Wegrzynowska-Drzymalska, K.; Grebicka, P.; Mlynarczyk, D.T.; Chelminiak-Dudkiewicz, D.; Kaczmarek, H.; Goslinski, T.; Ziegler-Borowska, M. Crosslinking of Chitosan with Dialdehyde Chitosan as a New Approach for Biomedical Applications. *Materials* **2020**, *13*, 3413. [CrossRef]

57. Chauhan, K.; Kaur, J.; Kumari, A.; Kumari, A.; Chauhan, G.S. Efficient method of starch functionalization to bis-quaternary structure unit. *Int. J. Biol. Macromol.* **2015**, *80*, 498–505. [CrossRef]

58. Zeng, L.; Qin, C.; Wang, L.; Li, W. Volatile compounds formed from the pyrolysis of chitosan. *Carbohydr. Polym.* **2011**, *83*, 1553–1557. [CrossRef]

59. Gierszewska, M.; Jakubowska, E.; Olewnik-Kruszkowska, E. Effect of chemical crosslinking on properties of chitosan-montmorillonite composites. *Polym. Test.* **2019**, *77*, 105872. [CrossRef]

60. Sionkowska, A.; Michalska-Sionkowska, M.; Walczak, M. Preparation and characterization of collagen/hyaluronic acid/chitosan film crosslinked with dialdehyde starch. *Int. J. Biol. Macromol.* **2020**, *149*, 290–295. [CrossRef] [PubMed]

61. Ezati, P.; Rhim, J.-W. pH-responsive chitosan-based film incorporated with alizarin for intelligent packaging applications. *Food Hydrocoll.* **2020**, *102*, 105629. [CrossRef]

62. Hoffmann, B.; Seitz, D.; Mencke, A.; Kokott, A.; Ziegler, G. Glutaraldehyde and oxidised dextran as crosslinker reagents for chitosan-based scaffolds for cartilage tissue engineering. *J. Mater. Sci. Mater. Med.* **2009**, *20*, 1495–1503. [CrossRef]

63. Ilkar Erdagi, S.; Asabuwa Ngwabebhoh, F.; Yildiz, U. Genipin crosslinked gelatin-diosgenin-nanocellulose hydrogels for potential wound dressing and healing applications. *Int. J. Biol. Macromol.* **2020**, *149*, 651–663. [CrossRef] [PubMed]

64. Wahlström, N.; Steinhagen, S.; Toth, G.; Pavia, H.; Edlund, U. Ulvan dialdehyde-gelatin hydrogels for removal of heavy metals and methylene blue from aqueous solution. *Carbohydr. Polym.* **2020**, *249*, 116841. [CrossRef] [PubMed]

65. Yuan, L.; Wu, Y.; Gu, Q.-s.; El-Hamshary, H.; El-Newehy, M.; Mo, X. Injectable photo crosslinked enhanced double-network hydrogels from modified sodium alginate and gelatin. *Int. J. Biol. Macromol.* **2017**, *96*, 569–577. [CrossRef]

66. Bin Ahmad, M.; Lim, J.J.; Shameli, K.; Ibrahim, N.A.; Tay, M.Y. Synthesis of Silver Nanoparticles in Chitosan, Gelatin and Chitosan/Gelatin Bionanocomposites by a Chemical Reducing Agent and Their Characterization. *Molecules* **2011**, *16*, 7237–7248. [CrossRef] [PubMed]

67. Osorio-Madrazo, A.; David, L.; Trombotto, S.; Lucas, J.-M.; Peniche-Covas, C.; Domard, A. Highly crystalline chitosan produced by multi-steps acid hydrolysis in the solid-state. *Carbohydr. Polym.* **2011**, *83*, 1730–1739. [CrossRef]

68. Yao, Y.; Wang, H.; Wang, R.; Chai, Y.; Ji, W. Fabrication and performance characterization of the membrane from self-dispersed gelatin-coupled cellulose microgels. *Cellulose* **2019**, *26*, 3255–3269. [CrossRef]

69. Quero, F.; Padilla, C.; Campos, V.; Luengo, J.; Caballero, L.; Melo, F.; Li, Q.; Eichhorn, S.J.; Enrione, J. Stress transfer and matrix-cohesive fracture mechanism in microfibrillated cellulose-gelatin nanocomposite films. *Carbohydr. Polym.* **2018**, *195*, 89–98. [CrossRef] [PubMed]

70. He, B.; Wang, X.; Xue, H. The performance of chitosan/gelatin composite microspheres in the wash-off procedure of reactive dyeing. *Color. Technol.* **2016**, *132*, 353–360. [CrossRef]

71. Ziegler-Borowska, M.; Chełminiak, D.; Kaczmarek, H.; Kaczmarek-Kędziera, A. Effect of side substituents on thermal stability of the modified chitosan and its nanocomposites with magnetite. *J. Therm. Anal. Calorim.* **2016**, *124*, 1267–1280. [CrossRef]

72. Dong, Y.; Zhao, S.; Lu, W.; Chen, N.; Zhu, D.; Li, Y. Preparation and characterization of enzymatically cross-linked gelatin/cellulose nanocrystal composite hydrogels. *RSC Adv.* **2021**, *11*, 10794–10803. [CrossRef]

73. Pulieri, E.; Chiono, V.; Ciardelli, G.; Vozzi, G.; Ahluwalia, A.; Domenici, C.; Vozzi, F.; Giusti, P. Chitosan/gelatin blends for biomedical applications. *J. Biomed. Mater. Res. A* **2008**, *86A*, 311–322. [CrossRef] [PubMed]

74. Sutirman, Z.A.; Sanagi, M.M.; Abd Karim, K.J.; Wan Ibrahim, W.A. Preparation of methacrylamide-functionalized crosslinked chitosan by free radical polymerization for the removal of lead ions. *Carbohydr. Polym.* **2016**, *151*, 1091–1099. [CrossRef] [PubMed]

75. Kaczmarek-Szczepańska, B.; Mazur, O.; Michalska-Sionkowska, M.; Łukowicz, K.; Osyczka, A.M. The Preparation and Characterization of Chitosan-Based Hydrogels Cross-Linked by Glyoxal. *Materials* **2021**, *14*, 2449. [CrossRef] [PubMed]
76. Zhao, Y.; Sun, Z. Effects of gelatin-polyphenol and gelatin–genipin cross-linking on the structure of gelatin hydrogels. *Int. J. Food Prop.* **2017**, *20* (Suppl. 3), S2822–S2832. [CrossRef]
77. Nieto-Suárez, M.; López-Quintela, M.A.; Lazzari, M. Preparation and characterization of crosslinked chitosan/gelatin scaffolds by ice segregation induced self-assembly. *Carbohydr. Polym.* **2016**, *141*, 175–183. [CrossRef]
78. Kim, S.; Nimni, M.E.; Yang, Z.; Han, B. Chitosan/gelatin–based films crosslinked by proanthocyanidin. *J. Biomed. Mater. Res. Part B Appl. Biomater.* **2005**, *75B*, 442–450. [CrossRef]
79. Yap, W.F.; Yunus, W.M.M.; Talib, Z.A.; Yusof, N.A. X-ray photoelectron spectroscopy and atomic force microscopy studies on crosslinked chitosan thin film. *Int. J. Phys. Sci.* **2011**, *6*, 2744–2749. [CrossRef]
80. Olewnik-Kruszkowska, E.; Gierszewska, M.; Grabska-Zielińska, S.; Skopińska-Wiśniewska, J.; Jakubowska, E. Examining the Impact of Squaric Acid as a Crosslinking Agent on the Properties of Chitosan-Based Films. *Int. J. Mol. Sci.* **2021**, *22*, 3329. [CrossRef]
81. Abbasi, A.; Eslamian, M.; Heyd, D.; Rousseau, D. Controlled Release of DSBP from Genipin-Crosslinked Gelatin Thin Films. *Pharm. Dev. Technol.* **2008**, *13*, 549–557. [CrossRef]
82. Bonilla, J.; Fortunati, E.; Atarés, L.; Chiralt, A.; Kenny, J.M. Physical, structural and antimicrobial properties of poly vinyl alcohol–chitosan biodegradable films. *Food Hydrocoll.* **2014**, *35*, 463–470. [CrossRef]
83. Khan, M.U.A.; Haider, S.; Raza, M.A.; Shah, S.A.; Razak, S.I.A.; Kadir, M.R.A.; Subhan, F.; Haider, A. Smart and pH-sensitive rGO/Arabinoxylan/chitosan composite for wound dressing: In-vitro drug delivery, antibacterial activity, and biological activities. *Int. J. Biol. Macromol.* **2021**, *192*, 820–831. [CrossRef] [PubMed]
84. Kudłacik-Kramarczyk, S.; Drabczyk, A.; Głąb, M.; Gajda, P.; Jaromin, A.; Czopek, A.; Zagórska, A.; Tyliszczak, B. Synthesis and Physicochemical Evaluation of Bees' Chitosan-Based Hydrogels Modified with Yellow Tea Extract. *Materials* **2021**, *14*, 3379. [CrossRef] [PubMed]
85. Skopinska-Wisniewska, J.; Tuszynska, M.; Olewnik-Kruszkowska, E. Comparative Study of Gelatin Hydrogels Modified by Various Cross-Linking Agents. *Materials* **2021**, *14*, 396. [CrossRef]
86. Kaczmarek, B.; Sionkowska, A.; Monteiro, F.J.; Carvalho, A.; Łukowicz, K.; Osyczka, A.M. Characterization of gelatin and chitosan scaffolds cross-linked by addition of dialdehyde starch. *Biomed. Mater.* **2017**, *13*, 015016. [CrossRef] [PubMed]
87. Wang, C.-H.; Liu, W.-S.; Sun, J.-F.; Hou, G.-G.; Chen, Q.; Cong, W.; Zhao, F. Non-toxic O-quaternized chitosan materials with better water solubility and antimicrobial function. *Int. J. Biol. Macromol.* **2016**, *84*, 418–427. [CrossRef] [PubMed]
88. Sajomsang, W. Synthetic methods and applications of chitosan containing pyridylmethyl moiety and its quaternized derivatives: A review. *Carbohydr. Polym.* **2010**, *80*, 631–647. [CrossRef]
89. Karami Juyani, A.; Saberi, M.; Kaka, G.; Jafari, M.; Riyahi, S.; Taghizadeh, M. Toxicity Study of Gelatin-chitosan Films with Bone Marrow Stromalcells Cultured in Rat. *Int. J. Pediatr.* **2014**, *2*, 87. [CrossRef]
90. Szurkowska, K.; Drobniewska, A.; Kolmas, J. Dual Doping of Silicon and Manganese in Hydroxyapatites: Physicochemical Properties and Preliminary Biological Studies. *Materials* **2019**, *12*, 2566. [CrossRef]
91. Tammaro, M.; Salluzzo, A.; Rimauro, J.; Schiavo, S.; Manzo, S. Experimental investigation to evaluate the potential environmental hazards of photovoltaic panels. *J. Hazard. Mater.* **2016**, *306*, 395–405. [CrossRef]
92. Isidori, M.; Ferrara, M.; Lavorgna, M.; Nardelli, A.; Parrella, A. In situ monitoring of urban air in Southern Italy with the tradescantia micronucleus bioassay and semipermeable membrane devices (SPMDs). *Chemosphere* **2003**, *52*, 121–126. [CrossRef]
93. Fahmy, A.; Kamoun, E.; El-Eisawy, R.; El-Fakharany, E.; Taha, T.; El-Damhougy, B.; Ahmed, F. Poly(vinyl alcohol)-hyaluronic Acid Membranes for Wound Dressing Applications: Synthesis and in vitro Bio-Evaluations. *J. Braz. Chem. Soc.* **2015**, *26*. [CrossRef]
94. Zhang, Y.; Lu, L.; Chen, Y.; Wang, J.; Chen, Y.; Mao, C.; Yang, M. Polydopamine modification of silk fibroin membranes significantly promotes their wound healing effect. *Biomater. Sci.* **2019**, *7*, 5232–5237. [CrossRef] [PubMed]
95. Singh, B.; Dhiman, A. Designing bio-mimetic moxifloxacin loaded hydrogel wound dressing to improve antioxidant and pharmacology properties. *RSC Adv.* **2015**, *5*, 44666–44678. [CrossRef]
96. Yang, T.; Zheng, J.; Zheng, B.-S.; Liu, F.; Wang, S.; Tang, C.-H. High internal phase emulsions stabilized by starch nanocrystals. *Food Hydrocoll.* **2018**, *82*, 230–238. [CrossRef]
97. Liu, Y.; Cai, Z.; Sheng, L.; Ma, M.; Xu, Q.; Jin, Y. Structure-property of crosslinked chitosan/silica composite films modified by genipin and glutaraldehyde under alkaline conditions. *Carbohydr. Polym.* **2019**, *215*, 348–357. [CrossRef]
98. Owens, D.K.; Wendt, R.C. Estimation of the surface free energy of polymers. *J. Appl. Polym. Sci.* **1969**, *13*, 1741–1747. [CrossRef]
99. Skopinska-Wisniewska, J.; Kuderko, J.; Bajek, A.; Maj, M.; Sionkowska, A.; Ziegler-Borowska, M. Collagen/elastin hydrogels cross-linked by squaric acid. *Mater. Sci. Eng. C* **2016**, *60*, 100–108. [CrossRef]
100. Johnson, B.T. Microtox® Acute Toxicity Test. In *Small-Scale Freshwater Toxicity Investigations: Toxicity Test Methods*; Blaise, C., Férard, J.-F., Eds.; Springer Netherlands: Dordrecht, The Netherlands, 2005; pp. 69–105.

International Journal of
Molecular Sciences

Article

Understanding Passive Membrane Permeation of Peptides: Physical Models and Sampling Methods Compared

Liuba Mazzanti and Tâp Ha-Duong *

BioCIS, CNRS, Université Paris-Saclay, 17 Avenue des Sciences, 91400 Orsay, France
* Correspondence: tap.ha-duong@universite-paris-saclay.fr

Abstract: The early characterization of drug membrane permeability is an important step in pharmaceutical developments to limit possible late failures in preclinical studies. This is particularly crucial for therapeutic peptides whose size generally prevents them from passively entering cells. However, a sequence-structure-dynamics-permeability relationship for peptides still needs further insight to help efficient therapeutic peptide design. In this perspective, we conducted here a computational study for estimating the permeability coefficient of a benchmark peptide by considering and comparing two different physical models: on the one hand, the inhomogeneous solubility–diffusion model, which requires umbrella–sampling simulations, and on the other hand, a chemical kinetics model which necessitates multiple unconstrained simulations. Notably, we assessed the accuracy of the two approaches in relation to their computational cost.

Keywords: peptide membrane permeability; molecular dynamics simulation; umbrella sampling; Markov State Model; free energy profile

Citation: Mazzanti, L.; Ha-Duong, T. Understanding Passive Membrane Permeation of Peptides: Physical Models and Sampling Methods Compared. *Int. J. Mol. Sci.* **2023**, 24, 5021. https://doi.org/10.3390/ijms24055021

Academic Editors: Geoffrey Brown, Enikö Kallay and Andrzej Kutner

Received: 1 February 2023
Revised: 21 February 2023
Accepted: 23 February 2023
Published: 6 March 2023

1. Introduction

To reach an intracellular target, a drug must cross the cellular membrane. This process can be performed by endocytosis, which is generally involved in internalizing macromolecules or drug nanocarriers [1], or by using membrane transporter proteins, such as the human ATP binding cassette (ABC) [2] and solute carrier (SLC) [3] families, or the outer membrane porins [4] and TonB-dependent transporters [5] of gram-negative bacteria. However, the predominant mechanism for xenobiotics is a passive diffusion across the lipid bilayer along a concentration gradient [6,7]. It is, thus, essential in pharmaceutical developments to characterize the drug membrane permeability to anticipate their pharmacokinetic properties. This is particularly crucial for peptide-based therapeutics. Indeed, peptides are very promising compounds for modulating protein-protein interactions (PPIs), especially those involved in intracellular signaling pathways [8]. This is partly due to their size, which is well-appropriate for covering the generally large molecular surfaces involved in PPIs. However, this size may turn out to be a disadvantage for membrane permeation.

Passive membrane permeability of drugs can be characterized experimentally by using the parallel artificial membrane permeability assay (PAMPA), which consists in measuring the number of compounds that have crossed a planar artificial membrane from donor to acceptor wells [9]. Alternative methods have been developed to monitor in real-time the entry of drugs into unilamellar liposomes using fluorescent probes [10,11]. In both approaches, the drug membrane permeability P is quantified as $P = J/\Delta C$, where J is the flux of the drug through the membrane and ΔC the difference between the donor (or outside) and acceptor (or inside) drug concentrations.

Throughout a membrane permeation process, a peptide can bind the lipid headgroup-water interface, fold into secondary structures, form aggregates, insert into the lipid hydrophobic tails, and perturb the bilayer organization [12,13]. These multiple possible events depend on the peptide apolar/polar/charged amino acid composition, its structure and flexibility, the solvent pH, the ionic strength, and the membrane content in lipids. All these factors make it challenging to predict the membrane permeability and deeply understand the mechanism of peptide translocation. To tackle this issue, molecular dynamics (MD) simulations of peptide crossing lipid bilayers are valuable tools to gain insight into the physical-chemical factors that govern the permeation process.

However, membrane permeation occurs on a millisecond to second timescale [10] which are generally out of reach for classical MD simulations. Thus, enhanced sampling methods, such as metadynamics or umbrella sampling (US), are needed to provides valuable information about the compound propensity to partition between water and membrane, the structures of the intermediate states within the bilayer, and about the mechanisms of permeation [14,15]. In particular, trajectories from US simulations allow to compute drug free energy profiles (FEP) across lipid bilayers and can be employed to estimate their membrane permeability by using the so-called inhomogeneous solubility-diffusion model (ISDM) proposed by Marrink and Berendsen [16]. This approach has been applied to several drugs over the past few years [14,17–19] and recently to peptides [20].

Since membrane permeability is a kinetic property, it can be naturally computed by using a kinetic model of the permeation process. This consists in describing the elementary steps of a compound membrane permeation and formulating each of the corresponding reaction rates as a function of the concentrations in the intermediate states. Then, the master differential equations are solved to yield the compound concentrations on both sides of the membrane as a function of time, from which the membrane permeability can be retrieved with Fick's law of diffusion [18]. Importantly, this method requires to preliminary determine the rate constants of the kinetic model. This can be done with Markov State Model (MSM) analyses which can extract long-time kinetic information of a molecular system from an ensemble of "short" MD trajectories starting from multiple initial configurations [21,22].

For small compounds, calculations based on kinetic models appear to provide membrane permeability values that better correlate with experiments than those based on ISDM [18]. Nevertheless, as far as we know, the former approach has never been applied to peptides. In the present study, we comparatively assessed the two computational methods to estimate the membrane permeability for a benchmark cyclic peptide of 10 amino acid residues. Notably, we scrutinized the accuracy of the two approaches in relation to their computational cost.

2. Results and Discussion

We applied the two computational approaches to a paradigmatic ten amino acids cyclic peptide for which PAMPA assays were performed [9]. Constrained or unconstrained MD trajectories of the peptide assembled with a planar lipid bilayer were collected to ultimately calculate the peptide permeability coefficient through Equation (4) or (8) for the ISDM or the MSM-based model, respectively. In both approaches, the free energy profile of the peptide perpendicularly crossing the planar bilayer membrane is crucial for estimating its permeability coefficient. In the ISDM method, the latter is directly related to the FEP $\Delta G(z)$ through Equations (2)–(4). On the other hand, in the MSM-based approach, without the knowledge of the energy barrier location, the Markov State Models could be biased since the phase space sampling by short trajectories depends on the starting points along the collective variable. Therefore, the initial configurations must correspond to maximum energy states for the trajectories not to be trapped in energy minima. Thus, we will present first the free energy analyses and, subsequently, the quantitative estimates for the permeability coefficients.

2.1. Free Energy Profiles

The weighted histogram analysis method (WHAM) [23] applied to the umbrella sampling trajectories for the cyclic decapeptide designated as CDP5 (Section 3.1) yields the free energy profile displayed in Figure 1. Moving from the bulk water to the membrane, the peptide encounters a first energy barrier of about ΔG_{in} = 3 kcal/mol located on the surface of the lipid headgroups in the water phase (z = 2.4 nm). Then the FEP exhibits a minimum on the other side of the lipid headgroup plane, inside the membrane (z = 1.4 nm), where the amphiphilic peptide is stabilized by both hydrophobic and polar interactions with the lipid tails and headgroups, respectively. We note that the free energy of this stable state is roughly equal to that one in the bulk water. Therefore, the free energy barrier for moving out from the interior of the membrane into the bulk water is also $\Delta G_{out} = \Delta G_{in} = 3$ kcal/mol. The major energy barrier for crossing the membrane is located at its center (z = 0.0 nm), indicating that the rate-limiting step during the peptide permeation is the flip-flop passage from one lipid layer to the other. The necessary energy to overcome this barrier is about ΔG_{flip} = 12 kcal/mol.

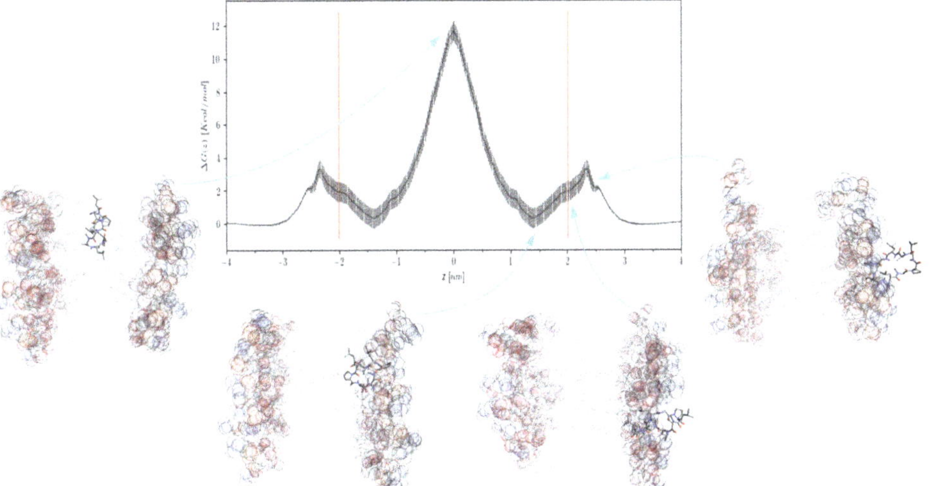

Figure 1. Symmetrized CDP5 free energy profile computed from umbrella sampling trajectories. Error bars were estimated using the `gmx wham` bootstrap algorithm [24]. The orange vertical line and shaded area indicate the mean and one σ of the lipid headgroup positions, respectively. Free energy difference $\Delta G(z) = G(z) - G(z_{max})$ was calculated with z_{max} = 3.8 nm in the water phase. Four representative structures of the peptide-membrane system are shown for the peptide positions z = 0.0, 1.4, 2.0, and 2.7 nm. Peptides, lipid headgroups, and lipid tails are displayed using sticks, semi-transparent spheres, and lines, respectively.

The FEP computed for the CDP5 peptide is very similar to those of amphiphilic cyclic hexapeptides reported by Sugita et al. [20], indicating that these cyclic peptides of comparable size should have the same mechanism of membrane translocation in which the flip-flop event within the bilayer is the rate-limiting step. This contrasts with the FEPs of amphiphilic small compounds computed by Dickson et al. [18], which exhibit two marked energy minima located at the inner lipid headgroup surfaces much lower than the free energy in the bulk water. For these small compounds, the energy barrier at the membrane center is generally small, depending on their hydrophobicity, indicating that the major rate-limiting step of their permeation process is moving from the interior of the membrane into the bulk water.

To compute the peptide membrane permeability using kinetic models, we built a first Markov State Model (*CDP5_2.4*) from unconstrained MD simulations of the peptide with initial positions at z = 2.4 nm and z = 0.0 nm corresponding to the two maxima of its FEP computed from US trajectories. Moreover, to assess the sensitivity of this approach to the initial configurations, we also built Markov State Models (*CDP5_2.3* and *CDP5_2.6*), respectively from trajectories starting at z = 2.3 nm and z = 0.0 nm on the one hand, and from trajectories starting from $z = 2.6$ nm and $z = 0.0$ nm, on the other hand. A fourth Markov State Model (*CDP5_all*) was also built using all the aforementioned trajectories (Section 3.3). From each of these sets of unconstrained MD trajectories, we computed the peptide stationary distributions π and the free energy profiles $\Delta G = -k_B T \log \pi$, which are shown in Figure 2.

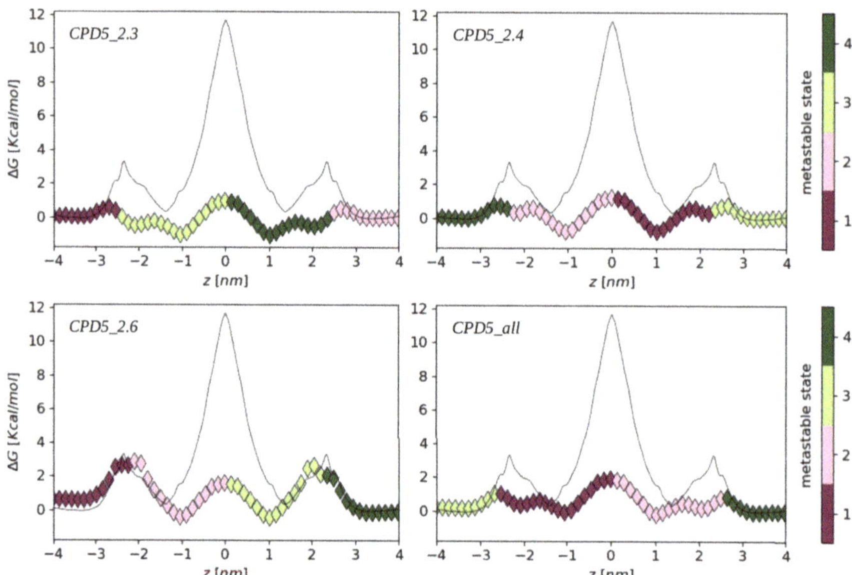

Figure 2. CDP5 free energy profiles computed from four different sets of unconstrained trajectories (Section 3.3). Diamonds indicating the discrete states are colored according to the metastable state to which they belong. FEPs were set to zero in the water phase, including the FEP computed from US trajectories (black lines).

FEPs derived from unconstrained trajectories show significant differences with respect to the profile computed from the US trajectories: First, the energy barriers to cross the lipid headgroups are smaller (by about 2 kcal/mol) than the US ones, except for the (*CDP5_2.6*) set of trajectories. Secondly, the positions of the energy minima inside the membrane are shifted toward the lipid tails ($z = 1.0$ nm) compared to the US FEP ($z = 1.4$ nm). However, the most striking difference is the free energy barrier corresponding to the flip-flop step: the unconstrained simulations yield a central free energy barrier of about 2 kcal/mol, while the US flip-flop barrier almost reaches 12 kcal/mol. This suggests that the peptide membrane permeability estimated with the kinetic models will be much higher than in the ISDM-based approach.

It is interesting to note that the four FEPs generated by the four sets of unconstrained simulations display different qualitative behaviors, implying that a slight change in their initial conditions could entail a sensible variation in the results. In particular, we notice that the free energy needed to translocate the peptide from inside the membrane into the water phase is larger than the energy barrier associated with the flip-flop step in the *CDP5_2.6* model, while it is smaller in the three other cases (Figure 2). The emergence of high in-out free energy barriers in the *CDP5_2.6* case is a consequence of the initial peptide positions, which were slightly shifted from the minor maxima of the US FEP ($z = 2.4$ nm) toward the water bulk. Therefore, fewer peptide trajectories could naturally sample this maximum energy state which increases the barrier height, compared to the three other cases (where it could even be observed a local shallow energy minimum at this location). This highlights the sensitivity of the peptide FEPs computed from unconstrained MD simulations upon their initial configurations.

2.2. Membrane Permeability Calculations

In the framework of the ISDM approach, an estimate of the peptide permeability coefficient can be directly obtained from the US trajectories by extracting the integrated autocorrelation times τ_{zz}, the collective variable variance σ_z^2, and the free energy profile $\Delta G(z)$ and applying Equations (2)–(4). For the CDP5 peptide, we get the value reported in Table 1. Compared to experiments, the ISDM-based estimation of the peptide membrane permeability is about one order of magnitude lower than the value found in PAMPA assays [25].

Table 1. Membrane permeability coefficient results for the cyclic decapeptide CDP5. PAMPA value was taken from [25].

Method	$\log[P(cm/s)] \pm \sigma_P$
PAMPA	-6.7
ISDM	-8.1 ± 0.4
MSM CDP5_2.3	-1.0 ± 0.1
MSM CDP5_2.4	-1.2 ± 0.1
MSM CDP5_2.6	-1.4 ± 0.1
MSM CDP5_all	-1.0 ± 0.1

On the other hand, further calculations and analysis are necessary to get an estimate of the permeability coefficient within the liposome kinetic model approach. Rate constants are derived as the inverse of the mean first passage times (MFPTs) and then converted to apparent rate constants for the liposome model using Equation (6). As detailed in Appendix A, integration of the kinetic model Equation (5) yields a three-exponential time evolution for the substrate in the inner aqueous compartment (Equation (A7)). By substituting k_I of Equation (8) with the fastest rate constants κ_n appearing in the substrate time evolution, we obtained the permeability coefficients reported in Table 1. Compared to the PAMPA value, it appears that the liposome kinetic models overestimate the peptide membrane permeability by about five to six orders of magnitude.

While comparing theoretical with experimental estimates of the peptide membrane permeability, we should bear in mind that PAMPA assays use hexadecane plates separating two water compartments [25] instead of a lipid bilayer or spherical liposomes. Thus, if we assume that crossing a hexadecane plate is easier than a lipid bilayer, this difference in the membrane nature might explain the lower permeability found by the ISDM-based method compared to experiments. On the other hand, the permeability estimated with liposome kinetic models being proportional to the liposome radius, it should be intuitively much lower than the permeability through a planar membrane. This is the opposite tendency that we can notice in Table 1 and, altogether, it appears that the MSM-based approach greatly overestimates the peptide membrane permeability compared to experiments.

The discrepancy between the ISDM-based and the MSM-based approaches can be understood as a consequence of the difference in the free energy barriers of the FEPs computed by the two methods. As the energy barrier for the flip-flop step is larger in the US FEP than in the unconstrained simulations-derived one, permeation through the membrane is intuitively less easily achieved within the ISDM framework than from the standpoint of the MSM-based liposome kinetic models. To a lesser extent, the slight variations of the energy barrier between the MSM-based FEPs can account for the slight differences in the permeability coefficient estimations (Table 1).

3. Methods and Materials

3.1. Building the Peptide-Membrane System

The peptide that we chose for the benchmark of the two methods that compute membrane permeability is a cyclic decapeptide named CDP5 (Figure 3). Four out of its ten residues were N-methylated to improve its cell permeability and oral bioavailability [26,27]. Its membrane permeability coefficient was measured using PAMPA experiments [9] with a hexadecane layer separating the donor and acceptor compartments [25]. It should be noted that this hexadecane layer differs from the lipid bilayer that will be considered in our simulations and which is a more realistic model of experimental liposomes. Nevertheless, as far as this study is concerned, the experimental value of CDP5 permeability coefficient reported in [25] ($\log P_{exp} = -6.7$) will be used as a reference for our theoretical results.

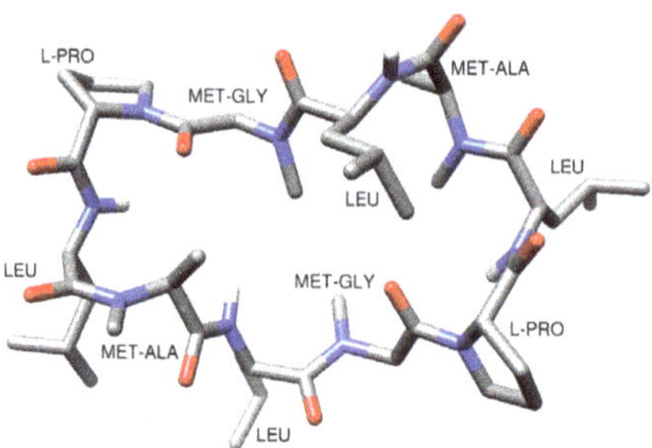

Figure 3. Sequence and structure of the cyclic decapeptide CDP5 with four N-methylated residues.

The modified peptide CDP5 was parameterized using the highly optimized chemical building blocks from CGenFF [28] to be compatible with the CHARMM36 force field for lipids [29]. An initial three-dimensional structure was generated from its sequence using OpenBabel [30] and PyMol [31]. After an energy minimization of 5000 steps and two short equilibration runs (1 ns in NVT and 2.5 ns in NPT ensemble), CDP5 was submitted to a 100 ns MD simulation in water with a sodium chloride concentration of 0.15 mol/L, using a V-rescale thermostat and a Parrinello-Raman barostat set at 300 K and 1 bar, respectively. Lennard-Jones potentials were cut off at 1.2 nm, electrostatic interactions were calculated using the smooth PME method [32], and covalent bonds with hydrogen were constrained using LINCS algorithm [33], for all simulations in this work. The resulting trajectory showed strong structural stability of the CDP5 peptide and was clustered based on RMSD with a cutoff value of 0.15 nm. The representative structure of the most populated cluster (Figure 3) provided the peptide conformation to be assembled with the membrane.

A lipid bilayer composed of 25 POPC (1-palmitoyl-2-oleoyl-sn-glycero-3-phos-phocholine) molecules per leaflet were built at the temperature of T = 293.15 K using CHARMM-GUI [34]. The membrane was energy minimized and equilibrated in water with 0.15 mol/L of NaCl by running a 250 ps NVT and a 125 ps NPT equilibration with a semi-isotropic pressure scaling, followed by three 500 ps NPT runs with decreasing position restraints on lipid phosphate groups from 1000 to 0 kJ/mol/Å^2. Then, the peptide and the lipid bilayer were assembled in a rectangular box, solvated, energy minimized, and equilibrated with 1 ns NVT and 1 ns NPT runs by decreasing the position restraints on peptide backbone heavy atoms and lipid phosphate groups from 4000 and 1000 to 50 and 0 kJ/mol/Å^2, respectively. The final conformation of the equilibration procedure was taken as the initial configuration for US simulations.

GROMACS 2021 [35] was used for all simulations, with CHARMM36 [29] force-field for lipids, and TIP3P [36] model for water.

3.2. ISDM-Based Calculation of Membrane Permeability

This subsection presents the calculations that were performed to estimate the peptide membrane permeability by using the inhomogeneous solubility-diffusion model (ISDM) [16].

Umbrella Sampling and WHAM Analysis

In the initial configuration of the peptide-membrane system, the solute is located in the bulk water at a distance z_{max} = 4 nm from the center of the lipid bilayer (the membrane plane being oriented perpendicular to the z-axis). Starting from this initial position, the solute was pulled toward and inside the lipid membrane by an increment of $\Delta z = 0.1$ nm until it reached the position $z_{min} = 0$ nm corresponding to the center of the bilayer. For each peptide position (or window), which is maintained with an umbrella potential of 1000 kJ/mol/Å^2, energy minimization was performed, followed by a 1 ns NVT and 1 ns NPT equilibration. Then, a production simulation of 250 ns was performed. The final conformation of the equilibration stage was pulled and used as a starting conformation for the next window.

The last 130 ns of the production trajectories were analyzed with the WHAM analysis [23] implemented in GROMACS [24]. More specifically, taking as input the values of the collective variable z for each window, gmx wham tool was used to compute both the free energy profile $\Delta G(z) = G(z) - G(z_{max})$ and the normalized integrated autocorrelation time τ_{zz} defined as:

$$\tau_{zz} = \frac{1}{\sigma_z^2} \int_0^\infty [z(0) - \langle z \rangle][z(t) - \langle z \rangle] dt \tag{1}$$

where σ_z^2 is the variance of the collective variable z defined as $\sigma_z^2 = \langle z^2 \rangle - \langle z \rangle^2$. It could be noted that gmx wham provides both averages and standard errors of the FEP and τ_{zz} by using bootstrap calculations [24].

Inhomogeneous Solubility-Diffusion Model

The ISDM considers that, due to the inhomogeneous nature of lipid membranes, the diffusion rate of a solute strongly depends on its position in the latter [16]. Moreover, it assumes that the diffusion process obeys the classical diffusion equation, implying that the dependence of $\langle z^2 \rangle$ on time is linear. Thus, the position-dependent diffusion coefficient can be calculated as the ratio of the variance σ_z^2 over the integrated autocorrelation time τ_{zz} provided by WHAM analyses of US simulations:

$$D(z) = \frac{\sigma_z^2}{\tau_{zz}} \tag{2}$$

Using the diffusion coefficient $D(z)$ and the free energy $\Delta G(z)$, the local resistance of the solute permeation through the membrane can be defined as [16]:

$$R(z) = \frac{\exp[\beta \Delta G(z)]}{D(z)} \tag{3}$$

where $\beta = 1/(k_B T)$, with T the temperature and k_B the Boltzmann's constant. The sum of all the $R(z)$ along the variable z gives the global resistance and its inverse yields the membrane permeability:

$$P_{ISDM} = \left[\int_{-z_{max}}^{z_{max}} R(z) dz \right]^{-1} \tag{4}$$

To estimate the errors on the permeability coefficient, 100 bootstrapped trajectories were generated from the original trajectory of each window using an in–house script. Then, for each bootstrapped trajectory, the permeability coefficient was calculated as described above. The final results were computed as the mean of these 100 values and errors were given by the standard deviation.

3.3. MSM-Based Calculation of Membrane Permeability

In this subsection, we present the different steps of the method for computing the peptide membrane permeability by using a kinetic Markov State Model (MSM) [21].

Multiple Unconstrained MD Simulations

A kinetic Markov State Model is essentially based on the transition probabilities between the model discretized states. These probabilities can be estimated by running multiple unbiased MD simulations, which make the system pass through its different minimum energy states. Nevertheless, to sufficiently sample the transition events, it is recommended to run these simulations from out-of-equilibrium initial configurations, which can "go down the hill" to the nearest metastable states in a short time [21,22]. The identification of these maximum energy states implies the preliminary determination of the system's free energy landscape. Regarding the permeation of CDP5 peptide, its FEP exhibits two main transition states, a major one at the center of the bilayer ($z = 0$ nm) and a minor one at the surface of the lipid headgroups (around $z = 2.4$ nm) (Figure 1). Consequently, we decided to run simulations of the peptide-membrane system, starting from four sets of 75 conformations each, extracted from the previous US simulations, with peptide positions $z = 0.0, 2.3, 2.4$, and 2.6 nm. Each of these 4×75 initial configurations were submitted to a 100 ns unconstrained MD production in the same conditions as the US simulations.

Building the Markov State Models

To study the sensitivity of the computed permeability to the initial configurations of the peptide-membrane system, we built four Markov models from four different sets of MD trajectories (Table 2).

Table 2. Markov State Models built from four different sets of 100 ns unconstrained trajectories of the peptide-membrane system.

MSM Name	Nb of Trajectories	Initial Positions
CDP5_2.3	2×75	$z = 0.0$ and 2.3 nm
CDP5_2.4	2×75	$z = 0.0$ and 2.4 nm
CDP5_2.6	2×75	$z = 0.0$ and 2.6 nm
CDP5_all	4×75	$z = 0.0, 2.3, 2.4$, and 2.6 nm

PyEMMA [22] was used to build and analyze the four MSMs. For each kinetic model, the estimated implied relaxation timescale was plotted as a function of different lag times τ (Figure 4). From these graphs, we considered that the implied relaxation timescales are approximately constant from 3 timesteps, i.e., $\tau = 300$ ps, and that this lag time is suitable for calculating accurate kinetic coefficients of all MSMs. These plots also indicate that the models are characterized by three implied slow relaxation timescales (blue, red, and green lines in Figure 4). The transition matrix was estimated using the above lag time value, and a Chapman–Kolmogorov test was performed to verify the Markovianity of each model [22].

Next, PCCA+ [37] clustering was applied to describe all trajectories of each MSM in terms of a few metastable states. Since three implied slow relaxation timescales could be identified, we asked PCCA+ to cluster the trajectories into four metastable states (Figure 5). Moreover, PCCA+ provides the values of the MFPTs between each pair of metastable states (Figure 5), which were used to derive the rate constants corresponding to the entry of the solute into the membrane k_{in}, the exit out of it k_{out}, and the flip-flop transitions from one lipid layer to the other k_{flip} (Figure 6A).

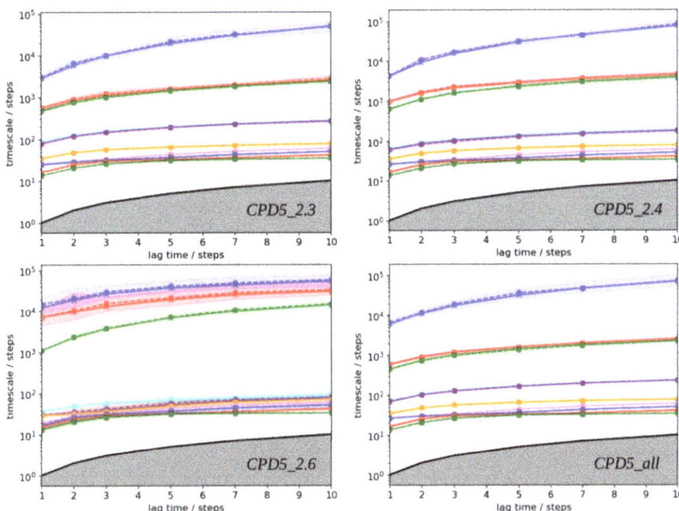

Figure 4. PyEMMA implied timescales as a function of lag time for each Markov State Model. Timescales are sorted in descending order using the color lines blue, red, green, cyan, violet, yellow, and pink. Black lines indicate timescale equal to lag time.

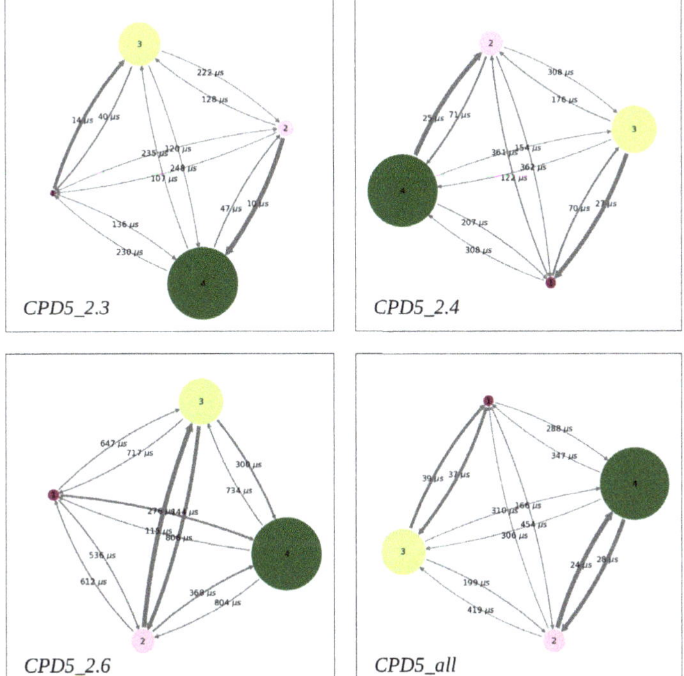

Figure 5. PyEMMA metastable states are represented by colored discs with a radius proportional to their population. The most populated states (labeled 3 and 4) always identify the water phase, while the smaller ones (labeled 1 and 2) represent the two lipid leaflets. Transitions are described with arrows whose thickness and label are associated to the rate constant and MFPT, respectively.

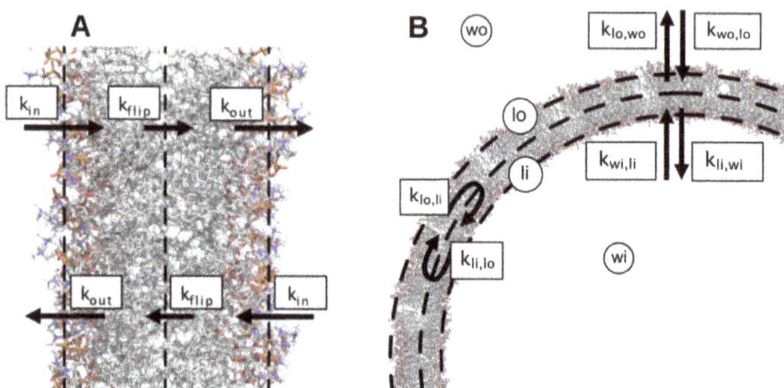

Figure 6. Designation of the kinetic rate constants in membrane planar (**A**) and spherical (**B**) geometry. In the liposome model, wo and lo stands for the outer water and lipid compartment, respectively, and similarly for the two inner compartments wi and li.

Kinetic Model Master Equations

Similarly to what was done by Dickson et al. [18], the master equations describing the passive diffusion of a solute from an outer water compartment into the water interior of a spherical liposome were solved by considering two lipid compartments (indicated as outer and inner lipid compartments) as schematically depicted in Figure 6B. The differential equations describing the time dependence of the solute in the four compartments read:

$$\frac{dS_I(t)}{dt} = \sum_J \left[-k_{IJ}S_I(t) + k_{JI}S_J(t) \right]$$

$$S_{wo}(0) = S_{tot}, \qquad S_{lo}(0) = S_{li}(0) = S_{wi}(0) = 0$$

(5)

where $I, J = wo, lo, li, wi$ are the indices that refer to the outer water, outer lipid, inner lipid, and inner water compartments, respectively, and the summation runs over the nearest compartments. At time $t = 0$, the solute is entirely in the outer water compartment (Equation (5)).

Knowing that the rate constant k of each kinetic step depends on the permeability coefficient P, the surface area A separating two compartments, and the volume V of the donor compartment, through the relation $k = P(A/V)$, it should be noted that the apparent rate constants $k_{wo,lo}, k_{lo,wo}, k_{lo,li}, k_{li,lo}, k_{li,wi}$, and $k_{wi,li}$ in the membrane spherical geometry are related to the rate constants k_{in}, k_{out}, and k_{flip} previously estimated by the Markov State Models in the membrane planar geometry as

$$k_{wo,lo} = \frac{V_W}{V_{wo}}\frac{A_l}{A_L}k_{in}, \quad k_{wi,li} = \frac{V_W}{V_{wi}}\frac{A_l}{A_L}k_{in}$$

$$k_{lo,wo} = \frac{V_L}{V_l}\frac{A_l}{A_L}k_{out}, \quad k_{li,wi} = \frac{V_L}{V_l}\frac{A_l}{A_L}k_{out}$$

$$k_{lo,li} = \frac{V_L}{V_l}\frac{A_l}{A_L}k_{flip}, \quad k_{li,lo} = \frac{V_L}{V_l}\frac{A_l}{A_L}k_{flip}$$

(6)

where A_L, V_L, and V_W are the lipid bilayer surface area, its interior volume, and the bulk water volume in the planar membrane systems simulated by MD, respectively. In the spherical liposome model, the surface area is considered equal on both outer and inner layers: $A_l \equiv A_{li} = A_{lo}$. Similarly, we assumed that the volumes of the two compartments in the lipid bilayer are equal: $V_l \equiv V_{li} = V_{lo}$. Quantitatively, we used a liposome radius $r = 100$ nm and a ratio of 1000:1:1:10 for the volumes of the four compartments $V_{wo} : V_{lo} : V_{li} : V_{wi}$, conforming to the experimental conditions of the liposome experiments reported in [10]. For the MD areas and volumes, we used the POPC area per lipid $A_L = 65.6$ Å^2 and the volume per lipid $V_L = 1.2$ nm^3 [18]. The bulk water volume V_W was calculated from the volume per molecule $30.5\,\text{Å}^3$ and the number of water per lipid (89 in our simulated systems).

Equations (5) are a system of four ordinary differential equations (three of which are independent) that we solved numerically using SciPy routines [38]. We also solved the system analytically, using the numerical values of the eigenvalues of the matrix associated with the system of differential equations (Appendix A).

Permeability Coefficient from Kinetic Models

Following reference [39] and subsequent papers [10,40], the permeation coefficient is extracted from experimental data by fitting the luminescence curve with a biexponential function as a function of time:

$$I(t) = I(\infty) + [I(\infty)_I - I(0)_I]e^{-k_I t} + [I(\infty)_{II} - I(0)_{II}]e^{-k_{II} t} \tag{7}$$

where $I(\infty) = I(\infty)_I + I(\infty)_{II}$ is the total maximal luminescence, $I(\infty)_I$ and $I(\infty)_{II}$ are the maximal luminescence of the fast and slow phase respectively, $I(0)_I$ and $I(0)_{II}$ are the initial luminescence for the two phases, and k_I and k_{II} are the corresponding rate constants.

For luminescence experiments with liposomes, the permeation coefficient has been related to the rate constant k_I of the fast exponential phase and the liposome radius as:

$$P_{lipo} = \frac{V_{wi}}{A_l}k_I = \frac{r}{3}k_I \tag{8}$$

From the point of view of the four-step model describing the process, where a peptide undergoes adsorption, flip-flop, and desorption through the liposome membrane bilayer, the relevant kinetics are expressed as three exponential functions, as in Equation (A7), rather than the two fitting functions for the luminescence intensity of Equation (7). We can, however, still use the rate constant associated with the fastest exponential phase in the liposome model and substitute its value in Equation (8) to obtain the membrane permeability coefficient, as it has been done in [18].

Errors on the permeability coefficient are derived using the Bayesian estimator in PyEMMA for the MSMs, which generates samples of transition matrices rather than one single transition matrix, as in the case of the maximum likelihood estimator. With this method, we obtained a sample of 100 transition matrices for each MSM, from which PyEMMA can compute samples for different quantities, such as the MFPT. The reported results were evaluated based on the mean value of the MFPT samples and their standard deviation, using propagation of uncertainties to compute the errors on the rate constants.

4. Conclusions

The ISDM approach is probably the most popular method used to estimate the membrane permeability of chemical compounds. However, Dickson et al. recently showed that an alternative method, based on Markov State Models of the permeation process, yielded permeability coefficients in better agreement with experimental data for seven small molecules [18]. In the present study, we applied and compared the two computational methods for a larger compound, the cyclic decapeptide CDP5. As opposed to the tendency reported by Dickson et al., we found that the ISDM-based and MSM-based method slightly underestimates and largely overestimates the peptide membrane permeability, respectively.

A probable explanation of the peptide permeability overestimation by our MSM-based calculations lies in the too-short duration of its unconstrained MD simulations required to build the Markov State Models. Indeed, due to its size and polarity, the CDP5 decapeptide cannot pass through the lipid headgroups as easily and rapidly as the small compounds simulated by Dickson et al. [18]. Thus, the peptide unconstrained simulations starting from maximum energy positions on the lipid headgroup surface in water require longer times to go down the hill and move towards the energy minimum inside the membrane. As a consequence, the peptide population π and the free energy $\Delta G = -k_B T \log \pi$ at these maximum energy positions are larger and lower than expected, respectively. Accordingly, the MSM-based peptide permeability coefficient is excessively large compared to experiments and ISDM calculations.

To remedy this default, it would probably be necessary to extend the duration of the unconstrained MD simulations and/or multiply their starting points. However, for the building of one Markov State Model, we have already performed a total of 15 µs MD simulations, whereas the ISDM approach has needed 10 µs US. In addition, the building of accurate MSMs requires running unconstrained MD simulations from maximum free energy states, which, very often, have to be predetermined by US calculations. Thus, predictions of the membrane permeability of peptides with the MSM-based method would require much more computer resources than the ISDM approach, which was able to yield reasonable estimations.

Author Contributions: Conceptualization, L.M. and T.H.-D.; Methodology, L.M. and T.H.-D.; Software, L.M. and T.H.-D.; Validation, L.M. and T.H.-D.; Formal analysis, L.M. and T.H.-D.; Investigation, L.M. and T.H.-D.; Resources, T.H.-D.; Data curation, L.M. and T.H.-D.; Writing—original draft, L.M. and T.H.-D.; Writing—review & editing, L.M. and T.H.-D.; Visualization, L.M. and T.H.-D.; Supervision, T.H.-D.; Project administration, T.H.-D.; Funding acquisition, T.H.-D. All authors have read and agreed to the published version of the manuscript.

Funding: This work was supported by HPC resources from GENCI-IDRIS (grants AD010713502 and A0130713807).

Data Availability Statement: All molecular dynamics trajectories supporting results reported in this study are available on request from the corresponding author. The raw data are not publicly available due to their size which prevent them from being uploaded to a publicly available repository.

Conflicts of Interest: The authors declare no conflict of interest regarding the publication of this article.

Appendix A. Analytical Resolution of the Kinetic Model Equations

Equation (5) can be rewritten in matrix form as

$$\frac{d\vec{S}(t)}{dt} = D \cdot \vec{S}(t) + \vec{B}$$
$$\vec{S}(0) = \vec{S}_0 \tag{A1}$$

where $\vec{S}$ (and similarly $\vec{S}_0$) is defined as

$$\vec{S} = \begin{pmatrix} S_{wo} \\ S_- \\ S_{wi} \end{pmatrix}, \quad S_- = \frac{1}{2}(S_i - S_o), \quad S_i = S_{wi} + S_{li} \tag{A2}$$

The matrix D, the coefficients vector $\vec{B}$, and the initial conditions $\vec{S}_0$ read:

$$D = \begin{pmatrix} -K_o & k_o & 0 \\ -k_f & -K_f & k'_f \\ 0 & k_i & -K_i \end{pmatrix} \tag{A3}$$

$$\vec{B} = \frac{S_{tot}}{2} \begin{pmatrix} k_o \\ k_- \\ k_i \end{pmatrix}, \quad \vec{S}_0 = \frac{S_{tot}}{2} \begin{pmatrix} 1 \\ -\frac{1}{2} \\ 0 \end{pmatrix} \tag{A4}$$

where S_{tot} indicates the constant total amount of solute in the system that is initially contained entirely in the outer aqueous compartment, as aforementioned. We have simplified notations by defining:

$$\begin{aligned} k_o &\equiv k_{lo,wo}, & K_o &\equiv k_o + k_{wo,lo} \\ k_i &\equiv k_{li,wi}, & K_i &\equiv k_i + k_{wi,li} \\ k_f &\equiv k_{li,lo}, & k'_f &\equiv k_{lo,li} \\ K_f &= k_f + k'_f, & k_- &= k'_f - k_f \end{aligned} \tag{A5}$$

Appendix A.1. Approximated Solution for the Inner Acqueous Compartment

Knowing that the volume of the outer aqueous compartment is much larger than the volumes of the other compartments, and for symmetry reasons, we infer that

$$K_o \simeq k_o, \quad k_i \simeq k_o, \quad K_f \simeq 2k'_f \simeq 2k_f \tag{A6}$$

The solution to the differential equations yields the time evolution for the solute in the aqueous inner compartment, which is given by the sum of three exponential functions plus constants, in terms of the rate constants and of the eigenvalues $\lambda_1, \lambda_2, \lambda_3$ of the matrix D, with $\lambda_n = -\kappa_n$ and $\kappa_n \geq 0$, for $n = 1, 2, 3$:

$$S_{wi}(t) = \sum_{n=1}^{3} \left[\frac{\beta_n}{\kappa_n} + \left(\nu_{0n} - \frac{\beta_n}{\kappa_n} \right) e^{-\kappa_n t} \right] \tag{A7}$$

where ν_{0n} are the initial conditions for the eigenvectors ν_n, and β_n are the components of $\vec{B}$ in the eigenvector basis:

$$\nu_{0n} \equiv \nu_n(t = 0) = \frac{S_{tot}}{2} k_i \frac{K_i - \sum_{m \neq n} \kappa_m}{\prod_{m \neq n} (\kappa_m - \kappa_n)} \tag{A8}$$

$$\beta_n = \frac{S_{tot}}{2} k_i \prod_{m \neq n} \frac{(K_i - \kappa_m)}{(\kappa_m - \kappa_n)}. \tag{A9}$$

The initial conditions ν_{0n} and the coefficients β_n are here derived in the approximations (A6), while the exact expressions are reported in Appendix A.2, together with the solutions for the solute in the remaining three compartments. In the limit $K_i \to k_i$, which implies $k_{wi,li} \ll k_{li,wi}$, the substrates are trivially constant as a function of time and equal to their initial values fixed in Equation (5):

$$S_{wo}(t) \equiv S_{tot}, \quad S_{lo}(t) \equiv S_{li}(t) \equiv S_{wi}(t) \equiv 0 \tag{A10}$$

The condition $K_i \simeq k_i$ would be satisfied if the inner aqueous compartment became much larger than the volume of the membrane, for instance, assuming that the rate constants are roughly of the same order of magnitude (as it will generally be the case for the current models). As we depart from this trivial limit, in other words, as the inner aqueous compartment volume becomes comparable to the membrane volume, we expect that the substrate would start decreasing in the outer aqueous compartment to partly fill the inner aqueous compartment once equilibrium is reached, as will be confirmed by our results.

Appendix A.2. Full Solutions for the Kinetic Model Equations

The exact solutions to the set of Equation (5) describing the time evolution of the solute inside the liposome model illustrated in Figure 6B are sums of three exponential functions plus constants:

$$S_{wo}(t) = \sum_{n=1}^{3} \left[\frac{k'_f}{k_f} - \frac{(K_i - \kappa_n)\left(k_f + k'_f - \kappa_n\right)}{k_i k_f} \right] \nu_n(t)$$

$$S_{lo}(t) = \sum_{n=1}^{3} \left[\frac{k'_f}{k_f} - \frac{(K_i - \kappa_n)\left(k'_f - \kappa_n\right)}{k_i k_f} \right] \nu_n(t) + \frac{S_{tot}}{2}$$

$$S_{li}(t) = \sum_{n=1}^{3} \left[\frac{k'_f}{k_f} - \frac{(K_i - \kappa_n)}{k_i} \right] \nu_n(t) + \frac{S_{tot}}{2} \tag{A11}$$

$$S_{wi}(t) = \sum_{n=1}^{3} \nu_n(t)$$

where $v_n(t)$ are the eigenvectors solutions to the system of differential equations:

$$v_n(t) = \frac{\beta_n}{\kappa_n} + \left(v_{0n} - \frac{\beta_n}{\kappa_n}\right)e^{-\kappa_n t}, \quad n = 1,2,3 \tag{A12}$$

If no specific approximation is applied for the membrane system, the initial conditions v_{0n} for the eigenvectors v_n yield:

$$v_{0n} = \frac{S_{tot}}{2}k_i \frac{K_i + k_- - \sum_{m \neq n}\kappa_m}{\prod_{m \neq n}(\kappa_m - \kappa_n)} \tag{A13}$$

and the constant coefficients β_n can be expressed as

$$\beta_j = \frac{S_{tot}}{2}k_i \frac{\prod_{m \neq n}(K_i - \kappa_m) + k_-\left(K_f + K_i - \sum_{m \neq n}\kappa_m\right) - k_f k_o + k'_f k_i}{\prod_{m \neq n}(\kappa_m - \kappa_n)} \tag{A14}$$

References

1. Ju, Y.; Guo, H.; Edman, M.; Hamm-Alvarez, S.F. Application of advances in endocytosis and membrane trafficking to drug delivery. *Adv. Drug Deliv. Rev.* **2020**, *157*, 118–141. [CrossRef] [PubMed]
2. Dean, M.; Moitra, K.; Allikmets, R. The human ATP-binding cassette (ABC) transporter superfamily. *Hum. Mutat.* **2022**, *43*, 1162–1182. [CrossRef] [PubMed]
3. Pizzagalli, M.D.; Bensimon, A.; Superti-Furga, G. A guide to plasma membrane solute carrier proteins. *FEBS J.* **2021**, *288*, 2784–2835. [CrossRef] [PubMed]
4. Vergalli, J.; Bodrenko, I.V.; Masi, M.; Moynié, L.; Acosta-Gutiérrez, S.; Naismith, J.H.; Davin-Regli, A.; Ceccarelli, M.; van den Berg, B.; Winterhalter, M.; et al. Porins and small-molecule translocation across the outer membrane of Gram-negative bacteria. *Nat. Rev. Microbiol.* **2020**, *18*, 164–176. [CrossRef]
5. Moynié, L.; Milenkovic, S.; Mislin, G.L.A.; Gasser, V.; Malloci, G.; Baco, E.; McCaughan, R.P.; Page, M.G.P.; Schalk, I.J.; Ceccarelli, M.; et al. The complex of ferric-enterobactin with its transporter from Pseudomonas aeruginosa suggests a two-site model. *Nat. Commun.* **2019**, *10*, 3673. [CrossRef]
6. Sugano, K.; Kansy, M.; Artursson, P.; Avdeef, A.; Bendels, S.; Di, L.; Ecker, G.F.; Faller, B.; Fischer, H.; Gerebtzoff, G.; et al. Coexistence of passive and carrier-mediated processes in drug transport. *Nat. Rev. Drug Discov.* **2010**, *9*, 597–614. [CrossRef]
7. Di, L.; Artursson, P.; Avdeef, A.; Ecker, G.F.; Faller, B.; Fischer, H.; Houston, J.B.; Kansy, M.; Kerns, E.H.; Krämer, S.D.; et al. Evidence-based approach to assess passive diffusion and carrier-mediated drug transport. *Drug Discov. Today* **2012**, *17*, 905–912. [CrossRef]
8. Lau, J.L.; Dunn, M.K. Therapeutic peptides: Historical perspectives, current development trends, and future directions. *Bioorg. Med. Chem.* **2018**, *26*, 2700–2707. [CrossRef]
9. Kansy, M.; Senner, F.; Gubernator, K. Physicochemical High Throughput Screening: Parallel Artificial Membrane Permeation Assay in the Description of Passive Absorption Processes. *J. Med. Chem.* **1998**, *41*, 1007–1010. [CrossRef]
10. Eyer, K.; Paech, F.; Schuler, F.; Kuhn, P.; Kissner, R.; Belli, S.; Dittrich, P.S.; Kramer, S.D. A liposomal fluorescence assay to study permeation kinetics of drug-like weak bases across the lipid bilayer. *J. Control. Release* **2014**, *173*, 102–109. [CrossRef]
11. Biedermann, F.; Ghale, G.; Hennig, A.; Nau, W.M. Fluorescent artificial receptor-based membrane assay (FARMA) for spatiotemporally resolved monitoring of biomembrane permeability. *Commun. Biol.* **2020**, *3*, 383. [CrossRef] [PubMed]
12. Di Pisa, M.; Chassaing, G.; Swiecicki, J.M. Translocation Mechanism(s) of Cell-Penetrating Peptides: Biophysical Studies Using Artificial Membrane Bilayers. *Biochemistry* **2015**, *54*, 194–207. [CrossRef]
13. Ulmschneider, J.P.; Ulmschneider, M.B. Molecular Dynamics Simulations Are Redefining Our View of Peptides Interacting with Biological Membranes. *Acc. Chem. Res.* **2018**, *51*, 1106–1116. [CrossRef] [PubMed]
14. Swift, R.V.; Amaro, R.E. Back to the Future: Can Physical Models of Passive Membrane Permeability Help Reduce Drug Candidate Attrition and Move Us Beyond QSPR? *Chem. Biol. Drug Des.* **2013**, *81*, 61–71. [CrossRef] [PubMed]
15. Kabelka, I.; Brozek, R.; Vacha, R. Selecting Collective Variables and Free-Energy Methods for Peptide Translocation across Membranes. *J. Chem. Inf. Model.* **2021**, *61*, 819–830. [CrossRef]
16. Marrink, S.J.; Berendsen, H.J.C. Simulation of water transport through a lipid membrane. *J. Phys. Chem.* **1994**, *98*, 4155–4168. [CrossRef]
17. Bennion, B.J.; Be, N.A.; McNerney, M.W.; Lao, V.; Carlson, E.M.; Valdez, C.A.; Malfatti, M.A.; Enright, H.A.; Nguyen, T.H.; Lightstone, F.C.; et al. Predicting a Drug's Membrane Permeability: A Computational Model Validated with in Vitro Permeability Assay Data. *J. Phys. Chem. B* **2017**, *121*, 5228–5237. [CrossRef]
18. Dickson, C.J.; Hornak, V.; Pearlstein, R.A.; Duca, J.S. Structure–Kinetic Relationships of Passive Membrane Permeation from Multiscale Modeling. *J. Am. Chem. Soc.* **2017**, *139*, 442–452. [CrossRef]

19. Faulkner, C.; de Leeuw, N.H. Predicting the Membrane Permeability of Fentanyl and Its Analogues by Molecular Dynamics Simulations. *J. Phys. Chem. B* **2021**, *125*, 8443–8449. [CrossRef]

20. Sugita, M.; Sugiyama, S.; Fujie, T.; Yoshikawa, Y.; Yanagisawa, K.; Ohue, M.; Akiyama, Y. Large-Scale Membrane Permeability Prediction of Cyclic Peptides Crossing a Lipid Bilayer Based on Enhanced Sampling Molecular Dynamics Simulations. *J. Chem. Inf. Model.* **2021**, *61*, 3681–3695. [CrossRef]

21. Prinz, J.H.; Wu, H.; Sarich, M.; Keller, B.; Senne, M.; Held, M.; Chodera, J.D.; Schutte, C.; Noé, F. Markov models of molecular kinetics: Generation and validation. *J. Chem. Phys.* **2011**, *134*, 174105. [CrossRef]

22. Scherer, M.K.; Trendelkamp-Schroer, B.; Paul, F.; Pérez-Hernández, G.; Hoffmann, M.; Plattner, N.; Wehmeyer, C.; Prinz, J.H.; Noe, F. PyEMMA 2: A Software Package for Estimation, Validation, and Analysis of Markov Models. *J. Chem. Theory Comput.* **2015**, *11*, 5525–5542. [CrossRef]

23. Kumar, S.; Rosenberg, J.M.; Bouzida, D.; Swendsen, R.H.; Kollman, P.A. The weighted histogram analysis method for free-energy calculations on biomolecules. I. The method. *J. Comput. Chem.* **1992**, *13*, 1011–1021. [CrossRef]

24. Hub, J.S.; de Groot, B.L.; van der Spoel, D. g_wham-A Free Weighted Histogram Analysis Implementation Including Robust Error and Autocorrelation Estimates. *J. Chem. Theory Comput.* **2010**, *6*, 3713–3720. [CrossRef]

25. Wang, S.; König, G.; Roth, H.J.; Fouché, M.; Rodde, S.; Riniker, S. Effect of Flexibility, Lipophilicity, and the Location of Polar Residues on the Passive Membrane Permeability of a Series of Cyclic Decapeptides. *J. Med. Chem.* **2021**, *64*, 12761–12773. [CrossRef]

26. Fouché, M.; Schäfer, M.; Berghausen, J.; Desrayaud, S.; Blatter, M.; Piéchon, P.; Dix, I.; Martin Garcia, A.; Roth, H.J. Design and Development of a Cyclic Decapeptide Scaffold with Suitable Properties for Bioavailability and Oral Exposure. *ChemMedChem* **2016**, *11*, 1048–1059. [CrossRef] [PubMed]

27. Fouché, M.; Schäfer, M.; Blatter, M.; Berghausen, J.; Desrayaud, S.; Roth, H.J. Pharmacokinetic Studies around the Mono- and Difunctionalization of a Bioavailable Cyclic Decapeptide Scaffold. *ChemMedChem* **2016**, *11*, 1060–1068. [CrossRef] [PubMed]

28. Vanommeslaeghe, K.; Hatcher, E.; Acharya, C.; Kundu, S.; Zhong, S.; Shim, J.; Darian, E.; Guvench, O.; Lopes, P.; Vorobyov, I.; et al. CHARMM general force field: A force field for drug-like molecules compatible with the CHARMM all-atom additive biological force fields. *J. Comput. Chem.* **2010**, *31*, 671–690. [CrossRef] [PubMed]

29. Huang, J.; MacKerell, A.D.J. CHARMM36 all-atom additive protein force field: Validation based on comparison to NMR data. *J. Comput. Chem.* **2013**, *34*, 2135–2145. [CrossRef]

30. O'Boyle, N.M.; Banck, M.; James, C.A.; Morley, C.; Vandermeersch, T.; Hutchison, G.R. Open Babel: An open chemical toolbox. *J. Cheminform.* **2011**, *3*, 33. [CrossRef]

31. Schrödinger, LLC. *The PyMOL Molecular Graphics System*, Version 1.8; Schrödinger, LLC.: Cambridge, MA, USA, 2015.

32. Essmann, U.; Perera, L.; Berkowitz, M.L.; Darden, T.; Lee, H.; Pedersen, L.G. A smooth particle mesh Ewald method. *J. Chem. Phys.* **1995**, *103*, 8577–8593. [CrossRef]

33. Hess, B.; Bekker, H.; Berendsen, H.J.C.; Fraaije, J.G.E.M. LINCS: A linear constraint solver for molecular simulations. *J. Comput. Chem.* **1997**, *18*, 1463–1472. [CrossRef]

34. Lee, J.; Cheng, X.; Swails, J.M.; Yeom, M.S.; Eastman, P.K.; Lemkul, J.A.; Wei, S.; Buckner, J.; Jeong, J.C.; Qi, Y.; et al. CHARMM-GUI Input Generator for NAMD, GROMACS, AMBER, OpenMM, and CHARMM/OpenMM Simulations Using the CHARMM36 Additive Force Field. *J. Chem. Theory Comput.* **2016**, *12*, 405–413. [CrossRef]

35. Abraham, M.J.; Murtola, T.; Schulz, R.; Páll, S.; Smith, J.C.; Hess, B.; Lindahl, E. GROMACS: High performance molecular simulations through multi-level parallelism from laptops to supercomputers. *SoftwareX* **2015**, *1–2*, 19–25. [CrossRef]

36. Jorgensen, W.L.; Chandrasekhar, J.; Madura, J.D.; Impey, R.W.; Klein, M.L. Comparison of simple potential functions for simulating liquid water. *J. Chem. Phys.* **1983**, *79*, 926–935. [CrossRef]

37. Röblitz, S.; Weber, M. Fuzzy spectral clustering by PCCA+: Application to Markov state models and data classification. *Adv. Data Anal. Classif.* **2013**, *7*, 147–179. [CrossRef]

38. Virtanen, P.; Gommers, R.; Oliphant, T.E.; Haberland, M.; Reddy, T.; Cournapeau, D.; Burovski, E.; Peterson, P.; Weckesser, W.; Bright, J.; et al. SciPy 1.0: Fundamental Algorithms for Scientific Computing in Python. *Nat. Methods* **2020**, *17*, 261–272. [CrossRef] [PubMed]

39. Thomae, A.V.; Wunderli-Allenspach, H.; Krämer, S.D. Permeation of aromatic carboxylic acids across lipid bilayers: The pH-partition hypothesis revisited. *Biophys. J.* **2005**, *89*, 1802–1811. [CrossRef]

40. Thomae, A.V.; Koch, T.; Panse, C.; Wunderli-Allenspach, H.; Krämer, S.D. Comparing the lipid membrane affinity and permeation of drug-like acids: The intriguing effects of cholesterol and charged lipids. *Pharm. Res.* **2007**, *24*, 1457–1472. [CrossRef]

International Journal of
Molecular Sciences

Article

Identifying Drug Candidates for COVID-19 with Large-Scale Drug Screening

Yifei Wu [1], Scott D. Pegan [2], David Crich [3], Lei Lou [1], Lauren Nicole Mullininx [1,4], Edward B. Starling [1,4], Carson Booth [1,4], Andrew Edward Chishom [1,4], Kuan Y. Chang [5,*] and Zhong-Ru Xie [1,*]

[1] School of Electrical and Computer Engineering, College of Engineering, University of Georgia, Athens, GA 30602, USA
[2] Division of Biomedical Sciences, School of Medicine, University of California Riverside, Riverside, CA 92521, USA
[3] Department of Pharmaceutical and Biomedical Sciences, College of Pharmacy, University of Georgia, Athens, GA 30602, USA
[4] Franklin College of Arts and Sciences, University of Georgia, Athens, GA 30602, USA
[5] Department of Computer Science and Engineering, National Taiwan Ocean University, Keelung 202, Taiwan
* Correspondence: kchang@ntou.edu.tw (K.Y.C.); paulxie@uga.edu (Z.-R.X.)

Abstract: Papain-like protease (PL^{pro}) is critical to COVID-19 infection. Therefore, it is a significant target protein for drug development. We virtually screened a 26,193 compound library against the PL^{pro} of SARS-CoV-2 and identified several drug candidates with convincing binding affinities. The three best compounds all had better estimated binding energy than those of the drug candidates proposed in previous studies. By analyzing the docking results for the drug candidates identified in this and previous studies, we demonstrate that the critical interactions between the compounds and PL^{pro} proposed by the computational approaches are consistent with those proposed by the biological experiments. In addition, the predicted binding energies of the compounds in the dataset showed a similar trend as their IC$_{50}$ values. The predicted ADME and drug-likeness properties also suggested that these identified compounds can be used for COVID-19 treatment.

Keywords: COVID-19; virtual screening; PL protease; MD simulations; drug discovery

Citation: Wu, Y.; Pegan, S.D.; Crich, D.; Lou, L.; Mullininx, L.N.; Starling, E.B.; Booth, C.; Chishom, A.E.; Chang, K.Y.; Xie, Z.-R. Identifying Drug Candidates for COVID-19 with Large-Scale Drug Screening. *Int. J. Mol. Sci.* **2023**, *24*, 4397. https://doi.org/10.3390/ijms24054397

Academic Editors: Geoffrey Brown, Enikö Kallay and Andrzej Kutner

Received: 3 January 2023
Revised: 11 February 2023
Accepted: 16 February 2023
Published: 23 February 2023

1. Introduction

With more than 756 million cases and 6.8 million deaths worldwide as of February 2023, the SARS-CoV-2 (COVID-19) pandemic is one of the most significant pandemics in recent history [1]. In the United States, deaths due to this virus have surpassed the 1918 H1N1 pandemic, the most lethal flu pandemic in the last century [2–4]. To curb these numbers, vaccines against COVID-19 have been in development since the sequencing of the virus in early 2020 [5,6]. As of the writing of this article, many vaccines have been approved worldwide and more are still in development: Pfizer-BioNTech and Moderna in the United States, Oxford-AstraZeneca in the United Kingdom, Covaxin in India, etc. [7]. Despite this success, cases continue to surge in the US, Japan, Germany, and many other high-income nations [8]. Although they have not experienced as many infections, low-income nations are disproportionally underserved by current vaccines, resulting in extended restrictions on their populations [7]. To ease the negative impact of these issues, continued drug development for COVID-19 is very important.

Drugs developed for COVID-19 vary in size and primarily target one of three viral proteins: RNA-dependent RNA polymerase (RdRp), 3C-like protease (3CL^{pro} or M^{pro}), and papain-like protease (PL^{pro}) [9,10]. Remdesivir, the first COVID-19 drug approved by the FDA, is an adenine analog nucleoside inhibitor that disables RdRp by incorporating itself into the new sequence and blocking transcription [11,12]. Molnupiravir (EIDD-2801) and Favipiravir are further drugs of this type in development [9–11]. The next most prominent drug in development is PF-07321332, developed by Pfizer, which binds to the active site of 3CL^{Pro} using a series of hydrogen bonds, with Glu166, Gln189, and His164 of 3CL^{Pro} being the most important residues anchoring and stabilizing the ligand throughout the binding process [10,13]. Its mechanism is similar to Boceprevir, which inhibits COVID-19 3CL^{Pro} and has been approved by the FDA to treat HCV [10,14]. On a separate note, many antibodies, such as REGEN-COV and Sotrovimab, approved by the FDA to treat mild COVID-19 have different targets from the aforementioned proteins [9].

PL^{Pro} is a well-known cysteine protease. PL^{Pro} not only cleaves non-structural proteins 1–3 (nsp1–3) from polyprotein 1ab, a major protein involved in early COVID-19 infection [10,15,16], but also aids in immune evasion by cleaving ubiquitin (Ub) and interferon (IFN)-stimulated gene 15 (ISG15), which are significant parts of the NF-κB pathway. Near the catalytic site, the last four residues of Ub bind to leucine, arginine, and two glycines, respectively, which are known as P1–P4. Most of the above substrates are anchored by P1–P4 before cleavage by the catalytic site [10]. Along with this, PL^{Pro} plays a role in attaching the replication/transcription complex (RTC) to the modified endoplasmic reticulum (ER) membranes [15].

This wide range of functions makes PL^{Pro} an attractive drug target. Most drugs against PL^{Pro} prevent Ub and ISG15 from reaching the catalytic site. GRL-0617, the most prominent drug candidate against PL^{Pro}, binds to P3 and P4 using H-bonds and hydrophobic interactions between Tyr265, Tyr 269, Pro248, and Pro249. GRL-0667, a more potent inhibitor (IC$_{50}$ = 0.32 μM), utilizes a similar mechanism, with its smaller size and conformation allowing for greater interaction with the binding site [10]. Some derivatives of GRL-0617's functional groups are more potent than GRL-0617, such as XR8-89 (IC$_{50}$ = 0.113 μM) and XR8-83 (IC$_{50}$ = 0.21 μM) [16]. Acting as a Zn^{2+} ejector and disrupting the interaction with Cys111, Ebselen also inhibits PL^{Pro} with an IC$_{50}$ of 0.67 $\pm$ 0.09 μM [10,17]. Regarding virtual screening studies, cyanobacterial metabolites, such as Cryptophycin 1 (-7.7 ± 0.09 kcal/mol), and compounds from natural sources, such as matcha green tea and Nigella sativa, have good binding affinity when docked to PL^{Pro}, indicating likely inhibitory activity as well [18–21].

To push drug development further, we screened a 26,193 compound library for potential candidates against PL^{Pro} using molecular docking and molecular mechanics with generalized Born and surface area (MM-GBSA) calculations. While the potential candidates were compared to GRL-0617 and our previous best compound, the top three candidates were selected to undergo further molecular dynamics (MD) simulations. Overall, this study identified three novel drug candidates (F3077-0136, F2883-0639, and F0514-5148) against PL^{Pro} that have lower binding affinities than the control compounds, potentially surpassing them in inhibitory power.

2. Results

2.1. Docking Analysis of Top Three Compounds against SARS-CoV-2 PL^{pro}

To develop effective drugs against SARS-CoV-2 PL^{Pro}, we docked 26,193 compounds on PL^{Pro} (PDB ID: 7LBR) by performing ligand–protein docking. Based on the docking results, the binding energies were calculated using Prime MM-GBSA in Maestro. As a result, the top three compounds with the best binding energies were found to be F3077-0136 (-91.92 kcal/mol), F2883-0639 (-90.96 kcal/mol), and F0514-5148 (-89.66 kcal/mol) (Table 1). Moreover, a reported PL^{Pro} inhibitor (GRL-0617) [22,23], the best drug candidate kaempferol 3-O-sophoroside 7-O-glucoside from our previous study [21], and the existing ligand XR8-89 in 7LBR were used as controls. From Table 1, it can be seen that these three compounds show better binding energies than the control compounds, indicating that these top three compounds may be drug candidates against SARS-CoV-2 PL^{Pro}.

Table 1. The estimated binding energies of the top three compounds and the control compounds.

Compound	Structure	Binding Energy (kcal/mol)
F3077-0136		−91.92
F2883-0639		−90.96
F0514-5148		−89.66
GRL-0617		−60.75 [21]
Kaempferol 3-O-sophoroside 7-O-glucoside		−87.97 [21]
XR8-89		−87.41

From 2D ligand–protein interaction diagrams (Figure 1), we found that the top three compounds could all interact with Tyr268. F3077-0136 and F0514-5148 interact with Tyr268 by each forming one pi–pi stacking. F2883-0639 interacts with Tyr268 by forming one hydrogen bond and one pi–pi stacking. Furthermore, both F2883-0639 and F0514-5148 interact with Asn267 by each forming one hydrogen bond. Notably, Asn267 and Tyr268 are the residues on blocking loop 2 (BL2), which plays an essential role in inhibitor binding [24]. This potentially explains why these top three compounds can tightly bind to PL^{pro}.

Additionally, F3077-0136 interacts with Arg166 by forming a hydrogen bond and a pi–cation interaction. F0514-5148 also interacts with Arg166 by forming a hydrogen bond. Both F3077-0136 and F2883-0639 interact with Tyr273 by each forming one hydrogen bond. All the interactions between the top three compounds and the residues in the binding pocket are summarized in Table 2. We also calculated the strength of the hydrogen bonds between the top three compounds and the residues (Table S1), and the data indicated that the hydrogen bonds between the top three compounds and the binding residues were strong.

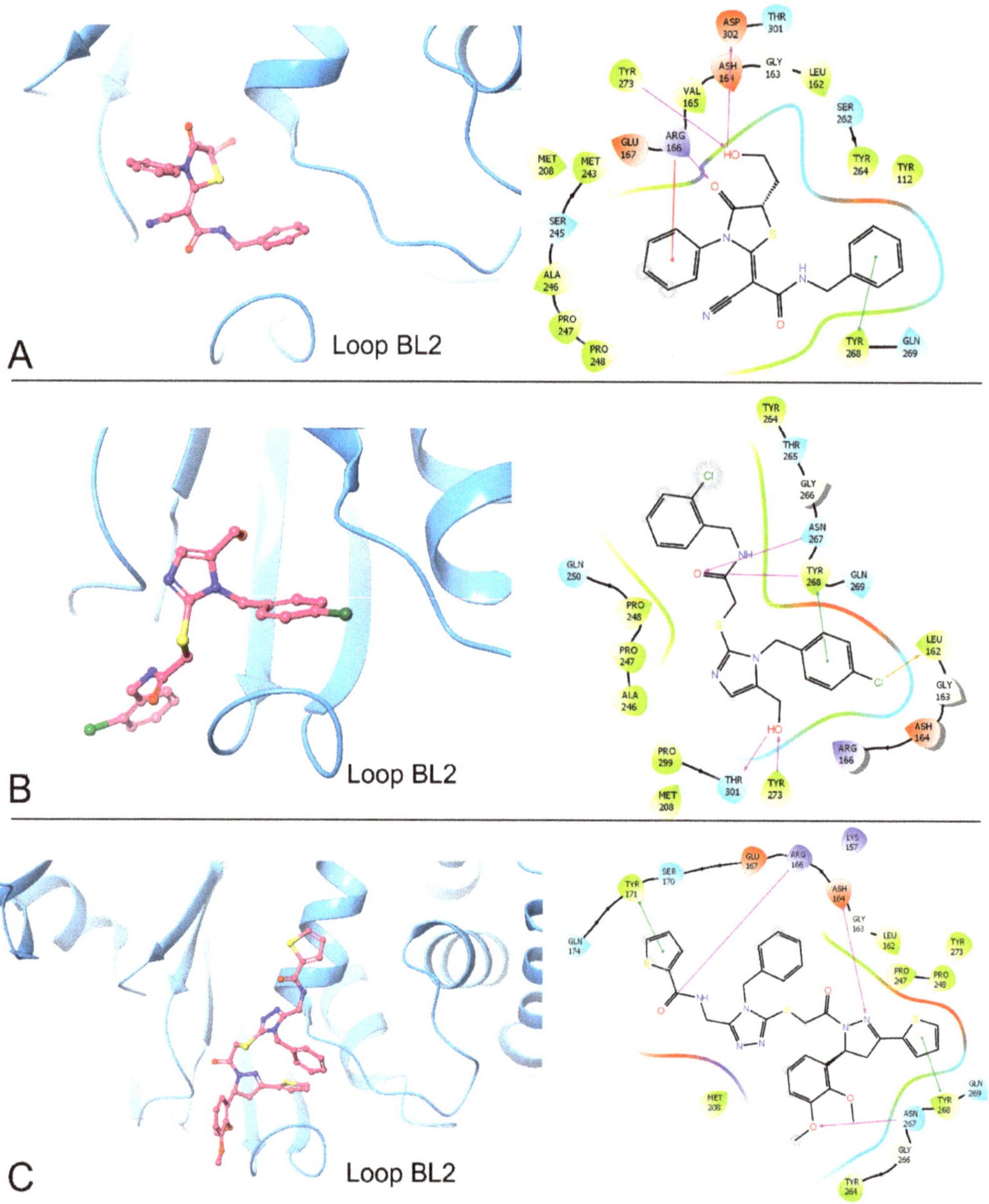

Figure 1. The docking poses and 2D ligand–protein interaction diagrams for 7LBR and the top three ligands: (**A**) F3077-0136; (**B**) F2883-0639; (**C**) F0514-5148. The pink arrow indicates the hydrogen bond; the green line represents pi–pi stacking; the red line represents pi–cation interaction; the yellow line represents the halogen bond.

Table 2. The interactions between the top three compounds and the residues of SARS-CoV-2 PL^{pro} and their distances (H-bond: hydrogen bond).

Residue	F3077-0136	F2883-0639	F0514-5148
Leu162		Halogen bond (3.33 Å)	
Asp164			H-bond (1.87 Å)
Arg166	H-bond (2.29 Å, 2.05 Å)		H-bond (2.31 Å)
	Pi–cation (4.58 Å)		
Tyr171			Pi–pi stacking (4.82 Å)
Asn267		H-bond (2.03 Å)	H-bond (2.02 Å)
Tyr268	Pi–pi stacking (4.64 Å)	H-bond (2.67 Å)	Pi–pi stacking (5.45 Å)
		Pi–pi stacking (4.78 Å)	
Tyr273	H-bond (1.85 Å)	H-bond (1.75 Å)	
Thr301		H-bond (2.71 Å)	
Asp302	H-bond (2.03 Å)		

2.2. Physicochemical Properties Prediction

To explore the physicochemical properties of the top three compounds, we predicted the absorption, distribution, metabolism, excretion (ADME), and the drug-likeness properties by using Qikprop in Maestro. The results are shown in Table 3. First, the molecular weights and the QPlogS of the top three compounds were in the recommended ranges of 130.0 to 725.0 and −6.5 to 0.5, respectively. F3077-0136 and F2883-0639 showed better solubility than F0514-5148 because the QPlogS value of F0514-5148 was close to −6.5. In addition, Lipinski's rule of five (RO5) and Jorgensen's rule of three (RO3) were used to evaluate the drug-likeness. Lipinski's rule of five (RO5) requires molecular weight < 500, an octanol–water partition coefficient (LogP) < 5, the number of hydrogen bond donors $\leq$ 5, and the number of hydrogen bond acceptors $\leq$ 10 [25]. Jorgensen's rule of three (RO3) requires aqueous solubility (LogS) > −5.7, the apparent Caco-2 cell permeability > 22 nm/s, and the number of primary metabolites < 7 [26]. All three compounds passed the RO5 and RO3 tests, while both F3077-0136 and F2883-0639 showed no RO5 and RO3 violations and F0514-5148 had two violations for each test. These results demonstrate that these three compounds are potential drug candidates. Therefore, F3077-0136, F2883-0639, and F0514-5148 were studied further.

Table 3. Selected Qikprop descriptors for the top three compounds and the control compounds.

Compound	mol_MW [1]	QPlogS [2]	RO5 [3]	RO3 [4]
F3077-0136	393.459	−5.404	0	0
F2883-0639	436.355	−5.440	0	0
F0514-5148	658.807	−6.367	2	2
GRL-0617	304.391	−4.952	0	0
F5367-0114	372.433	−4.230	0	0

[1] mol_MW represents the molecular weights of the molecules. The recommended range is 130.0–725.0. [2] QPlogS is the predicted aqueous solubility. The recommended range is −6.5~0.5. [3] RO5: number of violations of Lipinski's rule of five. The maximum is four. [4] RO3: number of violations of Jorgensen's rule of three. The maximum is three.

2.3. Molecular Dynamics (MD) Simulation Analysis

To further analyze the stability of PL^{pro} bound to the top three compounds, we conducted MD simulations to calculate the root-mean-square deviation (RMSD) for the proteins (Figure 2), the total energies (Figure 3), the H-bond (Figure S5), the radius of gyration (rGyr) (Figure S6), the solvent-accessible surface area (SASA) (Figure S7), and the root-mean-square fluctuation (RMSF) (Figure S8). RMSD can be used to evaluate the stability of a protein structure. As shown in Figure 2, the RMSD of 7LBR bound to F2883-0639 and F0514-5148 stabilized around 0.25 nm from 4 to 100 ns, similar to the RMSD for the cocrystal structure 7LBR-XR8-89. This suggested that the protein structures of 7LBR bound to F2883-0639 and F0514-5148 were stable. The RMSD of 7LBR bound to F3077-0136 ranged from 0.20 to 0.38 nm from 4 to 90 ns and then stabilized around 0.25 nm after 90 ns. Moreover, the complex 7LBR-F5367-0114, which showed worse binding energy (−29.81 kcal/mol), was applied as a negative control. The RMSD of 7LBR-F5367-0114 stabilized around 0.25 nm from 5 to 45 ns and then ranges from 0.3 to 0.4 nm until 100 nm. As the positive control, the RMSD of 7LBR bound to GRL-0617 ranged from 0.20 to 0.30 nm from 4 to 75 ns and then rose higher than 0.30 nm (Figure 2). Both the negative and positive controls showed higher RMSD values than the top three complexes. Therefore, we concluded that the proteins bound to the top three compounds were more stable than the proteins bound to F5367-0114 and GRL-0617. The total energies of the top three complexes and the control complexes are shown in Figure 3. The energies of the 7LBR–ligand complexes stabilized at around -1.49×10^6 KJ/mol, which also indicated the good stabilities of these systems. In addition, as depicted in Figure S5, 7LBR bound to F3077-0136 exhibited a higher number of hydrogen bonds than the other proposed complexes and was even better than the cocrystal structure 7LBR-XR8-89 from 0 to 60 ns, indicating the higher stability of 7LBR-F3077-0136. In Figure S6, the values for rGyr can be seen to stabilize around 2.35 to 2.43 nm, except for the negative control 7LBR-F5367-0114, indicating that the three proposed complexes have similar flexibilities compared to the positive control and the cocrystal structure. Furthermore, the SASA analysis plot (Figure S7) shows that the SASA values of the top three complexes were in the range of 159 nm^2 –172 nm^2, similar to the positive control and the cocrystal structure. This suggests that the top three complexes have similar solvent accessibilities as those of the positive control and the cocrystal structure. Notably, as shown in Figure S8, the top three complexes exhibited similar RMSF distributions compared to the positive control and the cocrystal structure, indicating their steady binding stabilities.

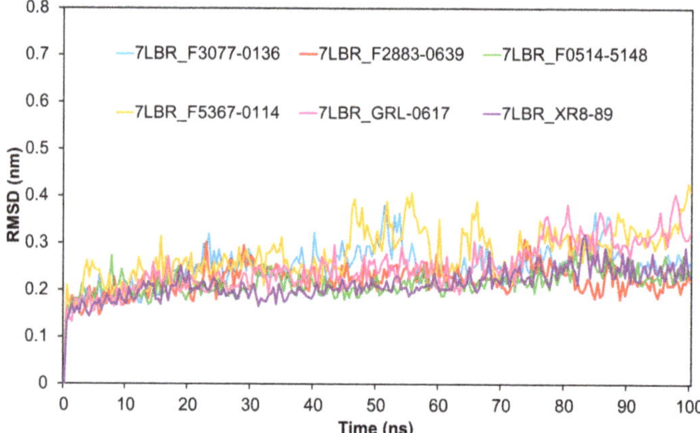

Figure 2. The comparison of the proteins' RMSD values. The blue, red, green, yellow, pink, and purple lines represent the RMSD values for 7LBR after binding to F3077-0136, F2883-0639, F0514-5148, F5367-0114 (negative control), GRL-0617 (drug candidate from a previous study), and the cocrystal structure 7LBR_XR8-89, respectively.

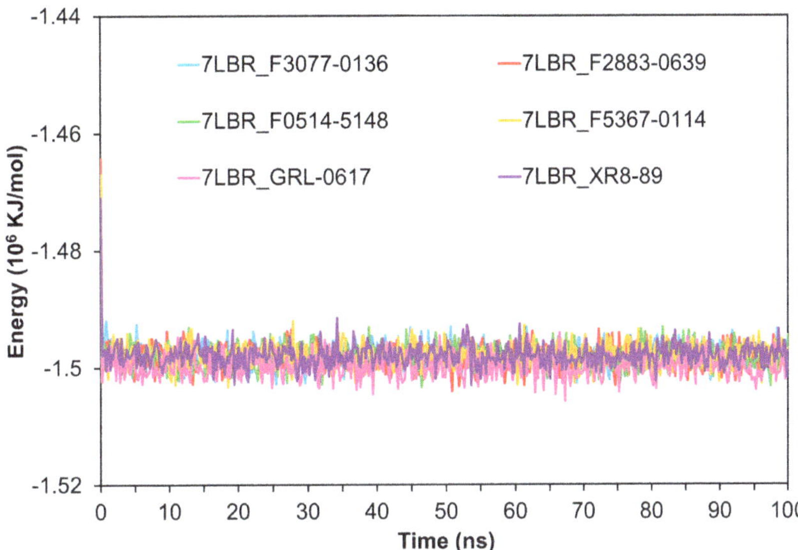

Figure 3. The comparison of the total energies of the protein–ligand complexes. The blue, red, green, yellow, pink, and purple lines represent the energies of the complexes 7LBR_F3077-0136, 7LBR_F2883-0639, 7LBR_F0514-5148, 7LBR_F5367-0114, and 7LBR_GRL-0617 and the cocrystal structure 7LBR_XR8-89, respectively.

3. Discussion

COVID-19 has led to the largest economic, educational, and societal disruptions of any pandemic seen throughout history, with an estimated USD 3.3 trillion deficit within the United States and USD 202.6 billion in lost revenue throughout American healthcare systems [27]. Hence, drug development for the treatment of COVID-19 has accelerated drastically. PL^{pro} is one of the critical therapeutic targets for COVID-19 treatment. Numerous studies have proposed PL^{pro} inhibitors, such as GRL-0617 and its derivative, XR8-89 [16,28]. GRL-0617 is not an ideal antiviral agent due to its insufficient potency [16]; however, it can be used as a control for virtual screening. The cocrystal structure (PDB ID: 7LBR) of XR8-89 with SARS-CoV PL^{pro} was applied for screening in this study. The existing ligand XR8-89 was used as another control. As one of the derivatives of GRL-0617, XR8-89 showed better binding energy than GRL-0617 (Table 1), which was consistent with their IC$_{50}$ data (the IC$_{50}$ of GRL-0617 is 1.61 μM and the IC$_{50}$ of XR8-89 is 0.113 μM) [16]. The binding energy of XR8-89 was close to those of the top three candidates, which indicates that the top three candidates should have similar or better inhibitory potency against PL^{pro}, as a previous study acknowledged that binding energy is a significant indicator of drug potency [29]. Furthermore, in addition to XR8-89, we tested the other compounds listed in the paper by Shen et al. [16] and compared the estimated binding energies with IC$_{50}$ values. As can be seen in Figure S1, our predicted binding energies showed a similar trend as the IC$_{50}$ values, suggesting the accuracy of the docking results.

In addition, there were six more candidates with better binding energies than the control group (Table S2). The binding energies of those candidates were around −88 kcal/mol. By comparing the protein–ligand interactions, we found that five of them interacted with Asn267, Tyr268, and Gln269 (Table S3). These three residues belong to the BL2 loop (Gly266-Gly271) which is an essential structure in PL^{pro} for inhibitor binding [24]. Therefore, F3222-1354, F1827-0078, F3166-0259, F3222-3821, and F1614-0151 may exert inhibitory effects on PL^{pro}. Among these five compounds, the Qikprop descriptors for F1827-0078, F3166-0259, and F1614-0151 fell within the recommended range, suggesting that these three compounds may be used for COVID-19 treatment (Table S4). To further validate our results, we are also testing these candidates using biochemical methods in our collaborative lab, and the outcomes will be published in the future once our collaborators complete the experiments.

4. Materials and Methods

4.1. Ligand Preparation

The structures of the 26,193 tested compounds were provided by the College of Pharmacy, University of Georgia. Before preparation, we analyzed this database using DataWarrior [30]. More than 90% of the compounds had molecular weights (MWs) below 450 (MW $\leq$ 450) (Figure S2), and around 96% of the compounds showed good hydrophilicity (cLogP $\leq$ 5) (Figure S3). After calculating the drug-likeness values in DataWarrior, around 70% of the compounds fell within the recommended range (drug-likeness value > 0) (Figure S4). All the tested compounds were prepared using Ligprep in Maestro 12.4 (Schrödinger). The process employed by Ligprep involves adding hydrogen molecules, computing the correct partial charges, and generating possible conformations. The force field used was OPLS3e by default [31].

4.2. Protein Preparation

The protein structure of SARS-CoV-2 PL^{pro} (PDB ID: 7LBR [16]) from RCSB's Protein Data Bank (https://www.rcsb.org/, accessed on 20 April 2022) [32] was prepared using Schrodinger Maestro. The protein preparation included three steps [33]. The first step was preprocessing, which included assigning bond orders, adding hydrogen, creating zero-order bonds to metals, creating disulfide, and generating het states using Epik [34]. The second step was optimization, which involved optimizing the hydrogen bond using PROPKA [35]. The third step was minimization, which was performed by using the OPLS3e force field [31].

4.3. Ligand–Protein Docking

To estimate the interactions between target proteins and the tested compounds, we conducted ligand–protein docking by using the Ligand Docking panel in Maestro. Before running docking jobs, a receptor grid box was generated based on the existing ligand XR8-89 (Ligand ID: XT7) in the structure 7LBR [16]. The size of the receptor grid box was set as the default (20 Å). Ligand–protein docking was performed as flexible docking in extra-precision (XP) mode.

4.4. MM-GBSA Calculation

To predict the binding energies of the tested compounds bound to PL^{pro}, we performed Prime molecular mechanics with generalized Born and surface area (MM-GBSA) calculations in Maestro. The pose-viewer files generated after dockings were uploaded into the Prime MM-GBSA panel to calculate binding energy. The force field used was OPLS3e [31].

4.5. ADME and Drug-Likeness Properties Prediction

After the calculation of binding energies, we applied the Qikprop module in Maestro to predict absorption, distribution, metabolism, excretion (ADME), and drug-likeness properties for further screening [36]. For Qikprop, the top three compounds were prepared using Ligprep. Finally, the descriptors, such as the rule of five (RO5) and rule of three (RO3), were applied to analyze the candidates.

4.6. Molecular Dynamics Simulations

To further investigate the dynamic interactions between PL^{pro} and the top three compounds, we conducted molecular dynamics (MD) simulations using GROMACS version 2030.4 and CHARMM36 force field [37,38]. The starting coordinates for the protein–ligand complexes were obtained from our ligand–protein docking studies. Then, we used

CHARMM-GUI to build the MD simulation solution box, which was a cubic box with a length of 112 Å, and filled it with water molecules [39]. Next, the minimized structures were equilibrated using a constant number of particles, volume, and temperature (NVT) ensemble and a number of particles, pressure, and temperature (NPT) ensemble. The target equilibration temperature was 303.15 K. Finally, MD simulations were performed for 100 ns and the conformations were calculated and recorded every 2 ps. During the simulations, the simulation solution box was isotropic and the periodic boundary condition was applied. After the MD simulations, we calculated the root-mean-square deviation (RMSD) for the protein and the energies of the system along the simulation trajectories.

5. Conclusions

This study demonstrated that F3077-0136, F2883-0639, and F0514-5148 were the best three drug candidates. In particular, F3077-0136 and F2883-0639 showed better drug-likeness properties. In this study, we applied several computational approaches to a 26,193 compound library provided by the College of Pharmacy, University of Georgia. All compounds were available for the ensuing assays. Before the biological assays, we docked all 26,193 compounds on PL^{pro}, predicted the binding affinities to identify the best inhibitors, predicted the drug-likeness, and performed MD simulations to validate the predicted binding. Overall, this study not only provides clues for bench research but may also speed up the process of COVID-19 drug development.

Supplementary Materials: The following supporting information can be downloaded at: https://www.mdpi.com/article/10.3390/ijms24054397/s1.

Author Contributions: Conceptualization, Z.-R.X.; methodology, Y.W. and L.L.; formal analysis, S.D.P. and D.C.; computational experiment: Y.W., L.L., L.N.M., E.B.S., C.B. and A.E.C.; writing—original draft preparation, L.N.M. and Y.W.; writing—review and editing, K.Y.C., S.D.P., D.C. and Z.-R.X.; project administration, K.Y.C. and Z.-R.X.; funding acquisition, K.Y.C. and Z.-R.X. All authors have read and agreed to the published version of the manuscript.

Funding: In this research, Y.W., L.L. and Z.-R.X. were supported by a faculty seed grant (2231464F22) from the University of Georgia and K.Y.C. was supported by the Ministry of Science and Technology, R.O.C., under MOST-108-2221-E-019-052.

Institutional Review Board Statement: Not applicable.

Informed Consent Statement: Not applicable.

Data Availability Statement: Publicly available data was generated in this study. This data can be found at: https://sourceforge.net/projects/msdock/.

Acknowledgments: We would like to acknowledge the Georgia Advanced Computing Resource Center (GACRC) and the College of Engineering's IT department from the UGA for technical support. This work used the Extreme Science and Engineering Discovery Environment (XSEDE) Bridges GPU at the Pittsburgh Supercomputing Center through allocation TGDPP180005.

Conflicts of Interest: The authors declare no conflict of interest.

References

1. WHO Coronavirus (COVID-19) Dashboard. Available online: https://covid19.who.int/ (accessed on 18 February 2023).
2. Johnson, N.P.A.S.; Mueller, J. Updating the accounts: Global mortality of the 1918-1920 "Spanish" influenza pandemic. *Bull. Hist. Med.* **2002**, *76*, 105–115. [CrossRef]
3. CDC COVID Data tracker. Prevention, Centers for Disease Control and Prevention. Available online: https://covid.cdc.gov/covid-data-tracker/#datatracker-home (accessed on 19 July 2022).
4. History of 1918 flu pandemic. Centers for Disease Control and Prevention. Available online: https://www.cdc.gov/flu/pandemic-resources/1918-commemoration/1918-pandemic-history.htm (accessed on 19 July 2022).
5. Wu, F.; Zhao, S.; Yu, B.; Chen, Y.-M.; Wang, W.; Song, Z.-G.; Hu, Y.; Tao, Z.-W.; Tian, J.-H.; Pei, Y.-Y.; et al. A new coronavirus associated with human respiratory disease in China. *Nature* **2020**, *579*, 265–269. [CrossRef]
6. Ozkan, K. How Close are We to a COVID-19 Vaccine? *J. Pure Appl. Microbiol.* **2020**, *14*, 893–902. [CrossRef]

7. Hassan, M.A.-K.; Aliyu, S. Delayed Access to COVID-19 Vaccines: A Perspective on Low-income Countries in Africa. *Int. J. Health Serv.* **2022**, *52*, 323–329. [CrossRef]

8. COVID-19 Map. Johns Hopkins Coronavirus Resource Center. 2022. Available online: https://coronavirus.jhu.edu/map.html (accessed on 19 July 2022).

9. Kim, S. COVID-19 Drug Development. *J. Microbiol. Biotechnol.* **2022**, *32*, 1–5. [CrossRef]

10. Lv, Z.; Cano, K.E.; Jia, L.; Drag, M.; Huang, T.T.; Olsen, S.K. Targeting SARS-CoV-2 Proteases for COVID-19 Antiviral Development. *Front. Chem.* **2021**, *9*, 819165. [CrossRef]

11. Tian, L.; Qiang, T.; Liang, C.; Ren, X.; Jia, M.; Zhang, J.; Li, J.; Wan, M.; YuWen, X.; Li, H.; et al. RNA-dependent RNA polymerase (RdRp) inhibitors: The current landscape and repurposing for the COVID-19 pandemic. *Eur. J. Med. Chem.* **2021**, *213*, 113201. [CrossRef]

12. Huang, L.; Chen, Y.; Xiao, J.; Luo, W.; Li, F.; Wang, Y.; Wang, Y.; Wang, Y. Progress in the Research and Development of Anti-COVID-19 Drugs. *Front. Public Health* **2020**, *8*, 365. [CrossRef]

13. Pavan, M.; Bolcato, G.; Bassani, D.; Sturlese, M.; Moro, S. Supervised Molecular Dynamics (SuMD) Insights into the mechanism of action of SARS-CoV-2 main protease inhibitor PF-07321332. *J. Enzyme Inhib. Med. Chem.* **2021**, *36*, 1646–1650. [CrossRef]

14. Ma, C.; Sacco, M.D.; Hurst, B.; Townsend, J.A.; Hu, Y.; Szeto, T.; Zhang, X.; Tarbet, B.; Marty, M.; Chen, Y.; et al. Boceprevir, GC-376, and calpain inhibitors II, XII inhibit SARS-CoV-2 viral replication by targeting the viral main protease. *Cell Res.* **2020**, *30*, 678–692. [CrossRef]

15. Lei, J.; Kusov, Y.; Hilgenfeld, R. Nsp3 of coronaviruses: Structures and functions of a large multi-domain protein. *Antivir. Res.* **2018**, *149*, 58–74. [CrossRef]

16. Shen, Z.; Ratia, K.; Cooper, L.; Kong, D.; Lee, H.; Kwon, Y.; Li, Y.; Alqarni, S.; Huang, F.; Dubrovskyi, O.; et al. Design of SARS-CoV-2 PLpro Inhibitors for COVID-19 Antiviral Therapy Leveraging Binding Cooperativity. *J. Med. Chem.* **2022**, *65*, 2940–2955. [CrossRef]

17. Weglarz-Tomczak, E.; Tomczak, J.M.; Talma, M.; Burda-Grabowska, M.; Giurg, M.; Brul, S. Identification of ebselen and its analogues as potent covalent inhibitors of papain-like protease from SARS-CoV-2. *Sci. Rep.* **2021**, *11*, 3640. [CrossRef]

18. Fatimawali Maulana, R.R.; Windah, A.L.L.; Wahongan, I.F.; Tumilaar, S.G.; Adam, A.A.; Kepel, B.J.; Bodhi, W.; Tallei, T.E. Data on the docking of phytoconstituents of betel plant and matcha green tea on SARS-CoV-2. *Data Br.* **2021**, *36*, 107049. [CrossRef]

19. Siddiqui, S.; Upadhyay, S.; Ahmad, R.; Gupta, A.; Srivastava, A.; Trivedi, A.; Husain, I.; Ahmad, B.; Ahamed, M.; Khan, M.A. Virtual screening of phytoconstituents from miracle herb nigella sativa targeting nucleocapsid protein and papain-like protease of SARS-CoV-2 for COVID-19 treatment. *J. Biomol. Struct. Dyn.* **2022**, *40*, 3928–3948. [CrossRef]

20. Naidoo, D.; Roy, A.; Kar, P.; Mutanda, T.; Anandraj, A. Cyanobacterial metabolites as promising drug leads against the M(pro) and PL(pro) of SARS-CoV-2, an in silico analysis. *J. Biomol. Struct. Dyn.* **2021**, *39*, 6218–6230. [CrossRef]

21. Wu, Y.; Pegan, S.D.; Crich, D.; Desrochers, E.; Starling, E.B.; Hansen, M.C.; Booth, C.; Mullininx, L.N.; Lou, L.; Chang, K.Y.; et al. Polyphenols as Alternative Treatments of COVID-19. *Comput. Struct. Biotechnol. J.* **2021**, *19*, 5371–5380. [CrossRef]

22. Hajbabaie, R.; Harper, M.T.; Rahman, T. Establishing an Analogue Based In Silico Pipeline in the Pursuit of Novel Inhibitory Scaffolds against the SARS Coronavirus 2 Papain-Like Protease. *Molecules* **2021**, *26*, 1134. [CrossRef]

23. Ghosh, A.K.; Takayama, J.; Aubin, Y.; Ratia, K.; Chaudhuri, R.; Baez, Y.; Sleeman, K.; Coughlin, M.; Nichols, D.B.; Mulhearn, D.C.; et al. Structure-Based Design, Synthesis, and Biological Evaluation of a Series of Novel and Reversible Inhibitors for the Severe Acute Respiratory Syndrome-Coronavirus Papain-Like Protease. *J. Med. Chem.* **2009**, *52*, 5228. [CrossRef]

24. Lee, H.; Lei, H.; Santarsiero, B.D.; Gatuz, J.L.; Cao, S.; Rice, A.J.; Patel, K.; Szypulinski, M.Z.; Ojeda, I.; Ghosh, A.K.; et al. Inhibitor Recognition Specificity of MERS-CoV Papain-like Protease May Differ from That of SARS-CoV. *ACS Chem. Biol.* **2015**, *10*, 1456–1465. [CrossRef]

25. Lipinski, C.A.; Lombardo, F.; Dominy, B.W.; Feeney, P.J. Experimental and computational approaches to estimate solubility and permeability in drug discovery and development settings. *Adv. Drug Deliv. Rev.* **2001**, *46*, 3–26. [CrossRef]

26. Jorgensen, W.L. The Many Roles of Computation in Drug Discovery. *Science* **2004**, *303*, 1813–1818. [CrossRef]

27. Kaye, A.D.; Okeagu, C.N.; Pham, A.D.; Silva, R.A.; Hurley, J.J.; Arron, B.L.; Sarfraz, N.; Lee, H.N.; Ghali, G.E.; Gamble, J.W.; et al. Economic impact of COVID-19 pandemic on healthcare facilities and systems: International perspectives. *Best Pract. Res. Clin. Anaesthesiol.* **2021**, *35*, 293–306. [CrossRef]

28. Calleja, D.J.; Lessene, G.; Komander, D. Inhibitors of SARS-CoV-2 PLpro. *Front. Chem.* **2022**, *10*, 1–20. [CrossRef]

29. Wan, S.; Bhati, A.P.; Zasada, S.J.; Coveney, P.V. Rapid, accurate, precise and reproducible ligand–protein binding free energy prediction. *Interface Focus* **2020**, *10*, 20200007. [CrossRef]

30. Sander, T.; Freyss, J.; Kor MVon Rufener, C. DataWarrior: An Open-Source Program for Chemistry Aware Data Visualization and Analysis. *J. Chem. Inf. Model.* **2015**, *55*, 460–473. [CrossRef]

31. Roos, K.; Wu, C.; Damm, W.; Reboul, M.; Stevenson, J.M.; Lu, C.; Dahlgren, M.K.; Mondal, S.; Chen, W.; Wang, L.; et al. OPLS3e: Extending Force Field Coverage for Drug-Like Small Molecules. *J. Chem. Theory Comput.* **2019**, *15*, 1863–1874. [CrossRef]

32. Berman, H.M.; Westbrook, J.; Feng, Z.; Gilliland, G.; Bhat, T.N.; Weissig, H.; Shindyalov, I.N.; Bourne, P.E. The Protein Data Bank. *Nucleic Acids Res.* **2000**, *28*, 235–242. [CrossRef]

33. Madhavi Sastry, G.; Adzhigirey, M.; Day, T.; Annabhimoju, R.; Sherman, W. Protein and ligand preparation: Parameters, protocols, and influence on virtual screening enrichments. *J. Comput. Aided. Mol. Des.* **2013**, *27*, 221–234. [CrossRef]

34. Shelley, J.C.; Cholleti, A.; Frye, L.L.; Greenwood, J.R.; Timlin, M.R.; Uchimaya, M. Epik: A software program for pK a prediction and protonation state generation for drug-like molecules. *J. Comput. Aided. Mol. Des.* **2007**, *21*, 681–691. [CrossRef]

35. Li, H.; Robertson, A.D.; Jensen, J.H. Very fast empirical prediction and rationalization of protein pKa values. *Proteins Struct. Funct. Bioinforma* **2005**, *61*, 704–721. [CrossRef]

36. *Schrödinger Release 2021-3, QikProp*; Schrödinger LLC: New York, NY, USA, 2021.

37. Huang, J.; Rauscher, S.; Nawrocki, G.; Ran, T.; Feig, M.; De Groot, B.L.; Grubmüller, H.; MacKerell, A.D. CHARMM36, An Improved Force Field for Folded and Intrinsically Disordered Proteins. *Biophys. J.* **2017**, *112*, 175a–176a. [CrossRef]

38. Vanommeslaeghe, K.; Hatcher, E.; Acharya, C.; Kundu, S.; Zhong, S.; Shim, J.; Darian, E.; Guvench, O.; Lopes, P.; Vorobyov, I.; et al. CHARMM general force field: A force field for drug-like molecules compatible with the CHARMM all-atom additive biological force fields. *J. Comput. Chem.* **2010**, *31*, 671–690. [CrossRef]

39. Jo, S.; Kim, T.; Iyer, V.G.; Im, W. CHARMM-GUI: A web-based graphical user interface for CHARMM. *J. Comput. Chem.* **2008**, *29*, 1859–1865. [CrossRef]

MDPI

St. Alban-Anlage 66

4052 Basel

Switzerland

www.mdpi.com

International Journal of Molecular Sciences Editorial Office

E-mail: ijms@mdpi.com

www.mdpi.com/journal/ijms